HUMAN STRUCTURE AND FUNCTION

NURSING APPLICATIONS IN CLINICAL PRACTICE

HUMAN STRUCTURE AND FUNCTION

NURSING APPLICATIONS IN CLINICAL PRACTICE

Christine G. Brooker
BSc, RGN, RM, RNT
formerly Senior Lecturer
in Nursing Studies
South Bank University
London

Artwork by
Susan E. McCormick, BA

MOSBY

Published by
Mosby-Year Book Europe, Ltd.
Brook House
2-16 Torrington Place
London WC1E 7LT

ISBN 0 7234 1846 2

Printed and bound in Great Britain by
BPCC Hazell Books Ltd

A CIP record for this book is available from the British Library.

For a full list of forthcoming titles and details of our nursing, midwifery and physiotherapy titles, please write to: Mosby Year Book Europe Limited, 2-16 Torrington Place, London WC1E 7LT, England.

Cover: Olympic medalist Matt Biondi. © Allsport U.K., Ltd./Simon Bruty. Cover border, TS yellow elastic ligament (Van Gieson). Reproduced with permission from Craigmyle, M., *A Colour Atlas of Histology.* 2nd edn. Wolfe Medical Publications, Ltd. 1986.

Cover design by Lara Last.

Foreword

For some time now, there has been a need for a comprehensive, up-to-date text for the new student beginning health studies. Many existing books assume a background knowledge of chemistry and fail to relate anatomy to physiology. *Human Structure and Function* avoids these difficulties by assuming no previous knowledge. Its strengths lie in the key definitions which open every chapter, the introduction of relevant chemistry at appropriate points within the text and the inclusion of anatomy. Students are constantly reminded of the relevance of theory to practice through the inclusion of **Nursing Practice Applications, Person-centred Studies**, and **Special Focus** sections. References for further reading and self-tests enable students to build on knowledge and monitor their own progress.

This book will be warmly welcomed by all those teaching the curricula on the new **Project 2000** diploma courses. Its clear and attractive format combining many diagrams and flow charts will make it a firm favourite with students of nursing and the many disciplines allied to health care.

Dinah Gould
Lecturer in Nursing Studies
King's College London
University of London

Acknowledgements

With many thanks to Dinah Gould and Janet Mardle for reviewing the text; Ann Osborne, Librarian at King's Lynn Health Science Library, The Queen Elizabeth Hospital; David Gardner for additional artwork; Roger Osborne, Faber and Faber; Eleanor Flood for editing; June Thompson, Griselda Campbell and Amy Salter, Mosby — Year Book Europe, Ltd.; the United Kingdom Central Council for Nursing, Midwifery and Health Visiting for supplying material on the new Code of Professional Conduct; and to my many friends and colleagues for their help and support. Special thanks to my partner David — without his computer skills, this book would not exist.

Dedication

For Darby,
a very special friend, who sadly died before this book
was completed.

Contents

PREFACE

The major aims of *Human Structure and Function* are to provide students of nursing with the knowledge they need for the biology and physiology topics covered within the **Common Foundation Programme (CFP)** of the **Project 2000** diploma courses, and to form a firm base upon which to build the appropriate aspects of the **branch programmes**. The content and level of knowledge included in the book have been influenced by comments from students, lecturers, clinical nurses and from several excellent textbooks. This knowledge of human structure (anatomy) and function (physiology) is provided in the context of nursing practice applications.

The principal theme is that of structure and *normal* function, and the close relationship between structure and function is emphasized within the text. The information is provided within a logical framework. Each of the eight sections of the book relates directly to functional areas of the body with overlap between sections as appropriate. The text is practice-based and, where appropriate, linked with relevant nursing research, health promotion and pathophysiology.

With assistance from learning aids incorporated throughout the text (see Introduction), students can use *Human Structure and Function* to:

- appreciate normal functioning of the body with its vital interactions that ensure homeostasis and health;
- understand how physiological dysfunction may lead to health deficiencies (illness);
- use their knowledge of body functioning during observation and assessment of people, clients and patients;
- apply this knowledge to appropriate areas of nursing practice with particular reference to current research into problems such as pain, incontinence or pressure sores; and
- integrate this knowledge with social and behavioural sciences in the promotion of health, prevention of ill-health, rehabilitation, and care relevant to the chosen branch programme.

No book or aid to learning is complete in itself. It succeeds only if it enables students to broaden and enhance their knowledge, and to apply this knowledge to daily nursing practice. As students prepare to pursue their chosen branch programme, *Human Structure and Function* will prepare them with a firm knowledge of anatomy and physiology, founded in holism and health.

Christine G. Brooker

INTRODUCTION

To the Student

Why study structure and function?

Why is a detailed knowledge of human anatomy and physiology relevant to today's nurses? Apart from the fact that the roles and responsibilities of nurses are expanding in concert with our increasing knowledge of health issues and medical science, there are several reasons why these topics are of particular relevance to nurses and students of nursing:

- Health and health needs are better understood when there is an appreciation of the relevant life sciences and an ability to integrate this knowledge with social and behavioural sciences.
- Breakdown in homeostasis can be more easily identified if normal functioning of physiological systems is clearly defined and there is an understanding of the biological factors that affect health.
- Physical care can be planned and implemented more effectively if the practitioner is familiar with relevant biological aspects of care. Knowledge of life sciences is considered, by students, to be important in physical care and in understanding reasons for a particular nursing action (Akinsanja 1985).
- Outcomes of care can be more accurately predicted and evaluated by the practitioner who has a broad knowledge of body function.
- Knowledge of homeostatic processes is essential for an understanding of the implications for care should homeostasis fail.
- When nurses advise people about maintaining health and avoiding problems it is important that they have an understanding of normal body function.
- Study of body structure and function will ensure that the nurse has a complete picture of the person who requires their care. This enables the practitioner to respond with the appropriate nursing skills.

Examples in the nursing literature support the view that the study of anatomy and physiology during a foundation course have relevance to nursing practice. Closs (1987), for example, concludes that education and research in biological sciences should aim to improve standards of practice in nursing.

Why emphasize healthy function?

A holistic approach to the study of nursing can be successful only if the structure and functioning of the body is included and integrated with other areas of knowledge required for practice.

Current nursing philosophy centres upon the *whole* person and emphasizes health and health maintenance. It is logical, therefore, to take a holistic view when learning about body structure and, more importantly, about the *healthy* functioning of its components. The holistic approach allows you to discover the close relationships between structure and function, and to appreciate the contribution that all body systems make towards homeostasis and health.

What is 'normal' function?

Body structures function as part of the integrated whole — no single organ or system can maintain homeostasis in isolation. At all levels — molecules, cells, tissues, organs and systems — there are interactions and essential relationships that affect the functioning of the whole person. A change in function in any major organ can affect the health of the individual.

Physiological function and health can be summarized as:

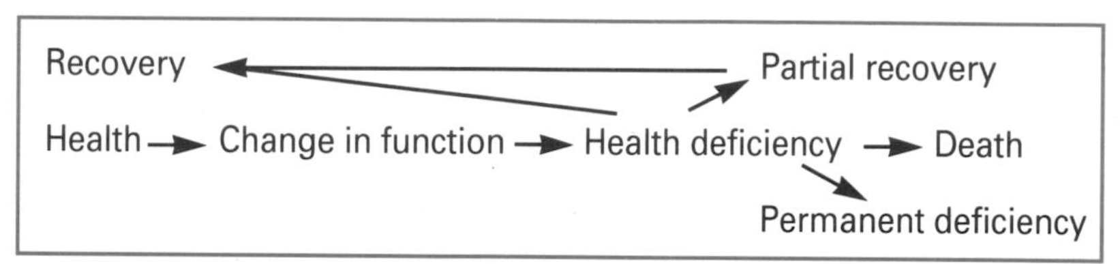

The table below lists factors required for physiological health and those that alter the homeostatic balance.

Criteria for normal physiological function:	**Factors associated with altered physiological function and homeostatic imbalance:**
■ genetic inheritance which programmes for normal function	■ inheritance of genes which cause altered function
■ normal chromosome complement	■ chromosomal abnormality
■ oxygen	■ lack of oxygen or poor utilization of oxygen
■ essential nutrients in the correct amounts	■ malnutrition: incorrect nutrients, malabsorption and loss of nutrients.
■ water	■ water imbalance
■ ability to regulate pH and electrolyte levels and to remove waste	■ electrolyte imbalance, and failure of excretory and homeostatic processes
■ protection from hostile environmental factors, injury and disease	■ lack of shelter, accidents, trauma, exposure to: toxic chemicals, radiation, contaminated water and food, pathogens (disease producing micro-organisms) and other stressors
■ movement and stability	■ immobility and instability of the body
■ body temperature within normal range	■ inability to maintain body temperature within normal range
■ cell growth, replication and repair	■ abnormal cell growth and division e.g. malignant disease
■ periods of inactivity and sleep	■ lack of adequate rest and sleep (this varies with individuals)
■ reproduction (not strictly essential for the individual but vital for the species)	

It is important that the image of a *healthy* person is retained during the study of body structure and function. As you pursue your nursing practice, you will find that most people within a community are not ill. They are normally functioning individuals who may become consumers of health care services at specific times, such as screening or during pregnancy. As illustrated in the health-illness continuum below, individuals become patients if they suffer health deficiencies (illness):

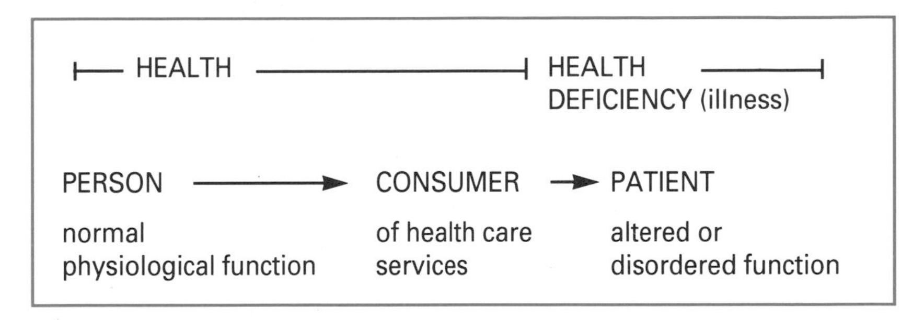

However, health is never static — it is a dynamic state which means different things to individual people and which may vary with their particular stage of life. The definition of 'health' for any individual, therefore, is never clear-cut.

Making this Book Work for You

A textbook can present a wealth of information; however, if that information is difficult to locate, to understand, and to apply to your daily activities as a nurse, then the textbook is of limited use. The learning aids featured in *Human Structure and Function* will enable you to quickly locate and utilize the information so that it benefits you. These learning aids include the following:

Overview located at the beginning of every chapter, gives you a brief outline of topics covered in the chapter.

Learning Outcomes listed at the beginning of every chapter, tell you what you should know after reading the chapter. The learning outcomes provide a framework for learning and the means to plan and take responsibility for your learning.

Key Words also listed at the beginning of each chapter, define specific terms you will encounter throughout that chapter. The keywords and other important terminology are also presented in **boldface** and defined within the chapter text. Additional terms related to the topic are presented in *italics* and defined throughout the chapter.

Person/family-centred Studies, Nursing Practice Applications and **Special Focus Topics** scattered throughout the chapters, help you relate your knowledge of structure and normal function to health and the problems which may occur when function is disordered. This information also demonstrates the relevance of this knowledge to the daily practice of nursing. (The people featured in these learning aids are fictitious and are not based on any person I have nursed or met whilst supervising students.)

Chapter Summary/Checklist located at the end of every chapter, enables you to quickly review main topics and plan revision.

Self Test presented after the chapter summary, enables you to test your knowledge of the main topics covered in the chapter and alerts you to topics you may wish to review again.

Further Reading and **References** list additional sources of information to enhance your understanding of particular topics.

Useful Addresses located at the end of some chapters, lists organizations concerned with particular illnesses or disabilities discussed in the chapter. You can contact these organizations to obtain additional information about particular topics. It is also useful to be aware of these organizations when you become a practising nurse.

In addition to these features, a detailed **index**, numerous **cross references**, and informative **illustrations** and **appendices** also help guide you through *Human Structure and Function* to make the most of the information it presents. We hope you also will use the blank spaces in your book to write important points in your own words (one of the best ways to check real understanding), and add related references and notes about how you used a particular topic during observation or practice situations.

References

Akinsanja, J. (1985) Learning about life. *Senior Nurse.* **2**, 5, 24-25.

Closs, J. (1987) Biological science and nurses. *Senior Nurse.* **7**, 5, 45.

Before You Begin

If you have not studied science subjects before, it might be helpful to read some basic texts prior to starting your exploration of human structure and function. A short list of appropriate books is provided but your tutors/lecturers will also offer advice and suggestions for your reading. If you have some knowledge of basic science, you may want to use these texts for a quick review:

Hole C.B. (1974) *An Introduction to Cell Biology.* London: Macmillan

Rose S. and Sanderson C. (1979) *The Chemistry of Life.* 2nd edition. Harmondsworth: Penguin Books.

Lewis M. and Waller G. (1980) *Thinking Chemistry.* Oxford: Oxford University Press.

Swift D. (1988) *Physics for GCSE.* Oxford: Basil Blackwell.

Minett P. Wayne D. Rubenstein D. (1989) *Human Form and Function.* London: Unwin Hyman Limited.

section one

living material and organization of the body

CHAPTER ONE

Outline of Living Material and Organization of the Body

Overview

- Cells – structure, transport, division, multiplication, genes, chromosomes, nucleic acids, protein synthesis, differentiation.
- Abnormalities of cell growth, including neoplasms.
- Tissues – overview of the main types within the body, including: glands, membranes.
- Organs, functional systems and body cavities.
- Anatomical terms, body planes.

Learning Outcomes

After studying Chapter 1 you should be able to:

- Describe an 'idealized' cell.
- Describe the main cellular organelles and their function.
- Explain the ways in which substances are transported across the cytoplasmic membrane.
- Describe mitosis.
- State the differences between mitosis and meiosis.
- Define the terms 'chromosome' and 'gene.'
- Explain the structure and role of nucleic acids in protein synthesis.
- Define 'cellular differentiation.'
- Describe the common abnormalities of cell growth.
- Discuss the factors which predispose cells to neoplastic changes.
- Describe ways of reducing the risk of malignancy.
- Discuss how cancers may be detected at an early and potentially treatable stage.
- Define 'tissue' and state the major types found in the body.
- Name the main types of glands and membranes.
- State the types of structure found in organs.
- Name the body cavities.
- Define the common anatomical terms used to describe body directions.
- Describe the anatomical position of the body.
- Describe body planes.

Key Words

Adenosine triphosphate (ATP) – a chemical which provides the energy for vital cellular activities.
Deoxyribonucleic acid (DNA) – a nucleic acid found in the cell nucleus, it has a complex helical structure which carries the genetic code.
Differentiation – the changes that occur in cells and tissues as they develop the ability to perform specialized functions.
Diffusion – the process by which gas and liquid molecules of different densities/concentrations mix when brought into contact with one another.
Diploid – used to describe a cell which has a full set of paired chromosomes (46 arranged in 23 pairs), seen in all cells except the gametes.
Gametes – the haploid (n) reproductive cells; ova and spermatozoa.
Gene – hereditary factors present on the chromosomes; consisting of DNA they are responsible for the transmission of inherited characteristics and the precise replication of proteins.
Glycoprotein – a compound of protein and carbohydrate, also known as a mucoprotein.
Haploid – used to describe a cell which has a half set of chromosomes, (i.e. 23 unpaired chromosomes), e.g. the gametes following meiosis.
Meiosis – a type of nuclear division which produces **gametes**. These special cells (ova and spermatozoa) are **haploid**.
Mitosis – a type of nuclear division in somatic (body) cells which results in the production of identical daughter cells with the same number of chromosomes as the original cell.
Organelles – small structures within the cell, e.g mitochondria, which perform specific functions.
Osmosis – the passage of a solvent, which is always water in a 'living system', from a dilute solution to a more concentrated one through a selectively permeable membrane.
Ribonucleic acid (RNA) – a nucleic acid found in the cell nucleus and in ribosomes, vital in the process of protein synthesis.
Transcription – the process by which genetic information is transferred from DNA to RNA, the first stage in the synthesis of proteins.
Translation – the process by which proteins are synthesized in the ribosomes from amino acids.

Introduction

The body comprises many billions of cells. These basic structural units form the great variety of tissues, organs and systems which perform the highly specialized, biochemical processes required to maintain the body in a stable state (**homeostasis**). The actual biochemical processes are carried out by the smaller subcellular structures or **organelles** found in individual cells. The specific functions performed by a cell are determined by the type and numbers of these organelles. Continuity of the species depends on cells and the genetic material found in the nucleus. At fertilization the spermatozoon and ovum unite to form the **zygote**, a single cell, which by cell division and **differentiation** eventually gives rise to all the cells of the body. Differentiation is the term used to describe the changes that occur in cells and tissues as they develop the ability to perform specialized functions. It is an incredible process when you consider the diversity of cells produced in this way, e.g. excitable cells found in the nervous system, a secretory cell lining the digestive tract, or a structural cell, such as bone. The study of cells is called **cytology**.

Structurally, cells consist of a gel substance called **protoplasm** which is a complex mixture of water, proteins, lipids, carbohydrates and electrolytes. Individual cells may vary in form (related to function) and may have different lifespans. Generally cells consist of an outer cytoplasmic membrane, which surrounds cytoplasm containing several organelles, and a nucleus with its genetic material. However, not all mature cells retain a nucleus e.g **erythrocytes**. (See Chapter 9.)

CELL STRUCTURE

The structure of an idealized cell is shown in *Figure 1.1*. The functions of the various components are described below.

Cytoplasm and cytoplasmic membrane

Antigens *– substances (molecule or part of a molecule) which stimulate the body defences to produce an antibody (a protein). See Chapter 19.*

Cytoplasm is the protoplasm within the cell, not including that within the nuclear membrane. The cell is surrounded by a complex protein/phospholipid membrane known as the **cytoplasmic membrane** which forms a boundary around individual cells, and separates intracellular and extracellular fluid. Certain substances can pass through this membrane by a variety of different processes (see page 7-8). On the surface of the cytoplasmic membrane are various carbohydrate-based **antigens**. One example are those on erythrocytes which determine blood group (see Chapter 9).

Organelles

Inside the cell, situated within the cytoplasm, are several smaller structures known generally as organelles; these perform a variety of metabolic functions.

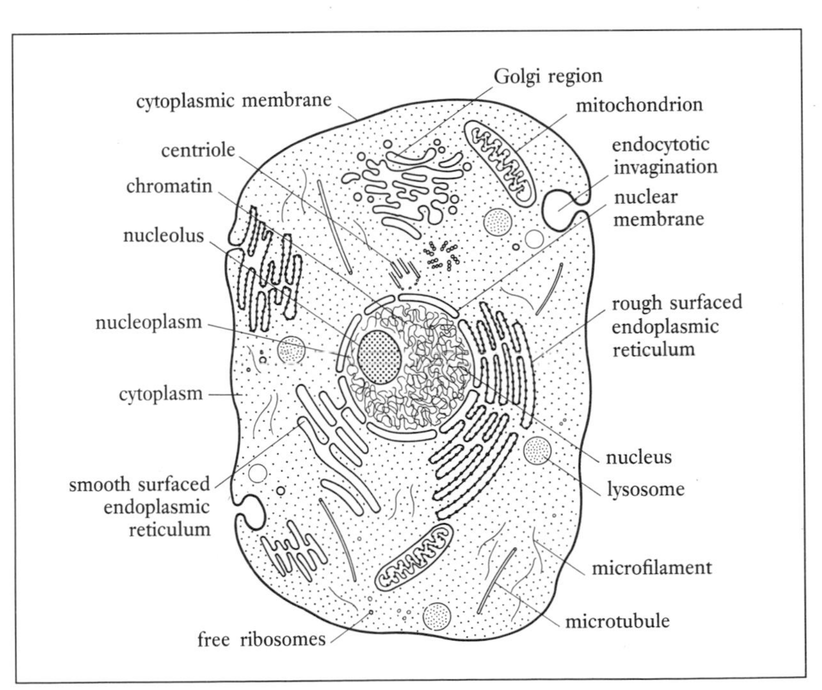

Figure 1.1. A cell (idealized).

Krebs' citric acid cycle *– a series of biochemical reactions where fuel molecules undergo changes in the presence of oxygen to produce ATP, carbon dioxide and water.*

Oxidative phosphorylation *– further biochemical reactions which produce large amounts of ATP.*

Mitochondria

Mitochondria are the 'cellular power stations' in the cell; the structures concerned with energy production from nutrients. The series of complex biochemical reactions known as the **Krebs' citric acid cycle** and **oxidative phosphorylation** occur in the mitochondria to produce energy in the form of **adenosine triphosphate (ATP)** (the chemical which provides energy for vital cellular activities). Cell function determines the number of mitochondria – greater numbers are present in liver and muscle cells which are metabolically more active (they have a large energy requirement).

Endoplasmic reticulum (rough and smooth)

Endoplasmic reticulum consists of a network of channels inside the cell which are involved with the production and movement of various substances, e.g. proteins. The rough endoplasmic reticulum is so called because of the presence of many ribosomes on its surface. In contrast, the smooth endoplasmic reticulum has no ribosomes attached.

Ribosomes

Ribosomes are very tiny organelles consisting of granules of protein and **ribonucleic acid (RNA)**. Ribonucleic acid is the nucleic acid, found in the ribosomes and nucleus, which is concerned with protein synthesis. Ribosomes may be free in the cytoplasm or associated with the rough endoplasmic reticulum.

Golgi region (body)

The **Golgi region** is a network of membranes in the cytoplasm situated close to the nucleus. These act as a storage area for cellular secretions and the modification of **glycoproteins** (a compound of protein and carbohydrates) prior to their use. They are larger in cells that have a secretory function.

Lysosomes

Within the cell are membranous sacs, derived from the Golgi region or the endoplasmic reticulum, containing *lytic* enzymes (capable of breaking down proteins, carbohydrates, etc). These sacs, known as **lysosomes**, are involved in the breakdown of substances or particles entering the cell, such as bacteria, and in the 'clearing up' processes initiated when cells are damaged. Lysosomes are found in *neutrophils* (a type of white blood cell). (See Chapter 9.)

Supporting structures – microfilaments and microtubules

Cells contain protein strands called **microfilaments**, which are concerned with maintaining cell shape, movement within cells and chromosomal migration during nuclear division (see description of mitosis pages 10-11). They are of particular importance in contractile cells such as skeletal muscle, where they are highly developed. Support is also achieved by the presence of **microtubules**, which are tubes arranged in bundles. These help to support the cell organelles and maintain cell shape. Microtubules form part of the **centrioles** (organelles which form the spindle during nuclear division). They also have a role in the intracellular transport of substances.

Motile structures – flagella and cilia

Certain cells have motile structures which project from their surface. They are formed from microtubules and include the microscopic hair-like **cilia**. The cilia are present on the cells that line the respiratory tract where they move mucus and debris out of the lungs. The uterine (fallopian) tubes also have cilia which aid the process of fertilization by moving the ovum along the tube. A much longer version of a cilia is known as a **flagellum**; the only human cell to have a flagellum is the spermatozoon, where it forms the 'tail' which allows the spermatozoon considerable motility (see Chapter 20).

The nucleus

The nucleus is concerned with cellular control through the synthesis of proteins and through cell division. It contains **chromatin,** a thread-like substance containing proteins and the genetic material of the cell. Within the chromatin are the 46 **chromosomes** which are made of **deoxyribonucleic acid (DNA)**, a complex helical structure which carries the genetic code. Each tiny segment of DNA is known as a **gene**. Genes are responsible for the transmission of inherited characteristics and the precise replication of proteins. The nucleus is enclosed by the nuclear membrane which separates nucleoplasm from cytoplasm; it allows the transfer of some substances by mechanisms yet to be fully understood. Also situated inside the nuclear membrane are the **nucleoli**, usually two in number; they contain both DNA and RNA and are involved in cell division.

Transport of Substances Across the Cytoplasmic Membrane

Amino acids *– building blocks which combine to form the many different proteins.*

Lipids *– fatty substances e.g. fatty acids.*

Electrolytes *– chemicals such as salts which dissociate in water to form electrically charged particles (ions) e.g. sodium (see Chapter 2).*

Urea *– a waste product of protein metabolism produced by the liver and excreted by the kidneys (see Chapters 13, 14, 15).*

For all the complex cellular processes required for health, various **organic** (contain carbon) and **inorganic** (contains no carbon) substances must be able to pass through the cytoplasmic membrane in both directions. These include glucose, amino acids, lipids, water, electrolytes, gases, urea, cellular secretions and in certain situations large particles for destruction, such as bacteria. The cytoplasmic membrane is **selectively permeable**, i.e. the movement of some substances is enabled while movement of others is restricted. The way these substances are moved is either by **passive**, e.g. osmosis, diffusion, or **active** means, e.g. active transport, ATP-requiring processes.

Osmosis

The method by which water moves in and out of the cell is known as **osmosis**. Water molecules pass from dilute to more concentrated solutions. It is an important mechanism in the maintenance of normal fluid balance within the body (see Chapter 2). Osmosis is a passive process in which the cell uses no energy.

Diffusion

Diffusion is the movement of molecules from areas of high molecular concentration to areas of low concentration (such as a sugar cube dissolving in a cup of tea). Substances pass through the cytoplasmic membrane by dissolving in its protein or phospholipid component or by utilizing specific carrier molecules. Substances are able to move in either direction, e.g. oxygen into the cell and waste carbon dioxide out of the cell. Certain cells, e.g. liver cells, are able to obtain glucose quickly by using a carrier molecule.

Diffusion is a passive process and, whether simple or involving a carrier molecule, it requires the presence of a concentration gradient.

Active transport

Active transport is a process requiring energy (ATP) to enable carrier molecules to move substances:

- Against a concentration gradient.
- Where no gradient exists.
- Where the substance is unable to diffuse through the cytoplasmic membrane.

Substances transported by this process include amino acids, glucose (some cells) and **electrolytes**, (electrically charged atoms), e.g. sodium, potassium.

Special ATP-requiring processes

Special transport processes requiring ATP are used in the bulk transport of substances. Particles may be engulfed by the cell, such as **phagocytosis** of bacteria by *leucocytes* (white blood cells) (see Chapter 9), or **pinocytosis**, where the cytoplasmic membrane surrounds a tiny drop of water which is then taken into the cell. The general term for these processes in which the material enters the cell is **endocytosis.** When the material is moved out of the cell the mechanism is called **exocytosis**, which occurs when hormones (see Chapter 8) or neurotransmitters (see Chapters 3 and 6) are released or when mucus is discharged from the cell.

Cell Replication

Cells replicate at different rates. Skin cells and those lining the digestive tract are continuously replaced, while others, such as liver cells, replicate only until the organ reaches full size, but they retain the ability to regenerate if the liver is damaged. Certain cells are not capable of replication after they mature, e.g. cells of the central nervous system (neurones) cannot be repaired or replaced. The result of this can be seen following a stroke, where damage to the brain leads to a permanent loss of function such as a weakness down one side of the body. However, it is sometimes possible for a nerve fibre to regenerate if the nerve cell body of the neurone is intact (see Chapter 3). This is also true for cardiac and skeletal muscle cells where damaged areas are replaced not by similar cells, but by inelastic fibrous tissue during repair.

Cell division

The way that cells replicate is through the division of existing cells. In order for cells to replicate nuclear division must take place. The two types of nuclear division are **mitosis,** where identical daughter cells are produced, and **meiosis**. Meiosis is a more complex reduction division resulting in the production of **gametes**, the male and female reproductive cells, (spermatazoa and ova) which have half the normal chromosome compliment (**haploid**). This ensures that both the female and the male each contribute half the genetic material when the male and female cells unite to form a zygote. The zygote then has the full compliment of chromosomes (**diploid**):

OVUM	+	SPERMATAZOON	=	ZYGOTE
(23 chromosomes)		(23 chromosomes)		(46 chromosomes)

A more detailed explanation can be found in Chapter 20.

Mitosis is the process by which the **somatic** (non-gamete) cells of the body divide to produce new cells during periods of growth or as replacement for cells which have reached the end of their lifespan. The events of somatic cell division, which include replication of the chromosomes, nuclear division and division of the cytoplasm, ensure that each new generation of cells contains the same genetic material.

The cell cycle

The cell cycle is the continuous process from one mitotic division to the next. The cycle includes all the processes necessary for nuclear and cell division. Very simply, the cell cycle can be divided into: *interphase* during which DNA replicates, *mitosis*, and *cytokinesis* (cytoplasmic division) (*Figure 1.2*). A knowledge of the cell cycle in 'normal' and malignant cells is used when planning radiotherapy for the treatment of some malignant diseases. During the cell cycle the cell is most susceptible to radiation, which disrupts DNA, during G_2 and mitosis. The dose and timing of radiation can be planned for maximum destruction of malignant cells whilst sparing 'normal' cells. Generally cells that divide frequently. e.g. those lining the digestive tract, are most sensitive to radiation (radiosensitive).

Interphase and DNA replication

Interphase is the stage during which all the normal day-to-day metabolic processes of cells occur. It is also the stage between two mitotic divisions when the cell grows and prepares the special structures and materials needed for division. The interphase can be subdivided into three stages: G_1, S and G_2 (see *Figure 1.2*).

The first part of interphase may last only hours in rapidly dividing cells, e.g. those of the digestive tract, but in other instances it may last longer. The cell grows and proteins are synthesized. Close to the nucleus there are two centrioles lying at 90° to each other. These

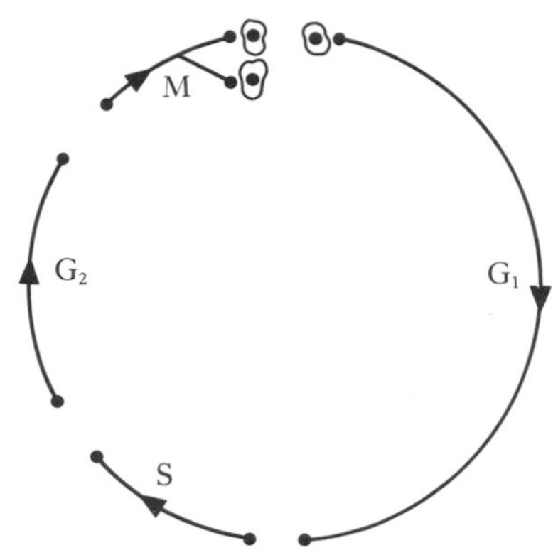

G_1 period of rapid growth and metabolic activity (variable length)
S DNA replication and growth
G_2 preparation for complete separation, growth and maturation
(interphase)
M mitosis and cytokinesis

Figure 1.2. The cell cycle.

replicate during interphase and each resultant pair will move to the opposite poles of the cell to commence the formation of the **mitotic spindle**. The spindle is formed from protein strands which consist of microtubules. During the next stage the DNA in the chromatin replicates. This means that the two new cells, which will result from mitosis, will have identical genetic material. The actual trigger for DNA replication and cell division is unknown; it is probable that there are several factors, e.g. cell volume/surface area ratio or the release of inhibition by protein hormones.

DNA is a complex chemical. Each strand of DNA is built from subunits formed from **nucleotides**, each composed of a nitrogenous base, a sugar and a phosphate. It has two strands coiled into a **double helix**. During replication the helical structure is uncoiled and the bonds between the two nucleotide strands are broken by **enzymes** (*chemical catalysts*) to produce two separated DNA strands. In an energy (ATP) requiring process, other enzymes cause two new nucleotide strands to be formed using the existing separate strands as a pattern (see *Figure 1.3*). Once replication is complete the double set of DNA joins with proteins to form new chromatin strands. The last part of interphase involves the metabolic processes which provide the enzymes and energy needed for mitosis.

Various factors such as chemicals, drugs and radiation may adversely affect this delicate process and prevent cell division. These are used therapeutically, to inhibit the growth and division of malignant cells, in the form of cytotoxic drugs (toxic to cells) or radiotherapy.

Mitosis

Mitosis is a continuous process but is usually described in four stages: *prophase, metaphase, anaphase* and *telophase. Figure 1.4* illustrates the stages of mitosis. The process of mitosis is usually followed by cytokinesis (see below). The whole process is completed within about two hours with some variation between different cell types.

Prophase (see Figure 1.4: 1a, 1b)

Prophase commences with a change in the chromatin which condenses and shortens to form visible chromosomes. (Remember that there is a double set of each chromosome resulting from DNA replication during interphase.) The double chromosomes are joined by the **centromere**. Each half of this double chromosome is known as a **chromatid**. The nucleoli start to breakdown and the nuclear membrane disappears. Each pair of centrioles moves to opposite poles of the cell where they commence the formation of the mitotic spindle which eventually reaches from one pair of centrioles to the other.

Metaphase (see Figure 1.4: 2)

During metaphase, the double chromosomes move towards the middle of the cell so that their centromeres are arranged precisely along the equator of the mitotic spindle.

Figure 1.3. DNA replication.

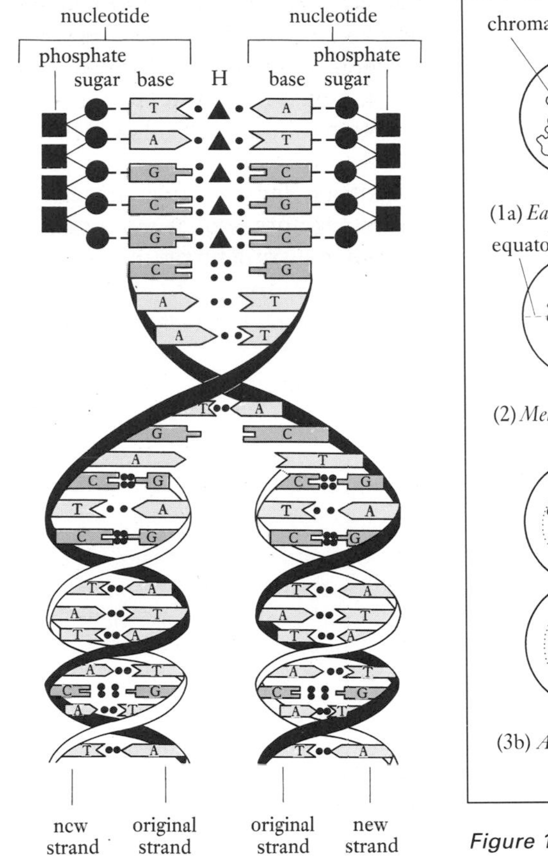

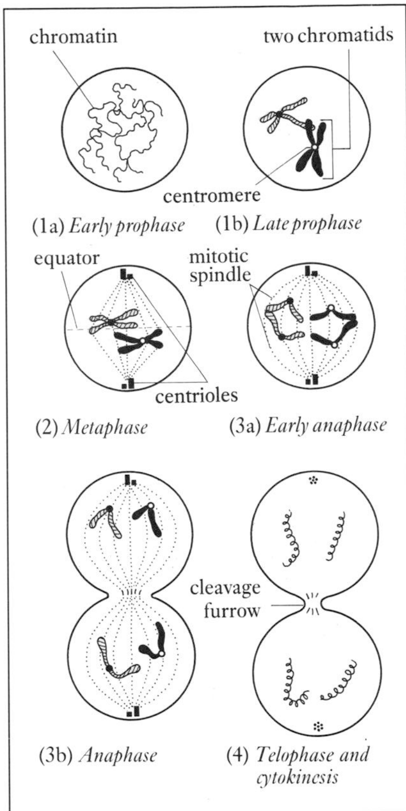

Figure 1.4. Stages of mitosis.

Anaphase (see Figure 1.4: 3a, b)

During anaphase the double chromosomes split at the centromere with each chromatid becoming a complete chromosome with its own centromere. The fibres of the mitotic spindle contract causing one chromosome from each new pair to be pulled to the opposite pole of the cell.

Telophase (see Figure 1.4: 4)

In telophase the set of chromosomes at each pole uncoils to form the thread-like chromatin. Other changes include a reversal of the events occurring during prophase: the nuclear membrane reforms, the mitotic spindle disappears and the nucleoli reform.

By the process of mitosis the body is able to grow and replace cells that sustain damage or reach the end of their allotted lifespan.

Cytokinesis

Cytokinesis (cytoplasmic division) is a separate process occurring after mitosis. The **cleavage furrow** which has formed round the cell during late anaphase continues to progress inwards in telophase. (*Figure 1.4: 3b, 4*) This process continues until the original cell has been pinched into two 'daughter' cells each with a nucleus containing identical genetic material.

Nuclear division is not always followed by cytokinesis. This means that some cells have more than one nucleus, such as may be seen in skeletal muscle. The two new cells, which are smaller than the original cell, now commence the cell cycle which starts with interphase and a period of growth before their own division.

Nucleic Acids and Protein Synthesis

Nucleic acids

The two nucleic acids found in cells are DNA (deoxyribonucleic acid) and RNA (ribonucleic acid).

DNA consists of the pentose sugar deoxyribose, phosphate groups, and four nitrogenous bases: **adenine (A)**, **thymine (T)**, **cytosine (C)** and **guanine (G)**. These three components form subunits or nucleotides which are arranged in a double helix with the two strands joined by hydrogen bonds. The bases are always paired A with T and C with G (see *Figure 1.3*). These are known as **complementary base pairs**. DNA is the genetic material present within the nucleus which is replicated prior to cell division. Apart from its role in cell division DNA acts as the master pattern for the precise synthesis of proteins. The term 'gene' is used to describe the tiny fragment of DNA which carries the information required to make a specific protein or subunit of a large protein molecule.

RNA is found in the nucleus, cytoplasm and ribosomes. Although it is derived from DNA, it differs from DNA in several ways *(Table 1.1)*. Its pentose sugar is ribose and the nitrogenous base sequence has **uracil** (**U**) substituting for thymine. Structurally it is a single strand which may

Table 1.1 . Summary of DNA and RNA details.

	DNA	***RNA***
Composition	Deoxyribose sugar Phosphate groups Nitrogenous bases: adenine, thymine, guanine and cytosine	Ribose sugar Phosphate groups Nitrogenous bases: adenine, uracil, guanine and cytosine
Structure	Two nucleotide strands coiled into a double helix	Single nucleotide strand which may be straight or folded
Sites	Nucleus	Nucleus, cytoplasm and ribosomes
Functions	Genetic material replicated in cell division acts as a pattern for protein synthesis	Undertakes the instructions needed for for protein synthesis

be folded or straight. There are three types of RNA which perform specific roles during protein synthesis: **messenger (mRNA)**, **transfer (tRNA)** and **ribosomal (rRNA).**

Protein synthesis

All physiological functions are dependent upon the ability of the body to make **proteins**. A protein is made up of many amino acids linked by peptide bonds to form **polypeptide chains**. We need to make many types of proteins: structural proteins, certain hormones and the multitude of enzymes required for controlling biochemical reactions. The sequence of nitrogenous bases in the DNA molecule is in fact a **genetic code** which is read and decoded by RNA. Three bases in the sequence form a **triplet** which codes for one amino acid in the new polypeptide chain, e.g. AAA codes for the amino acid phenylalanine. Several triplets arranged in a specific sequence in a gene will code for one complete polypeptide chain. The actual number of triplets needed will depend upon the length of the polypeptide chain (the size of the protein) to be made. Proteins are made by a complex process which occurs in the nucleus, the cytoplasm and inside the ribosomes. It has two distinct stages, **transcription** (the transfer of genetic information from DNA to RNA) and **translation** (the synthesis of proteins from amino acids within the ribosomes), and involves DNA and the three types of RNA. *Figure 1.5* illustrates the addition of the amino acid serine as part of a polypeptide chain.

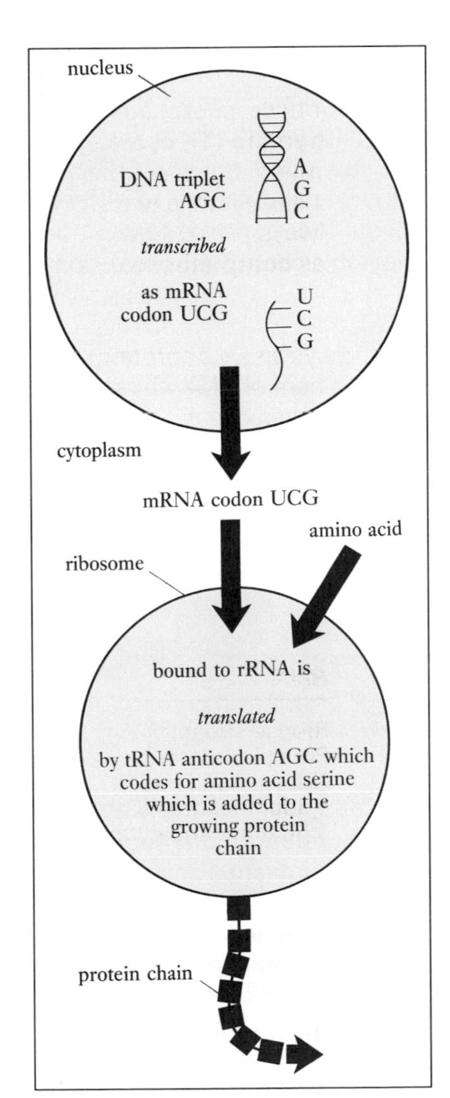

Figure 1.5. Transcription and translation.

Transcription

The first stage of transcription is the encoding of the DNA base sequence onto a molecule of mRNA. This is formed inside the nucleus by using one strand of the DNA as a pattern. The information from each DNA triplet is then carried as a **codon** of three complementary bases on the mRNA (remember the bases are different in RNA – U instead of T). By this process the genetic code has been transcribed. There is a total of 64 different codons as there are 64 different combinations of the four bases possible. Each codon either codes for an amino acid, switches on the process, or acts to turn off the process when the protein is complete (**terminal codon**). Once transcription is complete the mRNA leaves the nucleus and migrates through the cytoplasm to the ribosomes where it binds to rRNA ready for the next stage to start.

Translation

In this stage the tRNA is able to *translate* the base sequence in the codon of the mRNA into the amino acid required to build the protein. It is the **anticodons** (three bases) of tRNA which recognize the required amino acid, from the codon on the mRNA, and transfer this amino acid from the cytoplasm to the ribosome. Here the amino acid is added to the growing chain which will form the protein. The whole process is summarized in *Figure 1.5.*

The exact control mechanisms of protein synthesis are not fully understood but obviously the consequences of any small change in the process, such as the order of the bases, can have serious implications for the functioning of the individual. These changes, known as **mutations**, cause the production of abnormal proteins and result in genetic diseases. An example of a genetic disease involving an abnormal protein is sickle cell disease (see Chapter 9). Here an abnormal amino acid sequence in the protein part of the haemoglobin molecule results in defective erythrocytes.

Cell Development

Cell differentiation

The body has many different types of cells, each with a characteristic structure and well-defined function. When you think about the very start of life from only two cells, the ovum and spermatozoon, it becomes clear that some changes must occur in cells apart from the simple growth and division described in the cell cycle. The process by which a cell becomes more specialized and complex structurally and functionally is known as **differentiation**. *Figure 1.6* illustrates the structural and hence functional changes that occur as an erythrocyte matures. A cell that has become differentiated also loses the ability, that embyronic cells have, to develop into other cell types. The fact that cells with identical genetic material can become so different structurally and functionally, suggests that gene expression and protein synthesis are subject to considerable control and regulation at a cellular level.

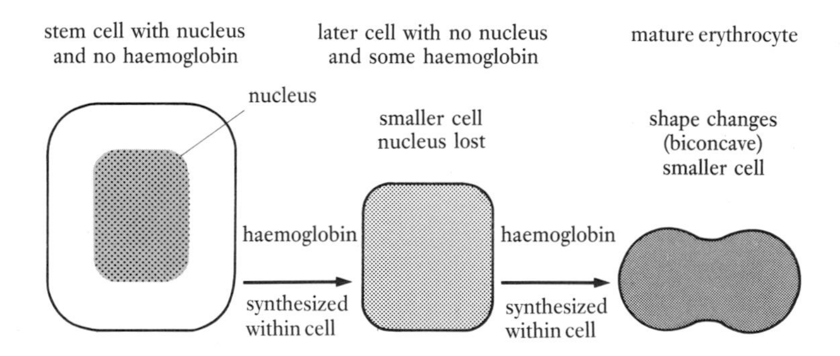

Figure 1.6. An example of differentiation – red blood cell differentiation.

Cell ageing

Cells, like the complete organism they form, will eventually age. The exact cause of this decline is not fully understood. Theories put forward include: (i) a failure of DNA repair mechanisms or simply running out of DNA as we grow older; (ii) repeated injuries to cells caused by toxins and chemicals, radiation, or lack of oxygen and nutrients over a period of many years; (iii) a breakdown or change in the immunological responses which prevent cell damage. It is probable that all three contribute to the process of cell ageing.

Abnormalities of cell growth

As with any complex process the scope for error is ever present. Amazingly, protein synthesis, and growth and division in human cells are generally accurate processes. When errors do occur they range from a failure of growth to a breakdown in control mechanisms which allows uncoordinated cell growth and division. This results in the development of malignant **tumours** (mass of tissues). Various cellular changes occur between the two extremes of no growth (**aplasia**) and tumour formation (**neoplasia**), including: atrophy, hypertrophy, hyperplasia, metaplasia and dysplasia. All of these are summarized in *Figure 1.7.*

Aplasia

Aplasia means 'absence of growth' and applies to a situation where an organ or structure fails to develop during uterine life. This particular abnormality of cell growth affects paired structures: only one of the pair develops, e.g. kidney. A less severe form is **hypoplasia** where the

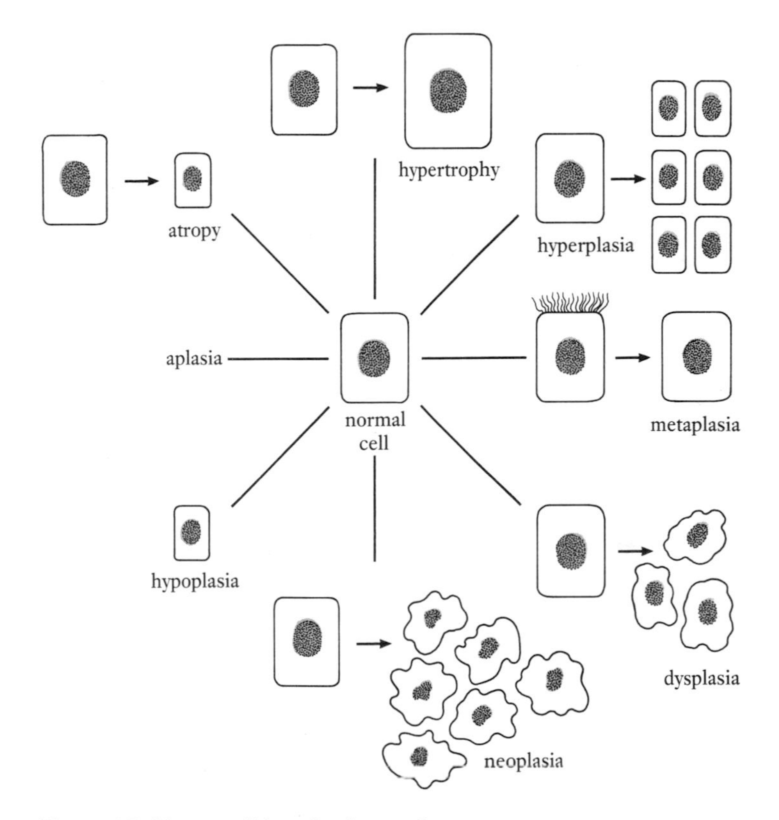

Figure 1.7. Abnormalities of cell growth.

development of an organ is incomplete resulting in a smaller than normal structure with consequent loss of function.

Atrophy

Atrophy is an acquired change that occurs when a previously normal structure becomes wasted and smaller. Causes of atrophy include:

- Normal deterioration, e.g. during the climacteric, period of time during which female reproductive function declines, (usually occurs between 45–55 years of age) when the ovaries become smaller.

- Abnormal deterioration, e.g. when a structure is starved of nutrients, its blood supply is impaired, it is subjected to constant pressure or after long-term disuse. This is of particular importance in nursing practice. People confined to bed due to illness or injury and those immobile for other reasons, e.g. severe arthritis, show evidence of muscle mass loss (see Chapter 17).

Hypertrophy

With **hypertrophy**, individual cells increase in size, resulting in an overall enlargement of the organ. This may occur where use is increased, e.g. the leg muscles in a runner. It can happen as part of the compensation mechanisms by which the body tries to minimize the effects of declining function, e.g. the myocardium (heart muscle) when the heart is failing.

Hyperplasia

An increase in actual cell numbers is known as **hyperplasia**. An example is the bone marrow hyperplasia which increases erythrocyte production in anaemia.

Metaplasia

Metaplasia is when there is a change in cell type, usually with the cells becoming less specialized. This is a reverse of the differentiation process and may be seen in areas such as the cervix or gallbladder. It

Table 1.2. Comparison of benign and malignant tumours.

	Benign	***Malignant***
Structure	Resembles its tissue of origin – well differentiated	May show any degree of differentiation; well to completely undifferentiated
Growth rate	Usually slow	Variable but usually rapid and uncoordinated; some malignant tumours are very slow growing
Enclosing capsule	Yes	No
Spread	No	Yes – locally by infiltration of tissues and by metastases which spread to other sites via blood, lymphatics, by 'seeding' along a natural channel and across body cavities
Effects	Pressure on a vital organ, e.g. brain. Production of hormones. Complications such as infection and bleeding. May become malignant.	Destroy vital organs. Cause weight loss and debility (**cachexia**) and eventual death in the absence of effective treatment. Some malignant tumours respond very well to treatment. (*See also* effects of benign tumours.)

Special Focus Cancer

Predisposing Factors
Geographical and environmental – e.g. breast cancer is uncommon in Japan but common in Europe and North America, even in Japanese women living in these areas.
Genetic – certain cancers show an increased incidence within some families, e.g. colonic polyps.
Chronic physical irritation – e.g. bladder cancer may be associated with urinary calculi.
Chemicals – there are numerous examples of carcinogenic chemicals: those contained in cigarette smoke (lung cancer), asbestos (lung cancer and mesothelioma), aniline dyes (bladder cancer), vinyl chloride (liver cancer and possibly lung and brain cancer).
Infective agents – viruses are implicated as the causative agent in several animal cancers. In humans there is a link between the papilloma virus (HPV) and some types of cervical cancer.
Radiation – excess exposure to ultra violet rays from sunlight causes skin cancers. Ionizing radiation causes an increased incidence where environmental levels increase, e.g. following accidental leaks from nuclear plants.

Reducing the risks – including measures which may be realistic
Reduce tobacco smoking.
Modify diet – reduce fat and increase fibre. Adequate intake of vitamin A may be protective against cancer.
Limit alcohol consumption – the recommended limit is 20 units/week for males, 13 units/week for non-pregnant females (see Chapter 14).
Avoid exposure to carcinogenic chemicals - in the workplace, follow safety procedures, e.g. wear protective clothing, and develop safer materials.
Practice safe sex – limit number of partners, consider using a barrier method of contraception, e.g. cap or condom. Think carefully about long-term use of oral contraceptive and use of hormone replacement therapy (current evidence to date is conflicting).

Common cancers (UK)
NB. Cancer causes nearly 25% (1 in 4) of deaths in the UK (HEA, 1988).
Lung/bronchus, gastro-intestinal tract (oesophagus, stomach, pancreas, bowel and rectum), breast, ovary, cervix and uterus, prostate and bladder, skin, leukaemia.

Early detection - what you can do
Be aware of and report early warning signs, such as a change in a skin mole or blood in the urine.
Practice breast awareness and testicular self-examination (see Chapter 20).
Have routine screenings and specialized tests – cervical smear tests, mammography (an X-ray of the breast), ultrasound and CAT scans to detect ovarian cancer (see Chapter 20). Monitoring of body fluids for cancer cells, e.g. urine for workers in some 'high risk' industries. Faecal examination for occult blood, (rectal and colonic cancers).

is due to longstanding irritation or infection. If the cause is removed the cells return to normal. If the cause remains, further more serious cell changes can occur, e.g. the bronchial epithelium in smokers.

Dysplasia

Dysplasia is a change in the size and shape of the cells that form the covering and lining tissues of the body (epithelial tissue, see pages 18-19). It results from chronic irritation and commonly affects the skin, cervix and oesophagus. These serious changes can lead to the development of a malignant tumour. Interestingly even at this stage spontaneous reversal can occur.

Neoplasia

Neoplasia is new growth that is characterized by very marked cell changes. This includes changes in DNA structure and abnormal, uncoordinated cell division which gives rise to a tumour. Tumours are usually classified according to the type of tissues from which they develop. They may be benign or malignant:

- Benign localised tumours which may cause problems from pressure or hormone production.

- Malignant tumours which are invasive both locally and by their ability to **metastasize**, that is spread to distant parts of the body via the blood or lymphatics (see *Table 1.2*). Cancer is the lay term used to describe any malignant condition.

Tumour formation and defence

Tumour cells develop as various DNA changes occur in the cells exposed to some **carcinogenic** (cancer-causing) agent. Sometimes these changes result in a benign tumour which may become malignant if an **oncogene** (cancer gene) already present in body cells becomes activated. At all stages the defence mechanisms of the body are attempting to prevent the formation of cancer cells by destroying carcinogens and repairing damaged DNA. Should this fail and a cell becomes malignant it can still be removed by specialized cells of the immune system. It is only in situations when these defences are completely overwhelmed that a malignant tumour develops. Malignant tumours occur most commonly in older people, which supports the immunological theory of cell ageing already described.

TISSUES

A tissue consists of a number of identical cells and the extracellular material or matrix which binds them together. The cells of a tissue have the same structure and perform the same functions in different parts of the body. The study of tissues is called **histology**. When techniques involve looking for abnormalities, such as the examination of **biopsies**, (when tissue samples are examined), it is known as **histopathology**.

Tissue origin and classification

The embryo has three basic types of tissue: **ectoderm**, **mesoderm** and **endoderm**. These are known as the *primary germ layers*. It is from these that the four groups of tissue (epithelium, connective, nervous and muscle) develop during early embryonic life.

Ectoderm —> some epithelium, nervous tissue.
Mesoderm —> some epithelium, muscle and connective tissue.
Endoderm —> some epithelium.

Epithelial tissues

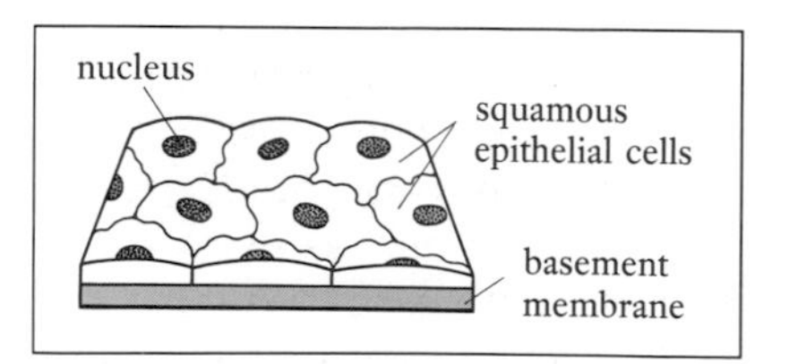

Figure 1.8. Squamous epithelium.

Epithelial tissues cover the body, line body cavities and form glands. The cells which make up epithelial tissue are close together and usually lie on a basement membrane. There is little extracellular material. Each type has developed to meet its basic function with further individual adaptations for special areas.

Simple epithelium
Epithelium may be **simple** consisting of a single cell layer. These cells may be squamous, cuboidal or columnar.

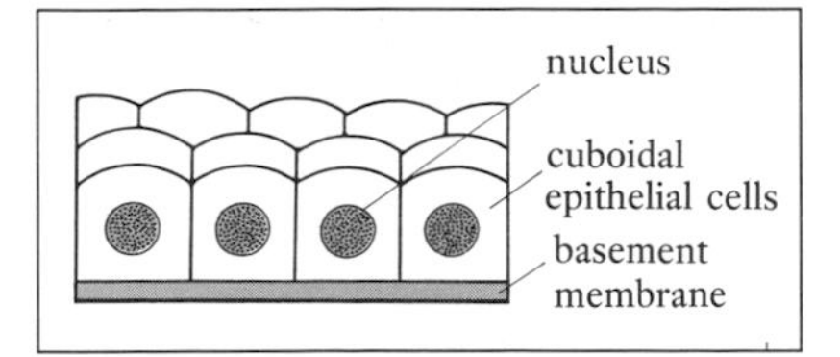

Figure 1.9. Cuboidal epithelium.

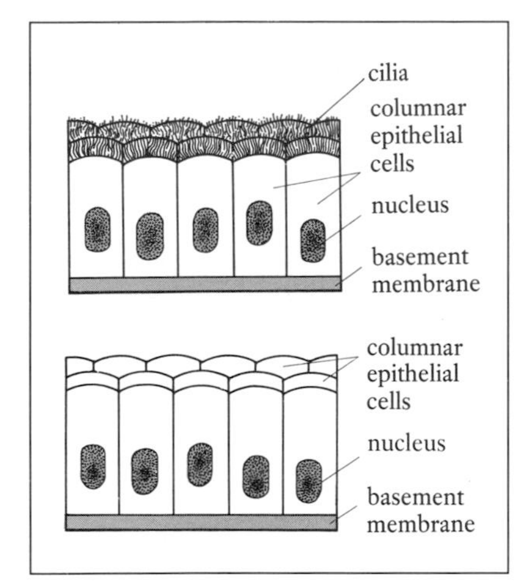

Figure 1.10. Columnar epithelium (*above* ciliated; *below* non–ciliated).

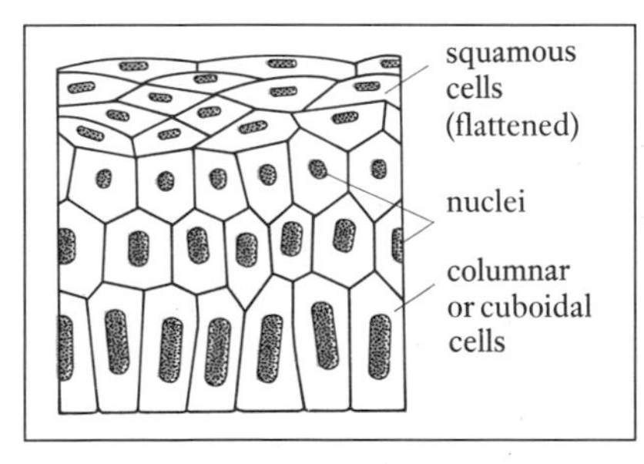

Figure 1.11. Stratified squamous epithelium.

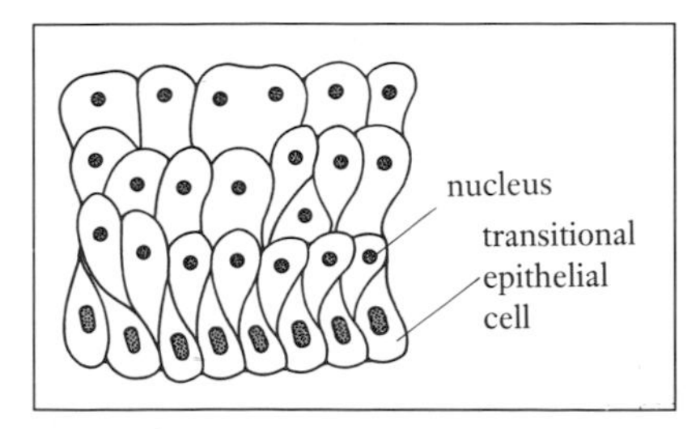

Figure 1.12. Transitional epithelium.

Squamous epithelium *(Figure 1.8)* – a single layer of flat cells found as a smooth lining in the heart and blood vessels (smooth tissue essential to prevent blood clotting), in the lymph vessels, in the glomeruli of the kidney and in the alveoli of the lungs. The single layer of fragile cells allows easy transfer of substances, e.g. by diffusion of gases between blood and alveoli, and by filtration in the glomeruli.

Cuboidal epithelium *(Figure 1.9)* is also a single layer but here the cells are cube shaped. This type of tissue is found in the kidney tubules where it is adapted for absorption (with many tiny projections called **microvilli** on its surface) and secretion. Cuboidal epithelium is also found in small glands and their ducts.

Columnar epithelium *(Figure 1.10)* – has cells which are tall. Many have microvilli, which increases the surface area available for absorption, and have **goblet cells** which produce mucus (goblet cells are illustrated in *Figure 12.3*). Columnar epithelium is found in the gastrointestinal tract and gallbladder, again concerned with absorption and secretion. Some types found in the respiratory tract and uterine tubes have cilia upon their surfaces. The ciliated surface of these delicate, highly specialized cells helps move mucus or reproductive cells.

Stratified epithelium

Stratified epithelium consists of many cell layers. These cells may be squamous, cuboidal and columnar. In addition there is the specialised transitional epithelium.

Stratified squamous epithelium *(Figure 1.11)* is composed of many layers of dividing columnar or cuboidal cells which flatten as they move to the surface. They are protective and are found as skin in dry areas where the outer layer of cells are keratinized (**keratin** is a tough fibrous protein) and dead. In moist areas, which are subject to wear and tear, a non-keratinized variety protects the mouth, oesophagus, pharynx, anus and vagina.

Transitional epithelium *(Figure 1.12)* is a type of epithelium found only in the urinary tract, where it lines the renal pelvis, ureter, bladder and urethra. Structurally it has cuboidal, columnar and dome-like surface cells which allow distension when the urinary structures fill with urine. Its surface cells are also able to withstand the effects of contact with urine.

Glands

Some epithelial tissues form specialized structures known as **glands.** These are groups of cells adapted to the production and secretion of various water-based secretions. Glands which secrete directly into the blood or lymph are known as **endocrine** glands and are ductless. With endocrine glands, secretions known as **hormones** are discharged directly into the extracellular spaces to enter the blood or lymph, e.g. thyroid gland (see Chapter 8). **Exocrine** glands secrete their products into ducts. Exocrine gland secretions leave via a duct into a body cavity or are discharged directly onto the surface skin, e.g. liver, mucous glands and sweat. Structurally, exocrine glands may

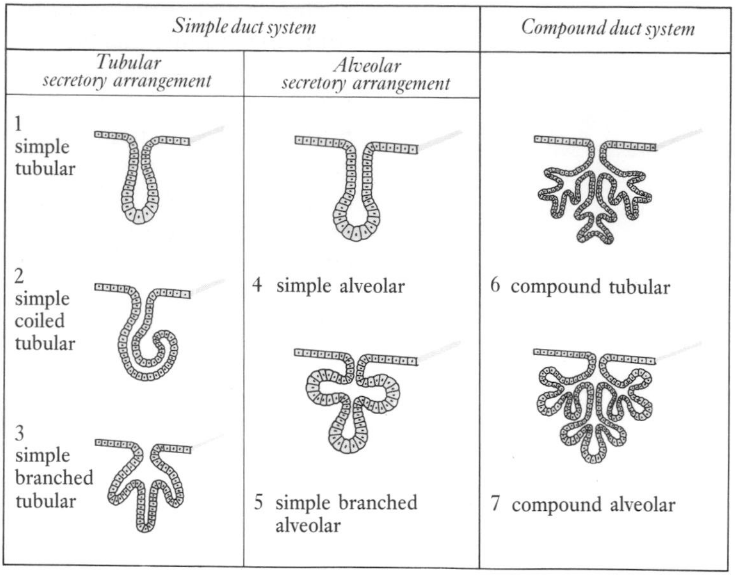

Simple duct system		*Compound duct system*
Tubular secretory arrangement	*Alveolar secretory arrangement*	
1 simple tubular	4 simple alveolar	6 compound tubular
2 simple coiled tubular	5 simple branched alveolar	7 compound alveolar
3 simple branched tubular		

Figure 1.13. Types of exocrine glands.

have a simple or compound duct system with a tubular or alveolar secretory arrangement (*Figure 1.13*). These secretions leave by endocytosis, cell rupture or a partial disruption where the portion containing the secretion is 'nipped off'.

Membranes

Specialized secretory sheets of epithelial tissue are called **membranes**. They cover organs, or line organs and body cavities. Here they provide support, reduce friction, provide nourishment and offer some protective functions. There are three types of membranes:

- **Mucous membranes** are found in the respiratory, genito-urinary and gastro-intestinal tracts and produce a sticky viscous fluid called mucus.
- **Serous membranes** are double layer membranes which line the major cavities of the body (*parietal layer*) and cover some organs (*visceral layer*). They form the *peritoneum* in the abdominal cavity, the *pleura* in the thoracic cavity, and the *pericardium* around the heart (see Chapters 13,12,10 respectively). Serous fluid produced between the two layers reduces friction during movement.
- **Synovial membrane** is found lining the cavity of freely movable joints. The fluid secreted by this membrane is **synovial** fluid, which acts as a lubricant.

Connective tissues

Connective tissues include a great variety of tissues. These range from loose connective tissue (adipose tissue), to supportive connective tissue (bone), to blood-forming tissue. Connective tissues are very widespread in the body and are concerned with much more than connecting and support. Connective tissue around organs offers protection and insulation. In the form of fat it provides an energy source. It is also involved with body protection as reticular tissue and with transportation as blood.

The hallmark of connective tissue is cells surrounded by a matrix which may be rigid (bone), gel-like (areolar) or fluid (blood). Fibres are also present as a network within the matrix. These special proteins may be collagen, elastic or reticular. Connective tissues are composed of the following cell types: *fibroblasts*, fat cells, *mast cells (inflammatory response)*, leucocytes, *plasma cells (immune response)* and *macrophages*. These cells are not all found in every type of connective tissue e.g. fat cells are primarily found in adipose connective tissue. The individual cells are discussed again in the relevant chapters, e.g. plasma cells (Chapter 19), but it might prove useful here to expand on the macrophage and its general importance in defence strategies. This large phagocytic cell is found fixed in certain organs and as a freely mobile cell in other areas. Apart from connective tissues they are seen in the liver as Kupffer cells (see Chapter 14), spleen, bone marrow, other lymphoid structures and as microglial cells (see Chapter 3) in the brain. Macrophages are also part of the wider immune processes involving other cell types.

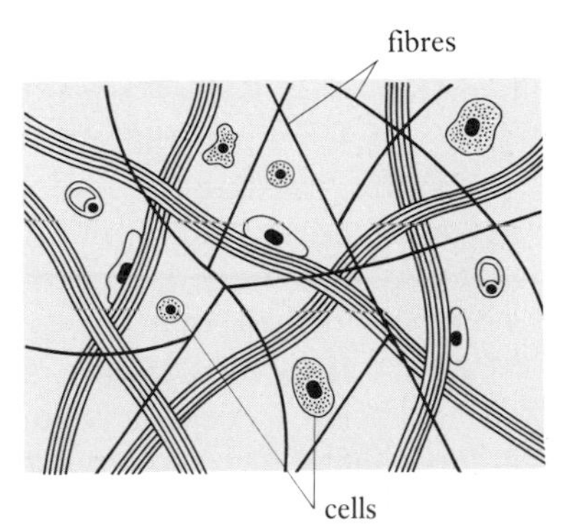

Figure 1.14. Areolar tissue.

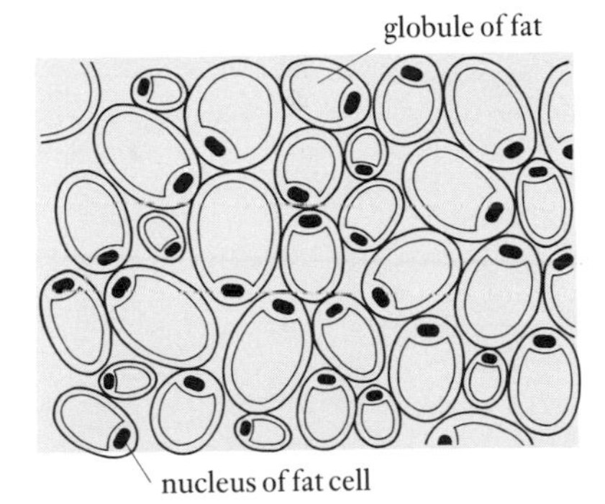

Figure 1.15. Adipose tissue.

Loose connective tissue

Areolar tissue *(Figure 1.14)* is a loose woven tissue with a semi-solid matrix containing collagen, elastic and reticular fibres. The most prominent cells are fibroblasts which are important in tissue repair. Also present are fat cells, mast cells and macrophages. Areolar tissue is found supporting vessels and nerves, around muscles, as part of glands and as the *subcutaneous* (under the skin) tissues. This open structure allows for considerable accumulation of water and salts. In certain situations this leads to tissue swelling (oedema, see Chapter 10).

Adipose tissue *(Figure 1.15)* consists of many fat cells in a basic areolar matrix. It is present in varying quantities, to some extent genetically determined, under the skin, around organs such as the kidneys, and between muscle fibres. Apart from support and insulation it provides the body with a valuable source of fuel.

Reticular tissue (*Figure 1.16*) consists of a network of reticular fibres. The matrix is loose and contains **reticular cells** (primitive cells

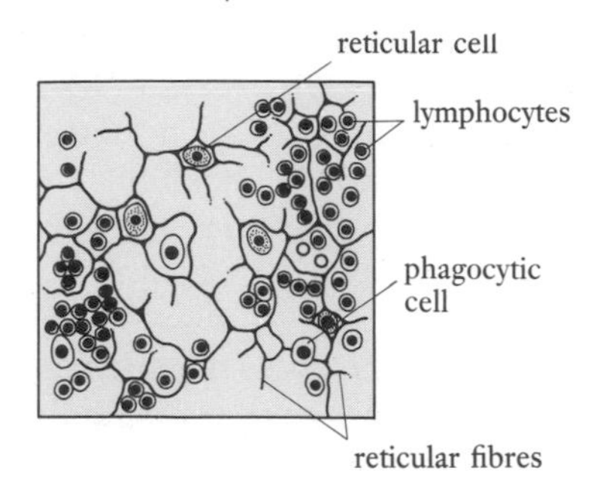

Figure 1.16. Reticular tissue.

similar to embryonic connective tissue), lymphocytes and phagocytic cells. It is found in the spleen, liver, lymph nodes and bone marrow where it provides support and has a role in protecting the body.

Dense connective tissue

The two types of dense connective tissue are fibrous and elastic tissue. **Fibrous tissue** *(Figure 1.17)* has regular and irregular fibres – it is a much tougher tissue consisting of mainly collagen fibres and a few fibroblasts within a scanty matrix. It forms ligaments, tendons, and the covering for bones and organs, such as the brain. This tissue is protective and assists with movement and stability by joining bones together and providing muscle attachments.

Elastic tissue *(Figure 1.18)*, as its name suggests, is a tough tissue with the ability to stretch and recoil due to the presence of elastic fibres within the matrix. It is an important component in structures required to distend and change shape: large arteries, epiglottis, vocal cords, trachea and the ear lobe.

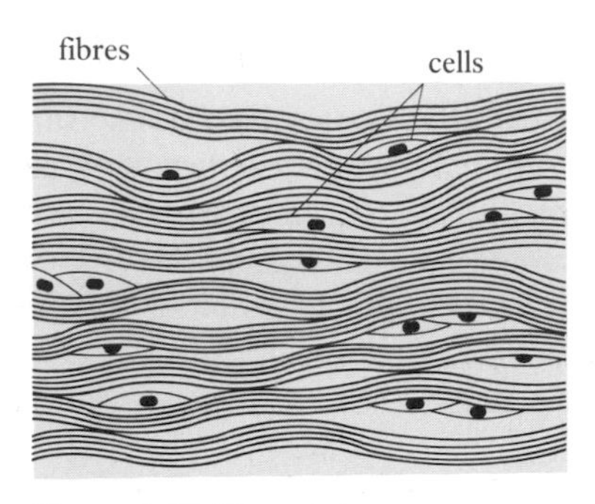

Figure 1.17. Fibrous tissue.

Cartilage

Cartilage has a solid matrix containing **chondrocytes** (cartilage cells) and collagen fibres which make the tissue firm and tough. The three types – hyaline, white fibrocartilage and elastic yellow cartilage (*Figure 1.19*) – perform quite different functions.

Hyaline cartilage, a smooth, shiny tissue found in the embryonic skeleton, covers the ends of long bones in joints and forms the cartilage of the nose, larynx, trachea, bronchi and the costal cartilages which join the ribs to the sternum.

White fibrocartilage is a strong flexible tissue able to absorb compressive forces. It is found as pads between the vertebrae (intervertebral discs), the pubic bones of the pelvis (symphysis pubis) and as the semilunar cartilages of the knee joint.

Yellow elastic cartilage helps structures maintain their shape by its great flexibility. It is found in the *pinna* of the ear (see Chapter 7) and in the *epiglottis* (see Chapter 12).

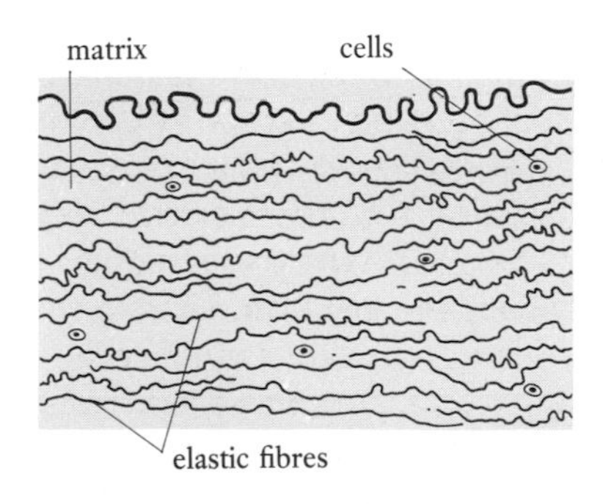

Figure 1.18. Elastic tissue.

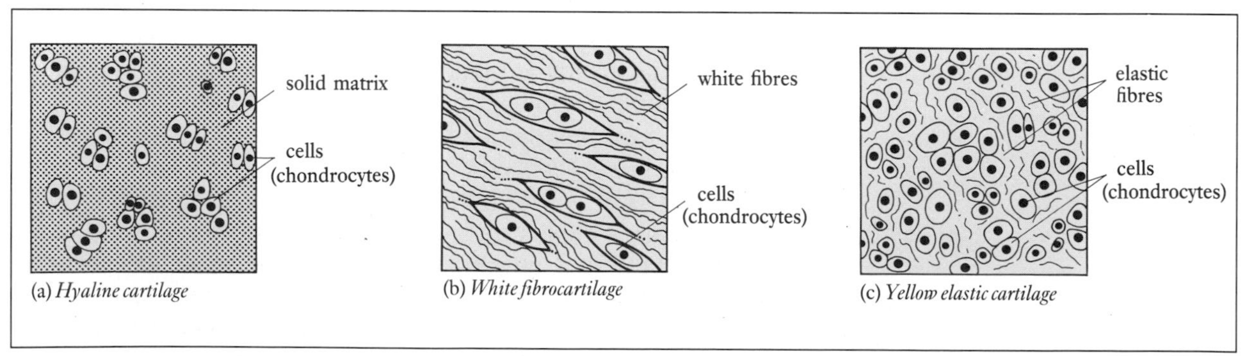

Figure 1.19. Cartilage.

Bone
There are two types of **bone** tissue, **compact** and **cancellous** (spongy) which are discussed in Chapter 16.

Blood and blood-forming tissues
These tissues are discussed in detail in Chapter 9.

Nervous tissue

Nervous tissue can be divided into: **neurones** (excitable cells which transmit the nerve impulse) and the non–excitable **neuroglia**, cells which support, insulate and protect. (See Chapter 3.)

Muscle tissue

There are three types of muscle tissue: **striated** which is voluntary (skeletal), **non-striated** which is involuntary (smooth), and **cardiac** (heart) (see Chapters 10,17).

Tissue repair

Physical or chemical injury to the body will initiate the inflammatory response and more specific immune reactions within the tissues (see Chapter 19). The way in which healing occurs after tissue injury will depend on the type of tissue and the degree of damage involved. This can be either: (i) *Regeneration* – when the tissue is repaired by the proliferation of identical cells to those lost. This occurs in relatively undifferentiated tissue, e.g. bone. (ii) *Fibrosis* – here the repair is effected by the production of fibrous tissue (scar) in place of the damaged tissue. This occurs in differentiated tissues, e.g. cardiac muscle. In reality most healing occurs through a combination of both processes.

Highly specialized tissue such as the nerve cells of the central nervous system and cardiac muscle cannot regenerate. N.B. It is sometimes possible for a nerve fibre to regenerate if the nerve cell body is intact. Obviously any serious damage to these tissues will result in loss of functional ability. On the other hand, blood cells are continuously regenerated and epithelial cells, e.g. those lining the gastro-intestinal tract, continually wear away and regenerate to maintain tissue integrity. Between these two extremes are tissues, such as liver and bone, that can regenerate, but do so more slowly. Healing also depends on many other factors such as tissue nutrition, presence of infection and age. Healing ability slows and becomes less efficient with increasing age. The question of healing is addressed much more fully in Chapter 19 where it is applied to nursing practice in a discussion of wound care and healing.

ORGANS

Tissues grouping together eventually become discrete functional units called **organs**. These may be hollow, e.g. stomach, or compact, e.g. liver. Several types of tissue may be represented in one organ; e.g. the stomach has an outer connective tissue and epithelium layer, middle muscle layer and a mucous membrane lining (epithelium). Compact organs consist of functional cells (**parenchymal**) supported by connective tissue and surrounded by a tough capsule.

Body Cavities

The body can be divided into cavities defined by the bony skeleton and muscles. The four body cavities are the **cranial, thoracic, abdominal** and **pelvic** cavities. Many of the tissues and organs which constitute the **functional systems** of the body are grouped within these cavities (e.g. *respiratory, digestive, nervous* and *reproductive structures*).

N.B. Not all the structures or organs mentioned in the text are shown in the illustrations (*Figure 1.20–1.24*). Some are situated behind other structures and some are too small to be shown with any clarity.

The cranial cavity

The cranial cavity is formed by the bones of the cranium which is the upper part of the skull. The bony box formed by the cranium surrounds and protects the brain. *Figure 1.20* shows the bones of the

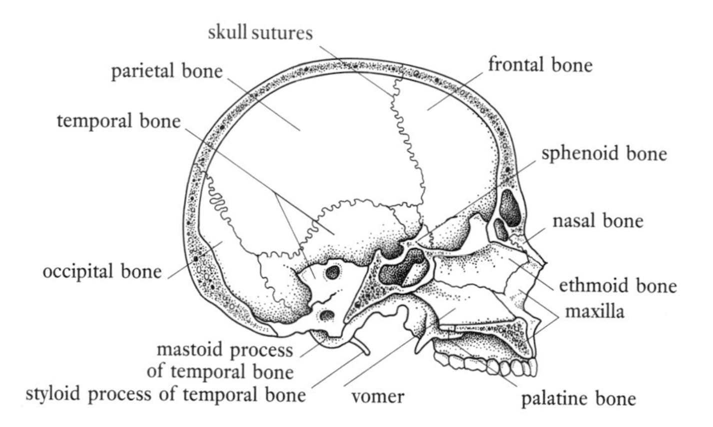

Figure 1.20. Cranial cavity.

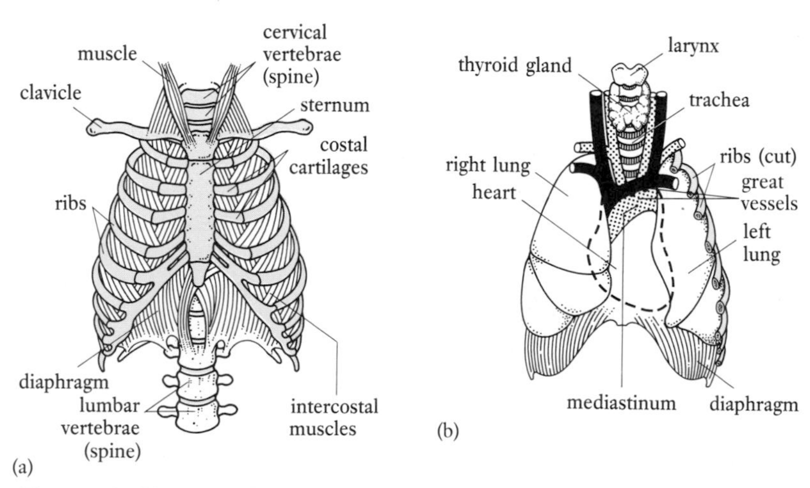

Figure 1.21. Thorax (a) External structures; (b) internal structures.

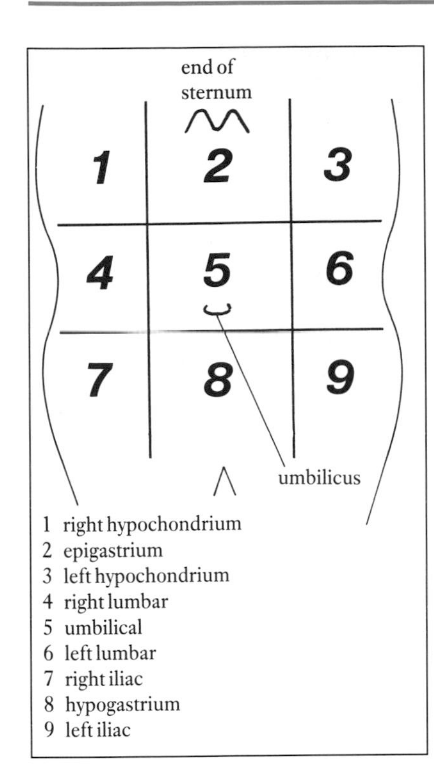

Figure 1.23. Regions of the abdomen.

cranium, (*frontal, parietal–2, temporal–2, occipital, sphenoid* and *ethmoid*) joined by fibrous joints called **sutures**. Also illustrated are some bones of the face (*nasal, maxilla, palatine, vomer*). All these bones are covered in detail in Chapter 18.

The thoracic cavity (thorax)

The cavity forming the upper portion of the trunk is known as the thoracic cavity (*Figure 1.21*). It is bounded by the structures of the root of neck, ribs, costal cartilages, intercostal muscles, sternum, spine and diaphragm (muscle which divides the thorax from the abdominal cavity). The thoracic cavity contains the trachea, two bronchi (Chapter 12), two lungs, heart and great vessels, the oesophagus (Chapter 13), nerves and lymphatics. The space between the lungs, occupied by the heart, is called the *mediastinum.*

The abdominal cavity

The large lower portion of the trunk is the abdominal cavity *(Figure 1.22)*. It is bounded by the spine, abdominal muscles (Chapter 18), lower ribs, diaphragm and the pelvic cavity below. Abdominal structures include the organs involved with digesting and absorbing nutrients, and the liver, kidneys and ureters, spleen, adrenal glands and associated blood vessels, nerves and lymphatics. *Figure 1.22(a)* shows the abdominal organs viewed from the front. The stomach, spleen, small bowel, caecum, appendix, colon, hepatic and splenic flexures, liver and gallbladder are shown. The *omentum* (fold of peritoneum) has been removed for clarity. *Figure 1.22(b)* illustrates the abdominal organs visible when most of the digestive organs have

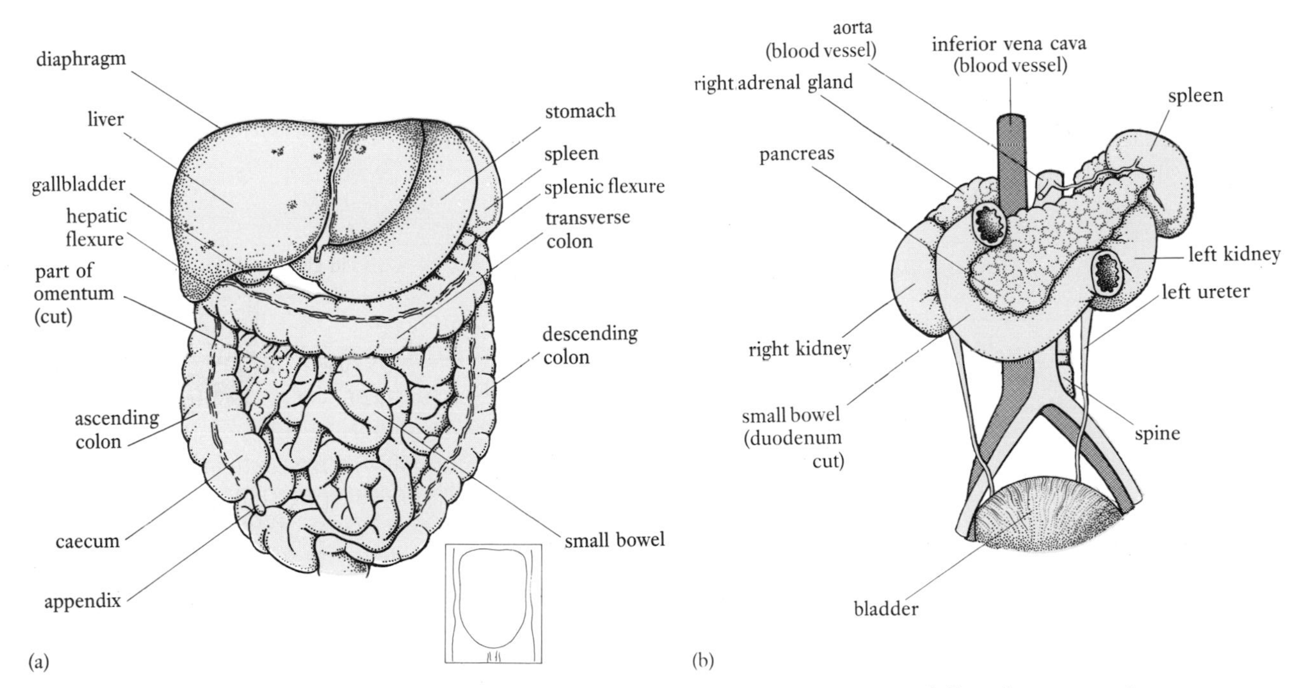

Figure 1.22. Abdominal cavity. (a) Anterior structures; (b) view with most of the organs of digestion removed.

been removed. The surface anatomy of the abdomen is divided into nine regions (*Figure 1.23*) a useful tool when describing the location of the various organs or of pain.

The pelvic cavity

The lowest portion of the trunk is the pelvic cavity (*Figure 1.24*). This is a continuation of the abdominal cavity. It is bounded by the bony pelvis, sacrum and muscles of the pelvic floor (Chapter 18), and contains openings for the urethra, vagina and anus. The pelvic cavity contains the female reproductive structures (ovaries, uterine tubes, uterus and vagina), some of the male reproductive structures (vas deferens, prostate gland, seminal vesicle but not the penis, testes and scrotum), the lower ureters, bladder, and urethra. Other pelvic organs are some coils of small bowel, the last part of the colon, rectum and anal canal.

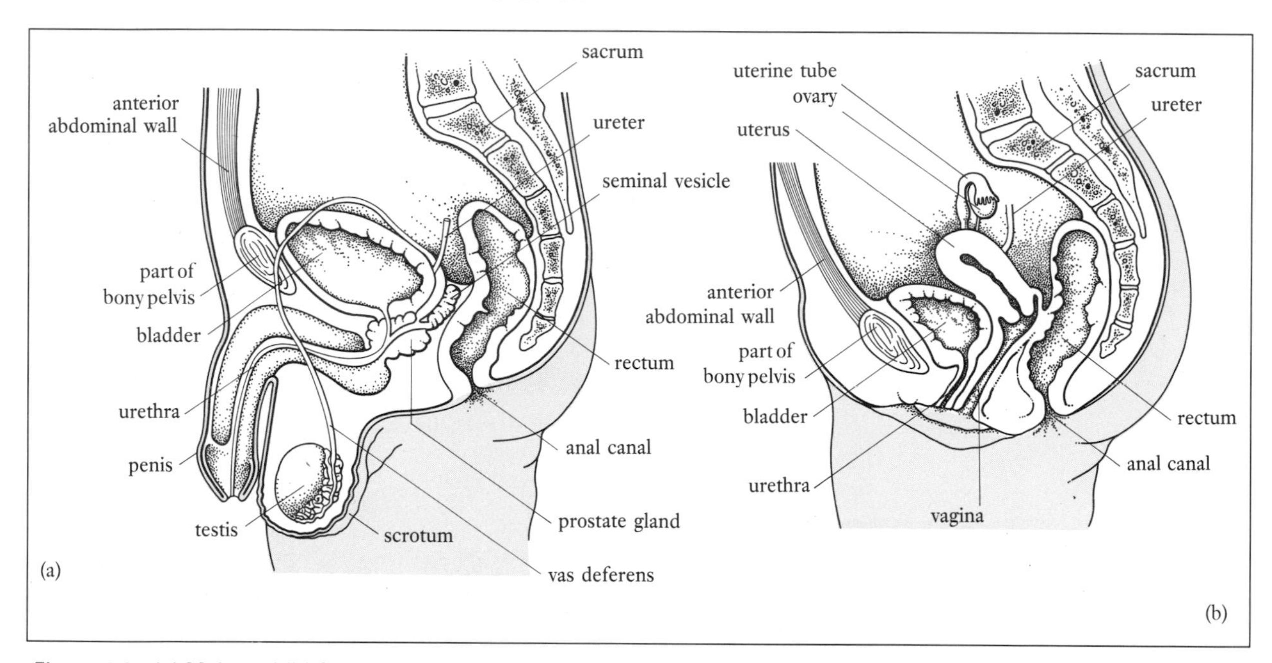

Figure 1.24. (a) Male and (b) female pelvis.

Anatomical Terminology

Anatomy is the study of body structures. A set of standard terms, accepted everywhere, is used to describe the position of body structures and their geographical relationships with each other. This ensures that the result will be both concise and understood.

The anatomical position and regional terms

The **anatomical position** of the body is used as a reference point when studying or describing the position of body structures. The person stands erect, faces forward, with arms at his or her side with palms uppermost. *Figure 1.25* shows the anatomical position. Each region of the body can be described using a specific term and many of

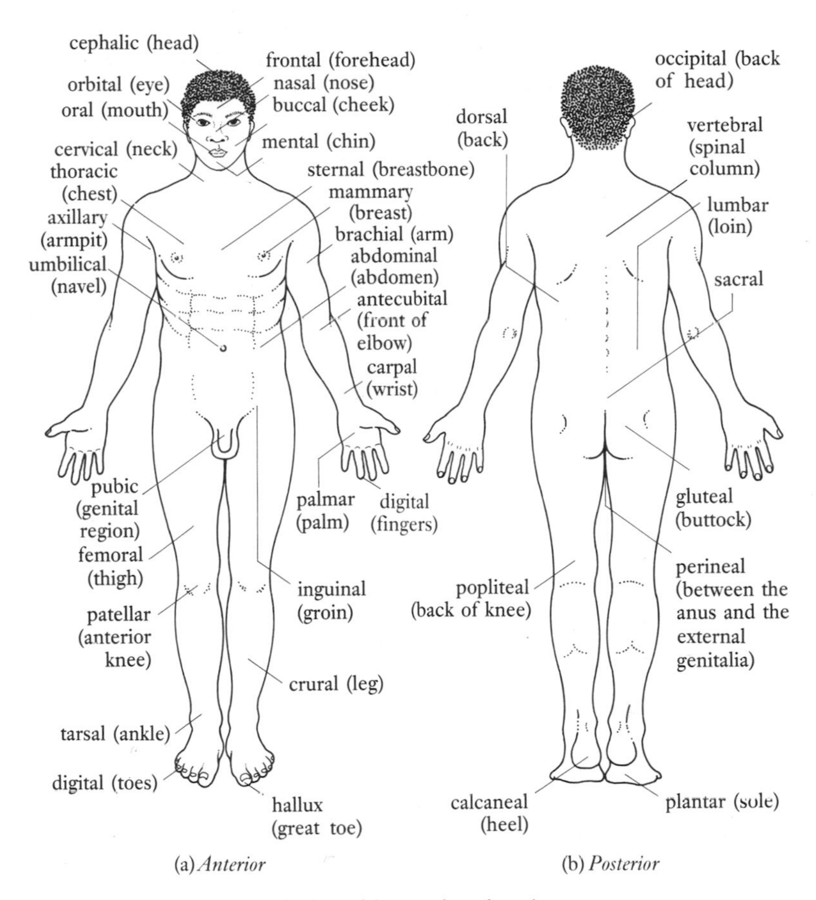

Figure 1.25. The anatomical position and regional terms.

these **regional terms** are illustrated in *Figure 1.25.*

Body planes

Body structures can be described in relation to three **planes** (imaginary lines) – **median** (midsagittal), **coronal** and **transverse** – which run through the body (*Figure 1.26*).

Directional terms

- *Superior* – above.
- *Inferior* – below.
- *Anterior (ventral)* – in front.
- *Posterior (dorsal)* – at the back. When describing the hands, the terms *palmar* and *dorsal* are used, with *plantar* and *dorsal* for the feet.
- *Afferent* – towards, e.g. sensory nerves going to the brain.
- *Efferent* – away from, e.g. motor nerves leaving the brain.
- *Peripheral* – at the edges of the body, e.g. fingers and toes.
- *Lateral* – away from the median line (middle), on the outer side.
- *Medial* – toward the median line, on the inner side.
- *Distal* – furthest away from a given point.
- *Proximal* – nearest the given point.

- *Internal* – towards the centre/inside of a cavity,
- *External* – towards the outside of a cavity.
- *Deep* – away from the surface of the body, e.g. deep veins in the legs.
- *Superficial* – near or on the surface of the body, e.g. superficial veins in the legs.

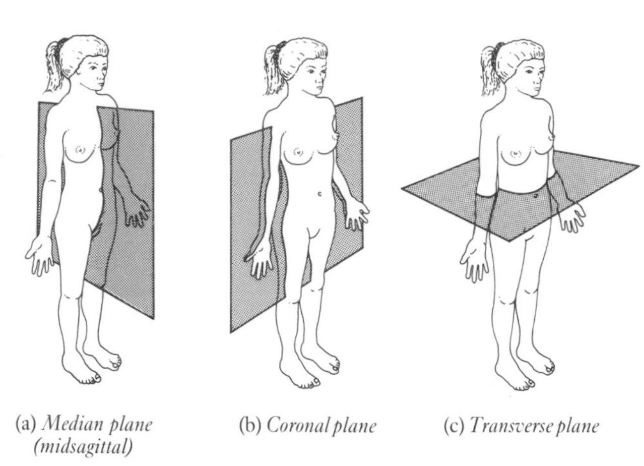

(a) *Median plane (midsagittal)* (b) *Coronal plane* (c) *Transverse plane*

Figure 1.26. Body planes.

SUMMARY/CHECK LIST

Cell structure – cytoplasm, cytoplasmic membrane, organelles, supporting structures, cilia, flagella, nucleus and related structures.
Transport across membranes – osmosis, diffusion, active transport, endocytosis.
Cell division – mitosis, meiosis, cell cycle (interphase, DNA replication, stages of mitosis and cytokinesis).
Protein synthesis and nucleic acids – nucleic acids (DNA and RNA), protein synthesis (transcription and translation).
Cell development – Cell differentiation, cell ageing, abnormalities of cell growth. Tumours – benign, malignant, formation and defence, *Special Focus* – Cancer.
Tissues – Tissue origin (endoderm, ectoderm, mesoderm). *Classification*: epithelium (simple, stratified), glands (endocrine, exocrine), membranes (mucous, serous, synovial); connective (loose, dense, cartilage, bone, blood); nervous (neurones, neuroglia); muscle (striated, smooth, cardiac). *Tissue repair* (regeneration, fibrosis).
Organs – hollow and compact.
Body cavities – cranium, thorax, abdomen, pelvis.
Anatomical terminology – position, regional terms, planes, directional terms.

?

Self Test

1 The basic structural unit of the body is:
(a) the cell organelle
(b) an organ
(c) the cell
(d) a tissue.

2 Which of the following statements are true?
(a) The simple diffusion of substances across the cytoplasmic membrane is an active process.
(b) Osmosis is concerned with the movement of water across a selectively permeable membrane.
(c) Cells such as leucocytes are able to engulf and destroy bacteria by a process known as phagocytosis.
(d) Active transport processes using carrier molecules are able to move substances against a concentration gradient.

3 Which of the following is the correct sequence of events when somatic cells divide?
(a) Interphase, mitosis, DNA replication and cytokinesis.
(b) DNA replication, mitosis, interphase and cytokinesis.
(c) Interphase and DNA replication, mitosis and cytokinesis.
(d) Mitosis, interphase, DNA replication and cytokinesis.

4 Describe four ways in which DNA and RNA differ.

5 Explain the term 'differentiation'.

6 Which of the following statements about cancer are true?
(a) Malignant tumours often spread in the body via the blood or lymph.
(b) Cancers cause about 30% of deaths in the UK.
(c) Infective agents are probably linked with the development of some human cancers.
(d) Prevention of cancer includes modifying diet and stopping smoking.

7 Which types of cell are not able to regenerate:
(a) Those of the central nervous system.
(b) Liver cells.
(c) Those lining the gastro-intestinal tract.
(d) Cardiac muscle cells.

8 What are the four basic tissues of the body? Give one example of each type.

9 Describe the boundaries and contents of the thoracic cavity and the female pelvic cavity.

10 Put the following directional terms in their correct pairs:

(a) distal	(b) afferent
(c) lateral	(d) superior
(e) efferent	(f) medial
(g) inferior	(h) proximal

Answers
1. c. 2. b, c, d. 3. c. 4. See Table 1.1. 5. See page 14. 6. a, c, d. 7. a, d. 8. Epithelial, connective, nervous and muscle. To check examples see pages 18-23. 9. See pages 25,26. 10. a-h, b-e, c-f, d-g.

References

Health Education Authority (1988) *Can You Avoid Cancer: a guide to reducing your risks.* London: Health Education Authority.

Further Reading

Alberts, B., Bray D., Lewis J., Raff M., Roberts K., Watson J,D. (1983), *Molecular Biology of The Cell* 1st edn. New York: Garland Publishing.
NB A comprehensive but complex coverage of cells and cell function.

Davies, I. (1983) *Studies in Biology Series No 151: Ageing* 1st edn. London: Edward Arnold.

Hole, C. B. (1974) *An Introduction to Cell Biology* 1st edn. London: Macmillan.
NB A general and simple account of cells and cell function.

Langelann, D. (1986) The biology of ageing: an introduction. *Geriatric Nursing and Home Care,* **6**, 2 16–19.

Rose, S. and Sanderson, C. (1979) *The Chemistry of Life,* 2nd edn. Harmondworth: Penguin Books.

Thomson, A.D. and Cotton, R.E. (1983) *Lecture Notes on Pathology,* 3rd edn. Oxford: Blackwell Scientific Publications.

Tschudin, V. (Ed). (1988) *Nursing the Patient with Cancer,* 1st edn. London: Prentice Hall.

Watson, J.D., Crick, F.H.C. (1953) Molecular structure of nucleic acid. A structure for deoxyribose nucleic acid, *Nature,* **171**, 737 – 738.

Wheater, P.R., Burkitt, H.G. and Daniels, V.G. (1979). *Functional Histology,* 1st edn. Edinburgh: Churchill Livingstone.

section two

the internal environment

CHAPTER TWO

THE INTERNAL ENVIRONMENT – BASIC CONCEPTS AND HOMEOSTASIS

OVERVIEW

- Homeostasis (what, why and how) – breakdown of homeostasis and implications for health.
- Related basic chemical concepts.
- Diffusion and osmosis (detailed).
- Fluid compartments – body fluids – electrolytes.
- Dehydration – fluid replacement.
- Hydrogen ion concentration (pH) – control mechanisms.

LEARNING OUTCOMES

After studying Chapter 2 you should be able to:

- Define homeostasis.
- Describe basic mechanisms by which homeostasis is maintained.
- Discuss the general effects on the body when these mechanisms fail.
- Discuss related basic chemical concepts: atom, element, molecule, compounds, ions, etc.
- Discuss the importance of diffusion and osmosis in maintaining the internal environment.
- Describe the fluid compartments and body fluids.
- State the important electrolytes and briefly discuss their functions in the body.
- Describe briefly how water balance is maintained.
- State how dehydration might occur and describe the effects on the body. Discuss groups at particular risk and describe how fluid balance can be restored.
- Describe the concept of pH and its regulation.

KEY WORDS

Acid – a substance which combines with alkalis to form salts. It contains an excess of hydrogen over hydroxyl ions, e.g. hydrochloric acid. Acids are hydrogen ion (proton) donors.
Alkali – a substance which combines with acids to form salts. Contains an excess of hydroxyl over hydrogen ions, e.g. sodium hydrogen carbonate (bicarbonate), sodium hydroxide. Alkalis are hydrogen ion (proton) acceptors.
Atom – the smallest stable part of an element that can exist and display the properties of that element.
Autoregulation – 'self-regulation' by mechanisms occurring within the body. Usually applied to local changes to blood flow or other physical or chemical feature which maintains optimum conditions within an organ in the event of changes to the internal or external environment.
Base – another term for alkali.
Buffers – chemicals which minimize changes in pH by donating or accepting hydrogen ions.
Electrolytes – ionic compounds that dissociate in water to form charged particles or ions. Able to conduct electricity.
Element – one of the unique substances (matter) that comprise all living and non-living things, e.g. oxygen, potassium, zinc, carbon.
Extracellular fluid – body fluid found outside the cells, e.g. blood, interstitial fluid.
Homeostasis – the maintenance of a stable physiological state by the autoregulatory processes of the body, e.g. temperature control.
Inorganic compounds – compounds which generally contain no carbon in their chemical structure, e.g. water, sodium hydroxide, sodium chloride. There are exceptions such as carbon dioxide which are classified as inorganic.
Interstitial fluid (tissue fluid) – fluid that surrounds the cells, which forms part of the extracellular fluid.
Intracellular fluid – fluid contained within the cell cytoplasm. Most body fluid is within the cells.
Ion – an atom with an electrical charge, divided into *cations,* which have a positive charge, and *anions,* which have a negative charge.
Molecule – two or more atoms joined together by a chemical bond, e.g. water.
Organic compounds – chemical compounds, such as glucose, which contain carbon in their chemical structure. The large biological molecules found in 'living' systems are organic, e.g. fats, proteins, nucleic acids, carbohydrates, steroids.
Reaction (chemical) – the energy consuming and releasing processes that change the structure of molecules. In the body they may be anabolic (building up), catabolic (breaking down) or involve the exchange of molecules, e.g. oxygen and carbon dioxide carriage by red blood cells. Some reactions can move in either direction and are said to be reversible:

$$A + B \rightleftharpoons AB$$

Solute – a substance which is dissolved in a solvent.
Solution – a liquid (solvent) containing a solid (solute) which has been dissolved in it.
Solvent – a liquid able to dissolve another substance.

INTRODUCTION

*The **fluid compartments** consist of the fluid inside the cells – intracellular and the fluid outside the cells – extracellular*

The chemical and physical internal environment of the body is determined by: the water content and its distribution within the fluid compartments, electrolyte levels, temperature, available nutrients, levels of waste products, and pH of various body fluids.

All cellular chemical reactions are dependent on the maintenance of these properties within a limited range. To this end, mechanisms operate in all body systems to maintain an optimum internal environment, thus ensuring that the complex interrelated functions of **metabolism** (all the biochemical processess occurring in the body) continue to occur. These mechanisms include:

- Excretion of carbon dioxide (CO_2) gas by the lungs. Retention of CO_2 will affect pH.
- Hydrogen carbonates (bicarbonates), phosphates and proteins, which, acting as **buffers**, are able to prevent quite small changes in blood pH. Buffers are chemicals which limit pH (acidity/alkalinity) changes (see page 48 for a fuller explanation).
- The kidneys, which help to regulate water balance, pH and electrolyte levels by their ability to excrete urine of variable composition in line with body needs.
- The skin, which is able to regulate temperature in several ways, e.g. via sweat glands.

Homeostasis

The term **homeostasis** means the maintenance of a stable but dynamic physiological state by **autoregulatory** mechanisms initiated from within the organism. For the individual cell homeostasis provides a constant internal environment in which to function. For the person it means optimal functioning of all body systems which is a good basis for health. The breakdown of homeostasis results in imbalance and disease.

Homeostatic control mechanisms

A basic homeostatic mechanism has three components: a **receptor** usually of nervous tissue, an overall **control area** and an **effector**. Communication between these three is via nerves (see Chapter 3) or endocrine structures (see Chapter 8).

The receptor samples the internal environment for changes, details of which are sent to the control area. The control area actually determines the normal value for a particular chemical or physical property; it interprets information received and then initiates an appropriate response. The last part involves the effector structures, e.g. kidney, glands, blood vessels, etc., which operate to reverse the changes. Most homeostatic mechanisms act by **negative feedback**, where an increase in the level of product from the process being monitored causes the process to be slowed or stopped.

Physiological processes regulated through receptors, control areas and effectors include temperature control, rate and depth of respiration, electrolyte levels and blood sugar. Let us consider the three components involved in temperature control:

- *Receptors– thermoreceptors* in the skin and brain (hypothalamus) monitor changes.
- *Control centre – thermoregulation centre* (in the hypothalamus) receives this information via nerves and initiates responses which will return the temperature to normal.
- *Effectors* – various structures in the skin (sweat glands, blood vessels) and skeletal muscles. If the temperature is too low (e.g. hypothermia) measures are taken to conserve or produce heat. Sweat is not produced, blood vessels constrict (**vasoconstriction**)

and muscles contract and relax to produce shivering. If temperature rises above normal (e.g. following vigorous exercise) the measures taken to cause heat loss are sweat production and dilation of blood vessels (**vasodilation**).

A few homeostatic mechanisms work through **positive feedback**, where an increase in the level of products from the process being monitored causes the process to be stimulated and increased. These usually involve reactions that occur more infrequently, e.g. normal blood clotting.

Breakdown in homeostasis

Any imbalance in homeostasis will have serious implications for the individual's ability to function normally. In the newborn the heat-regulating mechanisms are immature (Ho, 1989). This and other factors may lead to difficulties in maintaining a normal body temperature.

At the other end of the lifespan, when body cells and regulatory mechanisms are less efficient, there will be problems with homeostasis. Again this is well illustrated by lack of temperature control. In elderly people the thermoregulation mechanisms cope less well with extremes of temperature (Herbert, 1986); this may result in **hypothermia** (temperature below normal, see Chapter 19).

Most health deficiencies are linked to a breakdown in the normal homeostatic mechanisms, e.g. abnormal blood clotting or the loss of the trigger of high carbon dioxide levels for respiration in people with severe chronic lung disease.

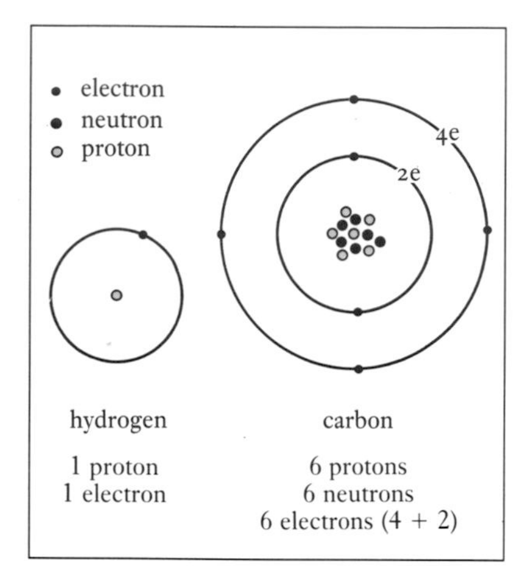

Figure 2.1. Atomic structure (diagrammatic – planetary model).

Basic Chemistry

It is appropriate here to explain some basic chemistry. You will find this helpful in understanding body processes and many practice applications. Only very basic facts have been included; these can form an introduction if science has not been studied before or as a revision of previous knowledge. The structure of proteins, carbohydrates and fats is covered in Chapter 13.

The composition of matter

All matter is made from substances known as **elements** (*Table 2.1*). There are 92 elements which occur naturally, e.g. carbon, hydrogen, oxygen and iron. Individual elements consist of tiny particles called **atoms** which display the chemical and physical properties of that element.

Elements can be identified by a chemical symbol, which is a short code that uses one or two letters, e.g. sodium is Na and hydrogen is H.

Atoms

Atoms consist of smaller *subatomic particles:* **protons** (positive charge), **neutrons** (no charge) and **electrons** (negative charge). The protons and neutrons cluster together to form the nucleus around which the electrons orbit (just as some satellites orbit the Earth),

Table 2.1. Elements of physiological importance

Element	*Symbol*	*Importance in the body*
Calcium	Ca	Forms structure of bones and teeth. Nerve impulse transmission, muscle contaction, and blood coagulation
Carbon	C	Present in all **organic** compounds, e.g. proteins, carbohydrates, lipids (fats)
Chlorine	Cl	An important negative ion (anion) in the extracellular fluid
Hydrogen	H	Concerned with body pH and as a constituent of organic compounds
Iodine	I	Required for the hormones of the thyroid gland
Iron	Fe	Constituent of haemoglobin, the oxygen-carrying pigment in the erythrocytes, and myoglobin in muscle
Magnesium	Mg	Found in bones and as a cofactor in many body reactions
Nitrogen	N	Constituent of proteins and nucleic acids
Oxygen	O	A constituent of many organic and **inorganic** compounds; needed for energy production
Phosphorus	P	As phosphates found in bone, teeth and nucleic acids
Potassium	K	An important positive ion (cation) found mostly in the intracellular fluid; required for nerve and muscle function
Sodium	Na	An important positive ion (cation) found mostly in the extracellular fluid
Sulphur	S	A constituent of proteins

In addition to these, the body needs other elements in extremely small amounts. These are known as *trace elements* and include:

Chromium Cr	Fluorine F	Selenium Se
Cobalt Co	Manganese Mn	Silicon Si
Copper Cu	Molybdenum Mo	Zinc Zn

Figure 2.1. The planetary model of atomic structure which we shall use describes electrons following fixed orbits within rings around the nucleus. More recent models describe electron clouds moving without fixed orbits within shells around the nucleus.

Atoms have equal numbers of positive protons and negative electrons, which means that overall the atom is neutral. Each element has a different number of subatomic particles within its atoms. This gives each element its unique properties: atomic number, mass number and atomic weight.

The atomic number

The **atomic number** is the number of protons within the nucleus, and the number of electrons in orbit, e.g. hydrogen, with one of each, has the atomic number 1.

The mass number

The total mass of the protons and neutrons is the **mass number**. The electrons with their extremely small mass are not significant within an atom. Rather confusingly most elements have variable mass atoms (different number of neutrons) known as **isotopes**; these have the same atomic number but a different mass number, e.g. hydrogen atoms exist in three forms (*hydrogen, deuterium* abd *tritium*).

Some elements produce heavy isotopes which are unstable and radioactive. These are known as **radioisotopes**; they consist of unstable matter which emits radiation as the nucleus disintegrates. The radiation may be of several types – e.g. alpha (α), beta (β), gamma (γ). Each radioisotope has a **half-life**, the time taken for a radioactive isotope to decay to half its original activity, e.g. radium's half-life is 1690 years. Radioisotopes are widely used in diagnostic tests, such as in certain scans and in radiotherapy, where the radiation is used to treat cancers and some other conditions (see Chapter 1).

Atomic Weight

Inorganic *– chemicals which generally contain no carbon in their structure e.g. water.*

Organic *– chemicals such as glucose which contain carbon in their structure.*

The **atomic weight** is the weight (mass number) of an average atom of a particular element; it takes account of the different isotopes of that element. Hydrogen has an atomic weight of 1.008 (the average of its three isotopes), but for our purposes it is taken to be 1. The relative unit of weight used for measuring atoms and subatomic particles is the *dalton* or *atomic mass unit (amu)*. The weight of a neutron and a proton have each been designated as being one atomic mass unit (1 amu). An example to illustrate this is sodium with its twelve neutrons and eleven protons has a mass number of 23 and hence an atomic weight of 23. The weight of other elements can be compared with that of hydrogen, e.g. oxygen with an atomic weight of 16 is sixteen times heavier than hydrogen.

Combinations: molecules and compounds

In general, atoms cannot exist separately – they combine with other atoms. A **molecule** is formed when two or more atoms are chemically bonded together. These atoms may be from the same element, e.g. two atoms of oxygen form one molecule of the element oxygen (O_2). When atoms from different elements combine they form molecules of a **compound**, e.g. two atoms of hydrogen combine with one atom of oxygen to form one molecule of the compound water H_2O.

A **compound** is a pure substance of which a molecule is the smallest unit which displays its individual properties. It is formed from the atoms of two or more elements but has quite different characteristics from its constituent atoms, e.g. a molecule of the compound sodium chloride is formed from atoms of sodium and chlorine.

Compounds are formed by atoms with an outer shell electron number that makes them unstable (reactive), i.e. not full. Atoms without a full outer shell will attempt to become stable by filling the shell. Conversely, when the outer electron shell is full or contains eight electrons the atom is very stable, these substances are described as being inert or unreactive, e.g. the gases helium, xenon, krypton, argon, etc.

Molecular weight

The molecular weight of a substance is the total atomic weights of its elements:

water H_2O		
2 hydrogen (atomic weight 1)	2 x 1 =	2
1 oxygen (atomic weight 16)	16 x 1 =	16
molecular weight of water	=	18
glucose $C_6H_{12}O_6$		
6 carbon (atomic weight 12)	6 x 12 =	72
12 hydrogen (atomic weight 1)	12 x 1 =	12
6 oxygen (atomic weight 16)	6 x 16 =	96
molecular weight of glucose	=	180

Molar concentration of solutions (molarity)

Molarity is very important and is used when expressing the concentration of substances in body fluids, e.g. glucose levels in the blood. The molar concentration of a solution (**molarity**) is measured by the number of moles of a substance present in a litre of the solution. It is part of the **Systeme Internationale (SI)** (see Appendix A: International System of Units), a system of measurement used for medical, scientific and technical purposes internationally.

The unit of measure is the **mole** (mol) which is the atomic weight or molecular weight in grams of a substance, e.g. 1 mole of glucose = 180 grams of glucose. One mole of any substance will contain the same number of atoms or molecules: 6.02×10^{23} (known as *Avogadro's* number). This means that the actual numbers of particles in a solution is always known, this being especially important in body fluids.

A **molar solution** is one in which 1 mole of the substance is dissolved in enough **solvent** (a liquid which is able to dissolve another substance, usually water in the body) to produce one litre of solution. Chemical substances are usually in very low concentration in the body and for this reason we divide moles by 1000 to obtain the **millimole** (mmol). This is a much more appropriate unit when expressing the concentration of substances in the body, e.g. serum calcium in the blood is 2.1–2.6 mmol/l. If an even smaller unit of measurement is needed the millimole can be divided by 1000 to give the **micromole** (µmol). The SI system of measurement can be applied to any substance of known molecular weight (electrolytes and non-electrolytes), and it has replaced the milliequivalent for physiological purposes.

Chemical bonds

Earlier we considered the combining of two or more atoms to form molecules and compounds. Obviously some kind of bond must hold them together. The actual bond is formed by the energy reaction between the outer shell electrons of the individual atoms and can be of three types: ionic, covalent and hydrogen bonds.

Ionic

One way in which atoms can bond together is through **ionic bonds**. The electrons in the outer shells move between the reacting atoms. This results in one atom shedding one or more electrons and becoming a positive ion (**cation**), and the other atom gaining one or more electrons and becoming a negative ion (**anion**). These electrically charged particles are known as **ions**. The attraction between these charged ions holds the two ions together forming an ionic bond. A good example of an ionic bond is the formation of sodium chloride (NaCl): sodium atoms give up electrons to become the positive cation (Na^+) and chlorine atoms receive electrons to form the negative anion (Cl^-) *(Figure 2.2)*. Again this illustrates an attempt by atoms to fill their outer electron shells. When water is not present these compounds usually form crystals, e.g. common salt (sodium chloride). Many ionic compounds belong to the group of chemicals called *salts*, formed when an **acid** (a substance which contains an excess of hydrogen ions over hydroxyl ions, and therefore donates hydrogen ions) and **alkali** (a substance which contains an excess of hydroxyl ions over hydrogen ions, and therefore accepts hydrogen ions) combine, e.g. hydrochloric acid (HCl) and sodium hydroxide (NaOH).

$$HCl + NaOH = NaCl + H_2O$$

Hydrochloric acid + Sodium hydroxide = Sodium chloride + water

Proteins, acids and bases are also ionic substances. When ionic compounds dissolve in water they dissociate or ionize into their ions, in which form they are known as **electrolytes**. The many electrolytes

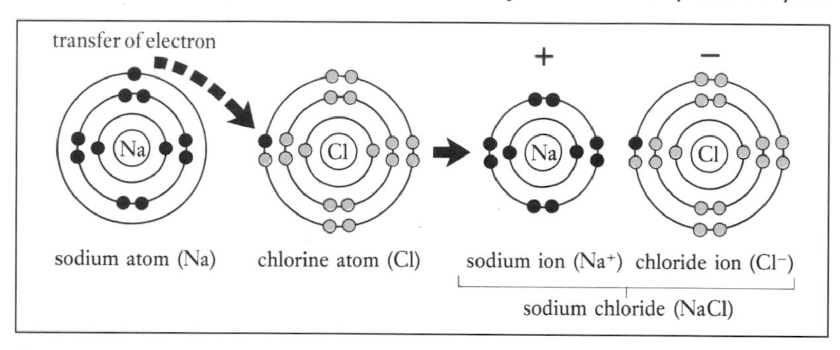

Figure 2.2. Ionic bond.

within the body are considered later in this chapter and in other chapters where appropriate.

Covalent

In a **covalent bond** the electrons are shared between the atoms, rather than the loss and gain situation of ionic bonds. In this way atoms can achieve partial stability. Examples of covalent bonds include hydrogen molecules, carbon dioxide and water.

Hydrogen bond

A **hydrogen bond** is a weak bond found between water molecules and in certain complex biological molecules, e.g. proteins and nucleic acids (see *Figure 1.3)*. The hydrogen atoms involved in the bond are already bonded covalently to other atoms, e.g. oxygen or nitrogen. Hydrogen bonds help to maintain the three-dimensional structure vital to the functioning of biological molecules such as enzymes.

Diffusion and Osmosis

A further account of diffusion and osmosis is now appropriate, although the concepts were discussed in Chapter 1 in terms of transport across cytoplasmic membranes. This further coverage will serve to illustrate the role of diffusion and osmosis in maintaining fluid compartments and the concentration of substances within body fluids.

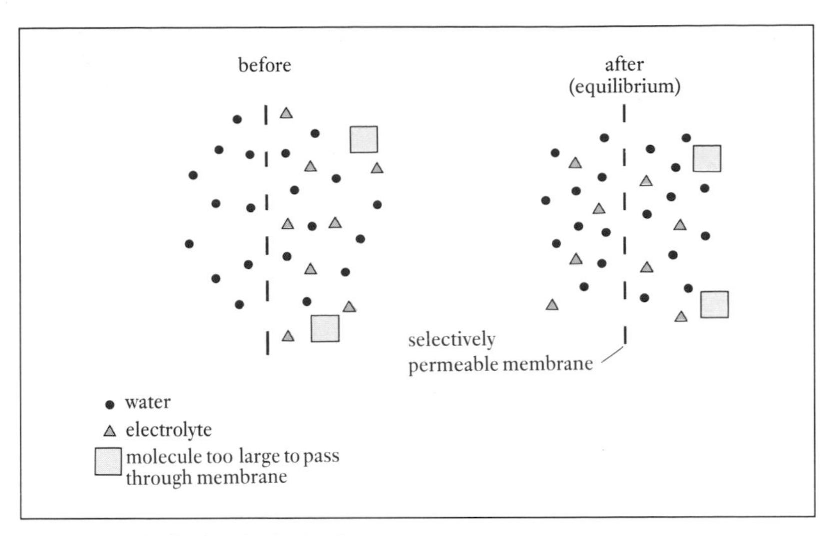

Figure 2.3. Diffusion in the body.

Diffusion

Diffusion involves the movement of substances from an area of high concentration to an area of lower concentration, i.e. down a concentration gradient. An everyday example of diffusion would be when an air freshner or perfume spray is used in one part of a room it takes some time for you to smell the spray in another part of the room as the spray molecules need to diffuse through the air. In the body the movement of substances takes place either across a selectively permeable membrane, which allows certain particles to pass through but holds back other larger particles, or down a concentration gradient within cells (*Figure 2.3*). Diffusion is a passive process which does not use body energy. The energy which powers diffusion is produced by the constant collision of molecules as they move randomly in solution. In a closed system the end result of diffusion will be the concentration on both sides of the membrane being equal, i.e. in equilibrium. For the body, diffusion is important in the movement of oxygen, carbon dioxide, fats, some hormones, electrolytes such as sodium, and the waste product urea. As already mentioned, the particles pass through the membrane by dissolving in either its protein *or* phospholipid component. A more complex type of diffusion known as **facilitated diffusion**, makes use of special carrier molecules situated in the membrane, e.g. glucose uses such a carrier molecule.

Osmosis

For physiological purposes the process of **osmosis** is the movement of water through a selectively permeable membrane. Pores in the membrane allow the movement of water molecules in either direction. Water will move when there is a difference in concentration on either side of the membrane. The direction of movement will be from high water to low water concentration. The water moves from a weak solution to a stronger, more concentrated solution. Most people will remember the experiment from school: a container of sucrose solution with a selectively permeable membrane at its base is placed in a beaker of water; after a while it becomes obvious that water has

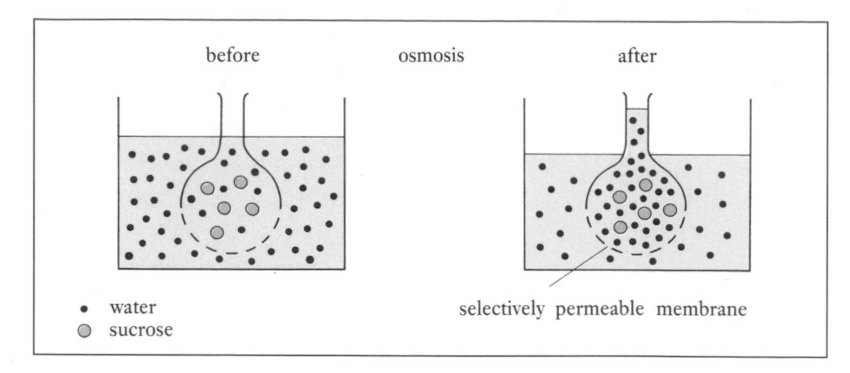

Figure 2.4. Osmosis.

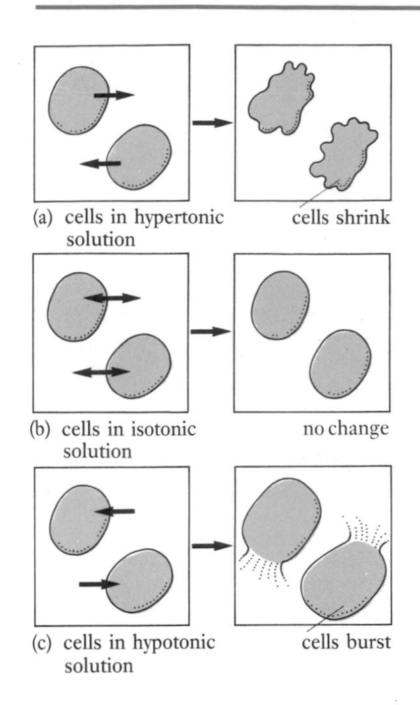

Figure 2.5. Effects of tonicity on cells.

moved into the sucrose as a rise in the fluid level of sucrose is observed *(Figure 2.4)*.

The force required to stop the movement of water by osmosis is known as the **osmotic pressure**; this increases with the concentration difference either side of the membrane. The concentration of dissolved osmotically significant particles (**solute**) in a **solution** (the liquid in which the solute has been dissolved) is termed its **osmolarity**. Osmolarity is measured in osmoles/litre (osmol/l) or milliosmoles/litre (mosmol/l), this tells us about a solution's osmotic properties, including osmotic pressure which depends upon the actual number of particles. N.B. Osmolarity and molarity will be the same for undissociated non-electrolyte substances but when substances dissociate into ions each ion has exactly the same osmotic action as an undissociated molecule. A solution of one molar sodium chloride which dissociates into sodium and chloride ions will contain two osmoles.

Tonicity is the osmotic concentration of a solution compared with that of body fluids. A solution with the same concentration as body fluids is called **isotonic**; cells placed in this solution show no water movement. A higher concentration solution is **hypertonic** *(Figure 2.5)*. Cells placed in it shrink as water leaves the cells. One with a lower concentration is **hypotonic**; here the cells swell and burst as they draw in water from the solution.

Osmosis is vital in maintaining the correct water distribution within the fluid compartments. To summarize it can be said that the two processes of diffusion and osmosis ensure that the state of the internal environment is always favourable to cellular processes.

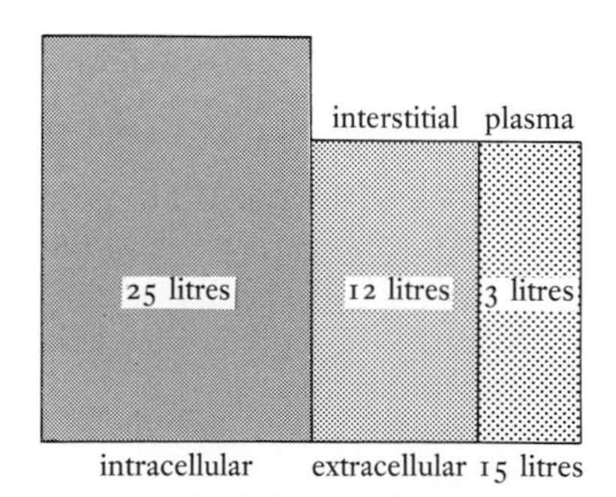

Figure 2.6. Fluid compartments.

Fluid Compartments and Body Fluids

In healthy young adults the body consists of approximately 50–65% water. Different types of tissue contain varying amounts of water, e.g. muscle and bone have a higher water content than adipose tiseue. Males, with their larger muscle mass, generally have a higher percentage of water than females who have more adipose tissue. Babies have a higher proportion of water (around 70%) which decreases during childhood. Individuals who are obese also have a lower percentage of water due to the proportionally higher amount of adipose tissue. The proportion of water decreases slightly in older people as part of normal ageing.

This water, with various non-electrolyte substances and electrolytes, is located within the intracellular and extracellular compartments of the body. Most of the water is actually within the cells as **intracellular fluid** (ICF). The **extracellular fluid** (ECF) consists of **interstitial fluid** (*tissue fluid*) which bathes the cells and the fluid part of blood (plasma). Other types of ECF include lymph, gastrointestinal juices and cerebrospinal fluid around the brain and spinal cord.

For example; a young adult male weighing 70 kg (11 stones) will have about 40 litres of water distributed around the fluid compartments as shown by *Figure 2.6*. This can be summarized as:

Intracellular fluid –	25 litres
Extracellular fluid –	15 litres (12 litres as interstitial fluid 3 litres in the plasma)
Total	40 litres

Electrolyte composition of fluid compartments

Each body fluid has its individual composition. Some fluids are very different from each other, e.g intracellular and extracellular fluids. Others, such as the two extracellular fluids – interstitial fluid and plasma – exhibit similarities. They do, however, differ in one important

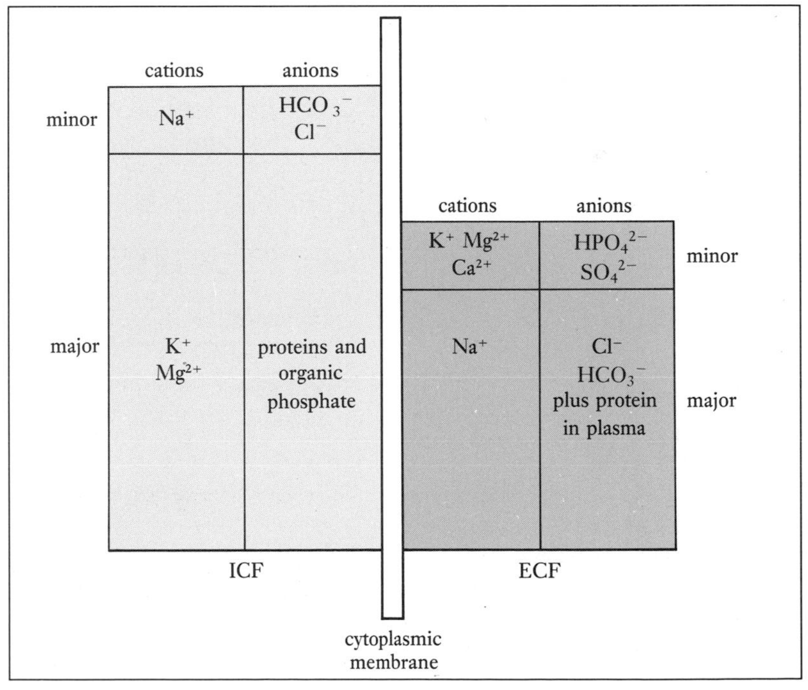

Figure 2.7. Comparision of the electrolyte composition of intracellular fluid (ICF) and extracellular fluid (ECF).

respect: the protein content is higher in plasma (important in the production of osmotic pressure).

NB The positively and negatively charged ions (electrolytes) are balanced within a compartment to give an overall electrical neutrality.

The ion concentration of intracellular and extracellular fluid is set out below (and see *Figure 2.7*). **Major anions** and **cations** are those which predominate and the **minor anions** and **cations** are those which are present in small quantities only.

Intracellular fluid contains:

- Major anions – organic phosphates and proteins.
- Minor anions – hydrogen carbonate, chloride.
- Major cations – potassium and magnesium.
- Minor cations – sodium.

Extracellular fluid contains:

- Major anions – hydrogen carbonate and chloride in interstitial fluid, and these plus protein in plasma.
- Minor anions – hydrogen phosphate, sulphate and organic ions.
- Major cation – sodium.
- Minor cations – potassium, magnesium and calcium.

A common investigation used to assess health is the measurement of the concentration of various ions in the serum (fluid part of blood after the sample has clotted); this test monitors the *serum electrolytes*. The ions measured include; sodium, chloride, potassium, hydrogen carbonate (bicarbonate), calcium and sometimes magnesium *(Table 2.2)*. The results can give a guide to the functioning of many homeostatic mechanisms, e.g. renal excretion of potassium, the level of which rapidly rises when the kidneys fail.

Table 2.2. Serum electrolytes – normal range.

Electrolyte	***Normal range (serum)***
Sodium (Na^+)	135–145 mmol/l
Chloride (Cl^-)	95–105 mmol/l
Potassium (K^+)	3.3–5.0 mmol/l
Hydrogen carbonate (bicarbonate) (HCO_3^-)	22–28 mmol/l
Calcium (Ca^{2+})	2.1–2.6 mmols/l
Magnesium (Mg^{2+})	0.75–1.0 mmols/l

*Normal ranges may vary slightly on a local basis.

Water balance

Normally the amount of water in the fluid compartments is determined by the amount of water: (i) taken in during eating and drinking; (ii) produced by metabolic processes; and (iii) leaving the body via the kidneys, skin, gastrointestinal tract and lungs:

IN	OUT
moist food	urine and faeces
fluids	sweat
metabolic processes	via the lungs

This is another example of a complex homeostatic mechanism which involves many functional systems. Overall control of this delicate balance is via the nervous and endocrine systems through *osmoreceptors* in the brain and the secretion of *antidiuretic hormone* (ADH) (see Chapters 8 and 15 for more detailed description).

Problems with water balance

Excess water (waterlogging of the tissues)
An increase in water in the interstitial fluid is called **oedema**. Areas so affected look puffy and form depressions or pits when pressure is applied. Oedema often occurs in the dependent parts such as the feet and legs. The skin over an oedematous area has an increased risk of developing a pressure sore (see Chapter 19). The formation of oedema is discussed in Chapter 10.

Special Focus Dehydration

Dehydration occurs when the waters lost over a period of time exceeds the water gained. Dehydration can, if it becomes severe, cause hypovolaemic shock due to reduced blood volume (see Chapter 10 for a discussion of this and other types of shock).

Causes
Excess loss. Diarrhoea. Vomiting. Haemorrhage. Severe burns or scalds due to loss of plasma. Conditions where urinary output volume is increased such as diabetes mellitus and diabetes insipidus. Severe sweating. Diuretic drugs which increase urinary output volume. *Inadequate intake.* Where suitable drinking water is unavailable. A physical factor exists such as unconsciousness. Psychological factors, e.g. severe depression, may prevent the individual taking an adequate intake.

People with an increased risk of dehydration
Babies and young children. Diarrhoea is an important cause of dehydration in this group, especially in the developing countries.
Elderly people. Those who do not respond to thirst, may not drink enough. Some may restrict their intake because they fear urinary incontinence.
Others. People with jobs that involve heavy work in hot conditions such as furnace workers. People exposed to high environment temperatures, especially those on holiday in hot climates. People whose homeostatic mechanisms are impaired by disordered body functions.

Water depletion (dehydration)
When water intake is inadequate or excess fluid is lost from the body **dehydration** (water depletion) occurs. It is often accompanied by a disturbance in electrolyte balance. Water is lost from the extracellular fluid and fluid moves out of the cells in an effort to equalize the concentration of the intracellular and extracellular fluids.

Person-centred Study Natalie

Natalie, aged 2, is the second member of her family to have gastroenteritis; her sister, aged 7, has also been ill. She had been 'poorly' with diarrhoea for 24 hours when her worried father, who cares for the girls alone, phoned the health centre for some advice.

Their family doctor visited Natalie at home and decided that she did not need to go to hospital although her negative water balance had caused dehydration.

Initially the fluid lost by Natalie was from the extracellular fluid (ECF). When the negative fluid balance continued an osmotic movement of fluid from intracellular fluid (ICF) to ECF occurred in an attempt to restore the balance between the fluid compartments. In Natalie's situation where negative water balance is due to diarrhoea, the electrolytes sodium and potassium would also be lost from the body.

After asking Natalie's father about her condition and an examination of Natalie the doctor noted the following physiological effects of the negative water balance and dehydration:

Dry, flushed skin – caused when fluid leaves the skin as ICF moves into the ECF. If fluid loss were to continue the skin would eventually lose its elasticity and the eyes become sunken.

NB Natalie may also feel hot and look flushed if her temperature was raised (see Chapter 19) in response to the infection causing her diarrhoea.

Dry mouth, sticky saliva, furred tongue – the salivary glands (see Chapter 13) produce less saliva in situations where the fluid volume of the body is reduced. The saliva that is produced tends to be sticky and viscous (thick). Without the normal cleansing action of saliva the mouth will feel dry and the tongue will be furred.

Thirst – Natalie will be thirsty. The loss of ECF will initially increase its solute concentration (until fluid moves from the ICF). This increase in concentration stimulates the osmoreceptor cells in the thirst centre (see Chapter 4) situated in the brain. This and the dry mouth will increase the desire to drink which eventually replaces fluid lost from the body.

Natalie is passing small amounts of *dark urine* – the dark urine is concentrated because it contains less water than normal. Homeostatic mechanisms concerned with fluid balance operate to conserve water within the body, e.g. osmoreceptors within the brain respond to changes in the concentration of solutes in the ECF. This initiates the release of antidiuretic hormone (ADH)from the posterior pituitary gland (see Chapter 8) which in turn increases kidney tubule permeability to water. This means that more water is reabsorbed rather than being excreted in the urine (see Chapter 15).

After discussion with Natalie's father, the doctor decided that, because Natalie was not vomiting, she could be rehydrated at home with oral fluids, and powders containing glucose and the electrolytes sodium and potassium. This treatment regimen will replace the lost fluid and restore the electrolyte balance to normal. Although this soon restored Natalie to her normal, happy, active self her father was still worried about protecting the family from future 'tummy upsets'. Next time he took Natalie to the clinic for a routine assessment, he discussed his worries with the health visitor who made some suggestions regarding food safety, which included: (i) wash hands after using the toilet and before preparing or eating food; (ii) select food which is fresh, in date and in good condition; (iii) store food at correct temperature, e.g. in refrigerator or freezer; (iv) store food properly, e.g. keep raw and cooked foods separate; (v) cook at temperatures high enough to kill most harmful bacteria; (vi) always follow the individual recommendations on cooking and 'stand time' for microwave ovens; (vii) only reheat food once; (viii) keep food preparation equipment and areas clean and dry; (ix) keep pets away from food and preparation areas; (x) if at all possible do not prepare food if you are unwell with diarrhoea and vomiting or have a septic lesion on your hands.

Fluid replacement in dehydration (rehydration)

Fluid replacement or **rehydration** can be achieved by:

- Oral fluids, if dehydration is not severe or when excessive vomiting is not a problem (see Person-centred Study, Natalie).
- Passing a tube into the gastrointestinal tract usually via the nose. This is a method used if the gastrointestinal tract is functioning (see Chapter 13), but when drinking is not possible, e.g. where the person is unconscious.
- Intravenous infusion, containing a combination of water, and appropriate electrolytes used when dehydration is severe, e.g. burns, or where the gastrointestinal tract is not absorbing fluid, such as after bowel surgery. Water and other substances in the

infusion fluid are transported through the capillary walls into the interstitial fluid. From here some will enter the cells to restore the equilibrium between the extracellular and intracellular fluids.

- Fluid infused subcutaneously (under the skin) or into the rectum (rarely administered in these ways).

Nursing Practice Application **Monitoring and evaluating fluid replacement**

It is important that the nurse is able to monitor and evaluate the adequacy of fluid replacement. The state of the skin and how the person feels are useful and simple assessments. A record of all intake and especially output in the form of a fluid balance chart is an important method of assessing hydration. When dehydration is severe it is necessary for the nurse to use more precise measures of the volume of cirulating fluid: blood pressure, pulse and central venous pressure (see *Figure 10.24*). The medical staff will also use blood tests to assess the efficacy of the treatment regimen.

Hydrogen Ion Concentration (Acid–Base Concentration)

The hydrogen ion concentration or pH is a method of expressing the acidity (concentration of hydrogen ions H^+) or alkalinity (concentration of hydroxyl ions OH^-) of a solution. pH is measured on a logarithmic scale of 0–14 (see *Figure 2.8*); these figures represent the indices of the concentration i.e. 10^0, 10^{-7},10^{-14} and are easier to use than the actual number of hydrogen ions involved (10^0 = 1 mole hydrogen ions/litre, 10^{-7} = 0.0000001 mole hydrogen ions/litre) *NB* 10^0 = 1 because the logarithm of 1 is 0. A solution with a pH of 0 has the greatest concentration of hydrogen ions (10^0 or 1 mole H^+/litre) and the lowest concentration of hydroxyl ions. It is therefore acidic.

pH 7 represents the point where hydrogen ions and hydroxyl ions are in the same concentration (10^{-7}); this solution is said to be neutral, that is neither acid or alkali, e.g. distilled water. At pH 14 (10^{-14}) the hydroxyl ions are at their greatest concentration and the hydrogen ions at their lowest. It is therefore alkaline. A solution with a pH of less than 7 is acidic and one where the pH is greater than 7 is alkaline. As hydrogen ion concentration decreases that of hydroxyl ions increases and vice versa. Since the pH scale is logarithmic, a change of one on the scale actually represents a tenfold change in the hydrogen ion concentration, e.g. a solution of pH 6 has ten times more hydrogen ions than a solution of pH 7. This is of particular importance when considering the pH of blood (normal range pH 7.35–7.45, i.e. slightly alkaline) as very small changes will greatly affect cell environment and function.

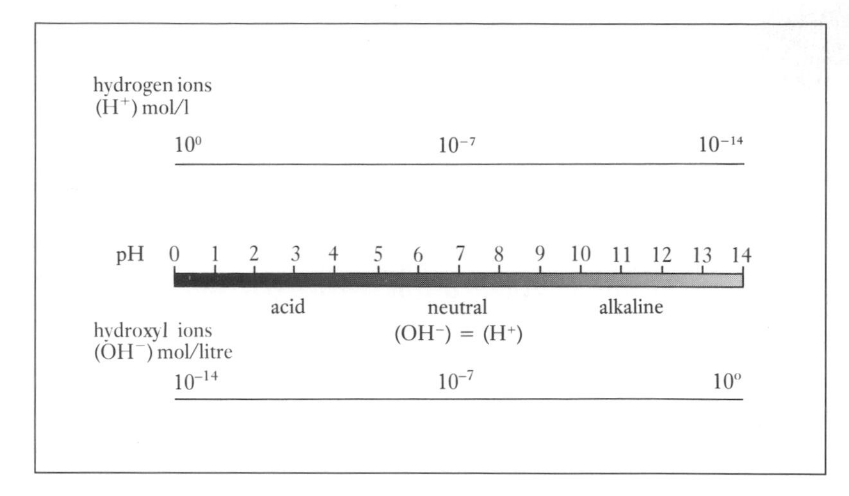

Figure 2.8. pH scale. *NB* At 37°C (body temperature) the neutral point on the pH scale is 6.8.

Regulation of pH

Acid–base balance in the body is maintained by homeostatic mechanisms occurring in the respiratory system, blood and kidneys. These mechanisms ensure that the blood pH stays within the range 7.35–7.45 during health. This is vital for enzyme activity, which is dependent upon environmental pH, and for other functional proteins, such as haemoglobin, which require the correct pH range for optimal function. Table 2.3 details the pH ranges of some body fluids.

Buffers

Buffers are substances which limit pH change by their ability to accept hydrogen ions from an acidic solution and donate hydrogen ions to an alkaline solution. In this way it is possible to prevent swings in pH which would inhibit cell function. The important buffer systems are:

Hydrogen carbonate (bicarbonate) system in the blood and ICF

The hydrogen carbonate system consists of hydrogen carbonate (HCO_3^-) and carbonic acid (H_2CO_3) in a ratio of 20:1. The hydrogen carbonates, either sodium hydrogen carbonate (in ECF) or potassium hydrogen carbonate (in ICF) act as a weak alkali to mop up excess hydrogen ions from a strong acid. Thus forming more carbonic acid, a weak acid, and a neutral salt, leaving pH unaffected:

HCl	+	$NaHCO_3$	→	H_2CO_3	+	$NaCl$
Hydrochloric acid	+	Sodium hydrogen carbonate		Carbonic acid	+	Sodium chloride
(strong acid)		(weak alkali)		(weak acid)		(salt)

The amount of hydrogen carbonate available in the blood for buffering pH is known as the **alkaline reserve**. It is regulated by the kidneys (see Chapter 15). Carbonic acid as a weak acid is able to donate hydrogen ions when the pH of a solution rises when a strong alkali is added:

$NaOH$	+	H_2CO_3	→	$NaHCO_3$	+	H_2O
Sodium hydroxide	+	Carbonic acid		Sodium hydrogen carbonate	+	water
(strong alkali)		(weak acid)		(weak alkali)		

Carbonic acid is derived from cellular metabolism, and its level in the blood is regulated by the lungs (see Chapter 12).

Hydrogen phosphates in the ICF, kidney and blood

The hydrogen phosphate system consists of two sodium salts: sodium dihydrogen phosphate (NaH_2PO_4), which acts as a weak acid, and disodium hydrogen phosphate (Na_2HPO_4), which acts as a weak alkali. These two act like the hydrogen carbonate system by donating and accepting hydrogen ions.

Protein buffers

Protein buffers, such as haemoglobin, limit pH changes in ICF and blood by their ability to ionize. Proteins are long chains consisting of many amino acids joined together. One end of the chain has a carboxyl group (COOH) which acts as a weak acid and donates hydrogen ions:

$$R-COOH \rightleftharpoons R-COO^- + H^+$$

The other end has an amino (amine) group (NH_2) which acts as a weak alkali and accepts hydrogen ions:

$$R-NH_2 + H^+ \rightleftharpoons R-NH_3^+$$

R = the different amino acid structures.

Table 2.3. pH range of some body fluids.

Fluid	***pH range***
Blood	7.35–7.45
Saliva	6.5–7.5
Gastric secretion	1.5–3.0
Pancreatic secretion	8.0–8.4
Urine	4.5–8.0

When homeostatic mechanisms fail and buffer systems are overwhelmed, changes occur in the pH of blood which may be life threatening. These changes can be (i) *acidosis* due to a fall in pH; (ii) *alkalosis* due to a rise in pH.

A fuller account of these departures from normal can be found in Chapters 12,13,15.

Summary/Check List

Homeostasis – control mechanisms, negative and positive feedback, breakdown in control.
Basic chemistry Matter – elements and those of physiological significance, atoms, atomic number, mass number, atomic weight, isotopes, radioisotopes, half-life. Molecules and compounds – molecular weight, molar concentration and molarity. Chemical bonds – ionic (ions and electrolytes), covalent, hydrogen bond.
Diffusion – simple, facilitated.
Osmosis – concepts, osmotic pressure, osmolarity, tonicity (isotonic, hypertonic and hypotonic solutions).
Fluid compartments and body fluids – intracellular fluid (ICF) and extracellular fluid (ECF) (interstitial fluid and plasma). Electrolyte composition of fluid compartments – major anions and cations in ICF and ECF, serum electrolytes. Water balance – water gain and loss. Problems with water balance. *Special Focus* – Dehydration. *Person-centred Study* – Natalie. Fluid replacement in dehydration. *Nursing Practice Application* – monitoring and evaluating fluid replacement.
Hydrogen ion concentration – the pH scale, regulation of pH, buffers (hydrogen carbonate system, hydrogen phosphates and proteins), pH range of some body fluids, acidosis and alkalosis.

?

Self Test

1 Which of the following statements about homeostasis are true?
(a) It is the self-regulatory process of the body which maintains the constancy of the internal environment.
(b) All homeostatic mechanisms are controlled by negative feedback.
(c) The only function of homeostatic mechanisms is to regulate the chemical concentration of body fluids.
(d) A breakdown in homeostasis leads to a decline in health.

2 Briefly describe the three components required for a homeostatic mechanism.

3 An atom with an electrical charge is called an:
(a) isotope, (b) electron, (c) element, (d) ion.

4 What are the chemical symbols for the following elements? sodium, calcium, potassium, iron, magnesium, chlorine, oxygen, hydrogen, nitrogen, carbon.

5 Describe the direction of water movement when cells are placed in the following solutions of sodium chloride:
(a) isotonic, (b) hypotonic, (c) hypertonic.

6 In healthy young adults (male) the amount of water in the body is approximately:
(a) 50%, (b) 70%, (c) 40%, (d) 60%.

7 A man has 40 litres of body water. How much of this will be found in the cells, around the cells and as blood plasma?

8 Which of the following statements are true?
(a) Sodium is the major cation of the ECF.
(b) Most potassium is intracellular.
(c) Chloride, calcium and hydrogen carbonate are all anions.
(d) Cations are ions with a positive charge.

9 Which groups have an increased risk of dehydration?

10 Mark on the pH scale: neutral point, blood pH, acidic range, pH of gastric secretions, pH of urine, alkaline range.

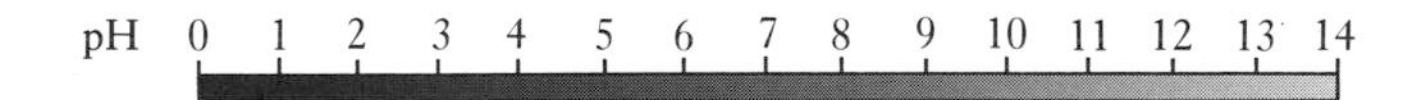

Answers

1. a, d. 2. See page 35. 3. d. 4. Na, Ca, K, Fe, Mg, Cl, O, H, N, C. 5. a. no movement, b. from solution to cells, c. from cells to solution 6. d. 7. 25 litres in cells, 12 litres as interstitial fluid and 3 litres as plasma. 8. a, b, d. 9. See page 45. 10. See *Figure 2.8 and Table 2.3.*

REFERENCES

Herbert, R. (1986) The biology of ageing: body temperature regulation. *Geriatric Nursing and Home Care*, **6** (4), 16–18.

Ho, E. (1989) The early days. *Nursing*, **3** (39), 12–14.

FURTHER READING

Foss, M. (1988) Acid-base balance, *The Professional Nurse*, **3** (12), 509, 511–513.

Herbert, R. (1986) The biology of ageing: maintenance of homeostasis, *Geriatric Nursing and Home Care*, **6** (3), 14– 16.

Lewis, M. and Waller, G. (1980). *Thinking Chemistry*, Oxford: Oxford University Press. *NB* Provides coverage of basic chemical concepts.

Miller, J.A. (1989) Intravenous therapy in fluid and electrolyte imbalance, *The Professional Nurse*, **4** (5), 237-241.

Robinson, J.R. (1975) *Fundamentals of Acid–Base Regulation*, 5th edn. Oxford: Blackwell Scientific Publications.

Willatts, S.M. (1982) *Lecture Notes on Fluid and Electrolyte Balance*, Oxford: Blackwell Scientific Publications.

section three

regulation, communication and control

CHAPTER THREE

Nervous Tissue and Basic Functions of the Nervous System

Overview

- Organization of nervous system.
- Nervous tissue.
- The nerve impulse.
- Reflex arc.

Learning Outcomes

After studying Chapter 3 you should be able to:

- Outline the organization of the nervous system.
- Describe the structure of a nerve cell.
- Describe the different types of neuroglial cells and their functions.
- Describe the main types of sensory receptors.
- Explain the process of adaptation that occurs in some sensory receptors.
- Discuss the transmission of nerve impulses.
- Describe a synapse and explain briefly how it functions.
- Outline the variety and role of neurotransmitters within the nervous system.
- Discuss common causes of disordered nerve function.
- Outline the effects of demyelination, such as in multiple sclerosis.
- Describe the reflex arc.

Introduction and Basic Organization

The nervous system works closely with the endocrine system (Chapter 8) to control and regulate all body functions and preserve homeostasis (the concept of homeostasis is discussed in Chapter 2). Communication within the body occurs via the two systems:

- The nervous system, which transmits nerve impulses as electrochemical energy (see pages 56-58).
- The endocrine system where, in a slower process, hormones produced by the endocrine glands are transported in the blood (see Chapter 8).

Key Words

Action potential – the change in electrical potential and charge across the cell membrane which occurs when a nerve conducts an impulse or when muscle fibres contract.
Autonomic nervous system – the part of the nervous system which controls the involuntary functions such as heart rate, glandular secretion and smooth muscle contraction.
Axon – the long extension of a nerve cell which conducts impulses away from the nerve cell body.
Cell body (nerve cell body) – the part of the neurone which contains its nucleus and other organelles (not centrioles).
Central nervous system (CNS) – the brain and spinal cord.
Dendrite – the branching processes of the nerve cell. They receive nerve impulses from other neurones and conduct them towards the nerve cell body.
Motor or **efferent** nerves – convey impulses from the central nervous system to muscles and glands.
Myelin – the fatty material which covers some nerves. It insulates the nerve fibres and is analogous with the plastic cover around electrical wires.
Neuroglia – support cells of the nervous system.
Neurone – a nerve cell consisting of a cell body, an axon and dendrites.
Neurotransmitter – a chemical which facilitates the transmission of a nerve impulse across a junction, e.g. acetylcholine.
Peripheral nervous system (PNS) – the part of the nervous system outside the brain and spinal cord. The motor and sensory nerves carrying impulses between the CNS and the rest of the body.
Potential difference – the voltage or potential energy in volts (V) or millivolts (mV) between two points, e.g. across a cell membrane.
Refractory period – the time for which a nerve or muscle fibre is unable to respond to a stimulus. It may be partial, when it may be overcome if the stimulus is sufficiently large, or absolute, when no response is possible.
Saltatory conduction – the transmission of a nerve impulse along a myelinated fibre where the impulse appears to leap from one node of Ranvier to the next.
Sensory or **afferent nerves** – convey impulses to the central nervous system from organs, skin, joints and skeletal muscles.
Summation – the process by which the excitatory and inhibitory effects of stimuli arriving at a nerve cell body are added together, sorted and integrated.
Synapse – the gap between the axon of one neurone and the dendrites of another, or the gap between an axon and an effector cell. Transmission of the nerve impulse across the gap depends upon the release of a chemical neurotransmitter to carry the impulse across the gap.
Threshold – the intensity of the stimulus required to initiate a response, eg an action potential.

The nervous system receives sensory data from external and internal sources, processes this data and initiates an appropriate response. The ability to sort **sensory input** and determine the required action is called **integration**. The response – the **motor output** – results in glandular or muscular activity. For example, you are strolling through a meadow when you see a bull grazing (sensory input), you realize the potential danger (integration) and remove yourself rapidly from the vicinity (motor output).

Our understanding of the nervous system is increasing rapidly but considerable research is necessary before its functioning is fully understood.

The nervous system consists of several different parts, which contribute to the integrated working of the whole system. For convenience it can be divided structurally into:

- **Central nervous system (CNS)** – the brain and spinal cord. It is here that sensory input is received and processed before an appropriate motor output is produced.

- **Peripheral nervous system (PNS)** – the cranial nerves (12 pairs) and spinal nerves (31 pairs), which link the brain and spinal cord with the rest of the body.

The PNS can be divided into two functional parts: the **sensory division (afferent nerves)**, which carries sensory data to the CNS from receptors in the skin, muscles and joints (*somatic*) or from the organs (*visceral*); and the **motor division (efferent nerves)**, which carries the motor output from the CNS. The motor division consists of the **voluntary** part of the PNS, which supplies skeletal muscle and can be controlled consciously, and the **autonomic nervous system (ANS)**, which supplies **involuntary** muscle and glands and controls such functions as heart rate and digestion. In turn, the ANS has two parts: The **sympathetic** ANS, which initiates the stress responses of the body, e.g. increased heart rate. The **parasympathetic** ANS, which controls the more peaceful functions of the body, e.g. digestion.

To summarize:

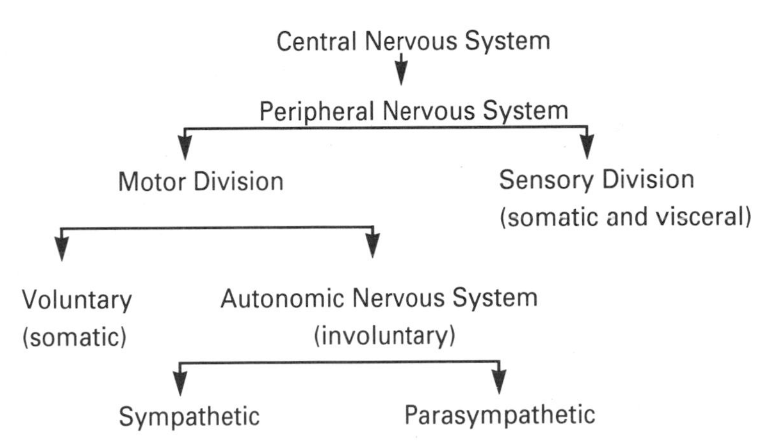

NERVE TISSUE

Nerve tissue consists of the excitable cells – **neurones** – and the **neuroglia** (non-excitable cells), which fulfil support functions within the nervous system.

Neurones

Neurones (nerve cells) are the highly differentiated units of the nervous system (*Figure 3.1*). They are very sensitive to changes in their environment and are easily damaged by exposure to chemicals, toxins, lack of oxygen, etc.; this is illustrated by the brain damage which results when the brain is deprived of oxygen for several minutes, such as may occur during a cardiac arrest.

As a result of their high degree of differentiation, neurones cannot replicate by mitosis (see Chapter 1), so damaged cells cannot be replaced. They compensate for this by being extremely long-lived. However, it is possible to repair limited damage to neurones outside the CNS, providing that the damage occurs some way from the cell

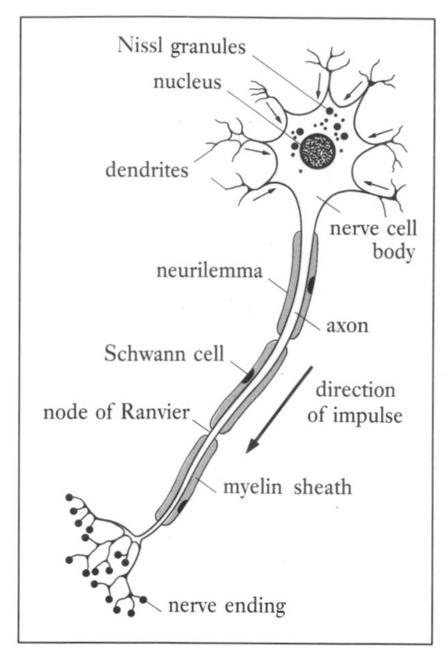

Figure 3.1. Neurone.

body – a peripheral axon which is cut or compressed may regenerate and regain function.

There are several types of neurones and all are adapted structurally for the transmission of electrical impulses. A neurone consists of:

- A **nerve cell body**, which contains a nucleus, other organelles and the **Nissl granules**, which are the rough endoplasmic reticulum concerned with protein synthesis. There are no centrioles, which reflects the neurone's inability to divide by mitosis.
- A long process, or fibre (from a few millimetres to a metre in length), called an **axon**, which transmits impulses away from the cell body. The **axoplasm** (cytoplasm) contains mitochondria and is enclosed by a membrane, the **axolemma**. The axon receives nutrients by the continuous movement of substances from the cell body. *Axonal transport* and *axoplasmic flow* carry enzymes and other substances from cell body to axon, where they are used to maintain structural integrity and metabolic functions.
- Several short branching processes, called **dendrites**, which transmit nerve impulses towards the cell body.

(Students who require a revision of cell structure, organelles and mitosis should have a quick look at Chapter One.)

Some axons are covered with a white fatty material called **myelin**, which protects and insulates the fibre. Myelin insulates the axon and prevents leakage of the electrical charge generated by the nerve impulse. Myelination thus allows very rapid impulse transmission, of up to 130 m/s (see **saltatory conduction**, page 61).

Myelin is produced by specialized neuroglial cells, known as **Schwann cells** in the PNS and as **oligodendrocytes** within the CNS. These cells use their cell membrane to enclose the axon (see below); the layer they form around the axon is known as the **neurilemma**. In myelinated axons the neurilemma is wrapped around the axon several times, and forms the **myelin sheath**; in non-myelinated axons there is neurilemma but no myelin sheath. The mainly myelinated fibres form the '**white matter**' of the CNS and the nerves of the PNS. The non–myelinated cell bodies form the '**grey matter**' of the nervous system. Gaps in the myelin sheath occur at intervals along the myelinated axon and are known as the **nodes of Ranvier**, which assist in rapid transmission of the impulse. These gaps represent the boundaries between individual Schwann cells.

Functionally, neurones may be motor (carrying impulses from the CNS to the muscles or glands), sensory (carrying incoming sensory information to the CNS) or serve as connections in the CNS, when they are called **interneurones**.

Loss of myelin (*demyelination*) in previously myelinated fibres will result in serious impulse conduction problems. A common example of a demyelinating condition is *multiple sclerosis*, which affects 1 in 2000 of the population in the United Kingdom (see *Family-centred Study*).

Family-centred Study Jill, Simon, Caroline and Kevin

Jill is 35 and lives with Simon and their two children (Caroline 9 and Kevin 5). Soon after Kevin was born Jill complained of double vision and weakness in her right arm. This soon cleared up and being busy with the new baby she forgot all about it. All was well for about a year until weakness and tingling in her legs took Jill to the doctor.

After some tests and a consultation at the hospital Jill and Simon were told that most probably she had multiple sclerosis.

Four years on and Jill is unable to walk and needs help with washing and dressing. Urinary incontinence (see Chapter 15) was a problem but an indwelling bladder catheter has made life much easier for Jill and the family. Preventing incontinence has also lessened the already considerable risk of skin breakdown and pressure sore development (see Chapter 19).

Simon has learnt about skin care, lifting (see Chapter 18) and bladder management and is able to care for Jill when he is home from work. Jill's complete dependence is a strain on their relationship and Jill worries that the children miss out on family outings and holidays. The family is well supported by Simon's parents who live nearby, social services and the health care team who give physical care, advice on practical matters and psychological support. Another help is the Multiple Sclerosis Society, which they joined soon after Jill's illness was diagnosed. They found that the chance to talk to people with similar problems was a big relief. They were able to talk about areas of anxiety such as sexuality and the effects of Jill's condition on the children.

The prognosis (outcome) for Jill is poor and she has a high risk of developing pressure sores, urinary infection and pneumonia. Multiple sclerosis has a very variable presentation and outcome, which depend upon the neurones affected. The signs and symptoms are due to demyelination and hardened areas on the myelin sheath. There is no curative treatment but problems such as muscle spasm can be relieved with drugs, e.g. diazepam. Some people have very little disability whilst others like Jill become totally dependent. The severe form can affect family relationships and cause financial problems, and presents health care professionals and social services with an enormous challenge if quality of life is to be maintained.

Supporting cells (non-excitable)

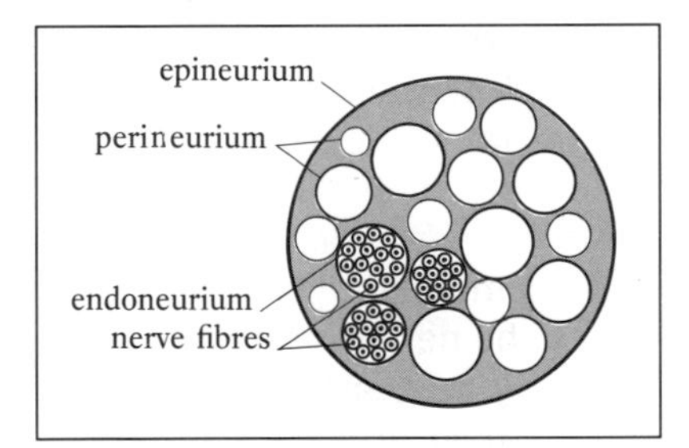

Figure 3.2. Nerve structure.

Supporting cells are known collectively as the neuroglia or **glial cells**. They are found in the CNS and PNS (when they are called Schwann cells) and they greatly outnumber the neurones. Neuroglial cells can replicate and, if replication becomes disordered, may form tumours called *gliomas*. Glial cells include:

- Oligodendrocytes and Schwann cells, whose function has already been discussed.
- **Astrocytes**, which are star-shaped cells found in the CNS. Astrocytes support neurones and surround blood vessels in the CNS to form an important component of the protective **'blood–brain barrier'** (see Chapter 4).
- **Microglial cells**, also found in the CNS, are a type of macrophage; these are phagocytic cells which are active in areas of inflammation or damage (see Chapters 1, 9 and 19).

The nerves of the PNS consist of many axons (nerve fibres) bound together to form a nerve (such as the many tiny wires/fibres in a telephone cable) (*Figure 3.2*).

Individual nerve fibres are surrounded by a delicate connective tissue, the **endoneurium**. Small bundles of nerve fibres are grouped together and enclosed in a sheath of connective tissue, the **perineurium**.

The entire nerve, which consists of many bundles of axons, is protected by an outer fibrous coat, the **epineurium**. Running between bundles of nerve fibres, and supported by connective tissue, are the blood (Chapter 10) and lymphatic (Chapter 11) vessels required by the nerve.

Nerves may be motor (efferent), sensory (afferent) or, more commonly, mixed (carrying both motor and sensory axons).

Sensory receptors

Sensory receptors are specialized structures which monitor changes in the body's environment by reacting to internal or external stimuli. Receptors are situated close to the body surface, e.g. in the skin or eye, and in 'deeper' structures, e.g. organs, joints and muscles. They range from a simple pressure receptor in the skin to very complex sense organs such as the eye (see Chapters 4 and 7).

The receptors send sensory data via the sensory neurones to the CNS, where it is integrated at different levels; this means that not all the sensory input reaches our consciousness but may be acted upon at a subconscious level, e.g. certain reflexes (instantaneous, involuntary response to a stimulus). The many types of receptors may be classified structurally or functionally as follows:

Structural classification

- **Free dendritic nerve endings** – found in the viscera, muscles, joints and skin, where they detect pain, touch and temperature.
- **Encapsulated receptors** (specialized nerve endings) – found in the skin, e.g. **Meissner's corpuscles**, which detect light pressure. Specialized receptors (**Golgi organs**) in the tendons and muscle spindles monitor stretching and with receptors in the joints, provide information about position (*proprioception*).
- **Non-nervous receptors** – the specialized cells found in the sense organs, e.g. taste buds on the tongue.

Functional classification (by type of stimulus)

- **Chemoreceptors** can monitor changes in the chemical environment. They are concerned with taste and smell plus the vital monitoring of carbon dioxide, oxygen and electrolyte levels in the blood.
- **Mechanoreceptors** are responsive to mechanical forces. They include pressure and touch receptors in the skin; **proprioceptors**, which monitor stretching, in muscles, joints and tendons, stretch receptors in the bladder, lung and gut; **baroreceptors**, which monitor pressure, in blood vessels and special receptors in the ear which respond to sound waves.
- **Thermoreceptors** situated in the skin respond to changes in temperature.
- **Nociceptors (pain) receptors** respond to many different stimuli, including intense pressure, extremes of temperature and the chemicals released during the inflammatory process (see Chapter 19). All of these are harmful to tissue and cause pain (see Chapter 4).
- **Photoreceptors** are special receptors in the eye which respond to light energy.

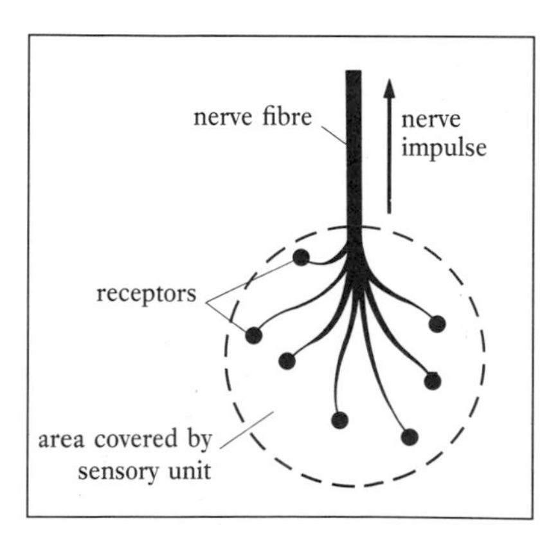

Figure 3.3. Sensory receptors and sensory unit.

Sensory receptors need a certain level of stimulation before they initiate a nerve impulse via the afferent fibres to the CNS; this is termed the **threshold**.

Some receptors can modify their response when a stimulus is repeated continuously – after a while they stop responding to small constant stimuli and respond only to changes. This is termed **adaptation** and explains why, for instance, we are not aware of our clothes touching our skin. Pain receptors, however, do not adapt, and continue to respond for as long as the potentially harmful stimulus is present.

Several sensory receptors channel their output into one afferent fibre; this is known as a **sensory unit** (see *Figure 3.3*).

The Nerve Impulse

Transmission

Nerve impulses are transmitted by nerve fibres in the form of **electrochemical energy** – energy that is produced by the electrical and concentration gradients that result from the movement of ions across the axon membrane. The chemical composition of the intracellular and extracellular fluids is vital to nerve impulse transmission (see Chapter 2). It is the difference in ion concentration between the intracellular and extracellular fluids that produces the electrical **potential difference** across the selectively permeable axon membrane.

The potential difference of –70 mV across the membrane is known as the **resting potential**, the negative sign indicates that the inside of the membrane is negative in respect to the outside. At this stage there is no impulse transmission and the membrane is said to be **polarized**.

For transmission of a nerve impulse to occur, a change in resting potential and membrane permeability (ie. what ions the membrane will allow through) is needed. This occurs when a stimulus reaches the axon. The movement of ions through specific ion channels in the axon membrane causes a change in the potential difference, and this allows the nerve impulse to move along the axon. This change in potential difference is known as the **action potential**. As sodium (Na^+) and chloride (Cl^-) ions move into the axon and potassium (K^+) ions move out, a potential difference of +40 mV is produced, a change of 110 mV. The membrane is now *depolarized* – the inside is electrically positive in respect to the outside. This process of **depolarization** occurs sequentially along the axon, allowing the propagation of the impulse as it regenerates in each segment. The increased speed of transmission in myelinated nerves is explained by the process of **saltatory conduction**, where impulses appear to leap from one node of Ranvier to the next.

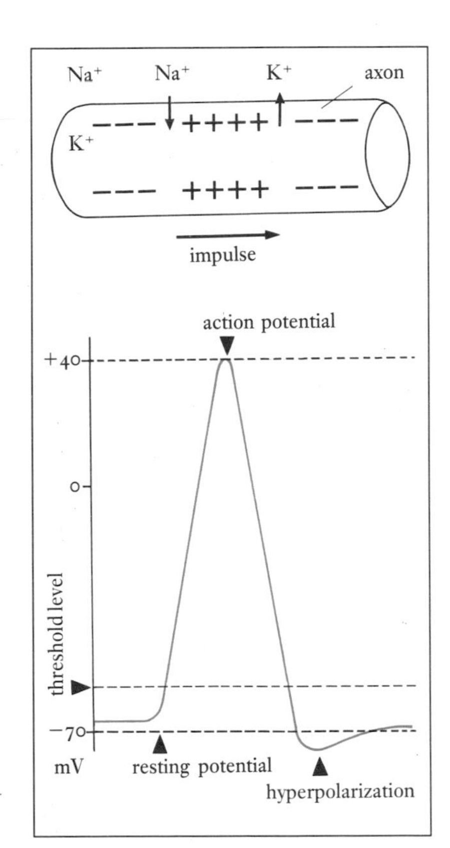

Figure 3.4. Transmission of the nerve impulse/action potential.

After the passage of the impulse the membrane is *repolarized* – returned to the resting potential. During **repolarization** there is an increase in the negative potential, which results in a short period of **hyperpolarization,** a situation which reduces excitability and prevents the transmission of more impulses (*Figure 3.4*). The ion concentrations of the resting potential are restored by an active transport sodium/potassium exchange pump and diffusion (Chapters 1 and 2).

The whole process takes very little time, only a few milliseconds (ms); this is an important feature in a system which depends upon rapid impulse transmission for communication. Speed of transmission depends not only on whether the fibre is myelinated but also on the diameter of the axon. Fibres can be divided into:

- The *'A' fibres* (alpha, α; beta, β; gamma, γ; and delta, δ, becoming progressively smaller and slower) are thick, heavily myelinated fibres capable of rapid transmission, e.g. motor and sensory fibres, some pain and thermal impulses.
- *'B' fibres* or intermediate size fibres which have some myelin and transmit more slowly, e.g. fibres in the autonomic nervous system.
- The *'C' fibres,* which are thin and have no myelin sheath. The lack of myelin means that they transmit at the slowest rate, e.g. autonomic system and some types of pain.

The ability to produce a potential difference across their membrane is a feature of many cells, but only the excitable cells – neurones and muscle cells (see Chapter 17) can produce an action potential by reversibly altering the potential difference.

Some rather special features of the action potential merit further consideration. An action potential in a particular neurone is always of the same size and duration, regardless of the type and intensity of the stimulus. This is the **all-or-none law** – either a stimulus produces a full action potential or it produces nothing; there is no partial response to smaller stimuli. The intensity of the stimulus and the degree of membrane depolarization required for the passage of an impulse is called the **threshold**. During and following the production of an action potential the axon membrane is in a **refractory** state, when its threshold increases and its response to stimuli is altered. When an action potential is being generated the axon is in an **absolute refractory period** which means that further stimuli, regardless of intensity, are incapable of producing another action potential. A **relative refractory** period, of around 5 ms follows the action potential, when only a very intense stimulus will cause a second action potential.

Synapse

A **synapse** is a junction or gap between two neurones or between a neurone and an effector cell, such as a gland or muscle (see also *neuromuscular junction*, Chapter 17). The processes that take place at a synapse allow the nerve impulse to continue across the gap, a feature vital to the continuity and integration of the nervous system. Basically the continuation of transmission can be achieved by electrical or chemical means.

Electrical synapse

In a few situations, where the gap between the cell membranes is extremely small and the electrical resistance (the resistance to the flow of electrical charge) is low, it is possible for an impulse generated in the first neurone to cross the gap to the second neurone. This type of **electrical synapse** is uncommon, but examples exist in the brain, where they allow rapid transmission.

Chemical synapse

Most synapses are of the **chemical** type, (*Figure 3.5*) where substances known as **neurotransmitters,** e.g. *acetylcholine,* are

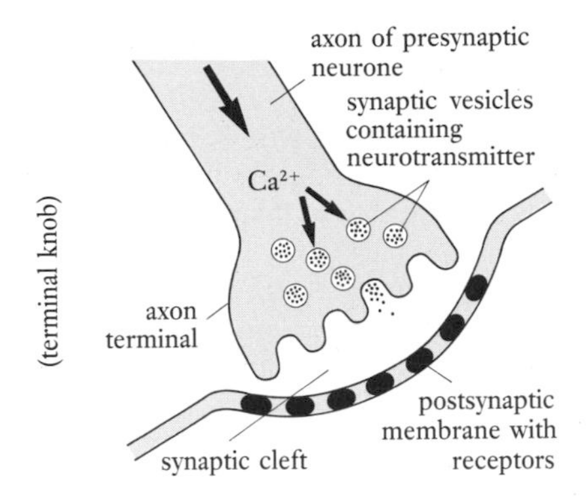

Figure 3.5. Synapse.

involved in the transmission of an impulse. A synapse consists of a presynaptic neurone, with a terminal knob containing vesicles (sacs) filled with the neurotransmitter, the synaptic cleft (gap) and a postsynaptic neurone, whose membrane contains receptor sites. These receptor sites are of different types; each type responds to a specific transmitter chemical, or to several.

The complex events occurring at a chemical synapse can be summarized as:

- The action potential reaches the terminal knob of the axon.
- The membrane allows calcium ions (from the extracellular fluid) into the axon knob. *NB* The calcium ions are later pumped out.
- The influx of calcium causes the synaptic vesicles to fuse with the presynaptic membrane and release neurotransmitter into the cleft, by the process of *exocytosis* (see Chapter 1).
- Released neurotransmitter diffuses across the gap, which accounts for the small delay, and attaches to special receptor sites on the postsynaptic membrane.
- The neurotransmitter alters the potential difference of the postsynaptic membrane by initiating the movement of sodium, potassium and chloride ions or by using a series of reactions requiring another chemical (a *second messenger*) to cause a membrane response. The postsynaptic membrane potential may be *excitatory*, when sufficient impulses will produce the threshold depolarization required to generate an action potential in the axon distal to the synapse, or *inhibitory*, when hyperpolarization occurs and no action potential in the axon is possible (see pages 61-62).

All this is rather a complex process involving hundreds of neurones; some of these contacts are excitatory and others inhibitory and, as you would expect, not all are active at any one time.

The activity of many axon terminals, or the rapid production of impulses from a few, is required before the threshold is reached and the postsynaptic neurone can produce an action potential. This process of **summation** sorts, adds and integrates the effects of excitatory and inhibitory synapse activity on a *temporal* (when one or more axon terminals are producing rapid impulses) or *spatial* (when many axon terminals are active at the same time) basis.

The total process takes a few milliseconds and concludes with the destruction of the neurotransmitter by an enzyme, e.g. **acetylcholinesterase**. This is necessary before the synapse can function again.

Examples of drug action at the synapse

Drugs such as atropine block the action of the chemical neurotransmitter, **acetylcholine**, at the postsynaptic receptor sites in neurones supplying the heart muscle, smooth muscle and exocrine glands, e.g. salivary glands. This is why atropine given before a general anaesthetic to dry up secretions will cause a very dry mouth. Atropine is also used to increase heart rate where *heart block*, which may occur after a *myocardial infarction* ('heart attack'), is causing an abnormally slow heart rate (*bradycardia*).

Table 3.1. Neurotransmitters.

Neurotransmitter and production sites	***Action: excitatory (E) or inhibitory (I)***
(1) MONOAMINES	
Acetylcholine (classified by some authorities as a monoamine) – CNS: brain PNS: skeletal muscle neuromuscular junctions, some synapses of the autonomic nerves	Both (E) and (I) depending on site
Noradrenaline – CNS: brain PNS: some synapses of the autonomic nerves; adrenal medulla also secretes noradrenaline as one of its hormones	Both (E) and (I) depending on site and type of receptor
Dopamine – CNS: basal nuclei and extrapyramidal tracts	(E) and (I)
5-hydroxytryptamine (5-HT) (serotonin) – CNS: brain and spinal cord	(I)
(2) AMINO ACIDS	
Gamma-aminobutyric acid (GABA) – CNS: brain and spinal cord	(I)
Glycine – CNS: brain and spinal cord	(I)
Glutamate – CNS: brain and spinal cord	(E)
(3) NEUROPEPTIDES	
Enkephalin – CNS: brain and spinal cord; pituitary gland	(I) Acts as a natural opiate (painkiller)
Endorphins – see Enkephalin	
Somatostatin – CNS: brain; also secreted by the pancreas	(I) Also inhibits growth hormone release
Substance P – CNS: brain PNS: sensory nerves carrying pain impulses	(E) Concerned with the transmission of pain

Pancuronium can be used to block the action of the neurotransmitter acetylcholine at receptor sites in neuromuscular junctions. It is used to relax skeletal muscle during surgery and in situations where ventilation is being provided by intermittent positive pressure respiration/ventilation (IPPR/V) as 'life support'.

Neurotransmitters

Detailed consideration of specific neurotransmitters can be found in *Table 3.1* and in other Chapters in Section 3. Neurotransmitters facilitate the passage of nerve impulses in all parts of the nervous system; they allow the vital connections between neurones and between neurones, muscles and glands.

Many different chemicals act as neurotransmitters, some have an excitatory effect and others are inhibitory. The major chemical groups of transmitters are: (i) *monoamines*; (ii) *amino acids*; and (iii) *neuropeptides*, a large group of substances that are thought to act as neurotransmitters. (*Peptides* are substances formed from short chains of amino acids.)

The list of neurotransmitters is certainly not complete – more examples are being discovered as research continues in this area.

Factors affecting nerve transmission

The transmission of the impulse depends upon adequate stimuli, the physical integrity of the neurone and functioning synapses.

A vast array of drugs affect nerve transmission; earlier we mentioned atropine and pancuronium, but other examples include sedatives and local anaesthetics.

A nerve subjected to temporary pressure or extreme cold will not have an adequate blood supply and the consequent lack of oxygen will reduce the transmission of impulses. When you get up after sitting on your leg, the tingling and 'pins and needles' you feel is due to resumption of impulse transmission as the pressure on the nerve is released. Numbness of the fingers and toes on a winter's day is due to the effect of cold on nerve function.

Apart from pressure and cold, nerve impulse transmission may be disordered by factors which include:

- Trauma, where the nerve is subjected to prolonged compression, is crushed or severed.
- Bacterial toxins, such as those produced by the bacteria causing *tetanus*.
- The effects of certain viruses e.g. *Herpes zoster* (shingles) and *human immunodeficiency virus (HIV)* (see Chapter 19).
- Metals such as lead and mercury, which are toxic to nerves.
- Problems with neurotransmitters, e.g. lack of *dopamine* causes the condition of *parkinsonism,* which is characterized by tremor, rigidity and walking difficulties.
- Demyelination, with its associated loss of impulse conduction in affected neurones, e.g. multiple sclerosis (see *Family-centred Study*, page 59). *Neuropathies* (nerve diseases) caused by lack of vitamin B_{12} and the *Guillain–Barre* syndrome are also due to demyelination.
- Nutritional problems, e.g. deficiency of the B vitamins, which cause peripheral neuropathy and degeneration in the spinal cord. This may be seen when the intake of alcohol is excessive, or in elderly people, who have an inadequate diet.
- Alcohol in excess is itself toxic to nerves, as well as causing problems with B vitamin intake and absorption.

Nursing Practice Application Pressure on nerves

Bandages, plaster casts, splints and crutches can cause pressure on nerves (see also Chapter 5). Care plans and preparation for discharge advice should reflect an awareness of this fact. Patient teaching should stress the importance of watching for altered nerve function caused by pressure, e.g. tingling 'pins and needles' or numbness, and reporting problems as soon as they are noticed. Tingling and numbness may also be caused by compression of blood vessels (see Chapter 17, Person-centred Study - Dave).

The Reflex Arc

Reflexes are rapid involuntary motor responses to sensory input. They illustrate the integration of the fundamental concepts considered in this chapter – sensory receptors, neurones, nerve impulse transmission, synapses and effectors (*Figure 3.6*).

There are several types of reflex (see Chapters 5 and 6) but an example of a simple reflex action would be the events (reflex arc) initiated when your finger touches a needle. Sensory receptors in the skin send impulses via the sensory neurone (PNS) to the spinal cord (CNS), where synapses and an interneurone are used to transmit the impulse to the motor neurone (PNS) connected to the muscle (effector). Contraction of the muscle causes withdrawal of your finger from the needle.

There are situations where reflexes can be modified by voluntary control. For example, your parents have a very old and valuable dinner service. One mealtime you pick up a very hot serving dish but, instead of dropping the dish (reflex) you hold on long enough to put it down safely. The reflex has been modified by knowing that the dish is irreplaceable!

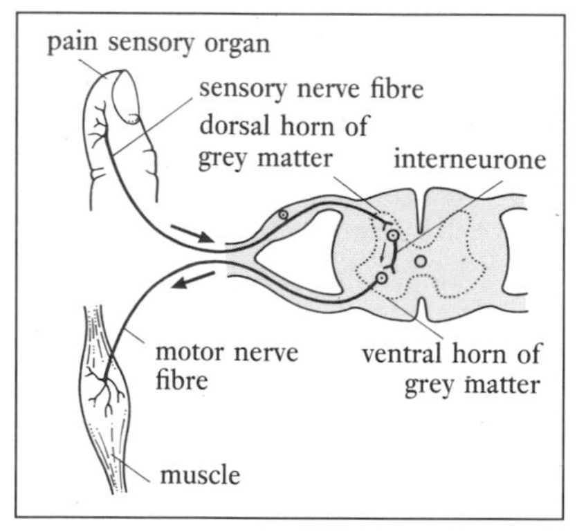

Figure 3.6. The reflex arc.

Summary/Check List

Introduction and basic organization – nervous system.
Nerve tissue – Neurones – cell body, axon, dendrites, myelination, *Family-centred Study* – Jill, Simon, Caroline and Kevin. Supporting cells (neuroglia) – Schwann cells, oligodendrocytes, astrocytes, microglia. Nerves – endoneurium, perineurium, epineurium. Sensory receptors – structural and functional classifications, adaptation, sensory unit.
Nerve impulse – Impulse transmission/conduction – resting potential, action potential, saltatory conduction, speed and size of nerve fibre, all-or-none law, thresholds, refractory periods. Synapses – electrical, chemical, summation, drugs and the synapse. Neurotransmitters – monoamines, amino acids and neuropeptides. Factors affecting nerve transmission – drugs, cold, pressure, other factors causing disordered transmission, e.g. alcohol. *Nursing Practice Application* – pressure on nerves.
Reflex arc – basic components, example of simple reflex, voluntary modification.

?

Self Test

1 The brain and spinal cord are known collectively as the:
(a) Peripheral nervous system.
(b) Central nervous system.
(c) Voluntary nervous system.
(d) Autonomic nervous system.
2 Briefly describe a typical neurone.
3 Which of the following statements are true?
(a) Neuroglial cells can replicate.
(b) Schwann cells are concerned with myelination in the CNS.
(c) Astrocytes are part of the 'blood–brain barrier'.
(d) The myelin sheath increases the speed of impulse transmission.
4 Put the following sensory receptors with their correct function:

(a) proprioceptor	(b) temperature
(c) photoreceptor	(d) chemical environment
(e) thermoreceptor	(f) pressure
(g) chemoreceptor	(h) light
(i) baroreceptor	(j) position sense

5 Briefly describe the events which cause an action potential.
6 What are the components of a chemical synapse?
7 Which of the following neurotransmitters is found at the neuromuscular junction in skeletal muscle:
(a) Dopamine
(b) Noradrenaline
(c) Acetylcholine
(d) GABA
8 Why does premedication which includes atropine cause a dry mouth?
9 Which of the following statements about demyelinating conditions are true?
(a) Physical trauma causes loss of myelin.
(b) Multiple sclerosis is a common demyelinating condition.
(c) Lack of vitamin B_{12} in the diet may lead to demyelination.
(d) The effects of demyelination are always the same.
10 Draw a simple reflex arc.

Answers
1. b. 2. See page 58. 3. a, c, d. 4. a-j; c-h; e-b; g-d; i-f. 5. See pages 61-62. 6. See page 63. 7. c. 8. See page 63. 9. b, c,. 10. See page 66 *(Figure 3.6)*.

Further Reading

Stein, J. F. (1982) *An Introduction to Neurophysiology*. Oxford: Blackwell Scientific Publications.
Ross Russell, R. W. and Wiles, C. M. (1985) *Neurology*. London: Heinemann.

Useful Addresses

Multiple Sclerosis Society
of Great Britain and Northern Ireland
25 Effie Road
London SW6 1EE

CHAPTER FOUR

The Central Nervous System

Overview

- Central nervous system (CNS) – development, protection, blood supply.
- Brain structure and function in detail.
- Spinal cord.
- Sensory and motor pathways.
- Pain.

Learning Outcomes

After studying Chapter 4 you should be able to:

- Outline the development of the brain and spinal cord.
- Name the meninges and describe their function.
- Describe the ventricular system and the formation, circulation and functions of cerebrospinal fluid (CSF).
- Explain the causes and effects of hydrocephaly.
- Explain what is meant by the 'blood–brain barrier'.
- Explain how an adequate blood supply to the brain is ensured at all times.
- Identify the main areas of the brain and briefly discuss their function.
- Describe the cerebral hemispheres.
- Discuss the functional areas of the cerebral cortex.
- Describe the location and role of the basal nuclei, thalamus, hypothalamus, pineal body and pituitary gland.
- Discuss the internal capsule, pyramidal and extrapyramidal motor tracts.
- Discuss the effects of damage caused by cerebrovascular accident (CVA) to the pyramidal tracts.
- Discuss the effects of extrapyramidal disorders such as parkinsonism.
- Identify the different areas which form the brain stem and discuss their role in the control of vital functions.
- Describe the structure and function of the cerebellum.

Learning Outcomes (continued)

- Describe the reticular formation and its role in levels of consciousness and sleep.
- Discuss the nursing observations used to assess neurological status. Describe how each is affected where intracranial pressure is raised.
- Describe the limbic system and its links with emotional state.
- Describe the structure of the spinal cord.
- Describe the major sensory and motor tracts in the spinal cord.
- Distinguish between flaccid and spastic paralysis.
- Describe pain transmission and perception, and discuss methods of pain relief.
- Describe the effects of spinal injury and spinal shock.

Key Words

Affective – having an emotional component, e.g. intonation of voice expresses emotion.
Basal nuclei (*ganglia*) – islands of grey matter situated deep within the white matter of the cerebral hemispheres.
Blood-brain barrier – the structural arrangement that ensures that brain capillary walls are relatively impermeable. This generally prevents the passage of harmful substances from blood to brain.
Brain stem – collective name for the midbrain, pons varolii and the medulla oblongata.
Cerebellum – part of the hindbrain situated below the cerebrum and behind the pons varolii.
Cerebral cortex – the thin layer of grey matter which covers the white matter of the cerebral hemispheres.
Cerebrospinal fluid (CSF) – fluid found in the ventricles of the brain and between the membranes that cover the brain and spinal cord. It provides a protective cushion around these vital structures.
Cerebrum – the forebrain; divided into hemispheres, it forms the largest part of the brain.
Choroid plexus – specialized blood capillaries within the ventricles of the brain, involved in the production of CSF.
Decussate – crossing over, i.e., of the motor nerves.
Meninges – three protective membranes (dura mater, arachnoid mater and pia mater) which cover the central nervous system (CNS).
Pyramidal cells (*Betz*) – large pyramid-shaped cells. They form the neurones of the precentral motor area of the cerebral cortex.
Ventricles – cavities in the brain which contain CSF.

Introduction

The **central nervous system (CNS)** consists of the brain and spinal cord (*Figure 4.1*). This has already been mentioned in Chapter 3 but now a more detailed approach will be taken. The CNS is the vital integration and communications network which controls and regulates every body function and every minute detail of our unique behaviour. As you would expect, the structure and functioning of the 'command centre' of the body are extremely complex. Many areas are as yet poorly understood, but our knowledge of the structure and, more significantly, function continues to increase. This is especially so since the advent of advanced scanning technology, which has increased

Ultrasound scan *– imaging technique where high frequency sound waves are used to produce an image.*

Computerized axial tomography *– imaging technique that uses computer-linked Xray equipment to provide cross-sectional views through various body planes.*

Magnetic resonance imaging (MRI) *– a technique that utilizes the response of hydrogen nuclei within tissues when exposed to a magnetic field. Used to determine biochemical function and to differentiate between different disease processes.*

Positron emission tomography (PET) *– a scanning technique which uses radioactive isotopes with a very short half life (Chapter 2). The radioactive isotopes gives off positrons (subatomic particle with a positive charge). When positrons meet electrons (negative charge) in the tissues there is an emission of gamma radiation. The areas where the gamma rays are detected will identify cells which are metabolically active. Useful for assessing function rather than structural changes.*

accessibility to the previously rather hidden CNS, e.g. *ultrasound (US), computerized axial tomography (CAT), magnetic resonance imaging (MRI)* – all used in the diagnosis of disease and *positron emission tomography (PET)* – used for research and for obtaining information about body function).

Development of the CNS

Development of the brain and spinal cord has started by the third week of pregnancy. A **neural plate** formed from the embryonic tissue ectoderm (see Chapter 1) eventually gives rise to all the neural tissue. This plate grows and folds in on itself (*invaginates*) to form the **neural tube**, the structure which, after many changes in shape, eventually becomes the brain and spinal cord. During pregnancy the developing nervous system is vulnerable to any damage which will affect formation and subsequent function. This is especially so in the early stages of development when several factors are associated with damage to the fetal nervous system, e.g. certain viruses, drugs and alcohol, lack of oxygen and lack of folate (a vitamin of the B complex).

Common developmental problems, such as *spina bifida* (which affects around 1 in 200 pregnancies), which may be associated with hydrocephaly (see page 74), occur when the neural tube is developing. These developmental problems are associated with structural defects in the vertebral column, which may leave the delicate neural tissue exposed and unprotected. This can result in varying degrees of abnormality. They may present no problems, cause disability or be incompatible with life.

After birth the CNS continues to develop; this development is especially important during the period from birth to 4 years, but continues until 13–15 years when adult brain size is reached (1300–1500 g) and myelination is complete. The male brain is larger than that of the female but brain size is proportional to overall body size.

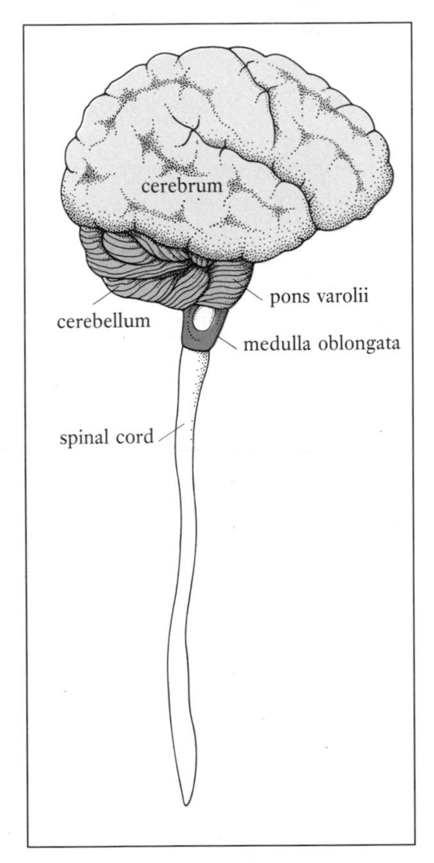

Figure 4.1. The central nervous system.

Protection for the CNS

The brain and spinal cord are well protected from external trauma and adverse internal changes by the bones of the skull, the vertebrae, the **meninges**, **ventricles** and **cerebrospinal fluid (CSF)**, an elaborate blood supply and the '**blood–brain barrier**' (see page 74).

Skull and vertebrae

The brain is entirely enclosed by the bones of the skull and the spinal cord runs within the bones of the *vertebral column* (see Chapter 18), deep inside the body.

Meninges

The three membranes of the meninges – the dura mater, the arachoid mater and the pia mater – cover and enclose the brain and spinal cord. They are formed from connective tissue (see *Figure 4.2*).

The outer membrane is the tough fibrous **dura mater**, which lines the skull and vertebral column and forms double folds of tissue – the **tentorium cerebelli** and **falx cerebri** – which divide areas in the brain and enclose the **venous sinuses** which drain blood from the brain. The space between the dura mater and the vertebra is called the **epidural space**. Local anaesthetic drugs are injected into this space at the lumbosacral region to relieve pain, especially during labour.

The middle membrane is the fragile, web-like **arachnoid mater**, which is separated from the dura mater by the **subdural space**. Beneath the arachnoid mater is the **subarachnoid space**, which is filled with cerebrospinal fluid. Bleeding into this space is known as **subarachnoid haemorrhage** and may result from the rupture of an aneurysm (an abnormal dilatation in the wall of an artery) or a cerebral bleed.

The inner membrane or **pia mater** is a fragile membrane containing many tiny blood vessels. It is in close contact with the brain and spinal cord where it follows every contour. *Meningitis* or inflammation of the meninges is a serious condition usually caused by bacterial or viral infection.

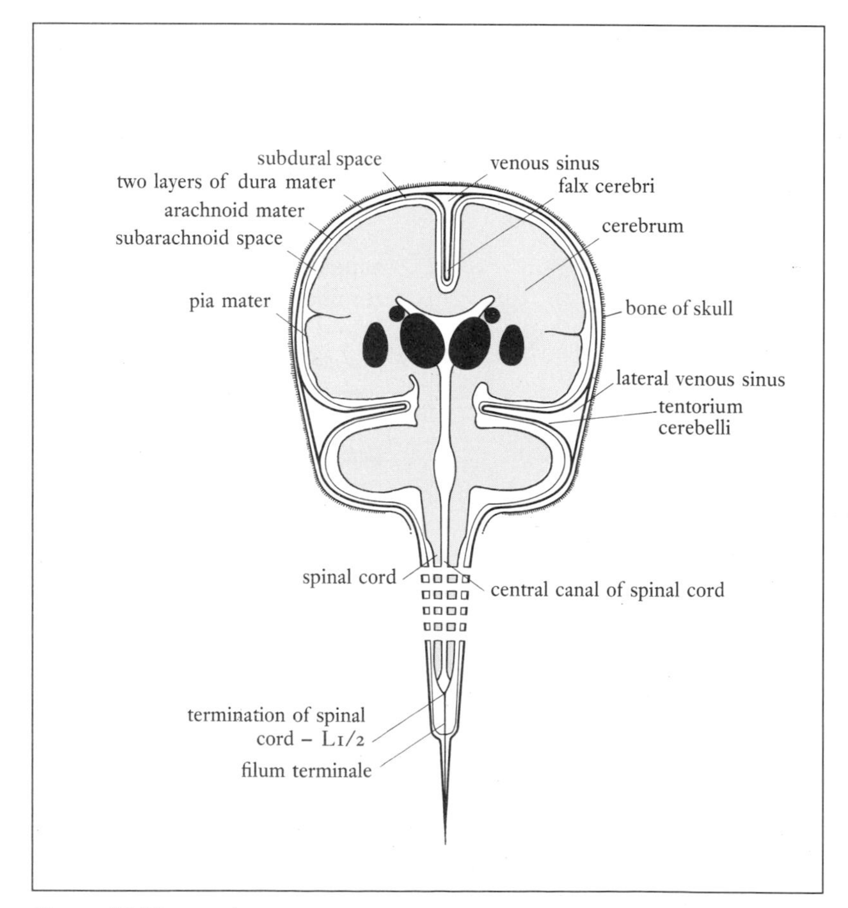

Figure 4.2. The meninges.

Ventricular system and cerebrospinal fluid

The ventricular system (*Figure 4.3(a)*) within the brain consists of the two **lateral ventricles**, located in the **cerebrum**, which communicate with the **third ventricle** via the **intraventricular foramina** (a foramen is a hole, or opening). The third ventricle is situated between the two parts of the **thalamus** (see page 81); it is connected to the **fourth ventricle** by the **cerebral aqueduct**. The location of the fourth ventricle is between the **cerebellum** and **pons varolii** (see *Figure 4.12*).

Cerebrospinal fluid (CSF) is produced by specialized capillaries known as **choroid plexuses** which line the ventricles (a plexus is a network of blood vessels or nerves). The CSF is a crystal-clear, slightly alkaline fluid derived from blood plasma (Chapter 9). As much as 300-500 ml per day is produced but only about 125–150 ml circulates within the system (it is being reabsorbed), where it exerts a pressure of 1.33–4 kPa (10–30 mmHg) depending on body position. Its specific gravity is approximately 1.005, which is similar to that of the brain tissue.

CSF consists of:

- Water.
- Proteins, e.g. globulins, albumin (at a lower concentration than in plasma).
- Glucose (at a lower concentration than in plasma).
- Electrolytes (more sodium and chloride and less potassium and calcium than plasma).
- Leucocytes (fewer than 5 x 10^6/litre).
- Small amounts of waste products.

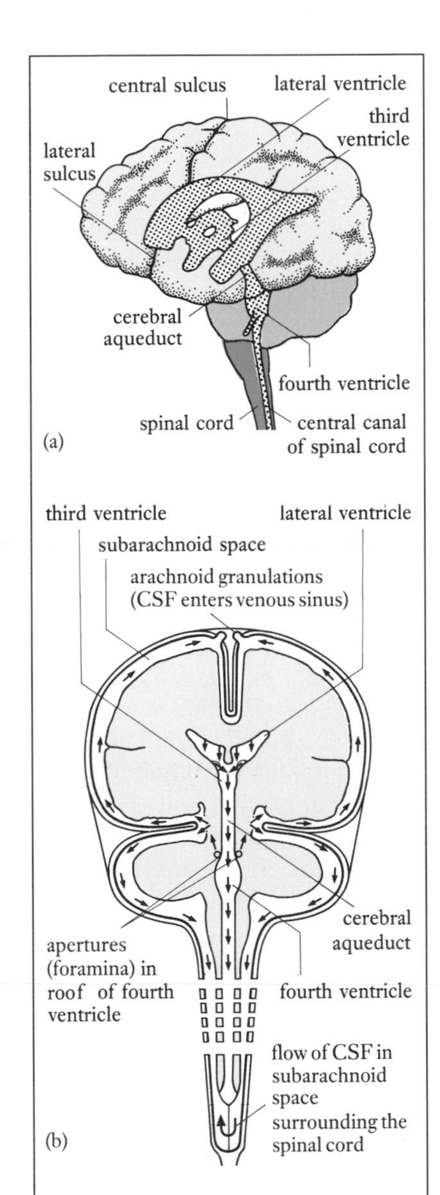

Figure 4.3. Position of the ventricular system: (a) lateral view (shown on intact brain); (b) circulation of cerebrospinal fluid (simplified).

The CSF enters the subarachnoid space through foramina in the fourth ventricle and circulates around the brain and spinal cord. Reabsorption of CSF occurs through the **arachnoid granulations** – portions of arachnoid mater which project into the venous blood sinuses (*Figure 4.3(b)*). The spinal cord ends at the level of the first/second lumbar vertebra and is secured to the vertebra by a modified portion of pia mater called the **filum terminale** (see *Figure 4.2*). The fluid-filled space below the spinal cord, which contains the last spinal nerves (the **cauda equina**) (see *Figure 4.15*), provides a safe area for obtaining specimens of CSF by *lumbar puncture.*

Functions of CSF are:

- To provide a cushion between the brain and skull which acts as a shock absorber in the event of trauma.
- To be involved in the metabolic processes of the brain by the exchange of substances between brain and CSF.
- To support the structures of the CNS by providing the correct external pressure.
- To form a protective barrier around the brain and spinal cord.

Hydrocephaly

Obstruction to the circulation of CSF or a failure in the reabsorption processes causes *hydrocephaly*, where excess CSF is present. This may be due to problems with nervous system development or be secondary to infections and tumours. It causes raised intracranial pressure and eventual damage to the ventricles and brain tissue. When hydrocephaly occurs in babies the bones of the skull, which have not yet fused together, are pushed apart and the head circumference increases (the structure of the skull in babies is discussed in Chapter 18). The management of hydrocephaly involves the use of specialized valves, e.g. *Spitz–Holter* valves, which drain excess CSF into the right side of the heart or peritoneal cavity from where it enters the circulation.

Blood-brain barrier

In common with other parts of the body the brain requires a stable internal environment for its proper functioning; it cannot cope with chemical variations in its extracelluar fluid. A special structural arrangement of **capillary endothelial (epithelium)** cells and **astrocytes** ensures that the brain capillary wall is relatively impermeable, and comprises the blood–brain barrier. This allows the passage of nutrients required by the neurones but generally prevents the entry of substances which may affect function and the loss of vital substances from the brain cells. There are, however, exceptions – as alcohol and certain drugs, e.g. anaesthetics, can cross the barrier and affect brain function. In babies and young children the barrier is less effective and allows the passage of substances, such as the bile pigment bilirubin (see Chapter 14) and lead, which may damage the brain.

Blood supply to the brain

It is essential that the brain receives sufficient oxygenated blood to maintain its complex metabolism. About 750-800 ml of blood circulates around the brain every minute, which means that the brain needs to receive approximately 15% of the cardiac output (the amount of blood pumped out by the heart in one minute – approximately 5 litres at rest). Any disruption in supply will rapidly cause changes in brain function; these range from the short period of unconsciousness experienced in fainting to the irreversible damage caused by a longer disruption, e.g. cardiac arrest.

To ensure the continuity of supply the brain has two pairs of major arteries (the *internal carotid arteries* and the *vertebral arteries*) feeding into a circular arrangement of arteries (*circle of Willis*) situated at the base of the brain (see *Figures 10.33, and 10.34*). Branches from this circle maintain the blood supply to different areas of the brain; this means that a problem in one branch does not cut off supply to the whole brain.

Blood flow can be increased in response to changes in blood oxygen, carbon dioxide and pH levels, which are continually monitored by the chemoreceptors; flow can also be diverted to areas with an increased need for oxygen. This is a good example of the

Nursing Practice Application Lumbar puncture

The diagnosis of conditions such as meningitis, multiple sclerosis and subarachnoid haemorrhage can be greatly enhanced by the examination of a specimen of CSF obtained by lumbar puncture. This involves the introduction of a hollow needle (after use of local anaesthetic) into the subarachnoid space between lumbar vertebrae 3 and 4 or 4 and 5; the use of these spaces ensures that the spinal cord is not damaged.

This investigation should not be undertaken when a person has raised intracranial pressure (an increase in pressure within the cranial cavity - Chapter 1). Release of pressure during the puncture could cause the **brain stem** to be forced through the *foramen magnum* (see pages 87-88) with serious results including unconsciousness.

It is important that the nurse explains the procedure and ensures that the person understands what will happen both during and after. To prevent the introduction of infection, sterile equipment is used and staff must adhere strictly to aseptic techniques. As well as obtaining specimens the pressure of the CSF can be measured by attaching a manometer to the needle.

Following lumbar puncture the person should stay in bed for up to 24 hours with only one pillow and be advised to sit up only very gradually. This prevents headache caused by loss of CSF and a reduction in its cushioning effect. The care plan should include: provision of adequate fluids to replace CSF volume, prescribed analgesia for headache or backache, neurological observations where approriate (see pages 87,88 for detailed discussion) and regular checks of puncture site for leakage of CSF.

ability of the brain to regulate its own homeostasis. Venous blood leaves the brain in blood plexuses and sinuses, which drain into the *internal jugular vein* and then into the *superior vena cava*, which carries blood to the right atrium of the heart (see *Figures 10.32, 10.35* and *10.36*).

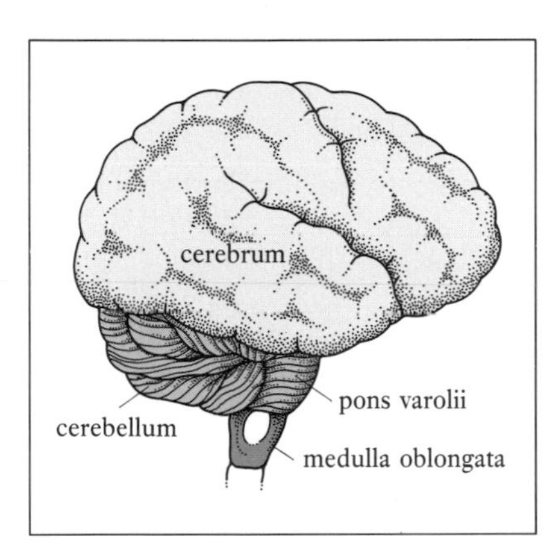

Figure 4.4. The brain.

The Brain

The brain is a semisolid (soft, with a consistency between a fluid and a solid) organ which occupies the cranial cavity (*Figure 4.4*). It consists of the **cerebrum** which forms from the embryonic forebrain (**cerebral hemispheres** and other structures such as the **thalamus**), **brain stem** (**midbrain, pons varolii** and **medulla oblongata**) and the **cerebellum,** which is part of the **hindbrain**.

Cerebrum

The cerebrum is the most highly developed and sophisticated part of the human brain. It consists of the two cerebral hemispheres, with the thalamus (see page 81), the **hypothalamus** (see pages 82-83) and the associated **pituitary gland** and **pineal body** (see pages 82-83).

Cerebral hemispheres

The cerebral hemispheres (*Figure 4.5*) are situated in the **anterior** and **middle cranial fossae** (a depression in a bone) and form the bulk of the brain mass. The **midsagittal** or **longitudinal fissure** (a fissure is a split or a cleft) divides the right and left hemispheres. They are also separated by a fold of dura mater, the falx cerebri, which dips down into the fissure to the level of the **corpus callosum.** The corpus callosum is a band of white matter which connects the two hemispheres deep within the brain. The nerve fibres in the corpus callosum are concerned with communication between the two

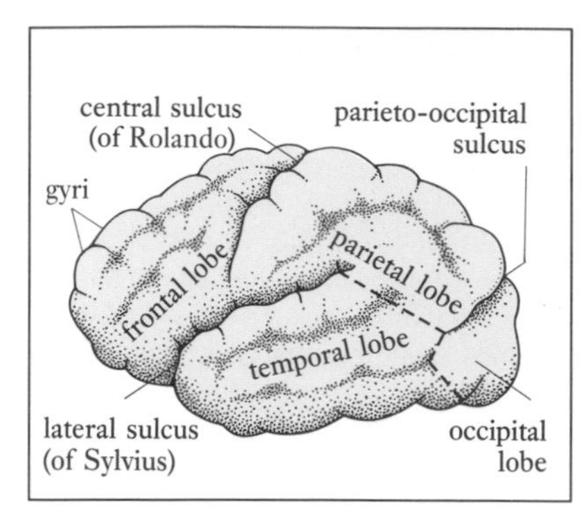

Figure 4.5. Cerebral hemispheres – gyri, sulci and lobes.

hemispheres. On the outer surface of the cerebral hemispheres is a thin layer of grey matter called the **cerebral cortex**. This cortex covers the white matter, which forms the major part of the hemispheres (although deep within the cortex there are isolated islands of grey matter called the **basal nuclei**; see *Figure 4.9*). The surface of the cerebral cortex has many folds and ridges. The raised areas of these convolutions are called **gyri** (singular, gyrus), they are separated by furrows – the **sulci** (singular, sulcus). This folding allows a much greater surface area of cerebral cortex to be fitted into the skull.

Each hemisphere is divided into lobes: **frontal, parietal, temporal** and **occipital**, which have the same names as the bones of the skull which overlie them. Deep sulci separate each lobe: the **central sulcus (sulcus of Rolando)** between the frontal and parietal lobes, the **lateral sulcus (sulcus of Sylvius)** separates the frontal from the temporal lobe and the **parieto–occipital sulcus** is found between the occipital and parietal lobes (*Figure 4.5*).

The gyri on either side of the central sulcus are very important functionally: the **precentral gyrus** is the main motor area of the cortex and the **postcentral gyrus** is the main sensory area. The white matter lying under the nerve cells of the cortex consists of three types of myelinated nerve fibres which form large tracts:

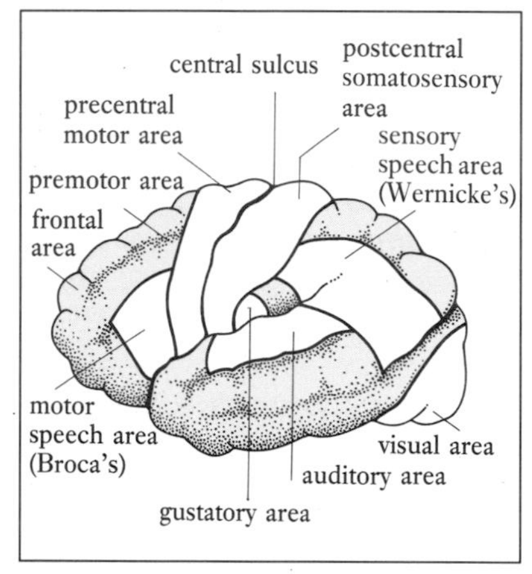

Figure 4.6. Functional areas of the cerebral cortex.

- **Commissures**, which connect the two cerebral hemispheres, the most important being the corpus callosum.
- **Association fibres**, which allow communication between individual gyri and between the lobes of the same hemisphere.
- **Projection fibres**, which are the fibres that connect the cerebral cortex with other areas of the brain, sensory receptors (afferents) and effectors (efferents).

The **internal capsule** (see *Figure 4.9*) is an area between the thalamus and the basal nuclei. It is formed by projection fibres massing together as they pass through the restricted space. Motor fibres leaving the cortex and sensory fibres travelling to the cortex are particularly vulnerable to damage in this area from a *stroke* or *cerebrovascular accident (CVA)* (see *Special Focus*, page 81).

Functional areas of the cerebral cortex

The cerebral cortex is the 'thinking' part of the brain; it is where you process sensory inputs, communicate, understand, make decisions and regulate voluntary muscle action. It is responsible for intellect, learning, memory, our sense of responsibility, and allows us to differentiate right from wrong (moral sense), i.e. the *higher functions*. Although we know a lot about cortical function it must be remembered that functional areas overlap, and any description is bound to suffer from some degree of simplification (*Figure 4.6*).

Motor areas

The main motor area is the **precentral primary motor area** situated in the frontal lobes of both hemispheres just anterior to the central sulcus. The neurones in the motor cortex are the pyramidal **Betz cells**

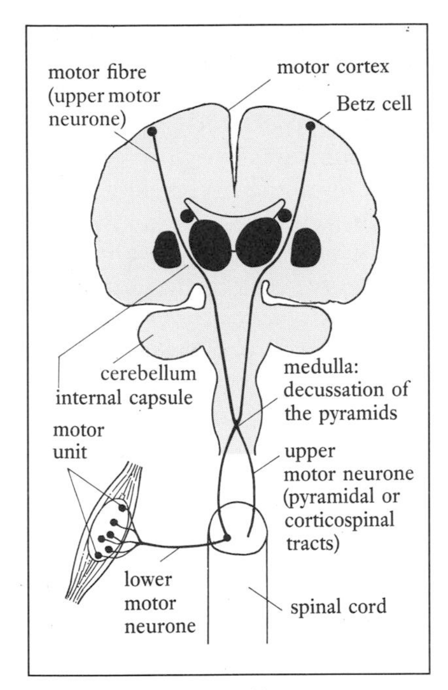

Figure 4.7. Upper and lower motor neurones.

which control voluntary (skeletal) muscle contraction, e.g. when you decide to flex your arm to pick up a glass. The fibres of the Betz cell bodies pass through the internal capsule to the medulla, where most cross over to the other side of the brain (**decussation**), with the result that the left motor cortex initiates voluntary movement on the right side of the body and vice versa.

The neurone from the cerebral cortex, known as the **upper motor neurone**, travels down the **pyramidal** or **corticospinal tracts** (see page 81) in the spinal cord to the appropriate level, where it synapses with the cell body of a second neurone, called the **lower motor neurone** (*Figure 4.7*). This second neurone, which arises in the spinal cord, carries the nerve impulse from the spinal cord to the neuromuscular junction in the voluntary muscle. One lower motor neurone supplies a number of muscle fibres known as a **motor unit**. Upper and lower motor neurones are discussed again on pages 91-92; a more specific knowledge is helpful in understanding how damage to the neurones causes the different effects (paralysis and muscle atrophy) seen in nursing practice.

The body is represented upside down on the motor cortex, with the toes being controlled from the medial part, and the head from the lowest part by the temporal lobe. In *Figure 4.8* the size of some body areas has been distorted, so that the size represents the amount of innervation and hence the movements possible, e.g. the tongue appears large because it has a large number of motor nerve fibres compared with other areas such as the trunk.

Anterior to the precentral motor area is the **premotor area** (see *Figure 4.6*) which exerts a general influence over the motor cortex and is responsible for learning motor skills that require manual dexterity. These skills involve several muscle groups in a repeated series or pattern of movements, such as those needed to use a computer keyboard, play the piano or tie a shoe lace.

Within the premotor area in one hemisphere is the motor speech area – **Broca's area**. The neurones here are responsible for the multitude of muscular movements needed for articulation. Broca's area is situated at the base of the frontal lobe just above the lateral sulcus. In right–handed individuals it is found in the left hemisphere and in left-handed people in the right hemisphere.

Sensory areas

The main sensory area is the **postcentral somatosensory cortex**, which is situated immediately posterior to the central sulcus in the parietal lobes of both hemispheres. This area receives sensory input from receptors in the skin (temperature, pressure, light touch and pain) and the proprioceptors in skeletal muscles. It can localize sensation to particular areas of the body, with the sensory input from the right side of the body being received in the left hemisphere and vice versa. Decussation of some sensory fibres occurs in the spinal cord, with others crossing over in the medulla. Sensory impulses are transmitted from receptor to sensory cortex via a three-neurone pathway; the first synapse is in the spinal cord or medulla and the second in the

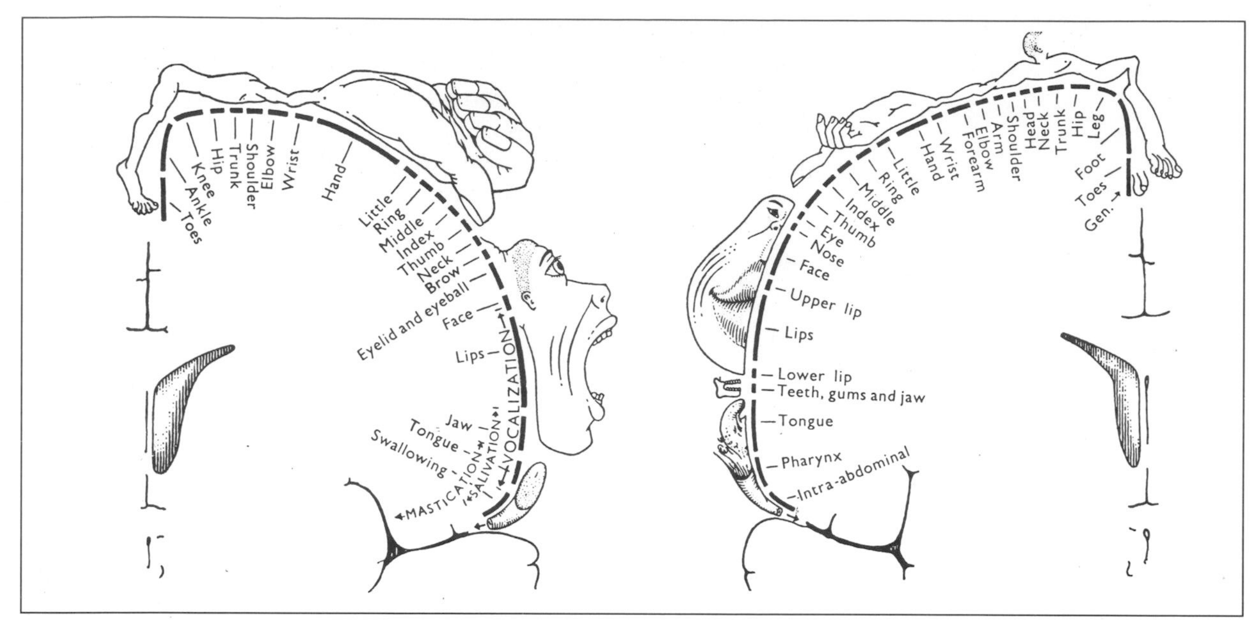

Figure 4.8. Representation of the motor and sensory areas (homunculus) (Adapted with the permission of Macmillan Publishing company from *The Cerebral Cortex of Man* by Wilder Penfield and Theodore Rassmussen. Copyright 1950 Macmillan Publishing company; copyright renewed © 1978 Theodore Rassmusen.)

thalamus. Again the parts of the body are represented upside down on the sensory cortex, with highly sensitive areas drawn much larger than their real size on the sensory homunculus (*Figure 4.8*), e.g. the face and finger tips have many more sensory receptors than the trunk.

The area just posterior to the postcentral sensory cortex is concerned with our ability to remember objects and to identify them by touch, e.g. with your eyes shut you can differentiate between a tennis ball and a golf ball placed one in each hand. This ability requires the integration of information about different factors such as size, shape and texture. There are connections between this parietal area and the postcentral sensory cortex which facilitates the integration of sensory input so that we can perceive specific sensations and know which area of the body is involved.

The sensory cortex has some highly specialized regions which receive sensory input from the sense organs. The sensory speech area, or **Wernicke's area**, is the part of the temporal lobe where the sounds of speech and visual impulses from written communications are received and understood. This area is situated with the motor speech area in the left hemisphere in right-handed individuals and in the right hemisphere in left-handed individuals; connection between the two areas allows language to be both understood and articulated.

Affective language areas in the opposite hemispheres to Broca's and Wernicke's areas provide an emotional element in communication, e.g. variations in voice intonation are enormous and can range from a gentle sympathetic tone when speaking to bereaved relatives to the short sharp tone which conveys displeasure.

Impulses from the retinae of the eyes are transmitted via the **optic nerves** for interpretation by the **visual cortex**, which is situated in the occipital lobes. Memories of previous experiences allows the recognition of familiar objects and people.

The **auditory cortex** (hearing) in the temporal lobes close to the lateral sulcus interprets impulses received via the **vestibulocochlear** (**auditory**) **nerves** from the cochlea of the inner ear.

Our ability to perceive taste is provided by the **gustatory cortex** in the parietal lobe, this receives impulses from the tongue via several different nerves.

The **olfactory cortex** (smell) is an area in the temporal lobes concerned with the sense of smell and the ability to differentiate odours. It receives impulses from the nose via the **olfactory nerves**. The functioning of this part of the brain is much more important in species that rely on smell for recognition than it is in humans.

Other cortical areas

Apart from motor and sensory aspects the cerebral cortex is concerned with a variety of functions which affect our behaviour. The anterior part of the frontal lobe, known as the **prefrontal cortex**, probably controls areas such as personality, anticipation of events and their effects and intelligence. This area of the human brain is more developed than in any other animal and accounts for the characteristics, such as moral sense, which are probably unique to humans.

Disorders arising in the prefrontal cortex can cause aggression or apathy. The person may lack concern about his or her personal grooming and cease to conform to socially acceptable behaviour or think out the consequences of his or her actions. These changes will present great difficulties within the family, especially for those involved as informal carers. Nurses can provide support for the family through explanation and information regarding help available, and by listening to fears and feelings of anger or isolation.

Other areas in the cortex are involved with our ability to interpret events: complex sensory inputs can be integrated so that you act quickly and appropriately. For example, a toddler pulls a bowl of tomato soup over his or her arm – the sensory input you receive includes hearing the child scream, seeing the red mark on the child's arm (and the stain spreading on the carpet) and, possibly, smelling the soup. You are probably unaware of these individual inputs as you pick up the child and hurry to the cold tap where immediate first aid with cold water will cool the area and reduce tissue damage.

Right brain versus left brain

Most functional areas are duplicated in both hemispheres but some, such as the speech areas, appear only on the *dominant* side of the brain, i.e. on the left side in right-handed individuals, and vice versa. There seem to be functional differences between the hemispheres with lateralization of some functions. One side is associated with particular functions, e.g. the left cerebral hemisphere in right-handed people tends to control language, numeracy and processes requiring logic and the right hemisphere, in this instance, will control the creative, imaginative, intuitive and artistic aspects.

There is communication between the two sides of the brain, with one side influencing the other. In this way a balance is created between 'sensible' behaviour and the more emotional side of our nature.

Individuals who lack a dominant side may have problems caused by the lack of a 'boss'; for example, Marieb (1989) suggests that the reading difficulty *dyslexia* may be linked to this lack of dominance.

Basal nuclei

The **basal nuclei** are isolated areas of motor nerve cell bodies (grey matter) deep within the white matter of the cerebral hemispheres (*Figure 4.9*). The basal nuclei are sometimes called the basal ganglia which is not strictly correct, as ganglia are structures of the peripheral nervous system.

The basal nuclei consist of several masses of tissue (the **caudate nucleus**, the **putamen** and the **globus pallidus**) situated close to the thalamus and internal capsule. When grouped together they are generally known as the *corpus striatum* (caudate, putamen) and the *lentiform nucleus* (putamen, globus pallidus) although there is some controversy over naming. They form part of the **extrapyramidal** (outside the pyramidal tracts) motor pathways and have connections with many areas of the brain, including the cerebral cortex, the thalamus and each other. Their exact functions are not well understood but they appear to be important in the initiation of slow and prolonged motor activities such as arm swinging during walking, posture and muscle tone.

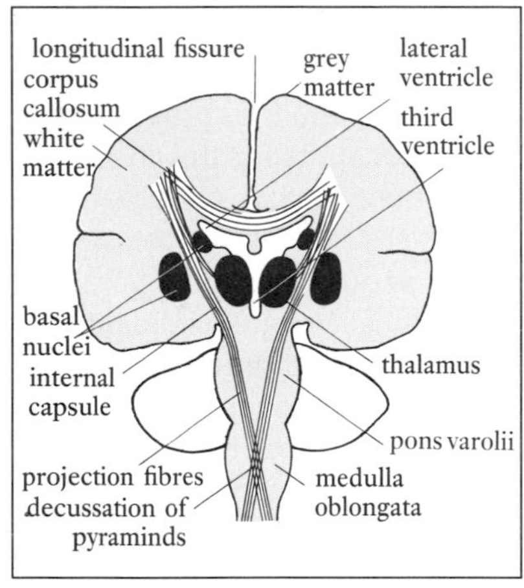

Figure 4.9. Cerebrum - white matter, basal nuclei, internal capsule, thalamus.

Their proper functioning is linked to the release of dopamine by another nucleus in the brain stem. When dopamine levels are reduced the basal nuclei no longer exert their inhibitory influences over motor activities, this leads to the development of the physical features of *Parkinson's disease* (parkinsonism). This is a common extrapyramidal condition which mainly affects older people. It is characterized by:

- Tremor, especially in the hands.
- Hesitation in starting movements, which causes problems with the initiation of walking – affected people tend to shuffle and have difficulty stopping once they are moving.
- Rigidity, which causes a mask-like facial expression.

Drugs such as L-dopa (a precursor (forerunner) of the neurotransmitter dopamine) are helpful in relieving the problems of parkinsonism, but may cease to be effective after prolonged use. Recent developments, involving the transplantation of dopamine-producing fetal tissue into the brains of affected individuals ('by stereotactic injection under local anaesthetic' – Fletcher 1992) have produced some encouraging results and the reduction of disability. There are, however, some very important ethical issues (the use of fetal tissue) and practical problems to be addressed before this technique becomes widely available.

Another example of an extrapyramidal disorder is *Huntington's disease (chorea)*, a hereditary condition which appears during adult life (usually during the 30s and 40s). Destruction of the basal nuclei causes involuntary jerky movements of the facial muscles and limbs. Later it progresses to dementia and eventual death. It is associated with a lack of GABA (see Chapter 3).

Internal capsule

The **internal capsule** is the area between the basal nuclei and the thalamus; it is formed by the efferent and afferent projection fibres which connect the cerebral cortex to other parts of the brain, muscles and sensory receptors(see *Figure 4.9*). The motor fibres which pass through the internal capsule form the **pyramidal tracts** (named for the pyramids in the medulla) or **corticospinal tracts**, which are the major motor pathways to the voluntary muscles. The motor activities controlled by these tracts are influenced (inhibited) by the extrapyramidal tracts which pass outside the internal capsule.

NB The tremor and rigidity of Parkinson's disease can be explained by the loss of basal nuclei extrapyramidal inhibition, which allows the pyramidal influences to become exaggerated.

Special Focus Cerebrovascular accident (CVA)

It is appropriate here to consider the very common problem of stroke or CVA. According to Macleod *et al.* (1987) every year 2 people in every 1000 of the population will have an initial stroke. This *pyramidal disorder*, which usually affects older people, can cause considerable disability in the 50% who survive the initial event. The various types of stroke are a leading cause of death (third most common) in developed countries. The disruption to cerebral blood supply and functioning may be due to haemorrhage, thrombus or embolism. Increased incidence is associated with arterial disease and high blood pressure. Frequently the damage occurs in the internal capsule and affects both the upper motor neurone fibres and the sensory fibres. The resultant damage and physical presentation can vary considerably and obviously depend on the location, the number of fibres and which fibres are affected.

Common problems include: (i) changes in consciousness – some people may be deeply unconscious following a stroke; (ii) *hemiplegia* (paralysis initially *flaccid* – muscles lack rigidity, decreased tone; becoming *spastic* – muscles are rigid, increased tone) or *hemiparesis* (weakness) on the side of the body opposite to the damage (remember motor fibres decussate in the medulla); (iii) problems with vision; which may include *diplopia* (double vision), visual field defects and blurred vision; (iv) communication difficulties may occur if the stroke affects the speech centre in the left side of the brain – the resultant problems with speech include *dysphasia*, which may be expressive (unable to express oneself) or receptive (unable to understand what is being said); (v) a decrease in emotional control and associated depression; and (vi) varying degrees of sensory loss.

Thalamus

The thalamus comprises two masses of grey matter situated in the cerebral hemispheres close to the basal nuclei and internal capsule. It forms part of the third ventricle and acts as a relay station for incoming impulses (see *Figure 4.9*). Some sensory neurones, you will remember, have their second synapse in the thalamus. The thalamus receives most of the sensory input from the sense organs, skin and viscera. It sorts this input and relays it to the sensory cerebral cortex. The thalamus also receives impulses from the basal nuclei and the cerebellum. It has an important role in the functioning of the cerebral cortex and memory.

Nursing Practice Application Care priorities (initial and long-term) following a stroke

- Maintenance of an adequate airway especially if unconscious. Altered consciousness will affect the cough reflex; this makes the inhalation of food, fluids or secretions more likely (see Chapter 12).
- Assessment of level of consciousness and other neurological observations (see pages 87,88). Any changes in condition will be noted, e.g. an extension of the damaged area.
- Careful positioning of limbs to aid mobility. This will help to prevent problems such as joint stiffness, foot drop and contractures.
- Frequent change of position to prevent pressure sore development (see Chapter 19). Remember that sensory loss may be present as well as motor loss. Change of position and breathing exercises will help to prevent chest infections.
- Maintenance of nutrition and hydration – may have difficulty with swallowing or feeding (see Chapter 13).
- Early involvement of nurse with medical officer, physiotherapist, speech therapist and occupational therapist to plan a rehabilitation programme which is realistic and involves the patient in the planning as appropriate.
- Involvement of family/friends in rehabilitation, education, advice and support for informal carers.
- Increase awareness of the affected side, i.e. encourage the person to 'remember' his or her affected limbs.
- Minimize problems with visual fields by approaching person from the unaffected side.
- Provide help with those activities that the individual is unable to manage, e.g. washing, dressing and using the toilet, but always encourage some independence.
- Encourage use of specially adapted equipment, e.g. velcro fastening on clothes and aids to mobility and continence.

Hypothalamus

Positioned below the thalamus, the group of nerve cells known as the **hypothalamus** forms part of the third ventricle and lie close to the point where the optic nerves decussate (**optic chiasma**). Hypothalamic tissue forms a stalk from which the **pituitary gland** or 'master' gland of the endocrine system (see Chapter 8) is suspended (*Figure 4.10*).

The hypothalamus is a vital component of homeostatic regulation; it contains some very important nuclei which influence most body tissues in some way. It regulates the action of the autonomic nervous system (see Chapter 6) which controls involuntary functions such as heart rate and digestion. Body temperature is controlled by the hypothalamus, which acts as a thermostat – the temperature of blood passing through the hypothalamus activates heat loss or heat conservation mechanisms as appropriate (see Chapter 19). Our appetite is regulated through the *hunger* and *satiety centres* in the hypothalamus, which control food intake; how useful it would be to have conscious control of these centres when trying to lose weight!

The hypothalamus is very important in the regulation of fluid balance and thirst; special cells called *osmoreceptors* respond to the osmolarity (see Chapter 2) of the blood and extracellular fluids. When the volume of body fluids is reduced these cells activate hypothalamic nulclei which cause antidiuretic hormone *(ADH)* to be released from the posterior pituitary gland. This hormone controls water loss in the kidney and reduces urinary volume (see Chapter 15). Other cells in the *thirst centre* cause you to feel thirsty and drink more fluid. In this way fluid balance is restored by reducing loss and increasing intake.

Through connections with other parts of the brain the hypothalamus

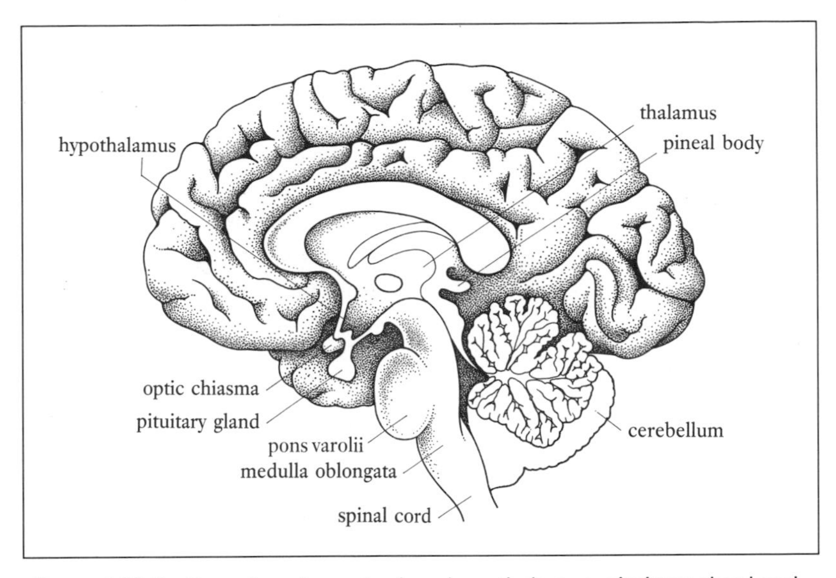

Figure 4.10. Section of cerebrum to show hypothalamus, pituitary gland and pineal body.

***5-HT**, also known as **serotonin**, is a monoamine (chemical with one amine group) derived from an amino acid. It is found in many body areas, e.g. brain, some blood cells, and intestinal tract. It is known to function as a neurotransmitter and is released when tissue damage occurs.*

is concerned with the expression of emotions such as rage, fear, pleasure and those involved in certain biological drives and rhythms, such as sexual behaviour. Certain cells in the hypothalamus form *pleasure* and *pain centres* which appear to influence our behaviour to a limited extent. The hypothalamus also influences our patterns of sleeping and waking, which are discussed on pages 86-87.

Earlier we mentioned that the pituitary gland is connected to the hypothalamus by a stalk, suggesting a close relationship between the two, which is definitely the case. Substances produced in the hypothalamus affect the release of anterior pituitary hormones by acting as releasing or inhibiting factors. The other connection is the fact that the two hormones released by the posterior pituitary (ADH and oxytocin) are actually produced by special hypothalamic nuclei. This link between hypothalamus and pituitary function is discussed more fully in Chapter 8.

Pineal body

The **pineal body** is situated close to the third ventricle and is part of the endocrine system (*Figure 4.10*). Its functions are not well understood but it is known to produce **melatonin** (a hormone derived from amino acid. It is known to inhibit the release of gonadotropin releasing hormone, which in turn controls the hormones which influence reproductive function) and **5-hydroxytryptamine**.

The pineal gland appears to be linked with biological rhythms, various diurnal cycles, sexual maturation and reproductive behaviour (see Chapters 8 and 20).

Brain stem

The brain stem is the lowest part of the brain and consists of the midbrain, pons varolii and medulla oblongata. This area of the brain controls the vital automatic functions, such as respiratory rate, contains the pathways which connect the cerebrum with the spinal cord and contains the nuclei of most of the cranial nerves (see *Figure 4.11*).

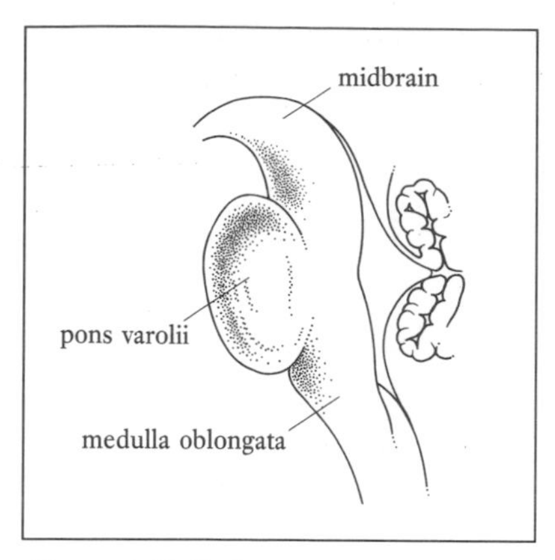

Figure 4.11. Brain stem.

Midbrain

The **midbrain** is the part of the brain stem between the cerebrum and pons varolii, which surrounds the cerebral aqueduct (page 73) connecting the third and fourth ventricles. It contains several nuclei, including:

- Those associated with the basal nuclei and the extrapyramidal motor tracts.
- Those involved with certain visual and auditory reflex functions.

The large descending and ascending nerve tracts in the midbrain are known as the **cerebral peduncles**.

Pons varolii

Situated between the midbrain and medulla the **pons varolii** has fibres which form a bridge (*pons*) between the two hemispheres of the cerebellum and connect the brain and spinal cord. It contains the nuclei of several cranial nerves and helps to control respiration.

Medulla oblongata

The **medulla oblongata** is the lowest part of the brain stem (and therefore of the brain). It extends from the pons to become the spinal cord, which leaves the cranial cavity via an opening in the skull known as the foramen magnum. It is the site of decussation of the major motor nerves of the corticospinal tracts, which form enlargements in the medulla called **pyramids** (which give their name to the pyramidal motor pathways). It receives impulses from sensory receptors in the skin and skeletal muscles; some of these sensory fibres decussate in the medulla prior to being relayed on to the thalamus. The medulla, which has connections with the cerebellum, contains the nuclei of several cranial nerves and special nuclei concerned with maintaining body equilibrium.

Many of the vital autonomic reflex functions required for homeostasis are controlled by centres in the medulla, including:

- The **respiratory centres**, which work with areas in the pons to control rate and depth of respiration (see Chapter 12). Different parts of the respiratory centres respond to increased carbon dioxide levels and, less importantly, to reduced oxygen levels in the blood; levels of these gases in the blood are monitored by chemoreceptors.
- The **cardiac centre**, which controls the strength and rate of cardiac contraction through the ANS (see Chapter 10).
- The **vasomotor centre (VMC)**, which responds to signals from the baroreceptors. It controls the diameter of the small peripheral arterioles (*peripheral resistance*) which regulates blood pressure. When the vessels constrict the blood pressure rises and vice versa (see Chapter 10).
- The **reflex centres**, concerned with sneezing, coughing, vomiting and hiccups.

Cerebellum

The **cerebellum** is situated in the **posterior cranial fossa**; it is located below the occipital lobe of the cerebrum and posterior to the pons and medulla *Figure 4.12*. It has two hemispheres, connected by the **vermis** (a worm-like structure), and consists of a cortex of grey matter covering the deeper white matter. The cerebellum communicates with the cerebral cortex, brain stem nuclei (see page 83), sensory receptors in the skin and the proprioceptors. These links allow the cerebellum to process sensory data and influence the voluntary motor activity required to maintain posture, balance and smooth coordinated motion. The functioning of the cerebellum is involuntary and we are not aware of its ability to correct and modify. However, when it becomes disordered we can appreciate the importance of its healthy functioning.

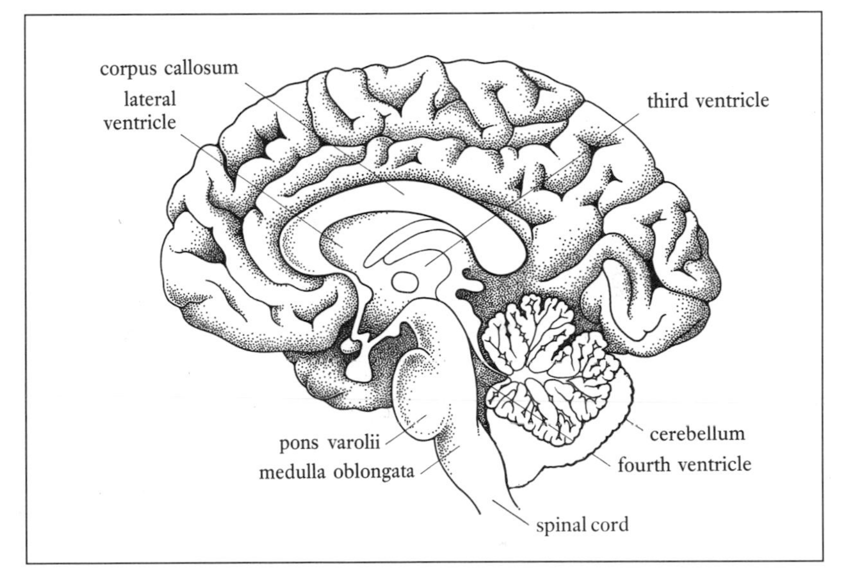

Figure 4.12. The cerebellum.

Nursing Practice Application Cerebellar dysfunction

People with cerebellar dysfunction, e.g. caused by alcohol excess, require a care plan that takes account of their clumsiness, lack of balance increasing the risk of falls, inability to perform tasks requiring precise movements and staggering gait which causes mobility problems. They may also have an intention tremor which occurs when they attempt to perform some task, such as lifting a cup, or communication problems in the form of slurred speech or *scanning-speech*, where syllables become separated.

Reticular formation

The **reticular formation** is a functional system of the brain stem which consists of neurones, neural pathways and some isolated nuclei. The reticular formation is important in the control of motor activity, some autonomic functions and the regulation of sensory inputs on route to the cerebral cortex. Through the **reticular activating system (RAS)** it is concerned with the state of cortical awareness and consciousness, modification of sensory inputs (familiar ones may be suppressed and unusual ones are allowed through), e.g. living on a busy road you are unaware of the usual continuous traffic noise, but you become aware of an isolated car travelling very fast. The RAS is inhibited by the sleep centres, alcohol, drugs that reduce anxiety, sedatives and severe trauma which may result in unconsciousness.

The reticular formation is involved with other structures in the regulation of *sleep–wake cycles*. Sleep takes up about a third of our lives but its purpose and mechanisms are poorly understood (see *Special Focus* – Sleep). We do know, however, that the proper amount of sleep is important in maintaining health and optimum functioning and for this reason it merits further discussion.

Special Focus Sleep

Sleep–wake cycles are thought to be controlled through the RAS, which keeps the cerebral cortex alert and awake, and sleep centres in the hypothalamus, which appear to control the timing of sleep. The process of sleep is much more complex than this suggests and involves the whole brain in ways which remain unclear. Sleep–wake patterns usually follow a circadian biorhythm of around 24 hours, where sleep occurs during the dark and wakefulness during the daylight hours.

The actual amount of sleep required by individuals varies considerably, with some people needing only 4–5 hours whilst others 'feel deprived' if they do not sleep 8–9 hours.

Modifications to the simple dark–light rhythm occur when routines are changed, such as when working at night or travelling through time zones (jet lag). Anxiety and depression may cause difficulty in getting to sleep or early waking. Sedatives, tranquillizers and alcohol all affect the type and patterns of sleep.

The functions of sleep are not fully understood but all of us can remember that dreadful feeling that accompanies lack of sleep. Sleep deprivation certainly affects coping mechanisms, emotional stability, restorative properties and our ability to perform skilled tasks.

One of the most common problems for people admitted to hospital is the disruption to their normal sleep–wake patterns. This is especially worrying when you consider that they are denied the restorative power of sleep during a period of increased physiological and psychological stress. In a study of 200 patients, Closs (1988) found that for most people sleep patterns changed after admission to hospital; 60% slept less well, 27% slept the same and 11% slept better. The common reasons given for sleep disruption were pain, noise, environmental temperature and uncomfortable beds.

Stages and types of sleep

In adults a typical night's sleep consists of alternating cycles of **non-rapid eye movement sleep (NREM)**, which has four stages (see below), and **rapid eye movement sleep (REM)**. These types and stages are identified by different patterns of brain waves recorded by the *electroencephalogram (EEG)*. Brain wave patterns are discussed more fully on pages 88-89. NREM sleep occurs first:

Stage 1 – (Transition stage) relaxation and drifting into sleep, vital signs (blood pressure, temperature, respiration and heart rate) are still normal and we are easily roused.

Stage 2 – more difficult to rouse.

Stage 3 – sleep is deeper and muscles relaxed. Vital signs decline. Dreaming occurs at this stage.

Stage 4 – the deepest level of sleep. Very difficult to rouse, often disorientated if woken. Vital signs at their lowest level. During this stage people may get out of bed or be incontinent of urine.

REM sleep occurs following a period of NREM sleep. It is characterized by a change in brain wave patterns, an increase in vital signs and inhibition of most muscles. Brain use of oxygen during REM exceeds the waking consumption. Dreaming also occurs during REM sleep.

Usually the cycles of NREM/REM sleep last for around 90 mins.

Lifespan and changing sleep needs

Small babies are asleep for approximately 16 hours a day. They have many sleep–wake cycles during the 24 hour period and these take little account of day and night. Gradually they adapt to sleeping mainly at night and being awake during the day.

By the mid-teens most individuals have reached the average adult sleeping pattern of 7–8 hours.

Sleep patterns may change during the middle years when a night's sleep may be broken by periods of wakefulness.

The elderly tend to have much lighter sleep with increasing periods of wakefulness. The lack of sleep at night is usually compensated for by

Special Focus **Sleep** (Continued)

frequent naps during the day. Older people appear to need more sleep and rest and will often plan for a sleep after lunch.

These different sleep needs have obvious implications for the nurse, who somehow must try to provide the environment in which individuals can achieve adequate sleep. This is vitally important to enhance the healing processes, and an individual's feeling of well–being but very difficult to organize within a busy ward or where accomodation problems exist, such as a teenager sharing a bedroom with an elderly grandparent.

Getting to sleep

Everyone has their bedtime routine, which probably helps prepare for sleep, e.g. warm bath, milky drink and changing into nightclothes. There are many factors which prevent proper sleep, such as pain, a strange hospital bed, being with other people, the ward light, being too hot or cold, a full bladder and the noise of other people and events. In a study, the two nursing actions which most helped people get back to sleep were pain relief and a hot drink (Closs, 1988).

Where individuals are still unable to sleep they might be prescribed a sedative drug, preferably for a limited period, as long-term use may lead to dependence and, paradoxically, to alteration in sleep patterns.

Consciousness/alterations in level of consciousness

Although sleep represents an altered state of consciousness it is not usually included when considering causes of an abnormal state of consciousness. Changes in consciousness result from some impairment of brain functioning, causes of which include fainting (*syncope*), accidental or intentional overdose of drugs or alcohol, general anaesthesia, head injury, stroke and other brain disorders (see *Figure 4.13* which shows an intracranial haemorrhage) and homeostatic

Person-centred Study **Billie**

Billie has come into hospital for observation after a fall from his pony Polly. He and some pals had been riding without hard hats when Polly took fright and Billie was thrown to the ground. Billie hit his head and, according to his friends, was 'knocked out' for a few minutes.

Neurological assessment and/or observation is important following head injury. There is a risk that bleeding (haematoma formation) – within the skull or swelling of the brain will cause an increase in intracranial pressure. Untreated, this *raised intracranial pressure (RIP)* may lead to further swelling, disruption to blood supply, altered consciousness, brain damage and eventually death if the brain stem (with all its vital centres) is forced down through the foramen magnum (*herniation or coning*).

During his hospital stay Billie is observed carefully for any changes that might indicate raised intracranial pressure. In an article concerning raised intracranial pressure, Allan (1986) recommends the use of standardized assessment techniques, e.g. Glasgow Coma Scale plus the other criteria mentioned on page 88.

The changes indicative of raised intracranial pressure are:

Conscious level – becoming less responsive, using the criteria of eye opening, verbal and motor response.

Blood pressure – the blood pressure increases (vasomotor centre involvement).

Pulse – the rate slows (cardiac centre involvement).

Respiration – changes in rate and depth (repiratory centre involvement).

Pupil size and reaction to light – if the brain is forced downwards it presses on the *oculomotor nerve (IIIrd cranial nerve,* see Chapter 5) which normally controls pupillary reflexes. If intracranial pressure is raised the pupils become unequal (dilate on affected side) and unresponsive to light, a fixed dilated pupil is an extremely grave prognostic sign.

Temperature – high body temperature may result from damage to the heat regulating centre in the hypothalamus.

Limb movement – spasticity or abnormal extension may occur as motor tracts and brain stem are compressed. This is another sign of poor prognosis.

Next morning Billie is seen by the medical officer who considers that he is fit to go home. Overnight all Billie's observation had been within normal limits and apart from the 'lump' on his head Billie feels very well. When his parents arrive Billie is discharged home with a reminder about wearing his hard hat (properly fitting and of the most recent safety standard). The nurse advises Billie's mother about contacting their doctor if Billie has headache, vomiting, visual problems or drowsiness, and the need for more rest and a gradual return to his normal energetic pastimes.

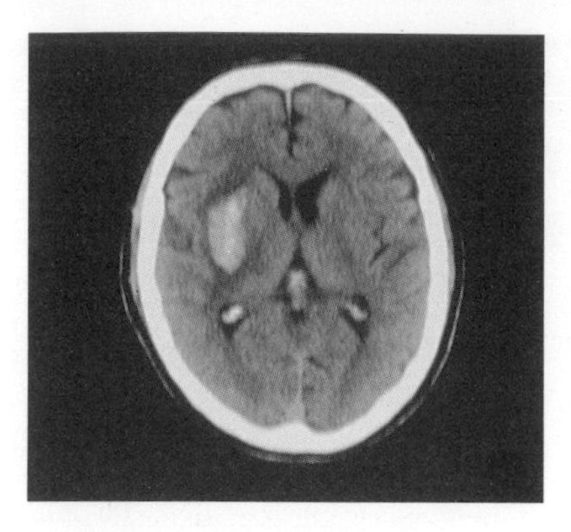

Figure 4.13. CAT scan showing intracranial haemorrage (pale area on left). (Simon Fraser/Science Photo Library. Reprinted with permission.)

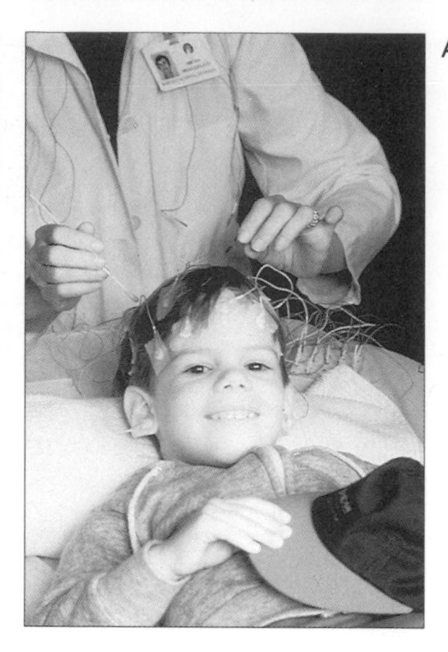

Figure 4.14. (a) Electroencephalograph electrodes in place (Larry Mulvehill/Science Photo Library. Reprinted with permission); (b) brain wave patterns. (Medical Photographic department, Guy's Hospital. Reprinted with permission).

breakdown, e.g. too much or too little sugar in the blood.

Changes in the state of consciousness can be assessed using a standardized instrument, such as the *Glasgow Coma Scale* (Allan, 1984) which tests verbal response, motor response and eye opening. These criteria – combined with vital signs (blood pressure, pulse, temperature and respiration), pupil size, their reaction to light and limb movements – provide a comprehensive neurological assessment (see *Person-centred Study* – Billie, page 87).

Brain wave patterns

The electrical activity of the many neurones in the brain can be recorded by the use of scalp electrodes and an electroencephalograph (*Figure 4.14*). The patterns of activity or **brain waves** are represented visually as an *electroencephalogram (EEG)*, which is a paper trace of the waves.

Brain wave patterns are very complex and unique to individuals; but there are four basic types:

- *Alpha* – low amplitude, slow waves associated with the relaxed awake state.
- *Beta* – higher frequency and more irregular, they occur when we are awake and 'thinking' or concentrating.
- *Theta* – even more irregular, these are seen in children and the first stages of sleep. Their presence in awake adults indicates adnormality.
- *Delta* – high amplitude waves which are recorded during deep sleep and in situations where the RAS is inhibited.

Brain wave patterns are useful in the diagnosis of many brain disorders, e.g. the *epilepsies.* These are characterized by abnormal electrical discharges in the brain that cause some alteration in consciousness, which may be accompanied by fits or seizures.

The EEG may be used in conjunction with other criteria to confirm *'brain stem death'*, although in the UK (and several other countries) it is not in routine use because some residual electrical activity may persist following brain stem death, and EEG measures cortical activity whereas tests for brain death are concerned with brain stem function (Allan, 1987). Diagnosis of 'brain stem death' is of particular importance when decisions are to be made regarding the withdrawal of life-support systems, especially where organ donation is anticipated.

Limbic system

The **limbic system** is not a well-defined system, but rather a diffuse collection of nuclei and fibres in the cerebral hemispheres. These areas, which are part of the primitive brain, influence feelings (affective) and emotions. The 'limbic system' is really the link between the **hippocampus** and **amygdaloid nucleus**, the hypothalamus (which you will remember is also concerned with our emotional state), the cerebral cortex (which puts a brake on our response to feelings) and several other areas.

Memory and learning

Memory is the ability to store past experiences which can be retrieved when required, very much like saving data on a computer disk to be used at some later date. As yet there is little information about the actual mechanisms involved in memory, but it would seem that we store facts in stages, that many areas of the brain are involved and that short-term memory is linked to the functioning of the hippocampus (area on the temporal lobe). Our ability to learn new facts and store them in the memory decreases with age. Many older people have perfect recall of memories of past experiences, but find their short-term memory does not function well, e.g. a lady of 85 can tell you all about food rationing during the Second World War but cannot recall what she had for supper that day.

Alzheimer's disease is a degenerative brain condition which causes dementia (it was previously called senile dementia). This very common condition affects memory, learning and other higher functions, especially in the elderly, but it can start in middle age. It is characterized by loss of short-term memory, an inability to learn new facts, confusion, disorientation and possible changes in personality. As you can imagine, this causes problems with everyday life for affected individuals and their families, who are often the only carers. Caring for and supporting people affected by Alzheimer's disease takes a considerable part of the resources of health authorities and social service departments, both in the community and in hospitals.

THE SPINAL CORD

The other part of the CNS is the **spinal cord** (*Figure 4.15*). This is continuous with the medulla oblongata and runs inside the vertebral canal (surrounded by the bones of the vertebral column), down the length of the back to the level of the first or second lumbar vertebra (L1/2) in adults but in young children it may reach L3. The cord is about 2.5 cm (1 in) thick and approximately 42 cm (17 in) long in an adult male. The spinal cord is protected by the three meningeal membranes and CSF (see pages 71-74). A modified portion of pia mater, the **filum terminale**, secures the end of the cord to the coccyx. Thirty-one pairs of spinal nerves (see Chapter 5) arise from the cord; these paired nerve roots enter and leave via apertures in the vertebra (**intervertebral foramina**). Below the level of L1/2 the nerve roots fan out to form a structure resembling a horse's tail, the **cauda equina**.

Several pathways in the spinal cord conduct sensory inputs to the brain and motor outputs from the brain. Various spinal reflexes occur within the spinal cord.

Structure of the spinal cord

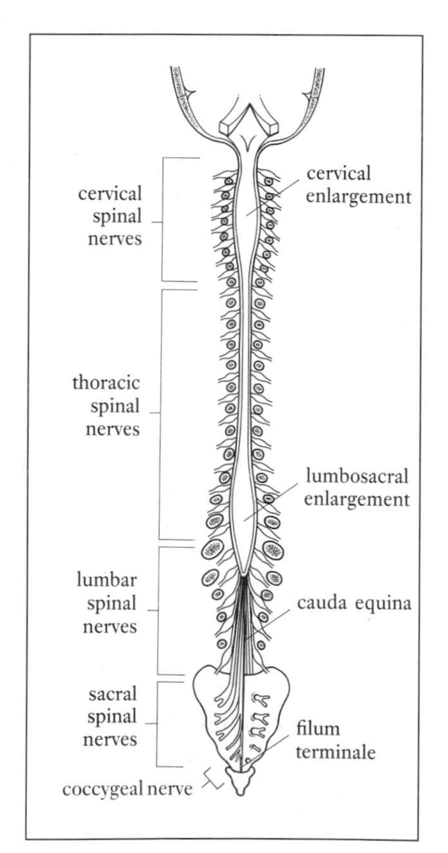

Figure 4.15. The spinal cord and spinal nerves.

The spinal cord consists of a butterfly–shaped area of grey matter with a central canal which is surrounded by white matter, neuroglia and blood vessels.

The grey matter can be divided into:

- Two anterior (ventral) horns or **motor horns**, which are largest at the level where nerves for the limbs leave the cord, they contain the cell bodies of the lower motor neurones whose axons supply skeletal muscles.
- Two posterior (dorsal) horns or **sensory horns**. The cell bodies for the sensory nerves entering the dorsal horns of the spinal cord are in the **dorsal** or **spinal ganglia** which are outside the spinal cord. These sensory nerves may synapse in the cord at the level of entry or travel up the ascending tracts of the spinal cord to synapses higher up in the cord or the brain. The dorsal horns also contain *interneurones* which link sensory and motor neurones (see simple reflex arc, Chapter 3).
- Lateral horns present in the thoracic and lumbar regions of the cord contain autonomic motor neurones that innervate the viscera. They leave the spinal cord with the somatic motor neurones of the ventral horns.

The dorsal and ventral nerve roots entering and leaving the spinal cord fuse just outside the cord to become the mixed spinal nerve (*Figure 4.16*).

The white matter of the spinal cord consists of the ascending sensory tracts, which carry afferent impulses to the brain, descending motor tracts, which carry efferent impulses from the brain or from higher levels in the cord, and the fibres, which connect one side of the cord with the other. The white matter is arranged into three columns or **funiculi** on each side of the cord: posterior, anterior and lateral.

Many ascending and descending tracts are contained within each column. Generally, the tracts consist of two or three neurones which synapse. Most tracts decussate at some point and they are paired, one on the right and one on the left side of the cord.

Motor pathways

The descending motor tracts are either pyramidal or extrapyramidal (see pages 77-80 and *Figures 4.7* and *4.17*).

Pyramidal tracts

The pyramidal tracts or corticospinal tracts originate in the Betz or pyramidal cells of the main motor area of the cerebral cortex. Their axons pass through the internal capsule and on to the medulla where most decussate. This upper motor neurone (UMN) descends in the spinal cord tract to the appropriate level where it synapses with the lower motor neurone (LMN), also referred to as the '*final common pathway*', which leaves the cord via the ventral nerve root to innervate the associated skeletal muscle fibres. The corticospinal tracts are concerned with voluntary movements.

At this point it is appropriate to consider the different effects of problems affecting UMN and LMN; these are summarized in *Table 4.1.*

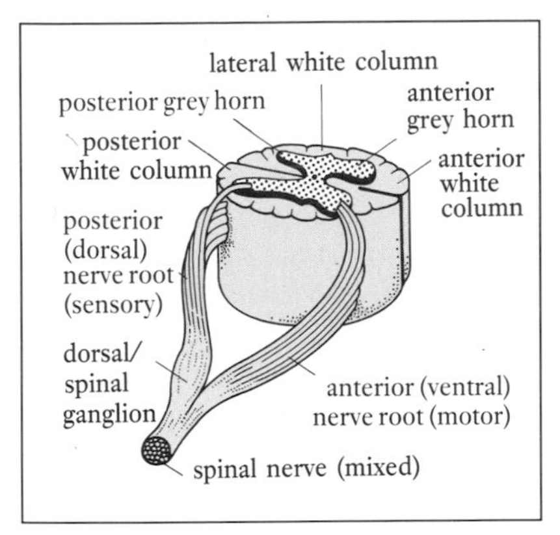

Figure 4.16. Section through the spinal cord showing structure and spinal nerve roots.

Table 4.1. Comparison of UMN and LMN disorders.

Site of problem	***Type of paralysis***	***Muscle atrophy***
Upper motor neurone, e.g. strokes	Spastic (increased tone), may be flaccid initially, affects one side of the body (hemiplegia)	No
Lower motor neurone, e.g. nerve trauma	Flaccid (decreased tone) affecting only the muscle group(s) innervated by that peripheral nerve	Yes

Damage to LMN may be accompanied by sensory disturbances such as an inability to detect temperature changes. This occurs because the peripheral nerves are usually mixed (both motor and sensory fibres).

NB Reflexes are also affected, see Chapter 5.

Extrapyramidal tracts

Extrapyramidal tracts include the **tectospinal**, **rubrospinal**, **reticulospinal** and **vestibulospinal tracts**. They arise in subcortical areas and travel in the spinal cord columns to the anterior horn. Their activity is influenced by links with the cerebellum, basal nuclei and brain stem nuclei. The extrapyramidal tracts modify coarse movements of voluntary muscles and affect posture and coordination:

- Tectospinal – fibres originate in the midbrain where they decussate.

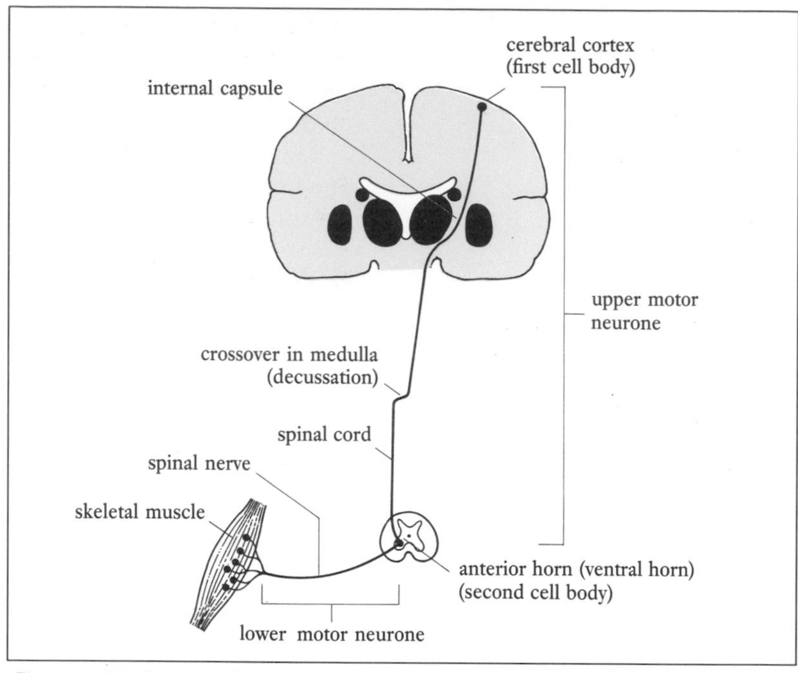

Figure 4.17. Motor pathways (pyramidal - two-neurone pathway).

They travel in the anterior spinal cord column and control posture, balance and muscle coordination.

- Rubrospinal – fibres originate in the red nucleus of the midbrain where they decussate. They travel in the lateral columns of the cord and control muscle tone and posture.
- Reticulospinal – fibres originate in the reticular formation; some decussate and others do not. They travel in the lateral and anterior columns to control muscle tone.
- Vestibulospinal – fibres originate in the vestibular nuclei in the medulla; some decussate and others do not. They travel in the anterior and lateral columns to control muscle tone.

Sensory pathways

Sensory pathways carry a variety of impulses from sensory receptors in the skin and proprioceptors in joints, muscles and tendons to the brain via a two- or three-neurone system (*Figure 4.18*). The pathways carry information about touch, pressure, temperature and pain to the sensory cortex, which in some instances can localize its site of origin. Information from the proprioceptors is carried to the cerebral cortex and to the cerebellum, where it is used to maintain posture and muscle tone. The major sensory pathways include: the **fasciculi cuneatus** and **gracilis** (a fasciculus is a small bundle), the **anterior/lateral spinothalamic tract** and the **anterior/posterior spinocerebellar tracts**.

Fasciculi cuneatus and gracilis

The fasciculi cuneatus and gracilis carry touch and pressure impulses from skin receptors, and position sense from proprioceptors, to the opposite side of the brain. *First-order sensory neurones* enter the cord and travel in the dorsal columns to synapse with *second-order neurones* in the nuclei cuneatus and gracilis in the medulla. The fibres decussate in the medulla before ascending to the thalamus where they synapse with *third-order neurones* which carry the impulses to the sensory cortex.

Anterior/lateral spinothalamic tracts
The anterior/lateral spinothalamic tracts carry poorly localized pressure and touch impulses, plus impulses for temperature and pain (in A and C fibres) to the opposite side of the brain. They synapse and decussate in the cord before ascending in the anterior and lateral columns to the thalamus, where they synapse with third-order neurones which carry the impulses to the sensory cortex.

Anterior/posterior spinocerebellar tracts
The anterior/posterior spinocerebellar tracts carry proprioceptor and touch impulses to the cerebellum, some fibres decussate and others do not. They synapse in the spinal cord and the second-order neurones ascend in the lateral columns to terminate in the cerebellum.

Spinal injury and spinal shock

The physiological effects of injury to the spinal cord obviously depend on the extent of the damage and the level at which it occurs. Common causes of spinal cord transection are traffic, industrial and sporting accidents, but diseases such as spinal tumours can cause similar effects through cord compression.

In the initial period after cord injury there is **spinal shock**, which is characterized by loss of bowel and bladder control, paralysis (flaccid

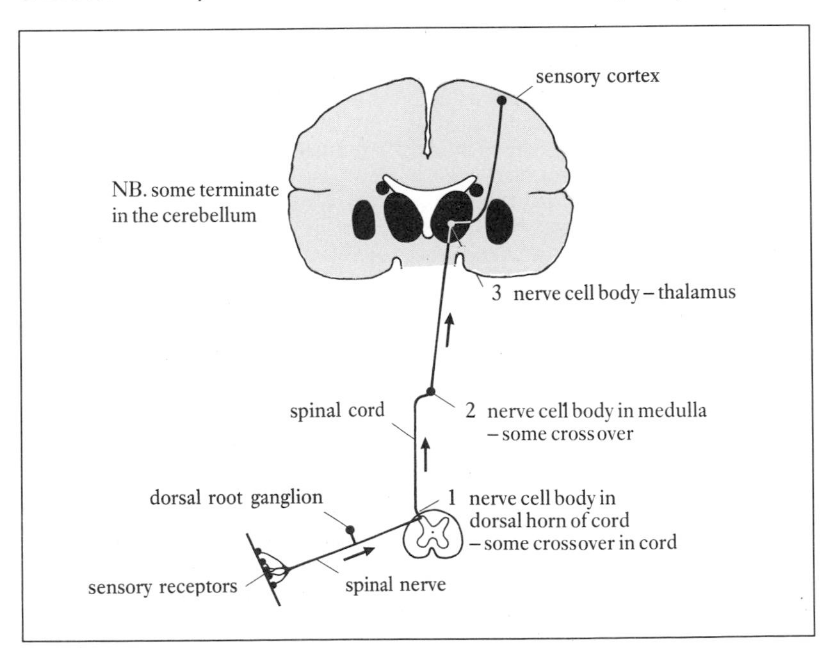

Figure 4.18. Sensory pathways (simplified two- or three-neurone pathway).

Nursing Practice Application Pain

Pain transmission and perception involve the whole nervous system but it seems both relevant and logical to look at pain now, having covered the brain and sensory tracts in some depth.

The management of pain is one of the most important and difficult issues in nursing practice. Pain is subjective and individuals all feel and describe painful sensations differently. The biggest problem is not being able to measure pain in units. Those working in health care need to accept what the individual has to say about his or her pain and act upon this information in a positive way.

Over recent years the concept of 'pain meters' has been developed; these allow people to grade their painful sensations on a scale, e.g. 0 (no pain) to 10 (the worst pain possible). There are many ways of recording pain intensity and Vandenbosch McCormick (1988) describes the use of 'pain flow sheets'.

Apart from what people say about pain there is a great deal of information to be gained from their non-verbal communication and physical appearance, e.g. a person in pain may keep very still, thrash around the bed or become sweaty. Changes in vital signs, e.g. increasing pulse, may be detected in a person with pain.

Some of the many factors which influence our response to pain include: (i) **pain threshold** and **tolerance** – some people appear not to feel pain so easily, and their pain threshold is said to be high; (ii) cultural differences, such as adopting a 'stiff upper lip' attitude to pain; (iii) factors which affect pain threshold, e.g. anxiety and information. Hayward (1975) found that information regarding pain and its duration given before surgery reduced the need for painkillers after operation; (iv) Latham (1993) suggests that the presence of several coexisting medical problems can amplify or confuse the picture of pain experienced and reported by older people.

Acute pain may have a physiological defence role and can be regarded in some situations as a warning that tissues are being damaged, e.g. acute inflammation. This is certainly not so for chronic pain, e.g. osteoarthritis (see Chapter 18) where joint damage has already occurred.

Physiology of Transmission

Pain impulses are carried to the CNS by the 'A' delta and 'C' nerve fibres (see Chapter 3), it may be that these fibres transmit different types of pain. Melzack and Wall (1965) put forward the *gate control theory* in which they suggested that a 'gate' operating at spinal cord level can block pain impulses from reaching a conscious level unless the input of impulses is large. Gate control theory is very complex, but basically works by fibres from touch receptors inhibiting the fibres carrying pain impulses. The descending spinal tracts are also involved. The gate control theory is important in our understanding of pain and we shall use it to explain other aspects of pain.

Various peptide chemicals produced in the body are involved in the transmission of pain impulses. These are the neurotransmitters (see Chapter 3): substance P, which enhances pain transmission and the opiate-like substances enkephalin and endorphin, which act as natural analgesics (painkillers) by inhibiting the synthesis of **prostaglandins,** substances (involved in the initiation of inflammatory effects when released from damaged tissue. NB Prostaglandins fufil many other roles in the body) which intensify pain. Other substances implicated in the production of pain are those produced by tissue damage and inflammation (see Chapter 19) such as **bradykinin,** 5–hydroxytryptamine and histamine, which all stimulate pain receptors.

Special types of pain

Referred pain

This is pain felt in a different location from which it originates, e.g. cardiac pain may be felt in the left arm, neck and jaw. This may occur because pain fibres from the affected structure enter the spinal cord at the same level as sensory fibres from the referred area.

Phantom limb syndrome

Following amputation of a limb the individual still experiences pain in the missing limb. This happens because the cut ends of sensory nerves continue to transmit pain impulses without inhibition (gate theory) to the brain, which localizes the pain incorrectly.

Chronic pain

As already discussed, chronic pain cannot be considered protective. Hanks and Hoskin (1986) state that chronic pain has no positive function. The perception of chronic pain, such as that caused by advanced malignancy, may also be influenced by the emotional state of a person facing the prospect of death (Waugh, 1988). Another important aspect of chronic pain is that the cause cannot be cured (Hockley, 1988).

Methods of pain relief

Just as pain perception is unique to the person, so its relief is equally individual. It is essential that care plans reflect a holistic approach to pain relief by considering the physical, emotional, social and cultural aspects. The management of acute and chronic pain differs in several ways. For instance, relief of acute pain is usually only required for a limited duration, such as following surgery, but relief of chronic pain may be required for life.

NB Interventions such as joint replacement, discussed in Chapter 18, may reduce pain substantially.

Physical

(i) Basic comfort such as an extra pillow; (ii) massage to enhance relaxation; (iii) heat or cold, e.g. ice pack (care is needed to prevent skin damage); (iv) transcutaneous electrical nerve stimulation (TENS), which probably works by stimulating the large fibres which inhibit the pain fibres (gate theory or the release of natural analgesic substances), see Chapter 20; (v) radiation; (vi) surgical division of a nerve; (vii) acupuncture, which may well stimulate the release of the natural opiates.

Pharmacological

(i) Analgesic drugs, such as morphine, aspirin etc.; (ii) drugs to lift mood, such as antidepressants; (iii) nerve blocks and epidural analgesia.

Psychological

(i) Distraction and diversion, e.g. TV, conversation, etc.; (ii) visualization techniques; (iii) relaxation and hypnosis; (iv) reducing anxiety and providing information.

becoming spastic) and loss of sensation and reflexes below the level of damage, reduction in blood pressure and problems with temperature control as sweating ceases below the damage level. If neurological function is still absent a few days after injury, it indicates that the cord damage and resultant disability is permanent.

Damage in the cervical region (see *Figure 4.15*) results in *quadriplegia,* which is paralysis of all four limbs, whereas damage in the thoracolumbar region results in *paraplegia*, where only the legs are affected. Various other effects include problems with sexual function and the enormous psychological and social problems associated with a sudden and permanent disability.

Individuals with spinal injuries require nursing interventions that ensure correct limb positioning, prevent skin breakdown, deal with bladder and bowel function, assist the person to express his or her sexuality and help all concerned to cope with permanent disability.

Summary/Check List

Development of CNS – embryonic – factors causing damage – development during infancy and childhood.

Protection for the CNS – Skull and vertebrae. Meninges. Ventricular system and CSF – composition of CSF, circulation and functions of CSF, *Nursing Practice Application* – lumbar puncture, hydrocephaly. Blood–brain barrier. Blood supply – continuity, regulation.

The brain – cerebrum – cerebral hemispheres (fissures, gyri, sulci, cortex, fibres). Functional areas of cerebral cortex – motor areas, sensory areas, other cortical areas. Right brain versus left brain. Basal nuclei – extrapyramidal tracts, parkinsonism, Huntington's disease. Internal capsule – pyramidal tracts. *Special Focus* – CVA, *Nursing Practice Application* – stroke. Thalamus. Hypothalamus – pituitary gland. Pineal body. Brain stem – midbrain, pons varolii, medulla oblongata. Cerebellum – *Nursing Practice Application* – cerebellar dysfunction. Reticular formation – *Special Focus* – sleep. Consciousness and alterations, *Person-centred Study* – Billie. Brain wave patterns – fits, brain stem death. Limbic system. Memory and learning – Alzheimer's disease.

Spinal cord – Structure (grey and white matter). Motor pathways – pyramidal, comparison of UMN and LMN disorders, extrapyramidal tracts. Sensory pathways. *Nursing Practice Application* – pain. Spinal injury and spinal shock.

?

Self Test

1 During pregnancy, development of the CNS has commenced by the:
(a) 3rd week (b) 12th week (c) 16th week (d) 20th week.

2 Which of the following statements are true?
(a) CSF is found between the dura and arachnoid mater.
(b) The composition of CSF is similar to that of plasma.
(c) The pia mater is the inner meningeal membrane.
(d) The 'blood–brain barrier' prevents brain damage due to exposure to lead in young children.

3 Draw and label a diagram of the brain to show the cerebrum, pons, medulla and cerebellum.

4 Put the following in their correct pairs:

(a) motor speech area	(e) extrapyramidal motor tracts
(b) basal nuclei	(f) visual cortex
(c) internal capsule	(g) occipital lobe
(d) Broca's area	(h) pyramidal motor tracts

5 Describe briefly how the hypothalamus influences fluid balance.

6 Why is the medulla oblongata so vital in maintaining homeostasis?

7 Which of the following indicate a rise in intracranial pressure?
- (a) Fall in blood pressure.
- (b) Decreasing verbal response.
- (c) Slowing pulse rate.
- (d) Pupils unresponsive to light.
- (e) Rise in blood pressure.
- (f) Rising pulse rate.

8 Which of the following statements are true?
- (a) The spinal cord terminates at the level of L1 or L2 in adults.
- (b) The grey matter of the cord surrounds the central white matter.
- (c) Sensory nerves entering the dorsal horn have their cell bodies outside the spinal cord.
- (d) Motor nerves leave the cord via the posterior horn.

9 Briefly describe the route taken by a nerve impulse from the Betz cells in the motor cortex until it reaches the motor unit.

10 Quadriplegia will follow spinal cord transection in which region:
(a) thoracic (b) lumbar (c) sacral (d) cervical.

Answers
1. a. 2. b, c. 3. See page 75. 4. a–d, b–e, c–h, g–f. 5. See page 82. 6. see page 84. 7. b, c, d, e. 8. a, c. 9. See pages 76-77, 91. 10. d.

References

Allan, D. (1984) Glasgow coma chart. *Nursing Mirror,* **158** (23), 32–34.

Allan, D. (1986) Raised intracranial pressure. *The Professional Nurse,* **2** (3), 78–80.

Allan, D. (1987) Criteria for brain stem death. *The Professional Nurse,* **2** (11), 357–359.

Closs, J. (1988) Sleep in surgical wards. *Nursing Times,* **84** (22), 52

Fletcher, S. (1992) Innovative treatment or ethical headache? Fetal transplantation in Parkinson's disease. *Professional Nurse,* **7** (9), 592-595.

Hanks, G.W. Hoskin, P.J. (1986) Pain control in advanced cancer: pharma-cological methods. *Journal of the Royal College of Physicians,* **20** (4), 276–281.

Hayward, J. (1975) *Information, a prescription against pain.* London: Royal College of Nursing.

Hockley, J. (1988) Setting standards for pain control. *Professional Nurse,* **3** (8), 310–313.

Latham, J. (1993) Treatment we can all believe in. Pain and its management in later life. *The Professional Nurse,* **8** (4), 212-220.

Macleod, J. , Edwards, C. and Bouchier, I. (1987) *Davidson's Principles and Practice of Medicine.* 15th edn. Edinburgh: Churchill Livingstone.

Marieb, E. N. (1989) *Human Anatomy and Physiology.* Redwood City, California: Benjamin/Cumming.

Melzack, R. and Wall, P. D. (1965) Pain mechanisms: a new theory. *Science* **150** (971–979).

Vandenbosch McCormick, T. (1988) How to use a pain flow sheet effectively. *Nursing 88 (USA),* **18** (8), 50–51.

Waugh, L. (1988) Psychological aspects of cancer pain. *The Professional Nurse,* **3** (12), 504–508.

Further Reading

Armstrong Ester, C. A and Hawkins, L. H. (1982) Day for night – circadian rhythms in the elderly. *Nursing Times,* **78** (30), 1263–1265.

Barasi, S. (1991) The physiology of pain. *Surgical Nurse,* **4** (5) 14–20.

Clifford Rose, F. and Capildeo, R. (1981) *Stroke – The Facts.* Oxford: Oxford University Press. *NB* Written for a lay audience but useful for health professionals.

Fordham, M. (1986) Neurophysical pain theories. *Nursing,* **3** (10), 365–372.

Horne, L. (1991) No more wakeful nights. *Professional Nurse,***6** (7), 383–385.

Jacques, A. (1992) Do you believe I'm in pain? Nurses' assessment of patient's pain. *Professional Nurse.* **7** (4), 249-251.

Kemp, J. (1984) Nursing at night. *Journal of Advanced Nursing,***9** (2), 217–223. *NB* A review of the literature.

Latham, J. (1985) Functional anatomy and physiology of pain. *The Professional Nurse,* **1** (2), 42–44.

Manion, R. (1988) Can I sleep now, nurse? *Nursing Standard,* **3**, 22–23.

McCaffery, M. (1983) *Nursing Management of the Patient in Pain.* London: Harper and Row.

Purchese, G. and Allan, D. (1984) *Neuromedical and Neurosurgical Nursing.* 2nd edn. London: Bailliere Tindall.

Ross Russell, R. W. and Wiles, C. M. (1985) *Neurology.* London: Heinemann.

Stead, W. (1985) One awake, all awake. *Nursing Mirror,* **160** (16), 20–21.

Stein, J. F. (1982) *An Introduction to Neurophysiology.* Oxford: Blackwell Scientific Publications.

Wilson-Barnett, J. and Batehup, L. (1988) *Patient problems: A Research Base for Nursing Care.* London: Scutari Press.

Useful Address

Disabled Living Foundation
346 Kensington High Street
London W 9 2HU

CHAPTER FIVE

THE PERIPHERAL NERVOUS SYSTEM

OVERVIEW

- Peripheral nervous system (PNS).
- Cranial nerves, spinal nerves, plexuses and peripheral nerves.
- Reflexes.

LEARNING OUTCOMES

After studying Chapter 5 you should be able to:

- Describe the different divisions of the PNS.
- Describe the cranial nerves and their functions.
- Describe the structure and formation of the spinal nerves.
- Identify the nerve plexuses and the major peripheral nerves arising from them.
- Outline the different types of reflex and be aware of their significance in neurological assessment.

KEY WORDS

Autonomic – meaning independent and self-governing, i.e. peripheral motor nerve fibres which innervate involuntary muscle and glands. Form the autonomic (involuntary) nervous system.
Cranial nerves – 12 pairs of peripheral nerves which originate from the brain.
Ganglion – a collection of nerve cell bodies in the PNS.
Plexus – a network or mass of nerve fibres.
Somatic – meaning of the body, i.e. peripheral motor nerve fibres which innervate voluntary muscles.
Spinal nerves – 31 pairs of peripheral nerves which originate from the spinal cord.

INTRODUCTION

The peripheral nervous system (PNS) is all that part of the nervous system outside the brain and spinal cord. It is of vital importance in the reception and transfer of sensory data, which is relayed to the CNS by *afferent* (meaning conducting inwards, towards the CNS) fibres, and in carrying motor impulses from the CNS, via *efferent* (meaning conducting outwards, away from the CNS) fibres, to effectors such as muscles and glands. Without this contact with the outside world all the complex functions of the CNS would be useless, rather like having a good idea which you keep to yourself!

The PNS consists of the peripheral nerves, of which most have both motor and sensory fibres (i.e. they are mixed nerves), ganglia and sensory receptors. The PNS has two types of ganglia: the dorsal root ganglia, which contain the cell bodies of the sensory nerves, and those which contain the cell bodies of autonomic motor neurones.

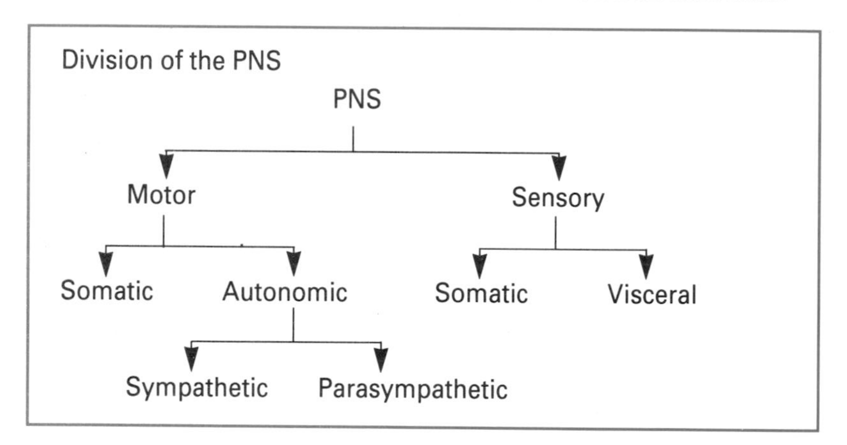

The peripheral nerves are either cranial or spinal, depending on whether they originate from the brain or spinal cord.

CRANIAL NERVES

The term **'cranial nerves'** is used for the 12 pairs of nerves which originate in the brain, i.e. pairs I and II from the forebrain and pairs III–XII from the brain stem *(Figure 5.1)*. The nerves are identified by names and Roman numerals. Most are concerned with the innervation of the head, neck and associated structures, but the vagus nerves (X) extend to structures in the thoracic and abdominal cavities.

Olfactory nerves (I) *(Figure 5.2)*. These are sensory and are concerned with smell. They carry impulses from the epithelium of the nasal cavity to the olfactory cortex in the temporal lobe of the cerebral hemispheres.

Optic nerves (II) *(Figure 5.3)*. These are sensory and are concerned

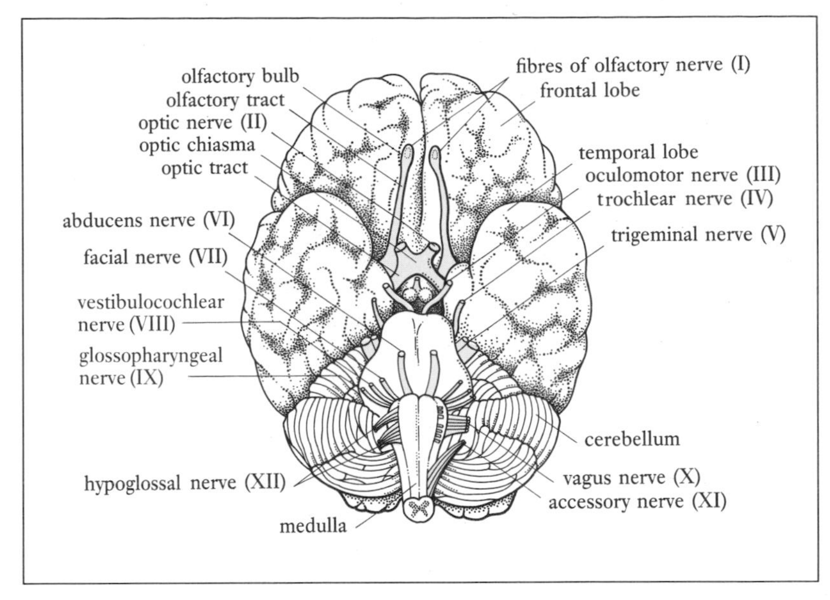

Figure 5.1. The origins of the cranial nerves.

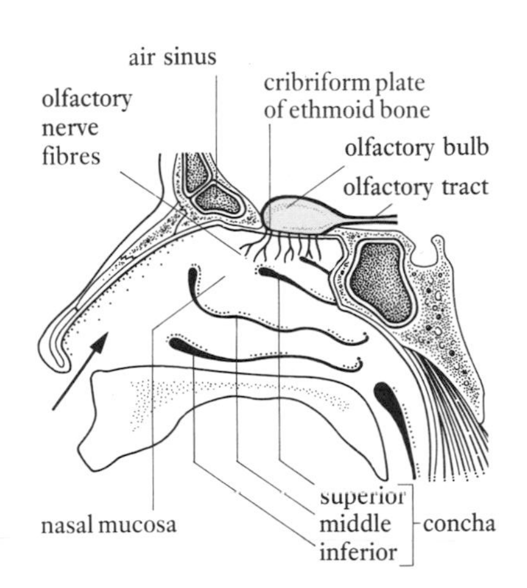

Figure 5.2. Olfactory nerves.

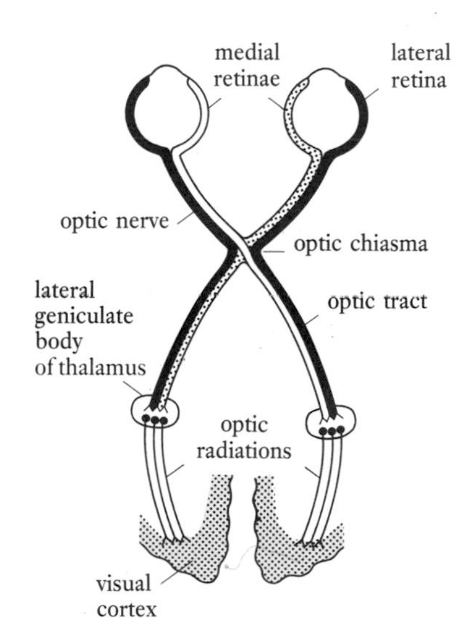

Figure 5.3. Optic nerves (visual pathways).

with sight. Impulses from the retina of the eye are conveyed to the visual cortex in the occipital lobe of the cerebral hemispheres. The optic nerves enter the cranial cavity and converge at the **optic chiasma,** where some fibres cross over. The **optic tracts** pass through the thalamus to terminate in the visual cortex. Damage to the optic nerves will cause blindness, whereas partial damage to the optic tracts (Chapter 7) distal to the chiasma causes visual field defects.

Oculomotor nerves (III). These are motor (somatic and parasympathetic) and originate in the midbrain. The motor somatic fibres innervate (supply nerves to) four of the extrinsic (external or outside) muscles that move the eyeball and the muscle that raises the upper eyelid. The parasympathetic fibres innervate the muscle that controls the size of the pupil (see Chapter 4) and the shape of the lens.

Disorders of the oculomotor nerves cause problems with eyeball movement – squints *(strabismus),* eyelid drooping *(ptosis),* focusing difficulties and double vision *(diplopia).*

Trochlear nerves (IV). These are motor, arise in the midbrain and innervate one extrinsic muscle of the eyeball.

Trigeminal nerves (V) *(Figure 5.4).* These are sensory and motor. The largest cranial nerve runs from the pons varolii (see Chapter 4) to supply the head and face. Its sensory fibres have their cell bodies in the trigeminal ganglion. The trigeminal nerve has three divisions: *(i) ophthalmic,* which carries sensory information from the anterior scalp, forehead, cornea, lacrimal glands, upper eyelids, nose and nasal mucosa; *(ii) maxillary,* which carries sensory information from

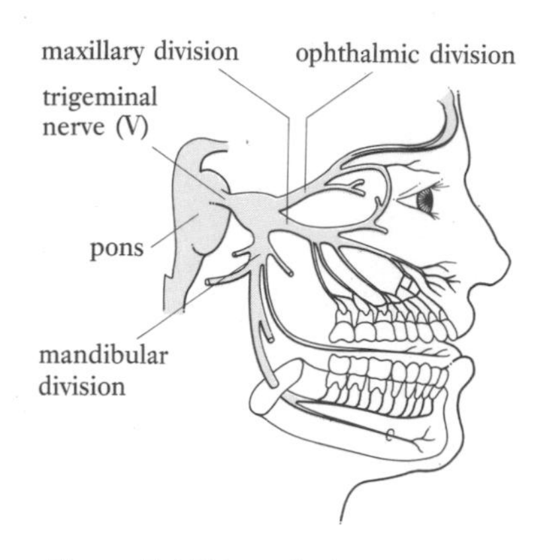

Figure 5.4. Trigeminal nerve.

the lower eyelids, the upper gums, teeth, lip and the cheeks; and *(iii) mandibular,* which carries sensory information from the lower gums, teeth and lip, chin, tongue and pinna of the ear. The mandibular division also contains motor fibres, which innervate the muscles involved in mastication (chewing).

A condition, of unknown cause, called *trigeminal neuralgia* results in agonizing pain felt in the face. The pain may be precipitated by simple stimuli, such as chewing and cold air. Pain relief with drugs is often inadequate and some people need more drastic methods, such as injections or cutting the trigeminal nerve.

The virus *herpes zoster* (shingles) can effect the trigeminal nerve producing a face rash, and corneal ulceration if the ophthalmic division is involved.

Abducens nerves (VI). These are motor and arise in the pons varolii. They innervate the remaining extrinsic muscle of the eyeball. Disorders cause strabismus.

Facial nerves (VII). These are sensory, motor and parasympathetic, and originate in the pons varolii. The sensory fibres carry impulses from the anterior part of the tongue (taste buds) to the cortex. The parasympathetic fibres supply the salivary glands, lacrimal glands and the glands of the nasal mucosa. The motor component innervates facial muscles and controls facial expression.

Disorders such as *Bell's palsy* cause paralysis of one side of the face, which leads to difficulty with eating and talking, loss of taste and an eye that waters constantly and will not close.

Vestibulocochlear/auditory nerves (VIII). These are sensory and carry impulses from structures in the inner ear. The **cochlear** branch carries impulses from the *organ of Corti* (involved in hearing) to the auditory cortex in the temporal lobe. The **vestibular** branch carries impulses from the *semicircular canals* (involved in balance) to the cerebellum. Damage can cause *tinnitus* (ringing/roaring noises in the ear), nerve deafness, loss of balance, dizziness and nausea.

Glossopharyngeal nerves (IX). These are motor, sensory and parasympathetic, and originate in the medulla oblongata. The motor fibres serve the tongue and pharynx and are concerned with swallowing and gag reflex. The sensory fibres transmit taste impulses from the back of the tongue to the cortex. The parasympathetic fibres stimulate the secretion of saliva from the parotid glands.

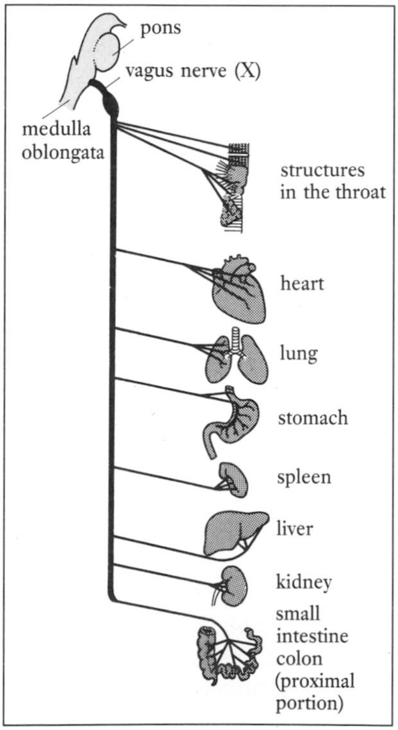

Figure 5.5. Vagus nerve.

Vagus nerves (X) *(Figure 5.5).* These are parasympathetic, motor and sensory. They originate in the medulla oblongata, extend down the neck to innervate the muscle and glands of the pharynx, larynx, trachea, lungs, heart and pass into the abdomen to supply the viscera (organs). The sensory fibres convey impulses to the brain from these structures. As you can imagine, the proper functioning of the vagus (X) nerve is vital to life and is discussed many times throughout the book.

Accessory nerves (XI). These are motor and originate in the medulla oblongata. They innervate the muscles of the pharynx, larynx, head, neck and shoulders. They control head movements and shrugging of the shoulders, a useful part of non-verbal communication.

Hypoglossal nerves (XII). These are motor and originate in the medulla oblongata. They provide motor fibres for fine tongue movements, which are required for moving food, swallowing and speaking.

SPINAL NERVES

The 31 pairs of mixed nerves arising from the spinal cord provide the connections between the CNS and the neck, body and limbs (*Figure 5.6*). They are classified and named by the level at which they leave the spinal cord:

- 8 cervical (C1–C8).
- 12 thoracic (T1–T12).
- 5 lumbar (L1–L5).
- 5 sacral (S1–S5).
- 1 coccygeal (C0).

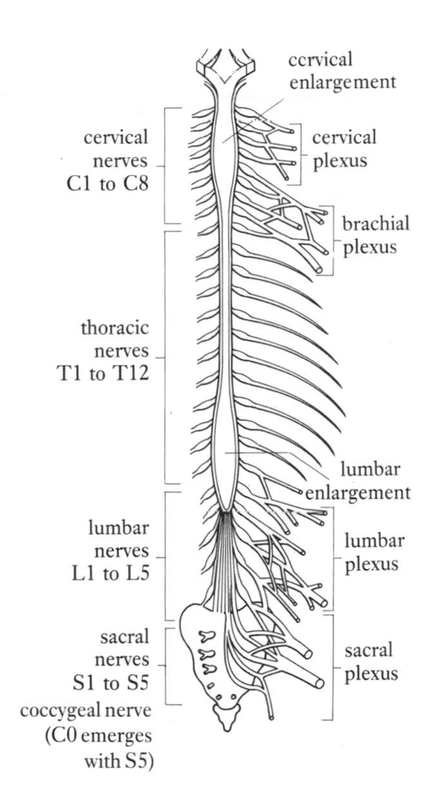

Figure 5.6. Spinal nerves. (N.B. for clarity the coccygeal plexus is shown in *Figure 5.12*).

There are eight cervical nerves but only seven vertebrae; the first pair of cervical nerves exits between the skull and the first cervical vertebra and the eighth pair exits below the last cervical vertebra. The other nerves are named for the vertebra immediately above their exit points. The nerve roots of the lumbar, sacral and coccygeal region extend into the vertebral canal as the cauda equina, before emerging from the vertebral column at the appropriate level.

Each spinal nerve has two connections with the spinal cord *(Figure 5.7)*. These are the **ventral (motor) root,** with its efferent fibres from the lower motor neurone (LMN), and the **dorsal (sensory) root,** with afferent fibres from the cell bodies in the **dorsal root ganglion.** These roots, covered with dura and arachnoid mater, fuse to form the mixed spinal nerve prior to emerging from the vertebral column via the intervertebral foramina.

Fibres of the autonomic system are associated with the ventral roots of the spinal nerves, sympathetic fibres in the thoracolumbar region and parasympathetic fibres in the sacral region (plus in the cranial nerves, as discussed earlier).

After leaving the vertebral column each spinal nerve divides into:

- A thin **dorsal ramus** (branch) of mixed fibres, which innervates

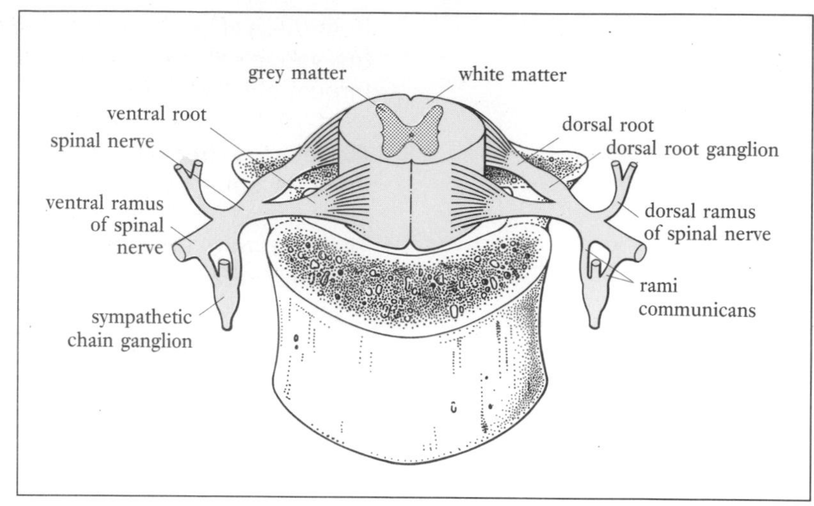

Figure 5.7. Formation of a spinal nerve.

the skin and muscles of the posterior part of the head, neck and trunk.

- A thicker **ventral ramus,** also of mixed fibres, which innervates the anterolateral aspects of the trunk and limbs.
- The **rami communicans** in the thoracic region, which form part of the sympathetic nervous system (see Chapter 6).

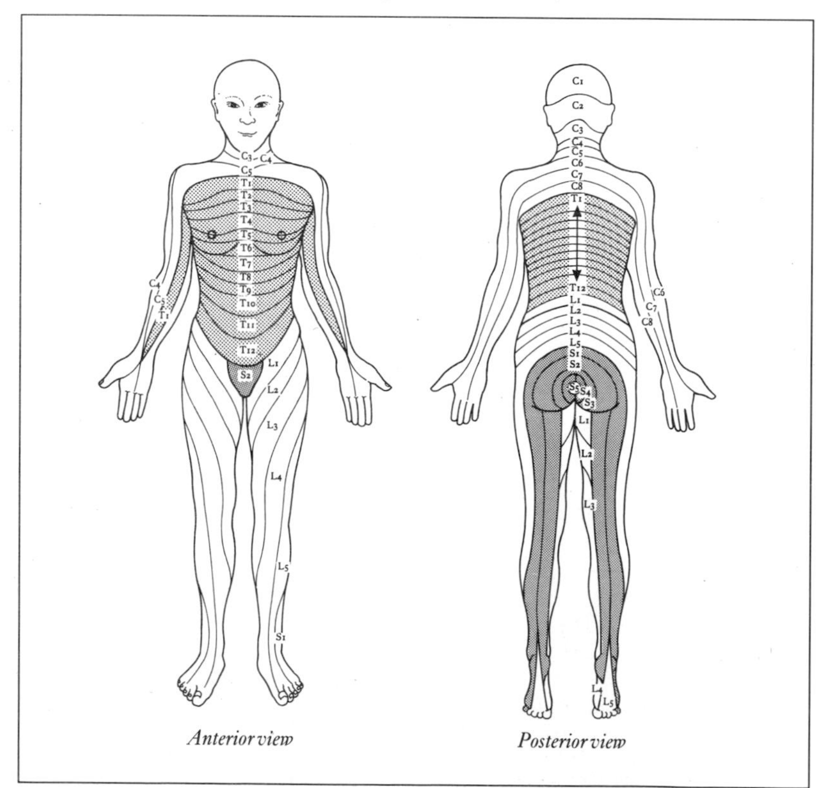

Figure 5.8. Sensory dermatomes.

In the cervical, lumbar and sacral/coccygeal regions the ventral rami mass together to form the **major plexuses**. The thoracic spinal nerves do not form a plexus but their ventral rami form the **intercostal nerves** that supply the ribs, intercostal muscles (respiratory muscles) and the skin and muscle of the thorax and abdominal wall.

Dermatomes

The cutaneous branches of a particular spinal nerve provide sensory innervation to an area of skin known as a **dermatome.** From *Figure 5.8* it can be seen that each nerve supplies a horizontal strip of trunk at the approximate level of its exit point. The situation is rather more complex for the limbs and there is considerable overlap of dermatomes.

During early embryonic development the body forms in segments, which are innervated by a corresponding cutaneous spinal nerve to form dermatomes. The overlap and size differences of adult dermatomes can be explained by embryonic limb development and the varying growth rates in different parts of the body. There is less overlap in the thoracic region than in the lower trunk, where dermatome boundaries are far from clear. In regions where considerable overlap exists a strip of skin may be innervated by two or more spinal nerves. It follows that malfunction of a single nerve, in these regions, may have only a limited adverse effect upon sensation.

Major plexuses

These are formed by the ventral rami of the spinal nerves (except the thoracic nerves) which mass together at the side of the vertebral column prior to forming branches that contain fibres from more than one nerve root. This adaptation serves to safeguard the nerve supply to individual parts of the body, e.g. a muscle has fibres from more than one spinal nerve (a good example of 'not putting all your eggs in one basket').

The identification of every individual nerve is beyond the scope of this book, which will discuss only the major nerves of each plexus where appropriate within the context of a nursing application. Readers requiring additional information should consult the reading list (e.g. Williams *et al.* (1989)).

Nursing Practice Application Hiccups

Hiccups (diaphragm spasm), are caused by irritation of the phrenic nerve. This usually causes only minor inconvenience for a short while in a healthy person but, when it is associated with established physiological dysfunction such as *uraemia* (high levels of urea in the blood), it may result in considerable distress and discomfort. The person affected soon becomes exhausted, eating and drinking become difficult and in some situations hiccups may cause pain, e.g. when abdominal muscles are strained or where the person has had recent surgery. Care plans should include: provision for pain relief and rest, help with eating/drinking and the administration of prescribed medication to stop the hiccups, such as chlorpromazine.

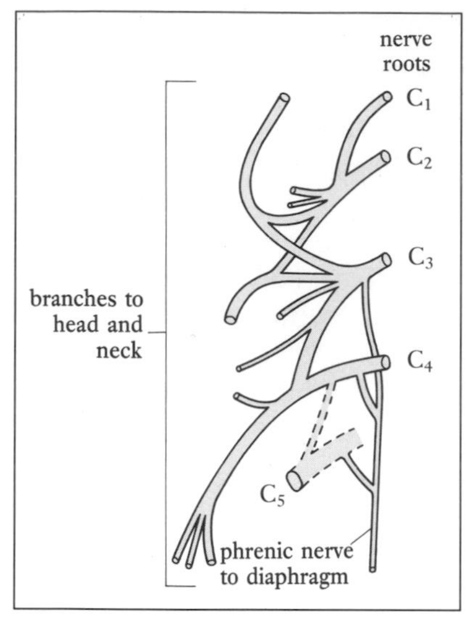

Figure. 5.9. Cervical plexus.

■ Cervical plexus *(Figure 5.9)*

The **cervical plexus** is situated in the neck under the *sternocleidomastoid* muscle (see Chapter 18) and is formed by spinal nerves C1–C4. Its branches supply the back of the head, neck and the *diaphragm* (respiratory muscle), which is innervated through a branch called the **phrenic nerve** (see *Nursing Practice Application* – Hiccups).

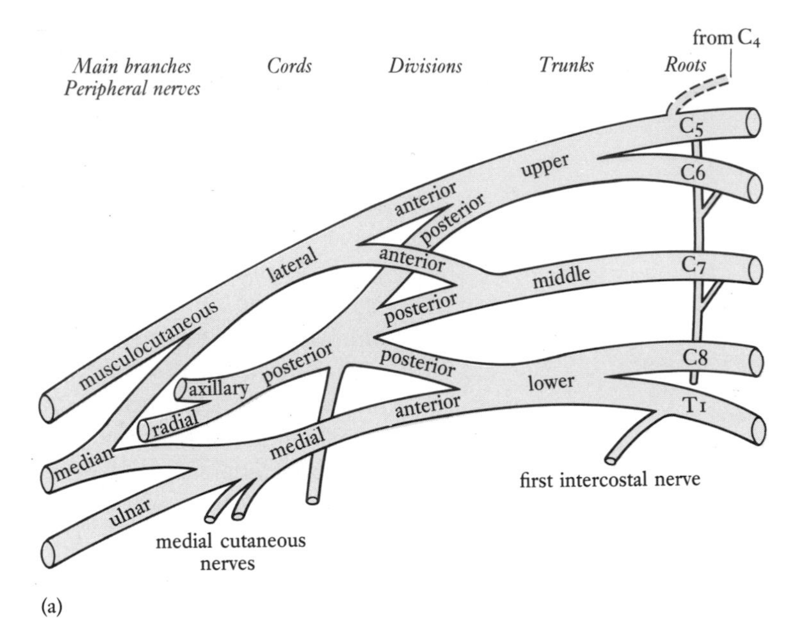

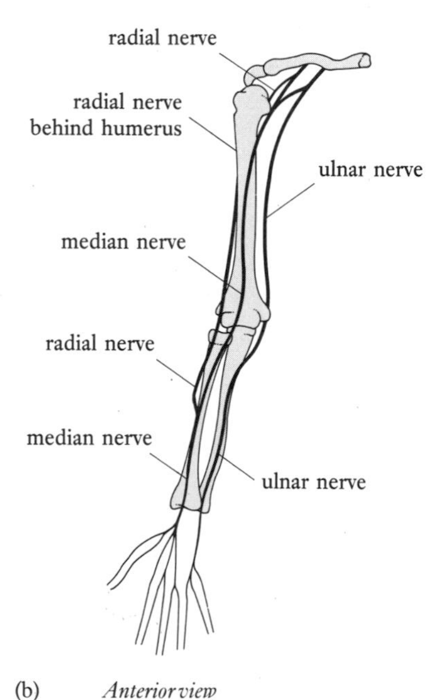

Figure 5.10. (a) Brachial plexus; (b) nerves of the arm (simplified).

Nursing Practice Application Brachial plexus damage

Damage to the brachial plexus can be caused by the incorrect use of crutches, such as leaning with all the pressure in the axilla or having crutches that are too long. Damage to the plexus may occur if a person's arms are not positioned carefully, as might be the case if consciousness is altered, e.g. during a general anaesthetic. The radial nerve may be affected by a condition known as *'Saturday night' palsy,* (Macleod et al 1987) which occurs if rather too much alcohol is consumed by an individual who then goes to sleep with his or her arm over a chair. They wake up with an arm that does not function properly, as well as a *'hangover'*.

■ Brachial plexus *(Figure 5.10)*

The **brachial plexus** is found in the neck and axilla and is formed from the spinal nerves C5–C8 and T1. Nerves from this plexus innervate the skin and muscles of the shoulder and arm and include the **axillary, radial, ulnar** and **median** nerves. When you hit your 'funny bone' it is the ulnar nerve that is responsible for the pain and tingling (See also *Nursing Practice Application,* above).

■ Lumbar plexus *(Figure 5.11)*

The **lumbar plexus** is situated in the *psoas* muscle of the back (see Chapter 18). The lumbar plexus is formed from the spinal nerves L1–L3 and part of L4. Nerves from this plexus innervate the psoas muscle, lower abdomen, hip, part of the genitalia, the thigh and the leg, and include: the **iliohypogastric, ilioinguinal, genitofemoral, lateral cutaneous nerve of the thigh, femoral** and **obturator nerves.**

■ Sacral plexus *(Figure 5.12)*

Situated in the posterior part of pelvic cavity, **the sacral plexus** is

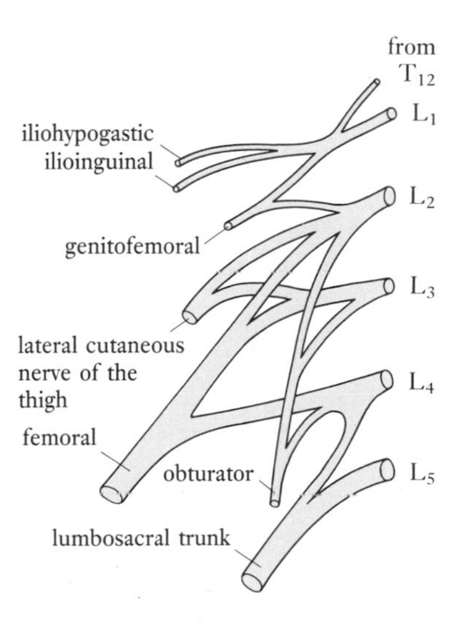

Figure 5.11. Lumbar plexus.

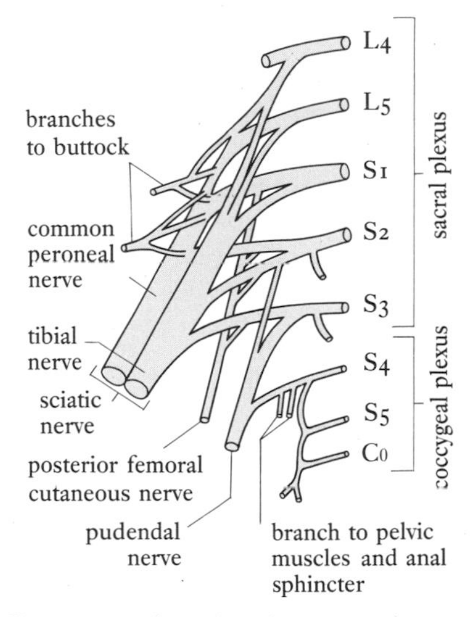

Figure 5.12. Sacral and coccygeal plexuses.

Nursing Practice Application Sciatic nerve

The position of the sciatic nerve has important implications for practice. When using the buttock as a site for intramuscular (IM) injections it must be remembered that the nerve passes through three quadrants of the buttock. This leaves only the outer upper quadrant as the area for injection, which avoids the risk of nerve damage (*Figure 5.13*). Back problems such as those leading to prolapse of an intervertebral disc in the lumbar region can cause considerable pain in the buttock and posterior aspect of the thigh (see Chapter 18).

formed from the spinal nerves L4/L5 **(lumbosacral trunk)** and S1–S3/4. Nerves from this plexus innervate the pelvic floor, pelvic structures, buttock and leg, and include: the **sciatic** (the largest nerve in the body), **posterior cutaneous nerve of the thigh, pudendal, tibial** and **peroneal nerves** (See *Nursing Practice Application,* above).

■ Coccygeal plexus (*Figure 5.12*)

The **coccygeal plexus** is formed from part of S4, S5 and the one coccygeal spinal nerve C0. It innervates the pelvic floor and the skin over the coccyx.

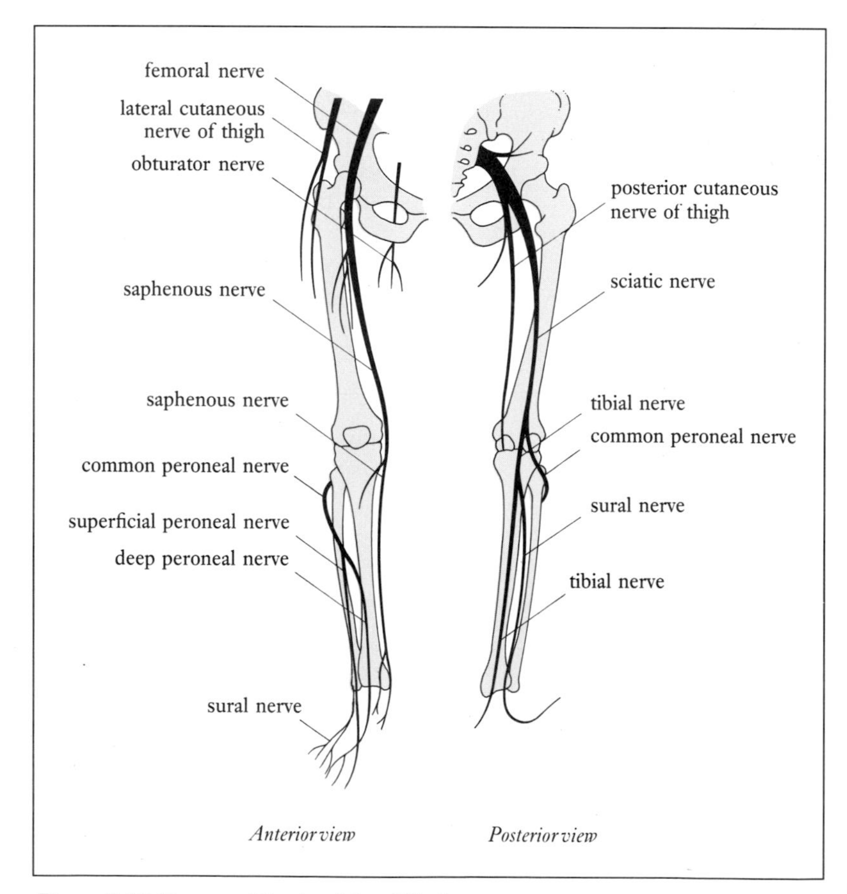

Figure 5.13. Nerves of the leg (simplified).

Reflexes

Reflexes, you will remember from Chapter 3, are the mainly protective, rapid involuntary motor responses to sensory inputs. The components of a reflex arc are discussed in Chapter 3. Functionally reflexes can be divided into:

- Those controlled by the autonomic nervous system (see Chapter 6), which initiate coughing, vomiting, secretion of saliva and maintenance of blood pressure, etc., by acting upon involuntary muscle and glands. Many of these reflexes operate through the medulla oblongata.
- Those that involve the contraction of voluntary muscle, which may be innate (inborn) or the result of learned behaviour that allows the automatic performance of some complex activity. Many of these postural or protective reflexes operate through the spinal cord and are known as **spinal reflexes.** In certain situations there is also some brain involvement, which will influence and modify the spinal reflexes.

Spinal reflexes

Tendon stretch reflexes are postural. They depend on the stretching of muscle fibres, which causes contraction of those muscle fibres, which in turn contributes to normal muscle tone. Tendon receptors known as *Golgi tendon organs* reverse this contraction if the tension in the muscle becomes too great.

Tendon stretch reflexes are simple *monosynaptic* reflexes involving sensory afferent fibres which synapse with the lower motor neurone (LMN) in the ventral horn of the spinal cord.

The tendon stretch reflexes can be tested where a tendon is stretched over a joint, e.g. the *knee-jerk reflex* (of the *patellar* tendon below the knee). When the patellar tendon is tapped sharply the quadriceps muscle of the thigh is stretched and contracts to cause the leg jerk *(Figure 5.14).*

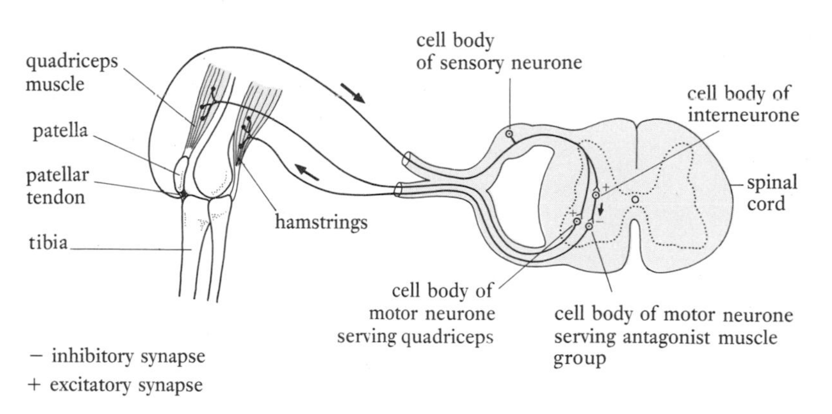

Figure 5.14. Knee-jerk reflex.

The stretch reflex is exaggerated where a disorder affects the upper motor neurone (UMN) because inhibitory influences from the UMN are lost. The reflex is absent where the LMN of the reflex arc is affected by disease or injury. It will also be reduced or absent if the function of the muscle groups involved is disordered (see *Table 4.1*).

Withdrawal or **flexion reflexes** are protective and involve more than one synapse *(polysynaptic).* They use interneurones to connect the sensory and motor neurones involved in the reflex, e.g. you withdraw your hand when you prick it on brambles whilst picking blackberries. These reflexes also initiate relaxation or even extension in other muscle groups so that posture is maintained as we move from pain or danger.

Another withdrawal/flexion reflex, the **plantar reflex,** occurs in response to painful/dangerous stimulation of the foot. It can be tested by stroking the sole of the foot: normally after infancy the toes curl up *(plantar flexor)*; in infants, prior to standing and walking the response is *plantar extensor*, where the toes extend and spread out. Disorders affecting the corticospinal tracts (UMN), such as a stroke, may cause the plantar response to be extensor, this sign of disordered function is known as a positive *Babinski response/sign.* The plantar response may be normal in disorders of the LMN but will be affected if the innervation to the relevant muscles is damaged.

Yet another reflex can be demonstrated by stroking the skin of the abdomen. Normally the underlying muscles will contract but the intensity of response varies in healthy individuals. The reflex is usually absent where UMN disease is present.

Reflexes, then, are altered in various ways by disordered function of the nervous system (see also spinal shock, Chapter 4); testing reflexes may help to localize certain disorders and is an important part of any neurological examination.

Summary/Check List

PNS Introduction – components, divisions.
Cranial nerves I–XII, function, disordered function.
Spinal nerves – cervical, thoracic, lumbar, sacral and coccygeal, structure and formation (roots, rami and plexuses). Dermatomes. Major plexuses – cervical plexus, *Nursing Practice Application* – hiccups. Brachial plexus, *Nursing Practice Application* – brachial plexus damage. Lumbar plexus. Sacral plexus, *Nursing Practice Application* – sciatic nerve. Coccygeal plexus.
Reflexes – autonomic reflexes. Spinal reflexes – tendon stretch, withdrawal/flexion, plantar (Babinski sign), abdominal, changes caused by disordered function.

?

Self Test

1 The PNS consists of the:
(a) Brain and peripheral nerves.
(b) Spinal cord and peripheral nerves.
(c) Muscles and lower motor neurones.
(d) Sensory receptors, peripheral nerves and ganglia.

2 Which of the following statements are true?
(a) All cranial nerves arise from the brain stem.
(b) There are twelve pairs of cranial nerves.
(c) The vagus nerve innervates structures in the chest and abdomen.
(d) The trigeminal is the largest cranial nerve.

3 Define: dorsal root, ventral root, mixed nerve, dorsal ramus and ventral ramus.

4 Correct the following statements:
(a) There are seven cervical spinal nerves.
(b) The thoracic spinal nerves form a plexus.
(c) The phrenic nerve forms part of the brachial plexus.
(d) The sciatic nerve runs through the outer upper quadrant of the buttock.

5 Describe the events occurring in the knee-jerk reflex.

Answers
1. d. 2. b, c and d. 3. See pages 103-104. 4. a. eight cervical nerves b. thoracic nerves do not form a plexus c. phrenic nerve is part of the cervical plexus d. the sciatic nerve does not pass through the outer upper quadrant but is found in the other three quadrants. 5. See page 109 and *Figure 5.14*.

References

Macleod, J., Edwards, C., Bouchier, I., Eds. (1987) *Davidson Principals and Practice of Medicine,* 15th edn. Edinburgh: Churchill Livingstone.

Further Reading

Ross Russell, R. W. and Wiles C. M. (1985) *Neurology.* London: Heinemann.
Williams, P.L., Warwick, R., Dyson, M. and Bannister, L.H. (1989) *Gray's Anatomy,* 37th edn. Edinburgh: Churchill Livingstone.

CHAPTER SIX

THE AUTONOMIC NERVOUS SYSTEM

OVERVIEW

- Autonomic nervous system (ANS).
- Sympathetic and parasympathetic divisions.
- Homeostatic role.
- Neurotransmitters.
- Stress.
- Pharmacology.

LEARNING OUTCOMES

After studying Chapter 6 you should be able to:

- Name the divisions of the ANS.
- Compare and contrast anatomical features, neurotransmitters and physiological functions of the divisions.
- Discuss the importance of the ANS in homeostatic mechanisms.
- Outline the effects of stress upon the ANS.
- Outline the ways in which drugs influence the ANS.

KEY WORDS

Acetylcholine – a neurotransmitter which functions in both divisions of the ANS.
Adrenaline – hormone released by the adrenal medulla. A catecholamine which augments the effects of the sympathetic nervous system.
Anticholinergic – a drug which inhibits the action of acetylcholine.
Catecholamine – important physiological substances such as adrenaline, noradrenaline and dopamine. They act as neutrotransmitters and as hormones. They are formed from the amino acid tyrosine in a series of chemical reactions.
Craniosacral – describes the outflow of the parasympathetic fibres from the CNS via the cranial nerves and the sacral part of the spinal cord.
Noradrenaline – the neurotransmitter functioning in part of the sympathetic division of the ANS. Also a hormone released by the adrenal medulla.
Parasympathetic – the division of the ANS concerned with resting functions of the body, e.g. digestion.
Parasympathomimetic (cholinergics) – a drug which stimulates parasympathetic action.
Preganglionic – refers to the autonomic fibre before the synapse in the autonomic ganglion.
Postganglionic – refers to the autonomic fibre after the synapse in the autonomic ganglion.
Sympathetic – the division of the ANS concerned with the body functions adapted to stressful situations.
Sympathomimetic (adrenergics) – a drug which stimulates sympathetic activity.
Thoracolumbar – describes the outflow of sympathetic fibres from the CNS via the thoracic and lumbar parts of the spinal cord.

INTRODUCTION

The **autonomic nervous system (ANS)** is the part of the peripheral nervous system (PNS) that controls involuntary functions vital to homeostatic regulation. It provides the motor innervation to smooth muscle, cardiac muscle and glands.

The ANS has two divisions: the **sympathetic** (concerned with stressful situations) and **parasympathetic** (concerned with resting functions). These are concerned with the many reflex actions which operate at subconscious levels in the CNS, e.g. in the medulla oblongata and spinal cord. We are, however, aware of some of its effects, e.g. increased secretion of saliva in anticipation of a favourite meal. The two divisions are mutually antagonistic, i.e. one opposes the effects of the other, but normally their combined activity produces a balanced situation.

The activities of the ANS are influenced by internal and external factors via sensory afferents from the viscera, higher centres, the hypothalamus and endocrine glands.

STRUCTURE AND FUNCTION OF THE ANS

A basic unit of the ANS consists of two neurones, ganglia outside the CNS (within which the neurones synapse) and neuroglia. The synapses outside the CNS in the ANS are in complete contrast to the voluntary motor system neurones which, you will remember, synapse in the spinal cord.

The two divisions of the ANS also show differences both in structural and functional features (see *Table 6.1*).

Some structures receive fibres from both divisions, e.g. heart, others, such as the sweat glands, have mostly sympathetic fibres and the remaining structures are innervated mainly with parasympathetic fibres, e.g. gastrointestinal glands.

Sympathetic division

The sympathetic division is concerned with the functions which enable the body to respond rapidly to stressful situations, the so-called *'flight or fight'* response.

Sympathetic fibres leave from the thoracic part of the spinal cord and the first part of the lumbar cord, i.e. they have **thoracolumbar** outflow *(Figure 6.1)*.

The **preganglionic fibres** (autonomic fibre before the synapse in the ganglion), which have their cell bodies in the lateral horn of the spinal cord, exit as part of the spinal nerve but very soon separate from the spinal nerve to form the **white rami communicans,** which is myelinated. The rami enter the chain of ganglia that runs either side of the vertebral column (the **sympathetic chain** or **paravertebral ganglia**). Here they may synapse with a second neurone, travel up or down the chain before synapsing or pass through the sympathetic chain to synapse in **prevertebral** or **collateral ganglia** which are situated in front of the vertebral column.

Table 6.1. Differences between the ANS divisions.

Features	***Sympathetic division***	***Parasympathetic division***
Outflow	Thoracolumbar	Craniosacral
General effects	Functions adapted for stressful situations Most are augmented by adrenal medulla	Functions adapted for resting situations
Neurotransmitters	Noradrenaline Acetylcholine	Acetylcholine
Position of ganglia	Outside but close to the CNS	Inside or close to effector structure
Length of fibres	Short preganglionic Long postganglionic	Long preganglionic Short postganglionic

The **postganglionic fibres** (autonomic fibre after the synapse in the ganglion) leaving the sympathetic chain are unmyelinated and form the **grey rami communicans**, which mostly rejoin the spinal nerves for their journey to the effector structures, which include sweat glands, smooth muscle and blood vessels.

Sympathetic ganglia

The sympathetic chain/paravertebral ganglia is a paired chain of ganglia joined by nerve fibres which extends from the cervical to the lumbar region. The **cervical ganglia** supply postganglionic fibres to the pupil of the eye, salivary glands, blood vessels in the head and the heart and lungs.

There are three prevertebral/collateral ganglia: the **coeliac, superior** and **inferior mesenteric ganglia,** which form plexuses that supply postganglionic fibres to abdominal and pelvic structures such as the gastrointestinal tract, kidney, bladder and reproductive organs.

The sympathetic division and the adrenal medulla

At this point we should consider the relationship between the sympathetic division and the **adrenal medulla** (see Chapter 8). Some fibres, which pass through the coeliac ganglia without synapsing, travel to the adrenal gland as very long preganglionic fibres (remember that preganglionic fibres are usually short). For simplicity we can regard the adrenal medulla as a sympathetic ganglion which has no postganglionic fibres. The adrenal medulla and sympathetic ganglia develop from the same embryonic tissue, which explains the close relationship. The adrenal medulla augments the functioning of the sympathetic division by releasing two **catecholamines** (physiologically important substances derived from tyrosine [amino acid]): **adrenaline** *(hormone*) and **noradrenaline** *(neurotransmitter/hormone*) directly into the blood.

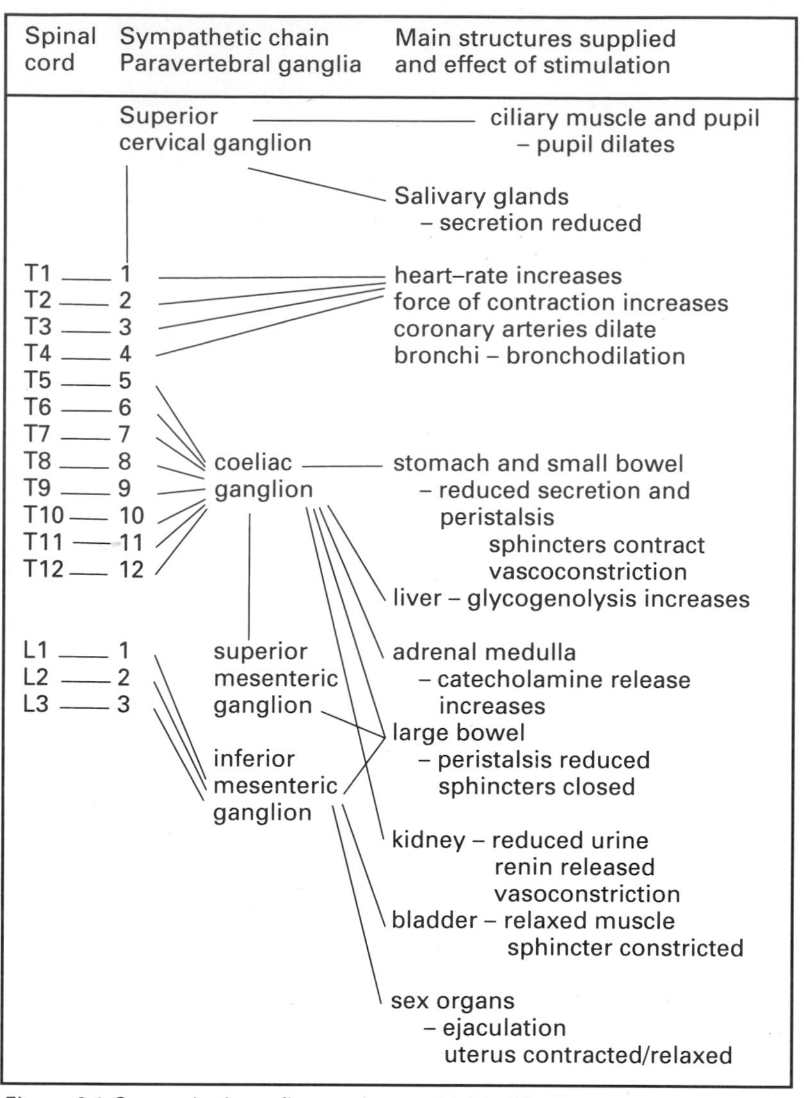

Figure 6.1. Sympathetic outflow and control (simplified).

Parasympathetic division

The parasympathetic division is the part of the ANS concerned with body functions associated with normal 'at rest' body functions. The cell bodies of its preganglionic fibres are situated in the brain stem and the lateral grey matter of the sacral part of the cord. Leaving the CNS with the cranial nerves and from the sacral region, the outflow is called **craniosacral** *(Figure 6.2, page 119).*

Generally the parasympathetic division is structurally less complex than the sympathetic division; most of its long preganglionic fibres synapse in ganglia situated within the effector structure. There are, however, four special ganglia for the parasympathetic fibres of the cranial nerves III, VII and IX: **the ciliary, sphenopalatine, submandibular** and **otic ganglia.** These fibres innervate structures in the head such as the salivary and lacrimal glands and the muscles inside the eye which control the shape of the lens and size of the pupil (see Chapter 4 for pupil reaction and neurological assessment). The other part of the cranial outflow is the very important **vagus** nerve (X) which provides most of the parasympathetic fibres in the body. The vagus innervates the oesophagus, heart, bronchi and lungs in the chest and passes through the diaphragm to supply the stomach, intestine, liver, gallbladder, pancreas and kidney (see *Figure 5.5*). Some preganglionic fibres of the vagus synapse with postganglionic fibres in plexuses, e.g. the supply to the heart, and in the ganglia within the walls of viscera such as the stomach. In the latter example a very short postganglionic fibre supplies the organ concerned.

The sacral parasympathetic outflow innervates the smooth muscle and glands of the large intestine, the bladder and the blood vessels of the genitalia. Preganglionic fibres synapse with post-ganglionic fibres in ganglia situated in the walls of these structures. These nerves are very important in the functioning of the anal and urinary sphincters and for sexual function, especially in the male. The pelvic structures involved also have sympathetic fibres and are influenced by voluntary nerves. We are therefore, able to exercise some voluntary control, e.g. we can overcome bladder emptying during an exciting television programme until the programme finishes.

Neurotransmitters and the ANS

Sympathetic division

The two neurotransmitters (see Chapter 3) of the sympathetic division synapses are **acetylcholine** and **noradrenaline.**

Acetylcholine is found at the preganglionic synapse and noradrenaline at the synapse between the postganglionic fibre and the effector structure.

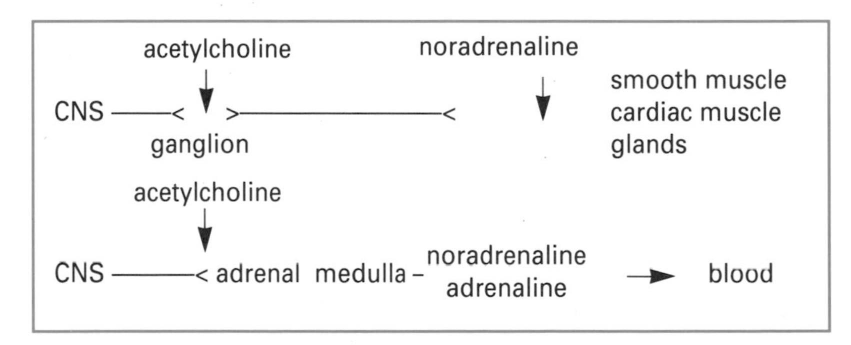

The postganglionic fibres using noradrenaline as their neurotransmitter are known as **adrenergic nerves.** There are two main types of adrenergic receptor: alpha α (divided into alpha 1 and 2), and beta β (divided into beta 1 and 2), which react differently to the noradrenaline and adrenaline produced by the adrenal medulla and have different responses to various drugs. A few postganglionic sympathetic nerves use acetylcholine, e.g. the sweat glands in the skin.

Parasympathetic division

Acetylcholine is the neurotransmitter at the synapse between pre- and postganglionic fibres and at the synapse with the effector structure in the parasympathetic division. It does, however, behave differently at the each of the two sites. The sites can be classified as being **nicotinic** or **muscarinic** which simply tells which of the two chemicals *nicotine* and *muscarine*, would bind to the receptor instead of acetylocholine which they mimic. This is an important distinction when we consider the effects of drugs upon the ANS. Fibres which use acetylcholine are known as **cholinergic nerves.**

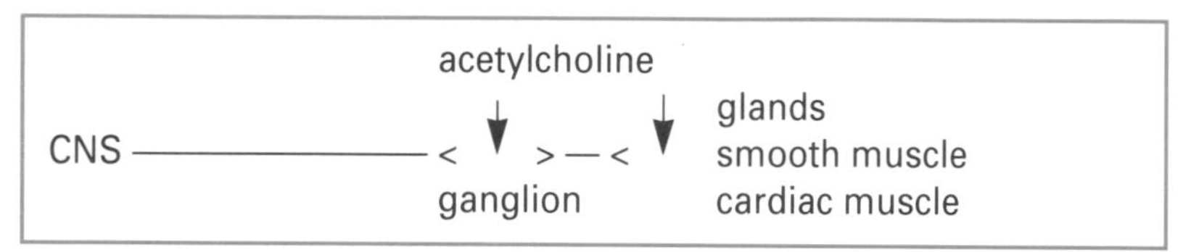

(See also Chapter 3.)

The ANS and homeostasis

It is worth restating the importance of the ANS to homeostatic regulation (*Table 6.2*). The ANS, which is an entirely motor system, is influenced by sensory nerve input from the muscle tissue and glands concerned, the hypothalamus (see Chapter 4), the cerebral cortex and endocrine glands. The technique of *biofeedback* may, for some people, increase their conscious control over certain autonomic functions, such as the ability to lower blood pressure.

Stress and the ANS

Biofeedback *– a technique where information (signals) gained from the monitoring of an autonomic function can be used to learn some conscious control over that function.*

A variety of physiological, psychological and social **stressors** (a term used by Selye (1978) to describe factors which cause stress responses) affect the ANS and hence the homeostatic balance of the body. Selye (1978) described the **general adaptation syndrome (GAS)**, which is a **triphasic** response to stressors: (i) **alarm** or initial reponse to the stressor; (ii) **resistance/adapation**, where the body structures adapt to continued contact with the stressor; and (iii) **exhaustion**, a situation where adaptation fails, and which may result in chronic illness or death.

Nursing Practice Application Stress in health care situations

Apart from the events seen in severe situations, e.g. shock due to haemorrhage, the nurse will often observe the physiological effects of anxiety and stress. These effects include: sweating, rapid pulse and restlessness and may be seen in people anxious about admission to hospital or the results of investigations or treatment. Another important group who may be affected by stress and anxiety are health care workers. It is vital that stress is recognized in both groups and appropriate coping strategies implemented to prevent or minimize its effects. Coping strategies, such as relaxation techniques, should help the person to regain his or her sense of control (Wallace, 1989).

Other stress-reducing measures include counselling and giving information. Boore (1978) found that stress indicators (urinary excretion of 17-hydroxycorticosteroid) in surgical patients could be reduced by giving information about care and treatment, and by teaching exercises pre-operatively (see Chapter 8).

Stress and pressure may affect standards of care and all nurses have a responsibility to notice if the health or safety of colleagues is being adversely affected. (UKCC Code of Professional Conduct, 1992). See Chapter 8 for a further discussion of stress.

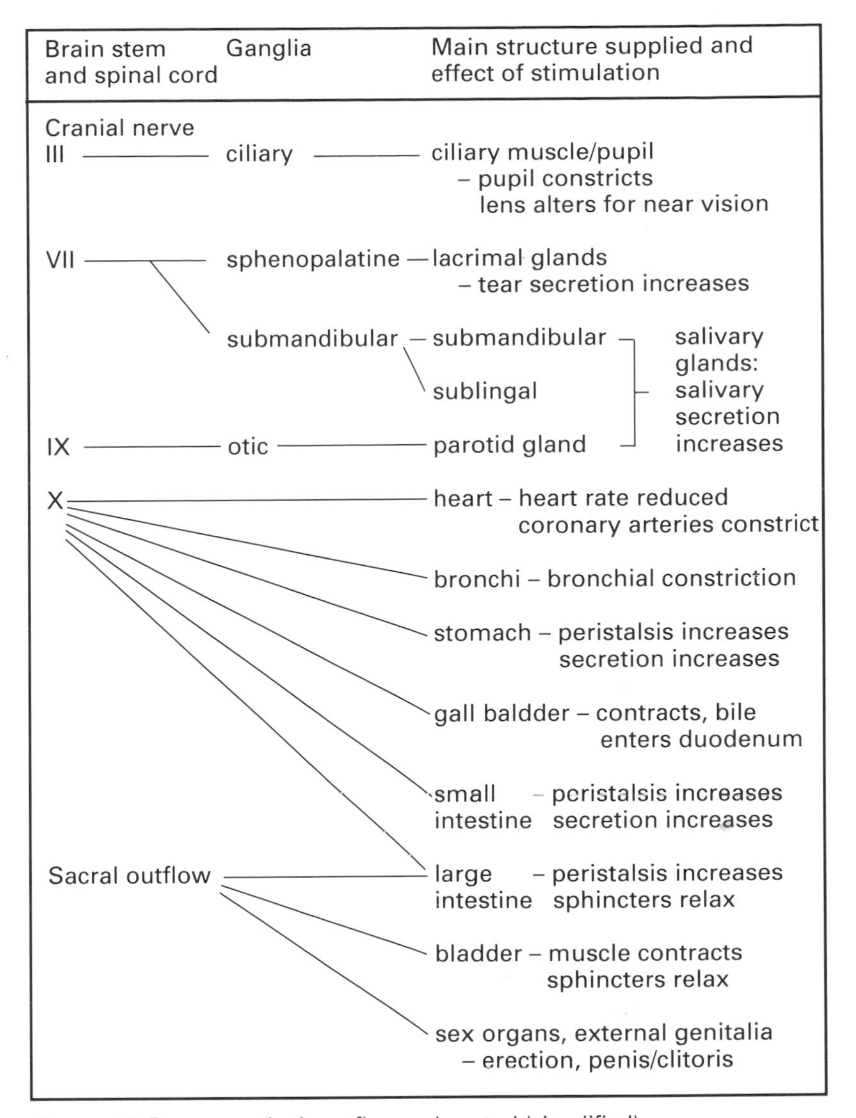

Figure 6.2. Parasympathetic outflow and control (simplified).

Basically the functioning of the sympathetic division, adrenal gland (medulla and cortex) and, to a much lesser extent, the parasympathetic division produces effects which allow the body to adapt to a specific stressor, e.g. extreme cold. Initially the adaptive heat-conserving mechanisms are stimulated by the sympathetic nerves, e.g. vasoconstriction (constriction of blood vessels) in the skin slows heat loss. If these coping mechanisms are unable to restore temperature homeostasis, the response to the stressor becomes non–adaptive, i.e. body temperature falls and metabolism slows (see hypothermia Chapter 19). At this stage the individual is unlikely to survive without medical intervention.

The list of stressors is enormous and includes fear, pain, cold or heat, trauma, blood loss, infection, hunger, low blood sugar and a

Table 6.2. Autonomic effects – a summary.

Effector structure	***Sympathetic action***	***Parasympathetic action***
Eye - pupil - ciliary muscle	Pupils dilate ——	Pupils constrict Changes shape of the lens (for close vision)
Lacrimal and nasal glands	Vasoconstriction No secretions	Secretion occurs; tears etc.
Salivary glands	Vasoconstriction No secretions	Saliva produced; watery with enzymes. Vasodilation
Oesophagus	Vasoconstriction	Secretion and increased motility/peristalsis
Stomach and small intestine	Vasoconstriction Reduced secretion and peristalsis Sphincters constricted	Secretion of digestive juices, peristalsis increased
Liver	Glucose released into the blood (glycogenolysis)	——
Gallbladder	Relaxed	Contracts; causing bile to enter the duodenum
Pancreas (endocrine function)	Influences insulin secretion	
Large intestine/rectum/anus	Reduced secretion and peristalsis Sphincters constricted	Secretion and peristalsis increased Sphincter relaxed - *defaecation*
Bronchi/lungs	Bronchodilation	Bronchial constriction
Heart	Heart rate and force of contraction increases Coronary arteries dilate	Heart rate and force of contraction reduced Coronary arteries constrict
Kidney	Vasoconstriction, reduced urine output Renin production	——
Bladder	Relaxed muscle, sphincter constricted	Muscle contracts, sphincter relaxed – *micturition*
Genitalia penis clitoris uterus	 Ejaculation —— Muscular contraction/relaxation	 Vasodilation- erection Vasodilation- erection ——
Adrenal medulla	Secretion - adrenaline and noradrenaline	——
Sweat glands - general - palms	Sweating (cholinergic fibres) Sweating (alpha receptors)	—— ——
Arrector pili muscles (hair follicles)	Muscles contract – hairs erect 'goose flesh'	——
Blood vessels	Most constrict so that heat is conserved, blood pressure increased and blood redirected to vital structures from skin and digestive organs. Vessels in skeletal muscle dilate during exercise (cholinergic fibres)	——
Adipose tissue	Lipolysis (fat breakdown for energy)	——

life crisis. The ability to cope with stressors varies between individuals. Coping with physiological stress is also dependent on hormonal, nutritional and immunological factors (Clark, 1984).

A very intense stressor which places you in some danger, such as the fear you feel on being followed along a dark road, produces the 'fight or flight' response when the sympathetic nervous system prepares you to cope by increasing heart rate, dilating vessels in skeletal muscle, dilating the bronchi and pupils and releasing glucose into the blood (see *Table 6.2*).

A less intense stressor causing anxiety over a period of time, such as a heavy workload, may result in maladaptive or 'unhealthy' physiological effects, which include hypertension, irregular heart rate, headaches, indigestion and high levels of glucose and fatty

Table 6.3. The effects of drugs on the ANS.

Drug group/example	***ANS division (S or P)/receptor***	***Effects/uses***	
Sympathomimetic	S	Effect	Sympathetic stimulation
adrenaline	beta (mainly)	Uses	Bronchodilation,allergy, anaphylaxis, cardiac arrest
dopamine (noradrenaline precursor)	beta 1	Uses	Cardiogenic shock to improve cardiac output
Selective adrenoceptor stimulant	S	Effect	Selective sympathetic stimulation
salbutamol	beta 2	Uses	Bronchodilation, inhibition of premature labour
Sympathetic blocking agents/ sympathetic inhibitors	S	Effect	Sympathetic blockade (inhibition)
alpha blocker – prazosin	alpha	Effect	Vasodilation, reduces blood pressure,
		Uses	In hypertension
beta blocker – atenolol	beta 1	Effects	Reduces heart rate and lowers blood pressure
		Uses	Hypertension, angina and arrhythmias
centrally acting drugs – methyldopa	Prevents the formation of noradrenaline	Effect	Reduces blood pressure
		Uses	Hypertension (not frequently)
Antidepressants			
tricyclics – imipramine	Prolongs the action of noradrenaline	Uses	Depression
monoamine oxidase inhibitors – phenelzine	Inhibits the enzymes which degrade monoamines, such as dopamine (CNS); leads to an increase in these neurotransmitters	Uses	Depression
Parasympathomimetics (cholinergics)	P	Effect	Enhance parasympathetic action
neostigmine	Inhibits the enzyme cholinesterase and prolongs the action of acetylcholine	Uses	Mainly for its action on voluntary motor end plates in myasthenia gravis or to reverse certain muscle relaxants used in surgery
carbachol	Acts like acetylcholine at muscarinic receptors	Effect	Contraction of bladder muscle
		Uses	In unobstructed retention of urine
Anticholinergics	P	Effect	Inhibits parasympathetic action
atropine	Inhibits muscarinic receptors	Effects	Relaxes smooth muscle and reduces salivary and bronchial secretions.
		Uses	As part of premedication, can be used for colic but side effects may make it unacceptable, used in heart block, used in ophthalmology to dilate the pupil

acids in the blood. If these effects are prolonged there appears to be an increased risk of developing conditions such as coronary heart disease.

Effect of drugs on the ANS

Very many drugs have an influence on ANS function. *Table 6.3, page 121* gives a summary of the more commonly used substances but further examples will be seen during your own observation and practice.

SUMMARY/CHECK LIST

ANS Introduction – divisions, influences, structure, basic unit, differences between sympathetic and parasympathetic divisions.
Sympathetic division – function, outflow, structure, ganglia, relationship with adrenal medulla.
Parasympathetic division – function, outflow, structure, ganglia.
Neurotransmitters – Sympathetic – noradrenaline, acetylcholine, adrenergic receptors. Parasympathetic – acetylcholine, cholinergic receptors.
ANS and its importance to homeostasis – influences on ANS, biofeedback, summary autonomic effects.
Stress and ANS – stressors, adaptive response, 'unhealthy' effects, *Nursing Practice Application* – stress.
Effects of drugs on the ANS.

?

SELF TEST

1 Outline briefly the structure and functions of the sympathetic and parasympathetic divisions of the ANS.
2 Explain where acetylcholine and noradrenaline are used in the ANS. Define the following terms: adrenergic, cholinergic, nicotinic, muscarinic.
3 Give two examples of vital functions controlled by the ANS.
4 Which of the following statements are true?
(a) Stressors are always physical.
(b) Most ANS response to stress is through the sympathetic division.
(c) The 'fight or flight' response results from an intense stressor.
(d) Prolonged exposure to stressors is linked to increased risk of coronary heart disease.
5 Why do beta blocker drugs such as atenolol slow the heart rate?

Answers
1. See *Table 6.1* 2. See pages 117-118. 3. Heart rate, blood pressure, temperature regulation, etc. 4. b, c, d. 5. They block beta 1 receptors in the heart which normally respond to the sympathetic division by increasing heart rate.

References

Boore, J.R.P. (1978) Prescription for recovery: The effect of pre-operative preparation of surgical patients on post–operative stress, recovery and infection. *RCN Research Report.* London: The Royal College of Nursing.

Clark, M. (1984) Stress and coping: constructs for nursing. *Journal of Advanced Nursing,* **9** (1), 3–13.

Selye, H. (1978) *The Stress of Life,* revised edn. New York: McGraw-Hill.

UKCC (1992) *Code of Professional Conduct for the Nurse, Midwife and Health Visitor,* 3rd edn. London: United Kingdom Central Council for Nursing, Midwifery and Health Visiting.

Wallace, A. (1989) An active role for patients in stress management. *The Professional Nurse,* **5** (2), 65–72.

Further Reading

British Medical Association and Royal Pharmaceutical Society of Great Britain (1992) *British National Formulary.* Number 24. London: BMA and The Pharmaceutical Press.

Holland, S. (1987) Stress in nursing. *Nursing Times,* **83** (21), 59–61.

Montague, S. E. (1980) The physiological basis of the stress reaction. *Nursing,* **1** (10), 422–425.

CHAPTER SEVEN

THE SPECIAL SENSES

OVERVIEW

- The eye and vision.
- The ear and hearing/balance.
- Olfactory epithelium and smell.
- Taste buds and taste.

LEARNING OUTCOMES

After studying Chapter 7 you should be able to:

- Describe how the eye is protected.
- Describe the structure of the eye and the functions of its component parts.
- Define refraction, accommodation and describe how light travels through the eye.
- Describe common problems with refraction and their correction.
- Describe how stereoscopic vision is produced.
- Explain how the retina adapts to dark and light.
- Outline how nerve impulses travel from the retina to the brain.
- Outline how visual acuity may be measured.
- Outline some of the changes in the eye associated with normal ageing.
- Describe the structure of the ear and the functions of its major components.
- Describe the passage of sound waves through the ear.
- Outline how nerve impulses travel from the organ of Corti to the brain.
- Outline the ability of the ear to differentiate between sounds of different intensity and pitch and to determine the direction of its source.
- Explain how the vestibular apparatus helps to maintain balance.
- Describe the changes in the ear associated with normal ageing.
- Describe the structure of the olfactory receptors and the mechanism of smell.
- Describe the structure of the taste buds and the mechanism of taste.
- Describe how smell and taste are affected by normal ageing.

Key Words

Accommodation – the ability of the eye to focus on near objects.
Choroid – the middle coat of the eye, contains pigment and blood vessels.
Cochlea – the spiral (snail-shaped) cavity of the inner ear, contains the organ of Corti and endings of the vestibulocochlear nerve.
Olfactory epithelium – chemoreceptor cells in the nose which respond to inhaled chemicals to give the sense of smell.
Organ of Corti – the organ of hearing within the cochlea.
Refraction – bending of light rays as they pass from one medium to another.
Retina – the inner coat of the eye, it contains rods and cones – the light-sensitive receptors.
Sclera – the fibrous outer coat of the eye.
Semicircular canals – three fluid-filled structures in the inner ear, concerned with equilibrium and balance.
Taste buds – the taste organs on the tongue containing chemoreceptor cells which respond to chemicals in food.

Introduction

The special senses are vision, hearing/balance, smell, taste, and touch or tactile sense (see Chapter 3, sensory receptors, and Chapter 19, skin). This chapter covers the structure and functioning of the eye, ear, olfactory epithelium and taste buds. These special sense organs are central to our ability to monitor and respond to environmental stimuli. Impulses from the sense organs are conveyed to the brain via the cranial nerves. These messages, along with those from other sensory receptors such as proprioceptors, allow the brain to form a comprehensive view of our immediate environment and to initiate appropriate responses.

As with other components of the nervous system the special senses function as part of the whole system to facilitate the communication between the outside world and the CNS, which is so essential for our protection, well-being and enjoyment. Try to imagine for a moment just how many complex neural processes are involved when, on seeing a brightly coloured flower, you bend closer to enjoy its scent.

Vision - The Eye

The eyes develop as outgrowths from the embryonic brain, starting around the fourth week of pregnancy. The developing eyes are vulnerable to damage during this early period from agents such as the *rubella* (German measles) virus.

The eye contains photoreceptors which respond to light from the environment. Light causes chemical changes in the photoreceptors which in turn initiate electrical potentials in the axons of retinal ganglion cells. These impulses are transmitted via the optic nerves (II) and tracts to the brain for processing.

Accessory structures

The eyeball is spherical and has a diameter of approximately 2.5 cm. Most of the eyeball is enclosed and protected within the bony orbit of the skull, only the very front portion is exposed; a pad of fat between eyeball and orbit also provides protection.

Other protective structures of the eyeball include the eyebrows, lashes, lids, conjunctiva and the lacrimal apparatus (*Figure 7.1*).

Eyebrows

The **eyebrows** grow obliquely over the superior orbital ridges of the frontal bone (see Chapter 18). They protect the eye from sweat, foreign bodies and from sunlight.

Eyelids and lashes

The **upper** and **lower lids** are thin, movable folds of tissue. They consist of an outer covering of skin, connective tissue the **tarsal plate** which supports two muscles – the *levator palpebrae superiorsis* and *orbicularis oculi,* which open and close the eye respectively (see Chapter 18) – and a lining of conjunctiva. The lids meet at the medial and lateral angles of the eye to form the **medial** and **lateral canthi** (singular, canthus; literally, corner). At the medial canthus there is a small raised, fleshy structure called the **caruncle**.

The **eyelashes** are curved hairs projecting from the lid borders. Glands along the inner edge of the lid include the **Meibomian glands**; these are modified sebaceous glands (see Chapter 19) which secrete an oily fluid which helps to keep the eye lubricated. The eyelids close reflexly every few seconds, this moves secretions across the eye to prevent drying. This *blink reflex* is also initiated by situations which threaten the eye: bright lights (such as when you put the main lights on after sitting with one small lamp) and if the lids or lashes are touched.

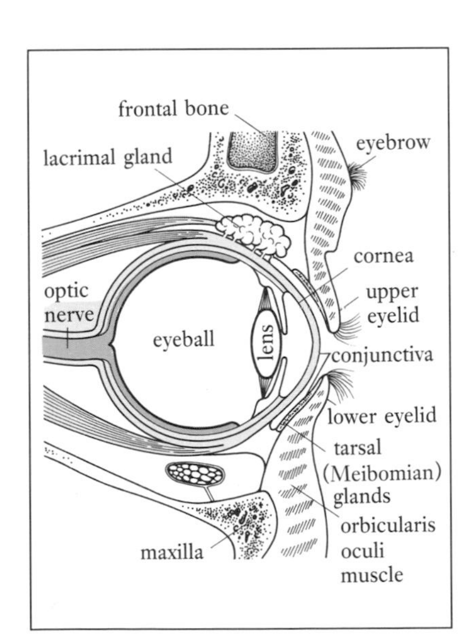

Figure 7.1. Eye and accessory structures.

Conjunctiva

The **conjunctiva** is a vascular transparent membrane which lines the lids and is reflected over the anterior part of the eyeball. Inferior and superior sacs are formed between the part of the conjunctiva linir g the lid and that covering the eye. The conjunctival covering prevents drying and damage to the eyeball.

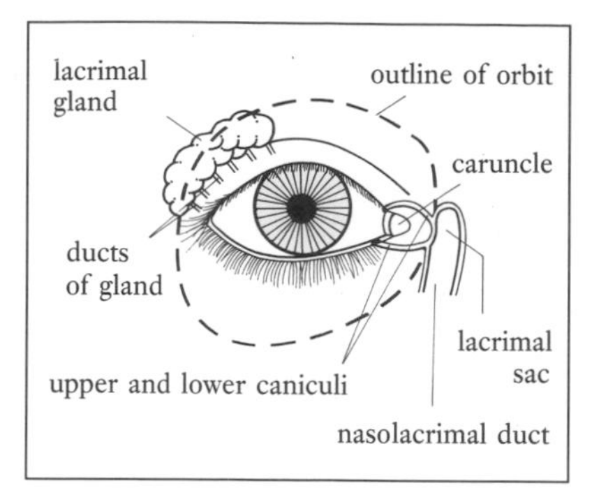

Figure 7.2. Lacrimal system.

Lacrimal apparatus

The **lacrimal gland** is situated within the orbit above the lateral aspect of each eye *(Figure 7.2).* The gland produces the watery fluid known as **tears**, which contains: water, mucus, salts and the *antibacterial* enzyme **lysozyme** (which is also found in other body fluids). Tears leave the gland via a series of small ducts to be moved across the eyeball towards the medial canthus by blinking. Tears enter the two **lacrimal canaliculi** through openings called **puncta** (singular, punctum) which drain the tears into the **lacrimal sac.** This sac is continuous with the **nasolacrimal duct**, which drains into the nasal cavity (you need to blow your nose whilst crying).

The continual washing of tears across the eye keeps it moist, clean and helps prevent bacterial infection. Excess tear production occurs when the eye is irritated by a foreign body or in response to some emotional situation. When the nasolacrimal ducts can no longer cope with the volume, tears overflow onto the face.

Tear production may diminish with age; affected individuals have dry eyes that are sore and susceptible to infection. Watery eye is caused by an obstruction in the drainage system. This may occur in newborn babies, during an upper respiratory tract infection, and in some older people.

Nursing Practice Application Eye Care

Where blinking or fluid secretion is impaired, such as in altered consciousness, measures to protect the eye and prevent corneal damage (see page 129) may include taping the lids shut, eye covers and the instillation of prescribed medication – 'artificial tears', e.g. hypromellose drops – to keep the eye moist, which are administered into the inferior conjunctival sac.

Extraocular (extrinsic) muscles of the eye

Each eye is moved by six muscles (see *Figure 7.3*). The muscles are attached to the orbit and the eyeball, and innervated (autonomically) by cranial nerves (see Chapter 5). This gives the precise control over eyeball movement and position that is essential for focusing on near or distant objects. Eyeball movement is also subject to voluntary control.

The six muscles fall into two groups:

- *Rectus muscles* (four):
  - **lateral rectus** moves eyeball outwards (VI nerve).
  - **medial rectus** moves eyeball inwards (III nerve).
  - **inferior rectus** moves eyeball downwards (III nerve).
  - **superior rectus** moves eyeball upwards (III nerve).
- *Oblique muscles* (two):
  - **superior oblique** moves the eye out and downwards (IV nerve).
  - **inferior oblique** moves the eye out and upwards (III nerve).

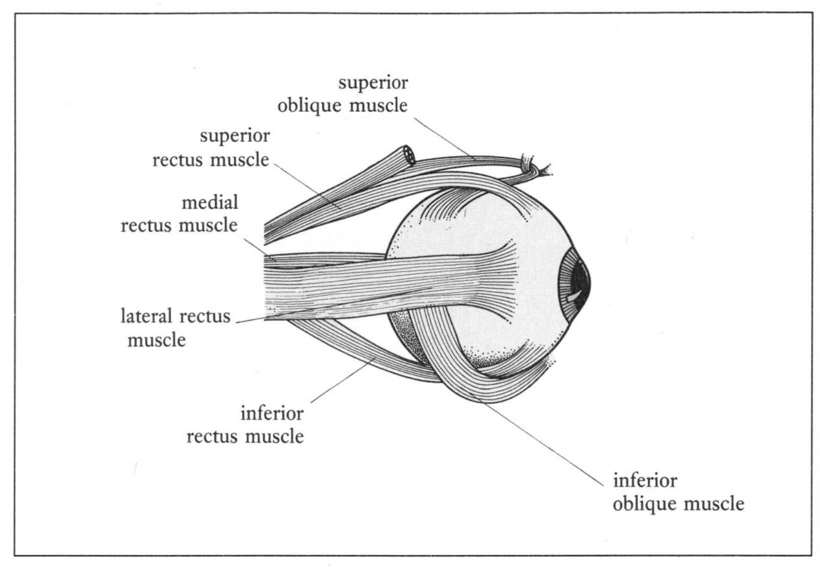

Figure 7.3. Extrinsic muscles.

The condition of *strabismus* (squint) occurs when the functioning of the six extrinsic (external) muscles and or their nerve supply is impaired (see Chapter 5).

THE EYEBALL

Layers of the eyeball

Outer layer

The eyeball has three layers *(Figure 7.4)*. An outer fibrous coat called the **sclera** forms the 'white of the eye'. The portion of outer coat over the very front of the eye is called the **cornea**. The cornea is a transparent avascular membrane which lets light into the eye. Its curved surface is the site of considerable light ray **refraction** (bending of light rays as they pass from one medium to another) (see page 132). The avascularity of the cornea allows for corneal grafts without the need for tissue/blood type matches required for other transplants (see Chapter 19).

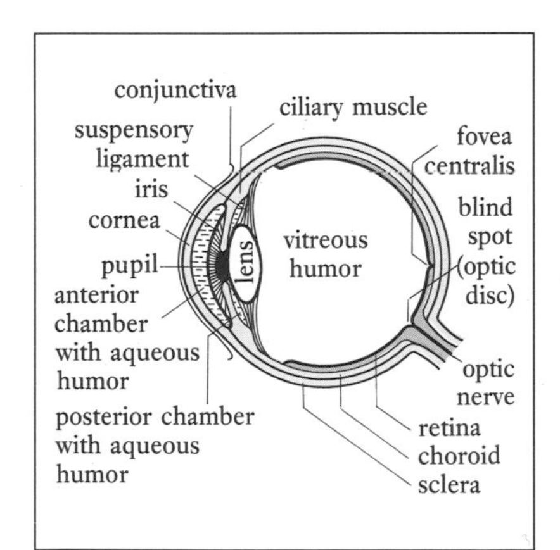

Figure 7.4. The eyeball.

Preventing eye damage

The cornea is very exposed and easily damaged (see *Nursing Practice Application,* page 128). Every possible effort should be made to provide information regarding eye safety. This will include using correct procedures for handling harmful chemicals and materials, wearing eye protection where appropriate and being aware of first aid measures in the event of an accident. Nurses need to think about areas of their own practice which could be hazardous such as preparing *cytotoxic* drugs. For eye protection a visor or goggles should be worn or the drug prepared within an enclosed cabinet (Health and Safety Executive, 1983). In addition, local safety policies e.g. in laboratories, should be known and followed.

Middle layer

The middle layer, or **uvea**, forms the **choroid**, **ciliary body** and **iris**. The choroid, found in the posterior five-sixths of the eye, is a highly vascular (many blood vessels) pigmented coat which absorbs light and prevents leakage or scattering of light. At the front of the eye the uvea forms the ciliary body, which contains the **ciliary muscles** that control **accommodation** (ability of the eye to focus on near objects). Extending from the ciliary body is the **suspensory ligament** which supports the **lens**.

Blood vessels in processes of the ciliary body produce a watery fluid, the **aqueous humor**, which fills the **anterior** and **posterior chambers** of the eye. Anterior to the ciliary body is the iris, which forms the coloured part of the eye (different amounts of brown pigment give blue, green, grey and brown eyes). The iris is a circular structure with muscle fibres (circular and radiating) which surround an opening, the **pupil**. The iris determines pupil size according to light intensity and emotional state – iris muscle fibres, you will remember, are reflexly controlled by fibres of the third (III) cranial nerve (see Chapters 5 and 6). Situated behind the cornea and in front of the lens, the iris forms the division between the anterior and posterior chambers.

Inner layer

The inner layer of the eyeball is the **retina** (*Figures 7.5 and 7.6*), a two-layer complex of pigment cells, photoreceptors and other neurones which receives light rays and converts them chemically to nerve impulses.

If these two layers become torn and separated, a *detatched retina* results, which leads to visual problems or blindness. Lasers are used to 'weld' the two layers together before vision is permanently impaired.

The retina covers the posterior part of the eye but only the part at the very back is photosensitive. There are two types of photoreceptor:

- The **rods**, which are more numerous and are responsible for peripheral and low light intensity vision, and contain the pigment **rhodopsin** (*visual purple*).
- The **cones**, which contain other pigments and give a high-definition colour vision in bright light.

There are distinct areas of the retina. The **blind spot** or **optic disc** (see *Figure 7.4*) is the point at which the optic nerve leaves the eye – as the name suggests there is an absence of photoreceptors. This area contains many blood vessels and can be visualized using an *ophthalmoscope* (*Figure 7.6*). This can aid the diagnosis of conditions such as raised intracranial pressure (see Chapter 4) where the optic disc bulges (*papilloedema*). At the lateral side of the optic disc is the **macula lutea** (yellow spot); this has a small depression, the **fovea centralis**, which contains only cones and is important in colour vision. The peripheral retina contains mainly rods.

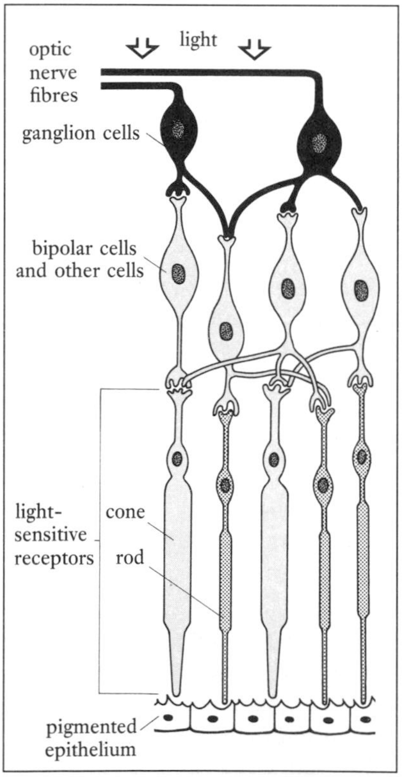

Figure 7.5. Cell layers of the retina (rods and cones).

The eye receives blood from a **central artery** and its venous blood drains via a **central vein**, both of which run in the middle of the optic nerve.

Inside the eye

The cavities/chambers

The anterior cavity of the eye is divided by the iris into the anterior and posterior chambers; both are filled with aqueous humor. Aqueous humor is produced by blood vessels in processes which form part of the ciliary body. The rate of production is determined by the intraocular (inside the eyeball) pressure. The correct volume is maintained by production equalling the amount which drains into the venous system via the **canal of Schlemm** (in the angle between the iris and cornea). The process of aqueous humor production and drainage can be compared with that of cerebrospinal fluid (see Chapter 4). The aqueous humor maintains intraocular pressure at a constant level during health; when fluid drainage is impaired the intraocular pressure rises, resulting in *glaucoma*, which causes visual disturbances and blindness if untreated.

Glaucoma, which may be primary or secondary to eye disease or trauma, is increasingly common after 40 years (it is estimated to affect as many as 2% of this older population). Glaucoma can be detected by examination of the eye and by measuring intraocular pressure. Many authorities recommend regular screening after 40 years. Management of glaucoma depends upon its severity but includes:

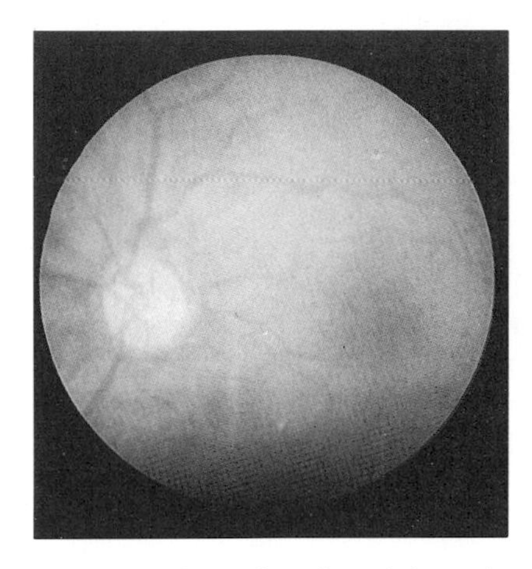

Figure 7.6. The retina viewed through an ophthalmoscope.
(Bedford, M. *A Colour Atlas of Opthalmological Diagnosis.* 2nd edn. Wolfe Medical Publications, Ltd. 1990. Reprinted with permission.)

- Miotic eye drops to constrict pupil and increase humor drainage, e.g. pilocarpine, which is *parasympathomimetic* (see Chapter 6).
- Surgery to improve drainage e.g. *iridectomy.*

The aqueous humor also provides nutrients and removes waste from the cornea and lens.

Posterior to the lens is the larger posterior cavity filled with the clear gel called **vitreous humor**. In contrast to the anterior cavity this is a closed system where the vitreous humor, formed during embryonic development, lasts for life. The vitreous humor contributes to intraocular pressure and helps to maintain eyeball shape and the position of lens and retina.

The lens

The lens, which separates the anterior and posterior cavities, is an avascular, biconvex, transparent structure (although it discolours with age) which alters shape to focus light onto the retina. The lens consists of epithelium on its anterior surface and fibres which form the rest of the lens. It is supported by the suspensory ligaments and is enclosed in a capsule. *Cataract* is an opacity or clouding of the lens; this is a common cause of impaired vision, but can be treated by removal/destruction of the lens. Following removal of the affected lens the ability to focus light is maintained by the implantation of a synthetic intraocular lens or the use of contact lenses or glasses of the correct prescription.

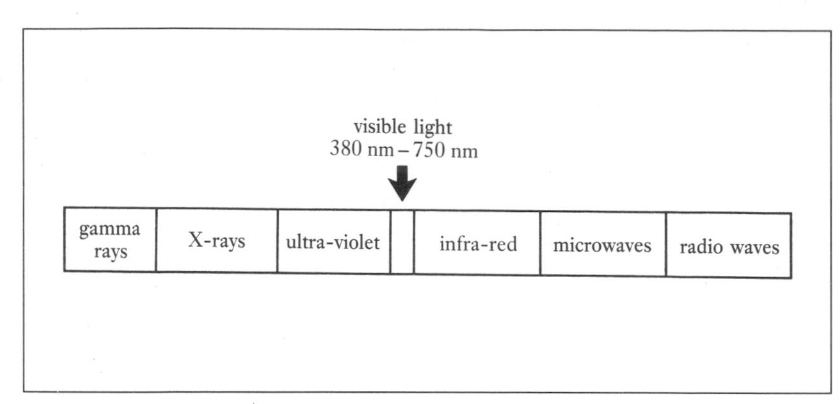

Figure 7.7. Electromagnetic spectrum.

colours of the visible spectrum
red
orange
yellow
green
blue
indigo
violet
white light
prism

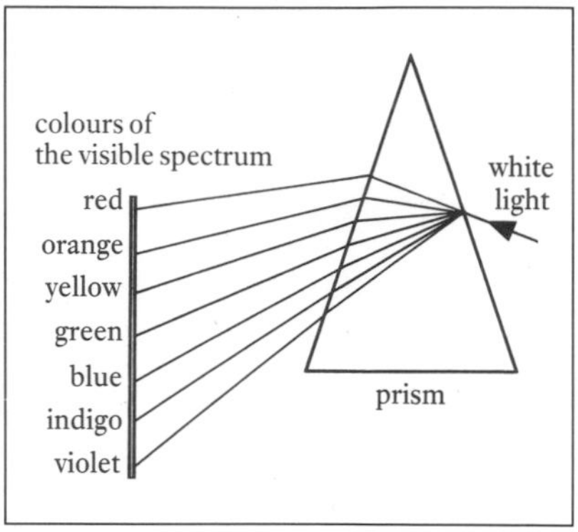

Figure 7.8. Visible spectrum.

Vision

The functioning of the eye can be compared with what happens when you take a photograph. Both the eye and the camera allow controlled amounts of light through a lens to be focused onto a photosensitive surface or film. The big difference comes when the image is processed; the image from the eye is interpreted and integrated by the CNS but the best you get from the camera is a set of photographs.

Light, refraction and the eye

It will be helpful at this stage if we remind ourselves about a few basic but vital facts concerning light energy. You may also want to consult the reading list for basic physics books.

Light travels in straight lines as waves of energy at the incredible rate of 300 000 km/s (186 000 miles/s). Our eyes can only respond to a range of wavelengths known as visible light, which forms a tiny part of the *electromagnetic spectrum* (*Figure 7.7*).

Visible light is split into the colours of the *visible spectrum* by passing it through a prism – the same effect is demonstrated when a rainbow is produced as sunlight passes through rain droplets (*Figure 7.8*).

We see objects as coloured when that particular colour wavelength is being reflected, e.g. grass reflects the green wavelength. White objects appear so because they are reflecting all wavelengths, in contrast, black objects absorb all wavelengths.

Another property of light is **refraction**, or bending when it passes from one medium to another, e.g. air to water. This happens when light enters the eye (*Figure 7.9*). Refraction can be demonstrated by placing a pen in a glass of water: the pen appears to bend as it enters the water. The eye uses refraction to bend the light so that it focuses correctly on the retina. Light is refracted by the cornea, aqueous humor, lens and vitreous humor. A **biconvex** lens causes convergence, like that in the eye, of the light so that it is focused onto a single point on the retina known as the **focal point** (*Figure 7.9*) (a **biconcave** lens causes divergence). The image produced on the

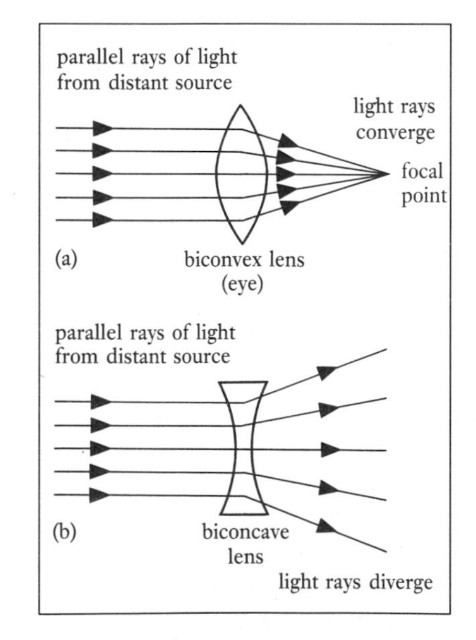

Figure 7.9. Refraction. (a) Eye (biconvex); (b) for comparison, a biconcave lens.

retina is actually upside down and reversed but is corrected by processing in the CNS. The distance between the lens and the focal point is called the focal length.

Just like the camera, we need to be able to focus light from both near and distant objects; more refraction occurs when focusing near objects.

Accommodation, distant and near vision

The ability of the lens to change shape (become more curved) and so increase refraction is known as **accommodation**. When we look at distant objects the ciliary muscle relaxes and the suspensory ligaments pull the lens flat; in this state there is little refraction at the lens (*Figure 7.10*). It is when we look at near objects that refraction is increased by contraction of the ciliary muscle which reduces the pull on the suspensory ligaments and so allows the lens to bulge and become more curved.

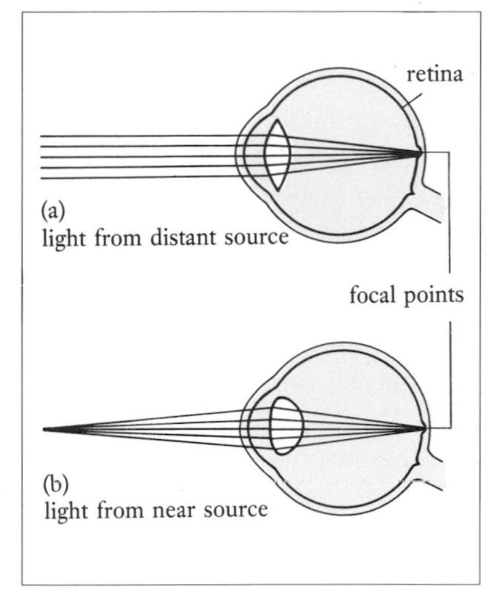

Figure 7.10. Distant and near vision

The **near point** is the closest point at which we can focus clearly on objects. It is directly related to the ability of the lens to increase refraction by accommodation. As age increases the lens loses elasticity and its ability to accommodate; this condition, known as *presbyopia,* moves the near point further from the eye as demonstrated by an older person reading at arm's length. The lens becomes less elastic and denser because the fibres increase.

Looking at near objects also causes the pupil to constrict reflexly, which reduces light entry and directs light through the most curved part of the lens. Also required for focusing of near objects is the autonomic convergence of the eyeballs (ask a friend to look at the end of their nose, note how the eyes converge); this prevents the production of two images which would cause double vision (*diplopia*). Obviously the eye is working hardest for near vision and prolonged close work can lead to 'tired eyes'.

Problems with refraction

Myopia (short sight/near sight)

The eyeball is elongated or the lens over-refracts the light entering the eye (see *Figure 7.11*). Near objects can be focused on the retina but distant objects are focused in front. Myopia is corrected by using a divergent biconcave lens.

Hypermetropia (long sight/far sight)

In this situation the eye is shortened or the lens cannot refract the light sufficiently (see *Figure 7.11*). Near objects are focused behind the retina and distant vision may require some lens accommodation. Correction is by the use of a convergent biconvex lens.

Astigmatism

Astigmatism is caused by defects in the curvature of the cornea, it is difficult to focus horizontal and vertical lines at the same time without blurring. Correction is with a cylindrical lens.

Retinal physiology

Sufficient amounts of visual pigments are required before the rods and cones can convert light energy to electrical impulses. The various pigments are formed from combinations of **retinene**, a light-sensitive molecule, and a protein (**opsin**). Retinene is derived from **vitamin A (retinol)**, lack of which can impair night vision and in severe cases cause blindness.

Rods contain the pigment **rhodopsin** which responds to dull light by changing shape and bleaching; this produces the receptor potential and triggers an impulse, which is transmitted to the optic nerve via bipolar and ganglion cells (see *Figure 7.5*). Cones have three pigments, which contain retinene but have different proteins. These three cone types respond to bright light of the red, green or blue wavelengths. Normal colour vision is inherited as a *dominant* gene (if present its characteristic is expressed) on the X chromosome. *Colour blindness* due to deficiency of one of these cone types often affects the red or green cones. Red-green colour blindness is inherited as a *recessive* gene (its characteristic is only expressed in the absence of a dominant gene on the homologus chromosome) on the X chromosome. Colour blindness is more common in males who only have one X chromosome, they will be affected if they inherit the recessive gene. Females, however, because they have two X chromosomes will only be colour blind if they inherit the recessive gene from both parents. Another example of this type of sex linked inheritance is haemophilia (Chapter 9).

X chromosome *- one of the sex chromosomes. Males have one X chromosome and a Y chromosome (XY) and females have two X chromosomes (XX).*

Impulses from the rods and cones (photoreceptors) are transmitted to the optic nerve fibres via bipolar and ganglion cells (see *Figure 7.5*).

Adaptation

When we enter a dark area from one of high light intensity and vice versa the photosensitive retinal pigments adapt to the new conditions:

■ Light ————> Dark (we see nothing but blackness).
The cones stop working, rhodopsin, which was bleached by the bright light, is formed and rod activity increases. Full adaptation may take some hours but adequate vision is restored in minutes.

■ Dark ————> Light (we are initially 'blinded' or dazzled).
Rhodopsin breakdown reduces retinal sensitivity, cone activity increases rapidly and within a few minutes we have high quality colour vision.

Binocular vision

Having two eyes which transmit slightly different images to the brain from different, but overlapping, visual fields gives us the advantage of three-dimensional or **stereoscopic vision**, which allows us to assess depth, speed and distance. You can test this by attempting to touch objects with one eye closed.

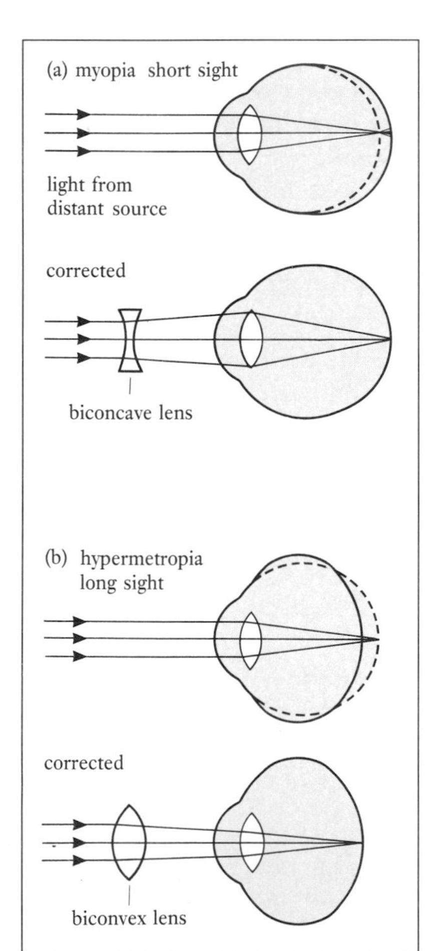

Figure 7.11. Problems with refraction and their correction.

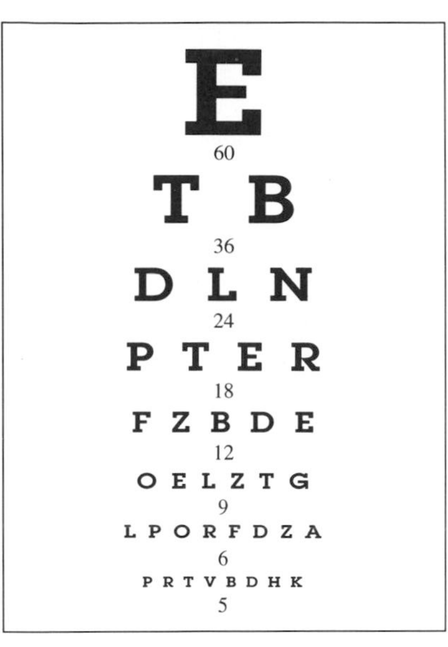

Figure 7.12. Snellen's type test chart.

Visual acuity

Visual acuity is the ability to see the difference between two points of light. This is what gives visual clarity, 'sharpness' and the ability to perceive detail. Individual acuity differs and can be tested by use of a *Snellen's type* test chart (see *Figure 7.12*) which has lines of different size letters. The letters are arranged in lines that can be read by a 'normal' eye at 60, 36, 24, 18, 12, 9, 6 and 5 m. Visual acuity for each eye is tested separately by asking the person to read the letters from a distance of 6 m with one eye covered. The visual acuity for each eye is expressed as 6 over the smallest line of letters that can be read, e.g. visual acuity of a person able to read the 6 m line at 6 m distance is 6/6.

As we age our acuity decreases with the amount of light entering the eye; the lens loses its transparency and the pupils tend to constrict as iris function declines.

Visual pathway

Axons from the ganglion cells in the retina leave via the **optic nerves** which merge and cross (decussate) at the **optic chiasma** *(Figure 7.13).* Although the nerves cross, it is important to note that only the fibres carrying impulses from the medial retina cross over; fibres from the lateral retina do not cross. This means that each side of the optic cortex receives impulses from both eyes. Loss of different parts of the visual field is known as *hemianopia*, the area lost depends on where the visual pathway is damaged.

Most impulses transmitted in the optic tracts synapse in the **lateral geniculate body** of the thalamus. Axons from the thalamus carry the impulses to the occipital lobes, which contain the visual cortex. It is here that the image is turned the right way up, reversed (right to left) and interpreted, all of which occurs almost instantly – you don't have to wait for these prints! Some fibres travel from the retina to midbrain to control visual reflexes such as the pupillary reflex.

Causes of visual impairment (blindness)

The causes of visual impairment worldwide include:

- Retinopathy (disease of the retina) associated with *diabetes mellitus* (see Chapter 8).
- Cataracts (see page 131) which may be congenital due to rubella or as part of ageing and disease, e.g. diabetes mellitus.
- Glaucoma (see page 131).
- Vitamin A deficiency (see page 134).
- Trauma, e.g. chemical injury (see page 129).
- Tumours.
- Infections such as *trachoma* (caused by the organism *Chlamydia trachomatis*) and *gonococcal ophthalmia* (see Chapter 20). Both of these infections may occur during birth if the infant has contact with infected vaginal discharge.

Person-centred study Anwar

Anwar has come into hospital to have planned abdominal surgery; he also has diabetes mellitus and, after a period of declining visual acuity, is now virtually without sight. Apart from needing care relating to the surgery and diabetes, his care plan should take account of his visual impairment.

The main problems for Anwar are communication and maintaining safety in a different environment. It is important that nurses speak to Anwar as they approach, describe every detail of the immediate environment and explain everything that happens in detail, e.g. the journey to theatre might involve going in a lift, which will need careful explanation to avoid anxiety. Maintaining safety involves everyone; all staff should be informed about Anwar's problems. Care must be taken to keep the immediate area as it has been described to Anwar. There should be no trailing cables or other objects which may cause a fall.

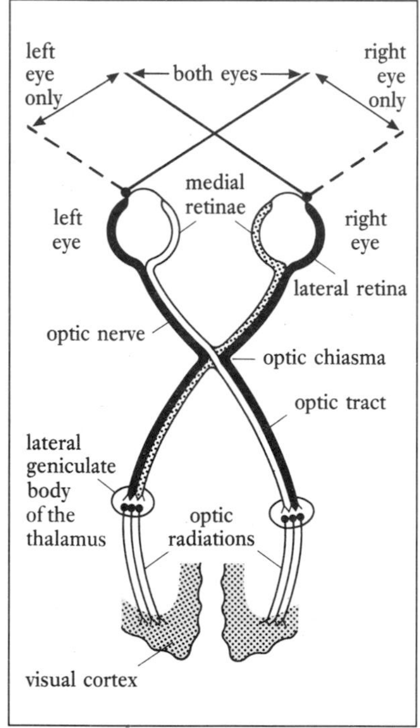

Figure 7.13. Visual pathways and fields.

Audition and Balance - The Ear

The ear develops from the embryonic brain starting around the fourth week of pregnancy. In common with the eye, the developing ear is vulnerable to damage during this early period from agents such as the *rubella* virus.

The ear has three main parts: the **outer** and **middle ear** which are concerned with the transmission of sound waves to the **inner ear**, which contains mechanoreceptors that convert sound energy to electrical energy. The inner ear also contains the **vestibular apparatus** which maintains balance/equilibrium with the help of mechanoreceptors that respond to changes in head position. Impulses travel via the **vestibulocochlear nerves** (VIII) to the brain for processing *(Figure 7.14)*.

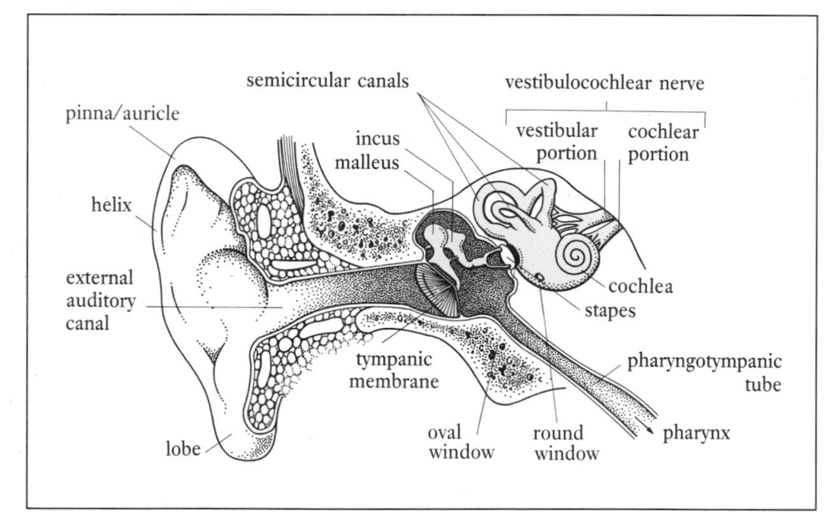

Figure 7.14. The ear.

Outer ear

The outer ear is formed from the **pinna (auricle)** and **external auditory canal (meatus)**. The pinna is formed from ridges of elastic cartilage (see *Figure 1.19*) covered with skin; the prominent outer ridge is called the **helix** and from this hangs the lobe, which has no cartilage. The pinna helps to funnel sound waves into the ear and offers some protection to the auditory canal. In humans the pinna does not move, which means that we must move our head towards a sound; many animals, e.g. horses, move their pinnae towards the sound source.

The external auditory canal is about 2.5 cm in length and initially runs upwards and backwards to end at the **tympanic membrane**. The first part of the canal is cartilaginous and the inner two-thirds is within the *temporal* bone (see Chapter 18). The canal is lined with skin (see Chapter 19) containing sebaceous glands, hairs and modified sweat glands known as **ceruminous glands**, which produce **cerumen** (wax). The sticky yellow–brown wax traps dust and other foreign bodies which enter the ear. An excess of wax which hardens can impede sound conduction and reduce hearing.

At the end of the canal the tympanic membrane (ear drum) forms a division between the outer and middle ear. This oval-shaped fibrous membrane is covered with skin on its outer surface and with mucous membrane on the internal surface, which is continuous with the middle ear lining. Sound waves cause the tympanic membrane to vibrate, which transmits sound energy to the middle ear structures.

Middle ear

The middle ear (**tympanic cavity**) is an air-filled cavity situated in the hard petrous ('stony') part of the temporal bone. The middle ear is lined with mucosa and bounded by the tympanic membrane, temporal bone and two openings: the **oval window**, which is occluded by the *stapes*, and the **round window** covered by fibrous tissue.

The middle ear communicates with the air cells (sinuses) of the **mastoid** process of the temporal bone and with the **nasopharynx** via a tube called the **pharyngotympanic (auditory) tube** (previously called the **Eustachian** tube). Air enters the middle ear through the pharyngotympanic tube to ensure that atmospheric pressure is maintained either side of the tympanic membrane; the equalization of pressure allows the tympanic membrane to vibrate correctly.

The pharyngotympanic tube is usually closed, but it opens during swallowing, yawning, etc. to equalize the pressure. The lining of the pharyngotympanic tube is continuous with that of the nasopharynx. During air travel the 'ear popping' noises are due to unequal pressures caused by an increase in altitude. To readjust the pressures it is necessary to swallow or suck a sweet. It is possible for micro-organisms to gain access to the ear from the respiratory tract via the tube and cause infection (*otitis media*), a common childhood infection.

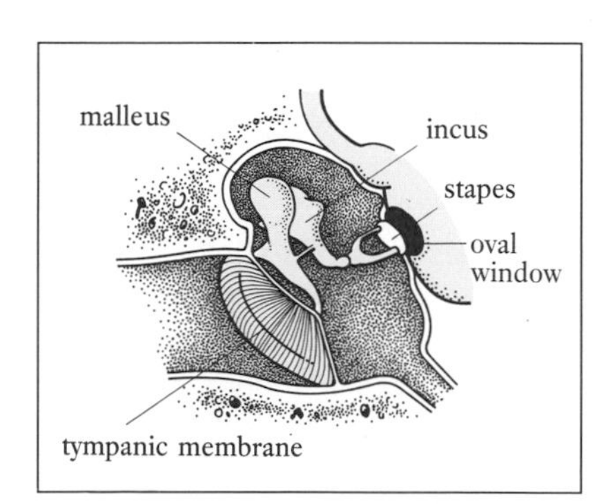

Figure 7.15. Middle ear ossicles.

Inside the middle ear are the three smallest bones of the body: **malleus (hammer)**, **incus (anvil)** and **stapes (stirrup)**, known collectively as the middle ear **ossicles** (*Figure 7.15*).

The malleus is attached to the tympanic membrane and the stapes is joined to the oval window; the incus, situated in the middle, articulates with both to complete the vibration system across the middle ear. When sound waves vibrate the tympanic membrane, the motion is transmitted through the ossicles to the oval window and hence to the fluid-filled inner ear. Sound conduction in the middle ear can be modified by two tiny muscles attached to the ossicles. These act reflexly to protect the ear from extremely loud noises by reducing conduction but, because of a time lag of a few milliseconds, damage may still occur.

Age changes in the middle ear include *otosclerosis* – the formation of new connective tissue causes the ossicles to adhere and the stapes to fuse to the oval window. This reduces the efficiency of sound conduction resulting in serious hearing impairment.

Inner Ear

The inner ear has two parts: (i) the bony labyrinth; and (ii) the membranous labyrinth (*Figure 7.16*). The **bony labyrinth** is formed by a complex arrangement of fluid-filled cavities within the temporal bone; it contains the auditory (hearing) and vestibular (balance) structures consisting of the **cochlea**, **vestibule** and **semicircular canals**. Inside the bony labyrinth is a layer of fluid (**perilymph**) which separates the bony labyrinth from the **membranous labyrinth** which fits inside the bony labyrinth rather like an inner tube inside a tyre. The membranous labyrinth contains the fluid **endolymph**. The fluids of the inner ear (perilymph and endolymph) are discussed again on pages 139 and 141.

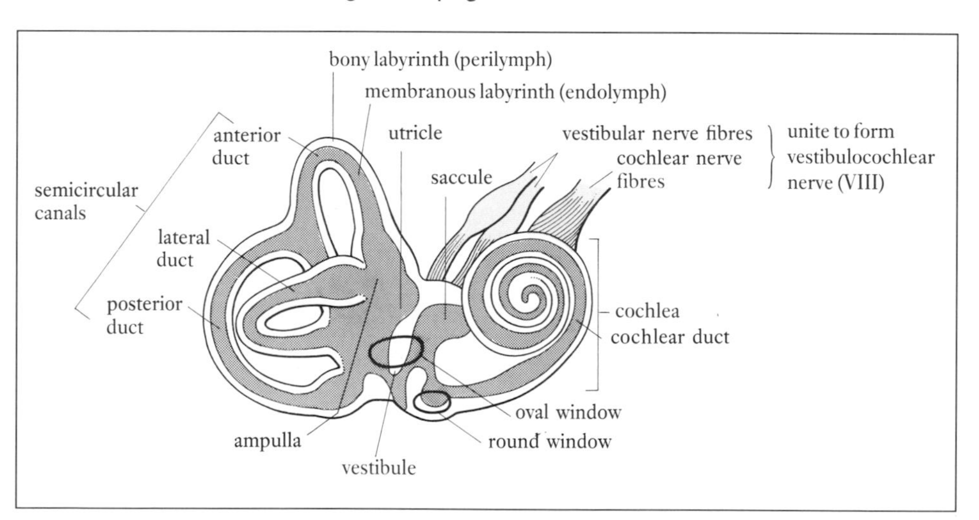

Figure 7.16. Inner ear.

The cochlea (*Figure 7.17*), which resembles a snail's shell, is a tube coiled around a central bony pillar. It is divided into two chambers by the **basilar membrane**; the superior chamber, the **scala vestibuli**, is in contact with the oval window, and the inferior chamber, the **scala tympani**, with the round window. Both chambers are filled with perilymph. The basilar membrane and another membrane (**Reissner's membrane**) enclose the **scala media** (**cochlear duct**) which

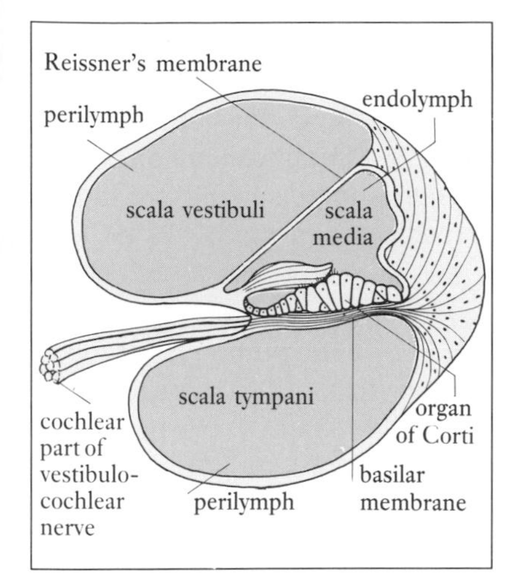

Figure 7.17. Cochlea and organ of Corti (section).

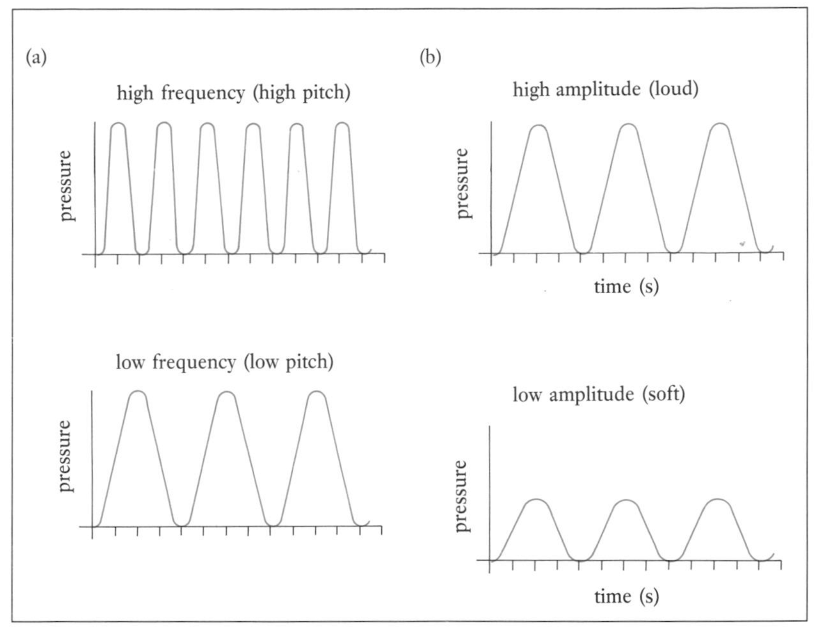

Figure 7.18. Sound – (a)Frequency/pitch; (b) amplitude/intensity.

contains endolymph and the auditory receptors ('hair cells') of the **organ of Corti** (*Figure 7.17*). Fibres from the auditory receptors eventually form the cochlear branch of the vestibulocochlear nerves (VIII).

The central area of the bony labyrinth between the cochlea and semicircular canals is called the vestibule; its lateral wall contains the oval and round windows. The vestibule is filled with perilymph and contains two communicating membranous sacs: the **saccule**, which connects with the membranous part of the cochlea, and the **utricle**, which extends into the **ampullae** of the semicircular canals. The saccule and utricle contain the **maculae** (see *Figure 7.20*), receptors concerned with equilibrium.

Projecting from the vestibule are three semicircular canals, which are orientated in the three planes of space: anterior (superior), posterior and lateral. Like the other parts of the inner ear they have a bony portion filled with perilymph and membranous ducts filled with endolymph. The membranous ducts have a distended area called the ampulla, which connects with the utricle and contains another equilibrium receptor; the **crista ampullaris** (see *Figure 7.20*). Nerve fibres from the ampullae, utricle and saccule eventually form the vestibular part of the vestibulocochlear nerves. To summarize, the inner ear has two parts – a bony labyrinth within which is fitted the membranous labyrinth. The space between the bony and membranous labyrinth is filled with perilymph and endolymph fills the membranous labyrinth.

AUDITION (HEARING)

Hearing depends on vibrations caused by sound waves reaching the inner ear, where receptors convert the mechanical energy to electrical potentials which stimulate nerve impulses. Before looking at hearing, it would be helpful to consult one of the books on basic physics given in Further Reading.

Sound and the ear

Sound is, in simple terms, the vibration of a medium, e.g. air, which causes a pressure change; the rate of vibration or **frequency** is measured in **hertz (Hz)** *(Figure 7.18(a))*. At 331 m/s, the passage of sound through air is quite slow (compare with light). This slowness accounts for the delay in hearing sounds, e.g. the sound of the starting gun of a race may reach the spectators after the runners have set off. We hear different sound frequencies as differences in **pitch**; high pitched sounds have a high frequency and vice versa. Humans can hear sounds between 20 and 20 000 Hz but we are most sensitive to frequencies between 1000 and 4000 Hz. Ability to hear high frequencies is lost with ageing, a condition known as *presbycusis*.

Sound also has **amplitude** – the degree of pressure change or vibration (*Figure 7.18 (b)*). This accounts for the loudness or **intensity** of the sound and is measured in **decibels (dB)**. Normal hearing covers a range of about 120 dB; this allows us to hear sounds which are only just audible, such as whispering (10–20 dB) to a level which causes pain, e.g. a jet engine (130 dB). We can also detect quite small changes in sound intensity. Localization of sound depends upon nuclei in the brain stem which interpret impulses which reflect the response of the ear to sound intensity and timing.

Problems with noise levels

Although the human ear can hear sounds at 130 dB it will sustain damage if exposed to sound levels greater than 90 dB. Loss of hearing will occur if the individual is exposed to noise over a prolonged period or at frequent intervals, e.g. machinery or loud music. The ear will also be damaged by an isolated exposure of very high intensity, e.g. an explosion. Other areas of behaviour affected by noise include:

- Loss of work efficiency; stress and irritability are linked to high levels.
- Communication difficulties, e.g. using a public telephone on a busy road (heavy traffic 85 dB).
- Sleep is disrupted by noise, e.g. the noise levels in a hospital ward (see Chapter 4).

Where possible we should avoid high intensity sounds or limit exposure by having frequent breaks. Proper ear protection should always be provided and worn for noisy jobs with regular hearing tests for those involved, e.g. people working in heavy engineering or construction industry.

Measures to reduce noise levels range from planning that removes heavy traffic from residential areas to working quietly on a ward at night.

Physiology of hearing

Sound waves directed by the pinna into the external auditory canal hit the tympanic membrane, which vibrates at that frequency. This movement is transferred via the ossicles to the oval window and hence to the inner ear. The fluids in the cochlea are displaced by the movement of stapes on the oval window. Sounds within the audible range cause waves in the perilymph to be transmitted to the endolymph of the scala media, which causes the movement of the many cochlear hair cells (receptors) of the organ of Corti. The movement of the hairs causes action potentials, which in turn produce action potentials in the fibres of the cochlear branch of the VIII nerves.

Summary of events

Sound → Pinna → External canal → Tympanic membrane
Ossicles → Oval window → Perilymph → Endolymph →
Organ of Corti → Nerve impulses.

Auditory pathway

Nerve impulses from the cochlea pass through the **spiral ganglion** and travel in the vestibulocochlear nerve to synapse in the auditory (cochlear) nuclei of the medulla. Here some cross the midline, so that impulses from both ears travel to each auditory cortex. From the medulla they travel to the auditory cortex situated in the temporal lobe, having synapsed in the midbrain or thalamus en route (*Figure 7.19*). This rather complex arrangement provides the hearing abilities discussed earlier, such as the localization of sound.

Auditory impairment (deafness)

Auditory impairment is usually partial and may be divided into two main types:

- *Conductive* – conduction of sound is impaired, e.g. excess hardened wax (see page 137), otitis media (see page 137), otosclerosis (see page 138).
- *Sensorineural (perceptive)* – caused by damage to hearing receptors, nerves or cortex. It may be due to maternal rubella during the first trimester (12 weeks) of pregnancy, loss of receptors with increasing age which leads to presbycusis (see page 140), noise damage (see page 140) and brain tumours.
- *Mixed deafness* – This is a combination of conductive and sensorineural deafness.

Tuning fork (512 Hz) tests *(Rinne's* and *Weber's)* may be helpful in differentiating between conductive and sensorineural hearing loss by testing air and bone sound conduction (see Further Reading, e.g. Bull 1985, for test details).

Hearing ability can also be tested by pure tone *audiometry*, where the individual is exposed to sounds of increasing frequency (Hz) and asked to indicate at what intensity (dB) the sound is heard. The test is performed in a soundproof area and each ear is tested separately. Sound conduction through air and bone can be tested during audiometry to determine the type of any hearing loss.

Nursing Practice Application **Auditory impairment**

Communication with a person with hearing loss can be enhanced by simple common sense measures which include: use of the 'good' ear, checking that any hearing aids are available and working, taking time and speaking clearly, ensuring that your mouth can be seen for lip reading, sign language and the use of other aids, such as printed information.

EQUILIBRIUM (BALANCE)

The vestibular apparatus in the inner ear responds to changes in the position of the head and to rotational movement. Its reflex functioning, with sensory input from the eye and proprioceptors, helps to control balance and eye movements.

Physiology of equilibrium

The vestibular apparatus consists of the semicircular canals and the vestibule (maculae in utricle and saccule) (*Figure 7.20(a)*). Changes in head position cause movement of the fluids within these structures, which affects the receptors. Activation of the receptors stimulates action potentials in the fibres of the vestibular branch of the VIII nerves.

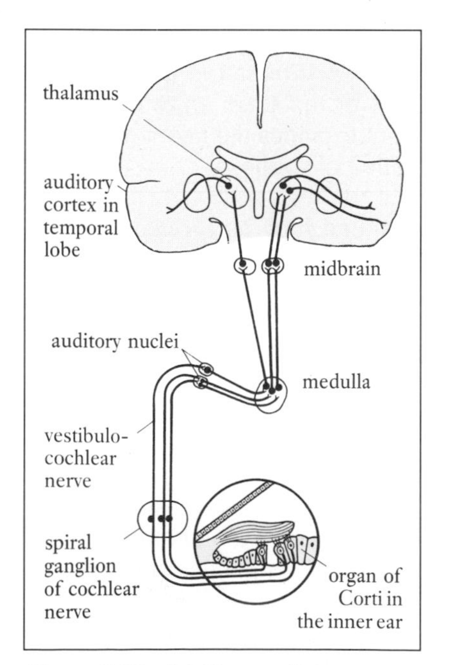

Figure 7.19. Auditory pathways (simplified).

Static equilibrium

The maculae situated in the saccule and utricle contain **hair cells**. These receptors for **static equilibrium** respond to head position relative to gravity and to linear changes in speed and direction. The hair cells project into a gelatinous mass or **cupula**, which contains chalky material called **otoliths** (*Figure 7.20(b)*). When the head is upright the maculae are vertical in the saccule and horizontal in the utricle, as the head tilts the hair orientation changes. Movement of the hairs activates the receptor cells which stimulate the vestibular nerve fibres. The otoliths increase the weight of the gelatinous mass over the hair cells.

Dynamic equilibrium

Dynamic equilibrium is controlled by receptors called **cristae**, which are situated in the ampulla of the semicircular canals. These respond to rotational movements of the head, such as nodding, shaking your head or riding a bicycle. A crista consists of **hair cells** which protrude into a gelatinous mass or **cupula**. As head position changes, the endolymph in one semicircular canal (the three canals cover the planes of space) moves over the cupula and reorientates the hairs. As before this movement of the hairs activates the receptor cell which stimulates the vestibular nerve fibres (*Figure 7.20(c)*).

If you are rotated quickly and then stop you will feel dizzy and nauseous, with loss of balance and *nystagmus* (lateral eyeball movements). This continues until your brain sorts out the incoming data from vestibular receptors, eyes and muscles.

A situation where the vestibular apparatus is at variance with the visual input may explain why some people suffer from the motion/travel sickness which can make any journey a nightmare. More seriously an excess of endolymph leads to a condition called *Meniere's disease*, which is characterized by *tinnitus* (ringing in the ears), hearing loss, nausea and dizziness.

Vestibular pathway

Nerve impulses from the vestibular apparatus reach the brain via the vestibular ganglia, which connect with the hair cells. The fibres travel in the VIII nerve to nuclei in the medulla. Some fibres travel to the cerebellum. Both these areas also receive impulses from the eyes and proprioceptors. All these impulses are integrated and the information is used in the reflex control of eye movements, muscle tone and posture.

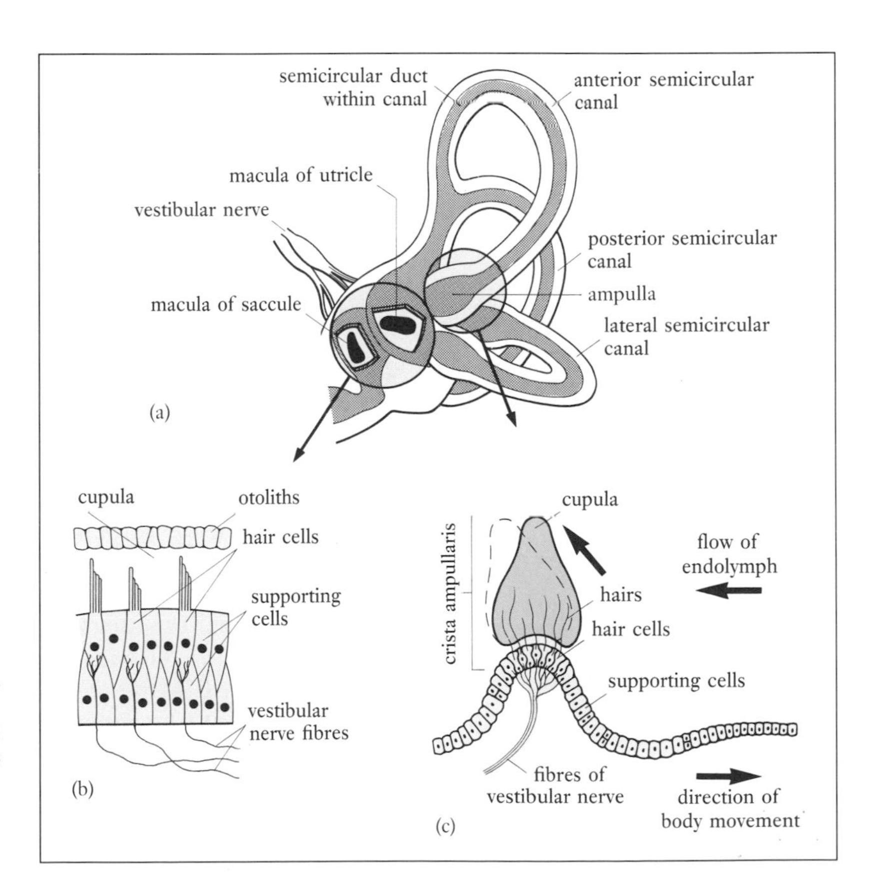

Figure 7.20. (a) Vestibular apparatus - vestibule and semicircular canals;
(b) detail of receptors in saccule and utricle of the vestibule (static equilibrium);
(c) detail of receptors in ampulla of the semicircular canals (dynamic equilibrium).

OLFACTION AND GUSTATION - CHEMICAL SENSES

Olfaction (smell)

A sense of smell is important as a protective mechanism. Many harmful agents have a characteristic odour, e.g. gas leaking from a domestic appliance. Odours also provide pleasurable experiences which enrich our lives, such as scented flowers in a city flat. In many species odours are a vital part of recognition between individuals and in reproductive behaviour. It may well be that our behaviour is also influenced by odours from chemicals known as *pheromones* contained within sweat and other body secretions (see Chapter 19).

There is a strong link between olfaction, appetite and taste. Appetite increases and food tends to taste better if it also smells good.

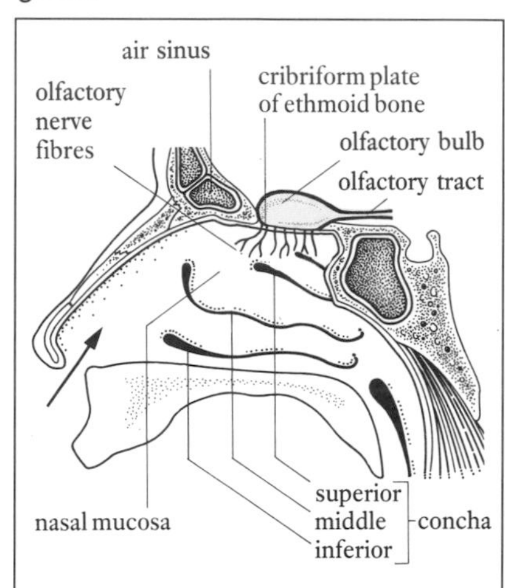

Figure 7.21. Olfactory structures.

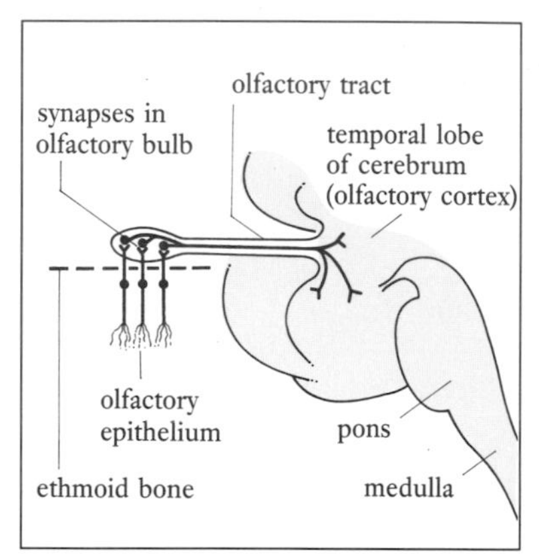

Figure 7.22. Olfactory pathway (simplified).

Physiology of olfaction

Specialized chemoreceptor cells are found within the **olfactory epithelium** which forms part of the mucous membrane lining the roof of the nasal cavity (*Figure 7.21*). The receptors respond to chemical vapours which enter the nose during inspiration (breathing in); the concentration of chemical in contact with the receptor is greatly enhanced by sniffing. Cilia (Chapter 1) projecting from each olfactory receptor increase the surface area for contact with the chemical odour, which dissolves in mucus prior to contact.

Axons from the olfactory receptors form the fibres of the **olfactory nerves** (I) (*Figure 7.22*) which pass through the *cribriform plate* of the *ethmoid bone* (see Chapter 18) prior to synapsing in the **olfactory bulbs**. The olfactory bulbs contain cells that integrate and modify impulses. **Olfactory tracts** transmit impulses to the olfactory cortex in the temporal lobes of each hemisphere for interpretation and to areas of the limbic system (see Chapter 4) concerned with the emotional aspects of smell. A specific odour may stimulate memories, e.g. the typical 'hospital smell' may remind you of a relative who died in hospital when you were a child.

Adaptation

The response to odours can be altered by processes in the brain. After a short exposure our awareness of an odour tends to decrease – a definite advantage when odours are unpleasant and of practical benefit by making way for new odours.

Gustation (taste)

The sense of taste is closely linked with that of olfaction. Being able to taste greatly enhances appetite and our enjoyment of food (compare food which is bland to a highly spiced curry). Taste stimulates digestive processes, such as the flow of saliva, and helps in homeostatic regulation by ensuring that we take in foods of different tastes which provide a balanced intake of minerals, vitamins and other nutrients. Rejection of harmful substances is assisted by their unpleasant taste/smell; we gag on 'bad food' or even vomit if swallowing does occur.

Physiology of gustation

Taste buds are situated on the tongue, although there are a few on the soft palate, cheeks, epiglottis and pharynx. They respond to chemicals dissolved in saliva and differentiate between four basic tastes: sweet, sour, salt and bitter. Taste buds are located in the tongue **papillae** and different areas of the tongue tend to detect a basic taste (*Figure 7.23*), although there is some overlap: front – sweet and salt, sides – sour and salt, back – bitter.

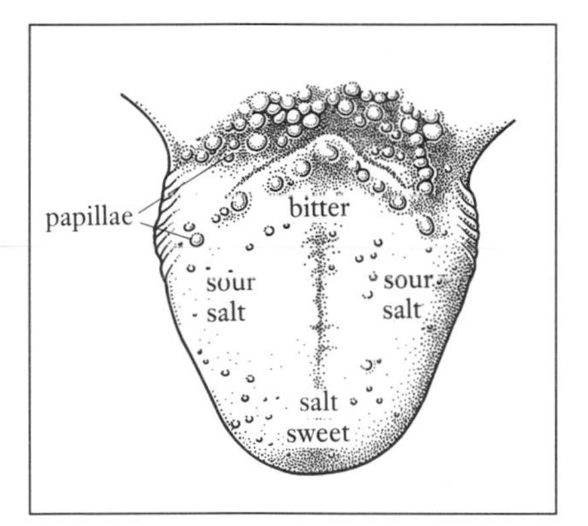

Figure 7.23. Taste regions of the tongue.

Each taste bud (*Figure 7.24*) is a cluster of cells within the epithelium of the tongue; its 'hairs' or **microvilli**, which project through the pore, contain the chemoreceptors. A receptor response of sufficient strength stimulates an action potential in sensory fibres leaving the taste bud. Taste depends upon the concentration of the chemical in contact with the receptors and the temperature, texture and smell of food. Adaptation to taste can occur within a few minutes.

Taste, and hence appetite, can be impaired if an upper respiratory tract infection affects the olfactory receptors in the nose. Taste is also affected where the mouth is dry and oral hygiene (see Chapter 13) declines such as through dehydration (see Chapter 2).

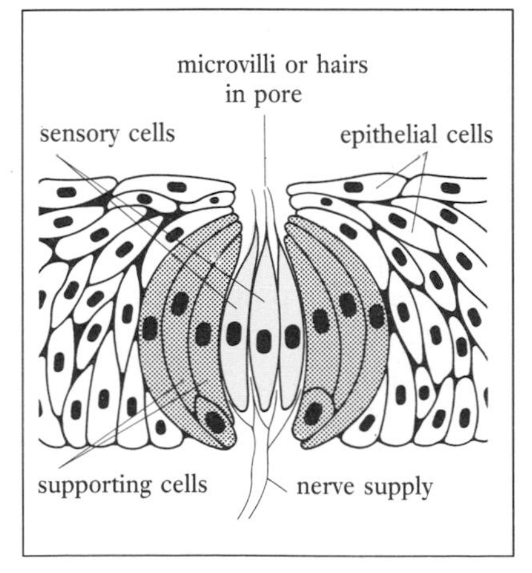

Figure 7.24. A taste bud.

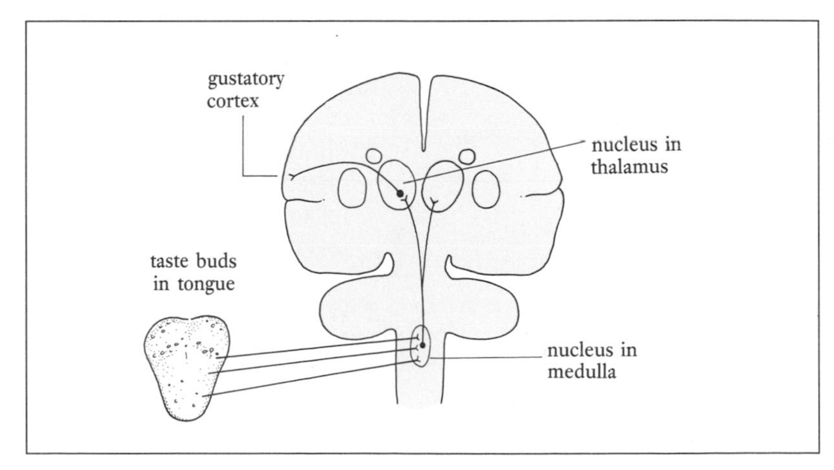

Figure 7.25. Gustatory pathways (simplified).

The nervous pathway (*Figure 7.25*) involves the *facial* (VII), *glossopharyngeal* (IX) and to a limited extent the *vagus* (X) nerves. The nerve fibres synapse within the medulla and pass to the thalamus prior to travelling to the gustatory cortex of both parietal lobes for interpretation.

The sense of taste and smell also change with normal ageing; gradual loss of taste and olfactory receptors, starting in the middle years, accounts for the greatly diminished sense of taste and smell in the older person. This may account for the common complaint voiced by older people that 'modern food' has no taste and contributes to a reduction in appetite. The loss of olfactory receptors impairs the ability to detect dangerous or unpleasant odours.

Disturbances in taste include the *aura* experienced by some people with epilepsy (see Chapter 4) who complain of a strange taste just prior to a fit. Taste may also change in some malignant conditions.

Summary/Check List

Introduction – special senses.
Vision – The eye (structure) – accessory structures, *Nursing Practice Application* – eye care. Extraocular muscles. Layers of the eyeball – sclera, uvea, retina (rods and cones). Inside the eye – cavities/chambers, lens. Vision (function) – Light, refraction, focal point. Accommodation, near point. Problems with refraction. Retinal physiology – pigments, adaptation, binocular and stereoscopic vision, acuity, visual pathway. Visual impairment – causes, *Person-centred Study* – Anwar.
Audition and balance – The ear (structure). Outer ear. Middle ear. Inner ear. Audition (function). Sound. Problems with noise levels. Physiology of hearing. Auditory pathway. Auditory impairment – types, *Nursing Practice Application* – auditory impairment. Equilibrium (balance) – vestibular apparatus. Physiology – static and dynamic equilibrium, vestibular pathway.
Olfaction (smell) – The nose, olfactory structures. Physiology of olfaction – olfactory pathway, adaptation.
Gustation (taste) – taste buds. Physiology of gustation – gustatory pathway. Changes in olfaction and gustation.

?

Self Test

1 Put the following in their correct pairs:
(a) tears
(b) posterior cavity
(c) aqueous humor
(d) lacrimal gland
(e) vitreous humor
(f) blood vessels in processes of ciliary body

2 Which of the following statements are true?
(a) The uvea consists of choroid, ciliary body and iris.
(b) The vascular cornea covers the front of the eye.
(c) Cones are colour vision receptors.
(d) The size of the pupil is controlled by the oculomotor nerve.

3 Describe briefly how light entering the eye is focused on the retina.

4 What changes are made to the image received by the visual cortex during interpretation?

5 Which of the following statements are true?
(a) The pinna helps to direct sound onto the tympanic membrane.
(b) The middle ear communicates with the nasopharynx via the pharyngotympanic tube.
(c) The stapes is attached to the round window.
(d) The membranous labyrinth is surrounded by perilymph and contains endolymph.

6 Put these in the correct order (moving from outside):
(a) ossicles (b) pinna (c) tympanic membrane (d) organ of Corti
(e) external auditory canal (f) cochlea (g) oval window.

7 What is the normal range of hearing and what characteristics of sound can we determine?

8 Explain briefly how the vestibular apparatus helps to maintain balance.

9 Outline the benefits of having the senses of smell and taste.

10 Which of the following statements are true?
(a) The taste buds on the back of the tongue are most sensitive to bitter tastes.
(b) Chemicals that stimulate the receptors for taste and smell are first dissolved in mucus or saliva.
(c) Impulses from the olfactory epithelium reach the brain in fibres of the second cranial nerve.
(d) Taste and smell receptors decrease with age.

Answers
1. a – d; c – f; e – b. 2. a, c, d 3. See pages 132-133. 4. See page 135. 5. a, b, d. 6. b,e,c,a,g,f,d. 7. See page 140. 8. See pages 142-143. 9. See pages 144,145. 10. a, b, d.

References

Health and Safety Executive (1983) *Precautions for the Safe Handling of Cytotoxic Drugs.* Guidance Note MS 21. London: HMSO.

Further Reading

Ballantyne, J. and Martin, J.A.M (1984) *Deafness,* 4th edn. Edinburgh: Churchill Livingstone.

Bull, P. D. (1985) *Lecture Notes on Diseases of the Ear, Nose and Throat,* 6th edn. Oxford: Blackwell Scientific Publications.

Dobree, J. H. and Boulter, E. (1982) *Blindness and Visual Handicap: The Facts.* Oxford: Oxford University Press.

Freeland, A. (1989) *Deafness: The Facts.* Oxford: Oxford Medical Publications.

Johnston, B. (1986) *Physics for GCSE.* London: Heinemann Educational Books.

Martin, F.N. (1986) *Introduction to Audiology,* 3rd edn. New Jersey: Prentice Hall.

Swift, D. (1988) *Physics for GCSE.* Oxford:Basil Blackwell.

Useful Addresses

Royal National Institute for the Blind
224 Great Portland Street
London W1N 6AA

Royal National Institute for the Deaf
105 Gower Street
London WC1E 6AH

CHAPTER EIGHT

THE ENDOCRINE SYSTEM: HORMONAL CONTROL AND REGULATION

OVERVIEW

- Introduction – endocrine structures and locations.
- Hormone structure, actions and control.
- Detail: hypothalamus, pituitary, thyroid, parathyroids, adrenals, pancreas.
- Control and regulation of: growth, general metabolism, calcium homeostasis, water/electrolyte balance, stress response, blood sugar.
- Thymus and pineal body.

LEARNING OUTCOMES

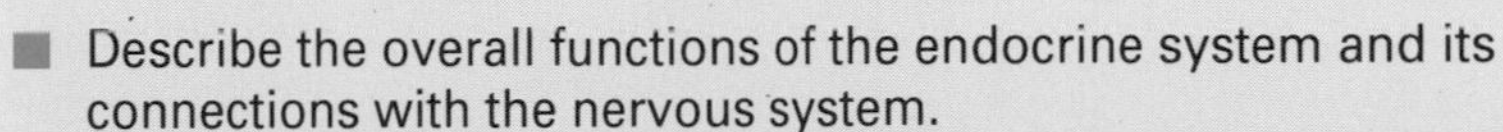

After studying Chapter 8 you should be able to:

- Describe the overall functions of the endocrine system and its connections with the nervous system.
- Describe the location of major endocrine glands and name other structures that produce hormones.
- Describe the structure of hormones and outline the ways in which they act on target cells.
- Explain how hormone secretion is regulated.
- Describe the relationship between the hypothalamus and the pituitary gland.
- List the hormones produced by the anterior pituitary and discuss their effects on growth and other endocrine structures.
- Describe the function of the posterior pituitary and the role of ADH in fluid balance.
- Describe how thyroid hormones control metabolism and outline their role in regulating serum calcium levels.
- Discuss the role of parathyroid hormone in calcium homeostasis.
- Describe the physiological effects of the hormones secreted by the adrenal cortex and medulla including fluid/electrolyte balance and the response to stress.
- Discuss the role of the pancreatic hormones – insulin and glucagon – in the control of blood sugar.
- Briefly outline the role of the thymus gland.
- Discuss the role of the pineal body.

KEY WORDS

Adenohypophysis – alternative name for anterior lobe of the pituitary.

Cyclic adenosine monophosphate (cyclic AMP) – the 'second messenger' substance formed from ATP, it initiates the effects of a hormone within its target cells in situations where the hormone does not enter the cell.

Endocrine gland-structure – a ductless glandular structure, its secretions are discharged directly into the extracellur spaces to enter the blood or lymph.

Hormones - the chemical messengers produced by endocrine structures which regulate the functions of distant organs or structures.

Hypophysis – alternative name for the pituitary gland.

Inhibiting hormone – a hypothalamic hormone that inhibits the secretion of a specific anterior pituitary hormone.

Metabolism – term used to describe all the biochemical processes occurring in the body.

Negative feedback – a regulatory mechanism by which hormone production is inhibited by high levels of that hormone in the blood. This mechanism regulates the release of most hormones.

Neurohypophysis – alternative name for posterior lobe of the pituitary gland.

Releasing factor/hormone – a hypothalamic hormone that stimulates the secretion of a specific anterior pituitary hormone.

INTRODUCTION

The endocrine structures work closely with the nervous system to regulate body processes and maintain homeostasis. They function through the release of **hormones** (chemical messengers produced by endocrine structures which regulate the functions of distant organs or structures) which travel in the blood to target cells. The onset of action of hormones is usually slower than the actions initiated by the nervous system but hormones usually control longer-term functions, e.g. growth (cf reflex action, Chapter 3). Although these two control systems are considered separately there are several examples where they are linked:

- The posterior lobe of the pituitary gland, which develops from the brain, stores and releases hormones made in nerve cells.
- Anterior pituitary hormone production is regulated by **releasing/inhibiting hormones** produced by the hypothalamus, which is part of the brain.
- The adrenal medulla develops from the same embryonic tissue as the sympathetic ganglia. Its action augments the sympathetic division of the ANS and it can be considered to be a sympathetic ganglion without a postganglionic fibre (Chapter 6).
- Chemicals which act as a hormone in one system may also function as a neurotransmitter, e.g. noradrenaline.

Endocrine structures (a ductless, glandular structure that secretes hormones directly into the extracellular spaces to enter the blood and lymph) and their hormone secretions regulate the **metabolic** processes of most cells and influence growth, nutrition, energy utilization, fluid and electrolyte balance, reproduction and stress responses.

ENDOCRINE STRUCTURES AND LOCATIONS

The major **endocrine glands** and their locations are (*Figure 8.1*):

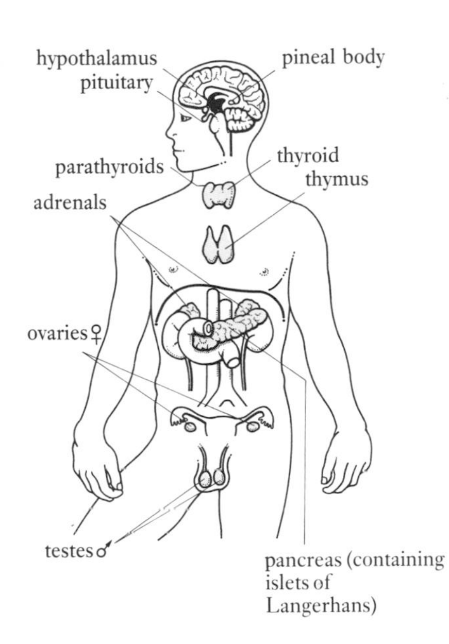

Figure 8.1. Location of major endocrine structures.

- Hypothalamus – brain.
- Pituitary gland – the pituitary fossa (depression) in the sphenoid bone of the skull.
- Thyroid gland – neck.
- Parathyroid glands (four) – posterior aspect of thyroid gland.
- Adrenal glands (two) – one on top of each kidney.
- Islets of Langerhans in pancreas – abdominal cavity. The pancreas has both endocrine function (islets of Langerhans, page 166) and exocrine (secretions leave the gland via a duct) function (Chapter 13).
- Gonads (two): ovaries (female) – pelvic cavity,
  testes (male) – scrotum (see Chapter 20).
- Pineal body – brain.
- Thymus – mediastinum (see Chapter 1).

Apart from the major endocrine glands there are structures which produce hormones in addition to their other functions (*Table 8.1*). These structures include the placenta, kidney and gastrointestinal tract, all of which are discussed in the appropriate chapters.

Certain tumours may also produce hormone-like chemicals, e.g. some

Table 8.1. Hormones produced by other structures.

Structure	*Hormone*
Trophoblast -> Placenta	Human chorionic gonadotrophin (hCG), oestrogens, progesterones
Kidneys	A factor which activates erythropoietin, 1,25-dihydroxycholecalciferol
Gastrointestinal tract	
stomach	Gastrin
small intestine	Secretin, cholecystokinin (CCK), enterogastrone and various other regulatory peptides
Heart	Atrial natriuretic peptides (ANP) (see Chapter 10)

lung cancers secrete substances which have the same effects as *antidiuretic hormone (ADH)* or *adrenocorticotrophic hormone (ACTH)*.

HORMONES

Structure

Hormones are divided into two groups (*Figure 8.2*): (i) those formed from amino acids (peptides and proteins), such as **thyroxine** and (ii) **steroids** derived from *cholesterol,* which include the sex hormones and those secreted by the adrenal cortex (*Figure 8.2(b)*).

Action

Hormones modify cell metabolism by changing the permeability of the cytoplasmic membrane, increasing enzyme synthesis or by altering enzyme activity (activation or deactivation). Some hormones act directly and others act indirectly through a 'second messenger' such as **cyclic AMP** (substance formed from ATP, initiates hormone effects in the target cell in situations where the hormone does not enter the cell).

The ways in which hormones cause these changes in their **target cells** include:

- Binding to specific receptors in the target cell membrane to alter permeability.
- Changing cyclic AMP levels, which in turn affects enzyme activity in the cell.
- Increasing production of mRNA or stimulating the ribosomes, both of which lead to an increase in protein (enzyme) synthesis (see Chapter 1).

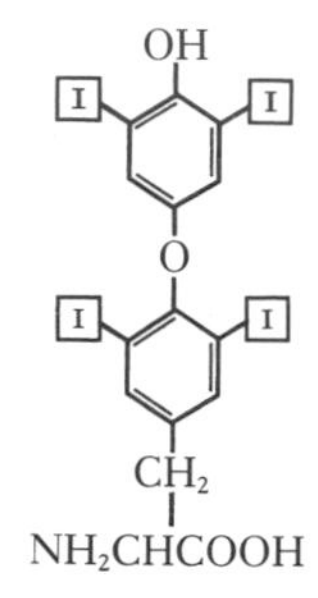

(a) thyroxine (T_4)
(amino acid based)

(b) cortisol
(steroid based)

Figure 8.2. Hormone structure.
(a) Amino-acid based;
(b) steroid-based.

The amount of intracellular calcium influences the storage and release of some hormones.

Substances known as **prostaglandins**, which are derived from fatty acids (see Chapter 13) such as *arachidonic acid,* influence cell metabolism by affecting the levels of cyclic AMP. Prostaglandins are produced in most tissues and produce hormone-like effects including fever and platelet aggregation. Prostaglandins are generally regarded as local hormones.

Hormone action may be rapid with immediate effects, e.g. adrenaline from the adrenal medulla, or not be apparent for hours or in some cases days, e.g. cortisol from the adrenal cortex.

Control of hormone secretion

Hormone secretion may be initiated by other hormones, the concentration of substances in the extracellular fluid or by the action of the nervous system. Most hormone release is regulated through **negative feedback/inhibition** (*Figure 8.3*), when hormone levels fall, release mechanisms are turned on and when the hormone level rises the mechanism is turned off (rather like cancelling the milk order when your fridge is full of milk). A similar mechanism occurs with hormones that respond to extracellular substances: high plasma levels of the substance inhibit hormone release. In this way hormone levels are

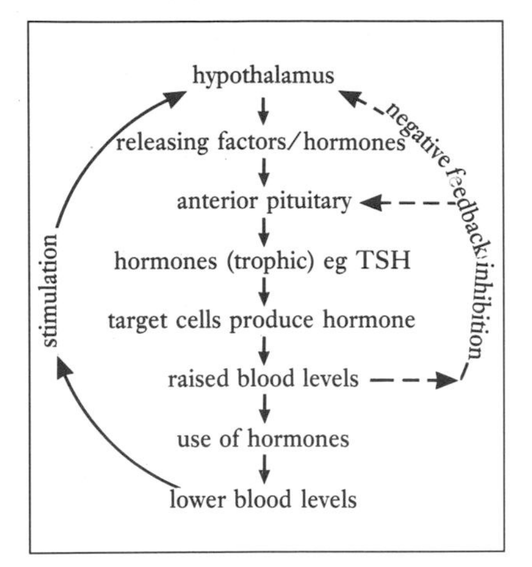

Figure 8.3. Negative feedback.

controlled, usually with very little variation in health.

Hormones have a specific *half-life* (time for removal of half the hormone from the plasma), after which they are mainly degraded by the liver into non-active metabolites to be excreted via bile or urine (see Chapters 14 and 15). Excretion of these metabolites by the kidneys can be measured in the urine and used to assess hormone levels.

Hypothalamus and Pituitary Gland

The **hypothalamus** (see Chapter 4) is of vital importance in the regulation of homeostasis. It controls pituitary function by: (i) producing the **releasing/inhibiting hormones** which regulate the hormone production of the anterior pituitary; and (ii) synthesizing of the two hormones stored and released by the posterior pituitary. The **pituitary gland** is often referred to as the *'master gland'*, because it stimulates other endocrine structures and metabolic processes.

Hypothalamic connections with other parts of the nervous system provide fine tuning for endocrine function. The tiny pituitary gland or **hypophysis,** weighing about 4 g, is really two quite separate parts: the anterior lobe (**adenohypophysis**), which develops from the embryonic pharynx, is glandular and the posterior lobe (**neurohypophysis**), which grows down from the brain, is composed of neural tissue.

The two lobes hang from the hypothalamus by a stalk containing blood vessels (*hypothalamohypophyseal portal system*) and nerve fibres. The pituitary rests within the pituitary fossa of the sphenoid bone very close to the optic chiasma (*Figure 8.4*). Tumours of the pituitary may cause visual problems if they press upon the optic nerves (Chapter 7).

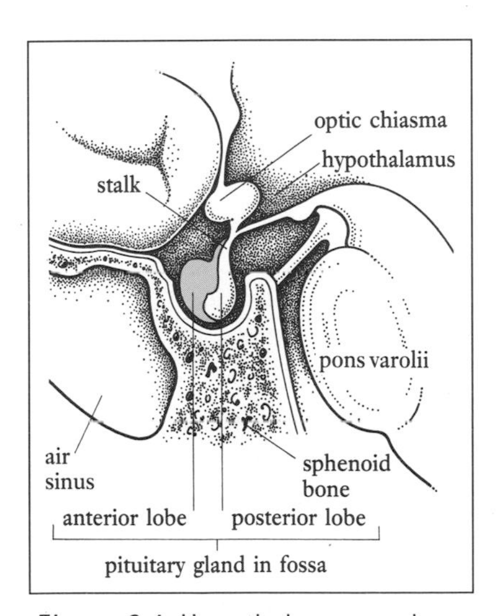

Figure 8.4. Hypothalamus and pituitary gland.

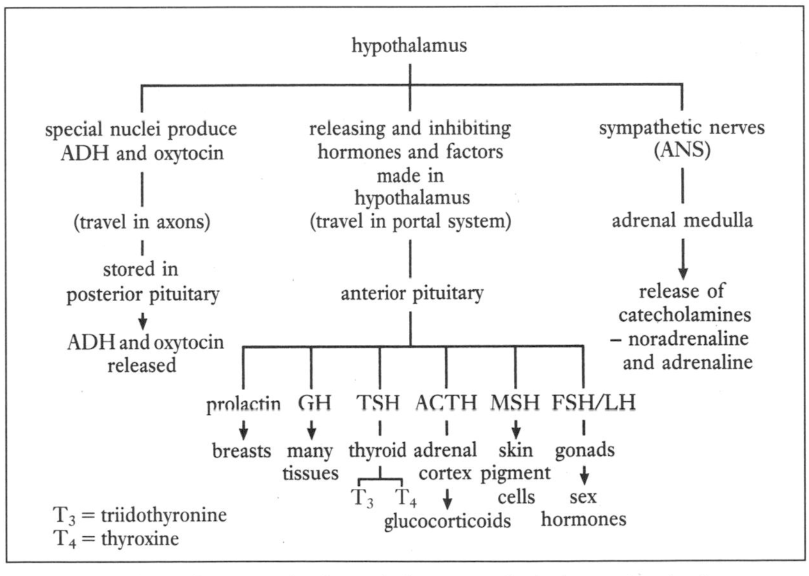

Figure 8.5. Flow diagram: the hypothalamus and pituitary gland – hormones and effects.

Anterior lobe (adenohypophysis)

The anterior lobe of the pituitary gland produces several hormones, which are controlled by hypothalamic releasing and inhibiting hormones that travel in the portal system of blood vessels in the stalk (*Figure 8.5*). The six anterior lobe hormones are the four **trophic** hormones (hormones that stimulate other endocrine glands): **adrenocorticotrophic hormone (ACTH)**, **thyroid stimulating hormone (TSH)** and the **gonadotrophins** – **follicle stimulating hormone (FSH)** and **luteinizing hormone (LH)** – and two that act on other tissues: **growth hormone (GH)** and **prolactin (PRL)**. The anterior lobe also produces a large precursor molecule which forms various substances including ACTH and **melanocyte stimulating hormone (MSH)** which acts on the skin pigment cells (see Chapter 19). Some authorities consider MSH to be produced by an area of distinct cells between the anterior and posterior lobes. This area is much more important to other species, e.g. those belonging to the amphibians.

Problems with anterior pituitary hormones are very diverse; there may be a complete failure *(panhypopituitarism)*, over- or undersecretion of some hormones or a breakdown in hypothalamic control.

Anterior lobe hormones

Growth hormone

Growth hormone is a protein (amino acid based) hormone with widespread effects on body tissues. It is regulated by two hypothalamic hormones: **growth hormone - releasing hormone (GHRH)** and **growth hormone - inhibiting hormone (GHIH)** or **somatostatin**, which is also produced by pancreatic and intestinal cells. Growth hormone stimulates the growth of bone, cartilage and muscle, and influences the metabolism of fats, proteins and carbohydrates (see Chapter 13). Growth hormone secretion increases during sleep and is affected by exercise, emotions and nutrition.

Hyposecretion (undersecretion) before the *epiphyseal plates* (see Chapter 16) ossify leads to *dwarfism* (*Figure 8.6 (a)*)and hypersecretion (oversecretion) to *gigantism* (*Figure 8.6 (b)*). Excess

A

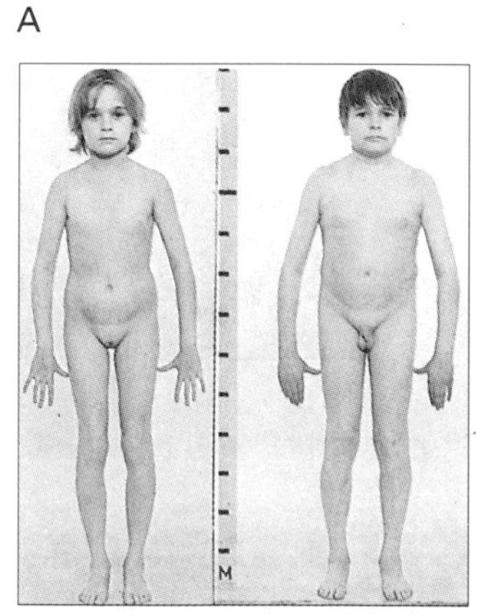

B

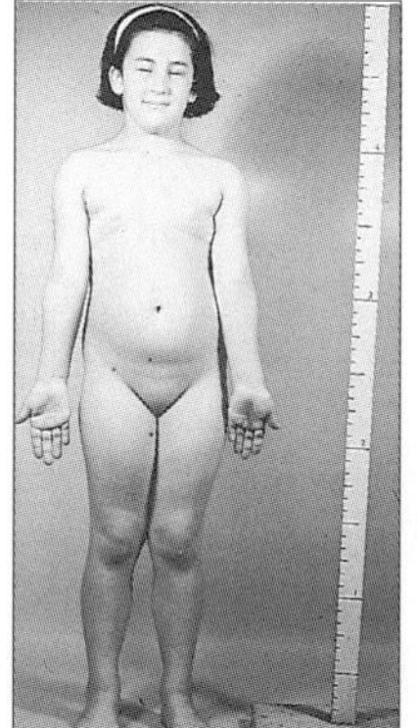

C

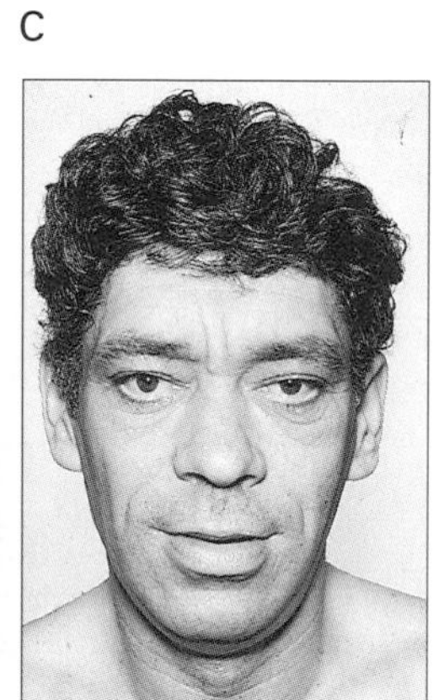

Figure 8.6. Growth problems: (a) dwarfism in two siblings: a 15-year-old boy and a 13-year-old girl; (b) gigantism in a 41/2- year-old girl; (c) acromegaly. (Hall, R., Evered, D.C. *A Colour Atlas of Endocrinology*. 2nd edn. Wolfe Medical Publications, Ltd. 1990. Reprinted with permission.)

secretion of GH after bone growth has ceased will cause the condition *acromegaly* which is characterized by enlargement of the bones of the feet, hands and face (*Figure 8.6(c)*).

Thyroid stimulating hormone

Thyroid stimulating hormone (TSH) is a protein hormone which stimulates thyroid growth and secretion. Its secretion is stimulated by the hypothalamic hormone **thyrotrophin releasing hormone (TRH)** and inhibited by high levels of thyroid hormones which act on the hypothalamus and pituitary.

Adrenocorticotrophic hormone

Adrenocorticotrophic hormone (ACTH) stimulates the secretion of *glucocorticoid* hormones from the adrenal cortex. Its release is stimulated by the hypothalamic factor – **corticotrophin releasing factor (CRF)** – and levels tend to be highest in the early morning but stressors (see Chapter 6) such as cold or anxiety can override this natural diurnal rhythm. Release of CRF and ACTH is inhibited by high levels of glucocorticoids.

Gonadotrophins (FSH and LH)

The **gonadotrophins** affect the functioning of the *gonads* (the ovaries and testes). **Follicle stimulating hormone (FSH)** stimulates ovum or spermatozoa production. **Luteinizing hormone (LH)** functions with FSH to stimulate ovulation and hormone release in the female and hormone production in the male (where it is known as **interstitial cell stimulating hormone – ICSH**). The hypothalamus starts to secrete **gonadotrophin releasing hormone (GnRH)** at puberty and the gonadotrophin secretion this stimulates brings about maturation of the gonads.

The complex pattern of gonadotrophin production is stimulated by GnRH and inhibited by the steroid hormones produced by the gonads (see Chapter 20 for a detailed discussion).

Prolactin

Prolactin is a protein hormone which stimulates *lactation* (milk production). In males and non-lactating females its regulation is mainly inhibitory through the hypothalamic hormone **prolactin inhibiting hormone (PIH)**, which is thought to be the neurotransmitter dopamine. During pregnancy, prolactin levels increase (possibly due to a hypothalamic releasing hormone, **prolactin releasing hormone – PRH**). The rise in prolactin leads to milk production, which will be further stimulated by suckling (see Chapter 20).

High levels of prolactin (*hyperprolactinaemia*) due to a pituitary tumour can cause female infertility and impotence in males.

Figure 8.7 illustrates the control of anterior lobe hormones.

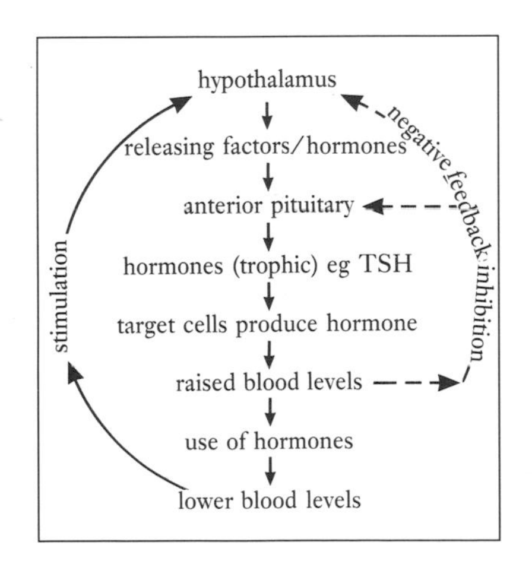

Figure 8.7. Control of anterior lobe hormones.

Posterior lobe (neurohypophysis)

The posterior lobe of the pituitary gland does not produce any hormones – it stores and secretes **oxytocin** and **antidiuretic hormone (ADH)** also known as *vasopressin*. Both these hormones are made up of nine amino acids; they differ by only two amino acids. The hormones are made in the **paraventricular** and **supraoptic nuclei** of the hypothalamus and travel in the nerve fibres of the stalk to the posterior lobe. Their release from the posterior pituitary is controlled by nerve impulses from the hypothalamus.

Hormones stored in posterior lobe

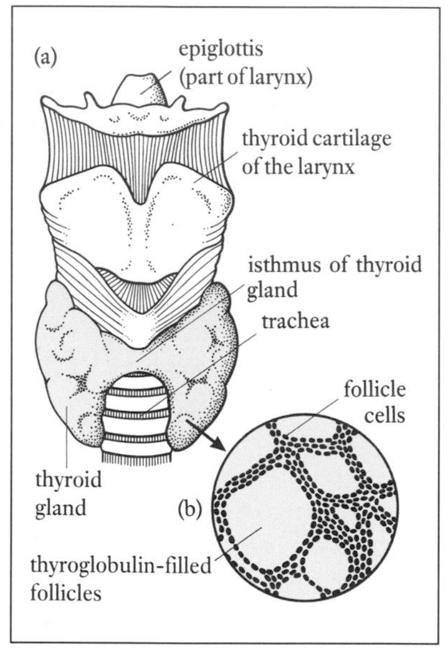

Figure 8.8. Thyroid gland: (a) Gross structure; (b) thyroglobulin – filled follicles.

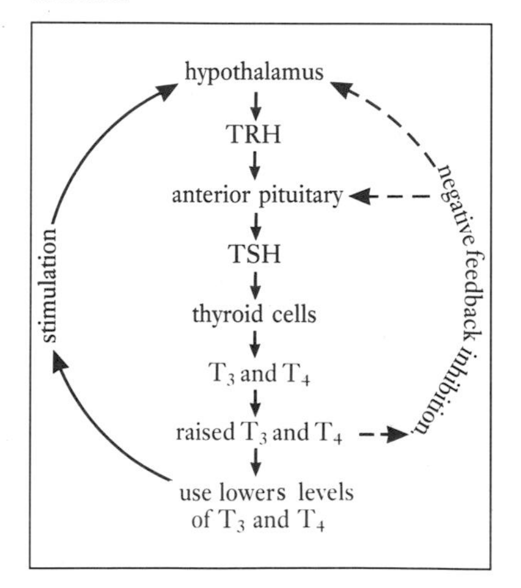

Figure 8.9. Negative feedback control of thyroid metabolic hormones.

Oxytocin

Oxytocin has an important role in parturition (childbirth). During labour the rising blood levels of oxytocin stimulate the expulsive contractions of the uterine muscle. Oxytocin is also important for the '*let down*' or ejection of milk from lactating breasts. Reflex release of oxytocin occurs in response to suckling, which results in more milk being ejected; this represents an example of **positive feedback.** Sometimes during breast feeding women experience cramp-like pains as the uterus contracts in response to the oxytocin release (see Chapter 20). Oxytocin and drugs (*oxytocics*) with similar properties are used to stimulate uterine contraction such as in the induction and active management of labour.

Antidiuretic hormone (ADH/vasopressin)

As its name suggests **antidiuretic hormone** reduces *diuresis* (urine production) in the kidneys. *Osmoreceptor* cells in the hypothalamus monitor the osmolarity (see Chapter 2) of the blood. When blood osmolarity is high, such as with dehydration, the posterior lobe releases ADH which causes the kidney tubules to become more permeable to water (see Chapter 15). More water is reabsorbed by the kidney, urinary volume is reduced, and the volume and osmolarity of the blood returns to normal.

When blood osmolarity is low, such as after a large fluid intake, no ADH is released and excess water is excreted by the kidney. The ADH mechanism, with processes which include the action of the thirst centre, help to maintain fluid balance homeostasis (see Chapter 4).

High concentrations of ADH raise blood pressure by increasing its effect on the smooth muscle in the vessel walls; this pressor (vasoconstriction) effect accounts for its other name, vasopressin.

Hyposecretion of ADH leads to the development of a condition known as *diabetes insipidus* which results in the production of large volumes of urine (as much as 20 litres in 24 hours). Homeostasis can still be maintained if fluid intake increases to match the loss. However, life will be somewhat inconvenient if you consider that the unfortunate person will need to drink copiously and spend a considerable amount of time *micturating* (urinating). Treatment with intranasal vasopressin analogues (similar substances which may have advantages such as longer duration of action or lack of pressor effects) and other drugs is usually effective.

Thyroid Gland

The **thyroid gland** is situated in the anterior part of the neck at the level of C5–T1. The thyroid has two lobes (joined by an **isthmus** of tissue) which lie one either side of the trachea and below the larynx (*Figure 8.8(a)*). The thyroid is extremely well supplied with arterial blood from branches of the external carotid and subclavian arteries (see *Figure 10.33*). The resultant vascularity increases the need for careful haemostasis during thyroid surgery. The venous blood returns to the heart via the *jugular veins* (see *Figure 10.35*).

Structure

Each lobe of the thyroid consists of many **follicles** with walls of cuboidal epithelial cells, which produce **thyroglobulin** (*Figure 8.8(b)*), a *colloidal* (particles suspended in a liquid which do not settle out or pass through membranes) substance which forms two of the thyroid hormones (see below). The thyroglobulin is stored within the follicles until required. A third hormone (calcitonin) is secreted by **parafollicular** cells found in the tissue between the follicles.

Thyroid hormones

A

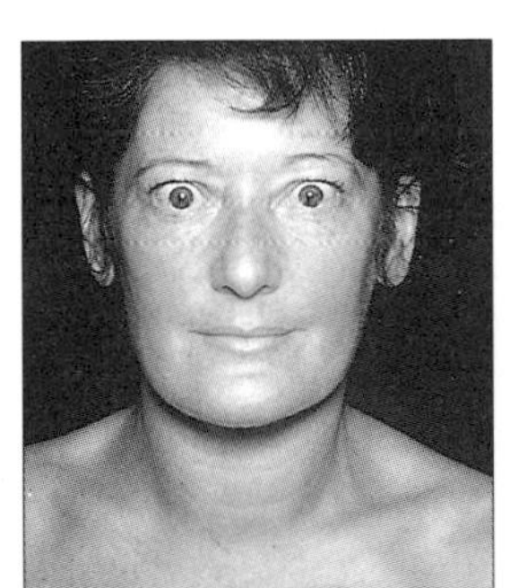

B

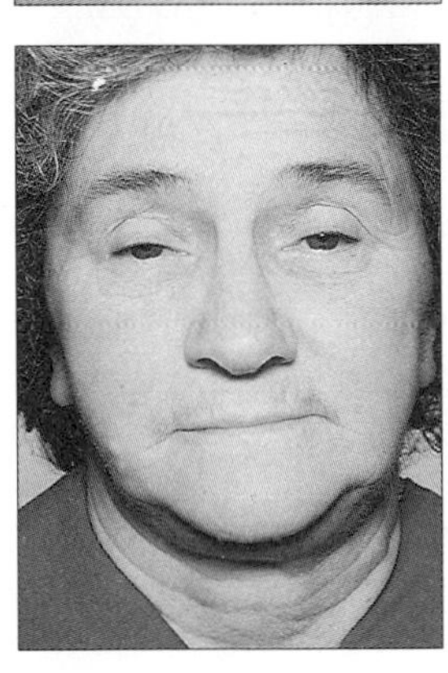

Figure 8.10. Problems with thyroid hormones. (a) Hypothyroidism; (b) hyperthyroidism. (Hall, R., Evered, D.C. *A Colour Atlas of Endocrinology*. 2nd edn. Wolfe Medical Publications, Ltd. 1990. Reprinted with permission.)

The thyroid produces three hormones:

- **Thyroxine** (T_4) and **triiodothyronine** (T_3), which regulate metabolism.
- **Calcitonin**, which lowers serum calcium levels.

Thyroxine (T_4) and triiodothyronine (T_3)

Thyroxine and **triiodothyronine** are formed from the amino acid *tyrosine* and possess either four (T_4; thyroxine) or three (T_3; triiodothyronine) atoms of iodine. Thyroxine is formed from thyroglobulin by the addition of iodine from the diet and is secreted from the follicles following action by enzymes. Some of this thyroxine is converted to triiodothyronine (a more active form) in the follicles, although this usually occurs at the target cells. Thyroid hormones can be stored as colloid for several weeks. T_4 and T_3 travel in the blood to their target cells bound to a special protein, **thyroxine binding globulin (TBG)**.

T_4 and T_3 are the major hormones controlling metabolism. Their effects include control and regulation of the:

- Basal metabolic rate (BMR) – the rate at which oxygen is consumed by cells.
- Metabolism of carbohydrates and fats to produce energy and protein synthesis.
- Growth and development, e.g. of the skeletal and nervous systems.
- Proper functioning of reproductive and nervous systems.

The synthesis and release of T_4 and T_3 is stimulated by TSH from the anterior pituitary, which in turn depends on the release of TRH from the hypothalamus. High levels of T_4/T_3 normally switch off the TSH/TRH (see *Figure 8.9*) and reducing levels switch it on. Other factors influence thyroid hormone release, e.g. pregnancy, where extra energy is required, and decreased by other hormones, e.g. corticosteroids.

Calcitonin

Calcitonin is a polypeptide hormone secreted by the parafollicular cells (or C cells) of the thyroid. As an antagonist (opposing) of the parathyroid hormone (see page 159) it lowers serum calcium and phosphate levels.

Calcitonin acts on bone and the kidneys. It inhibits the reabsorption of calcium from bone and increases excretion of calcium and phosphates in the urine.

The release of calcitonin depends upon the level of calcium in the blood: if calcium levels rise above normal (2.1–2.6 mmol/l), calcitonin is released. Release of calcitonin is also stimulated by various gastrointestinal hormones, e.g. gastrin and secretin; this helps to reduce intestinal absorption of calcium. Calcitonin, with its short half-life (around 10 minutes) has a fine-tuning role in controlling calcium metabolism (see parathyroid glands, pages 159-160).

Problems with thyroid function

Underactivity of the thyroid, which reduces the metabolic rate, is known as *hypothyroidism (myxoedema)* in adults and *cretinism* when it occurs in children (*Figure 8.10(a)*). It may be due to:

- Severe iodine deficiency with neck swelling or *goitre* – common in isolated, mountainous regions some distance from the sea where the soil and plants are low in iodine and fish (which contains iodine) is difficult to obtain. Previously common in Derbyshire, hence the term *'Derbyshire neck'*.
- Lack of TSH or TRH.
- Autoimmune conditions where antibodies destroy the thyroid (see Chapter 19).
- After surgery if too much of the gland is removed.

The effects of hypothyroidism include low body temperature, puffy eyes and face, bradycardia (slow pulse rate), weight gain, hair changes, dry, coarse skin and mental slowness. Hypothyroidism in children leads to stunted growth and learning difficulties. Treatment involves the replacement of thyroid hormones or correction of iodine deficiency.

Overactivity will have the opposite effects, with an increase in metabolic rate – this is known as *hyperthyroidism (thyrotoxicosis)* (*Figure 8.10 (b)*). Hyperthyroid states may be caused by:

- Excess TSH from a pituitary tumour.
- *Grave's disease* or diffuse hyperplasia (see Chapter 1) caused by antibodies that abnormally stimulate thyroid hormones.
- Isolated hormone-secreting thyroid nodules.

Hyperthyroidism causes raised temperature and sweating, tachycardia (rapid pulse rate) and cardiac failure, weight loss (although the appetite is good) and restlessness. *Exophthalmos* (eyeball protrusion), probably due to changes in the fat behind the eye and the extraocular muscles (see Chapter 7), may occur in Graves' disease. This can lead to eye damage. Hyperthyroid states may be treated with antithyroid drugs, radioactive iodine (which destroys the gland) or surgery.

Swellings of the thyroid with or without excess hormone production may cause pressure effects such as *dysphagia* (difficult swallowing) due to pressure on the oesophagus or pressure on the trachea causing *dyspnoea* (difficult breathing).

Parathyroid Glands

These minute structures are only some 6 mm in length. There are usually four situated on the posterior aspect of the thyroid (*Figure 8.11*). Some people, however, have more than four, the extra glands may be located elsewhere in the neck or mediastinum.

Structure

Each parathyroid gland is enclosed in a connective tissue capsule and consists of columns of round cells interspaced with vascular channels.

Parathyroid hormone

The parathyroid glands produce only one hormone, **parathyroid hormone (PTH)** or **parathormone**. This protein hormone maintains serum calcium levels within the homeostatic range of 2.1–2.6 mmol/l. This range of serum calcium is required for proper muscle contraction, nerve impulse transmission and blood clotting.

PTH acts in a variety of ways to raise serum calcium levels and reduce phosphorus (phosphate) levels (*Figure 8.12*):

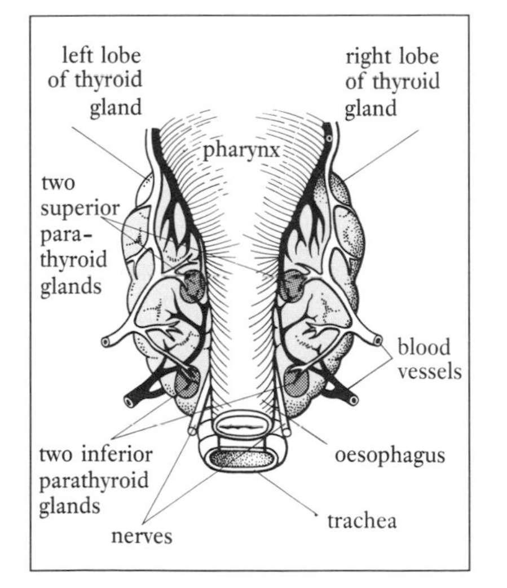

Figure 8.11. Parathyroid glands – gross structure and location (viewed from behind).

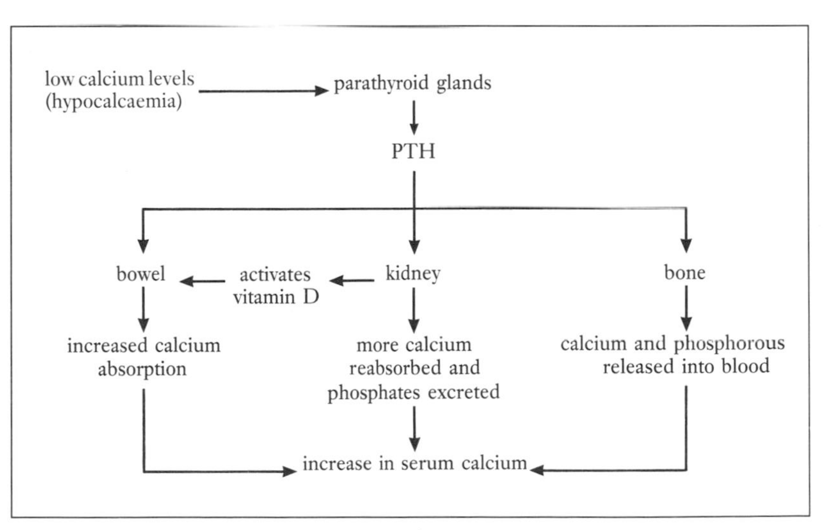

Figure 8.12. Parathyroid hormone – calcium and vitamin D metabolism.

Nursing Practice Application **Tetany following thyroid surgery**

The postoperative care plan for a person having thyroid surgery should include observations for tetany caused by hypocalcaemia. Initially the increase in nerve excitability will cause the person to complain of tingling around the mouth and in the hands and feet. There may also be painful *carpopedal* spasm (hands and feet) and more rarely laryngeal spasm and fits. Latent tetany may be detected by *Trousseau's sign*, where a cuff inflated around the upper arm induces forearm muscle spasm. Emergency medical management of tetany includes the administration of intravenous calcium gluconate and monitoring of serum calcium levels.

- The kidney reabsorbs calcium and excretes phosphate. PTH also stimulates the kidney to convert vitamin D into its active form, 1,25-dihydroxycholecalciferol.
- 1,25-dihydroxycholecalciferol acts upon intestinal cells causing increased absorption of calcium from food.
- PTH stimulates *osteoclasts* (see Chapter 16) in bone which reabsorb bone matrix releasing calcium and phosphate into the blood.

PTH release is stimulated by low levels of calcium in the blood (*hypocalcaemia*) and inhibited by levels above normal (*hypercalcaemia*). For further discussion of calcium metabolism, see Chapter 16.

Problems with parathyroid function

Parathyroid tumours cause hypersecretion of PTH – *hyperparathyroidism*. This results in hypercalcaemia, muscle weakness, nerve conduction problems, weakened bones and kidney problems such as stones and renal failure.

Undersecretion of PTH, or *hypoparathyroidism*, may be caused by trauma or removal of parathyroid glands during thyroid surgery, autoimmune disease and congenital defects. The resultant hypocalcaemia leads to increased nerve excitability and muscle spasm known as *tetany*. Other causes of hypocalcaemia include inadequate calcium intake which occurs in rickets, and renal failure where the failing kidneys do not convert vitamin D to the active form required for calcium absorption. *NB Alkalosis* (see Chapter 2) produced by *hyperventilation* (see Chapter 12), persistent vomiting and the ingestion of alkalis (see Chapter 13) can also produce tetany as the change in blood pH reduces the amount of available ionized calcium.

Adrenal Glands

The two triangular adrenal glands are situated one on the upper pole of each kidney (see *Figure 8.13(a)*). Their position on top of the kidneys accounts for their alternative name, the *suprarenal glands*. The adrenal glands derive arterial blood from a branch of the aorta and from the renal and phrenic arteries. Venous blood drains directly into the inferior vena cava from the right gland and into the renal vein from the left gland (see *Figures 10.41* and *10.43*).

The adrenal glands are divided into a middle part, the *medulla,* and the *cortex* around the outside (*Figure 8.13(b)*). The medulla and cortex are really two separate endocrine structures which develop from different embryonic tissues; the medulla from the ectoderm of the neural crest and the cortex from the mesoderm.

Adrenal cortex

The **cortex** has an outer capsule and three distinct layers, (*Figure 8.13(b)*) which produce a large group of steroid hormones which are derived from cholesterol and known collectively as **corticosteroids**.

From the outside working in, the layers are:

Zona glomerulosa, where clumps of cells secrete *mineralocorticoids* which regulate electrolyte/fluid balance.
Zona fasciculata, the widest layer, in which cells arranged in parallel columns secrete *glucocorticoids* which are important metabolic hormones.
Zona reticularis, which consists of a network of cells which produce small amounts of glucocorticoids and sex hormones.

There is considerable overlap of secretion and the different hormone groups share similarities of function, e.g. glucocorticoids have a slight mineralocorticoid effect.

Adrenal cortex hormones

The adrenal cortex secretes three hormone groups, glucocorticoids, mineralocorticoids and sex hormones.

Glucocorticoids

The **glucocorticoids** form a large group of hormones which are essential to life, they include **cortisol (hydrocortisone),**

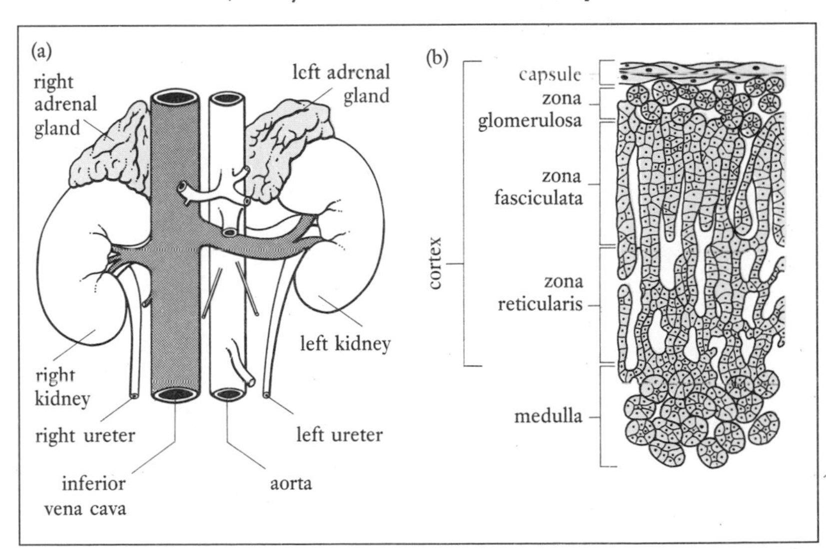

Figure 8.13. Adrenal glands. (a) Position; (b) layers through the adrenal gland.

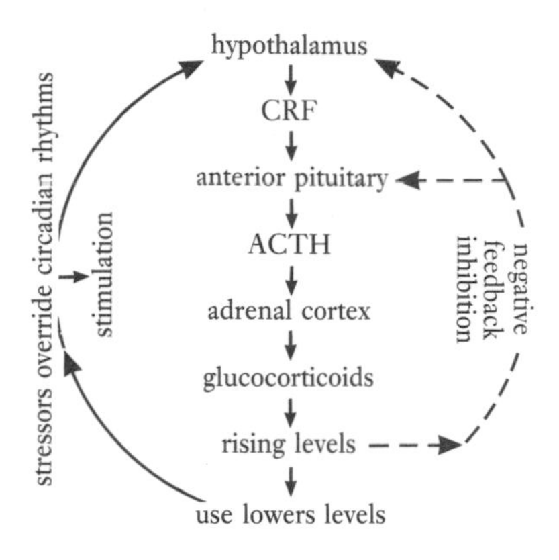

Figure 8.14. Control of adrenal cortex glucocorticoid secretion.

corticosterone and **cortisone**. They are concerned with many metabolic processes and responses to stress. Their actions can be summarized as follows:

- Control and modification of carbohydrate metabolism by the stimulation of *gluconeogenesis* in the liver. This is the production of glucose from non-carbohydrate sources such as amino acids and glycerol. Some of the glucose is stored by the liver as *glycogen* (storage carbohydrate). These processes are important in the regulation of blood glucose levels.
- Increases protein breakdown into amino acids for the synthesis of new proteins.
- Increases the release of fatty acids and their oxidation for energy.
- Causes some sodium and water reabsorption in the kidney tubules.
- Suppresses the inflammatory response and allergy.
- Depresses the immune processes.
- Increases gastric secretion of hydrochloric acid and enzymes.

The release of glucocorticoids is controlled by a negative feedback mechanism; CRF from the hypothalamus stimulates the anterior pituitary to secrete ACTH which in turn causes the release of glucocorticoids, e.g. cortisol from the cortex. Rising levels of cortisol will inhibit both the hypothalamus and pituitary (*Figure 8.14*). The release of glucocorticoids is not constant, a circadian rhythm exists with an early morning peak and a low level in the evening.

Stressors such as pain, cold and *hypoglycaemia* (low blood glucose) override the circadian rhythm allowing the secretion of the extra glucocorticoids required to cope with stress (see page 166 and Chapter 6). Glucocorticoids are mostly excreted in the urine after being changed chemically within the liver. Boore (1978) used the urinary excretion of *17-hydroxycorticosteroids* as an indicator of physiological stress in her study of surgical patients (see also Chapter 6).

Mineralocorticoids

The **mineralocorticoids** form a group of hormones of which **aldosterone** is the most important. Aldosterone helps to regulate

Nursing Practice Application Therapeutic use of glucocorticoids

Glucocorticoids, e.g. prednisolone, are used widely in the management of inflammatory conditions such as asthma and rheumatoid arthritis. Nurses need to plan an education programme which includes: (i) the importance of carrying a 'steroid' card which provides information for health care professionals, e.g. in an emergency; (ii) the need to inform others, e.g. dentist, that they are taking 'steroids'; (iii) an understanding of side-effects which should be reported, such as indigestion (which may indicate peptic ulcer), signs of Cushing's syndrome (see page 164) and signs of insufficient dose, e.g. fatigue, anorexia (loss of appetite) and weakness; (iv) ending the medication only with medical supervision (usually reduced gradually) because of adrenal function suppression. This is caused by long-term use of glucocorticoids leading to adrenal atrophy (see Chapter 1).

electrolyte/fluid homeostasis by causing the distal kidney tubules to reabsorb sodium and water and eliminate potassium or hydrogen ions (see Chapter 15). The release of aldosterone has little to do with the hypothalamus or pituitary although ACTH has some effect in stress situations; it depends upon the concentration of ions in the blood and the blood pressure. Aldosterone secretion increases when sodium and chloride levels are reduced or when potassium levels increase in the blood, and when the volume of extracellular fluid (see Chapter 2), and hence the blood pressure, is low.

Special cells in the kidney, known as the *juxtaglomerular apparatus (JGA),* respond to these changes by releasing the enzyme **renin**, which activates the plasma protein *angiotensinogen* by converting it to **angiotensin** (*Figure 8.15*). The angiotensin stimulates the adrenal cortex to produce aldosterone, which acts upon the distal tubules. Angiotensin also causes vasoconstriction, which raises the blood pressure. The **renin–angiotensin–aldosterone** response does not operate in isolation; it is closely linked to the osmoreceptors and ADH secretion (see page 156) which inhibits renin release. Renin secretion is also stimulated by sympathetic nerves. Reduced sodium or increased potassium levels in the plasma stimulate the adrenal cortex directly.

Sex hormones

Compared with the *gonads* (ovaries and testes) the amounts of **sex hormones** secreted by the adrenal cortex are insignificant; they are mainly **androgens** (male hormones) with very small amounts of the female hormones, **oestrogens** and **progesterones.** Their role is somewhat unclear but levels are high during fetal life and puberty. Inappropriate secretion can lead to masculinization in females.

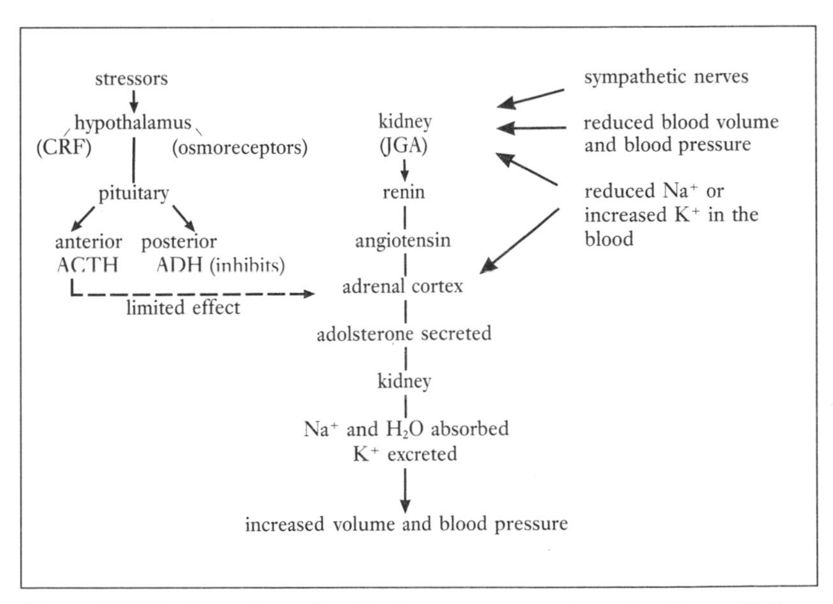

Figure 8.15. Renin–angiotensin–aldosterone response (see also *Figure 15.6*).

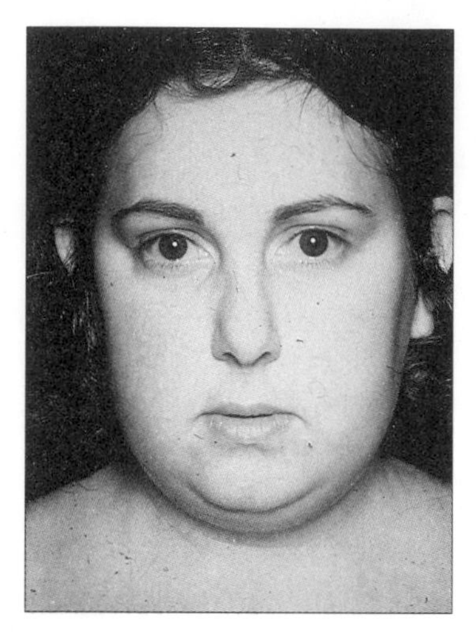

Figure 8.16. Cushing's syndrome. (Hall, R., Evered, D.C. *A Colour Atlas of Endocrinology.* 2nd edn. Wolfe Medical Publications, Ltd. 1990. Reprinted with permission.)

Hypersecretion of corticosteroids or Cushing's syndrome

The causes of *Cushing's syndrome* (*Figure 8.16*) include ACTH-producing pituitary tumour, adrenal cortex hyperplasia or tumour, administration of ACTH and ectopic sources of ACTH, such as bronchial tumours. Similar effects are seen after high-dose corticosteroid drugs, e.g. prednisolone, administered over a long period. The effects of hypersecretion are many and varied:

- *Hyperglycaemia* (high blood sugar), due to excessive gluconeogenesis, and development of diabetes mellitus (see page 168).
- Sodium and water retention leading to hypertension (high blood pressure) and oedema, muscle weakness due to potassium loss.
- Changes in fat and protein metabolism leading to typical 'moon' face, fat redistribution to the abdomen and back ('buffalo hump') and muscle wastage in the limbs.
- The suppression of inflammatory response delays wound healing and masks signs of infection.
- The skin becomes thin, fragile and easily damaged. Excessive bruising occurs and *striae* (stretch marks) appear on abdomen and thighs. *NB* Special care is needed after surgery and in the prevention of skin breakdown (see Chapter 19).
- *Osteoporosis* (see Chapters 16 and 20) and spontaneous fractures, especially vertebral collapse.
- *Peptic ulcer* development linked to an increase in gastric acid secretion.
- Serious psychotic illness – depression, paranoia, euphoria.

Adrenal cortex insufficiency

This usually results in the hyposecretion of both glucocorticoids and mineralocorticoids. It may be due to disease in the adrenals (*Addison's disease*) caused by autoimmune failure or tuberculosis. Pituitary or hypothalamic defects are a secondary cause of hyposecretion.

Loss of cortical hormones leads to the serious problems of dehydration, electrolyte imbalance, hypotension and hypoglycaemia. Affected individuals have pigmented skin (as ACTH levels rise more MSH is produced), lose weight and complain of gastrointestinal disturbances and weakness.

Aldosteronism

An excess secretion of aldosterone may result from an adrenal tumour (*Conn's syndrome*) or secondary to conditions such as cardiac failure (see Chapter 10). Aldosteronism leads to retention of sodium and water which cause hypertension. Other problems include alkalosis, muscle weakness and cardiac arrhythmias which are due to *hypokalaemia* caused by the loss of potassium in the urine.

Adrenal medulla

The medulla forms the smaller part of the adrenal gland. It consists of **chromaffin** cells situated around blood vessels and sinusoids. The medullary cells produce two **catecholamine** hormones which augment the effects of the sympathetic nervous system in the 'fight or flight' response to stress (see Chapter 6).

Adrenal medulla hormones

The medullary hormones are **adrenaline** and **noradrenaline** which, you will remember, is the postganglionic sympathetic neurotransmitter. Both are *monoamines* (substance having one amine group) derived from the amino acid tyrosine. More adrenaline than noradrenaline is produced (around 80% of production is adrenaline). Their metabolic effects are similar and include increase in heart rate, vasoconstriction, rise in blood pressure, diversion of blood to the vital organs, increase in blood sugar, stimulation of respiration and bronchodilation. Adrenaline has most effect on the heart and noradrenaline on vasoconstriction and blood pressure. Release of medullary hormones is through sympathetic nerve stimulation in response to a stressor (see Chapter 6). The catecholamines have a short half-life, they are degraded by enzymes and their metabolites excreted in the urine.

Problems with adrenal medulla secretion

Hyposecretion is not a problem because the medulla is functionally part of the sympathetic nervous system. It is possible to live without the adrenal medulla, which is demonstrated by the fact that after removal of the adrenal glands it is only necessary to replace the cortical hormones.

Excess production of catecholamines produces a prolonged 'fight and flight' state with rapid heart rate and hypertension. This may result from prolonged exposure to a stressor or from a rare tumour of the medulla or sympathetic chain known as a

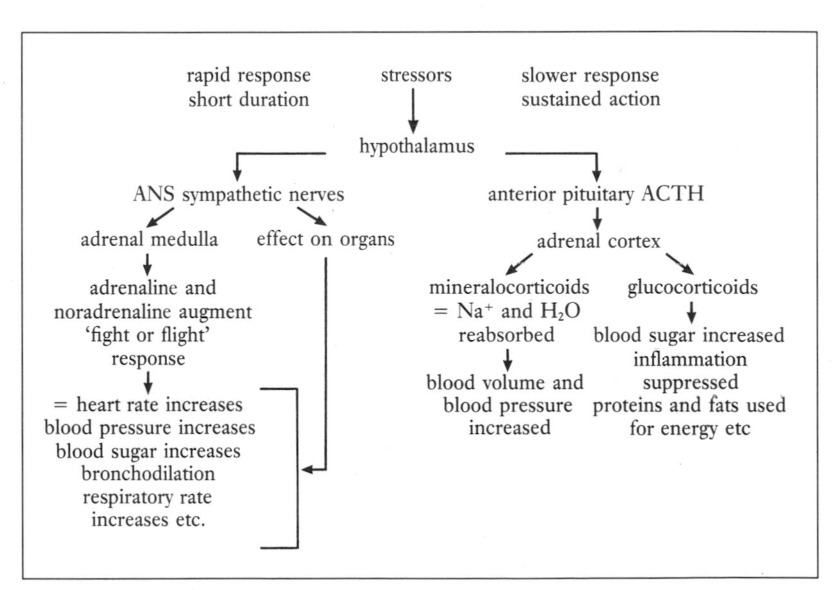

Figure 8.17. Adrenal gland and stress.

phaeochromocytoma. Tumours produce surges of catecholamines with resultant swings in blood pressure, heart rate and sweating. Diagnosis of phaeochromocytoma may be confirmed by the presence of excess urinary metabolites.

Stress and adrenal response

Physiological responses to stressors have been mentioned in respect of the ANS (see Chapter 6) and both the adrenal cortex and medulla. The ANS and medulla respond rapidly, usually for a short time, whereas the cortex responds more slowly but its effects are more sustained (*Figure 8.17*).

The Pancreas

Structure

The pancreas, a soft tapering gland situated in the left upper abdomen, is around 23 cm in length and lies partly behind the stomach. The pancreas is a mixed gland, having both endocrine and exocrine functions (see Chapter 13). It receives arterial blood from the mesenteric and splenic arteries and its venous blood returns to the circulation via the hepatic portal vein and the liver (see *Figures 10.42(a)* and *10.44*).

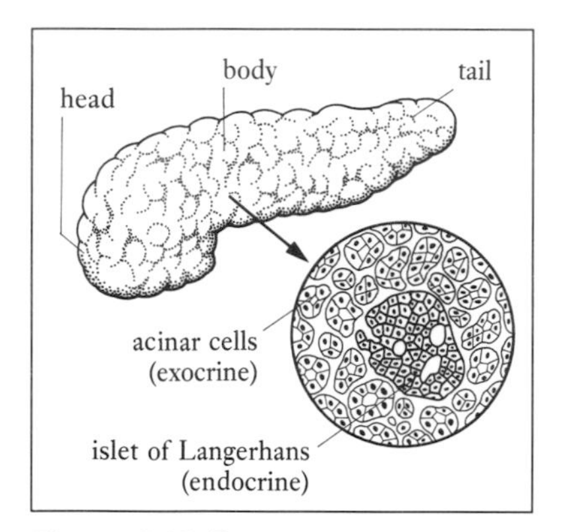

Figure 8.18. Pancreas – gross and microscopic structure.

The pancreas consists of (*Figure 8.18*):

- Exocrine *acinar* cells that produce digestive enzymes which leave the gland via ducts.
- Clusters of endocrine cells, known as the **islets of Langerhans**, scattered throughout the acinar cells like currants in a cake.

The pancreatic islets contain three types of hormone-producing cells: *alpha* α cells, which secrete *glucagon*, the more common *beta* β cells which secrete *insulin* and *delta* δ cells which produce several substances including *somatostatin* (see page 154), also called growth hormone release-inhibiting hormone).

Pancreatic hormones

Glucagon and insulin are very important in the metabolism of carbohydrates and the control of blood glucose levels within the normal range: 3.9–5.8 mmol/l (fasting) to 9 mmol/l after food. The brain and other tissues rely on a constant supply of glucose for energy. Permanent damage can occur if brain tissue is deprived of glucose, such as may happen when blood glucose levels are low (hypoglycaemia).

Glucagon

Glucagon is a polypeptide hormone which causes an elevation in blood sugar. It acts on the liver, where stored glycogen is broken down (*glycogenolysis*) to release glucose. Also in the liver, the process of *gluconeogenesis* which produces glucose from amino acids and glycerol is stimulated by glucagon. The most important stimulus to glucagon release is a reduction in blood glucose levels, such as during a period of fasting. It is also stimulated by certain amino acids and the hormone adrenaline. Glucagon release is inhibited by high blood glucose levels, fatty acids and somatostatin.

Insulin

Insulin is a polypeptide hormone which consists of two amino acid chains joined by chemical bonds. It is formed as *proinsulin* and converted to insulin within the pancreatic cells. The major physiological effect of insulin is a reduction in blood glucose levels but it also influences the metabolism of fats and protein.

The reduction in blood glucose is mainly achieved by the effect of insulin on glucose transport across some cell membranes – it speeds up the entry of glucose into muscle and adipose. Insulin also stimulates glucose metabolism within cells. There is increased *glycolysis* (breakdown of glucose) for energy with the conversion of surplus glucose to glycogen in liver and muscle cells. Insulin also enhances the entry of substances such as amino acids and potassium into the cell. Glycogenolysis and gluconeogenesis, both of which would increase blood glucose, are inhibited.

Insulin stimulates the storage of *triglycerides* (a *lipid* or fatty substance which consists of glycerol and three fatty acids – see Chapter 13) in adipose tissue and inhibits the breakdown of fatty acids. Protein synthesis from amino acids increases and the breakdown of proteins is inhibited. Insulin release is largely in response to a rise in blood glucose but vagal stimulation and gastrointestinal hormones also cause its release. Low blood glucose and the hormones adrenaline and somatostatin inhibit insulin secretion.

Somatostatin

Somatostatin is the same substance as GHIH produced by the hypothalamus. Its secretion from the pancreas inhibits the release of glucagon, insulin and the pancreatic enzymes.

Blood glucose regulation – a summary

Having considered the major hormonal controls of blood glucose homeostasis, it is worth putting these together with other influences which include GH, glucocorticoids, adrenaline and thyroid hormones (*Figure 8.19*).

Problems with blood glucose regulation

The most common problem is that of *diabetes mellitus*, where an absolute or relative lack of insulin causes serious homeostatic imbalance.

There are two main types of diabetes mellitus:

- *Type I* or *insulin-dependent diabetes mellitus (IDDM)*, where there is usually an absolute lack of insulin. This type of diabetes usually occurs first in people under 50. Management is with insulin injections and a diet modified to individual requirements. Recent

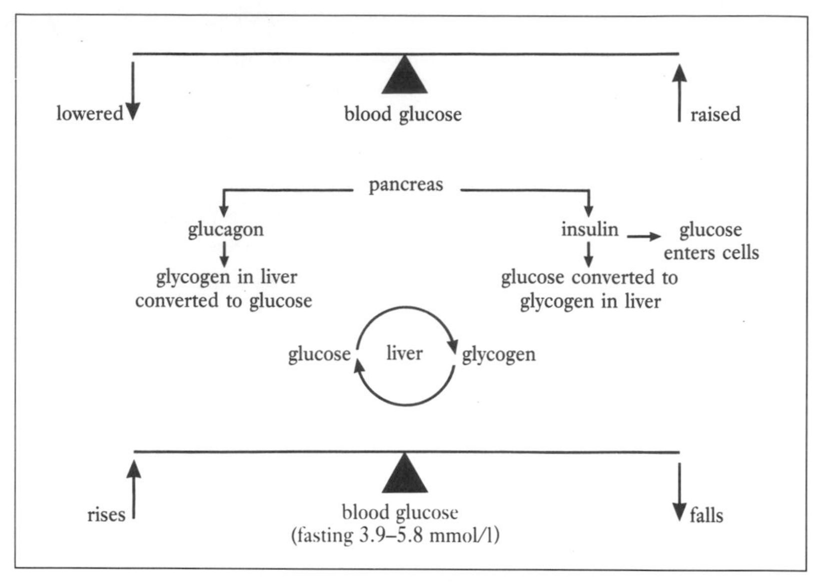

Figure 8.19. Regulation of blood glucose (other influences on blood glucose include diet, GH, somatostatin, gastrointestinal hormones, thyroid hormones, glucocorticoids and adrenaline).

developments have made it possible for donor islet cells to be transplanted into some people with diabetes. The cells, once injected, take root in the liver where they will hopefully start producing insulin. This type of management is not suitable for all and still requires development and evaluation.

- *Type II* or *non-insulin-dependent diabetes mellitus (NIDDM)* tends to be milder. With this form there is usually a relative lack of insulin due to insulin resistance and it is seen in older, overweight people. This type of diabetes may be controlled by diet and drugs (oral hypoglycaemics) to reduce the blood glucose. In some individuals the diabetes will disappear with a return to their normal body weight.

  Secondary diabetic states may follow *pancreatitis* (inflammation of the pancreas), corticosteroid therapy (see page 162), thiazide diuretics (see Chapter 15) and be associated with other endocrine problems such as Cushing's syndrome (see page 164).

Some physiological effects of diabetes mellitus

The physiological effects of diabetes mellitus vary in severity between types and individuals but include (see also *Figure 8.20*):

- Hyperglycaemia, when glucose levels exceed the *renal threshold* (see Chapter 15) it is excreted in the urine (*glycosuria*). The glycosuria may result in *pruritus* (itching) due to fungal infections, e.g. *candidiasis* (thrush).
- *Polyuria* (increased urinary volume). As more glucose and hence water is excreted this leads to serious dehydration and electrolyte imbalance.
- *Polydipsia* (intense thirst) and high fluid intake linked to the polyuria.

Person-centred Study **Andrew**

Andrew is 19 and has just started an engineering course at a university 100 miles from home. He has felt tired and 'run down' which he thought might be due to the work and a few late nights! More recently Andrew has noticed that he has lost weight, is always thirsty and seems to pass a lot of urine. When he visited the health centre the medical officer suspected a diagnosis of diabetes from the history plus the fact that Andrew had glucose in his urine and a blood test showed that glucose levels were raised (15 mmol/l).

The diagnosis was confirmed by the consultant when Andrew and his mother attended the local hospital. Obviously the priority for Andrew is the stabilization of his diabetes but this will require the identification and careful consideration of his lifestyle and individual needs. Initially Andrew may be admitted to hospital but stabilization is more easily achieved at home when the person is undertaking his normal activities. Andrew and his family will be involved with a multidisciplinary team whose aim will be to help him to understand and cope with the diabetes. The team includes ward nurses, specialist diabetic nurse, medical staff, dietitians and the staff at the university health centre.

Andrew and his family will need information and education covering the following aspects of his management: (i) physiology of blood glucose control and the changes in diabetes; (ii) monitoring blood glucose and regular contact with the 'team'; (iii) insulin injections: technique, care of equipment and storage of insulin; (iv) diet modification: a high fibre/unrefined carbohydrate, reduced fat diet which provides adequate energy and the correct nutrients. High fibre diets are useful in the control of diabetes because fibre ensures a slow and even absorption of nutrients from food which avoids sudden swings in blood glucose; (v) sport and exercise: effects on energy requirements; (vi) recognition and management of hypoglycaemia; (vii) recognition of ketoacidosis and hyperglycaemia, which may be precipitated by stressors such as infections; (viii) support from groups such as British Diabetic Association.

Nurses should ensure that the education programme is designed to enable Andrew (and his family) to resume a normal healthy lifestyle in spite of his chronic condition. The long-term complications of vascular problems, increased infection risk and neuropathy should be discussed with particular reference to healthy eating, hygiene, foot and skin care with chiropody when required.

Vascular disease such as atherosclerosis occurs more frequently in people with diabetes, who have a consequent increase in the incidence of peripheral vascular disease and gangrene, myocardial infarction (see Chapter 10), cerebrovascular accident (see Chapter 4), retinopathy (see Chapter 7) leading to blindness and nephropathy with renal failure (see Chapter 15). Abnormal lipid (fat) metabolism with an elevation in cholesterol levels is linked to the arterial deterioration, especially in the smaller vessels supplying the feet, retina and nephrons.

Sensory problems caused by peripheral neuropathy may lead to the diabetic being unaware of fairly minor foot damage which, because of a poor blood supply and infection, could lead to gangrene.

The importance of good blood glucose control should be stressed as the major way to reduce the risks.

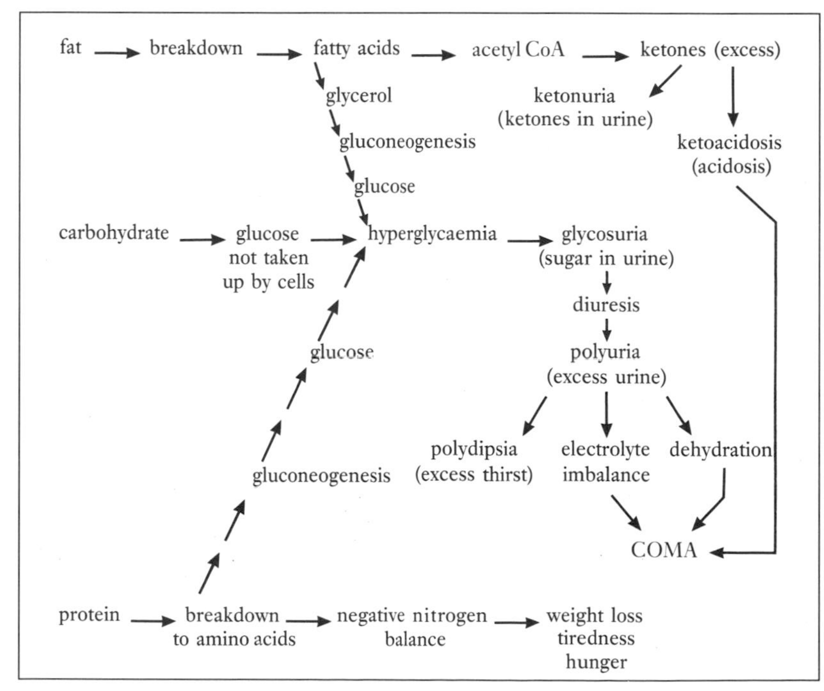

Figure 8.20. Physiological changes associated with diabetes.

- Protein is broken down and converted into glucose by gluconeogenesis. This further increases blood glucose levels. It is important to note that, although blood glucose is high, without insulin it is unavailable for cell use. If glucose is not available the body will turn to other fuel molecules, e.g. fats.
- Fat breakdown releases fatty acids which are metabolized to acetyl-CoA (important metabolic molecule, see Chapter 13). Acidic *ketones* are formed from excess acetyl-CoA and in high quantities cause a fall in blood pH. This leads to *ketoacidosis*, (acidosis due to acidic ketones) which is more commonly seen in type I diabetes. Ketones are excreted by the lungs and in the urine to produce a characteristic odour of acetone. *NB* In type II diabetes there is less risk of ketoacidosis but individuals may present with hyperglycaemia and dehydration (*hyperosmolar coma*).
- Altered consciousness and deep sighing respirations 'air hunger' as homeostatic mechanisms break down. Increased respiration is an attempt to remove more carbon dioxide and correct the metabolic acidosis (see Chapter 12).

Thymus Gland

The **thymus gland**, which has two lobes, is situated in the mediastinum. The lobes consist of many lobules, each with a medulla and cortex. The thymus is fairly large in infants and grows during childhood to reach a maximum during puberty. It becomes smaller during adult life and is gradually replaced by connective tissue in old age.

Structurally the thymus is similar to the spleen and other lymphatic tissue (see Chapter 11). It contains many lymphocytes in the cortex and **thymic corpuscles (Hassal's corpuscles)** in the medulla.

The thymus produces peptide hormones including **thymosin** and **thymopoietin**, which stimulate the proper development of *T lymphocytes* which form part of the immune system (see Chapter 19).

Pineal Body

The pineal body is a small reddish structure situated close to the third ventricle of the brain. It consists of nervous and secretory cells. In humans its function is poorly understood but it is known to produce **melatonin**. Other substances found in the pineal body include *histamine* and *5-hydroxytryptamine* (see Chapter 4).

Release of melatonin is linked to the amount of light entering the eye; levels of melatonin are low during the day and high at night. Melatonin appears to inhibit the release of gonadotrophins which affects the timing of puberty. Certainly the amount of light (day length) is known to influence mating and reproduction in other animal species. The link between melatonin and the timing of puberty in humans is further strengthened by the fact that some secreting tumours of the pineal gland delay the onset of puberty while disease processes that destroy the pineal may lead to abnormally early puberty.

Other aspects of behaviour, such as mood, may be linked to the secretion of melatonin; some types of depression – *SAD (seasonal affective disorder)* – become more frequent as the days shorten and winter approaches. Some people affected in this way have been helped by exposure to special lights which presumably inhibit melatonin.

The secretion of melatonin may also be linked with body processes that follow diurnal rhythms, such as sleep, appetite and hormone release. This would account for the general physiological upsets associated with 'jet lag', shift work and working or living in areas with no natural light.

Summary/Check List

Introduction and links with nervous system.
Major endocrine gland structures – location, other hormone-producing structures.
Hormones – structure. Action. Control of secretion/negative feedback. Half-life.
Hypothalamus and pituitary – releasing and inhibiting hormones. Anterior lobe — anterior lobe hormones, growth problems. Posterior lobe — posterior lobe hormones, problems with ADH.
Thyroid – structure, hormones (metabolism and calcium control), problems with secretion.
Parathyroids – structure. Hormone. Calcium metabolism. Problems with function, *Nursing Practice Application* – tetany.
Adrenals – Cortex. Hormones - glucocorticoids, mineralocorticoids (renin–angiotensin–aldosterone), sex hormones, *Nursing Practice Application* – glucocorticoid therapy. Problems with cortical secretion. Medulla. Hormones (catecholamines). Problems with medullary secretion. Stress and adrenal responses.
Pancreas – Structure (endocrine). Hormones — glucagon, insulin, somatostatin. Blood glucose regulation. Problems with pancreatic hormones - diabetes mellitus, *Person-centred Study* – Andrew.
Thymus – immune response.
Pineal – melatonin, puberty, mood, biological rhythms.

Self Test

1 Outline the ways in which the hypothalamus controls pituitary function.

2 Which of the following statements are true?
(a) ADH and oxytocin are made in the neurohypophysis.
(b) Prostaglandins are derived from fatty acids.
(c) ACTH is controlled by positive feedback.
(d) GHIH is also known as somatostatin.

3 Which of the following would result from an excess of T_3 and T_4?
(a) sweating (b) weight loss (c) mental slowness
(d) bradycardia (e) tetany (f) restlessness.

4 Outline how PTH raises the serum calcium.

5 (a) Which adrenal hormones regulate metabolism of carbohydrates?

(b) Which 'fight or flight' hormone has most effect upon the heart?
(c) Which adrenal hormone release is controlled by ion concentration and blood pressure?

6 Define the following: gluconeogenesis, glycogenolysis, glycolysis.

7 Explain how diabetes mellitus causes dehydration and ketoacidosis.

8 Which of the following do not affect blood glucose levels?
(a) prolactin (b) glucagon (c) cortisol
(d) calcitonin (e) insulin.

Answers
1. See page 153. 2. b, d. 3. a, b, f. 4. See pages159-160. 5. a. glucocorticoids; b. adrenaline; c. aldosterone. 6. See pages 166,167. 7. See pages 168-170. 8. a, d.

References

Boore, J.R.P. (1978) Prescription for recovery: The effect of pre-operative preparation of surgical patients on post-operative stress, recovery and infection. *RCN Research Report*. London: Royal College of Nursing.

Further Reading

Edwards, C. R. W. (1986) *Endocrinology (Integrated Clinical Science)*. London: Heinemann.

Jennings, P. E. (1988) Diabetes: Is self-help the answer? *The Professional Nurse,* **3** (8), 284–289.

O'Riordan, J. L. H., Malan, P. G. and Gould, R. P. (Eds) (1982) *Essentials of Endocrinology*. Oxford: Blackwell Scientific Publications.

Reading, S. (1989) Blood glucose monitoring in diabetic control. *The Professional Nurse,* **4** (7), 354-355.

Siddons, H., McAughey, D. (1992) Professional development brings specialist knowledge. The role of the diabetes specialist nurse: the Manchester model. *Professional Nurse,* **7** (5), 321–324.

Useful Addresses

British Diabetic Association
10 Queen Anne Street
London W1M 0BD.

section four

body transport systems

CHAPTER NINE

Blood

Overview

- Composition and functions of the blood.
- Plasma.
- Blood cells – erythrocytes, leucocytes, thrombocytes.
- Haemostasis.
- Blood groups and transfusion.

Learning Outcomes

After studying Chapter 9 you should be able to:

- Describe the major components and physical properties of blood.
- List the major functions of blood.
- Outline the precautions required for the safe handling of blood and body fluids.
- Describe the composition and functions of plasma.
- Outline the formation of blood cells.
- Describe the structure and functions of erythrocytes.
- Discuss the substances required for healthy erythrocyte production.
- Describe the structure and functions of haemoglobin and the mechanism for its breakdown.
- Outline the main types of anaemia and discuss its effects.
- Describe the haemoglobinopathies: sickle-cell disease and thalassaemia.
- Describe the various types of leucocyte, their production and functions.
- Discuss alterations in leucocyte production and the problems which result from abnormal white blood cell counts.
- Describe the production and functions of platelets (thrombocytes).
- Outline the factors and processes involved in haemostasis.
- Describe abnormal haemostasis: thrombosis, bleeding disorders, etc.

Learning Outcomes (continued)

- Describe ABO and rhesus blood groups.
- Discuss rhesus incompatibility and haemolytic disease of the newborn, and their prophylaxis.
- Outline the indications for transfusion of blood products and describe the measures taken to ensure safe transfusion.
- Describe the complications associated with blood transfusion and the appropriate nursing observations..

Key Words

Agglutination – an abnormal 'clumping' or 'sticking together' of cells, which may occur with a mismatched blood transfusion.
Anticoagulant – substance which prevents or delays coagulation (clotting).
Clotting factors – substances which control the process of blood clotting/coagulation.
Coagulation – blood clotting, the last stage of haemostasis.
Erythrocyte (red blood cell) – blood cells containing haemoglobin; they carry gases and buffer pH change.
Erythropoiesis – the formation of erythrocytes in the bone marrow.
Haematology – the science dealing with the blood.
Haemoglobin – complex molecule of iron-containing pigment and protein. Found in red cells.
Haemopoiesis – formation of blood cells.
Haemostasis – the processes which prevent inappropriate bleeding.
Interstitial fluid (tissue fluid) – fluid surrounding the cells, forms part of extracellular fluid.
Leucocyte (white blood cell) – the general name given to the different blood cells concerned with immunity and protection against infection.
Leucopoiesis – formation of leucocytes.
Macrocyte – an abnormally large erythrocyte.
Microcyte – an abnormally small erythrocyte.
Phagocytosis – process of enveloping and destroying bacteria and other particles by certain leucocytes, e.g. polymorphonuclear and monocytes, and other cells.
Plasma – the fluid part of blood in which the cells are suspended.
Platelet (thrombocyte) – the non-nucleated (without a nucleus), disc-shaped cellular fragment derived from large multinucleate (more than one nucleus) cells. Platelets are concerned with blood clotting.
Serum – clear fluid produced after blood has coagulated; plasma – clotting factors = serum.
Thrombosis – inappropriate intravascular clotting; formation of a thrombus (clot) in the heart or blood vessels.

Introduction

__Arteries__ – blood vessels which convey blood away from the heart.
__Veins__ – blood vessels which convey blood to the heart.
__Capillaries__ – extremely thin-walled vessels which connect arteries and veins. The thin capillary wall allows the transfer of substances between blood and interstitial fluid.

The blood (which flows within a network of vessels) and the heart (which maintains the circulation by pumping a steady flow of blood) form the main transport system of the body. Blood is carried around the body by a branching system of vessels (**arteries, veins** and **capillaries**). The substances and oxygen carried by the blood can cross the capillary wall and enter the **interstitial fluid** (fluid surrounding the cells); conversely, waste and other molecules can leave the interstitial fluid and enter the capillary. This molecular exchange is a vital part of homeostatic regulation.

Excess interstitial fluid is collected by the vessels of the **lymphatic system** (see Chapter 11), which return it to the circulation (see Chapter 10) so that a constant blood volume is maintained. The fluid within the *lymphatic vessels*, known as **lymph,** is also concerned with the transport of substances around the body. The vital role of the lymphatic system in body defence and immunity is discussed in Chapter 19.

Composition and Functions of Blood

Composition

Blood, a fluid connective tissue (see Chapter 1), is a sticky, red liquid which appears uniform but really consists of many cellular components suspended in fluid. Blood is slightly alkaline, with a normal pH range of 7.35–7.45 (see Chapter 2).

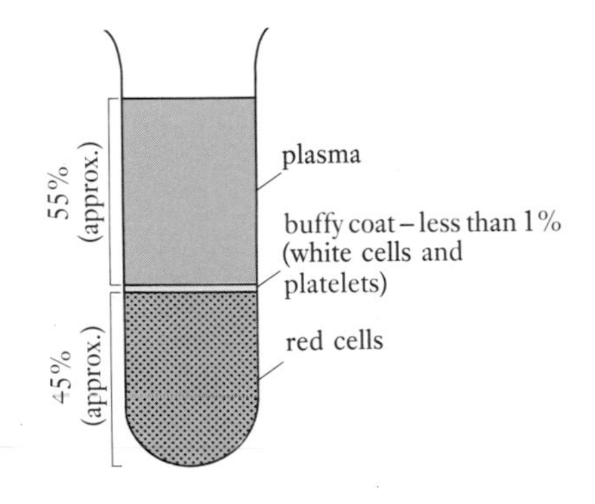

Figure 9.1. Blood components.

The blood cells – **erythrocytes** (red cells – the most numerous type of blood cell), **leucocytes** (white cells) and **thrombocytes** (**platelets**) form about 45% of blood volume, with **plasma** the fluid part forming the remaining 55%. The proportion of erythrocytes to plasma is termed the *packed cell volume (PCV)* or *haematocrit*. A sample of blood in a test tube will normally form a clot surrounded by **serum**. However, a sample of blood spun in a centrifuge or exposed to **anticoagulants** (substances which prevent or delay clotting) separates into three layers: the erythrocytes form the thick bottom layer, there is then a thin layer of leucocytes and platelets called the *buffy coat* and finally the plasma (*Figure 9.1*).

In adults blood forms approximately 7–8% of the body weight, males have 5–6 litres and females 4–5 litres. The percentage of body weight is higher in children and lower in overweight individuals (adipose tissue does not contain much blood).

Functions

Blood is important in maintaining the internal environment and homeostasis. It:

- Carries oxygen and nutrients to cells.
- Removes waste such as carbon dioxide to the lungs and

Nursing Practice Application **Safe handling of blood**

Micro-organisms, such as *human immunodeficiency virus (HIV)*, which causes *acquired immune deficiency syndrome (AIDS)* and the *hepatitis B* virus are present in the blood and body fluids of affected individuals. (A vaccine for hepatitis B is available for high-risk groups, which includes health care workers.)

Health care workers need to exercise care when handling blood, other body fluids and contaminated items, such as used needles, syringes, intravenous equipment and linen. This care should include, the use of gloves and plastic apron (RCN, 1986) as appropriate, correct disposal of used items, especially 'sharps', dealing safely with spillages, and following proper procedures when accidents occur, especially those involving dirty needles.

---

nitrogenous waste to the kidneys.
- Maintains body temperature by heat distribution.
- Transports enzymes and hormones to their areas of action.
- Maintains pH, fluid and electrolyte balance.
- Protects against infection with leucocytes and antibodies.
- Prevents severe haemorrhage by clot production (**coagulation** the last stage in haemostasis).

Unfortunately, blood may also provide the means by which harmful agents, e.g. parasites, micro-organisms and to a lesser extent malignant cells, travel around the body.

PLASMA

Plasma is the straw-coloured, slightly alkaline fluid that forms approximately 55% of blood volume. It consists of water, plasma proteins, inorganic ions and the substances normally transported via the blood. Plasma is:

- 90–91% water.
- 7–8% plasma proteins, e.g. *albumin, globulin, fibrinogen* and many more (see below).
- 1–2% inorganic ions, hormones, enzymes, nutrients and waste, e.g. urea.

Plasma proteins

There are between 60 and 80 grams of plasma proteins in every litre of blood. They are concerned with providing the osmotic pressures that maintain fluid compartment volumes, contributing to blood viscosity and blood pressure, carriage of substances (e.g. hormones), buffering pH changes, inflammatory and immune responses, blood coagulation and providing a protein reserve.

Albumin (35–50 g/l)

Albumin is made in the liver. It has a molecular weight of 69 000 (albumin is therefore a large molecule which is too large to pass through the capillaries of the healthy kidney; its presence in the urine indicates that renal physiology is disordered, as may occur in kidney inflammation, see Chapter 15). Albumin is important in maintaining the osmotic pressure of plasma, which ensures that water mostly stays in the blood (some water and protein do leak into the interstitial fluid and are returned to the blood via the lymphatics). Reduced albumin levels, which may be due to liver malfunction, results in oedema (see Chapter 10) as excess water leaks out of the blood vessels and 'waterlogs' the interstitial spaces. Sodium ions also contribute to osmotic pressure.

Albumin makes plasma viscous, or sticky. Blood is definitely 'thicker than water' by a factor of four and therefore contributes to blood pressure (see Chapter 10). Protein molecules such as albumin act as buffers to limit pH changes (see Chapter 2). Albumin also carries the substances *bilirubin, bile acids*, calcium and drugs, e.g. aspirin. These substances are bound chemically to albumin prior to transportation.

Globulins (23–35 g/l)

The **globulins** can be divided into three fractions; *alpha* α, *beta* β and *gamma* γ. Alpha and beta globulins are made in the liver but gamma globulin is produced by cells of the immune system. Alpha globulins transport many hormones, including cortisol and thyroxine (see Chapter 8). The beta globulins are important in the transport of lipids including cholesterol, fat-soluble vitamins (A, D and K), insulin and iron. Our ability to fight infection depends on the gamma globulins (*immunoglobulins*) (see Chapter 19).

Fibrinogen (1.5–4 g/l)

Fibrinogen, made in the liver, is one of the **clotting factors** (substances which control the process of blood clotting/coagulation). During **coagulation** soluble fibrinogen is converted to insoluble *fibrin* to form a clot (see page 192). Fibrin also 'seals off' areas of inflammation and prevents bacteria spreading.

Other plasma proteins

These include:

- Other clotting factors, e.g. *prothrombin.*
- **Complement**, a complex of several proteins involved in the immune and inflammatory responses (see Chapter 19).
- **Kinins**, e.g. *bradykinin* (cause pain by stimulating nerve endings), which are proteins associated with inflammation.

Inorganic ions/electrolytes

The main plasma ions are sodium, chloride, potassium, calcium, hydrogen carbonate, phosphate and magnesium (see Chapter 2). The ions assist in maintaining blood pH and osmotic pressure.

Gases

Oxygen is carried mainly by the erythrocytes, although a small amount is dissolved in the plasma. Carbon dioxide is much more soluble than oxygen and most is carried as hydrogen carbonate (bicarbonate) ions in the plasma, although erythrocytes carry a small amount.

Hormones and enzymes

Hormones and enzymes form a variable component of the plasma, depending on what substances are actually in transit. Hormones are secreted into the blood by endocrine glands (see Chapter 8) and transported to their target tissues.

Some enzymes are part of the functioning of the blood, e.g. clotting; while others, which are produced by cell breakdown, e.g. *aspartate aminotransferase* from damaged cardiac muscle after a myocardial infarction (see Chapter 10), are merely being transported.

Nutrients

Glucose, amino acids, vitamins and some lipids are absorbed from the gastrointestinal tract and distributed around the body.

Metabolic waste

The plasma carries nitrogenous waste – urea, uric acid and creatinine – to the kidneys for excretion. Other waste products include bilirubin (bound to albumin) and the metabolites of drugs and hormones.

Blood Cells

Earlier we mentioned that cells form approximately 45% of the blood volume. The majority are erythrocytes, followed by platelets, with the least numerous being the leucocytes.

Blood cell formation - haemopoiesis

All blood cells develop from a primitive cell found in the bone marrow. This **stem cell** or **haemocytoblast** gives rise to intermediate cells (proerythroblast, megakaryoblast, myeloid stem cells – myeloblast and monoblast, lymphoid stem cells) that eventually become erythrocytes, platelets and several varieties of leucocytes (see *Figure 9.2* and page 181). Some leucocytes are further processed by lymphoid tissues and thymus (see Chapter 19 and *Figure 19.6*).

During embryonic life the blood cells are made in the yolk sac, and then by the liver and spleen. At the time of birth all blood production has shifted to the red marrow present in the medullary cavities of the bones (see Chapter 16). Red marrow is gradually replaced by fatty yellow marrow during childhood, until in adults the only haemopoietic sites are the skull, ribs, sternum, vertebrae, pelvis and the ends of the long bones. At times of extra demand, such as severe anaemia, the body compensates by extending haemopoietic marrow and reviving hepatic (liver) **haemopoiesis**. Detailed discussion of haemopoiesis is included with individual cell types.

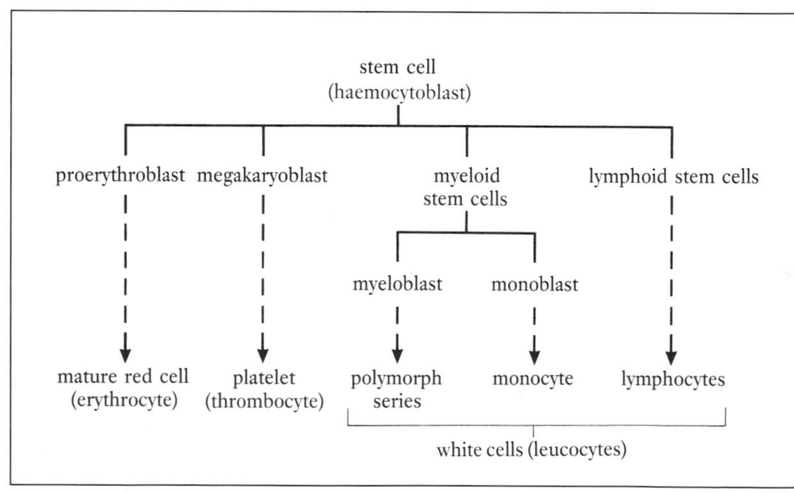

Figure 9.2. Formation of blood cells.

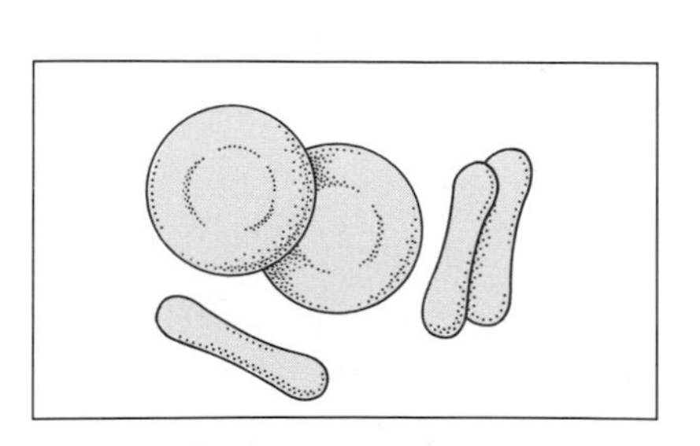

Figure 9.3. Erythrocytes.

Erythrocytes (Red Cells)

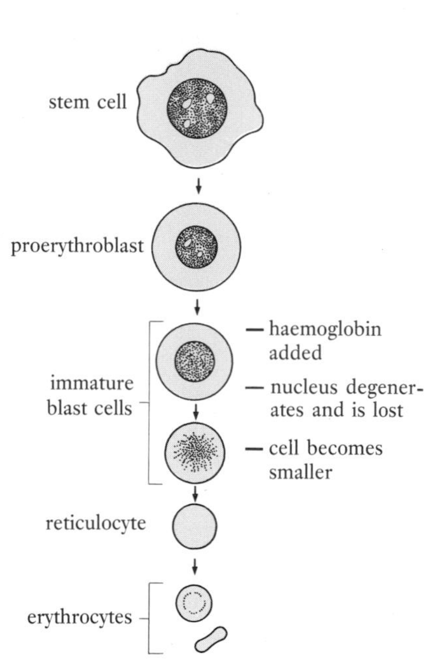

Figure 9.4. Formation of erythrocytes.

Normal mature erythrocytes are non-nucleated biconcave discs which contain **haemoglobin** (*Figure 9.3*). They measure 7–8 μm in diameter. Differentiation (see Chapter 1) occurring during development ensures that the mature erythrocyte is well adapted to its primary function of oxygen carriage. Erythrocytes have very few organelles. A cell membrane encloses the haemoglobin. The *glycoproteins* that determine blood group (see page 198) are situated on the membrane surface. The shape and large surface area of red cells means that all areas are near the surface and can engage in gaseous exchange. Erythrocytes can distort to travel along narrow capillaries. The ability to distort is due to the surface area:size ratio and the presence of special proteins, e.g. *spectrin,* in the membrane. Spectrin and other proteins are able to control and stabilize erythrocyte shape.

The normal range for erythrocyte numbers is 3.8–5.8 x 10^{12}/l in females and 4.5–6.5 x 10^{12}/l in males. Gender differences in erythrocyte numbers and haemoglobin concentration may be explained by menstrual blood loss and stimulation of erythrocyte production by *androgens* (male hormones).

Erythropoiesis - erythrocryte formation

Erythrocytes are produced in the red marrow from **proerythroblasts**. These are committed cells (i.e. their only purpose is to produce erythrocytes), which develop from stem cells (haemocytoblasts). The developing erythrocytes pass through several stages where immature blast cells lose organelles and nucleus, fill with haemoglobin and become smaller. The 'young' erythrocyte or **reticulocyte** is produced in about 3–4 days. Reticulocytes mature after they have been released into the blood (*Figure 9.4*). In adults reticulocytes normally form less than 2% of the circulating red cells, rising where demand for erythrocytes increases, e.g. following severe bleeding.

Erythrocyte numbers remain fairly constant; cells at the end of their lifespan (remember, erythrocytes cannot divide) are replaced by new cells. Replacement involves the production of many millions of cells every minute.

The red marrow forms erythrocytes as a result of direct stimulation by the hormone **erythropoietin**, which is produced in response to *hypoxia* (lack of oxygen). **Renal erythropoietic factor (REF)**, produced by the hypoxic kidney acts upon a plasma protein to form erythropoietin. Other hormones known to influence **erythropoiesis** include corticosteroids, androgens, oestrogens, thyroid hormones and growth hormone.

Erythropoietin production and hence erythropoiesis is increased at high altitudes where the partial pressure of atmospheric oxygen is low. Shortage of oxygen with chronic respiratory diseases will increase erythropoiesis as the body compensates. These situations lead to the production of extra erythrocytes, or *secondary polycythaemia*.

Erythropoietin production can be affected by renal disease, which explains why people with renal failure are usually *anaemic.*

Dietary requirements for healthy erythrocytes

To keep pace with erythrocyte production the body needs a constant supply of dietary raw materials including:

- Iron for the *haem* part of haemoglobin.
- Protein for the *globin* part of haemoglobin and other erythrocyte proteins.
- Vitamin B_{12}(hydroxycobalamin) and folic acid (folate) for DNA synthesis (see Chapter 1) and erythrocyte growth.
- Vitamin C for the absorption and utilization of iron and folate.
- Trace elements, e.g. copper.

Iron metabolism

The importance of iron in erythropoiesis merits further consideration. An adult body contains about 4 g of iron. Most is in haemoglobin and some is stored in the liver and spleen. Iron is toxic and for this reason is combined with proteins to form **ferritin** and **haemosiderin**, which are storage complexes. Whilst in transit around the body, iron is bound to the protein **transferrin**.

Iron absorption from food is enhanced by:

- Intake of foods rich in available iron, e.g. meat, eggs and green vegetables.
- Adequate vitamin C (often destroyed in cooking).
- Gastric acid, which converts ferric iron into the absorbable ferrous state.

An adult male needs to absorb about 1 mg iron daily to replace the losses in urine, faeces and sweat. Females require 2 mg daily to replace that lost during menstruation and the extra demands of pregnancy. During childhood the requirement for iron is also 2 mg daily.

Absorption of iron can be inhibited by *phytic acid* and phosphates in cereals. Vegetarians and those with a high fibre intake should ensure that they have adequate iron.

Iron deficiency due to inadequate intake, malabsorption or excess loss causes anaemia. This and the effects of B vitamin deficiency are considered later.

Conversely, excess amounts of iron may cause problems. Normal function is disrupted as iron is deposited in the liver, heart and other organs. Excess iron storage, known as *siderosis*, may be caused by repeated blood transfusion, high iron intake and abnormal absorption.

Acute iron poisoning

Accidental poisoning with iron is common in small children – the similarity of iron preparations to brightly coloured sweets makes them attractive to children.

All medicines should be kept in 'childproof' containers, out of reach of a resourceful toddler, and any left over should be disposed of safely. If an accident occurs, the child needs immediate emergency care in hospital where *gastric lavage* (stomach washout) is performed and the iron chelating agent, desferrioxamine, administered. Chelating agents are molecules which bind with metals to form 'complexes' which can then be eliminated safely.

Haemoglobin (Hb)

Haemoglobin is the red iron-containing pigment–protein complex contained in erythrocytes. A molecule of haemoglobin (molecular weight 68 000) consists of four **haem** groups each with an atom of ferrous iron (Fe^{2+}) and, four **globin chains** (*Figure 9.5*). Several forms of haemoglobin exist, they differ in their globin chain composition. Fetal haemoglobin (HbF) has two alpha (α) chains and two gamma (γ) chains. There are two forms of adult haemoglobin (HbA and HbA_2). The major form HbA has two alpha (α) chains and two beta (β) chains, and the minor form HbA_2 has two alpha chains and two delta (δ) chains.

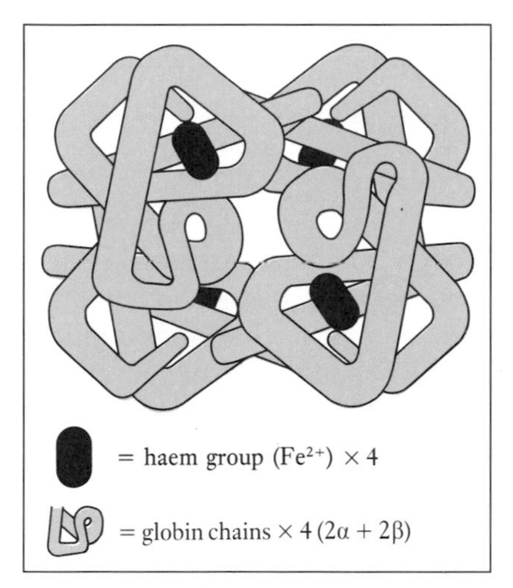

Figure 9.5. Structure of Haemoglobin

Each molecule of haemoglobin combines reversibly with four molecules of oxygen to form **oxyhaemoglobin** which is bright red (see Chapter 12). The affinity of haemoglobin for oxygen changes as the four molecules of oxygen combine sequentially, giving rise to the *sigmoid* shape of haemoglobin's oxygen dissociation curve (see *Figure 12.18*).

Dark red reduced haemoglobin, having given up its oxygen in the tissues, carries some carbon dioxide bound to the globin as **carbaminohaemoglobin** (see Chapter 12).

Haemoglobin is an important component of buffer systems which limit pH changes in the blood (see Chapters 2 and 12).

In the fetus the special haemoglobin (fetal haemoglobin; HbF) has a high oxygen affinity which overcomes the low partial pressure of oxygen in the placenta (the fetus is receiving 'second-hand' oxygen via the maternal circulation). By early childhood HbF has been mostly replaced by the adult haemoglobins (HbA, HbA_2). Normally adults have 11.5–18 g of haemoglobin in every 100 ml of blood (11.5–18 g/dl). Males have a higher concentration for reasons already discussed. For haematological indices see *Table 9.1*.

Table 9.1. Haematological values.

Measurement	***Male***	***Female***
Red cell count	4.5-6.5 x 10^{12}/l	3.8-5.8 x 10^{12}/l
Packed cell volume (PCV)	40-54%	35-47%
Haemoglobin (Hb)	13-18 g/dl	11.5-16.5 g/dl
Mean cell volume (MCV)	← 78-95 fl →	
Mean cell haemoglobin concentration (MCHC)	← 30-35 g/dl →	
Mean cell haemoglobin (MCH)	← 27-32 pg →	
Erythrocyte sedimentation	normal	0-7 mm/h
rate(ESR) – results depend	uncertain	8-20 mm/h
on age, gender and method	abnormal	>20 mm/h

Summary - erythrocyte functions

- Carriage of oxygen.
- Buffering pH changes.
- Carriage of some carbon dioxide.

Destruction of erythrocytes/haemoglobin

Erythrocytes have a lifespan of 100–120 days in the blood. Without a nucleus they are unable to synthesize proteins or divide. Old or damaged erythrocytes are destroyed by **macrophages**, the **phagocytic** cells found in the spleen and liver. This liberates iron and protein, which are recycled, and haem, which is excreted by an unusual route. The degradation of haem produces the pigments **bilirubin** and **biliverdin**. These are bound *(conjugated)* to glucuronic acid in the liver and then pass into the intestine via the bile (see Chapter 14).

Disordered erythrocyte function

Anaemia

Anaemia means a reduction in erythrocyte numbers, a reduction in the amount of haemoglobin, or both, which leads to diminished oxygen-carrying capacity. It is important to remember that anaemia is a 'symptom' of underlying disease. Causes of anaemia may be:

- *Haemorrhagic* – acute blood loss, e.g. following a road accident, or more insidious loss due to chronic bleeding, e.g. peptic ulcer or heavy menstrual loss. Management involves treating the cause, and possibly blood transfusion.
- *Deficiency* – of iron may result from inadequate intake, poor absorption or chronic blood loss. Iron deficiency results in small red cells (**microcytes**) with reduced haemoglobin. Iron supplements are given and transfusion may be necessary if the anaemia is severe.

  Lack of vitamin B_{12} and folate give rise to a *macrocytic* or *megaloblastic anaemia* (see *Figure 9.6*), so called because *megaloblasts* (abnormal erythrocyte precursors) are produced by the marrow. The erythrocytes, in vitamin B_{12} and folate deficiency, are large (**macrocytes**). Vitamin B_{12} may be lacking in the diet of strict vegans but deficiency is usually due to problems with absorption which requires the *intrinsic factor* (gastric glycoprotein, a factor that is essential for absorption of vitamin B_{12}). This is missing in *pernicious anaemia,* where the stomach lining atrophies, or following gastrectomy (removal of part or all of the stomach). There may be problems with absorption in the small bowel.

  Folate may be lacking in the diet when intake does not meet demand, such as during pregnancy, and some drugs prevent the use of folate, e.g. methotrexate (cytotoxic antimetabolite), anti-epileptic drugs.

  Where vitamin B_{12} is deficient there may also be nerve changes, e.g. demyelination (see Chapter 3). The effects of deficiency may take time to develop as the liver stores reserves of folate (sufficient to last some months) and vitamin B_{12} (enough for 3–5 years). Treatment is based on supplements of vitamin B_{12} (injection if the intrinsic factor is absent) and oral folate.
- *Haemolytic* – some anaemias are due to the excessive breakdown

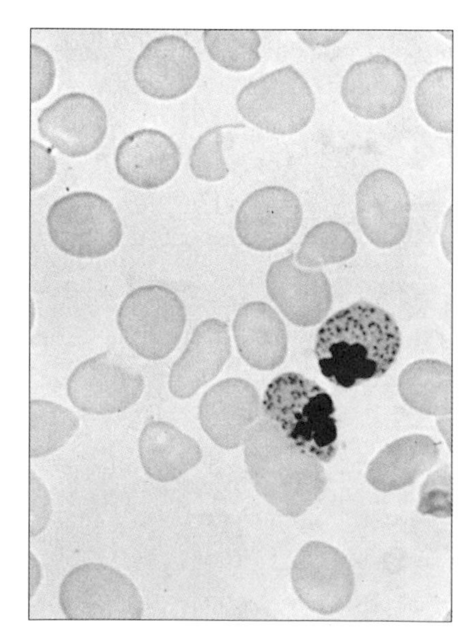

Figure 9.6. Abnormal erythrocytes – macrocytic anaemia with megaloblasts. (Hayhoe, F. G.J., Flemans, R.J. *A Colour Atlas of Haematological Cytology.* 3rd edn. Wolfe Medical Publications, Ltd. 1991. Reprinted with permission.)

Person-centred Study **Kate**

Kate is retired and has lived alone since her friend died six months ago. They were always busy with the house and garden but Kate can't be bothered any more. She feels so tired and gets breathless after the slightest exertion. Cooking is a real chore and she tends to make do with toast.

Kate decided to visit her GP when she became dizzy and had palpitations. Kate's doctor noted her pallor and felt that her history suggested anaemia. The signs and symptoms experienced by Kate can be explained by the diminished oxygen-carrying capacity of her blood and the reduced supply of oxygen reaching the tissues: (i) *tiredness* – reduced oxygen reaching the muscles causes fatigue; (ii) *breathlessness* – the respiratory rate increases on slight exertion in an attempt to supply more oxygen to hypoxic tissues; (iii) *dizziness* – some degree of cerebral hypoxia as the brain receives less oxygen; (iv) *palpitations* – heart rate increases and the person becomes aware of his or her heart beating as the body tries to circulate more oxygenated blood to the tissues; (v) *pallor* – the skin is paler than normal. This is because haemoglobin, which normally contributes to the 'pinkish' skin colour (in Caucasians), may be reduced.

A blood test (see *Table 9.1*) confirmed that Kate had iron deficiency anaemia: (i) *Hb* – 9.0 g/dl; (ii) *MCV, MCHC and MCH* – all reduced; (iii) blood film – erythrocytes are microcytic and hypochromic (pale); (iv) *serum iron* – reduced.

Several factors have contributed to Kate's anaemia: inadequate iron intake, depression, which has caused Kate to neglect her needs, and slight bleeding from her haemorrhoids.

Kate is prescribed some iron tablets, the practice nurse arranges to visit and discuss nutrition, and she is referred to a surgeon regarding her haemorrhoids. Recently the practice has employed a counsellor and Kate is referred for help regarding her bereavement-linked depression.

(*haemolysis*) of red cells. Causes include drugs, abnormal erythrocytes, overactive spleen, severe infections, haemoglobinopathies (see pages 185-187) and mismatched blood transfusion (see pages 199 and 201-202).

- *Aplastic* – the bone marrow is destroyed and ceases to produce red cells, white cells and platelets. This increases infection risk and bleeding tendency in addition to anaemia. Causes of marrow aplasia include cytotoxic drugs, e.g. busulphan, radiation and bone cancers.

Each type of anaemia is characterized by specific physiological effects but all cause changes due to reduced oxygen-carrying capacity (see *Person-centred Study* – Kate).

Polycythaemia

In addition to secondary polcythaemia (page 181), there is a condition known as *primary polycythaemia vera* where erythrocytes increase without any physiological stimulus. It is a serious condition with increased PCV and blood viscosity leading to abnormal clotting. Primary polycythaemia is one of a group of *myeloproliferative disorders* and can progress to leukaemia. Management includes venesection, radiotherapy and cytotoxic drugs.

Haemoglobinopathies

Of the many abnormal haemoglobins, *sickle-cell disease* and *thalassaemia* are the most common. These are genetic conditions

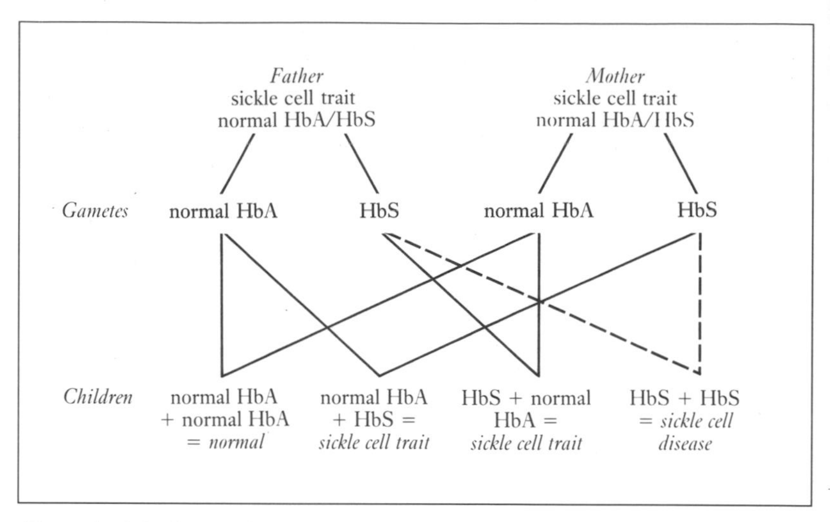

Figure 9.7. Inheritance of sickle-cell disease) HbA, adult haemoglobin; HbS, sickle-cell haemoglobin).

where the amino acid sequence in the globin chains (see page 183) is abnormal or globin production is reduced. The haemoglobin structure is abnormal, which in turn changes red cell shape and functioning.

Sickle-cell disease affects people from areas where *falciparum malaria* is common (equatorial Africa, part of India and part of the Eastern Mediterranean) and their descendants in the USA, West Indies and Europe. People who are *heterozygous* (see below) for the sickle-cell gene have an increased ability to withstand the malaria parasite.

People with sickle-cell disease produce a different type of haemoglobin (HbS, for sickle-cell) than normal adult haemoglobin. All people in at-risk groups should be screened for HbS prior to procedures which may cause hypoxia, e.g. an anaesthetic. The condition occurs in individuals who are *homozygous* for the sickle gene, i.e. inherited from both parents (see *Figure 9.7*). Their erythrocytes become 'sickle-shaped', leading to reduced oxygen-carrying capacity, blockage of blood vessels with pain and tissue death (*infarction*) and rapid erythrocyte destruction by the spleen, causing haemolytic anaemia (see page 184-185). During sickle-cell crisis the immediate need is for hydration and pain control.

Sickle-cell trait occurs in *heterozygous* individuals who have inherited one abnormal gene (compare with homozygous, above). People with sickle-cell trait can pass the abnormal gene to their children and may themselves show symptoms when hypoxic.

Thalassaemia affects populations around the Mediterranean, in the Middle East and Far East. Abnormal globin production results in fragile erythrocytes with reduced oxygen carriage. Excessive erythrocyte breakdown in the spleen results in haemolytic anaemia and iron overload. Homozyygous individuals have the severe *thalassaemia major* whereas heterozygotes have the much milder *thalassaemia minor.*

Haemoglobinopathy management depends on type and severity but

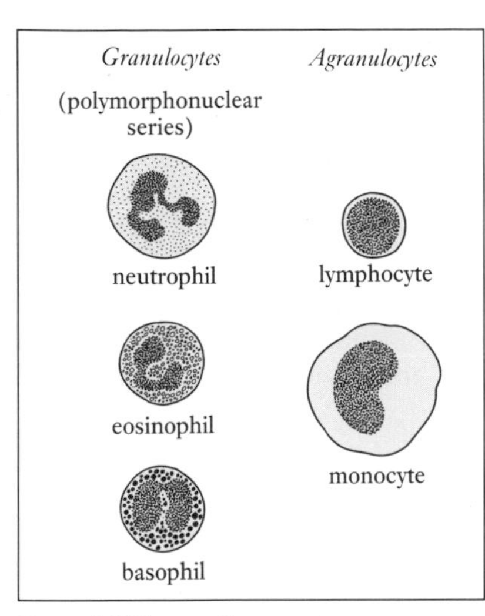

Figure 9.8. Types of leucocytes.

may include avoidance of stressors, e.g. cold and hypoxia, folate supplements, blood transfusion, iron chelating agents, splenectomy (removal of the spleen), symptom control, genetic counselling and antenatal diagnosis, with termination of pregnancy offered if the fetus is affected.

Leucocytes (White Cells)

All leucocytes are concerned in some way with body defences. Leucocytes are the least numerous of the blood cells; normally there are 4–11 x 10^9/l (see *Table 9.2)*. Unlike erythrocytes they contain nuclei and other organelles and are mobile. Some have the ability to pass through blood vessels into the tissues, a process called **diapedesis**. This means that leucocytes can locate and move to areas of inflammation. The movement, known as **chemotaxis**, is a response to chemicals released at the site of inflammation. Bacterial or viral infections cause the levels of certain leucocytes to increase rapidly in the blood. This rise above normal is called *leucocytosis*. Inflammation caused by tissue damage has a similar effect.

The main types of leucocytes are:

- **Polymorphonuclear cells (polymorphs)**, or **granulocytes**, which have a many-lobed nucleus and granules in their cytoplasm, e.g. **neutrophils, basophils** and **eosinophils**.
- **Agranulocytes**, which have no granules, e.g. **monocytes** and **lymphocytes**.

The nomenclature is somewhat complex, especially where the same cell gets renamed when it moves into the tissues. The functioning of different leucocytes is interrelated.

Polymorphonuclear white cells (granulocytes)

Development in the marrow takes a few days. The mature polymorph is approximately 12 μm in diameter. Following release they have a lifespan of a few hours to several days.

Neutrophils

Neutrophils are the most abundant of the leucocytes and, as their name suggests, they take up neutral dye in the laboratory to stain violet. Some of their granules store antibacterial lytic enzymes. Neutrophils are **phagocytic** (*Figure 9.9*) and their main function is to engulf and destroy foreign particles, e.g. bacteria. They are attracted to inflamed areas by chemicals and, after engulfing the bacteria, destroy them with enzymes. The effectiveness of phagocytosis is enhanced by the presence of immunoglobulin (IgG) (see also Chapters 1 and 19). Pus resulting from bacterial infection is a mixture of dead neutrophils, debris, fluid and bacteria.

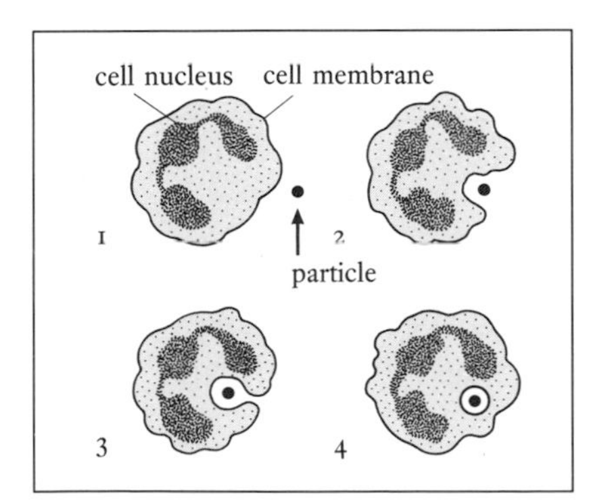

Figure 9.9. Phagocytosis.

Basophils

Basophils take up basic dye in the laboratory to stain blue/black, and are the least abundant of the leucocytes. Their granules contain **histamine** and **heparin**. Histamine causes smooth muscle contraction, vasodilation and increased vessel permeability, all of which attract leucocytes to the area. Heparin is an anticoagulant but its role in inflammation is unclear. Basophils which migrate to the tissues are known as **mast cells**. They are found around small blood vessels where they bind to antibodies (immunoglobulin [IgE]) prior to releasing chemicals involved in anaphylactic reactions and inflammation (see Chapter 19).

Eosinophils

Eosinophils are mobile, phagocytic cells which take up acid dye in the laboratory to stain red. They are present in the blood, respiratory and intestinal mucosa, and skin. Eosinophils are important in immune processes involving immunoglobulin (IgE) and allergic reactions. They reduce the effects of some inflammatory chemicals, e.g bradykinin, and destroy foreign proteins by phagocytosis. Blood levels increase in parasitic conditions and allergies such as asthma. Glucocorticoid hormones which suppress inflammatory and allergic responses, reduce eosinophil numbers (see Chapter 8).

Agranulocytes

Monocytes

Monocytes are large phagocytic cells up to 18 μm in diameter. They develop in the marrow over 2–3 days and, when released into the blood, have a large kidney-shaped nucleus. Monocytes move into the tissues where they differentiate into **macrophages** (also called **histiocytes**) which may remain active for months. Macrophages ('big eaters') live up to their name as the phagocytic cells of the liver, spleen, lymph nodes and marrow, which form the **mononuclear phagocytic** (*reticuloendothelial*) system concerned with body defences (see Chapters 1 and 19). Macrophages have an important

Table 9.2 Leucocyte count and differential

Measurement	***Value***
Total leucocyte (white blood cell) count	4-11 x 10^9/l
Polymorphonuclear cells (granulocytes):	
neutrophils	2.5-7.5 x 10^9/l (40-75%)
basophils	0.01-0.1 x 10^9/l (0-1%)
eosinophils	0.04-0.4 x 10^9/l (1-6%)
Agranulocytes:	
monocytes	0.2-0.8 x 10^9/l (2-10%)
lymphocytes	1.0-3.5 x 10^9/l (20-45%)

defensive role, are responsible for phagocytosis and, in the immune response, are required for lymphocyte function.

Lymphocytes

Lymphocytes have a large spherical nucleus. There are two basic types, large (10–14 μm) and small (5–10 μm). Their development in the marrow takes only 1 or 2 days. Immature lymphocytes released into the blood migrate to lymphoid tissue and the thymus for further development (see Chapter 19). This produces two distinct functional groups of lymphocytes: **T lymphocytes**, which are involved in *cell-mediated immune responses*, e.g. rejection of transplants, and **B lymphocytes**, which are part of *humoral immunity*, developing into **plasma cells** which produce antibodies (immunoglobins). During their lifespan, which may be days or years, lymphocytes spend most of their time in the lymphoid tissues, moving from lymphoid tissues to blood via lymphatic vessels (see Chapters 11 and 19).

Summary – leucocyte functions

- Phagocytosis (polymorphonuclear cells and monocytes).
- Involvement in anaphylaxis and inflammation (basophils).
- Reduce effects of some inflammatory chemicals (eosinophils).
- Processing of antigens (monocytes).
- Humoral immunity through antibodies (B lymphocytes).
- Cell-mediated immunity (T lymphocytes).
  (See also Chapter 19.)

Leucopoiesis – leucocyte formation

Leucocytes

In common with erythrocytes and thrombocytes, the leucocytes develop in the marrow from uncommitted haemocytoblasts. **Myeloid stem cells** give rise to the polymorph series (through **myeloblast**, **promyelocyte** and **myelocytes**) and monocytes (through **monoblast** and **promonocyte**) and **lymphoid stem cells** develop (through **lymphoblast** and **prolymphocyte**) into lymphocytes (see *Figure 9.10*). The stimuli of **leucopoiesis** are not as well understood as those of erythropoiesis, but certain hormonal agents are known to be involved. These are produced by a variety of cells, including leucocytes, possibly as part of immune responses.

Problems with leucocytes

Leucocytosis

Leucocytosis is a physiological increase in leucocytes in response to infection. It can, however, be abnormal; for example, in *infectious mononucleosis* (glandular fever) or *leukaemia* where atypical and immature cells are produced.

Leukaemia

A malignant condition of leucopoietic tissue, leukaemia is characterized by proliferation of immature and abnormal white cells. It may be acute, usually affecting children, or chronic, where older adults are affected. Both types are named for the white cell or precursor involved, e.g. acute lymphoblastic leukaemia (ALL), where lymphoblasts proliferate and chronic granulocytic leukaemia, where

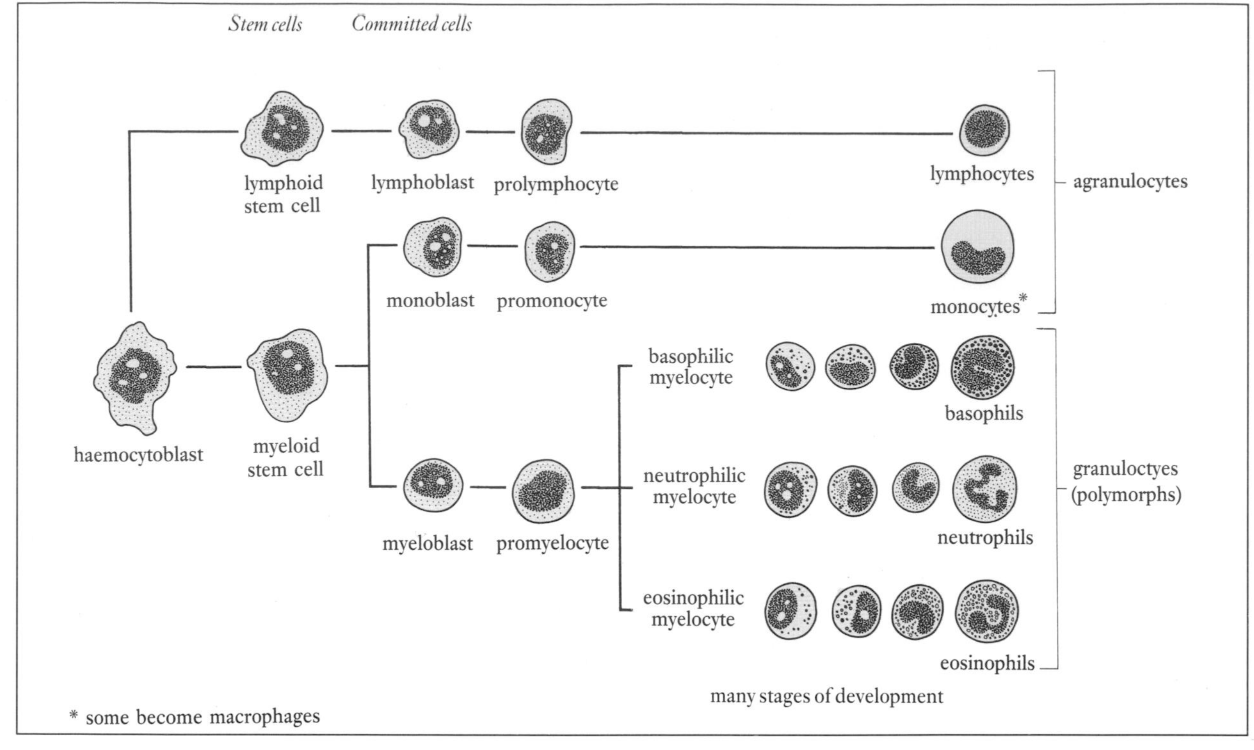

Figure 9.10. Formation of leucocytes.

granulocytes are predominant. Overproduction of leucocytes disrupts other haemopoiesis resulting in anaemia and a bleeding tendency. The huge numbers of abnormal leucocytes 'crowd out' the erythrocytes and platelets. Although leucocyte numbers are often very high (200 x 10^9/l) the affected person is vulnerable to infection as immature white cells do not function correctly.

Leukaemia is diagnosed by examination of the blood and marrow (if the product is faulty it makes sense to check production). This is obtained from a suitable site, e.g. ilium or sternum, by *bone marrow puncture.*

Leukaemia management depends on type and stage, but includes chemotherapy with cytotoxic drugs, radiotherapy, bone marrow transplant, protection from infection, antibiotics, blood and platelet transfusion and drugs to reduce uric acid levels (which rise as cells are destroyed by cytotoxic drugs).

It is now possible to produce a remission of several years in some types of leukaemia but for others treatment is palliative (i.e. alleviates the symptoms but does not cure the disease). The prognosis in children with certain types of leukaemia is often very good, reflecting the successful development of effective chemotherapy.

Leucopenia

Leucopenia is the term given to a reduction in leucocyte numbers which is associated with increased susceptibility to infection. Causes include radiation, cytotoxic and other drugs, chemicals or general marrow suppression *(pancytopenia)* (see aplastic anaemia, page 185).

Nursing Practice Application Leukaemia

Nursing interventions should be planned to minimize the effects of disordered haemopoiesis, infection risk, anaemia and bleeding, e.g. possible isolation and observation for infection, care planned to prevent fatigue and oral or intravenous medication to avoid intramuscular injections.

The aggressive treatment of leukaemia can cause distressing side-effects, e.g. vomiting and alopecia (hair loss); nurses should ensure that this is fully explained and, where possible, satisfactory solutions found for the problems. The person, and his or her family and friends will need support and information at all stages.

PLATELETS (THROMBOCYTES)

Platelets are cellular fragments vital in **haemostasis**. They are formed in the bone marrow from the large multinucleate cells known as **megakaryocytes** which develop from the **megakaryoblast** (see *Figure 9.2*). Non-nucleated disc-shaped platelets break from the cytoplasm of the megakaryocytes and are released into the blood (*Figure 9.11*).

Platelets measure 2-4 μm in diameter. Their lifespan is about 10 days, after which they are destroyed by macrophages in the spleen and liver. The platelet count in adults is normally 150-400 x 10^9/l. This range alters very little in health which suggests a homeostatic mechanism controlling **thrombopoiesis** (production of thrombocytes). Production is probably controlled by platelet numbers and the hormone **thrombopoietin**.

Platelets contain a contractile protein in microtubules and microfilaments which cause clot retraction. The platelet membrane and cytoplasmic granules contain substances, e.g. enzymes, inflammatory chemicals and *phospholipids* involved in platelet aggregation and clotting.

The process of haemostasis depends on platelet aggregation, plug formation and release of chemicals required during clotting.

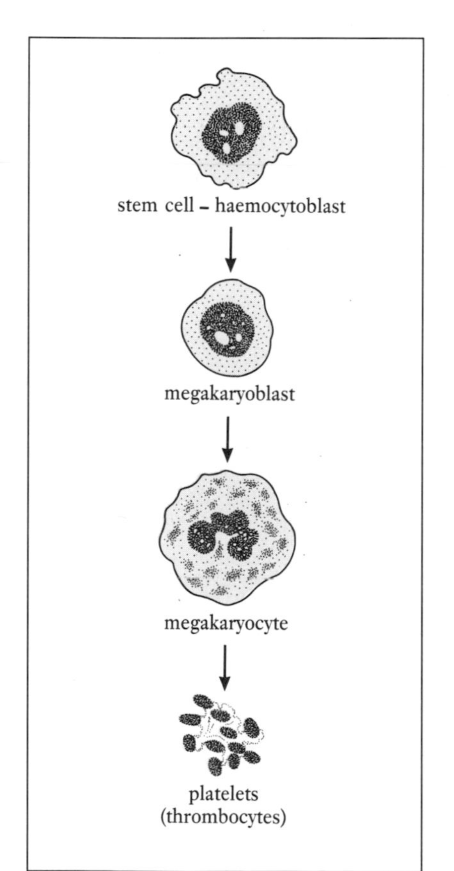

Figure 9.11. Megakaryocytes and the formation of platelets.

HAEMOSTASIS

Haemostasis is the physiological process by which bleeding from small blood vessels is controlled. If the endothelial lining of a blood vessel is damaged, a complex series of events is initiated by substances found in the blood and others that are released by tissue damage and platelets. Haemostasis involves four overlapping processes: **vasoconstriction, platelet plug formation, clotting (coagulation)** and **fibrinolysis**.

Vasoconstriction (vasospasm)

A few seconds after an injury the blood vessel constricts in response to chemicals released by platelets, e.g. 5-hydroxytryptamine (5-HT) and the prostaglandin derivative **thromboxane A_2**. Vasoconstriction, which can last for 20–30 minutes, limits blood flow and loss while a platelet plug forms and clotting occurs.

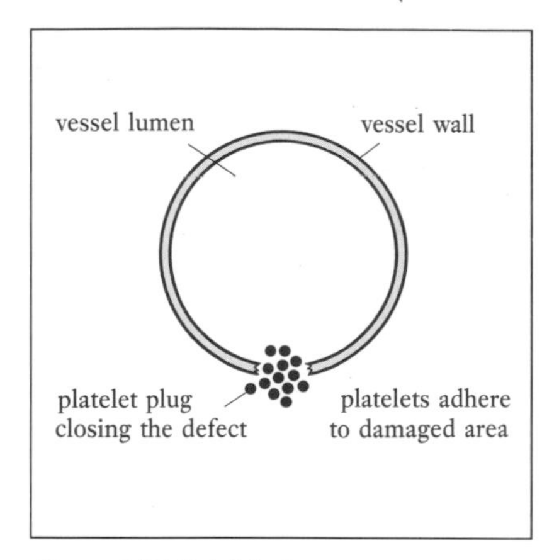

Figure 9.12. Platelet plug.

Platelet plug formation

Platelets do not normally adhere to undamaged blood vessels. Substances such as **collagen,** exposed when damage occurs, cause platelets to adhere at the site. When platelets adhere to the vessel wall they cause the release of **adenosine diphosphate (ADP)** from the vessel wall itself, from granules in the platelets and from other blood cells. Platelet aggregation is increased by this release of ADP. The presence of thromboxane A_2 and 5-HT, which are released by platelet degranulation (break up of granules with the release of chemicals), further increases aggregation and vasoconstriction. Platelets attracted to the area form a plug which closes the vessel defect on a temporary basis (*Figure 9.12*).

Degranulation also produces factors which initiate **thrombin** formation (thrombin also encourages platelet aggregation) and clotting.

Another prostaglandin derivative, **prostacylin**, produced by cells in the vessel, inhibits platelet aggregation and causes vasodilation. Opposing thromboxane A_2, it is probably important in preventing intravascular clotting (clotting inside a blood vessel).

Aspirin and other substances known to inhibit the synthesis of prostaglandins are prescribed to reduce platelet adhesion and **thrombus** (clot) formation after myocardial infarction. There is, however, a risk of bleeding, especially from the gastrointestinal tract. Drugs such as dipyridamole, which augment the action of prostacyclin and inhibit thromboxane synthesis, can also be used to minimize the risk of thrombus formation in a variety of situations, e.g. following the insertion of prosthetic heart valves.

Clotting/coagulation (fibrin clot formation)

The coagulation of blood is an extremely complex process. It relies on **enzyme cascade amplification** (where the product of one reaction triggers the next reaction) to produce a **fibrin** clot. Information regarding the twelve clotting/coagulation factors is provided in *Table 9.3.*

There are two coagulation pathways (see flow chart, page 194): (i) **intrinsic**, which is initiated by the exposure of **factor XII** and platelets to collagen and platelet breakdown; and (ii) **extrinsic**, which depends on tissue damage and **thromboplastin** release. Both pathways are required for normal clotting and in the body they are usually both activated by the event initiating clotting. The extrinsic pathway has fewer stages and is quicker, but both mechanisms converge to follow a final common pathway (*Figure 9.13*).

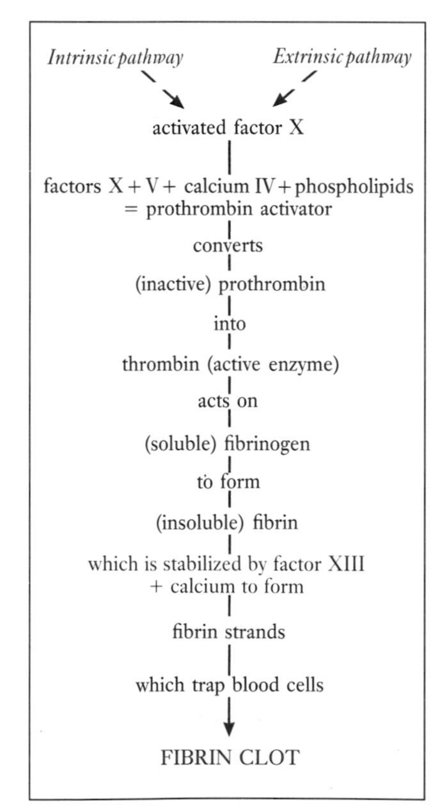

Figure 9.13. Final common pathway of coagulation (X → fibrin clot).

Table 9.3. Coagulation factors.

Number	Name	Origin	Action
I	Fibrinogen	Plasma protein made by liver	Converted to fibrin
II	Prothrombin*	Plasma protein made by liver	Converted to thrombin
III	Thromboplastin	Released by damaged tissues	Joins with other factors to activate X
IV	Calcium ions	Obtained from diet or released from bone	Required for most stages of clotting
V	Labile factor (proaccelerin)	Plasma protein made by liver	Required by intrinsic and extrinsic pathways to form prothrombin activator
VI	No longer in use		
VII	Stable factor* (proconvertin)	Plasma protein made by liver	Extrinsic pathway
VIII	Antihaemophilic globulin (AHG), Antihaemophilic factor A	Globulin made by liver	Intrinsic pathway Lack causes *haemophilia A*
IX	Christmas factor* (Antihaemphilic factor B), Plasma thromboplastin component	Plasma protein made by liver	Intrinsic pathway Lack causes *haemophilia B* (*Christmas disease*)
X	Stuart-Power* factor	Plasma protein made by liver	Vital for both pathways
XI	Plasma thromboplastin antecedent (PTA)	Plasma protein made by liver	Intrinsic pathway
XII	Hageman factor	Made by liver	Intrinsic pathway. Activation of other enzyme systems in the plasma
XIII	Fibrin stabilizing factor (FSF)	Made by liver	Forms insoluble fibrin

NB Although numbered the factors do not react in this sequence.
* needs Vitamin K.

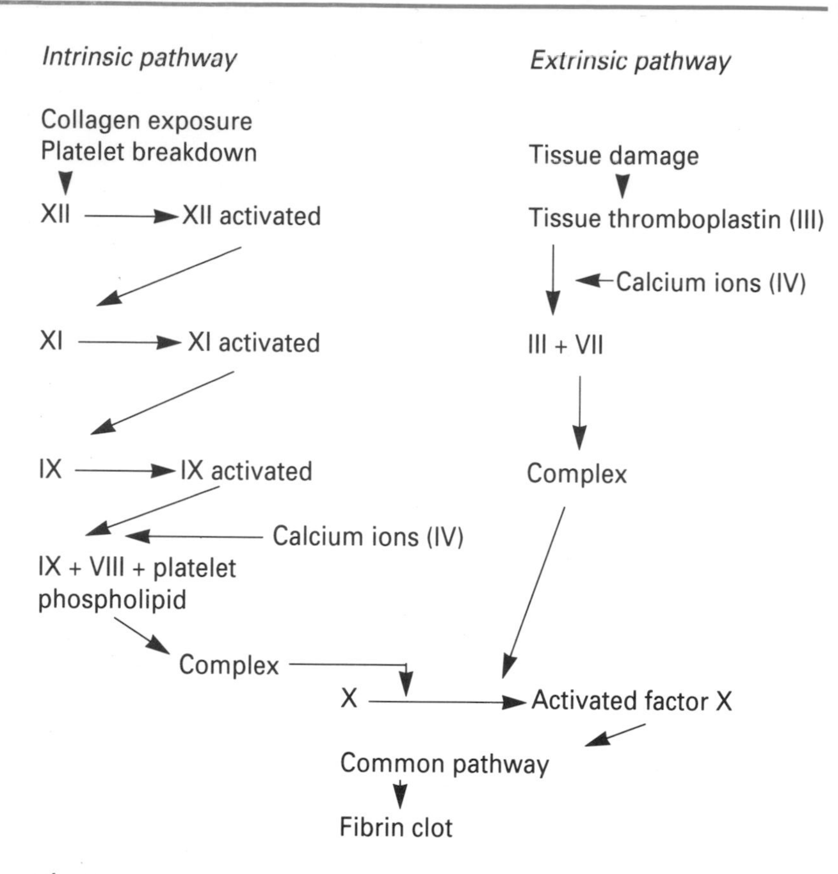

Detailed discussion of these separate mechanisms is not included and readers requiring more information are directed to Further Reading, e.g. Thompson and Cotton (1983).

The stages of both pathways culminate in the activation of factor X. **Prothrombin activator** is formed when factor X complexes with factor V, platelet phospholidids and calcium. In the next step prothrombin activator converts inactive **prothrombin** into the active enzyme **thrombin**. This proteolytic enzyme (one that breaks down protein) changes **fibrinogen**, the soluble plasma protein, into a network of **fibrin** strands. Platelets and other blood cells become trapped in the mesh to form a fibrin clot which is stabilized by **factor XIII**. The stages of clotting normally take 3–8 minutes to complete.

In health, blood clots only when a blood vessel is damaged. Anticoagulant substances normally present in the blood, called **antithrombins**, limit the action of clotting factors to areas where coagulation is needed.

Fibrinolysis

Discussion of haemostasis would be incomplete without details of how the body deals with clots after healing. A new clot is rather soft and sticky but within an hour it has become smaller and firmer. This is **clot retraction**, a process involving the contractile protein **thrombosthenin** in platelets. The edges of the damaged area are held more closely together promoting healing.

When healing is complete the proteolytic enzyme **plasmin** breaks down the fibrin clot and phagocytic cells remove the debris. Plasmin is derived from the activation of **plasminogen**, a plasma protein made by the liver. Sources of plasminogen activators include clotting factors, blood cells at the clot site, and tissues, e.g. lung, kidney and endothelium. In this way the homeostatic balance between clotting and vessel patency (vessel remains open) is maintained. During menstruation the uterine lining produces anticoagulants which prevent the coagulation of blood within the uterus (see Chapter 20).

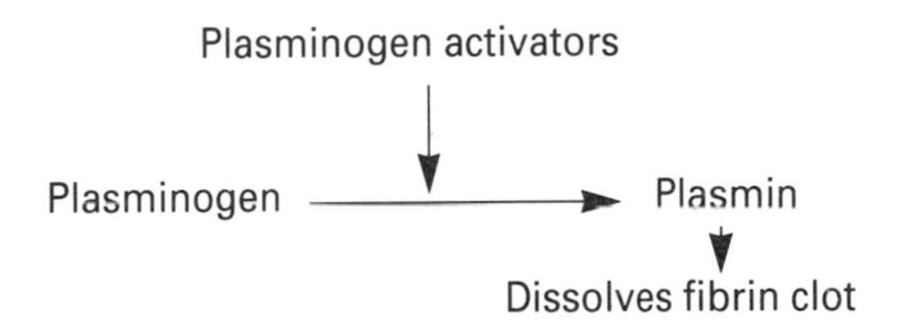

Problems with haemostasis

Thrombocytopenia

Thrombocytopenia is a reduction in the number of platelets in the blood. Causes include marrow depression (due to drugs, chemicals, radiation, cancer or infection), leukaemia and hypersplenism. Many cases, however, are *idiopathic* (cause unknown). Affected people bleed easily; they bruise and develop small purple spots on the skin (*petechiae*). Platelet transfusions may be required.

Reduced clotting factors

Production of clotting factors depends largely upon healthy liver function and the absorption of vitamin K. Severe liver malfunction or malabsorption of vitamin K will both cause clotting factor deficiency. Bile is needed for vitamin K absorption by the gut (see Chapters 13 and 14).

Haemophilias

The two bleeding disorders, *haemophilia A* (lack of factor VIII) and *Christmas disease* (lack of factor IX), are inherited as sex-linked (X-chromosome) recessive genes (*Figure 9.14*). The inheritance of this type of sex-linked recessive condition is also discussed on page 134 in relation to red-green colour blindness.

Asymptomatic females carry the gene which they transmit to some of their offspring; theoretically 50% of males will have haemophilia and 50% of females will be carriers. All the daughters of an affected male will be carriers but his sons will be unaffected. Genetic counselling and antenatal diagnosis are available for affected families.

NB It is possible for a female to have haemophilia – if a female carrier and a male haemophiliac have a female child she could inherit the defective gene from both parents. For further details of haemophilia inheritance, readers are directed to the genetics text in Further Reading (Emery and Mueller, 1988).

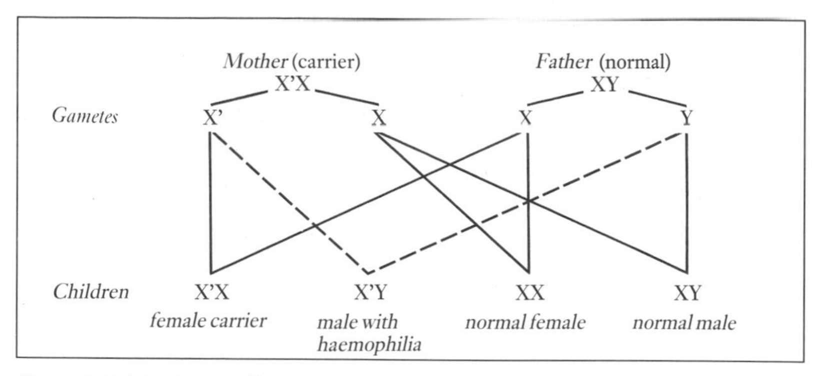

Figure 9.14. Inheritance of haemophilia.

The haemophilias cause bleeding of variable severity and are managed with intravenous injections of the missing coagulation factor, blood transfusion and lifestyle modification, e.g. avoiding hazardous activities. In the UK preparations of factors VIII and IX are now heat-treated to destroy harmful agents, e.g. HIV. Unfortunately in the early 1980s many people were supplied with infected blood products and have become HIV-positive and since developed AIDS (see Chapter 19).

Disseminated intravascular coagulation (DIC)

Disseminated intravascular coagulation (DIC) is a complication of various other conditions, e.g. intrauterine death and severe trauma. Widespread clotting occurs in small vessels with consequent depletion of fibrinogen and platelets. This leads to bleeding, which is controlled by transfusion of fibrinogen, platelets and fresh blood in conjunction with heparin to inhibit further clotting.

Thromboembolic conditions – due to intravascular clotting

The body normally controls clotting by limiting it to areas of vessel damage and fibrinolysis. Inappropriate coagulation leads to *thrombus* (plural, thrombi) formation. If the thrombus detaches from the vessel to travel in the circulation it is known as an *embolus* (plural, emboli).

Problems arise when a thrombus or embolus is large enough to occlude (block) a blood vessel. Tissue deprived of blood cannot function and eventually dies. The area of devitalized tissue is called an *infarction*. The majority of adult deaths in developed countries are caused by thromboembolic events affecting the brain (cerebral thrombosis/ embolism – stroke), heart (coronary thrombosis) or lungs (pulmonary embolism).

The conditions that predispose to inappropriate coagulation include:

- Blood stasis, e.g. associated with immobility or heart failure. It may cause *deep vein thrombosis (DVT)* with clots in leg and pelvic veins. Prevention is achieved through early ambulation, leg exercises and deep breathing, prophylactic heparin and elastic compression stockings which must be selected and fitted with

Nursing Practice Application Oral anticoagulant therapy

It is important that people taking anticoagulants are monitored carefully for bleeding and have regular prothrombin time blood tests. Many drugs, including aspirin, should not be taken unless specifically prescribed. Preparation for discharge from hospital should include education about the drug and its effects, especially bleeding, and the need to carry an anticoagulant card.

care. In a study of 15 types of below-the-knee compression hosiery only five were found to be effective, i.e. had significant linear trend graduated compression (Cornwall *et al.* 1987). Nurses can improve compliance by educating people about the purpose of compression hosiery (Dale and Gibson 1992). The danger with DVT is a clot breaking off, which then travels in the circulation through the heart (right side), to the lungs. Here it causes pulmonary infarction or death if a major vessel is blocked.

- Vessel wall disease and/or damage, e.g. *atheroma* (see Chapter 10), trauma and infection.
- Changes in the blood, e.g. polycythaemia vera (see page 185) and excess platelets which increase coagulation.

Drugs used in the prophylaxis and treatment of thromboembolic conditions include anticoagulant drugs (*Table 9.4*) and plasminogen activators (fibrinolytic drugs). Plasminogen activators, such as *alteplase, urokinase* (produced by kidney) and the bacterial *streptokinase,* increase fibrinolysis and restore vessel patency. The plasminogen activators are effective if commenced within an hour of a pulmonary embolism or within a few (3–12) hours for myocardial infarction.

Table 9.4. Anticoagulants.

Drug	***Action***	***Antidote***
Heparin sulphate	Inhibits thrombin, active at once	Protamine
Warfarin (coumarins)	Vitamin K inhibition effective after 36–48 hours	Vitamin K
Ancrod formation	Reduces fibrin Antivenom	(derived from pit-viper venom)

Blood Groups

The surface membrane of erythrocytes and other blood cells contain glycoprotein antigens which determine blood groups. Individuals inherit many different antigens but for most purposes only ABO and rhesus groups are clinically significant.

ABO groups

Antigen *– a molecule or part of molecule which is recognised by the body defences and stimulates an immune response.*

Antibody *– a protein which binds to a specific antigen.*

The presence of two **antigens** (**A** and **B**), the basis of **ABO groups**, was discovered by Landsteiner in 1900. Blood group depends on whether these antigens are present or not. The presence of A antigen denotes group A, whereas group B has B antigen, group O has neither antigen and group AB has both antigens (*Table 9.5*). Prior to Landsteiner's work, blood transfusion carried a high risk of fatal incompatibility. Safety improved to some extent, but transfusion reactions still occurred until **rhesus groups** were identified some forty years later.

Table 9.5. Antigens and antibodies.

Group	*Erythrocyte (antigen)*	*Plasma (antibody)*
A	A	Anti-B
B	B	Anti-A
O	None	Anti-A and anti-B
AB	A and B	None

We inherit one blood group gene from each parent (*Table 9.6*). The genes for antigens A and B are co-dominant over the gene for O. The inheritance of two A genes results in group A and two B genes in group B. If the A or B gene is inherited with an O gene the blood group will still be A or B. This occurs because A and B, being co-dominant will determine the blood group in both homozygous (A+A, B+B) and heterozygous states (A+O, B+O). If a person inherits an A and B gene they will be group AB. The commonest group, O, will only result if an individual inherits two O genes – they must be homozygous for the O gene (O+O).

Table 9.6. Inheritance of ABO groups and distribution in Caucasian population in United Kingdom.

Genes inherited	*Genotype*	*Group (phenotype)*	*%*
O+O	OO	O	46
A+A or A+O	AA, AO	A	42
B+B or B+O	BB, BO	B	9

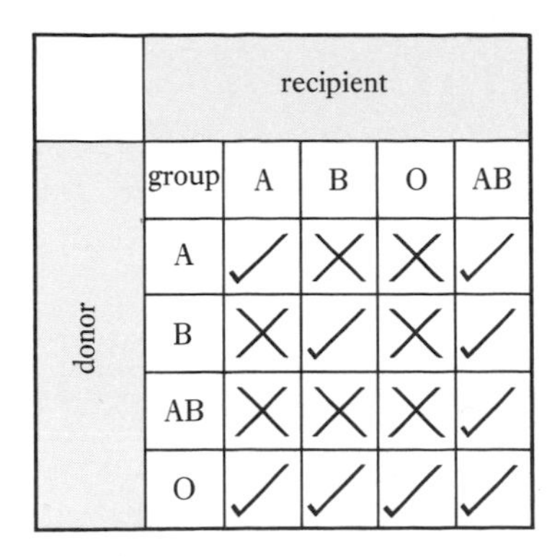

	recipient			
donor group	A	B	O	AB
A	✓	✗	✗	✓
B	✗	✓	✗	✓
AB	✗	✗	✗	✓
O	✓	✓	✓	✓

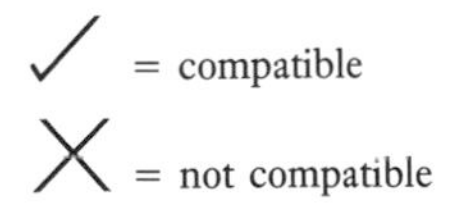

Figure 9.15. ABO compatibility for transfusion.

ABO antigens are present on the erythrocyte at birth. They are also known as **agglutinogens** because they cause erythrocyte **agglutination** (clumping together) if exposed to the specific **antibody**. The antibody or **agglutinin** is not present at birth but is produced during the first months (*Table 9.5*). An antibody that acts against the antigen not present on the individual's erythrocyte (it would be crazy to have an antibody against your own erythrocyte) forms in the plasma. This production of preformed antibodies is unusual in that it occurs without exposure to the appropriate antigen. Considerable debate continues but antibody production is probably stimulated by micro-organisms in the gut.

Before a transfusion of blood it is vital to know that the donor blood is compatible with that of the recipient. In the vast majority of cases (except in an extreme emergency) a blood sample is taken from the recipient for *group* and *cross-match*. This involves ascertaining the recipient's group (ABO/rhesus) and ensuring compatibility by mixing donor erythrocytes with recipient serum to check for agglutination (direct cross-match). Obviously only blood that shows no agglutination is transfused (see *Figure 9.15*).

Generally, group O, having no antigens, can be given to any group, and group AB, with no antibodies, can receive blood from any group, providing that other groups, e.g. rhesus, are compatible. Individuals from groups A and B can receive their own group and O. The terms 'universal donor and recipient', which only apply to ABO groups, have little relevance and should be discarded.

In most transfusions it is not necessary to consider reactions between recipient erythrocyte antigens and donor antibodies because they are sufficiently diluted by the recipient's blood.

If ABO incompatible blood is transfused, antibody in the recipient's plasma will agglutinate the donated erythrocytes with disastrous results. Red cells clump together and block small blood vessels. Later the clumps undergo haemolysis and haemoglobin is released. Organs such as the kidney are damaged by the haemoglobin resulting in renal failure (see *Special Focus* – transfusion of blood and blood components, page 201).

Rhesus group

The other important antigen on red cells is the **rhesus factor (Rh)**, identified in 1940 by Landsteiner and Weiner. There are several genes for **Rh antigens** which include those designated **CDE/cde**. CDE are dominant over cde. However, only D/d is really important for clinical purposes. The presence of D antigen (genes DD or Dd) makes the individual **Rh-positive** (85% in UK are positive) and those without the antigen (genes dd) are **Rh-negative**. The inheritance of rhesus factor is shown in *Figure 9.16*.

In contrast to the ABO system there are no preformed antibodies (agglutinins) to D. Formation of **anti-D** requires the exposure of a Rh-negative person to Rh-positive blood. This will occur if Rh-positive blood is transfused to an Rh-negative person or an Rh-negative woman is pregnant with an Rh-positive fetus. Antibodies are formed

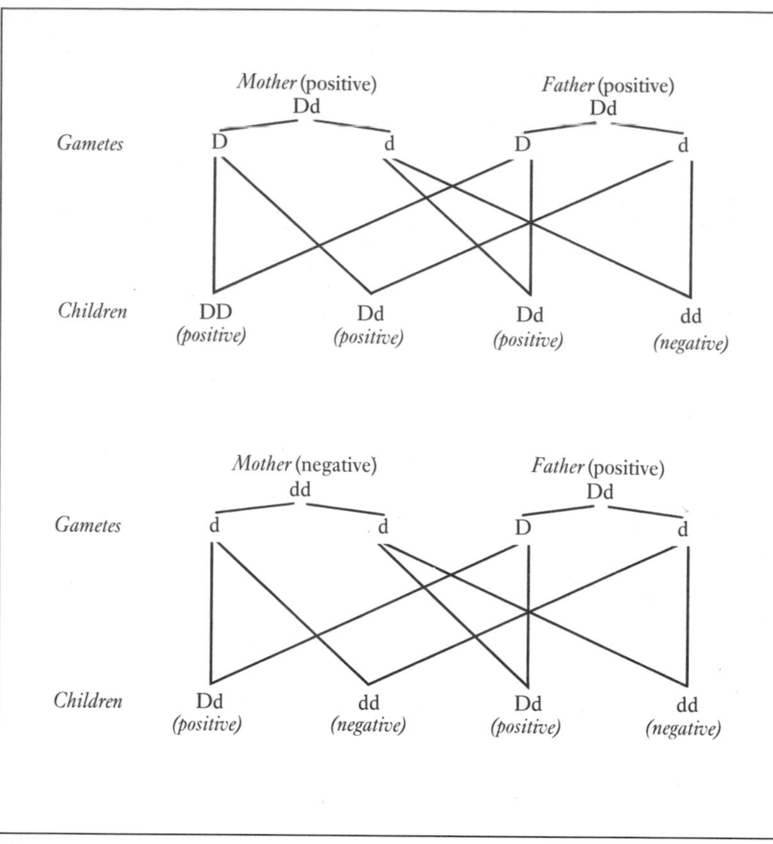

Figure 9.16. Inheritance of rhesus factor. Dd, positive (heterozygous); DD, positive (homozygous); dd, negative.

after the first exposure and these cause reactions at second or subsequent exposures (see rhesus incompatibility, below).

We now have eight groups, as each ABO group may be rhesus-positive or negative (A Rh-positive, A Rh-negative, B Rh-positive, B Rh-negative, O Rh-positive, O Rh-negative, AB Rh-positive and AB Rh-negative). Tests for Rh-compatibility prior to transfusion are similar to those for ABO groups. Rh-positive individuals may receive Rh-positive or Rh-negative (remember no preformed antibodies) but Rh-negative people must only receive Rh-negative blood.

Rhesus incompatibility – haemolytic disease of the newborn

Problems occur when fetal Rh-positive erythrocytes enter the maternal (Rh-negative) circulation via the placenta. This occurs with a full term pregnancy, during childbirth or when a pregnancy ends in spontaneous miscarriage or termination. The woman makes anti-D in response to the Rh-positive cells. Next time she is pregnant with an Rh-positive fetus anti-D crosses the placenta to cause haemolysis of fetal blood. The fetus will be anaemic, hypoxic and jaundiced (bilirubin release from the erythrocytes). If severe this will lead to brain damage or death (the baby will be stillborn or will die soon after birth). This is haemolytic disease of the newborn or *erythroblastosis fetalis.* Intrauterine transfusion and exchange transfusions after birth are used for affected babies.

Special Focus **Transfusion of blood and blood components**

Blood transfusion is a safe procedure, but potential hazards must always be anticipated and prevented where possible.

Transfusion of blood or blood products is indicated in the following situations:

Haemorrhage: whole blood replacing volume and cells.

Anaemia: plasma-reduced blood (packed cells) replacing cells rather than volume.

Thrombocytopenia: platelets to treat bleeding.

Leucopenia: granulocytes to combat infection in marrow suppression.

Plasma: used to replace that lost from major burns. Plasma substitutes, e.g. gelatin and dextran, are used short term to expand the blood volume. They also reduce the risk of plasma transmitted disease.

Plasma components: e.g. factors used in haemophilia (Factor VIII), fibrinogen and albumin.

Blood transfusion safety *(see page 177 for risks to health care workers)*

Precautions: (i) donors screened for disease, e.g. HIV, and asked not to donate; (ii) sterile, single-use equipment is used; (iii) blood anticoagulated and stored at 4°C for a maximum of 35 days (whole blood); (iv) group and cross-match tests; (v) vigorous checking procedures to ensure the person receives the correct blood, which is in good condition and in date; (vi) blood should be removed from refrigeration 30 minutes prior to use and allowed to reach room temperature. When large volumes are to be transfused quickly it can be warmed in a special blood warmer. A transfusion of cold blood could reduce body temperature and cause cardiac problems; (vii) complete records kept of blood used, blood bags retained in case of reactions; (viii) giving sets changed at regular intervals to minimize infection risk – at least every 24 hours and on completion of transfusion; (ix) filters used where appropriate to prevent micro-aggregates being transfused.

Prophylaxis and anti-D gamma globulin

Anti-D immunoglobulin (an intramuscular injection) given to Rh-negative women within 72 hours of delivery, miscarriage or termination destroys any fetal erythrocytes in the circulation before her immune system is stimulated to produce anti-D.

At risk Rh-negative women of child bearing age require education regarding Rh-incompatibility and the need for anti-D immunoglobulin after each pregnancy. Nurses should be aware of at-risk women in their care and ensure that they receive the correct information and treatment (see Chapter 19).

Selected transfusion complications/problems

ABO incompatibility

ABO incompatibility causes an extremely serious reaction produced by the agglutination and haemolysis of mismatched erythrocytes (see page 199). Soon after incompatible blood is transfused the following occur: loin pain, breathing difficulties, discomfort at the infusion site, rapid pulse, hypotension, pyrexia (fever) and rigors, haemoglobin in the urine, reduced urine output and, possibly, renal failure. Other manifestations may include rashes and diarrhoea.

Allergic/pyrogenic reaction

Allergic reactions occurring in response to donated blood may cause effects ranging from minor *urticarial* rashes and pyrexia to laryngeal swelling with airway obstruction (see Chapter 12). *Pyrogens* (which increase temperature) present in the blood may cause a rise in temperature.

Circulatory overload

Circulatory overload may occur if whole blood is used to treat anaemia, especially in the elderly or those with cardiac problems. It

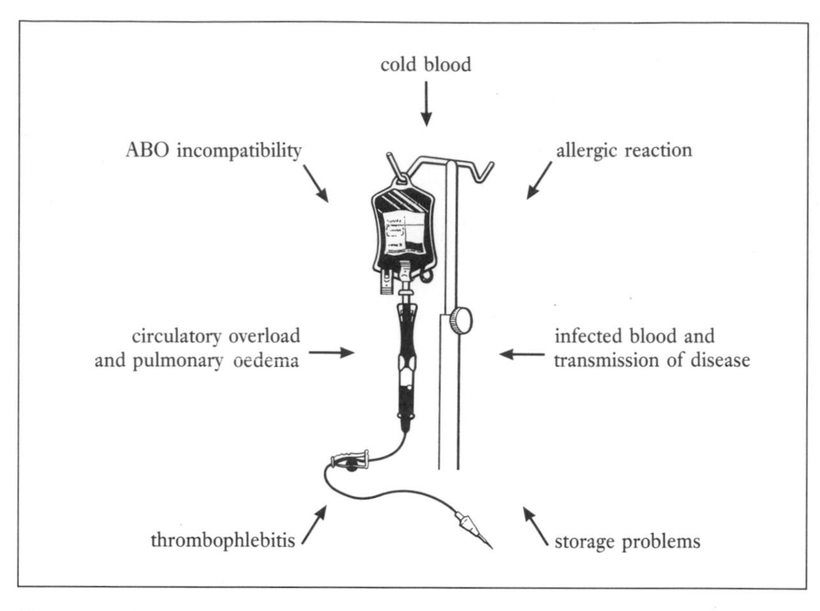

Figure 9.17. Complications of blood transfusion.

> *Nursing Practice Application*
> **Observations and blood transfusion**
> *(see Figure 9.17)*
>
> Baseline vital signs should be available for comparison during transfusion. Nurses should monitor the patient's temperature, pulse, blood pressure, colour, respiration and general condition at regular intervals, at least hourly, during transfusion. These observations should be more frequent (every 15 minutes) during the early stages when reactions are common. It is important to record fluid balance, test urine for blood and observe for rashes, rigors, pain (chest, loin or transfusion site), inflammation at the site, jaundice, signs of bleeding and any other abnormalities. Early detection and prompt treatment could save life.

causes *pulmonary oedema* (fluid in the lung) with rapid respiration, frothy sputum, chest pain and tachycardia.

Infected blood and transmission of specific diseases
(see Special Focus, Transfusion of blood and blood components).
Bacteria may contaminate the blood or equipment. This results in *septicaemia* with pyrexia, rigors and shock (hypotension), which may be fatal. Specific infections such as HIV and hepatitis B have been transmitted via infected blood.

Thrombophlebitis
Inflammation of vein with thrombus formation. Phlebitis is the most common complication of intravenous infusion/blood transfusion (Francombe, 1988).

Cold blood
Cold blood may cause cardiac problems and may lower temperature.

Storage problems
Platelets, leucocytes and clotting factors do not store well. People having large transfusions may have problems with haemostasis.

NB A fresh transfusion is needed if leucocytes are required. During storage, potassium leaks from the erythrocytes, which may cause hyperkalaemia (raised levels of potassium in the blood) and a risk of cardiac arrhythmias (abnormal heart rhythm for which cardiac monitoring may be required). Large transfusions may produce hypocalcaemia and tetany (see Chapter 8 and *Nursing Practice Application* for discussion of nursing observations during blood transfusion).

Summary/Check List

Introduction
Composition and functions of blood – composition, functions, *Nursing Practice Application* - safe handling of blood.
Plasma – Contents – proteins, ions, gases, hormones, enzymes, nutrients, waste.
Blood cells – haemopoiesis.
Erythrocytes – structure, erythropoiesis, dietary requirements, iron metabolism, acute iron poisoning, haemoglobin, destruction of erythrocytes, disordered erythrocyte function – anaemia, *Person-centred Study* – Kate, polycythaemia, haemoglobinopathies – sickle-cell disease and thalassaemia.
Leucocytes – Polymorphonuclear cells (granulocytes) – neutrophils, basophils, eosinophils. Agranulocytes – monocytes, lymphocytes. Leucopoiesis. Problems with leucocytes – leucocytosis, leukaemia, *Nursing Practice Application* – leukaemia, leucopenia.
Platelets – thrombopoiesis, structure, function.
Haemostasis – vasoconstriction, platelet plug formation, clotting, fibrinolysis. Problems with haemostasis – thrombocytopenia, reduced clotting factors, haemophilias, DIC, thromboembolic conditions, *Nursing Practice Application* – anticoagulants.
Blood groups – ABO – agglutinogens, inheritance and distribution, agglutinins, cross-match/compatibility. Rhesus group – inheritance, rhesus incompatibility, prophylaxis and anti-D. *Special Focus* – blood transfusion. Complications of transfusion, *Nursing Practice Application* – observations and blood transfusion.

?

Self Test

1 Which of the following describes plasma?
(a) Red alkaline fluid forming 55% of blood volume.
(b) Straw-coloured neutral fluid forming 45% of blood volume.
(c) Red neutral fluid forming 45% of blood volume.
(d) Straw-coloured alkaline fluid forming 55% of blood volume.
2 Describe how erythrocyte structure is adapted to oxygen carriage.
3 Which of the following statements are true?
(a) Leucocyte is the general name for all white cells.
(b) Neutrophils are phagocytic.
(c) White cells are not produced in the marrow.
(d) The normal white count is 4–11 x 10^9/l.
4 Outline the role of platelets in haemostasis.
5 Put the following in the correct order:
(a) clotting (b) vasoconstriction (c) fibrinolysis (d) platelet plug.
6 Explain how a man (A Rh-positive) and a woman (B Rh-positive) can have children who are group O Rh-positive.
7 What would cause an Rh-negative woman to produce anti-D?
8 Which of the following transfusions are compatible?

Donor	Recipient
(a) A Rh-negative	AB Rh-positive
(b) O Rh-positive	O Rh-negative
(c) B Rh-positive	A Rh-positive
(d) O Rh-positive	B Rh-positive

Answers.
1. d. 2. See page 181.
3. a, b, d. 4. See pages 191-194. 5. b, d, a, c. 6. See pages 198-200. 7. Transfusion of Rh-positive blood, pregnant with Rh-positive fetus. 8. a, d

References

Cornwall, J. V., Dore, C. J. and Lewis, D. (1987) Graduated compression and its relation to venous filling times. *British Medical Journal*, **295**, 1087–1090.

Dale, J.J., Gibson, B. (1992) Information will enhance compliance: Informing clients about compression hosiery. *Professional Nurse*, **7** (11) 755-760.

Francombe, P. (1988) Intravenous filters and phlebitis. *Nursing Times* **84** (26), 34–35.

RCN (1986) *AIDS – Nursing Guidelines*. Second report of the RCN AIDS working party. London: RCN.

Further Reading

Adler, M. W. (Ed.) (1988) *ABC of AIDS*. London: British Medical Journal Publications.

Cluroe, S. (1989) Blood transfusions. *Nursing*, **3** (40), 8–11

Elliot, J. (1988) Nursing care (safety), In Adler, M.W. (Ed.), *ABC of AIDS*. London: British Medical Journal Publications, p.42.

Emery, A.E.H., Mueller, R.F. (1988) *Elements of Medical Genetics*, 7th edn. Edinburgh: Churchill Livingstone.

Jeffries, D.J. (1988) Control of infection policies. In Adler, M.W. (Ed.) *ABC of AIDS*. London: British Medical Journal Publications, pp. 44,46.

Keele, C. A., Neil, E. and Joels, N. (1982; Reprinted with corrections 1984) *Samson Wright's Applied Physiology*. (Part I), 13th edn. Oxford: Oxford University Press.

Love, C. (1990) Deep vein thrombosis – threat to recovery. *Nursing Times*, **86** (5) 40–43.

Love, C. (1990) Deep vein thrombosis – methods of prevention. *Nursing Times*, **86** (6), 52–55.

Morgan, G. (1989) Nursing the patient with leukaemia. *Nursing*, **3** (40), 19–22.

RCN Oncology Nursing Society (1989) *Safe Practice with Cytotoxics*. Harrow: Scutari Press.

Thompson, A. D. and Cotton, R. E. (1983) *Lecture Notes on Pathology*, 3rd edn. Oxford: Blackwell Scientific, pp. 513–535.

CHAPTER TEN

CARDIOVASCULAR SYSTEM: HEART, VESSELS AND CIRCULATION

OVERVIEW

- Structure of the heart.
- Cardiac physiology.
- Blood vessel structure.
- Physiology of the circulation.
- Circulatory pathways.

LEARNING OUTCOMES

After studying Chapter 10 you should be able to:

- Describe the location, size and shape of the heart.
- Describe the layers of the heart, relating structure to function.
- Outline the special properties of cardiac muscle and compare it with skeletal muscle.
- Describe the chambers of the heart.
- Describe heart valves, their position and function.
- Describe the coronary circulation.
- Discuss the multifactorial causation of coronary heart disease and its effects on cardiac function.
- Describe blood flow through the heart.
- Explain the conduction pathway in the heart.
- Explain the normal electrocardiogram and briefly describe some common disturbances of heart rhythm.
- Describe the events of the cardiac cycle.
- Outline the regulation of heart rate and stroke volume.
- Explain how the ANS regulates cardiac function.
- Describe the structure of arteries, veins and capillaries, and explain how each is adapted to its function.
- Explain the terms blood flow, peripheral resistance and blood pressure.
- Outline ways in which the body controls blood vessel size and blood flow.
- Describe the factors which influence and control blood pressure.

Learning Outcomes (continued)

- Outline the physiological changes occurring in shock and describe the different causes.
- Explain capillary bed exchange of water and molecules between blood and the cells.
- Describe the pulmonary circulation and outline the process of gaseous exchange.
- Describe the major arteries and veins of the systemic circulation.
- Describe the hepatic portal circulation and explain its role.

Key Words

Artery – vessel carrying blood away from the heart.
Atria – two thin-walled upper chambers of the heart which receive blood.
Blood pressure – force exerted by the blood on the vessel walls.
Capillaries – network of tiny blood vessels linking arteries with veins. Their walls are only one cell thick, allow exchange of molecules between blood and cells.
Cardiac output – amount of blood pumped out of the left ventricle in one minute.
Cardiovascular system (CVS) – the heart and blood vessels.
Diastole – relaxation phase of cardiac cycle when the heart fills with blood.
Endocardium – smooth layer of endothelium (epithelium) lining the heart and blood vessels.
Mediastinum – the space in the chest, between the lungs, which contains the heart and blood vessels.
Myocardium – specialized cardiac muscle forming the middle layer of the heart.
Pericardium – double serous/fibrous membrane which encloses the heart.
Peripheral resistance – friction or resistance between the blood and the vessel (arteriolar) walls.
Stroke volume – amount of blood pumped out by each ventricular contraction.
Systole – contraction phase of the cardiac cycle when the heart pumps blood into the arteries.
Vein – vessel carrying blood to the heart.
Ventricles – two thick-walled lower chambers of the heart which pump blood out into the circulation.

Introduction

The heart and the circulation of blood within the network of blood vessels is vital in homeostatic control. Cells and tissues constantly need oxygen and nutrients for metabolism and the waste produced must be removed. Homeostasis is maintained by the ability of the **cardiovascular system** (CVS) (the heart and blood vessels) to adapt to changes in physiological conditions and needs. This feature of the CVS is easy to take for granted, but without this response an individual would faint on getting out of bed, would have no chance of running for a bus and would have difficulty digesting a large meal.

HEART

Popular opinion perceives the heart to be the location of emotion, as illustrated by expression such as 'sweetheart', 'heartbreak' and 'heartache'. Not so, the bounding heart before examination or interview is due to hormones and the ANS. The heart is simply a double pump which circulates fluid around a system of pipes (*Figure 10.1*).

The heart is two extremely efficient mechanical pumps which act in unison: one circulates blood around the body in the systemic circulation and the other pumps blood through the lungs. Most of the time we are not aware of its efforts or the scale of these labours; during a lifetime of 70 years the heart will beat some 3000 million times at an average rate of 80/min.

Early development

During early embryonic life the heart develops from a simple tube, which initially forms a single chamber. As early as the fourth week after fertilization this primitive structure is pumping blood.

During the next 3–4 weeks the single-chamber heart rotates and undergoes the structural changes which produce the double-pump heart with four chambers. There are, however, several structural modifications to the fetal heart and circulation (see page 209 and Chapter 20).

Cardiac development during the vital early weeks may be adversely affected by various factors, e.g. maternal infection with the *rubella virus*. Damage during development may result in structural problems, such as *septal defects* ('hole in the heart', see page 209).

Location, shape and size of the heart

The heart is a hollow organ, situated in the **mediastinum**, (a space between the lungs), and behind the sternum. It is roughly cone-shaped with its base uppermost and the apex inclined to the left. The

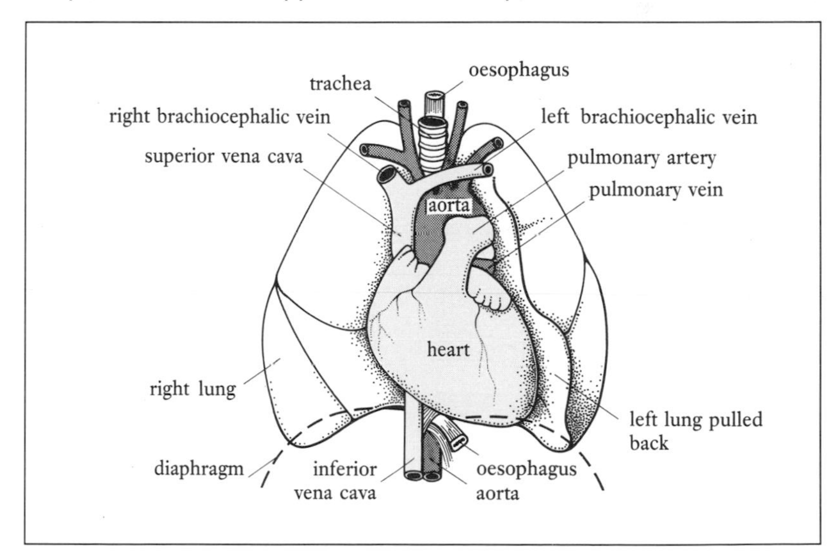

Figure 10.1. The heart, great vessels and lungs.

proportions of a person's fist, it measures around 10 cm from base to apex and weighs approximately 300 g. The base is level with the second costal cartilage and the apex can normally be located just medial to the midclavicular line in the fifth intercostal space (*Figure 10.2*).

Functions of the heart

- Circulates oxygenated blood to the tissues via the high pressure (systemic) general circulation.
- Pumps deoxygenated blood to the lungs via the low pressure pulmonary circulation, where the blood loses carbon dioxide and picks up oxygen (see Chapters 9 and 12).

Structure of the heart wall

Endocardium

The **endocardium** (see *Figure 10.5*) is an *endothelial* tissue which lines the chambers of the heart and covers the valves. It also forms the lining of the vessels entering and leaving the heart. This smooth, glistening tissue allows blood flow without turbulence which would damage the vessel walls (an eroded river bank is further damaged by the turbulent flow of water).

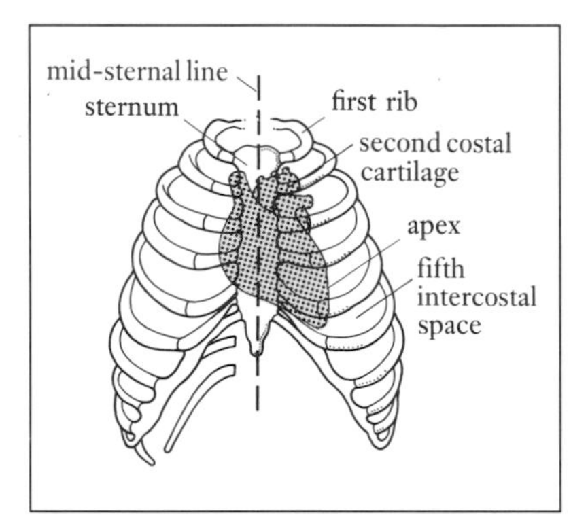

Figure 10.2. Position of the heart.

Myocardium

The middle layer of the heart wall consists of highly specialized muscular tissue known as the **myocardium**. Cardiac muscle has features in common with both skeletal and smooth muscle (see Chapter 17). It has striations, is involuntary and has an internal control system or *pacemaker*.

The fibres are short, have one or two nuclei and the many mitochondria reflect high levels of metabolic activity. Boundaries between individual cells are not well defined, and darkly staining **intercalated discs** join the cells together (see *Figure 10.3*). This means that ions can exchange freely across the muscle cell walls, allowing the action potential (see Chapter 3) and hence the wave of contraction to pass easily across the myocardium, which behaves like a single unit or **syncitium**.

The myocardium depends on *aerobic* (oxygen-requiring) *respiration* for its energy requirements and its performance is rapidly impaired by an oxygen insufficiency. Skeletal muscle can build up an '*oxygen debt*' and switch to *anaerobic respiration* (see Chapters 12 and 13), e.g. when you decide to sprint for the train. Not so the heart, which needs a constant supply of oxygen to use fuel molecules (e.g. glucose, lactic acid, fatty acids) to produce ATP. It is protected from fatigue (you can't really have a 'tired' heart) and *tetanic* contraction (see Chapter 17), where pumping would cease, by a refractory period (see Chapter 3). This may be absolute, when contraction is not possible, or relative, where a sufficiently large stimulus will cause contraction.

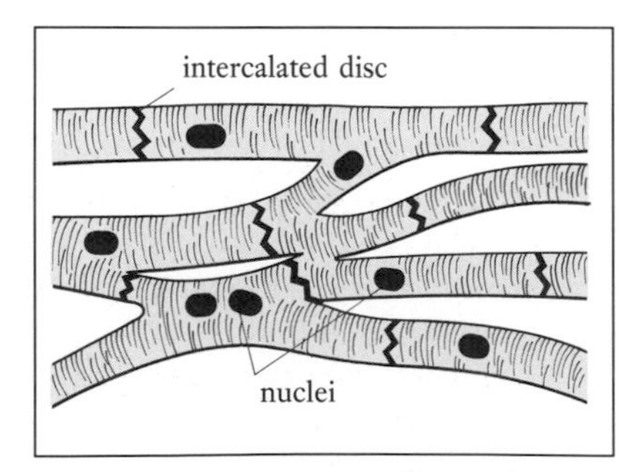

Figure 10.3. Cardiac muscle.

The myocardium reaches maximal thickness over the apex (ventricles), which has to exert sufficient power to pump blood to all areas of the body. The force of the myocardial contraction is proportional to the degree of stretching of the ventricular muscle fibres (within physiological limits). As the heart fills with blood during **diastole** (relaxation phase) the fibres are stretched and the next contraction is more powerful. This relationship between length and tension, known as **Starling's law of the heart**, helps to ensure that output from both ventricles is equal.

Pericardium

The outer sac enclosing the heart is known as the **parietal pericardium**. It has an outer fibrous and an inner serous layer. The **visceral pericardium (epicardium)** is a second serous layer found as part of the cardiac wall. A potential space between the two serous layers contains *serous fluid,* which lubricates and prevents friction as the heart contracts (see Chapter 1). The heart is protected by the outer fibrous layer, which also attaches it to the diaphragm, vessels and chest wall.

Chambers of the heart

The heart consists of four chambers: two upper chambers called **atria** (which receive blood) and two lower chambers called **ventricles** (which pump blood into the circulation). Between the atria and ventricles are valves, and the right and left sides of the heart are divided by a **septum** (see *Figure 10.5*). Before birth an opening between the atria, the **foramen ovale**, allows blood to bypass the fetal lungs (see Chapter 20). There is normally no communication between right and left sides after birth but defects in the atrial septum (ASD) or ventricular septum (VSD) present after birth can be successfully repaired by surgery.

The atria receive deoxygenated blood returning from the tissues (right atrium) and oxygenated blood from the lungs (left atrium). Each atrium has a small appendage known as the *auricular appendage* or *auricle*. The ventricles are much larger chambers with thick muscle which contracts to pump blood out into the general (left ventricle) and pulmonary circulation (right ventricle).

Heart valves

Blood flow through the heart is controlled by valves derived from endothelium. They allow flow in one direction only, by preventing backflow. **Atrioventicular (AV) valves** between atria and ventricles stop backflow of blood into the atria when the ventricles contract. The valve between the right atrium and right ventricle is the **tricuspid**, which has three cusps or flaps. On the left side the **mitral** (shaped like a bishop's mitre) or **bicuspid valve** has two cusps. Both valves are stabilized by cords (**chordae tendineae**) which fix the cusps to **papillary muscle** in the ventricular wall (*Figure 10.4(a)*). This arrangment prevents the valve being blown inside out. Rising pressure in the ventricles during **systole** (contraction phase) causes the AV valve to close.

Two **semilunar** (half-moon shape) valves guard the entrance to the great vessels from the ventricles (*Figure 10.4(b)*); the **aortic** valve, between left ventricle and aorta, and the **pulmonary valve**, between right ventricle and pulmonary artery. They open during

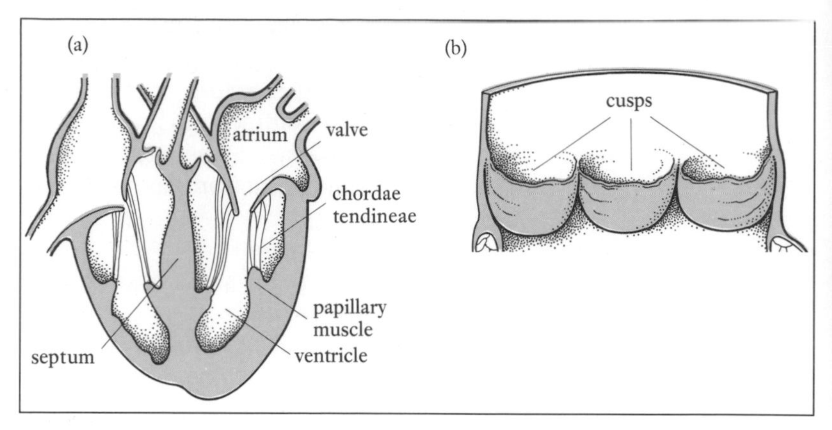

Figure 10.4. Heart valves. (a) AV valve (left); (b) semilunar valve.

ventricular contraction, to allow blood into the circulation, and close as the pressure in the aorta and pulmonary artery rises.

The **heart sounds**, heard when a stethoscope is placed against the chest wall, are the sounds of the valves closing (similarly, the sound is heard when you slam a door, not when it is opened). The **first sound** (Lub) is the AV valves closing and the **second sound** (Dup) is heard as the semilunar valves close.

The action of heart valves is important for the efficiency of the heart. Damage to the valves resulting from diseases such as *rheumatic fever* or *endocarditis* (endocardial inflammation) may cause *incompetence* where the valve is 'leaky' or *stenosis* where the valve is narrow and stiff. This valvular damage may produce abnormal heart sounds or *murmurs* caused by turbulence, as blood leaks back into the heart or is forced through the narrowed valve. Both incompetent and stenosed valves lead to increased cardiac

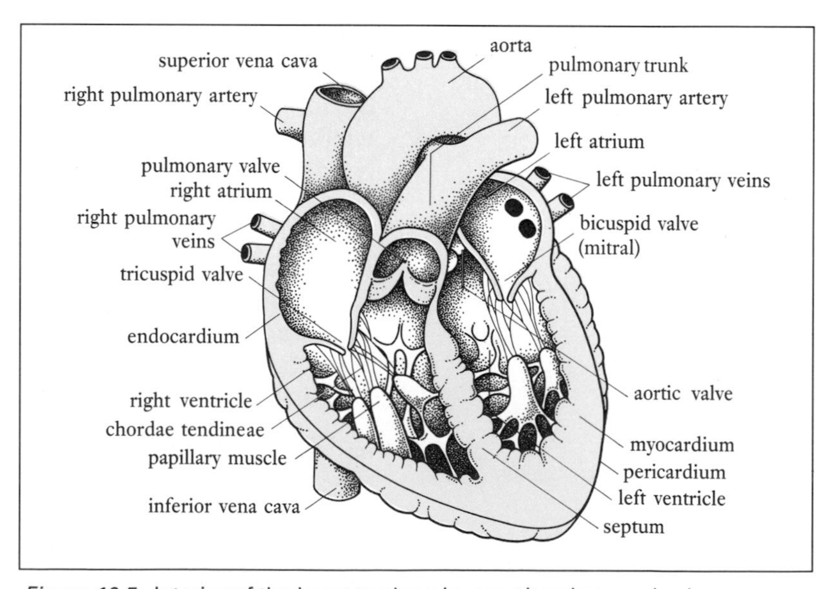

Figure 10.5. Interior of the heart to show layers, chambers and valves.

work, loss of efficiency and eventual heart failure. The mitral and aortic valves are most often affected but can be treated surgically.

Coronary circulation

The myocardium receives its blood supply from the right and left **coronary arteries**, which are the first branches from the aorta. These two arteries divide into smaller branches which run in grooves on the surface of the heart. The **left coronary artery** forms the **circumflex** and **anterior interventricular arteries** and the **right coronary artery** the **marginal** and **posterior interventicular arteries** (*Figure 10.6*).

This arrangement of arteries, which encloses the heart like a crown (hence coronary), ensures that myocardial cells are supplied with adequate oxygen and nutrients for their continuous functioning. With the body at rest, the myocardium receives about 250 ml of blood every minute, which is quite a lot for an organ weighing less than half a kilogram. Myocardial cells extract more oxygen from the blood than other body structures. Most of the blood reaches the myocardium during diastole as the coronary vessels are compressed during systole. When the myocardium needs more oxygen, such as during exercise, the coronary vessels dilate and blood flow increases. Various factors influence blood flow changes, including lack of oxygen, high levels of carbon dioxide, the autonomic nervous system (see Chapter 6) and the pressure in the aorta, which depends on myocardial contraction.

Deoxygenated blood leaves the myocardium via **cardiac veins**. Some of these form the **coronary sinus** which returns most of the blood to the right atrium. The remainder is drained directly into the right atrium via cardiac veins.

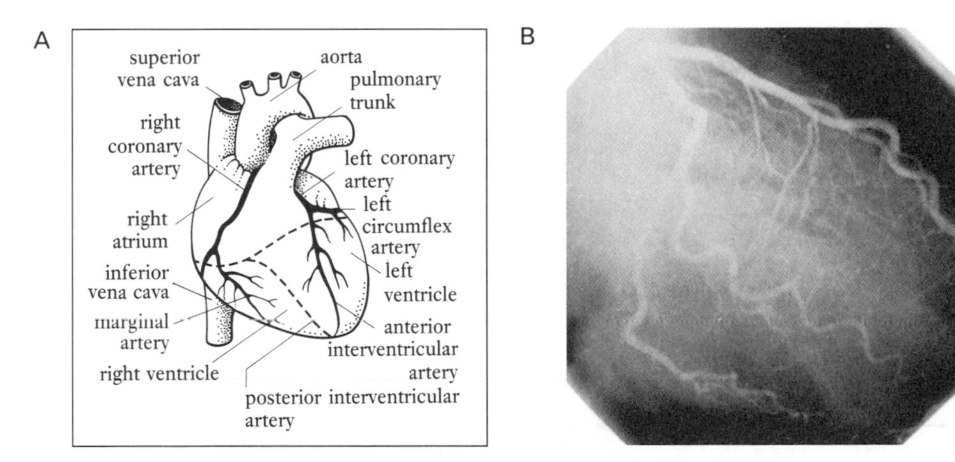

Figure 10.6. (a) Coronary circulation; (b) coronary angiograph – injection of left coronary artery.(Walton, S.,Underwood, S.R.Hunter, G.J *A Colour Atlas of Diagnostic Investigations in Cardiology*. Wolfe Medical Publications, Ltd. 1992. Reprinted with permission.)

Blood flow through the heart

Venous blood, low in oxygen and rich in carbon dioxide, returns to the heart via the **superior** and **inferior venae cavae** and the **coronary sinus** (*Figure 10.8*). The blood enters the right atrium and

Special Focus **Coronary heart/artery disease (CHD)**

Cardiac muscle is totally dependent on an adequate supply of oxygen for its proper function (remember it relies on aerobic respiration). When blood flow and oxygen levels are inadequate the myocardial fibres become *ischaemic* (lacking blood). This, if severe, will cause cell death or *infarction*.

Myocardial blood flow will be reduced if the coronary arteries are narrowed by the *atheromatous* plaques of *atherosclerosis* (see Chapter 9 and arterial disease, page 225) or arterial spasm. The results of this narrowing may be *angina pectoris* characterized by chest pain on exertion, a *myocardial infarction* or sudden death if a major coronary vessel is occluded by a *thrombosis*.

Coronary heart disease (CHD) is the 'plague' of developed economies where it most commonly affects people in the least affluent social groups. The UK has the unenviable distinction of having one of the highest death rates due to CHD, with over 160 000 deaths in 1985 (Ball and Mann, 1988).

The risk factors linked to the development of atherosclerosis, which include smoking, hypertension and elevated blood lipid (fat) levels, are discussed on page 225 within the context of arterial disease. Associated health promotion issues are discussed in relation to the *Person-centred Study* – Derek (page 213).

Angina pectoris – is due to a gradual build up of atheroma within the coronary arteries. It gives rise to pain in the chest, neck and left arm (see Chapter 4, referred pain). The pain is experienced in stressful situations, such as exertion or emotion, when myocardial demand for extra oxygen is not met, but usually disappears after rest. A person with angina is usually treated with drugs: (i) *nitrates* (e.g. glyceryl trinitrate), which dilate the coronary vessels and can be used when pain occurs or is anticipated; (ii) *beta blockers* (e.g atenolol, see Chapter 6), which reduce cardiac work and hence oxygen demands; (iii) *calcium channel blockers* (e.g. nifedipine), which relax smooth muscle and dilate the coronary vessels.

Nurses should ensure that the person has sufficient information regarding his or her drug regimen and understands the importance of compliance, safe storage and common side-effects. Simple measures, such as avoiding sudden exertion, especially after a meal, and keeping warm, will help to reduce pain attacks.

Severe angina may be treated surgically by *coronary artery bypass graft* operations. Recently techniques involving the use of an inflated balloon *angioplasty* or *laser* to open up the narrowed vessel have been developed.

Iindividuals with angina need help and encouragement to change certain aspects of their lifestyles: they should stop or reduce smoking, take regular gentle exercise, reduce their intake of saturated fat, increase their fibre intake, maintain normal weight, reduce alcohol intake and take active measures to manage stress, e.g. relaxation (see Chapter 6).

Myocardial infarction (*Figure 10.7*) – This is the 'heart attack' or 'coronary' of lay terminology. It usually occurs when a coronary artery is occluded by extensive atheroma, a thrombosis which forms on the damaged vessel wall, or an embolus. Men are more commonly affected than women, until after the menopause when this gender difference disappears. The area of myocardium normally supplied by that vessel becomes ischaemic, cells die and form an infarction. The wedge-shaped infarcted region eventually becomes fibrous scar tissue because cardiac muscle fibres do not regenerate.

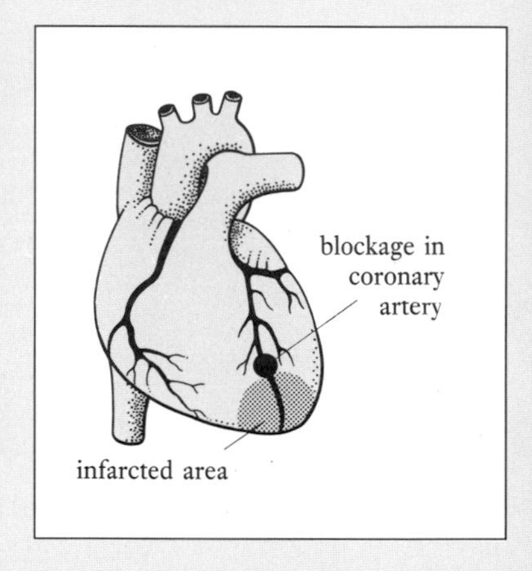

Figure 10.7. Myocardial infarction.

The prognosis following myocardial infarction depends on the size and location of the area affected. Many people recover completely as *collateral vessels* develop to carry blood to the muscle fibres, but some will have permanent loss of function which may lead to heart failure (see pages 221-222). Where a large area of myocardium is affected the person may die at once or within a few hours due to alterations in heart rhythm.

Person-centred Study **Derek**

Derek, who is 51 years old, has a pig farm which he manages with his wife and son. He has had angina pectoris for about 3 years, it is controlled by nitrates and beta blockers. Today, whilst doing the accounts, he suddenly had severe pain in his chest and left arm. When Derek gained no relief from the usual regimen of a glyceryl trinitrate tablet under his tongue and a rest on the bed his wife called their GP, who provisionally diagnosed a myocardial infarction, gave Derek an injection of morphine for the pain and arranged for his admission to hospital. When he arrived in the Accident and Emergency department the staff noted the following clinical features: (i) *crushing chest pain* resulting from the ischaemic changes in the muscle; (ii) *pallor* due to poor skin perfusion; (iii) *sweating* due to the effects of pain and anxiety on the autonomic nervous system (see Chapter 6); (iv) *nausea and vomiting* occurring as blood is diverted from the digestive tract to the vital structures; (v) *tachycardia* (rapid heart rate) caused in part by pain and anxiety, but also increased in an attempt to maintain an adequate circulation; (vi) *hypotension* (low blood pressure), which occurs because the infarcted left ventricle is unable to maintain an adequate **cardiac output** (the amount of blood pumped out of the left ventricle in one minute); (vii) *breathlessnesss* caused by lack of oxygen, which may be exacerbated by some degree of left ventricular failure causing a build up of blood in the lungs and *pulmonary oedema* (see page 222).

Derek had an *electrocardiogram* (ECG) which showed changes that indicated an infarction (see page 215-216) and blood was taken to estimate cardiac enzyme levels. Damaged cardiac muscle (and other tissues) release enzymes into the blood, these can be measured and used with clinical features and ECG changes to confirm the diagnosis. The enzymes usually measured are: *creatine kinase (CK)* and its specific myocardial isoenzyme (CK-MB), *aspartate aminotransferase (AST)* and *lactate dehydrogenase (LD)*. All these rise and peak at different intervals following an infarction or some other muscle damage which can complicate interpretation so measurements are usually taken over consecutive days. In addition Derek's leucocytes and erythrocycte sedimentation rate (ESR) will increase in response to the inflammation associated with the infarcted area.

Prior to transfer to the cororary care unit Derek is given further pain relief and an antiemetic for the vomiting. An intravenous infusion is sited and fibrinolytic treatment commenced. Plasminogen activators (e.g. alteplase) which increase fibrinolysis and thrombus degradation are given intravenously within hours of the coronary occlusion (see Chapter 9).

The initial nursing responsibilities are; (i) pain relief; (ii) information and explanation for Derek and his family to reduce anxiety. A recent study suggests that counselling after the first infarction will reduce anxiety and depression in the individual and anxiety in the individual's partners (Thompson, 1989); (iii) planning care that minimizes cardiac work and ensures rest; (iv) monitoring of: blood pressure, pulse, colour, respirations, temperature, urinary output and heart rhythm to detect potentially life-threatening *arrhythmias* (see *Figure 10.12*); (v) administration of prescribed drugs and oxygen.

Later the nurse, with Derek and other members of the care team, will have a vital role in planning a rehabilitation programme and preparing Derek and his family for discharge. Increased activity, driving, travel, return to work and resumption of sexual activity all depend upon the individual concerned and the severity of the infarct.

The programme will also include provision of relevant health education information (see below) and support for the family in making lifestyle changes (Deans and Hoskins, 1987).

Longer-term medical management and treatment may include: (i) lifestyle adaptations (see reducing the risk); (ii) continued treatment with beta blockers; (iii) drugs to lower serum cholesterol where *hyperlipidaemia* (high levels of lipids in the blood) exists; (iv) antiplatelet drugs, such as aspirin, which appear to reduce the risk of further myocardial infarctions; (v) coronary artery bypass graft using a vein from the leg; (vi) omega-3 fatty acids (certain fish oils), which may have a beneficial effect in individuals who have suffered a myocardial infarction (Siess *et al.*, 1980; Hay *et al.*, 1982). The results of a trial by Burr *et al.* (1989) also suggested that a modest intake of fatty fish (2-3 portions/week) reduces the mortality of men who have had a myocardial infarction.

Reducing the risk of further infarction for Derek and the risk of CHD generally.

The aetiology of arterial disease including that affecting the coronary arteries is multifactorial. Some risk factors such as smoking are proven but for others the link is more tenuous (see page 225). A study of people following myocardial infarction revealed that the main causes of heart attack were thought, by the subjects, to be overwork and worry rather than physical causes (Murray, 1989). Sensible measures and changes in lifestyle which reduce the risks are: (i) stop or reduce smoking; (ii) regular exercise; (iii) avoid obesity; (iv) reduce the intake of saturated fats, these tend to increase blood cholesterol levels and alter the ratio between high and low density lipids (see page 225); (v) increase unrefined carbohydrate intake (fibre), especially the gel-forming type; (vi) reduce alcohol to recommended levels (see Chapter 14); (vii) some authorities recommend that salt intake be restricted especially if the person has hypertension; (viii) development of stress management strategies; (ix) regular blood pressure checks – early detection of hypertension and initiation of treatment.

passes through the tricuspid valve into the right ventricle. Blood is pumped from the ventricle, through the semilunar valve into the **pulmonary artery**, which transports blood to the lungs. In the pulmonary circulation the close contact between the **capillaries** (network of tiny blood vessels linking arteries with veins – their walls are only one cell thick, allow exchange of molecules between blood and cells) and *alveoli* (the air sacs in the lung) allows gaseous exchange; carbon dioxide from the blood moves into the alveoli and oxygen moves into the blood (see Chapter 12).

Blood rich in oxygen and low in carbon dioxide returns to the left atrium via four **pulmonary veins** (two from each lung) and passes through the bicuspid valve into the left ventricle. Contraction of the ventricle moves blood through the semilunar valve into the **aorta** and to all parts of the body via the systemic circulation.

Cardiac conduction system

Heart muscle has the ability to contract without stimulation from the nervous system. The coordinated contraction of the heart depends upon a conduction system formed from specialized non-contractile muscle cells and is facilitated the structure of myocardium (see page 208-209). This allows an impulse to spread from atria to ventricles in an orderly manner. The intrinsic conduction system consists of a **sinoatrial (SA) node, atrioventricular (AV) node, atrioventricular bundle (bundle of His),** the **bundle branches** and **Purkinje fibres** (see *Figure 10.9*).

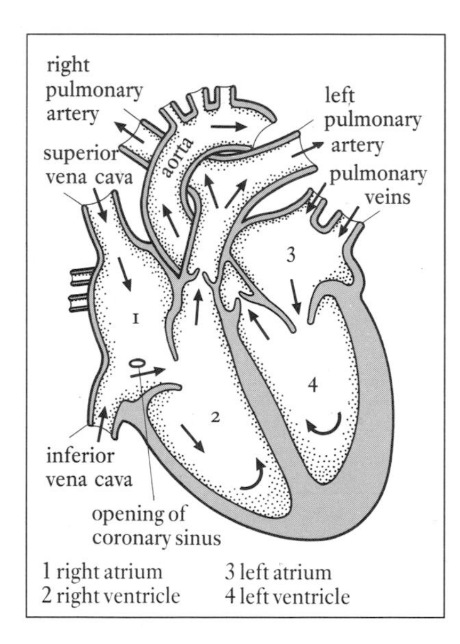

Figure 10.8. Blood flow direction.

The impulse is developed by the SA node situated in the right atrium, close to the opening of the superior venae cava. The cells of the SA node are **autorhythmic**, i.e. changes in membrane ion permeability cause depolarization and production of an action potential without innervation. Cells of the SA node depolarize more frequently than those in other areas. This sets the heart rate and for this reason is known as the '**pacemaker**'. The impulse from the SA node spreads through both atria causing depolarization and contraction.

The next area to receive the impulse is the AV node at the bottom of the right atrium. Conduction through the AV node is delayed by 0.1 s, allowing time for the completion of atrial systole. The impulse passes from the AV node down the AV bundle and right and left bundles to the apex of each ventricle. It travels rapidly via the Purkinje fibres to the ventricular muscle cells which contract stimultaneously to produce a coordinated ventricular systole.

We have looked in detail at the intrinsic control of cardiac contraction. However, the autonomic nerves also influence heart rate (see Chapter 6). The parasympathetic vagus nerve acts as a 'brake' on heart rate and the sympathetic nerves, with adrenaline, accelerate heart rate.

Problems with the conducting system lead to **arrhythmias** (without proper rhythm – irregular or abnormal heart rhythm) such as *ectopic beats* (where a new pacemaker 'starts up'), *fibrillation* and various degrees of *heart block* (where the rate is slow, e.g. 40/min in

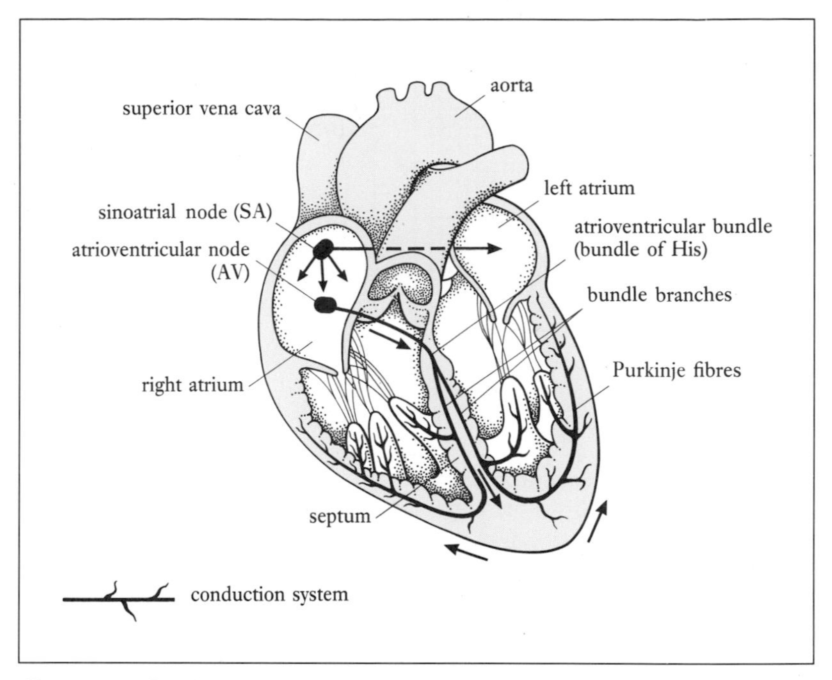

Figure 10.9. Conduction through the heart.

complete *heart block* (see *Figure 10.12,* page 217)). Some types of heart block are treated with drugs such as atropine (see Chapters 3 and 6) but others need the insertion of an electronic pacemaker. This may be temporary, where wires attached to an external pulse generator are introduced into the right ventricle via the veins. It is also possible to 'pace' the heart indirectly with wires placed in the oesophagus. These methods are unsuitable for long-term use because of the external pulse generator. When a permanent pacemaker is required the pacemaking device is sited under the skin of the chest wall.

Electrocardiography (ECG)

It is possible to record the electrical events that occur in the myocardium during conduction and contraction using electrodes placed on the skin (*Figure 10.10*).

The electrical impulses are picked up using an **electrocardiograph**, which produces the **electrocardiogram (ECG)**, displayed graphically on an oscilloscope or as a paper tracing. The normal heart produces a typical waveform, **sinus rhythm**, which consists of five deflection waves, known universally as P, Q, R, S and T (*Figure 10.11*). The **PQRST complex**, which lasts about 0.8 s, represents a complete **cardiac cycle**.

The small **P wave** represents the atrial depolarization (see Chapter 3) caused by conduction of the impulse from the SA node across the atria. After a short pause (0.1 s) the atria contract. The next three waves, which come close together, are the large **QRS complex**, this represents ventricular depolarization which immediately precedes contraction. The **T wave** represents ventricular repolarization. Atrial

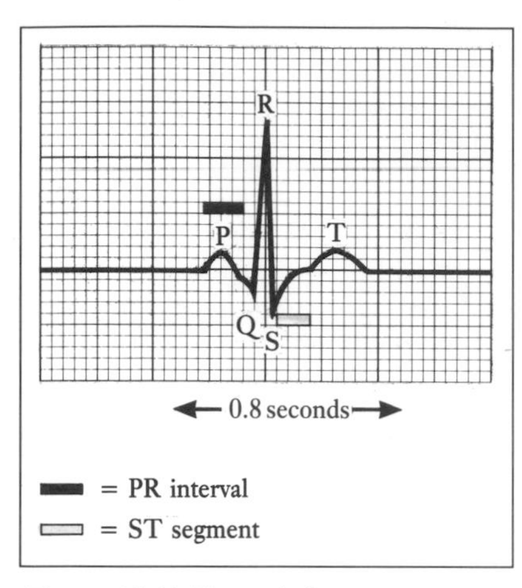

Figure 10.11. Normal electrocardiogram tracing (PQRST compex).

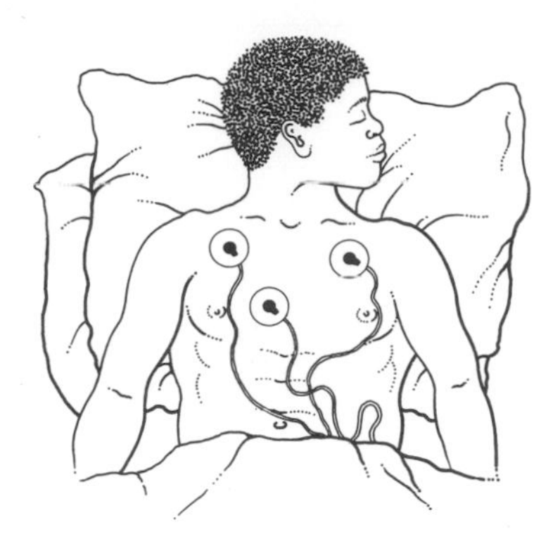

Figure 10.10. Chest leads in place.

repolarization is not usually represented in wave form as it is hidden by the much larger QRS complex. The **PR interval** is the time taken for impulse conduction from the SA node, through the atria to the AV node and on to the ventricles. The **ST segment** represents a period of inactivity and ventricular depolarization.

Normal changes in the sinus rate may occur with respiration; decrease on expiration and increase on inspiration. This is *sinus arrhythmia.*

Alterations in the waveform pattern or time intervals may indicate problems with the conduction system or myocardial function (*Figure 10.12*). The 12-lead ECG is a very useful diagnostic aid following myocardial infarction, heart failure, drug toxicity and many other situations.

Cardiac cycle

The **cardiac cycle** (see *Figure 10.13* page 219) describes the events occurring during one heart beat. It lasts around 0.8 s when the heart rate is 70/min. It is divided into diastole and systole and usually refers to ventricular activity. Although we will describe events in stages, you should remember that it is a continuous process.

Late diastole

The heart is completely relaxed and pressures are low. Blood enters from the circulation, and flows through the atria and into the ventricles. At this stage, the AV valves are open and the semilunar valves are closed. Near the end of diastole the SA node 'fires', atrial systole occurs and the ventricles receive the remaining 30% of blood.

Systole

The atria are relaxed and the pressure rises in the ventricles until the AV valves close. For a brief time, the ventricles are closed chambers, a period called **isovolumetric contraction**. Soon the venticular pressure is greater than that in the aorta and pulmonary artery, and the semilunar valves open to allow blood to leave the ventricles – **ventricular ejection phase**. This is the period of maximum pressure in the left ventricle and aorta (see systolic blood pressure, page 230).

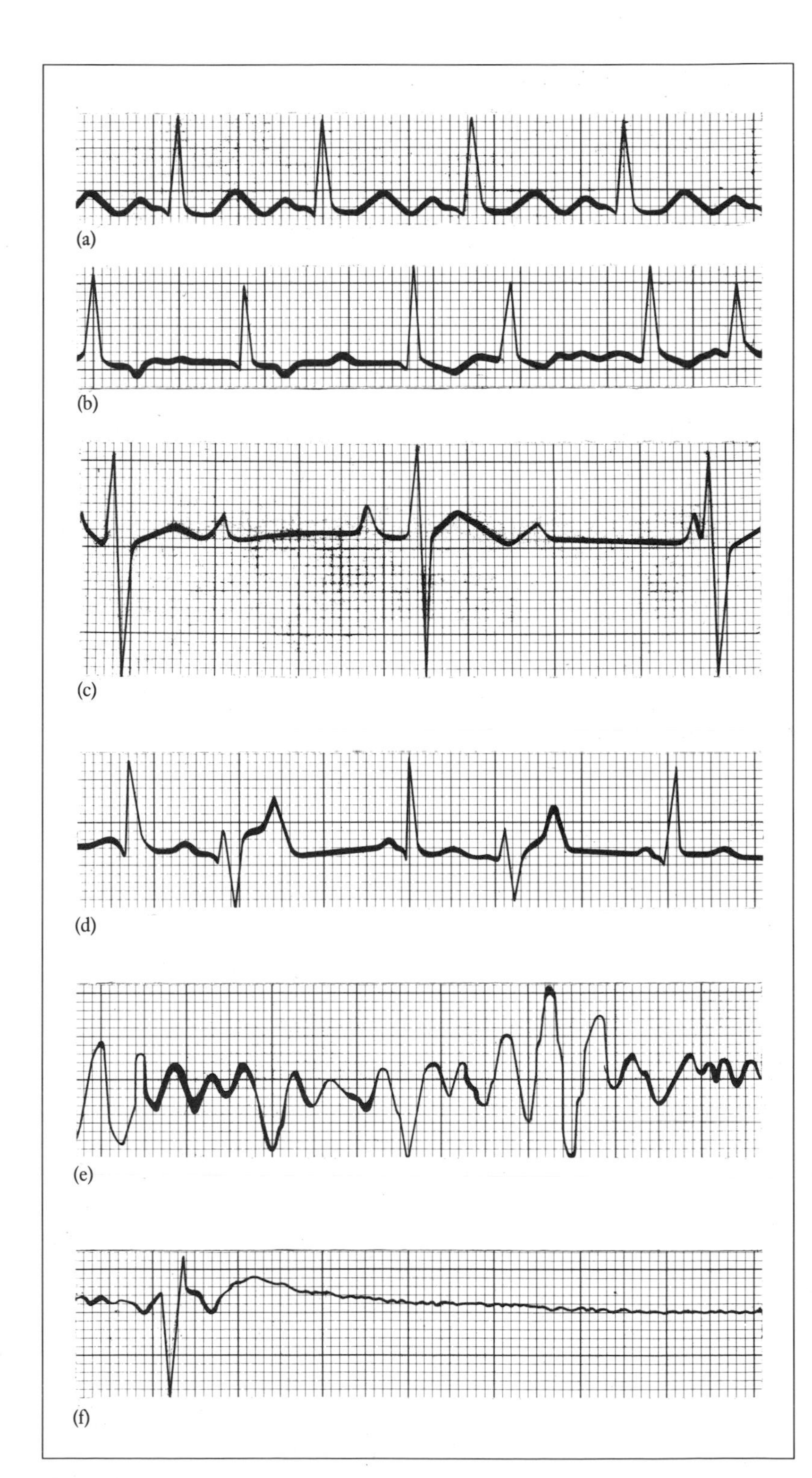

Figure 10.12. Normal and abnormal heart rhythms. (a) *Sinus rhythm* (b) *Atrial fibrillation* – atrial rate is rapid and irregular with absence of proper P wave. Ventricular rate is irregular and produces an irregular pulse. (c) *Complete heart block* (third-degree AV block) – no impulses from the AV node reach the ventricles. Atrial contraction is regular but the slower ventricular contraction is initiated by fibres in the AV bundle or ventricle. (d) *Ventricular ectopic beats* – there are wide QRS complexes without a P wave. These can occur in healthy people when they are associated with an excess of tea, coffee or alcohol. They may also be a feature of disease such as myocardial infarction. (e) *Ventricular fibrillation* – a completely disordered pattern of irregular QRS complexes which produce no cardiac output. This is the usual reason for cardiac arrest and is the commonest cause of sudden death. It requires the commencement of immediate resuscitation. (f) *Asystole* – no QRS complexes seen. Again this is a cardiac arrest situation which requires immediate resuscitation.

Nursing Practice Application **Cardiac arrest and resuscitation**

Cardiac arrest is the cessation of an effective heart beat, either because the ventricles are fibrillating or because of ventricular asystole. Causes include myocardial ischaemia, electrocution and alterations in serum potassium. If an effective circulation is not restored within 2–3 minutes the brain will be irreversibly damaged.

Cardiac arrest is characterized by unconsciousness, absent major pulse such as the *carotid*, gasping or absent respiration, dilated pupils and cyanosis or pallor.

Immediate resuscitation must be initiated by the person discovering the arrest. Nurses need to be competent in *cardiopulmonary* resuscitation and be fully conversant with local emergency procedures as they will be required to start treatment before the arrival of the specialist cardiac arrest team ('crash team').

Resuscitation: (i) *Airway* – must be opened and cleared of any obstruction, e.g. vomit, displaced dentures; (ii) *Breathing* – expired air mouth-to-mouth or mouth-to-nose ventilation. Where available a Brook's airway or airway with a bag and face mask is used. Later when the team arrive ventilation is maintained by endotracheal intubation; (iii) *Circulation* – is maintained by external cardiac massage, where the heart is compressed between the sternum and spine to produce an output of blood. In some situations a *precordial* blow to the lower end of the sternum may be sufficient to restart the heart.

Specialist measures – when the team arrives: (i) *Drugs and defibrillation* – an intravenous line is sited for the administration of drugs and the correction of the *metabolic acidosis* (see Chapters 2 and 12) with sodium hydrogen carbonate solution. The specific management depends on the cause of the arrest, e.g in ventricular fibrillation the heart is defibrillated by passing direct current electricity through the ventricles via paddles applied to the chest wall. Drugs used depend on the ECG (see below) and include adrenaline (possibly intracardiac), atropine, calcium chloride, isoprenaline and lignocaine; (ii) *Electrocardiogram* – is needed to determine the cause – ventricular fibrillation or asystole.

Readers requiring further information regarding resuscitation, are directed to Further Reading, e.g. Macleod *et al.* (1987) and Wynne (1990).

Early diastole

The ventricles relax, ventricular pressure falls and back-pressure in the great vessels causes the semilunar valves to close. The ventricles are again closed chambers – the **isovolumetric relaxation phase**. Meanwhile the the atria have been filling with blood (atrial diastole). The rising pressure eventually forces open the AV valves and blood flows into the ventricles, which is where we started. The cycle is complete.

Earlier we mentioned that the cardiac cycle lasts for 0.8 s when the heart rate is around 70/min. Diastole lasts 0.5 s and ventricular systole 0.3 s. When the heart rate increases, diastole becomes shorter, which may become a problem where the rate is very fast and the ventricles have no time to fill between systoles. All the events of the cardiac cycle: pressures, volumes, valves open/shut, heart sounds and ECG are shown in *Figure 10.13.*

Cardiac output

Cardiac output (CO) is the amount of blood pumped by the left ventricle in 1 minute. It depends upon **heart rate (HR)** and **stroke volume** (SV, the amount of blood pumped out with each ventricular contraction) CO = HR x SV, e.g. at rest:

$$70/\text{min} \times V\ 70\ \text{ml} = 4900\ \text{ml}\ \therefore\ CO = 4900\ \text{ml/min}$$

The cardiac output is altered by changes in rate and stroke volume.

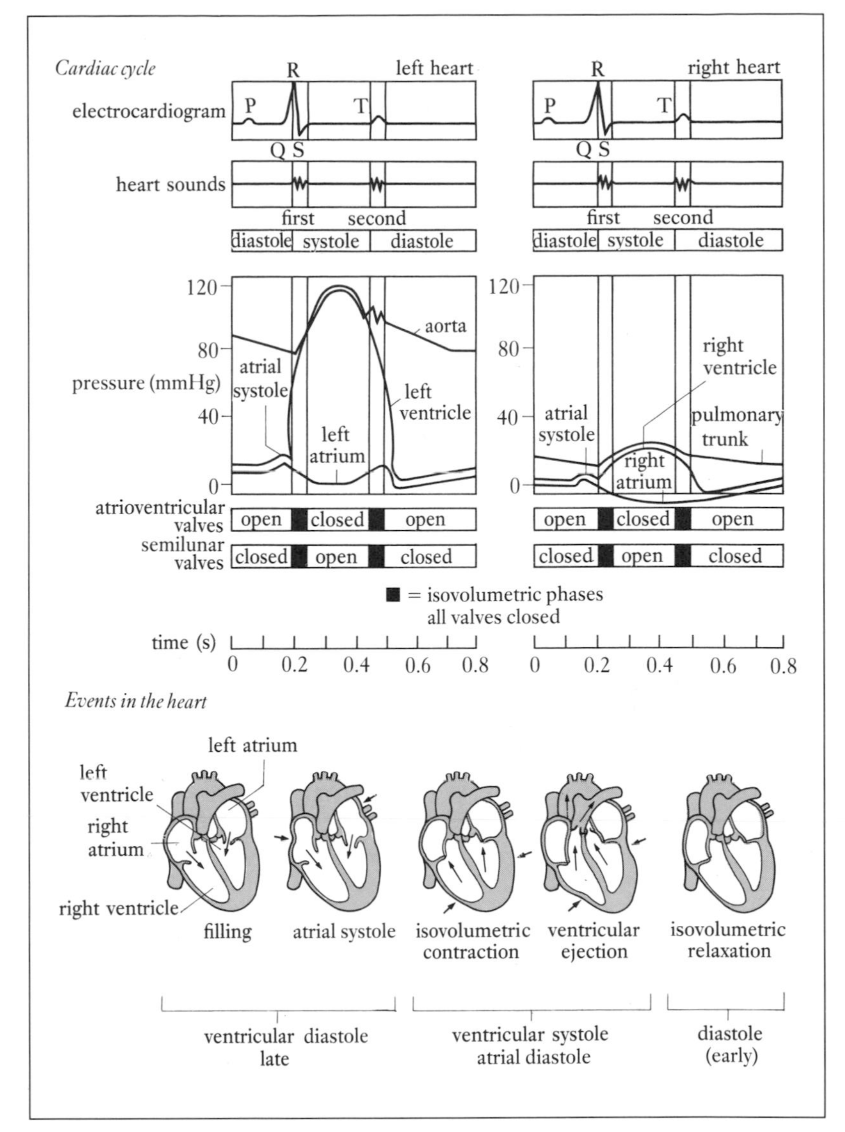

Figure 10.13. Events of the cardiac cycle.

During exercise an average person can increase his or her output to 20–25 l/min. The heart's facility to do this is called the **cardiac reserve**. In very fit people cardiac output may increase to 35 l/min, which is seven times the resting value! It might be helpful to consider the regulation of stroke volume and heart rate.

Stroke volume

Stroke volume is the difference between ventricular end diastolic volume (EDV) and end systolic volume (ESV) SV = EDV – ESV e.g.;

120 ml – 50 ml = 70 ml ∴ SV = 70 ml

Earlier we discussed the relationship between myocardial fibre stretch and the force of contraction, Starling's law of the heart (see page 209). When venous return increases, in exercise or with slow heart rate, the myocardial fibres stretch and the next contraction is more powerful, resulting in an increased output. If the venous return is reduced, e.g. after haemorrhage, there is a reduction in stroke volume and hence output. In health the balance between right and left is ensured: if the output of one side of the heart increases, the other side will receive more blood causing its output to increase.

Stroke volume is also influenced by sympathetic nerves and the adrenal hormones; adrenaline and noradrenaline (see Chapters 6 and 8). These affect myocardial contractility, enhancing the force of ventricular contraction. Stroke volume increases as the ventricles empty more efficiently. Some drugs (*inotropic*) used to treat heart disease affect myocardial contractility; contractility is decreased by beta blockers such as atenolol and increased by digoxin (see page 222).

Heart rate

The adult **heart rate** is normally 60–80 beats/min but can be altered by several factors in response to changing body needs (*Figure 10.14*). Clearly you need different effort from the heart whilst sleeping and during a game of squash. The heart rate, which is much faster at birth (120–140/min), slows during childhood to reach the adult level and then slows in old age. Women tend to have faster heart rates than men. A resting rate greater than 100/min is termed *tachycardia* and a rate less than 60/min is known as *bradycardia*.

The **cardiovascular centre (CVC)**, situated in the medulla and pons, consists of the **cardiac centre** which controls rate and force of contraction and the vasomotor centre which controls blood vessels and blood pressure. There is some overlap of function. The CVC controls heart rate via the ANS; the parasympathetic vagus nerves act as a brake on the SA node to slow heart rate and stimulation of sympathetic nerves (beta receptors) increases heart rate. At rest both parts of the ANS influence the heart but inhibition by parasympathetic nerves is the more important.

The many factors influencing the CVC include peripheral chemoreceptors (monitoring oxygen levels), baroreceptors (monitoring blood pressure) and data from the cerebral cortex and hypothalamus. This well coordinated control mechanism allows us to respond to changes in oxygen demand by increasing heart rate and cardiac output.

When needed, heart rate can increase as we respond to posture, exercise and stressors, such as cold, pain, anxiety, fever and anger or fear (adrenal hormones and the 'fight or flight' response). Abnormal blood electrolyte levels can have serious effects on cardiac contraction, especially calcium and potassium levels, which may lead to life-threatening arrhythmias. A cardiac mechanism called the *Bainbridge reflex*, which involves stretch receptors in the right atrium, can increase heart rate by 15% through sympathetic stimulation.

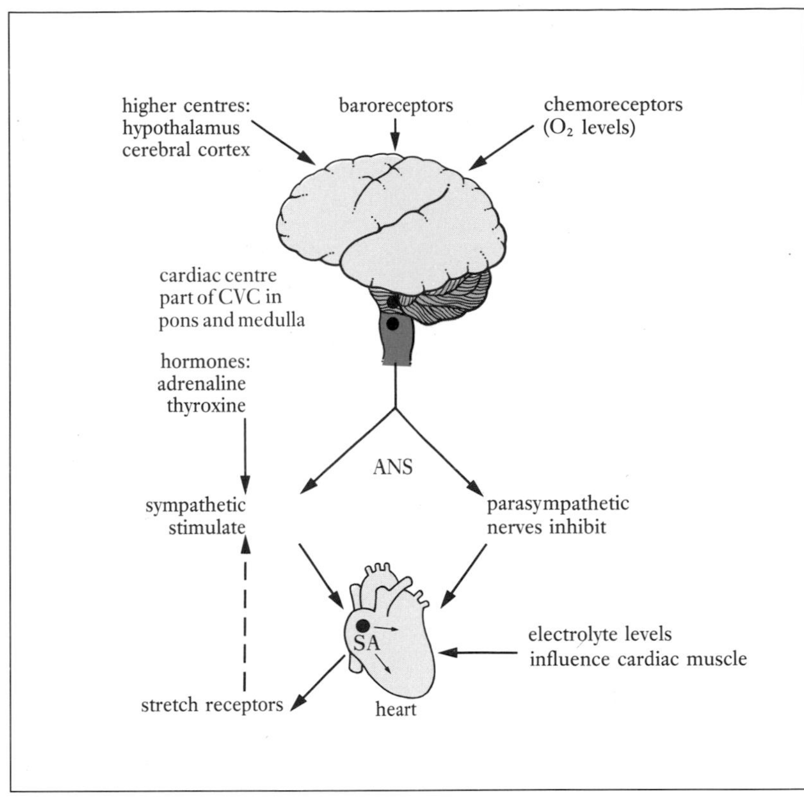

Figure 10.14. Regulation of heart rate. (SA, sinoatrial node; ANS, autonomic nervous system; CVC, cardiovascular centre).

It is important to remember that cardiac output, stroke volume and heart rate are interrelated components which should be viewed as a whole unit. Associated topics of blood pressure and peripheral resistance are covered later in this chapter.

Problems with cardiac output

The breakdown of homeostatic controls on cardiac output will have profound effects on cells and their functioning. The results of poor tissue oxygenation will eventually affect the efficiency of every body system:

- *Acute circulatory failure* – cardiac output is insufficient to maintain blood pressure and tissue perfusion. It is seen in various types of shock (see pages 232-233).
- *Heart failure* – a chronic inability of either the left or right side of the heart to pump effectively and produce a cardiac output sufficient for the person's needs. One side may fail initially but both sides may eventually be affected to produce *congestive cardiac failure* (CCF). If the output of the heart is reduced whilst blood is still returning obviously there will be congestion and a backlog of blood in the vessels trying to empty blood into the heart. A useful analogy is what happens when a motorway accident causes traffic congestion which eventually spreads to smaller roads leading to the motorway, and to the towns and villages nearby.

Nursing Practice Application Apex beat

Heart rate is generally determined by counting the radial pulse but in some situations (e.g. atrial fibrillation) it is necessary to listen to the *apex beat*. This is best heard in the mid-clavicular line at the level of the fifth intercostal space. Apex beat is required in atrial fibrillation because radial pulse may be inaccurate, as some beats will be too weak to reach the wrist. It is also important during treatment with certain drugs e.g. digoxin. Two nurses should perform simultaneous recordings of apex and radial pulse prior to drug administration (see below). The simultaneous recordings remove any difficulties in interpretation caused by changes in heart rate between separate recordings.

At first the heart tries to *compensate*, by *hypertrophy* and enlargement (see Chapter 1), in an attempt to increase output. If cardiac function continues to deteriorate a point is reached when the failure is *decompensated*, a serious state where overstretching causes further output reduction and death.

Left heart failure will lead to high pressure in the pulmonary veins, pulmonary congestion and *pulmonary oedema* where fluid from the pulmonary capillaries leaks out into the alveoli. As you can imagine these are not the ideal conditions for gaseous exchange and result in *orthopnoea* (breathlessness relieved by an upright position) and *cyanosis* (blue coloration). Another feature is *paroxysmal nocturnal dyspnoea* where the person suffers sudden breathless attacks in the night. The causes of left-sided failure include hypertension, infarction affecting the left ventricle and mitral or aortic valve malfunction.

If the right side fails there will be widespread venous congestion which will eventually affect most tissues. Its effects include oedema (see page 235 and Chapter 11) in dependent areas such as the ankles or sacral area, anorexia, nausea, constipation and reduced urinary volume. Right-side failure may be due to problems in the heart such as pulmonary or tricuspid valve malfunction, but more commonly it follows left-sided failure. Here the right ventricle is exhausted trying to pump blood into the congested lungs. Chronic lung diseases may also increase the work required by the right ventricle to the point where it no longer achieves an adequate output; this type of heart failure is called *cor pulmonale*.

Nursing Practice Application Cardiac glycosides - digoxin

Many people with cardiac conditions (e.g. heart failure, some arrhythmias) are treated with digoxin (a drug derived from the foxglove). Digoxin slows heart rate, increases contractility and improves cardiac output. Digoxin is a very useful therapeutic agent, but has serious side-effects which must be anticipated.

An important nursing role, apart from teaching people about their drugs and the need for compliance, is careful observation for side-effects. Digoxin may cause bradycardia and care plans should include determination of apex beat/radial pulse prior to administration. Should the apical rate fall below 60/min the dose is withheld. Digoxin may cause other serious arrhythmias such as ventricular tachycardia. Other toxic effects include confusion and disorientation, easy to mistakenly attribute to another cause, and gastrointestinal problems such as anorexia, nausea, vomiting and abdominal pain.

Vessels and Circulation

Having discussed the heart in detail we now need to look at the 'plumbing' (the blood vessels) and the many mechanisms that control their size, blood flow and pressure (circulation).

Blood vessels

Blood vessels form a closed network. Within this network blood circulates around the body. There are three basic types of vessel: **arteries,** which carry blood away from heart; **veins,** which carry blood back to the heart and **capillaries,** where exchange between blood and cells occurs. Capillaries form the link between the **arterioles** (small arteries) and **venules** (small veins).

Vessel structure

Capillaries have the least complicated structure. They comprise one layer of endothelial cells, which is ideally suited to their role in molecular exchange (*Figure 10.15*).

Arteries and veins are more complex, but both have the same basic three-layer structure. The outside layer or **tunica adventia,** made of fibrous tissue and collagen fibres, covers and protects the vessel. A network of tiny blood vessels, the **vasa vasorum**, supplies blood to the outer layer of larger veins and arteries. The variable middle layer or **tunica media** consists of elastic fibres and involuntary muscle innervated by sympathetic nerve fibres. This layer, thickest in the arteries, is responsible for alterations in vessel lumen size. When the muscle is relaxed the vessel opens (**vasodilation**) and when it contracts the vessel narrows (**vasoconstriction**). The inner layer or **tunica intima** comprises a very smooth endothelium and a basement membrane, which reduces friction. The endothelium is continuous with that lining the heart.

The variations in structure between vessels of different size or type reflect their specific functions, e.g. large arteries such as the aorta are well supplied with elastic tissue which allows for expansion as the left ventricle contracts.

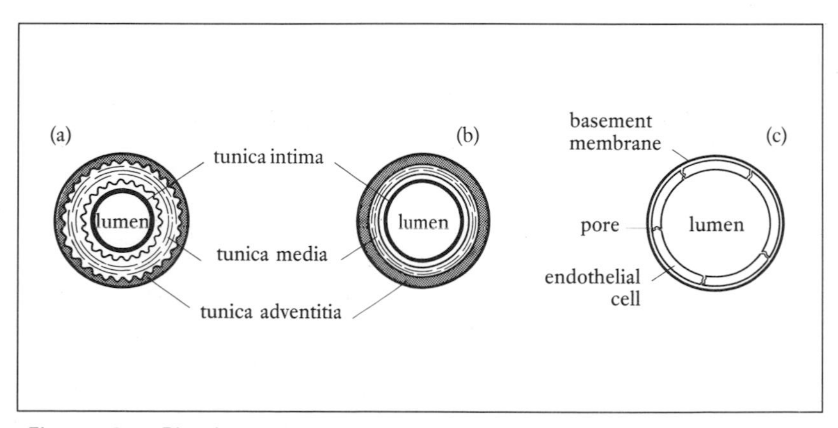

Figure 10.15. Blood vessel structure. (a) Artery; (b) vein; (c) capillary.

Arteries

The arteries, which carry blood away from the heart, include the large elastic arteries, medium arteries and the tiny arterioles which feed into the capillary network.

Many structures receive blood from more than one artery; these may **anastomose** (communicate) to form a network of vessels, e.g. the *circle of Willis* that supplies the brain. This has an obvious advantage as, when one vessel is occluded the rest of the structure still receives blood through a collateral circulation.

Some arteries are **end arteries**, e.g *central artery of the retina*. If these are occluded there are no alternative vessels and the structure normally supplied with blood is irreversibly damaged.

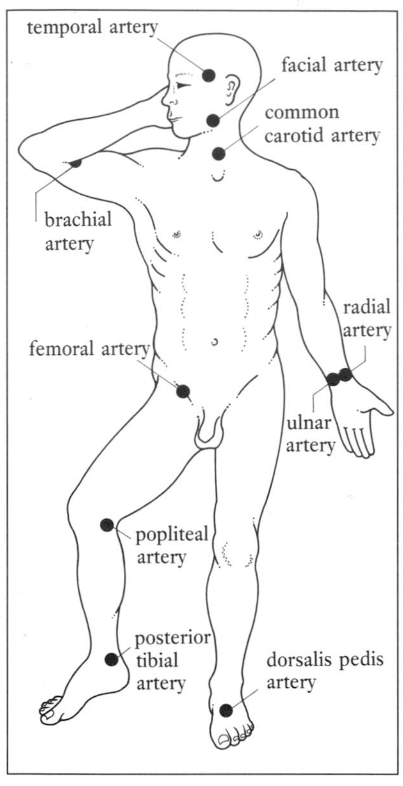

Figure 10.16. Arterial pulses and pressure points for the arrest of haemorrhage.

Large elastic arteries such as the aorta and its main branches have elastic fibres within the tunica media. The aorta has a diameter of 2.5 cm which offers a low resistance to flow. It distends readily as the elastic fibres stretch as blood is ejected during ventricular systole. It maintains blood flow during diastole by the recoil of the elastic fibres. These properties help to produce a steady arterial flow, rather than sudden gushes of blood. As age increases these vessels lose their elasticity. This results in increased resistance and blood pressure. Large arteries have considerable smooth muscle but are not subject to much nervous vasoconstriction.

The medium arteries have more muscle and less elastic tissue in their tunica media. These arteries, which generally supply specific organs, are subject to more vasoconstriction.

The smallest arteries, the arterioles, have a thick layer of smooth muscle. Resistance to flow in these vessels is high and the arterioles are important in determining overall **peripheral resistance** (friction or resistance between blood and the vessel walls). Their structure is well related to function. Vasoconstriction and vasodilation change vessel diameter in response to nervous and local chemical stimuli. This ability to constrict or dilate, and the presence of smooth muscle **precapillary sphincters** between the terminal arterioles and capillaries, is vital in controlling blood flow through the capillary network.

Arterial pulse

The pulse is the wave of expansion felt in the arteries as the left ventricle contracts. It is most conveniently felt where an artery passes over a bone, e.g. *radial artery* felt at the wrist (*Figure 10.16*).

Capillaries

Capillaries are minute vessels that form a dense network found in most tissues, but some connective tissues, such as tendons, have very few capillaries and the lens of the eye has none. Capillaries, which on average measure 750 μm in length and 7 μm in diameter, provide a delivery and waste collection service for individual cells. They consist of a single layer of endothelial cells, the tunica intima, which allows oxygen and small molecular-weight nutrients to pass from blood to cell, and waste such as carbon dioxide to move in the opposite direction. Some capillaries have pores or *fenestrations* in their wall which increase permeability. This structural feature is

Special Focus Arterial diseases

Atherosclerosis (see thrombo-embolic conditions Chapter 9, *Special Focus*, page 212, and Person-centred Study, page 213) is a degenerative process which results in structural changes in arteries. Most adults over 40 have some degree of arterial change and, as already mentioned, atherosclerotic disease is a major cause of death in developed countries.

Stages in development are: (i) the appearance of fatty streaks on the tunica intima; (ii) formation of a fatty plaque of atheroma containing *cholesterol* and *low density lipoproteins (LDL)* with thickening of the tunica intima; (iii) the atheromatous plaque enlarges with fibrosis and calcification, the media is disrupted and ulceration of the endothelium may occur. This severe later stage results in vessel narrowing and ischaemia. The roughened lining gives rise to turbulent flow, abnormal haemostasis and thrombus/embolus formation. Damage to the vessel wall may eventually lead to the development of an aneurysm or dilatation of the wall (*Figure 10.17*).

There is considerable information available regarding factors that may predispose to arterial change but some of the evidence fails to prove direct causal relationships. Unfortunately, media coverage of isolated and unsubstantiated work leads to confusion and conflict about how we should promote healthy cardio-vascular function.

It is generally accepted that atherosclerosis does not have a single cause but results from the damaging effects of several factors operating together (multifactorial aetiology). Atherosclerosis causes CHD (see page 212), strokes (see Chapter 4) and peripheral vascular disease necessitating amputation.

Factors linked to the development of atherosclerosis: (i) increasing age; (ii) more common in males than females, until after the menopause when this difference disappears (see Chapter 20); (iii) hereditary factors – certain families have an increased incidence, e.g. genetically determined familial *hypercholesterolaemia*; (iv) ethnic origin, e.g. in the UK people of Asian origin have an increased mortality compared with Caucasians; (v) hypertension because of damage to the arterial lining; (vi) high blood levels of LDLs and cholesterol are linked to the development of atherosclerosis. This relationship between lipids and arterial disease is complex. For reduced risk it is important to have a high ratio of *high density lipoproteins (HDL)* to LDL. The HDLs, such as omega-3 fatty acids (see page 213), are thought to have a protective role in reducing thrombus formation. Intake of polyunsaturated fats tends to lower blood cholesterol/LDL levels, but a high intake of saturated fats tends to increase cholesterol/LDLs. More information about the structure, absorption and metabolism of lipids can be found in Chapter 13. Still not definite however, is the link between dietary saturated fat and cholesterol levels and risk. Before you become agitated about your cholesterol intake remember that you need it as a precursor for the steroid hormones and the body is able to make it as well as ingest it; (vii) chemicals in cigarette smoke – carbon monoxide affects the endothelium, nicotine increases platelet adhesiveness and indirectly causes vasoconstriction; (viii) sedentary lifestyle, lack of exercise and obesity all lead to an unfavourable HDL:LDL ratio; (ix) stressful lifestyle or job may be linked with atherosclerosis. Personality type may also have some influence. Type A personalities, characterized by aggressive restless behaviour, appear to have a greater risk than the calmer type B personality. The evidence for personality influences is contradictory and is disputed by many authorities; (x) diabetes mellitus (see Chapter 8).

Aneurysm: a permanent dilatation of an artery with disruption of the inner and middle coats. Those occurring in the aorta are often due to atherosclerosis and are associated with hypertension. Where the defect has been diagnosed, elective surgical treatment can be very effective. Sometimes, however, the aneurysm 'dissects' and bleeding occurs between the vessel layers; this may require surgery. If an aortic aneurysm ruptures, the result, as you can imagine, will be disastrous unless immediate specialist surgery is available.

The 'berry' aneurysms which affect vessels of the circle of Willis result from a congenital vessel defect. Rupture, which leads to subarachnoid or intracerebral haemorrhage (see Chapter 4), is also associated with hypertension.

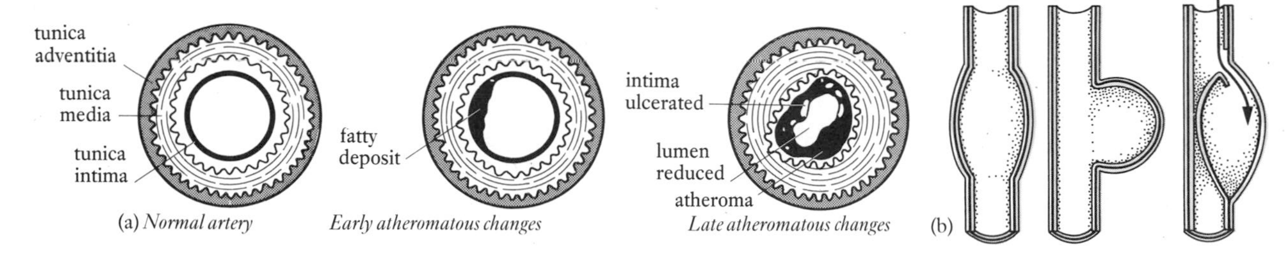

Figure 10.17. Arterial disease. (a) Atherosclerosis; (b) aneurysm.

Nursing Practice Application Pulse

Counting the pulse is used to determine the heart rate, with which the pulse should correspond. Resting rate in adults is normally 60–80 beats/min. An abnormally rapid pulse (tachycardia) may be caused by anxiety, pain (see Chapter 4), hyperthyroidism (see Chapter 8), infections, heart disease or severe haemorrhage. Abnormal slowing of rate (bradycardia) may be due to heart block, drug side-effects, hypothyroidism (see Chapter 8) or raised intracranial pressure (see Chapter 4).

Characteristics other than rate should be noted when monitoring pulse rate. These include the rhythm (normally regular), volume (which may be bounding or thready) and the elasticity of the artery.

Although accessibility of the radial pulse makes it the first choice for routine use in adults it may be necessary to use other arteries or record the apex beat (see page 222). Other arteries used include the temporal (small babies), carotid, facial, brachial (children), ulnar, femoral, popliteal, posterior tibial and dorsalis pedis (see *Figure 10.16*). These alternative sites might be required, for instance, if both wrists were injured or to check vessel patency after surgery to clear an occlusion, e.g. dorsalis pedis used after femoral artery surgery. Knowledge of these arteries is also important as they form the *pressure points* used to arrest haemorrhage, e.g. the posterior tibial or dorsalis pedis are compressed to stop bleeding from the foot.

useful in the gut for the absorption of nutrients, and in the kidneys where waste is removed from the blood.

As already mentioned, blood flow into the capillaries is controlled by the radius of the arterioles and by precapillary sphincters in the arterioles feeding into the capillary network (see *Figure 10.18*). Not all the capillaries are open and functional at the same time. When they are shut blood passes through **arteriovenous shunts** or **anastomoses** and takes no part in molecular exchange. These shunts provide a direct link between arterioles and venules which bypasses the capillary network. These mechanisms are important in diverting blood from one area to another as physiological needs change. After a meal the capillaries of the gastrointestinal tract are full of blood but between meals they close. Capillaries in skeletal muscle open during exercise and those in the skin open when heat loss is required.

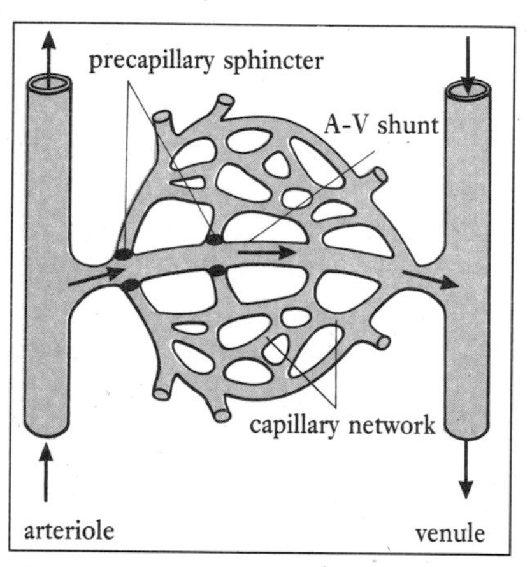

Figure 10.18. Capillary network and arteriovenous shunt.

Larger channels or **sinusoids**, lined with phagocytic macrophages, connect arterioles and venules in organs such as the liver, spleen and some endocrine structures. Here blood flow is slower, allowing time for processes such as phagocytosis of micro-organisms.

The capillaries/sinusoids eventually join up with the tiny venules which form the first part of the venous system.

Veins

Blood leaves the capillary network via venules which unite to form veins. These eventually unite to become the venae cavae. Veins have the basic three coats, but with less elastic tissue, a thinner muscle layer and a thicker tunica adventitia than arteries. They are easily distended and, because they have a large diameter, offer very little resistance to blood flow. Veins are capacity (able to hold a lot of blood) vessels and generally contain 60–65% of the blood volume. This capacity is variable because vessel size can be changed by the

venomotor tone, which is produced as the muscle layer contracts or relaxes in response to sympathetic activity. When veins are partially empty they flatten and become elliptical in shape.

Blood flow in the veins depends on small pressure gradients, but these require help, apart from venomotor tone, to provide a **venous return** sufficient for the heart to produce an adequate cardiac output. In the veins of the limbs the endothelial lining is modified to form valves which allow flow in one direction only (*Figure 10.19(a)*). These **valves** allow us to overcome the effects of gravity whilst standing upright, by helping to prevent pooling of blood in the leg veins. Contraction of skeletal muscles in the leg squeezes blood from valve to valve. When standing for any length of time it is prudent to operate this '**skeletal muscle pump**' by wiggling your toes to aid venous return and prevent pooling of blood in the leg veins and fainting. Numerous anastomoses between veins provide alternative routes for blood; skeletal muscle contracts to push blood from the superficial veins through these communicating channels into the deep veins of the legs and on to the heart. Pressure changes in the thorax and abdomen during breathing, known as the '**respiratory pump'**, also aid venous return. On inspiration, rising abdominal pressure squeezes the veins and the fall in thoracic and right atrial pressure 'sucks' blood back to the heart.

Specialized venous structures called **venous sinuses** are present in the heart and brain. In the heart they drain venous blood from the myocardium. Those in the brain drain venous blood and are involved in the absorption of CSF (see Chapter 4). They are thin-walled channels supported by external connective tissue.

Varicose veins

These dilated and tortuous veins occur when the valves are damaged by high pressure which builds up if the venous return from the legs is impeded (*Figure 10.19(b)*). Standing for long periods at work, pregnancy, abdominal tumours and obesity can all cause this high pressure. Usually it is the superficial leg veins which are affected. Varicose veins are unsightly, and cause swelling and discomfort. They bleed if traumatized and in severe cases will lead to ulcers which are notoriously resistant to treatment.

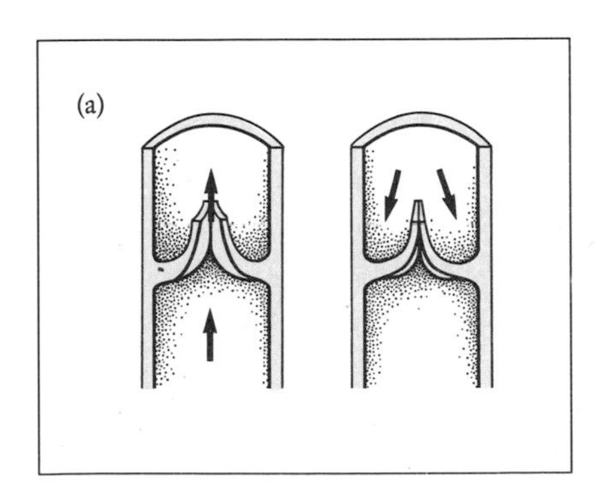

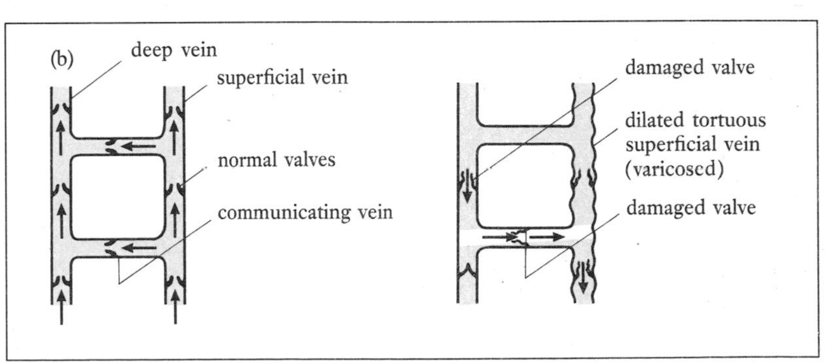

Figure 10.19. (a) Valves in veins; (b) varicosities.

Nursing Practice Application Venous return

During periods of immobility or after surgery it is important to maintain venous return and prevent stagnation of blood which could lead to the development of deep vein thrombosis (see Chapter 9). Nursing care plans should make use of physiological mechanisms by encouraging deep breathing, leg exercises and early ambulation.

When mobilizing a person after prolonged bed rest it should be remembered that lack of gravity (whilst horizontal) will reduce venomotor tone. If people stand up too quickly they will feel giddy and faint because their venous return is inadequate to provide the increased cardiac output required to supply the brain when upright.

Circulatory Physiology

The circulation of blood is a continuous and dynamic process influenced by many factors that enable it to respond appropriately to changing needs. Circulation of blood around the closed cardiovascular system depends upon flow, pressure and resistance.

Blood flow

The **blood flow** is the amount of blood flowing within the circulation (which equals cardiac output) or through an individual structure. The flow rate of any fluid (not just blood) depends on resistance to flow and pressure differences along the tube/pipe (vessel).

The flow in blood vessels is usually smooth or **laminar** but damage to the vessel lining will affect flow: atheromatous plaques, for instance, produce turbulence.

Obviously flow rates alter as body activities and needs change, e.g. at rest the skeletal muscles receive about 20% of the resting cardiac output but during strenuous exercise most of the increased cardiac output is diverted to the muscles. Many structures regulate their own blood flow by controlling arterioles, sphincters and flow through capillary networks, e.g. heart, kidney, skeletal muscle. This *autoregulation* is based on local chemical conditions, such as pH, potassium, carbon dioxide and oxygen levels, and the release of substances from cells, e.g. lactic acid. This causes vasodilation and increased flow, which is known as **active hyperaemia**. In this way, metabolically active tissues can receive more blood and therefore more oxygen and nutrients, and waste can be removed.

Pressure

All fluids exert pressure on the walls of their container (**hydrostatic pressure**). The force exerted by blood on the vessel walls is called **blood pressure**. Blood only flows around the circulation if pressure gradients exist within the CVS. The heart initiates the pressure which powers the circulation, but pressure needs to fall between the left ventricle and the tissues and again between the tissues and the right atrium. Pressure also falls between the right ventricle, the pulmonary vessels and the left atrium. The pressure within different vessels varies but when we talk about blood pressure it is generally taken to mean the pressure within large arteries close to the heart (see page 230).

Resistance

Resistance is the amount of friction experienced by the blood as it flows through the vessels. Most of this resistance occurs in the peripheral vessels and is known as the **peripheral resistance** (PR).

The factors which produce resistance are: vessel diameter, blood viscosity and vessel length. Resistance to flow increases with the length of the vessel, as vessels narrow (vasocontriction) and when blood becomes more viscous (sticky). Vessel length and blood viscosity are relatively stable in health and alterations in peripheral resistance are mainly due to changes in vessel diameter. Chemical stimuli, hormones, inflammatory chemicals and sympathetic nerves cause the smooth muscle of arterioles to contract (vasoconstrict) to increase peripheral resistance or to relax (vasodilate) to decrease peripheral resistance.

The nervous control is through the **vasomotor centre (VMC)** which is part of the cardiovascular centre in the brain (see page 220). It transmits nerve impulses which constrict the blood vessels (see *Figure 10.20*). The level of activity in the VMC depends on the amount of inhibition by the **baroreceptors**. These are sensory receptors responding to pressure changes in the **carotid sinus** and **aortic arch**. When pressure is normal the baroreceptors partially inhibit the VMC which sends sufficient impulses to maintain normal vasomotor tone (partial constriction). Low pressure reduces baroreceptor inhibition of the VMC, which sends more impulses to the vessels, which constrict. High pressure results in complete

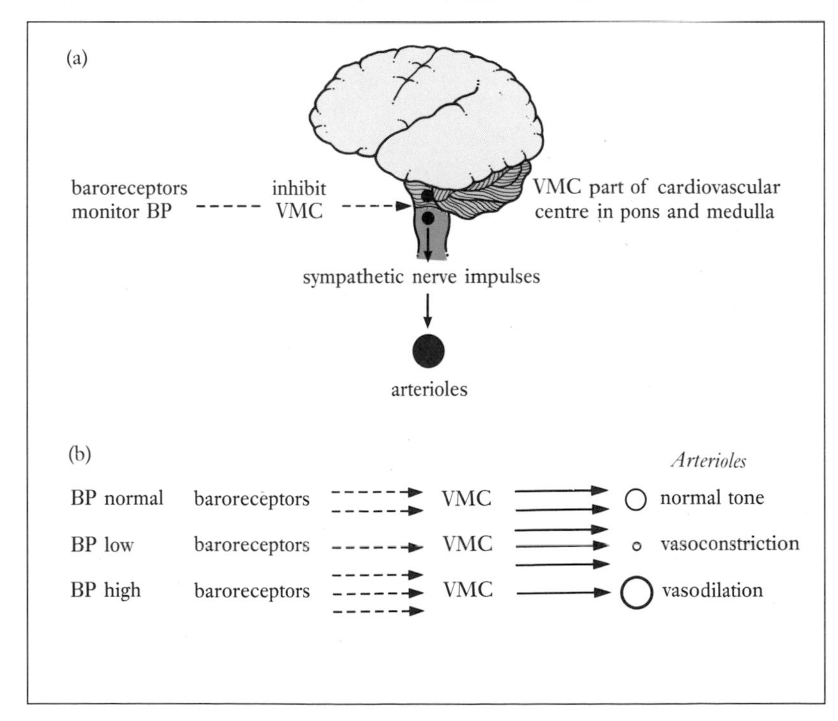

Figure 10.20. (a) Vasomotor centre (VMC); (b) control of vessel size in response to alterations in blood pressure (BP).

inhibition of the VMC and vasodilation. Baroreceptor/VMC mechanisms are vital in the regulation of arterial blood pressure. They also affect heart rate and contractility through functional overlap with the cardiac centre.

Arterial blood pressure

Pressure is exerted by the blood on the arterial walls as the left ventricle pumps blood into the aorta. The pressure is produced when flow meets resistance, which can be summarized as:

Blood pressure (BP) = Peripheral resistance (PR) x Cardiac output (CO)

Blood pressure, clinically, is generally measured in millimetres of mercury (mmHg) although the SI unit for pressure is the pascal (Pa) or kilopascal (kPa). Blood pressure has two readings: **systolic** which represents the highest pressure reached in the arteries after ventricular systole, and **diastolic**, which is the lowest pressure reached during ventricular filling. A 'typical' blood pressure for a young adult would be 120/70 mmHg (16/9.3kPa). You might be wondering why the diastolic pressure does not fall to zero – this is because the arteries are never quite empty and the next systole occurs before the pressure can fall further. The difference between systolic and diastolic pressures is termed the **pulse pressure**.
Factors contributing to blood pressure:

- Peripheral resistance (page 229).
- Cardiac output (dependent on rate and stroke volume) (pages 218, 219).
- Blood volume, a low volume reduces blood pressure, e.g. after haemorrhage, and an increased volume raises blood pressure (page 232).
- Venous return (page 227).
- Blood viscosity which affects peripheral resistance (Chapter 9 and page 229).
- Elasticity of large arteries; arterial disease will cause blood pressure to rise as blood is forced into non-distending vessels (page 224 and Special Focus).

Variations in blood pressure occur in healthy individuals as their situation changes, e.g. posture, exercise, strong emotions or stress causing the release of adrenal hormones and general level of activity.

An increased blood pressure or volume stretches the atria, which produce *atrial natriuretic peptides* (ANP, see Chapter 8). These hormones prevent renal absorption of water and sodium and inhibit the secretion of renin and ADH. This, with general vasodilation, reduces the blood pressure.

Blood pressure also varies between races and in Western countries it rises with age. In general, men have higher blood pressure than women.

Problems with blood pressure

Hypertension

Blood pressure (BP) sustained above the 'normal' level for age and gender is known as *hypertension*. The general concensus is that a diastolic pressure of 90–95 mmHg is abnormal. Drug therapy with beta blockers, antihypertensives, calcium channel blockers and diuretics (see Chapter 15) can be effective in reducing BP and associated cardiovascular disease such as stroke. Affected individuals should also be encouraged to adopt a healthy lifestyle (see page 213).

Around 20–30% of adults in Western countries have an elevated blood pressure. In many cases this will be discovered by screening before problems arise. Hypertension may be *essential* (primary), where the cause is unknown, or *secondary* to conditions such as renal disease, coarctation (compression) of the aorta or the rare adrenal medulla tumour (*phaeochromocytoma,* Chapter 8). High blood pressure may result in cardiac failure, renal failure and arterial disease leading to stroke and myocardial infarction.

Nursing Practice Application **Recording BP**

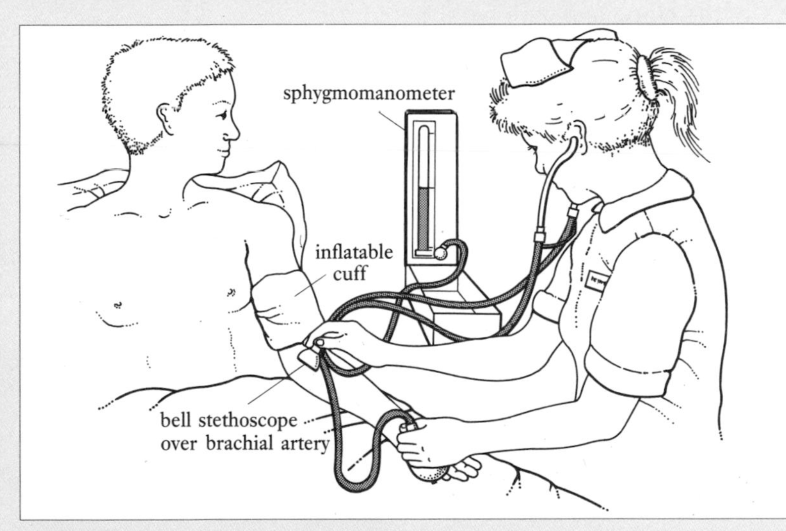

Figure 10.21. Measuring blood pressure.

Measuring BP is a skill and nurses should be aware of and take account of the many factors which may affect accuracy and consistency. The person should be at rest, with his or her arm supported at heart level and without constricting clothing around the upper arm (*Figure 10.21*). It seems rather obvious that the equipment should be in working order, but a study in a general hospital found only a third of the sphygmomanometers (sample 110) to be in proper working order (Bell and Siklos, 1984). Where a portable boxed sphygmomanometer is used, it should be on a flat surface and the observer's eye should be level with the column of mercury. The cuff should be of the correct size and applied correctly over the brachial artery.

There should be a definite nursing policy regarding which phase of *Korotkov sounds* (sounds heard whilst recording blood pressure), *phase 4* or *phase 5*, will represent diastolic pressure (*Figure 10.22*). More consistent results are obtained if one nurse records blood pressure for an individual patient over the whole duty shift, this is of particular importance where frequent observations are required for assessment, e.g. following head injury (see Chapter 4). The environment should be quiet if possible.

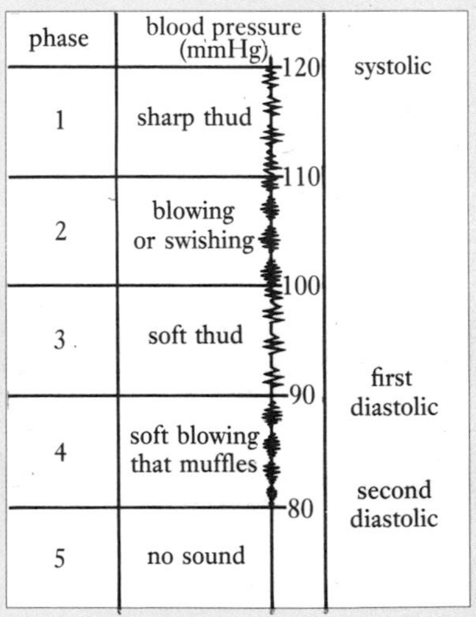

Figure 10.22. Korotkov sounds.

Hypotension

Many people function perfectly well on a blood pressure lower than the average values. Systolic pressure below 100 mmHg is usually accepted as *hypotension*. A chronically low BP may bc caused by adrenal failure (*Addison's disease*, see Chapter 8). Sudden falls in BP cause fainting if cerebral blood flow is inadequate. This may be a transitory event associated with, rising quickly, standing in the heat, lack of food or intense excitment. Much more serious is a situation, such as shock, where the blood pressure is too low to maintain circulation and organ perfusion, e.g. after major haemorrhage.

Shock

Shock (*Figure 10.23*) is a syndrome (collection of signs and symptoms) where there is a reduction in the effective circulating blood volume. It is characterized by clinical features which include:

- *Hypotension* – due to reduced cardiac output and blood volume.
- *Tachycardia/weak pulse* – heart rate increases in an attempt to increase output.
- *Pallor* – peripheral vasoconstriction as blood flow is diverted to vital organs, e.g. heart and brain.
- *Cyanosis* (seen first in the nail beds and lips) – poor oxygenation of the blood and peripheral 'shutdown'.
- *Oliguria* – reduced renal perfusion initiates renin–angiotensin–aldosterone response and the release of ADH (see Chapters 8 and 15) which reduce fluid loss in urine.
- *Altered consciousness/confusion* – inadequate oxygenated blood reaching the brain.
- *Sweating* – due to the action of the sympathetic nervous system.

The body attempts to minimize the effects of shock through baroreceptor and sympathetic activity which increase: vasoconstriction and peripheral resistance, blood volume (drawn from tissue fluid), venous return, heart rate and cardiac output. Adrenaline and noradrenaline enhance sympathetic effects and the release of corticosteroids ensures a supply of energy. Fluid-conserving mechanisms involving ADH and renin–angiotensin (also a powerful vasoconstrictor) are initiated. Autoregulatory processes help to ensure that 'vital' organs are perfused.

If the state of shock is not reversed the cells become *hypoxic* (deficient in oxygen), dilation of the capillary network occurs and the vessels become permeable, allowing fluid to leak from the blood. This loss of fluid increases the degree of shock by reducing the circulating blood volume and blood pressure. Irreversible cell damage and eventual death will occur if this 'vicious circle' of events remains unbroken. The events occurring in shock can be described by using the stages of the triphasic stress response (see Chapter 6).

Types and causes of shock

Hypovolaemic (decreased blood volume)

The causes of hypovolaemia include :

- Haemorrhage.

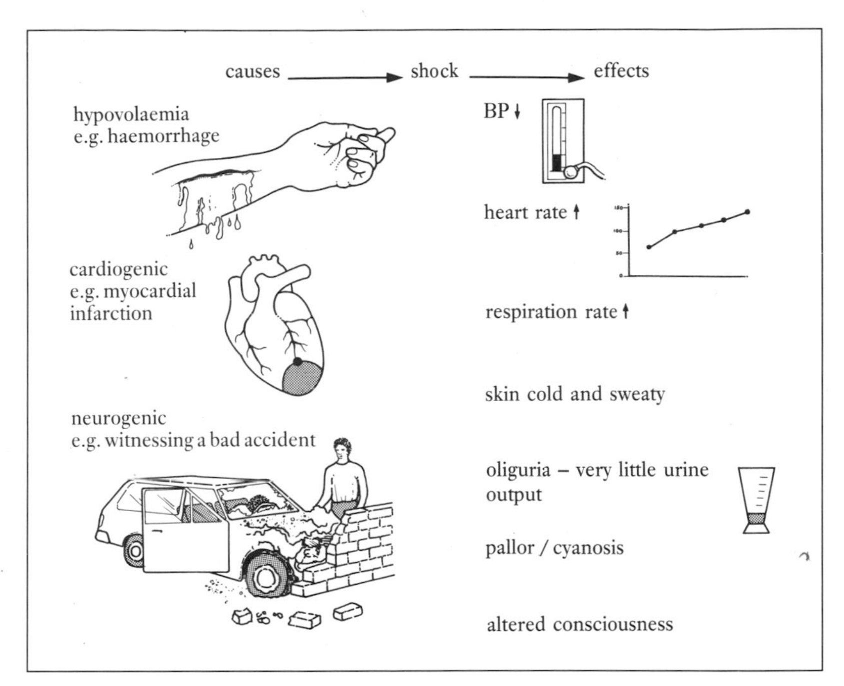

Figure 10.23. Causes and effects of shock.

- Loss of other fluids from the body, e.g. burns when plasma is lost, vomiting or diarrhoea.
- *Septicaemia* (circulation and multiplication of micro-organisms within the blood) or *anaphylaxis* (a state of shock caused by an intense antigen-antibody reaction) (some authorities classify these separately; see also Chapter 19) causing vasodilation and eventual movement of fluid from the vascular compartment as vessel permeability increases. Vasodilation increases the capacity of the circulation and movement of fluid from the blood results in hypovolaemia.

Cardiogenic (pump failure)
Causes of cardiogenic shock include extensive myocardial infarction, pericardial tamponade (collection of blood in the pericardium which obstructs the movements of the heart) and pulmonary embolus.

Neurogenic
It is possible that nervous pathways may operate in some situations (Thomson and Cotton, 1983). The shock results from changes in vasomotor tone and may follow events such as witnessing a major road accident or receiving very bad news. This mechanism of shock is not universally accepted.

Capillary bed and exchange of molecules

It is within the capillary network that oxygen and nutrients diffuse from the blood through the interstitial fluid to the cells. Waste products of cell metabolism make the reverse journey. All this depends on the delicate balance between **hydrostatic** and **osmotic pressure** (*Figure 10.25*).

Nursing Practice Application **Monitoring condition in shock**

The vital signs normally used to monitor an individual's condition may be inadequate in shock. Blood pressure may only be affected in the late stages and may be inaccurate (see page 232). *Central venous pressure* (CVP) which measures pressure in the right atrium is a useful guide to condition and fluid requirements. A catheter passed into the right atrium via the subclavian or jugular vein is attached to an intravenous (IV) administration set with manometer and three-way tap (*Figure 10.24*).

This arrangement with a three-way tap allows direct communication between catheter and manometer. The level of fluid in the manometer in cmH_2O pressure represents the central venous pressure. When measurements are not being taken the tap position is changed and the IV fluid infusion resumed.

Where an individual has serious heart or lung malfunction (e.g. after extensive myocardial infarction or massive pulmonary embolus) the CVP is not always a dependable indicator of left heart function. Where it is helpful to have accurate information about the left side of the heart a *Swan-Gantz* catheter passed via the veins into the right atrium is allowed to float through the ventricle to wedge in a small pulmonary artery. Inflation of a small balloon at its tip wedges the catheter and blocks off the pulmonary artery behind the balloon. Now the pressure in the pulmonary capillaries, and hence the left atrium and ventricle, is measured. The catheter, which is attached to a pressure transducer, records *pulmonary capillary wedge pressure* (PCWP), which gives information regarding left ventricular pressure at the end of diastole. PCWP is recorded at intervals and between measurements the balloon is deflated. When the balloon is deflated the catheter records pulmonary artery pressure.

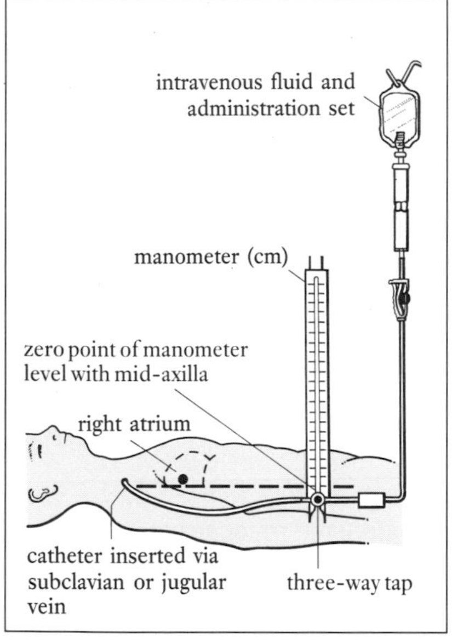

Figure 10.24. Central venous pressure.

The hydrostatic pressure (see page 228) at the arterial end of the capillary is sufficiently high to overcome the osmotic pressure (see Chapter 2) exerted by the non-diffusible plasma proteins (mainly albumin) and some inorganic ions (see Chapter 9). Fluid and nutrients consequently move through the permeable capillary wall into the interstitial spaces. Hydrostatic pressure falls (see page 228) across the capillary network and at the venous end the hydrostatic pressure is lower than the osmotic pressure. This results in fluid and waste being drawn back into the capillary. The minimal amount of fluid remaining within the interstitial compartment drains into lymphatic vessels and is returned to the circulation (see Chapter 11).

Molecules enter and leave cells by various processes which allow

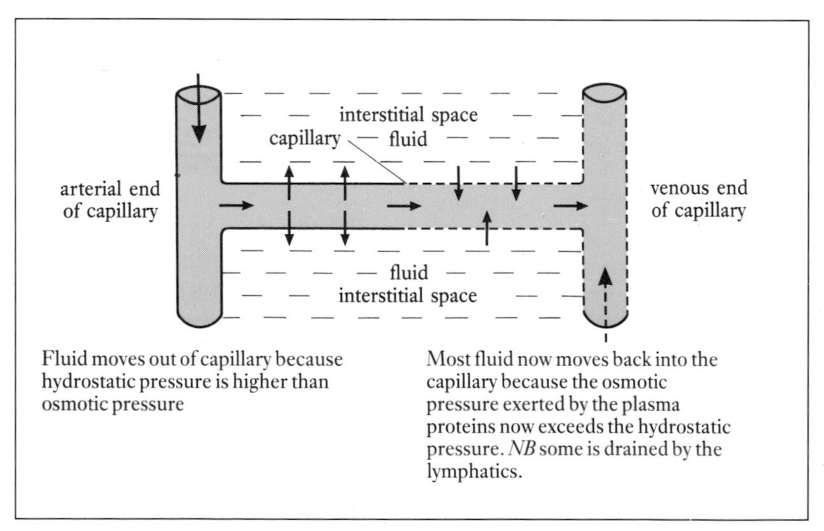

Figure 10.25. Formation of interstitial fluid (tissue fluid) – hydrostatic and osmotic pressure.

them to cross the cytoplasmic membrane (*Figure 10.26*). These processes include; osmosis, diffusion, active transport and bulk transport (see Chapters 1 and 2).

Slow blood flow through the capillaries gives time for these processes to occur. Autoregulatory mechanisms (see page 228) ensure that capillary flow is appropriate for local cellular needs.

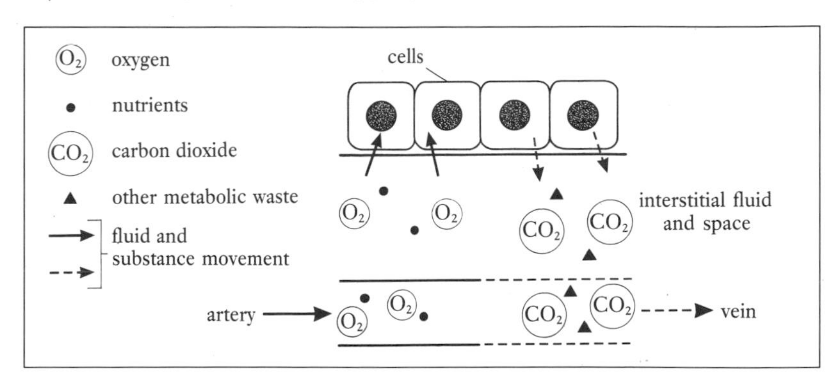

Figure 10.26. Exchange of molecules between blood and cells.

Homeostatic imbalance – oedema

Oedema is swelling caused by excess interstitial fluid collecting in the tissues. The swelling is usually most evident in the dependent areas, such as feet, ankles and legs when standing, or the sacral area if the person is confined to bed. Fluid may also collect abnormally in the peritoneal cavity (*ascites*), between the pleurae (*pleural effusion*) or pericardium (*pericardial effusion*). Problems affecting the pulmonary capillaries give rise to *pulmonary oedema* where the fluid collects within the alveoli.

Oedema will result from a failure of any part of the homeostatic mechanisms involved in maintaining the fluid compartments. Oedematous tissue becomes hypoxic and metabolites accumulate, which renders the tissues more susceptible to injury. This has serious implications for nursing practice as this waterlogged, poorly oxygenated tissue is more likely to break down and develop pressure sores (see Chapter 19).

Main causes of oedema

Figure 10.27 illustrates the possible causes of oedema:

- Reduced osmotic pressure in the blood caused by a decrease in serum protein levels (*hypoproteinaemia*), this may be due to liver disease, starvation or protein loss from the kidneys.
- Increased venous hydrostatic pressure such as in congestive cardiac failure (see pages 221-222), increased blood volume of pregnancy, fluid or sodium retention and varicose veins.
- Increased capillary permeability such as that caused by inflammatory chemicals.
- Obstruction in the lymphatic vessels, e.g. malignant cells from the breast may block the lymphatics in the arm, causing oedema. This happens because excess fluid in the interstitial space is unable to drain away.

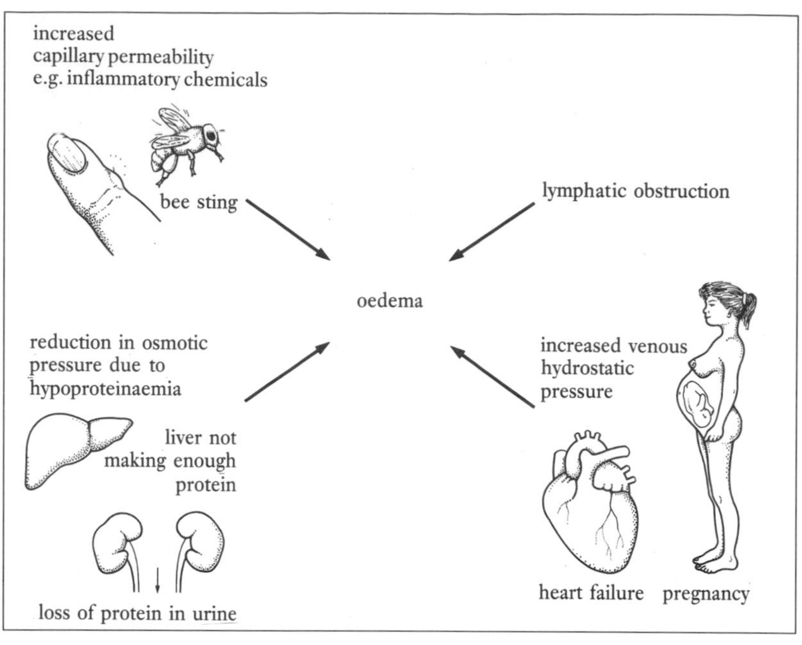

Figure 10.27. The causes of oedema.

CIRCULATION

Circulation consists of two main systems – the low-pressure pulmonary circulation through the lungs and the higher pressure systemic circulation, which supplies blood to all cells and tissues. A detailed account of every vessel and its location is not within the scope of this book. With this in mind, a brief coverage of major vessels is included, which should be sufficient for everyday nursing practice. Where readers require more specific information they are directed to a detailed anatomical work, e.g. Williams *et al.* (1989).

Pulmonary circulation

The **pulmonary circulation** is concerned with transporting blood low in oxygen and high in carbon dioxide to the lungs, where it gains oxygen and loses carbon dioxide. Dark red venous blood returning to the right side of the heart is pumped into the **pulmonary trunk** by ventricular systole. This large artery divides into the right and left **pulmonary arteries** which serve the different lungs. These divide into smaller branches which eventually become arterioles feeding into a capillary network. It is here that the intimate contact between capillaries and lung alveoli allows the process of gaseous exchange (*Figure 10.28*). Gases diffuse down pressure gradients. This means that oxygen passes from alveolar air into the blood and carbon dioxide moves from the blood to the alveolar air, a process known as *external respiration* (see Chapter 12).

The blood, now high in oxygen and low in carbon dioxide, is transported by venules to larger veins which eventually form four **pulmonary veins** (two from each lung). The pulmonary veins return

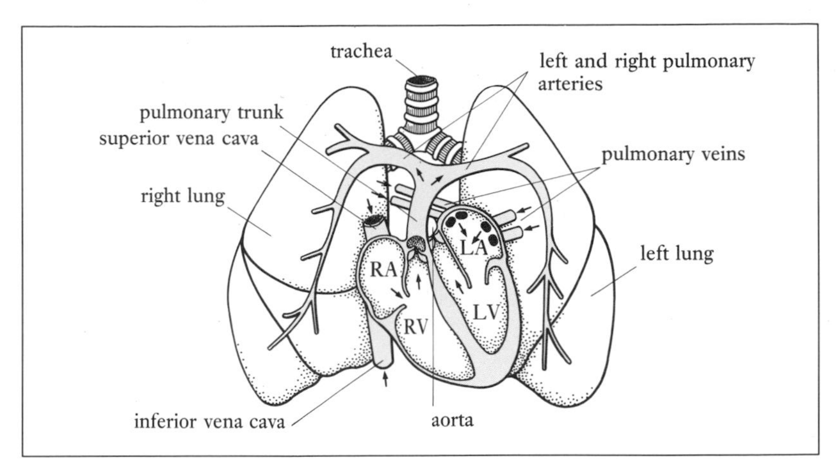

Figure 10.28. Pulmonary circulation.

the bright red oxgenated blood to the left atrium completing the pulmonary circuit.

NB It is worth remembering that the pulmonary circulation does not supply the lungs with arterial blood (see page 241).

Systemic or general circulation

The systemic circulation circulates oxygen-rich blood via the main artery (aorta) and its branches (*Figure 10.29*).

These branches eventually form the arterioles, which supply the capillary networks in individual organs and tissues. *Internal respiration* or gaseous exchange between blood and cells ensures that there is oxygen for *cellular respiration* (energy production) and that carbon dioxide is removed from the tissues. Venules collecting the oxygen-depleted blood become veins that form the venae cavae, which drain into the right atrium to complete the circuit (*Figure 10.30*).

The general circulation is rather more complex than this and, as an aid to understanding, we shall first consider the arterial supply to a particular area and where appropriate the corresponding venous return.

Aorta

The **aorta** ascends from the left ventricle, arches over and descends behind the heart to pass through the thorax and abdomen. Branches of paired or single arteries exit from the aorta as it descends. Many of these are named for the organ they supply. The first branches from the ascending aorta are the **coronary arteries** (see *Figure 10.6*) which supply the myocardium. Venous blood from the myocardium returns to the right atrium via the **coronary sinus** and **cardiac veins** (see page 211). Baroreceptors concerned with regulation of blood pressure are situated in the aortic arch (see pages 229,230 and *Figure 10.20*).

Aortic arch

Three arteries branch from the aortic arch; **left common carotid, left subclavian** and the **brachiocephalic (innominate)** which becomes the **right subclavian** and **right common carotid** (*Figure 10.31*). These arteries supply the neck, head and upper extremities.

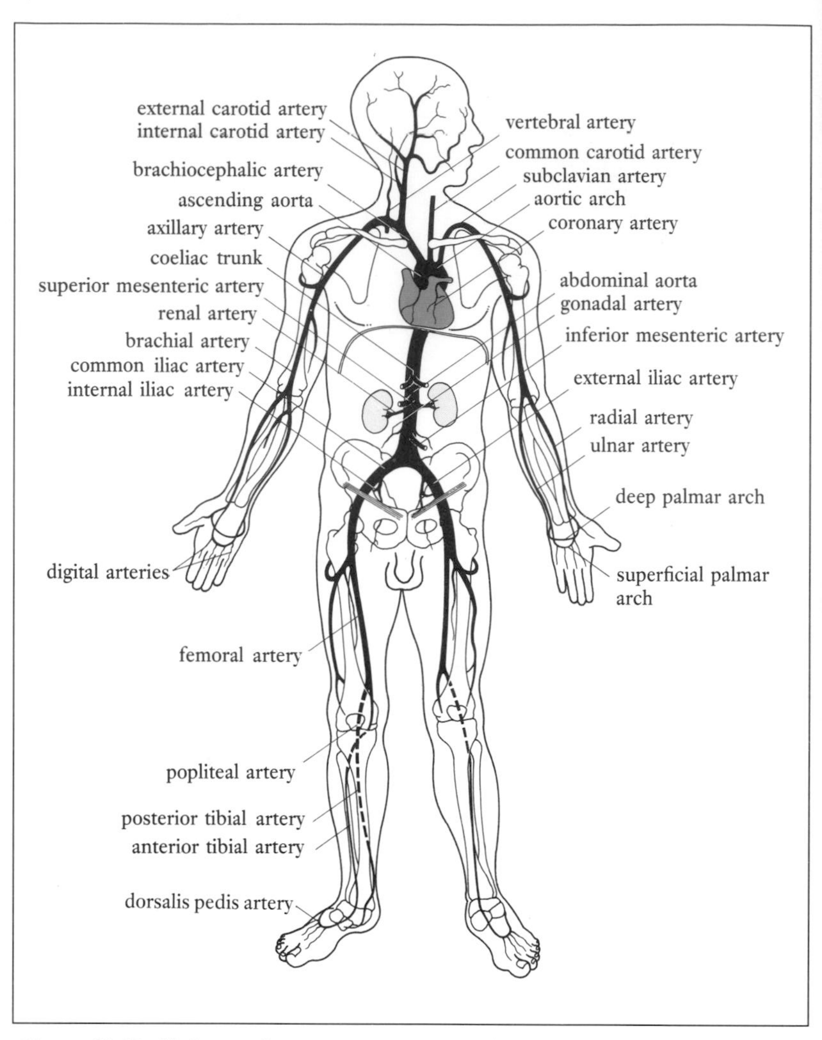

Figure 10.29. Main arteries.

Superior vena cava and venous returna

Venous return is via the superior vena cava formed from the **right** and **left brachiocephalic veins** which are derived from the **subclavian** and **internal jugular veins** (*Figure 10.32*).

Blood vessels of the head and neck-arteries

The main arteries supplying the head and neck are the two common carotids which run one up each side of the neck. Baroreceptors are found where the arteries divide into **internal** and **external carotid arteries** in an area known as the **carotid sinus**. Close by are the *carotid bodies* containing chemoreceptors which regulate respiratory rate (see Chapter 12).

The external carotid forms arteries to the face (**facial**), tongue (**lingual**), scalp (**temporal, occipital**), muscles of mastication (**maxillary**), bones of the skull and dura (**middle meningeal**), and neck and associated structures, e.g. thyroid (*Figure 10.33*).

The **internal carotid arteries** enter the skull through the temporal bones. They contribute to the **circle of Willis** (*Figure*

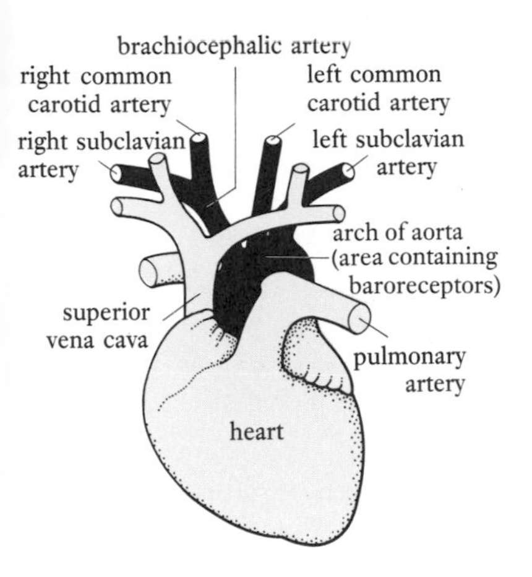

Figure 10.31. Aortic arch.

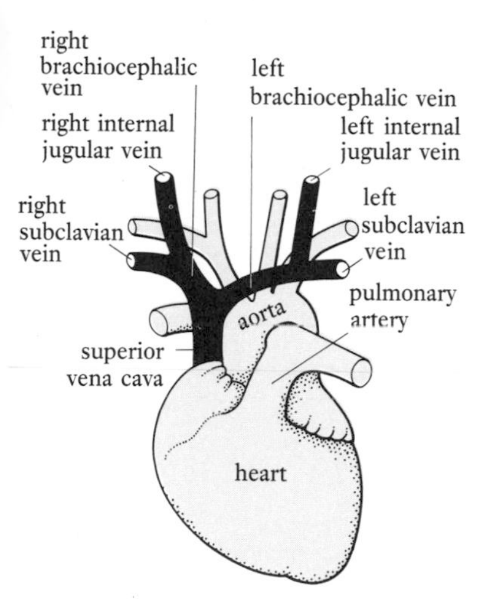

Figure 10.32. Superior vena cava.

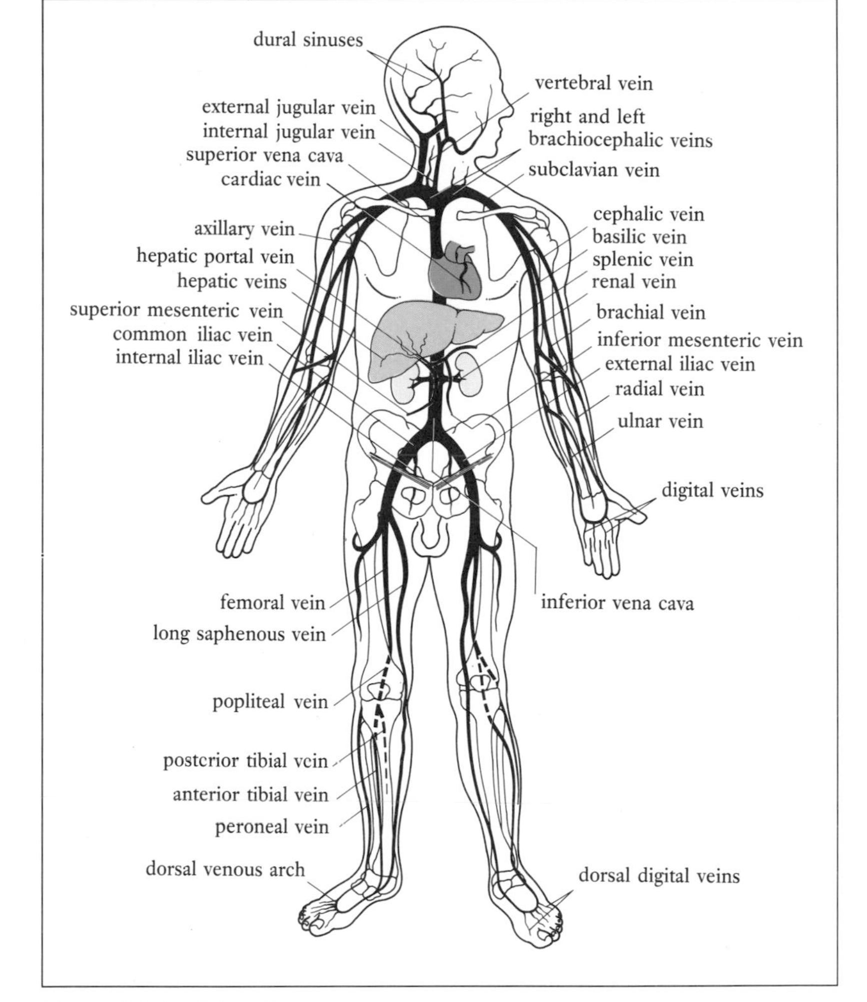

Figure 10.30. Main veins.

10.34) and supply the eye (**ophthalmic artery**). The circle of Willis (an *arterial anastomosis* – see Chapter 4) supplying the brain is formed from the **anterior cerebral arteries** (internal carotids), **communicating arteries** and the **posterior cerebral arteries** formed from the **basilar artery**. The basilar artery is formed from a unification of two **vertebral arteries** (derived from the subclavian arteries) which enter the skull through the foramen magnum. The brain stem is also supplied by the basilar artery.

Blood vessels of the head and neck – venous return

Most blood from superficial areas drains via veins (the names of which correspond with the arteries branching from the external carotid) which form the **external jugular veins** in the neck. The external jugular veins pass over the *sternocleidomastoid muscles* (see Chapter 18), behind the clavicles to empty into the subclavian veins (*Figure 10.35*).

Venous blood from the brain empties into **venous sinuses**

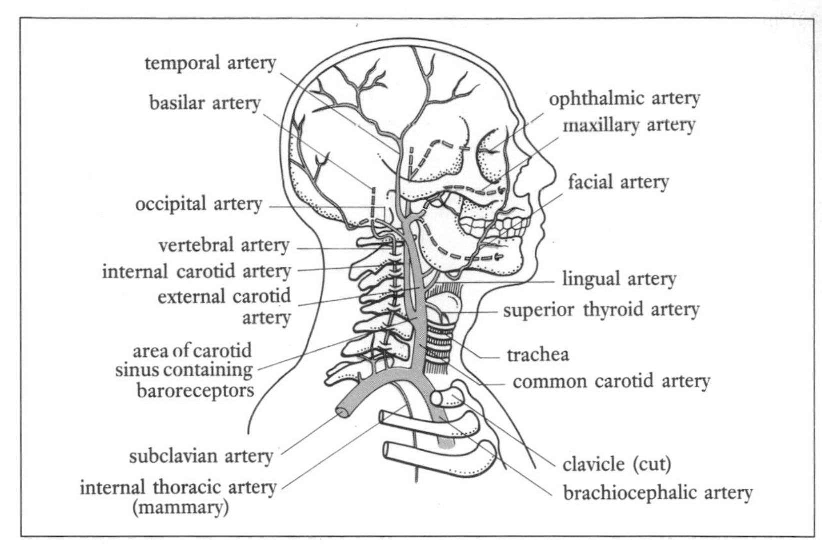

Figure 10.33. Arteries - head and neck.

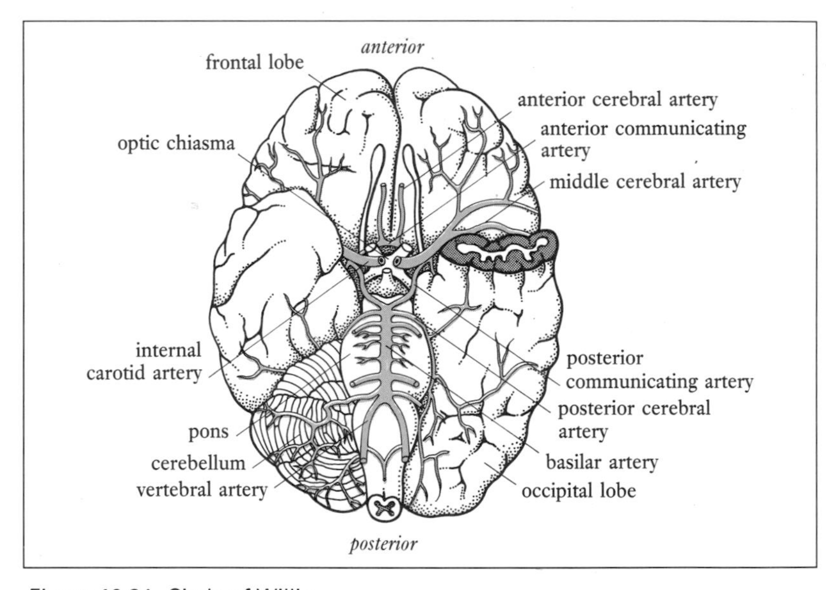

Figure 10.34. Circle of Willis.

(*Figure 10.36*) which are enclosed in the fibrous *dura mater* (Chapter 4). The main venous sinuses include the **superior** and **inferior sagittal sinuses**, **straight sinus**, two **transverse sinuses**, the **sigmoid** and **cavernous sinuses**, which drain into two **internal jugular veins**. *Arachnoid granulations* projecting into the superior sagittal sinus return cerebrospinal fluid to the circulation (see Chapter 4).

The internal jugular veins also receive blood from the deep structures of the face and neck as they pass behind the sternocleidomastoid muscles before entering the subclavian veins to form the right and left brachiocephalic veins.

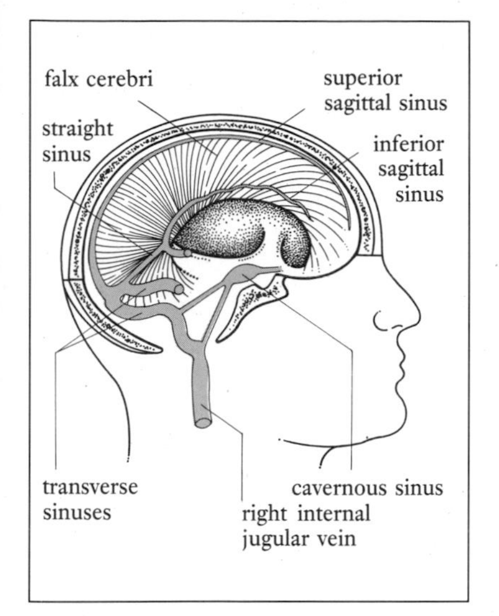

Figure 10.36. Venous sinuses of the brain.

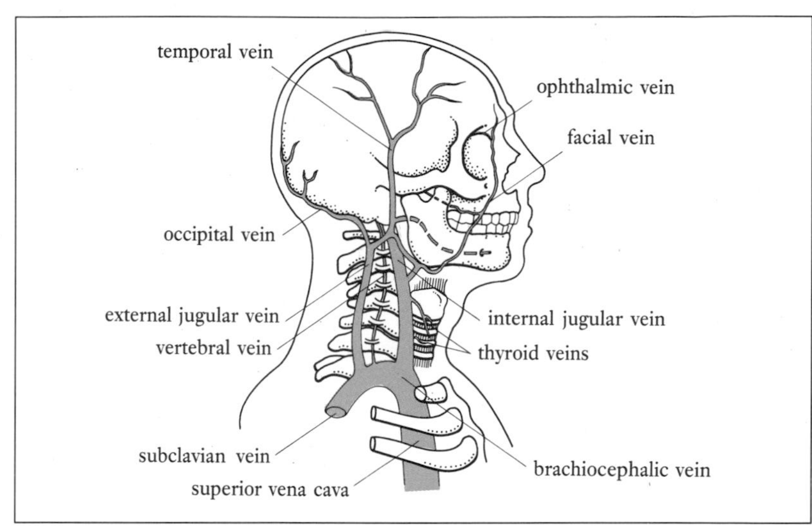

Figure 10.35. Veins - head and neck.

Blood vessels of the upper extremity – arteries

The subclavian artery supplies branches to the neck, thoracic cavity and breast (**internal mammary artery**) before passing between the clavicles and first rib to enter the axilla. Now called the **axillary artery** it provides branches to the shoulder joint, axilla and muscles of the chest wall. On entering the arm it is known as the **brachial artery**, which supplies the upper arm. As it passes over the inner aspect of the elbow joint it provides a convenient point for the measurement of arterial blood pressure (see *Figure 10.21*). Below the elbow the brachial artery divides into the **radial** and **ulnar arteries** which supply the lateral and medial aspects of the forearm. Where the radial artery passes close to the surface in the wrist it can be used to count pulse rate (see page 224). The radial and ulnar arteries anastomose to form the arteries supplying the hand: **superficial** and **deep palmar arches** and the **digital arteries** (*Figure 10.37*).

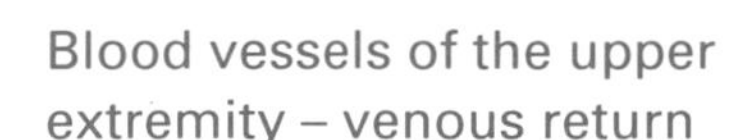

Blood vessels of the upper extremity – venous return

There are two groups of veins within the arm (*Figure 10.38*):

- A system of **deep veins** which take the name of the arteries and follow the same route – **deep** and **superficial palmar arches**, **ulnar**, **radial**, **brachial** and **axillary veins**, which empty in the subclavian veins.
- A system of **superficial veins** – **median**, **cephalic**, **basilic** and **median cubital veins**. These run close to the surface and are used to take samples of blood and for intravenous infusion or injection.

Thoracic blood vessels and descending aorta

The aorta passes behind the heart and travels through the thoracic cavity giving off branches to the chest wall **(posterior intercostal arteries**). Many branches leave the thoracic aorta and include those supplying the oesophagus (**oesophageal arteries**), lungs/bronchi and pleura (**bronchial arteries**) and pericardium (**pericardial artery**) (*Figure 10.39*).

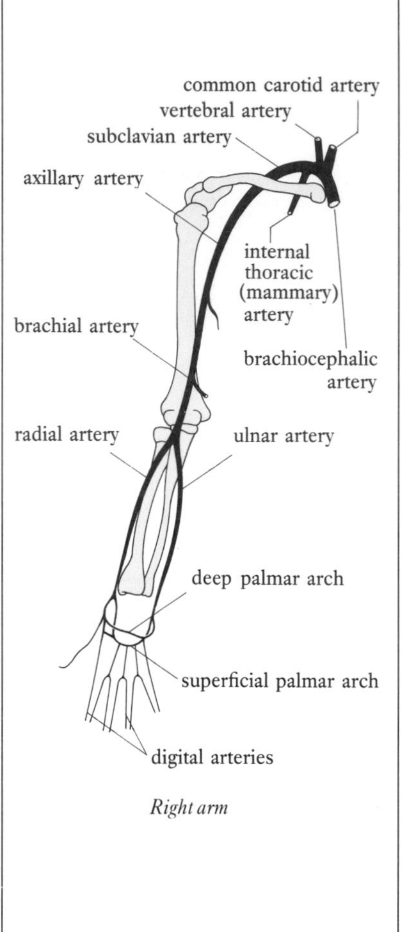

Figure 10.37. Arteries of the upper extremity.

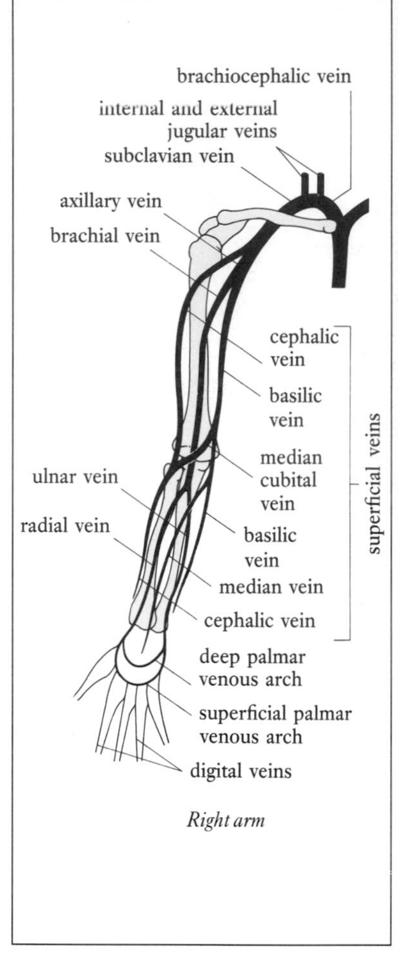

Figure 10.38. Veins of the upper extremity.

Venous return from the thorax

Major veins, which include **intercostal, oesophageal,** and **bronchial veins** (not shown on *Figure 10.40*) empty into the **azygos system** of veins which run up the sides of the vertebral column. The **hemiazygous vein** joins the **azygos vein**, which empties into the superior vena cava. An **accessory hemiazygos vein** also drains the thorax. In addition the azygos system of veins receives blood from the abdominal cavity via the **ascending lumbar veins** (*Figure 10.40*).

NB The lower oesophageal veins drain via the hepatic portal system.

Abdominal aorta and branches

The thoracic aorta passes through the diaphragm to become the **abdominal aorta**. It travels down the posterior wall of the abdomen to the level of the 4th lumbar vertebra where it divides into the two **common iliac arteries**. Branches from the abdominal aorta include (see *Figure 10.41*):

- Unpaired **coeliac, superior** and **inferior mesenteric arteries** (see page 243).

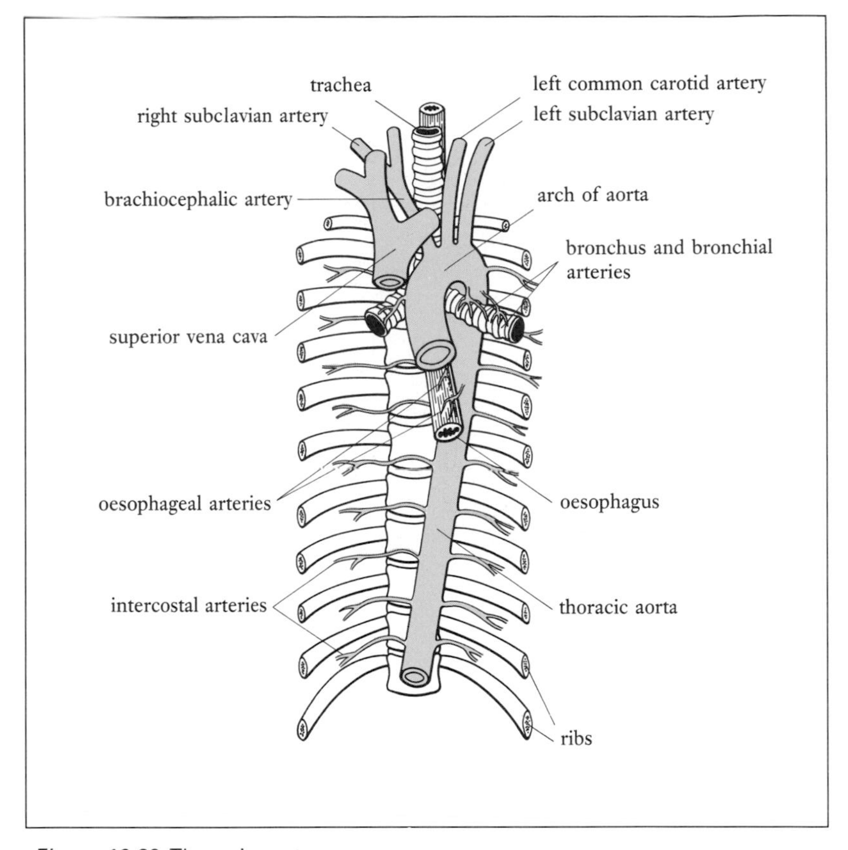

Figure 10.39. Thoracic aorta.

- Paired arteries supplying the diaphragm (**inferior phrenic arteries**) and the posterior abdominal wall (four **lumbar arteries**).
- Paired visceral arteries supplying organs: adrenals (**suprarenal arteries**), kidneys (**renal arteries**) and the gonads (**ovarian** or **testicular arteries**).

Arterial supply to the gastrointestinal tract and liver

The coeliac artery (trunk) divides into three branches (*Figure 10.42(a), page 246*).

- The **left gastric artery** supplying the stomach and lower oesophagus.
- A **common hepatic artery** which sends branches to the stomach, duodenum and pancreas before dividing into **right** and **left hepatic arteries** supplying the liver and gallbladder (see Chapter 14).
- The **splenic artery** which branches to supply the stomach, pancreas and spleen.

The superior mesenteric artery and its branches supply the pancreas, entire small bowel and the proximal part of the large bowel.

The inferior mesenteric artery and its branches supply the distal large bowel and most of the rectum (*Figure 10.42(b)*).

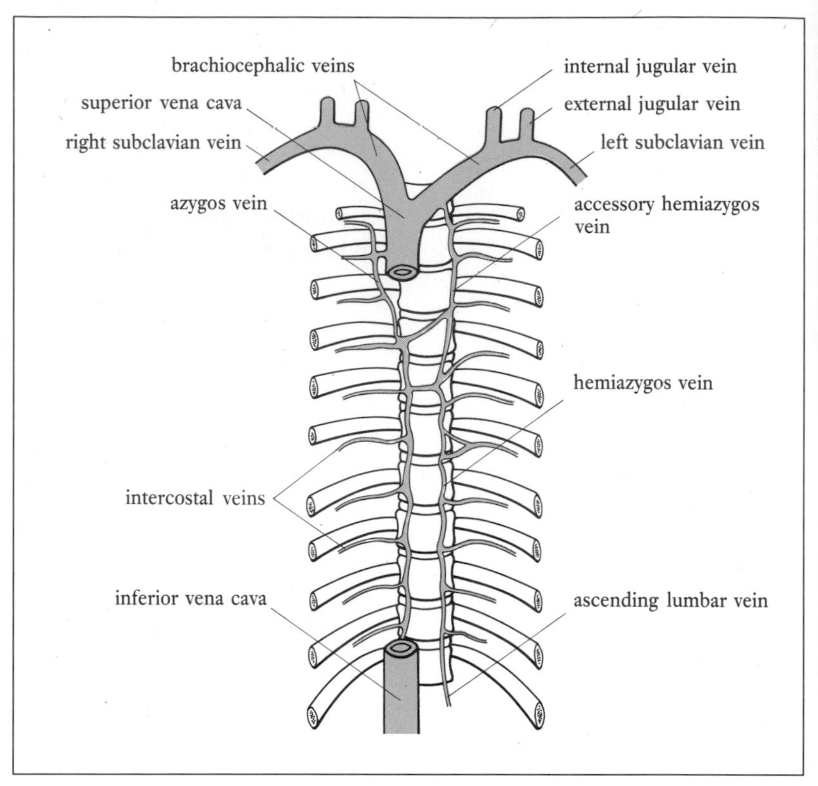

Figure 10.40. Venous return from the thorax.

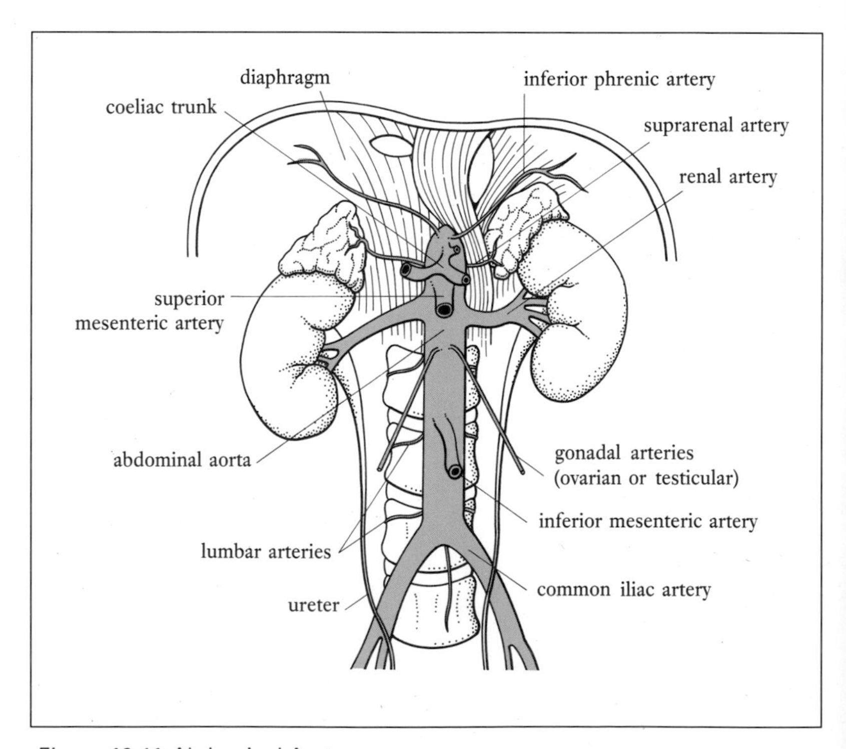

Figure 10.41. Abdominal Aorta.

Venous return from the abdominal cavity

Most of the venous blood from the structures supplied by paired arteries drains via veins of the same name emptying into the **inferior vena cava (IVC)** (see *Figure 10.43*). The IVC, formed by the two **common iliac veins,** carries blood from the lower part of the body. It passes through the central tendon of the diaphragm to the right atrium. The **renal**, **lumbar**, **right suprarenal** and **right gonadal veins** all drain directly into the IVC but the **left suprarenal** and **gonadal veins** drain via the left renal vein.

The **right** and **left hepatic veins** drain venous blood from the liver directly into the IVC.

N.B. Venous blood from structures supplied by the coeliac and mesenteric arteries travels through the liver before entering the IVC (see hepatic portal circulation).

Hepatic portal circulation

The veins carrying blood from the lower oesophagus, stomach, spleen, pancreas, small bowel and large bowel/rectum drain into a common vessel called the **hepatic portal vein** (*Figure 10.44)*. The major vessels which form the hepatic portal vein include:

- **Oesophageal veins** – lower oesophagus (drain eventually into gastric veins).
- **Gastric veins** – stomach.
- **Splenic vein** – spleen and part of stomach and pancreas.
- **Superior mesenteric vein** – small bowel and proximal large bowel.
- **Inferior mesenteric vein** – distal large bowel and rectum.

This portal system ensures that venous blood passes through a second capillary bed (liver sinusoids) before returning to the general circulation. This rather complicated arrangement makes considerable sense, as nutrients absorbed from the gastrointestinal tract and hormones from the pancreas are conveyed directly to the liver for processing (see Chapters 8, 13 and 14).

Hepatic portal dysfunction

Problems occur when the pressure of blood in the hepatic portal vein is high (*portal hypertension*). This is often caused by *cirrhosis* (see Chapter 14) of the liver and leads to back pressure being transmitted to areas where the hepatic portal and systemic circulations anastomose.

Varicosities develop at these points, which include the gastro-oesophageal junction (*oesophageal varices*) and the rectum (*haemorrhoids*). Massive haemorrhage from oesophageal varices is a common feature of portal hypertension.

Blood vessels of the pelvis and lower extremities – arteries

The aorta divides to form the **right** and **left common iliac arteries**, which supply blood to the lower abdomen, pelvic structures and legs. At the level of the *sacroiliac joints* (see Chapter 18) the common iliac arteries split into the **internal** and **external iliac arteries** (see *Figure 10.45*).

Branches of the paired internal iliac arteries supply the pelvis, lower rectum, bladder, uterus, vagina, prostate and external genitalia.

The right and left external iliac arteries supply the lower limbs. In each leg, the external iliac arteries pass behind the *inguinal ligament* in the groin to become the **femoral artery** supplying groin and thigh. The femoral artery passes down the anterior thigh until, near the knee, it moves posteriorly to enter the *popliteal fossa*. Now

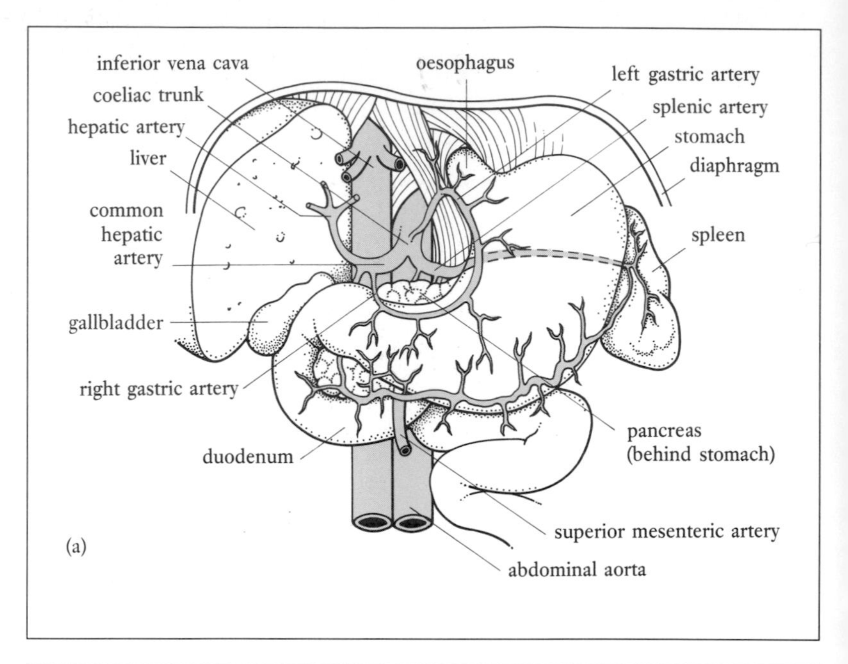

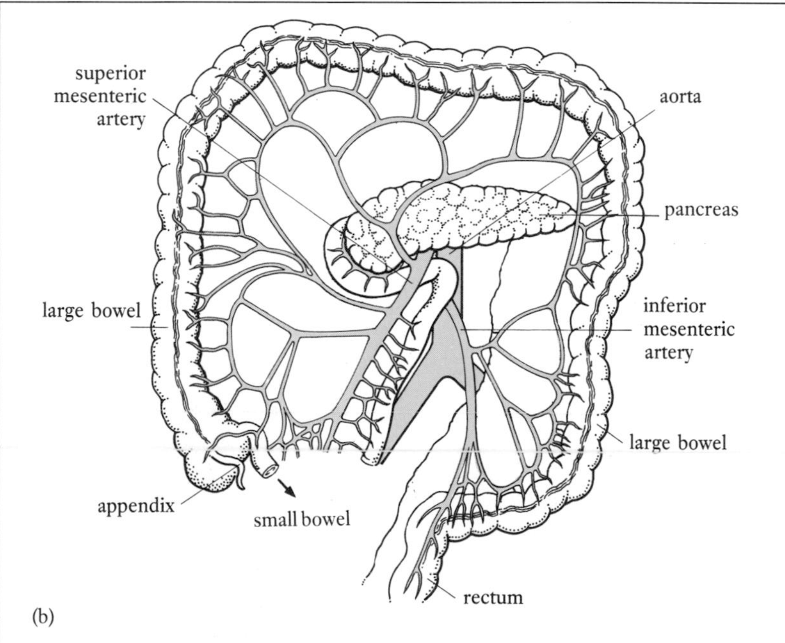

Figure 10.42. Arteries of the gastrointestinal organs. (a) Coeliac artery; (b) mesenteric arteries.

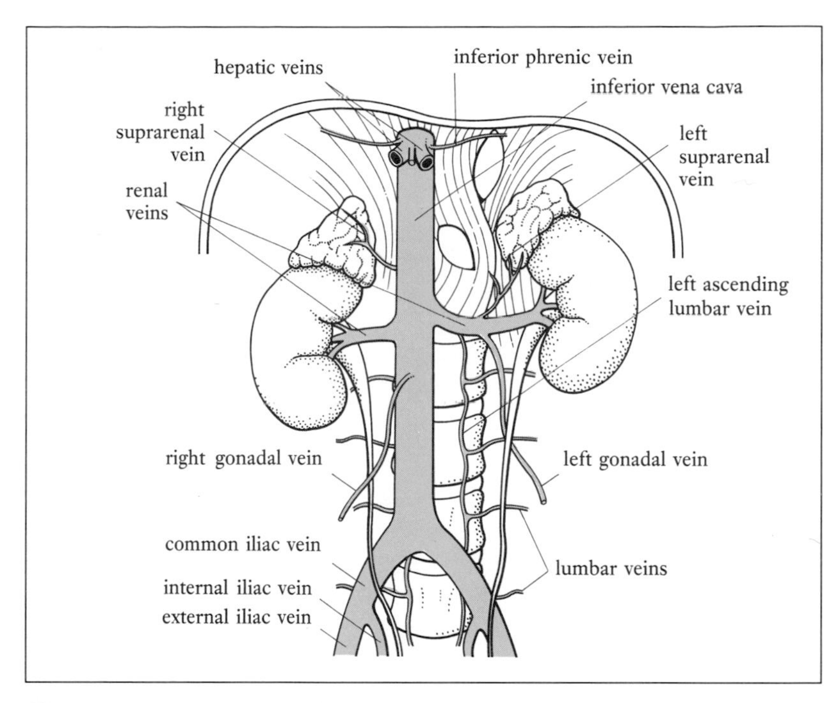

Figure 10.43. Venous return from the abdominal cavity.

known as the **popliteal artery** it divides into **posterior** and **anterior tibial arteries**.

The posterior tibial artery supplies the back of the leg and branches to form the **peroneal artery** which supplies the lateral aspects. It becomes the **plantar** and **digital arteries** supplying sole and toes. The anterior tibial artery supplies the front of the leg and at the ankle forms the **dorsalis pedis artery** (see pulses, page 226) which supplies the dorsal aspect of the foot. With the plantar arteries it forms the **plantar arch**.

Blood vessels of the lower extremities – venous return

The veins of the leg are in two groups, deep and superficial. The **deep veins** of the leg generally accompany and have the same names as the arteries (*Figure 10.46*). **Digital** and **plantar veins** drain the foot. They form the **posterior tibial vein** which travels deep within the muscles of the calf. An **anterior tibial** vein runs up the front of the leg to the knee where it joins the posterior tibial vein to become the **popliteal vein**. It leaves the knee to become the **femoral vein**, which travels up the thigh and passes behind the inguinal ligament to form the **external iliac vein**. **Internal iliac veins** drain the pelvic structures. The external and internal iliac veins join to become common iliac veins which unite to form the IVC. These deep veins of the leg and pelvis are a frequent site for the formation of deep vein thrombosis (see Chapter 9).

The two **superficial vessels** – **long (great)** and **short (small) saphenous veins** – run from the dorsal venous arch up the leg draining the superficial structures. Many communicating vessels

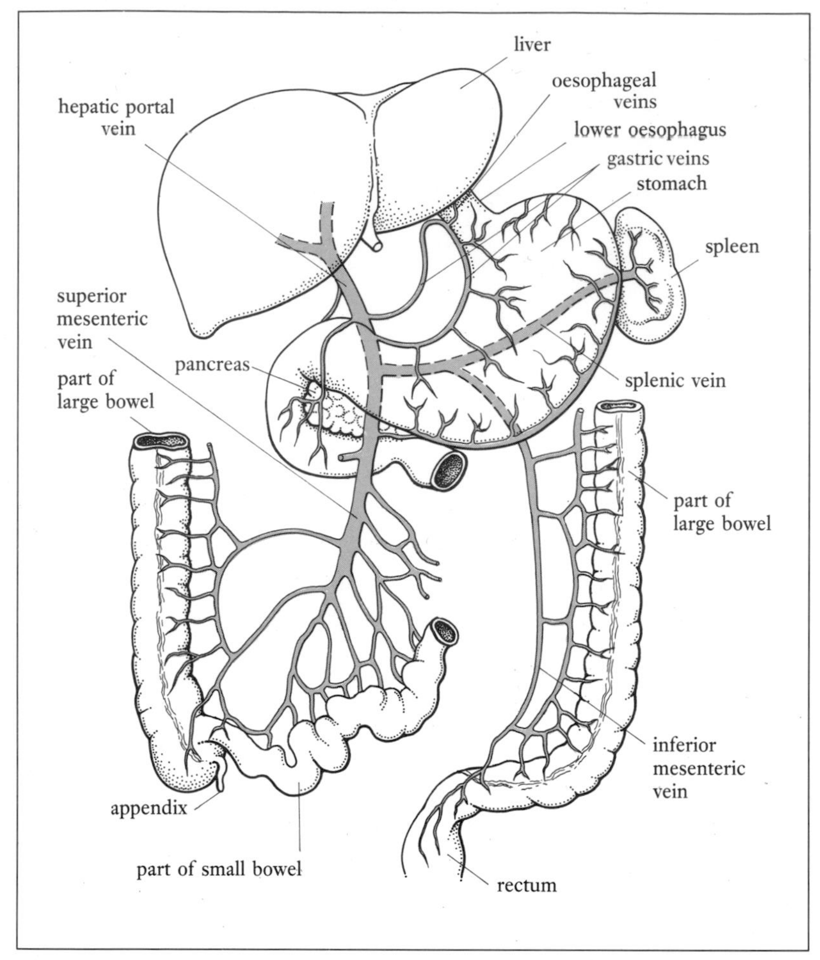

Figure 10.44. Venous drainage from gastrointestinal tract and the hepatic portal vein.

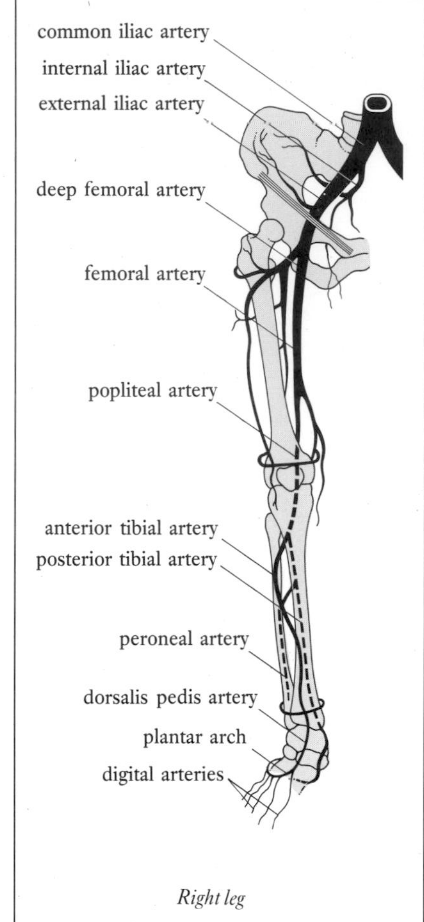

Figure 10.45. Arteries of the pelvis and lower extremities.

allow blood to pass between superficial and deep systems. The long saphenous, which is the longest vein in the body, joins the femoral vein just before the inguinal ligament. The many valves present in the saphenous veins are easily damaged by high pressure resulting in varicosities (see page 227). The long saphenous vein is used in coronary artery vein grafts where diseased coronary arteries are bypassed using a length of vein (see pages 212 and 213). The short saphenous vein drains into the deep system at the popliteal vein.

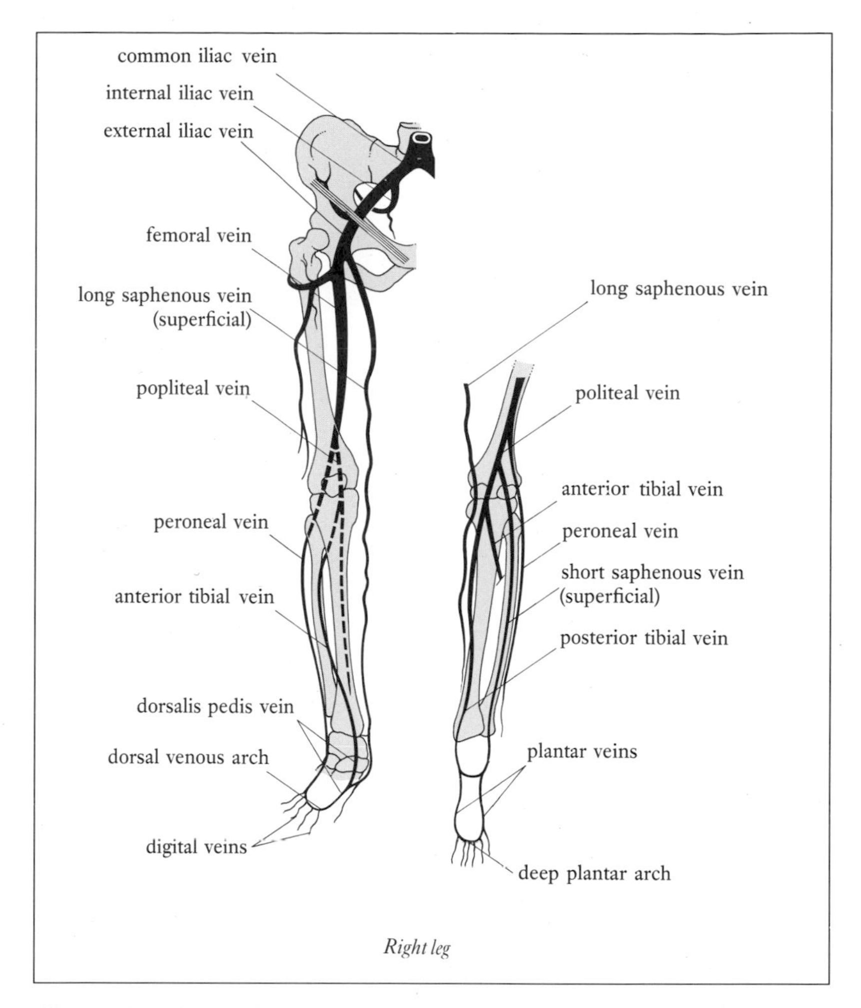

Figure 10.46. Veins of the lower extremities and pelvis.

Summary/Check List

Introduction: CVS.
Heart – Early development. Location. Shape. Size. Functions. Structure – wall, chambers, valves. Coronary circulation, *Special Focus* – coronary heart disease (angina pectoris, myocardial infarction), *Person-centred Study* – Derek. Blood flow direction. Conduction system – problems, pacemakers. ECG – normal and abnormal, *Nursing Practice Application* – cardiac arrest and resuscitation. Cardiac cycle. Cardiac output, Stroke volume. Heart rate, *Nursing Practice Application* – apex beat. Problems with cardiac output, *Nursing Practice Application* – cardiac glycosides.
Vessels – general structure. Arteries – arterial pulse, *Nursing Practice Application* – pulse. *Special Focus* – arterial disease (atherosclerosis, aneurysm). Capillaries, AV shunts. Veins, valves, *Nursing Practice Application* – venous return, varicose veins.
Circulatory physiology – Blood flow. Pressure. Resistance. Arterial blood pressure, *Nursing Practice Application* – recording blood pressure, problems with blood pressure. Shock, *Nursing Practice Application* – monitoring condition in shock. Capillary bed and exchange of molecules. Oedema.
Circulation – Pulmonary. Systemic. Hepatic portal.

?

Self Test

1 Complete the following;
(a) The heart is situated in the space between the lungs known as the ___________.
(b) In the heart the upper chambers are called the _____ and the lower chambers the _________.
(c) The middle layer of the heart is known as the ___________.
(d) The left atrioventricular valve has two names; ______ or ________.

2 Describe how heart muscle receives an adequate blood supply.

3 Draw a diagram to show the direction of blood flow through the heart.

4 Put the following components of the conduction system in the correct sequence:
(a) Bundle of His (b) SA node (c) Purkinje fibres
(d) AV node (e) Bundle branches.

5 Complete the following:
(a) CO = ? x SV
(b) EDV - ESV = ?

6 Arteries have the same basic structure as veins – true or false?

7 Describe ways in which blood flow through capillary beds is controlled.

8 Which factors contribute to arterial blood pressure?

9 Define 'shock'.

10 Explain how a low serum albumin level causes oedema.

11 Which of the following statements are true?
(a) Arteries only carry oxygenated blood.
(b) The pulmonary circulation is a low pressure system.
(c) The pulmonary artery supplies blood to the lung tissue.
(d) The first branches from the aorta are the coronary arteries.

12 Describe the structure(s) supplied/drained by the following vessels:
(a) vertebral arteries (b) internal iliac artery
(c) coeliac artery (d) renal vein
(e) azygos veins (f) long saphenous vein

Answers
1. a. mediastinum; b. atria, venticles; c. myocardium; d. mitral, bicuspid. 2. Page 211. 3. Page 211, 214. 4. b, d, a, e, c. 5. a. HR; b. SV. 6. True. 7. Pages 224,226. 8. Page 230.9. Page 232. 10. Page 234-235. 11. b, d. 12. a. brain; b. pelvic structures; c. oesophagus, stomach, duodenum, liver, gallbladder, pancreas and spleen; d. kidney; e. thoracic structures; f. superficial tissues of the leg.

References

Ball, M. and Mann, J. (1988) *Lipids and Heart Disease: A Practical Approach* Oxford: Oxford University Press.

Bell, M. J. and Siklos, P. (1984) Recording blood pressure: accuracy of staff and equipment. *Nursing Times*, **80** (26), 32–34.

Burr, M.L., Gilbert, J.F., Holliday, R.M., Elwood, P.C., Fehily, A.M., Rogers, S., Sweetnam, P.M. and Deadman, N.M. (1989) Effects of changes in fat, fish and fibre intakes on death and myocardial reinfarction: Diet And Reinfarction Trial (DART). *Lancet*, **ii**, 757–761.

Deans, W. and Hoskins, R. (1987) Preventing coronary heart disease. *The Professional Nurse*, **2** (10), 328–329.

Hay, C.R.M, Durber, A.P, and Saynor, R. (1982) Effect of fish oil on platelet kinetics in patients with ischaemic heart disease. *Lancet*, **i**, 1269–1272.

Murray, P. J. (1989) Rehabilitation information and health beliefs in post-coronary patients: do we meet their information needs? *Journal of Advanced Nursing*, **14** (8), 686–693.

Siess, W., Roth, P., Scherer, B., Kurzman, I., Bohlig, B. and Weber, P.C. (1980) Platelet-membrane fatty acids, platelet aggregation and thromboxane formation during a mackeral diet. *Lancet*, **i**, 441–444.

Thomson, A.D. and Cotton, R.E. (1983) *Lecture notes on pathology*. 3rd edn. Oxford: Blackwell Scientific Publications.

Thompson, D. (1989) A randomized controlled trial of in-hospital nursing support for first time myocardial infarction patients and their partners: effects on anxiety and depression. *Journal of Advanced Nursing*. **14** (4), 291–297.

Further Reading

Darbyshire, P. (1988) Making sense of central venous pressure monitoring *Nursing Times*, **84** (6), 36–38.

Department of Health (1991) *Dietary Reference Values for Food Energy and Nutrients for the United Kingdom*. Report on Health and Social Subjects, no. 41. Report of the Panel on Dietary Reference values of the Committee on Medical Aspects of Food Policy. London: HMSO.

DHSS (1984). *Diet and Cardiovascular Disease*. Report on Health and Social Subjects, no. 28. Committee on Medical Aspects of Food Policy. London: HMSO.

Holmes, S. (1989) Cardiac nursing: diet and heart disease. *Nursing*, **3** (36), 9–11.

Keele, C. A., Neil, E. and Joels, N. (1982, reprinted with corrections 1984) *Samson Wright's Applied Physiology (part II)*, 13th edn. Oxford: Oxford University Press.

Macleod, J., Edwards, C. and Bouchier, I. (Eds) (1987) *Davidson's Principles and Practice of Medicine*, 15th edn, pp. 134–136. Edinburgh: Churchill Livingstone

Miller, J. A. (1988) Recording central venous pressure. *The Professional Nurse*, **3** (6), 188–189.

National Advisory Committee on Nutrition Education (1983). *Proposals for Nutritional Guidelines for Health Education in Britain*. London: Health Education Council.

Russell, P. (1989) Cardio-pulmonary resuscitation – our benefit or the patient's ? *Senior Nurse*, **9** (6), 28–29.

Thom, A. (1989) Who decides? *Nursing Times*, **85** (2), 35–37.

Turton, P. (1986) Relaxation techniques. *Nursing*, **3** (9), 348–351.

Williams, P.L., Warwick, R., Dyson, M. and Bannister, L.H. (1989) *Gray's Anatomy*, 37th edn. Edinburgh: Churchill Livingstone.

Wynne, G. (1990) Resuscitation: revised guidelines for life support. *Nursing Times*, **86** (3), 70–75.

CHAPTER ELEVEN

LYMPHATICS: VESSELS, NODES AND TISSUES

OVERVIEW

- Lymphatic vessels, lymph and transport.
- Lymph nodes.
- Lymphoid tissues/organs – spleen.

LEARNING OUTCOMES

After studying Chapter 11 you should be able to:

- Describe the structure of lymphatic vessels.
- Name the major lymphatic vessels.
- Outline the formation and transport of lymph.
- Describe lymph node structure.
- Describe the location of lymph nodes.
- Outline the functions of the lymph nodes.
- Give examples of lymphoid tissues and their location.
- Outline lymphoid tissue functions.
- Describe the structure and functions of the spleen.

KEY WORDS

Lymph – fluid within the lymph vessels. It is derived from interstitial fluid.
Lymphatics – a term which includes the network of lymph vessels and nodes.
Lymph capillaries – vessels which convey fluid (lymph) from the interstitial spaces to the larger lymphatic vessels and ducts.
Lymph node – small masses of lymphoid tissue found along lymph vessels with collections at strategic points, e.g. axilla.

Introduction

Our look at body transport systems is completed by consideration of the **lymphatics** (network of lymph vessels and nodes) and associated lymphoid tissues and organs. Often overlooked, the lymphatic system is vital. It returns interstitial fluid at a rate of 3 litres/24 hours and proteins to the cardiovascular system, and forms an important component of our defence strategies (see Chapter 19).

Lymphatic Vessels

Tiny blind-ended **lymph capillaries** drain **lymph** (fluid derived from interstitial fluid) from the interstitial spaces (see *Figure 11.2*). They are found in most parts of the body except some connective tissues, e.g. bone, and the central nervous system. Their endothelial walls are highly permeable and allow the movement of water, waste and some proteins. The lymph capillaries in the small bowel mucosa are called **lacteals** (see Chapter 13). The lacteals absorb the products of digested fats and carry the creamy lymph, known as **chyle**, to the blood. The network of lymph capillaries unites to form the larger lymph vessels, which have valves to prevent back flow. Lymph vessels are similar in structure to veins (see Chapter 10), but with many more valves.

The lymphatic vessels convey the lymph to two main ducts. Lymph vessels from the right arm, right side of the head and chest drain into the **right lymphatic duct** (*Figure 11.1(b)*) which empties into the great veins in the neck, right subclavian and internal jugular veins (*Figure 11.1(a)*). The rest of the body drains lymph via the **thoracic duct** (*Figure 11.1(b)*) which commences at the level of the

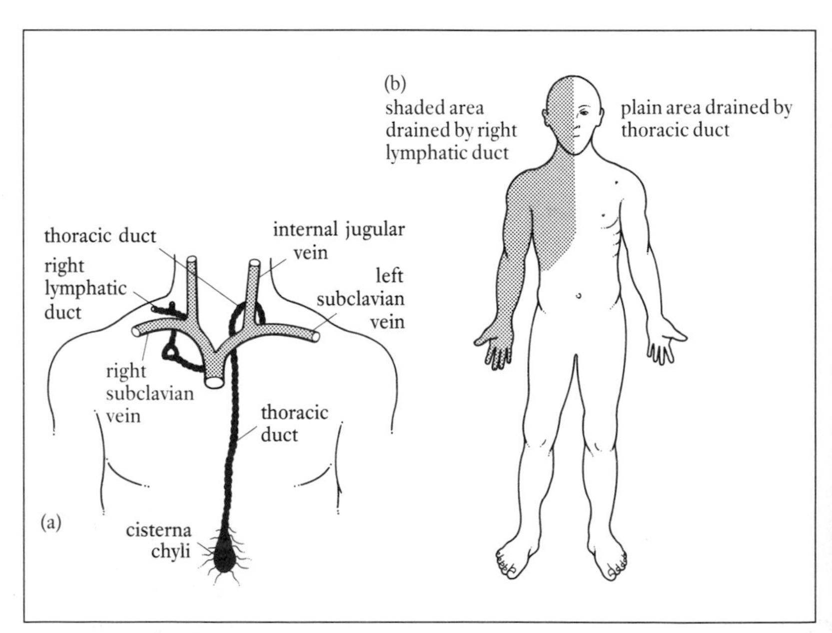

Figure 11.1. (a) Right lymphatic and thoracic ducts; (b) areas drained by each duct.

L2–T12 vertebrae with a dilation called the **cisterna chyli**, which receives lymph from the legs and digestive tract. The thoracic duct passes through the diaphragm accepting lymph from the left thorax, arm and head prior to emptying into the junction between left subclavian and internal jugular veins (*Figure 11.1(a)*).

Lymph: formation and transportation

In Chapter 10 we considered the formation of interstitial fluid from plasma. Most is drawn back at the venous end of the capillary by the osmotic pressure of the blood, but the 3 litres remaining daily in the interstitial spaces forms the **lymph**.

Lymph, which consists of water, proteins and cellular waste, may on occassions contain lipids, possibly hormones, dead cells and bacteria.

Lymph flows in one direction only – away from the periphery, towards the heart. This flow is achieved without a pump such as the heart – muscular contraction squeezes the vessels and the '*thoracic pump*', which aids venous return, also operates for lymph. Pulsation in adjacent arteries helps to move lymph, as does contraction of the smooth muscle coat of the larger lymph vessels.

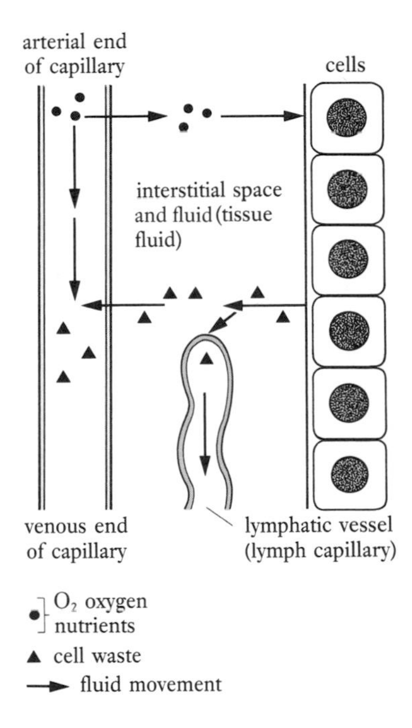

Figure 11.2. Formation of lymph.

Lymph Nodes

Structure and location

Lymph nodes are divided by connective tissue **trabeculae** (fibrous bands) into segments containing a network of reticular fibres, blood vessels, **lymphocytes** (which move between blood and lymph nodes) and phagocytic macrophages (*Figure 11.3*) (see Chapter 9). Enclosed

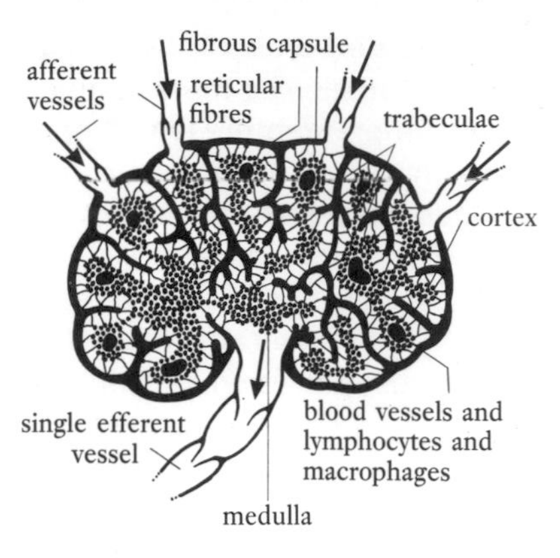

Figure 11.3. A lymph node.

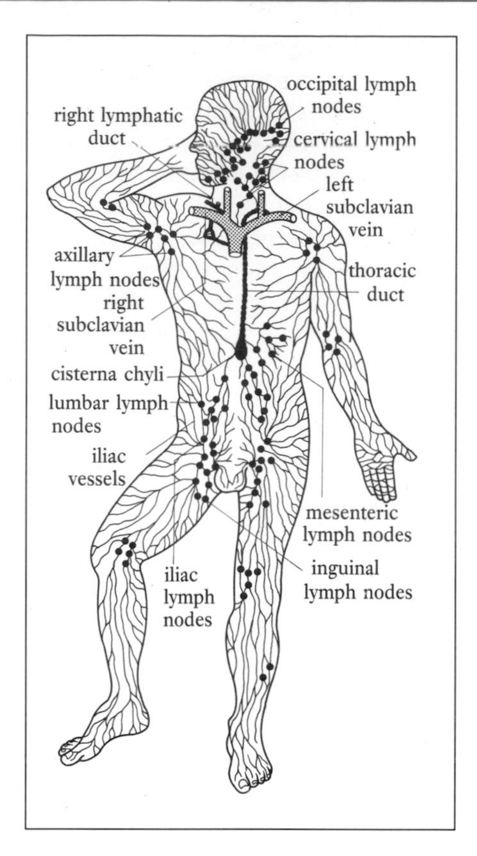

Figure 11.4. Lymphatic system: location of vessels and nodes.

within a fibrous capsule, the lymph nodes are kidney-shaped, with a cortex and medulla. They vary from a few millimetres to 2.5 cm in diameter.

Several afferent lymphatic vessels lead into each node but lymph leaves the node in a single efferent vessel.

Many hundreds of lymph nodes are located throughout the system and lymph passes through several nodes (*Figure 11.4*). There are collections of nodes at certain strategic points, which include the deep nodes, found in the abdomen, thorax and neck, and the superficial inguinal, axillary, occipital and cervical nodes. These strategically placed lymph nodes provide points where extraneous particles, such as bacteria, can be intercepted before reaching the blood. When you have a sore throat and 'cold' the cervical nodes enlarge in response to the infection, which is why they can be felt in the neck.

Lymph node functions

As part of the surveillance system of the body, lymph nodes 'filter' lymph and remove extraneous particles such as bacteria, viruses and malignant cells. Hopefully, extraneous particles are destroyed by the macrophages. If the node is overwhelmed, bacteria can infect the node and spread along the vessel, which becomes inflamed (*lymphangitis*). This shows as a red line tracking to the next set of nodes. Malignant cells are destroyed in small numbers but the lymph nodes can themselves become the site of metastatic cancer (see Chapter 1). The lymph vessels can also provide a route by which malignancy spreads from the primary site to other organs, e.g. breast cancer (see Chapter 20) frequently shows lymphatic spread.

Lymph nodes provide a site for the proliferation of T and B lymphocytes and the production of antibodies (see Chapter 19).

Problems: lymph vessels and nodes

Vessel obstruction

If lymph vessels become blocked the interstitial fluid is unable to drain away as lymph. Its increased protein content and osmotic pressure causes more fluid to collect in the extracellular spaces as oedema (see Chapter 10). Causes of lymphatic obstruction include metastatic malignant cells and parasitic diseases, such as *filariasis* (infestation with tiny parasitic thread-like worms).

Lymphomas

Lymphomas are a group of primary malignant diseases affecting the lymph nodes. One of the most common is *Hodgkin's disease*, which is characterized by painless lymph node enlargement, *hepato-splenomegaly* (enlargement of liver and spleen), pyrexia (fever), weight loss and anaemia. The prognosis has been greatly improved by developments in diagnosis, by determining the extent of the disease (known as *staging*) and selecting the treatment modalities most effective for the stage identified. Treatment is based on radiotherapy, chemotherapy with cytotoxic drugs, removal of the spleen (splenectomy) in some centres and supportive measures, such as blood transfusion. For a person with very early-stage disease (stage IA) the 5-year survival, with treatment, is over 90% (Macleod *et al.* 1987).

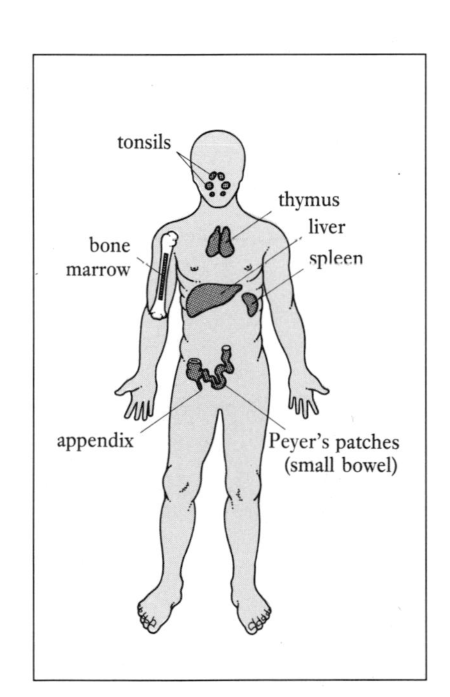

Figure 11.5. Lymphoid tissues: location.

Lymphoid Tissues/Organs

Other collections of lymphoid tissue similar in structure to the lymph nodes are found in sites other than the lymphatics. These include the **spleen, tonsils, thymus, bone marrow, liver, small bowel (Peyer's patches)** and **appendix** (see *Figure 11.5*).

In common with lymph nodes these lymphoid accumulations form part of the body defences but they do not filter lymph. Some structures have other specialized roles, e.g. spleen degrades old blood cells. More detailed accounts of some structures can be found in the relevant chapters, (e.g. thymus, Chapter 8; tonsils, Chapter 12; bowel, Chapter 13; liver, Chapter 14; and in Chapter 19 which deals with body defences). The spleen will, however, be covered within this chapter.

The spleen

The **spleen**, situated just below the diaphragm in the left upper abdomen, is the most extensive lymphoid organ (*Figure 11.6*). It is a vascular structure weighing around 200 g. It receives blood from the splenic artery and its venous blood drains via the splenic vein into the hepatic portal vein (see Chapter 10). These vessels, plus autonomic

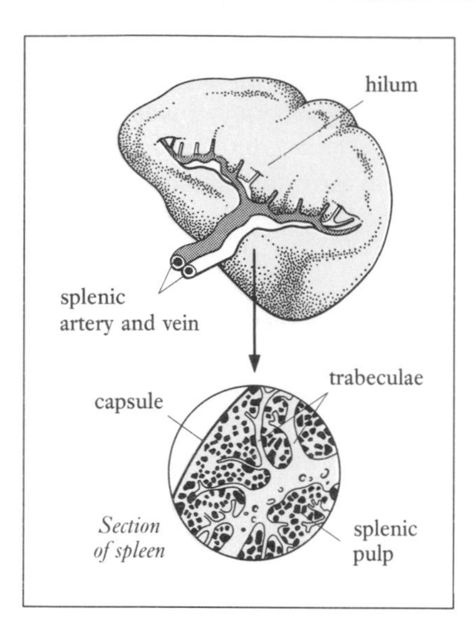

Figure 11.6. Spleen.

nerves and lymphatic vessels, enter or leave the spleen at the **hilum**. A fibrous capsule encloses the **splenic pulp** which contains lymphocytes, macrophages and reticular fibres divided into compartments by trabeculae. Sinusoids (see Chapter 10) within the spleen allow intimate contact between blood and pulp.

Functions of the spleen

- Destruction of old or defective erythrocytes and platelets (see Chapter 9).
- Filters blood and destroys by phagocytosis any micro-organisms or other debris.
- Is the site of lymphocyte proliferation and antibody production.
- Is the site of fetal blood cell production and produces erythrocytes in times of increased need.
- Possible role as a reservoir for blood. This is important in some mammals but is of limited significance in humans.

All this sounds rather impressive but it is entirely possible to function without a spleen. Its removal may be required to treat conditions such as haemolytic anaemia or following trauma. The spleen is often damaged during road accidents, with severe internal haemorrhage and shock. Following splenectomy other structures take over its role.

Summary/Check List

Introduction – lymphatics.
Lymphatic vessels – lymph: vessels and main ducts. formation and transportation. lymph capillaries.
Lymph nodes – structure and location. Functions.
Lymphatic problems – lymphatic obstruction and lymphomas.
Lymphoid – tissues/organs – structure and location. Spleen.

Self Test

1 Which of the following statements are true?
(a) The proper functioning of the cardiovascular system depends upon the lymphatics.
(b) Lacteals are the lymph capillaries of the large bowel.
(c) Lymphatic vessels from the legs and digestive tract empty into cisterna chyli.
(d) The thoracic duct empties into the veins in the left side of the neck.

2 List the mechanisms involved in lymph transport.

3 Briefly describe the role of the lymph nodes.

4 Relate the structure of the spleen to its functions.

Answers
1. a,c,d. 2. See page 255. 3. See page 256-257. 4. See pages 257-258.

References

Macleod, J., Edwards, C., and Bouchier, I. (Eds), (1987) *Davidson's Principles and Practice of Medicine*, 15th edn. Edinburgh: Churchill Livingstone.

Further Reading

Guyton, A. C. (1981) *Textbook of Medical Physiology* (Ch. 31), 6th edn. Philadelphia: W B Saunders.
Roitt, I. M., Brostoff, J. and Male, D. K. (1985) *Immunology* (Section 3). Edinburgh: Churchill Livingstone.

section five

obtaining and using raw material for metabolism and excreting waste

CHAPTER TWELVE

RESPIRATION

OVERVIEW

- Respiratory tract structure and function.
- Breathing/ventilation.
- Behaviour of gases.
- Gaseous exchange.
- Transport of gases.
- Control of ventilation.
- Respiratory role in pH regulation.

LEARNING OUTCOMES

After studying Chapter 12 you should be able to:

- Describe the respiratory tract, relating structure to function.
- Discuss the effects of smoking upon the respiratory tract.
- Explain the mechanisms of breathing.
- Briefly describe the role of the respiratory muscles.
- Describe intrathoracic pressure changes during breathing.
- Use your physiological knowledge in the assessment of breathing and indentification of common breathing problems.
- Outline factors which influence airflow and pulmonary ventilation.
- Describe lung volumes and pulmonary function tests.
- Define alveolar ventilation.
- Outline the behaviour of gases.
- State the composition of air (atmospheric, alveolar and expired).
- Explain how gaseous exchange occurs.
- Explain the differences in the amount of oxygen and carbon dioxide in arterial and venous blood.
- Outline the importance of the ventilation–perfusion relationship.
- Describe how oxygen is transported.
- Explain the oxygen dissociation curve.
- Describe how carbon dioxide is transported.
- Explain how ventilation is controlled.
- Outline the effects of exercise and altitude upon ventilation.
- Describe briefly how respiration helps to regulate blood pH.

KEY WORDS

Expiration – breathing out.
External respiration – gaseous exchange between alveolar air and the blood.
Gaseous exchange – the interchange/movement of gases in the body. It occurs either by bulk flow or by diffusion through tissues.
Inspiration – breathing in.
Internal respiration (cellular) – gaseous exchange between blood and the cells.
Oxygenation – combine or saturate with oxygen, e.g. when all four haem groups are combined with oxygen the fully saturated haemoglobin molecule is known as oxyhaemoglobin.
PCO_2 – the partial pressure (*P*) of carbon dioxide (CO_2). $PaCO_2$ = the partial pressure of carbon dioxide in arterial blood.
PO_2 – the partial pressure (*P*) of oxygen (O_2). PaO_2 = the partial pressure of oxygen in arterial blood.
Pulmonary – relating to the lungs.
Respiration – the process which supplies oxygen to the tissues and removes carbon dioxide.
Ventilation – mechanical process of breathing.

INTRODUCTION

The functions of the respiratory tract are to provide oxygen for cellular metabolism and to remove waste carbon dioxide. This is achieved by **ventilation** (mechanical process of breathing), **external respiration** (gaseous exchange between alveolar air and the blood), gaseous transport and **internal respiration** (gaseous exchange between blood and the cells). These processes, vital to life, involve close links with the cardiovascular system and the blood (see Chapters 9 and 10). Reduced functional efficiency, e.g. in disease, leads to insufficient oxygen reaching the tissues, a build-up of carbon dioxide in the body and possibly pH disturbance. We survive for only a few minutes if no oxygen reaches the brain.

RESPIRATORY STRUCTURES

The respiratory structures include the *nasal cavity* (*nose*), *pharynx, larynx, trachea, bronchi* and smaller *bronchioles* which warm, moisten and filter air en route to the lungs (see *Figure 12.1*). **Gaseous exchange** (the interchange/movement of gases in the body, occurring either by bulk flow or by diffusion through tissues) occurs in the system of tiny ducts and *alveoli* (air sacs; singular, alveolus) within the two lungs.

Early development

Development of the respiratory structures commences during the second month of embryonic life as a groove in the wall of the pharynx, which soon forms an enclosed tube. This tube divides to form two tubes, one of which will become the larynx and trachea while the other forms the oesophagus. This close developmental relationship means that a failure in oesophageal development can result in an abnormal opening (*fistula*) forming between the trachea and the oesophagus *(tracheo-oesophageal fistula).*

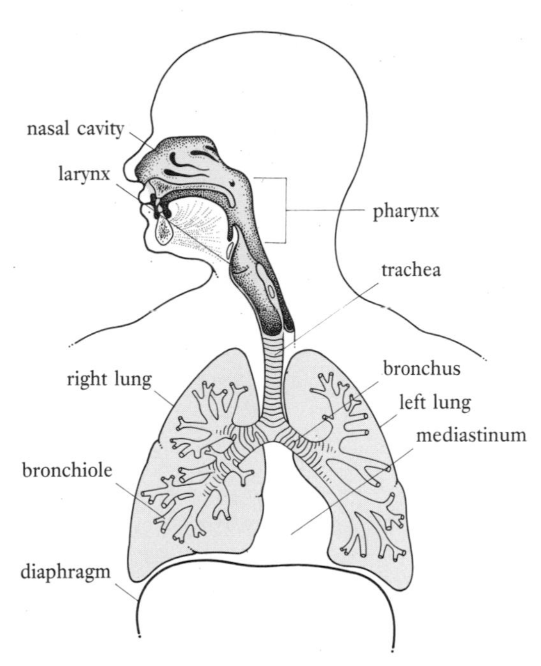

Figure 12.1. Respiratory structures.

The tiny *lung buds* or outgrowths, which form on the lower end of the trachea will eventually develop into the lungs, bronchi and alveoli. Before birth the fluid-filled fetal lungs play no part in gaseous exchange. Prior to lung inflation, which occurs as the infant takes its first breath, all gaseous exchange takes place through the placenta (see Chapter 20). Although the respiratory structures may be formed, a lack of *surfactant* (fluid that reduces surface tension in the alveoli) in the lungs of preterm babies may cause problems with lung inflation (see page 283 and respiratory distress syndrome, page 274).

Nose

Much more than a convenient place to balance your 'specs', the **nose** is a respiratory structure that helps to produce speech and that contains the olfactory receptors for smell (see Chapter 7). It is the only part of the respiratory tract visible externally.

The **nasal cavity** is formed from the bones of the face and skull, the nasal bone and hyaline cartilage (see *Figure 18.10b,c*). It is irregular in shape and is divided in the midline by a **septum** formed by the vomer and ethmoid bones (posteriorly) and nasal cartilage (anteriorly). The roof of the cavity consists of the **cribriform plate** (part of the ethmoid bone), sphenoid bone, frontal bone and nasal bone; its lateral walls are the ethmoid, maxilla and conchae, and inferiorly it is separated from the oral cavity by the hard and soft palates (see Chapter 13).

Air enters via the two **anterior nares** (the nostrils), which open

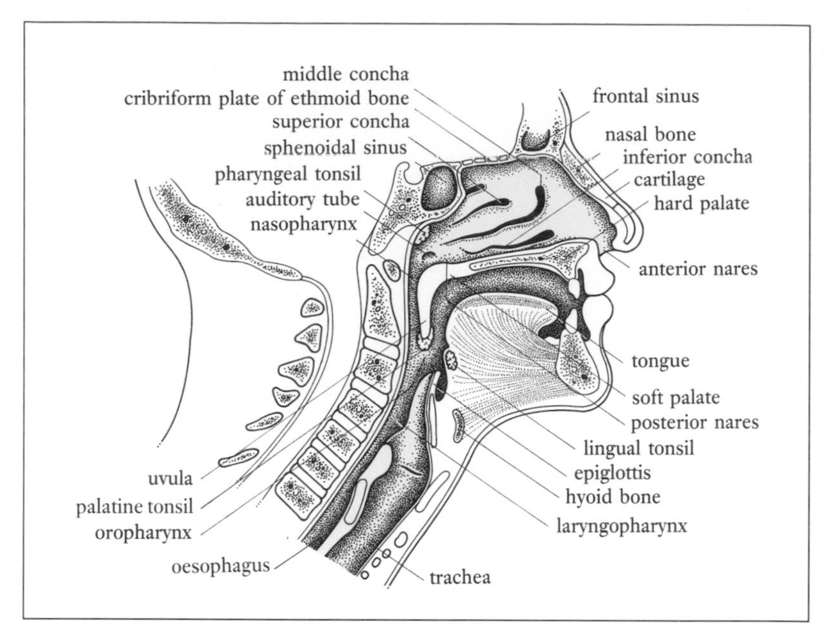

Figure 12.2. Upper respiratory tract.

into the nasal **vestibule**. This part of the nose is lined with skin and contains hairs which trap large foreign particles entering the nose. The internal part of the cavity is lined with highly vascular **ciliated columnar epithelium** (see *Figures 1.10* and *12.3*) containing many mucus-producing **goblet** cells. This lining is continuous with the **sinuses** (see below). Three **conchae** or **turbinate** bones project into the nasal cavity increasing mucosal surface area and causing air turbulance.

The specialized respiratory mucosa warms and moistens air, and smaller particles of dust carrying micro-organisms adhere to the mucus (which contains an antibacterial enzyme – lysozyme) and are later moved by the cilia (see Chapter 1) to the pharynx where the mucus is swallowed or expectorated (coughed up). An irritation or upper respiratory tract infection can cause considerable swelling of the nasal mucosa resulting in nasal obstruction. Most of us cope by mouth breathing while this lasts, but for small babies who normally breathe through the nose it is important to keep this route open.

NB Air entering the respiratory tract via the mouth during mouth breathing has the disadvantage of bypassing the modification and protective functions of the nose.

Another type of epithelium found in the nasal cavity roof is that concerned with olfaction, which is discussed more fully in Chapter 7.

The paranasal sinuses (see Chapter 18), found in the bones of the face and skull are air-filled and function in voice production and lighten the skull. They produce mucus which drains into the nasal cavity – the frontal and sphenoidal sinuses drain through the roof and the ethmoidal and maxillary sinuses drain via the lateral walls. This close relationship explains why upper respiratory infections can so easily spread to the sinuses (*sinusitis*).

Also opening into the nasal cavity are the *nasolacrimal ducts*, which carry secretions (tears) from the lacrimal glands (see Chapter 7). Excess tear production during 'crying' will result in discharge of fluid from the nose. At the back of the nasal cavity the **posterior nares** (see *Figure 12.2*) communicate with the pharynx.

The Pharynx

A funnel-shaped structure, running from the base of the skull to the level of the lower cervical vertebrae, the **pharynx** is divided into three parts: **nasopharynx** (behind the nose), **oropharynx** (behind the mouth) and **laryngopharynx**, which leads into the larynx and oesophagus (see *Figure 12.2*). The nasopharynx is exclusively respiratory but the two lower parts provide a common passageway for air and food which enter the oesophagus during swallowing (see pages 268-269 and Chapter 13).

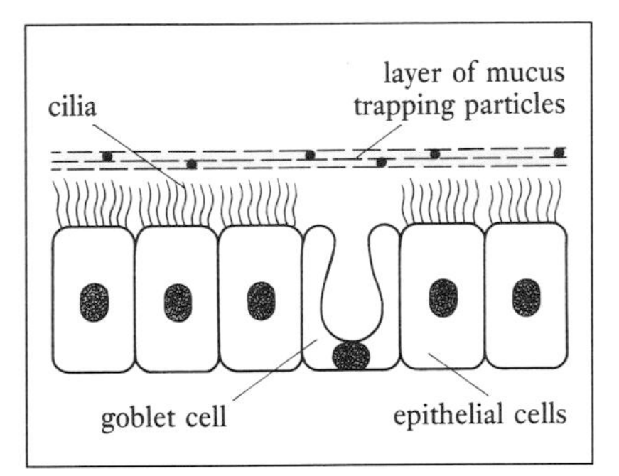

Figure 12.3. Epithelial lining of respiratory passages.

The pharynx receives blood via several vessels, including the pharyngeal artery and branches from the facial and laryngeal arteries. Venous blood returns to the heart by a pharyngeal plexus and the internal jugular vein.

The pharynx has constrictor muscles concerned with swallowing, which are mostly innervated by the vagus and glossopharyngeal nerves, a fibrous layer and an epithelial lining, which changes along its length. The complexity of the pharynx means that many problems may occur which affect function, e.g. difficulty swallowing (*dysphagia*).

The nasopharyngeal portion is lined with respiratory mucosa which continues to warm, moisten and filter the air. The oropharynx and laryngopharynx are lined with stratified squamous epithelium (see Chapter 1) continuous with that of the oesophagus. This variation protects the tissues from friction caused by food.

Nasopharynx

The **nasopharynx** lies at the level of the sphenoid bone and above the muscular soft palate. The *pharyngotympanic* (auditory) tubes open into the nasopharynx (see Chapter 7). These communicate with the middle ear and equalize pressure either side of the tympanic membrane during yawning and swallowing. They also provide a route by which respiratory infections spread to the middle ear causing *otitis media*. Lymphoid tissue known as the **pharyngeal tonsils** (called *adenoids* when enlarged) are located on the posterior wall of the nasopharynx; these provide defence against some micro-organisms entering the respiratory tract (see Chapters 11 and 19). Enlargement of the pharyngeal tonsils (adenoids) due to chronic infection will lead to obstruction of the nasal airway and mouth breathing.

During swallowing the soft palate and **uvula** (the small fleshy body which hangs down from the soft palate) move upwards to isolate the nasopharynx and stop food entering it.

Oropharynx

The **oropharynx** lies behind the mouth and extends from the soft palate to the *hyoid* bone (see *Figures 12.2* and *12.4*). As part of both respiratory and digestive tracts it conveys air and food. Two further paired masses of lymphoid tissue are located within the oropharynx: the **palatine tonsils** in folds formed by the soft palate and oropharynx, and the **lingual tonsils** at the base of the tongue.

Laryngopharynx

The last part of the pharynx extends from the hyoid bone to the level at which the respiratory and digestive tracts separate. The next part of the respiratory tract is the larynx, and the **laryngopharynx** becomes the oesophagus, which conveys food to the stomach. When we swallow the larynx is closed off to ensure that food does not 'go the wrong way' and enter the respiratory tract.

Larynx

The **larynx** lies anterior to the laryngopharynx and opens into the trachea; it extends between the third and sixth cervical vertebrae. Androgens released during puberty cause the male larynx to enlarge. This gives rise to a deeper voice and the *'Adam's apple'* prominence on the throat. The three functions of the larynx are to provide an open airway between the pharynx and trachea, to direct food into the oesophagus during swallowing and voice production.

Arterial blood is supplied by laryngeal branches of the thyroid arteries and venous blood drains via laryngeal veins into the thyroid veins and internal jugular veins.

Innervation of the larynx is through the superior and recurrent laryngeal nerves. These supply parasympathetic sensory and motor fibres from the vagus and accessory nerves to the epithelium and laryngeal muscles. Sympathetic fibres from the cervical ganglion supply the larynx.

The larynx is formed from cartilages joined by ligaments and membranes. The main laryngeal cartilages, which also provide attachment for several muscles, are the **epiglottis** (elastic cartilage) and the **thyroid**, **cricoid** and two **arytenoid cartilages**, which are hyaline cartilage (see *Figure 12.4*). The thyroid cartilage consists of two curved portions which fuse in the midline to form the **laryngeal prominence** or Adam's apple. Two pyramidal-shaped arytenoid cartilages form the lateral and posterior walls and anchor the **vocal cords** within the interior of the larynx. Below the thyroid cartilage is the complete ring of the cricoid cartilage which joins the larynx to the trachea. The leaf-shaped epiglottis, joined to the upper border of the thyroid cartilage, is involved in the mechanism which protects the respiratory tract from food; it contains some taste buds (see gustation, Chapter 7).

Swallowing

Food entering the pharynx initiates a series of reflex mechanisms: the nasopharynx is closed off by the soft palate, a sphincter closes the inlet into the larynx, and an upward movement of the larynx

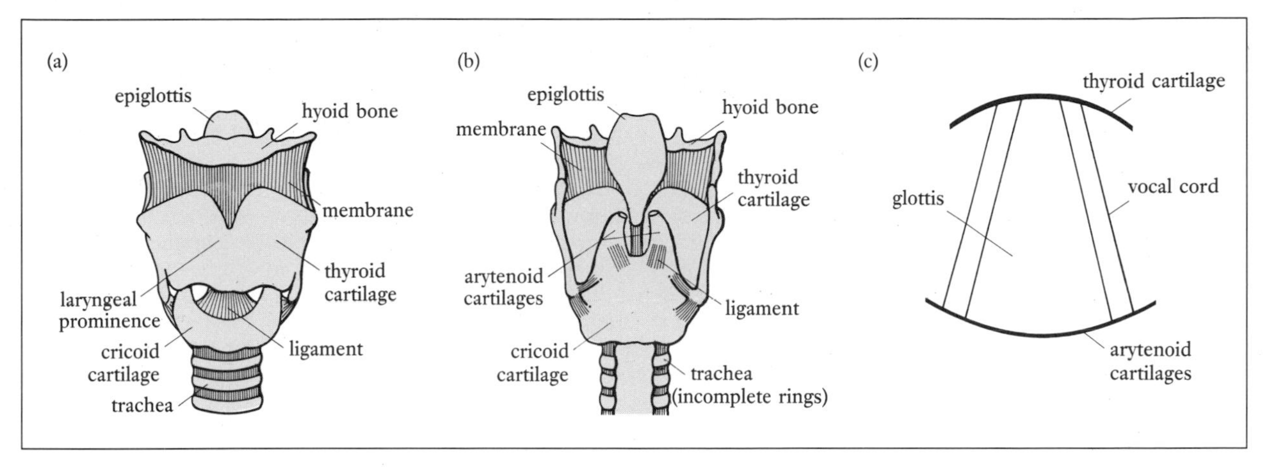

Figure 12.4. Larynx. (a) anterior view; (b) posterior view; (c) vocal cords.

(you can confirm this by feeling your throat whilst drinking) causes the epiglottis to cover the opening. If these mechanisms fail to operate properly and food does 'go the wrong way' the cough reflex (see page 271) expels the food and usually prevents it from entering the respiratory passages (see *Nursing Practice Application*, page 270).

Lining and vocal cords

The upper part of the larynx is lined with stratified squamous epithelium (see Chapter 1 for a detailed account of epithelial tissues) and the lower part with ciliated columnar epithelium (see *Figure 12.3*) which warms, moistens and filters the air. Mucus and debris are moved upwards by the cilia for expulsion. The lining is modified to form folds or **vocal cords** which stretch from front to back inside the larynx. The vocal cords are separated by a gap known as the **glottis**, through which air passes (*Figure 12.4*).

Valsalva's manoeuvre

In certain situations, such as straining to defaecate when constipated or when lifting heavy weights, air is forced out of the respiratory passages with the glottis closed. This increases pressure in the thorax and abdomen, and reduces venous return to the heart.

Voice production

Contraction and relaxation of the laryngeal muscles move the arytenoid cartilages which, in turn, alter the position and tension of the vocal cords, either closer together, narrowing the glottis, or further apart to increase the opening. Sound is produced as expired (breathed out) air is forced through the larynx to vibrate the vocal cords. The pitch or frequency of the voice changes with the length (tension) of the vocal cords and loudness depends on the force of air moving through the glottis. Males have deeper voices because their vocal cords become longer and thicker, due to laryngeal enlargement in response to male sex hormones released during puberty. Turning sounds into recognizable speech is achieved by the muscles of the lips, tongue, soft palate and pharynx (try talking clearly without moving your tongue or lips). The paranasal sinuses,

Nursing Practice Application Airway maintenance

Obviously a patent (open) airway is vital to life and we have already discussed ways in which this is achieved, e.g. the epiglottis. In situations which include altered consciousness, with loss of swallowing and cough reflexes, there is risk of airway obstruction if the tongue falls back or food/debris enters the larynx. Care plans should include: (i) careful observations of colour and respiration, (ii) placing in a position (*Figure 12.5a*) to minimize risk of inhaling vomit (N.B. A new 'recovery position' has recently been adopted by first aid organizations, such as St. Johns Ambulance); (iii) ensuring that people without the swallowing and cough reflexes are never given anything orally; and (iv) the provision of specialist equipment such as plastic airways and suction.

In certain situations the *Heimlich manoeuvre*, a first aid measure, can be used to clear the larynx or trachea of a foreign body such as food (Heimlich, 1975). Compression over the upper abdomen causes the diaphragm to rise and the resulting expulsion of air from the lungs hopefully dislodges the obstruction (*Figure 12.5b*).

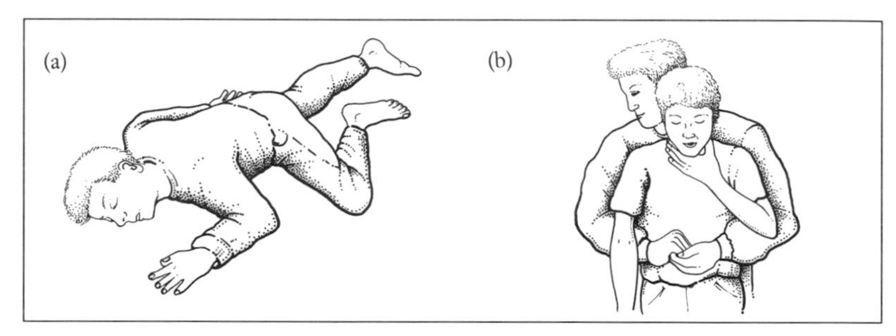

Figure 12.5. Maintaining an airway. (a) A position for an unconscious person; (b) Heimlich manoeuvre.

nasal and oral cavities, and pharynx give the voice resonance, which explains why a 'cold' and blocked sinuses make your voice sound dull and flat. Voice changes result from a variety of causes including *laryngitis, laryngeal tumours* and damage to the *recurrent laryngeal nerve* during thyroid surgery.

Trachea

The **trachea** (windpipe) extends from the larynx to the level of the fifth thoracic vertebra. Here, it bifurcates (divides) into the right and left main bronchi (see *Figure 12.6a*). Between 10–12 cm in length, the trachea passes through the mediastinum anterior to the oesophagus. The trachea is formed by 16 to 20 'C'-shaped rings of hyaline cartilage, which keep it patent, smooth muscle and fibro-elastic tissue. The cartilage rings are open at the back to allow oesophageal distention during swallowing.

The lining mucosa is ciliated columnar epithelium containing numerous goblet cells. In common with the other respiratory structures the tracheal mucosa warms, moistens and filters the air. Once the mucus has trapped particles it is moved upwards by the beating cilia to be swallowed or expectorated.

The trachea receives blood via the thyroid arteries and its venous blood drains into the brachiocephalic veins. It is innervated by parasympathetic sensory and motor fibres of the vagus nerve (recurrent laryngeal nerve) and sympathetic fibres from the cervical ganglion.

Intermittent positive pressure respiration/ventilation *– involves the use of a machine (ventilator) which delivers gases to the patient to inflate the lungs. This assisted ventilation is used in situations where breathing is inadequate or is not occurring spontaneously e.g. altered consciousness following severe head injury. Assisted ventilation can be achieved by tracheostomy or via a tube passed into the trachea via the mouth (endotracheal).*

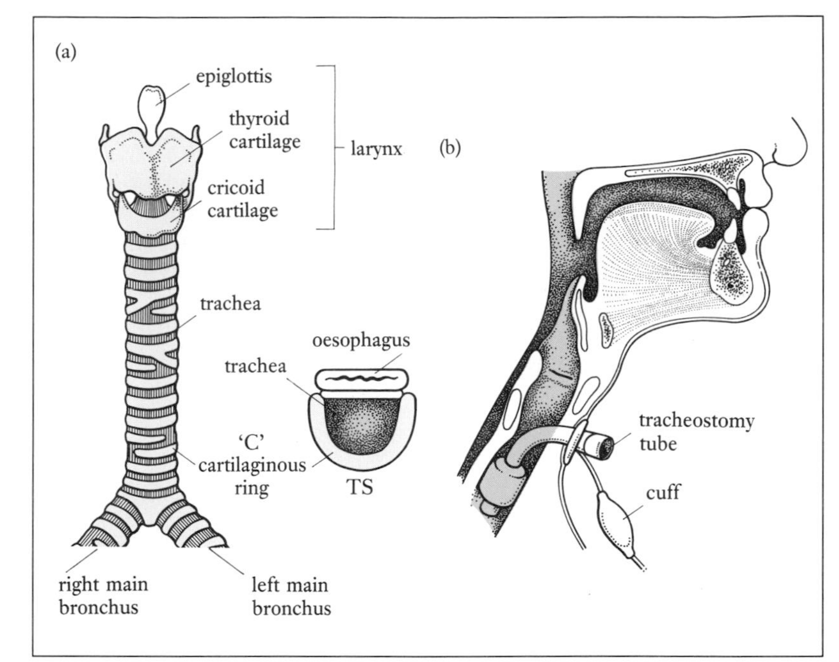

Figure 12.6. (a) Trachea; (b) tracheostomy.

Tracheostomy

An opening in the trachea known as a *tracheostomy* (*Figure 12.6b*) may be made for a variety of reasons which include:

- Long-term intermittent positive pressure respiration/ventilation (IPPR/V).
- To remove copious sticky secretions by suction.
- Dead space reduction (see page 272) and improved ventilation.
- Obstruction in the upper airways or vocal cord paralysis.

A plastic or metal tube inserted into the tracheal incision keeps it patent. Tracheostomy does, however, have certain drawbacks: micro-organisms can easily enter the lungs, the individual has communication difficulties, as speech is affected; and air no longer passes through the upper airways for warming and humidification.

Cough and sneeze

Coughing and sneezing are protective mechanisms that enable the respiratory tract to expel excess mucus and foreign bodies. Irritation of the mucosa stimulates a cough, which is a forced expiration of air from the mouth. Pressure build-up is achieved by a closed glottis during the start of the expiration. Once released, the air leaves at great speed. Sneezing is similar but the forced expiration is via the nose.

Bronchial tree and alveoli

At the level of the fifth thoracic vertebra the trachea divides into the **right** and **left primary (main) bronchi**. Each primary bronchus runs through the mediastinum to enter the lung at the **hilum**. The primary bronchus on the right is shorter, wider, and slopes more vertically than does the one on the left. This arrangement means that foreign bodies which 'slip past' the elaborate defences already discussed are most often inhaled into the right primary bronchus (see page 273). After entering the lungs the primary bronchi divide into smaller **secondary (lobar) bronchi** *(Figurc 12.7)*, three on the right and two on the left. These secondary bronchi divide into **tertiary (segmental) bronchi** which eventually branch to become tiny airways known as **bronchioles**. The bronchioles themselves divide to become **terminal bronchioles**, **respiratory bronchioles** and the minute **alveolar ducts** leading into the **alveoli**.

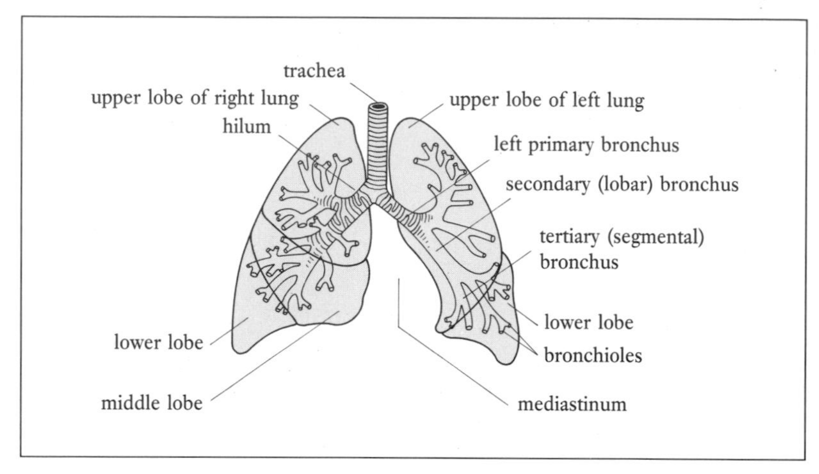

Figure 12.7. Bronchial tree.

The larger bronchi are similar in structure to the trachea, with cartilage to keep the airways open, a smooth muscle layer and a ciliated mucous membrane lining, which traps particles in the mucus (moved upwards by ciliary action) and warms and moistens air. This structural arrangement means that no gaseous exchange occurs – the bronchi with the nasal cavities, pharynx, larynx and trachea form the **conducting airways**. This area, with a volume of around 150 ml constitutes the **anatomical dead space**, so-called because none of this air takes part in gaseous exchange. The various modifications to structure occurring as the airways divide and become smaller reflect the change in function from conduction to gas exchange:

- Cartilage thins and is absent from the bronchioles. Without the rigid cartilage the diameter of the airways can be altered by changes in smooth muscle tone controlled by the autonomic nerves. Sympathetic stimulation of the beta-receptors (see Chapter 6) relaxes the muscle causing *bronchodilation*. Parasympathetic activity has the opposite effect to cause *bronchoconstriction* (see *Table 12.1*).
- The ciliated columnar epithelium becomes a single layer of squamous epithelium which forms part of the **respiratory membrane** found in the respiratory bronchioles, alveolar ducts and alveoli. This area of the bronchial tree, structually adapted for gaseous exchange, is known as the **respiratory zone**.

Table 12.1. Factors affecting bronchial muscle tone.

Factor	*Effects*
Bronchodilation	
(i) Sympathetic stimulation of beta-receptors: a. Exercise b. Drugs – Salbutamol Adrenaline	Relaxes bronchial muscle, causes bronchodilation and increased air entry
(ii) Inhibition of parasympathetic nerves by anticholinergics: a. Drugs – Ipratropium bromide	Prevents parasympathetic effects of muscle contraction and bronchoconstriction
Bronchoconstriction	
(i) Parasympathetic activity: a. During rest b. Irritant chemicals, e.g. cigarette smoke c. Allergic conditions, e.g. some types of asthma d. Infections	Contracts bronchial muscle, causes bronchoconstriction and reduced air entry

Arterial blood reaches the bronchial tree via bronchial arteries which branch from the aorta to provide a good blood supply. The bronchial veins drain venous blood into the azygos veins which empty into the superior vena cava.

Inhalation of foreign bodies in young children

Previously we mentioned that inhaled foreign bodies usually lodge in the right main bronchus. Young children are at particular risk and it is very easy for small particles to be aspirated (breathed in). For this reason young children should not be allowed access to food such as peanuts and toys should be suitable for their age group and meet the most stringent safety standards.

Alveoli (detail)

The many millions of **alveoli** which cluster around the respiratory bronchioles and alveolar ducts form gas(air)-filled cavities within the lungs (*Figure 12.8a*). The alveoli are in very close contact with the networks of pulmonary capillaries (see Chapter 10). The capillary endothelium and the alveolar walls form the extremely thin (less than 0.5 μm thick) **respiratory membrane** (see *Figure 12.8b*) which permits the diffusion of gases between alveolar air and blood – **external respiration**. The branching structure of the respiratory zone and the many clusters of alveoli result in a very large surface area available for gas exchange (in adults the alveolar surface is similar in size to a tennis court).

Nursing Practice Application Inhalers and nebulizers (see *Table 6.3)*

Many people with respiratory conditions such as asthma, causing bronchoconstriction, need bronchodilator drugs, e.g. salbutamol (adrenergic stimulator) or ipratropium bromide (anticholinergic). These drugs are most effective when administered via an inhaler or nebulizer because they are delivered to the problem area.

An important nursing responsibility is to ensure that the individual has sufficient knowledge to use both drug and appliance safely and efficiently. People using inhaled drugs for asthma will only have satisfactory asthma control where their inhaler technique is good (Ellis,1990).

A programme of drug education, commenced before discharge and continued in the community, should include: how the drug works, technique for using the appliance, correct dosage, associated problems, e.g. common side-effects, care of the appliance and drug storage.

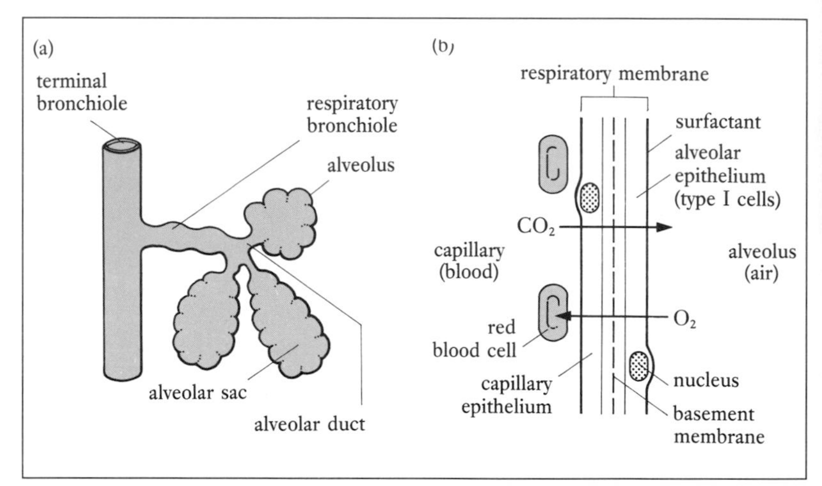

Figure 12.8. (a) Alveoli; (b) alveolar–capillary membrane. NB For clarity, cuboidal (type II) cells and macrophages are not shown.

The alveolar wall contains three cell types:

- Phagocytic macrophages (see Chapters 1, 9 and 19) which engulf and destroy micro-organisms and debris reaching the alveoli.
- Squamous epithelial cells (type I), which form the wall.
- Cuboidal epithelial cells (type II), which secrete a phospholipid fluid known as **surfactant**. Surfactant moistens the respiratory membrane, reduces surface tension and prevents alveolar collapse, and is essential for efficient gaseous exchange (see page 283).

Preterm infants who have immature lungs and a lack of surfactant develop *respiratory distress syndrome* caused by 'stiff lungs' and alveolar collapse.

Lungs

The two lungs almost completely fill the thoracic cavity during life. Lying one on each side, they are separated by the mediastinum, a space containing the heart, vessels, trachea, bronchi, oesophagus and other structures (see *Figure 12.9*).

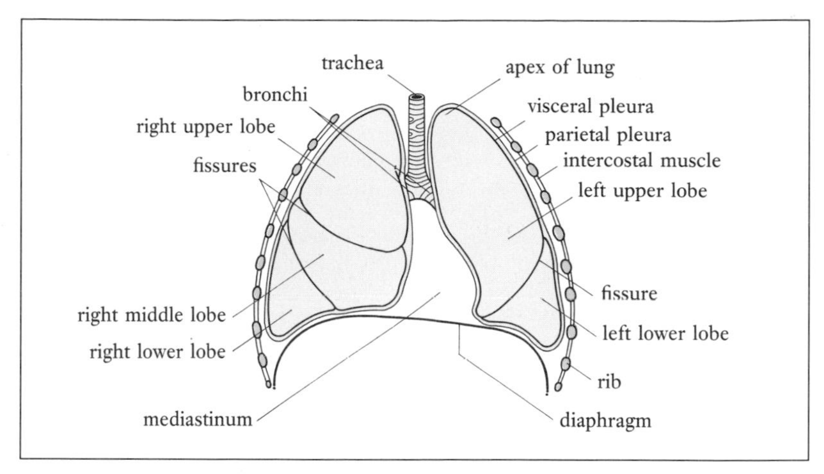

Figure 12.9. Lungs and pleura.

Each cone-shaped lung has:

- A narrow *apex* level with the clavicle.
- A convex *costal surface* which comprises the anterior, lateral and posterior surfaces in contact with the ribs, *intercostal muscles* and *costal cartilages*.
- The concave base which rests upon the *diaphragm*.
- A **hilum** or depression on its medial surface where the primary bronchi, nerves, blood vessels and a very good supply of lymphatics enter or leave.

The lungs are divided by **fissures** into **lobes**. The left lung, which has to make room for the heart, has two lobes (upper and lower) and the right lung has three (upper, middle and lower lobes). Each lobe is further subdivided into smaller **bronchopulmonary segments**. As the main bronchus enters the lung it divides to form the smaller branches within the lobes and segments.

Lung tissue comprises a connective framework which helps to prevent the spread of infection, e.g. *tuberculosis,* and supports the airways, alveoli and pulmonary capillaries. Being filled with air it is spongy, stretchy (elastic) and weighs very little.

Lung tissue receives its arterial supply from the bronchial arteries and venous blood leaves via bronchial veins (see page 273). This, however, is only part of the story; blood for **oxygenation** (combine or saturate with oxygen) arrives from the right side of the heart via the pulmonary artery and, after gaseous exchange, returns to the left side of the heart via four pulmonary veins (see Chapter 10).

The lungs are innervated by parasympathetic fibres of the vagus nerves which are concerned with bronchoconstriction and mucus secretion, and sympathetic fibres which initiate bronchodilation (see *Table 12.1*).

Nursing Practice Application Pleural/chest aspiration (thoracocentesis)

Pleural effusion or excess pleural fluid may result from a variety of causes, e.g. pulmonary infarction (see Chapter 9), malignancy or pneumonia. Fluid may be removed for diagnosis or to ease dyspnoea and discomfort. The fluid is withdrawn, by the medical officer, using a syringe with needle inserted between the ribs. When caring for a person who requires this procedure it is important that the nurse takes time to give an explanation and information which meets the patient's needs and reduces anxiety. They should explain what is done, stressing that local anaesthetic is used and that the nurse will stay throughout. The position (*Figure 12.10*) should be explained, as well as the importance of the patient saying if they need to cough or move during the aspiration, as this may cause the needle to penetrate the visceral pleura and lungs.

Pleura

The **pleurae** or double *serous membranes* (see Chapter 1) form three closed compartments, one for each lung and one for the mediastinal contents. The **visceral pleura** (inside layer) covers the lungs and the **parietal pleura** (outside layer) lines the thoracic cavity and mediastinum. Between the two layers is a potential space lubricated by pleural fluid, a serous secretion produced by the pleura. Pleural fluid reduces friction as the lungs move during breathing and causes the two layers of the pleura to adhere together by surface tension, helping to create a slight negative pressure within the pleural space (see page 279). Consequently the pleurae hold the lungs to the chest wall. This is vital to lung function as it means that lung inflation and deflation can occur as the chest wall moves.

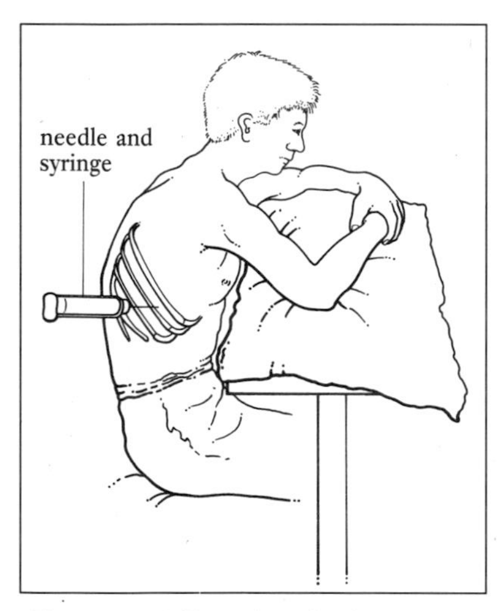

Figure 12.10 Pleural aspiration.

Pleurisy

Inflammation of the pleura may lead to a decrease in pleural fluid resulting in friction between the two layers. A '*pleural rub*' may be heard with a stethoscope when the person is asked to take a deep breath The affected individual suffers sharp, stabbing pain when they take a breath which inevitably leads to a reluctance to breathe normally. Shallow breathing is inefficient (see *Table 12.3*). Adequate pain relief is essential if the characteristic shallow respirations are to be replaced by breathing of normal depth which ensures adequate alveolar ventilation (see page 285).

Other types of pleurisy result in excess fluid production (*effusion*) which may prevent normal lung expansion and cause *dyspnoea* (difficult respiration).

Environmental and occupational hazards to respiratory health

Dust, fumes and chemicals present in the workplace or environment are implicated in the development of chronic respiratory disease (bronchial tumours, COAD (page 277), *pneumoconioses* and *fibrosis*). Noxious agents include asbestos, coal dust, arsenic, nickel, chromium, vinyl chloride, silica, cadmium, radiation and vegetable matter, e.g. dust from mouldy hay or mushroom compost.

Atmospheric pollution also appears to have some influence – many chronic respiratory conditions are more common in industrial and urban areas.

Special Focus Smoking and the respiratory tract

Smoking causes as many as 100 000 premature deaths in the UK every year. Most of these deaths are due to *bronchial tumours* (*Figure 12.11*), chronic airways limitation, which includes *chronic obstructive airways disease* (*COAD*), and heart disease (see Chapter 10). Apart from the deaths there is increased morbidity (being diseased) from heart, respiratory and peripheral vascular disease.

Carcinogens (see Chapter 1) in tobacco are known to cause malignant changes in the bronchi/lungs. The incidence in smokers is some forty times greater than in non-smokers (Health Education Authority, 1988). Bronchial cancer is the commonest malignancy in men and is becoming increasingly common in women.

Unfortunately most bronchial tumours are diagnosed too late for surgery to be effective and the average life expectancy following diagnosis is less than a year. Cytotoxic drugs and radiotherapy may be used to treat certain tumours and provide palliation in others.

Smoking also causes changes in the respiratory tract leading to *chronic bronchitis* and *emphysema* (over-distention and destruction of alveoli) which are two types of COAD. The chemicals in the smoke cause increased goblet cell activity, increased mucus and reduction of ciliary action. The excess mucus, inflammation, mucosal oedema and fibrosis results in narrowing and blockage of small airways, disruption of alveoli and the eventual reduction in the surface area available for gaseous exchange (see *Person-centred Study* – Sid, page 278).

There is no doubt that smoking is a health hazard to those who smoke and for those people with whom they live or work (passive smoking). Throughout this book reference is made to smoking-related diseases, but in particular see Chapter 10. Nurses are ideally placed to provide information and advice about smoking for people such as Sid (page 278). This help will only be effective if nurses have a sound knowledge of the physiological effects of smoking and the necessary communication skills to 'sell the idea'. In an evaluation of the Health Education Council (now the Health Education Authority) package (*Nurses and Smoking*, 1983) it was found that both these aspects were lacking (Haverty *et al.*, 1987).

Various organizations disseminate information about the effects of smoking and offer advice to health professionals and people wanting to give up, e.g. Health Education Authority, Action on Smoking and Health (see page 301 for addresses).

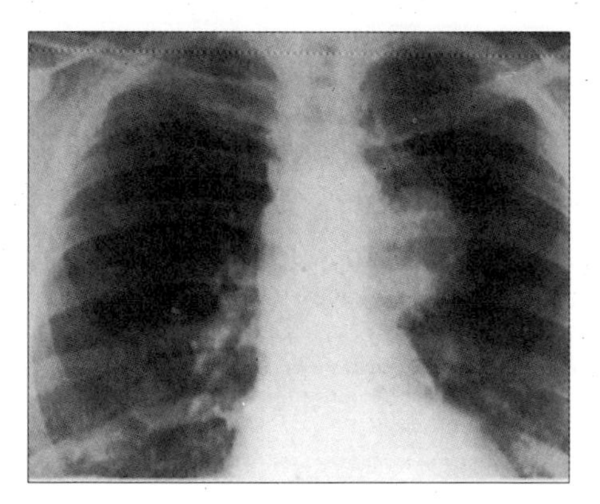

Figure 12.11. Bronchial tumour (an X-ray). (Studdy, P.R. *Diagnostic Picture Tests in Respiratory Medicine.* Wolfe Medical Publications, Ltd. 1990. Reprinted with permission).

Breathing/ Ventilation

Breathing (**ventilation**) is the mechanical process by which air moves in and out of the lungs. It consists of **inspiration** (breathing in) where the lungs expand to fill with air and **expiration** (breathing out) where they recoil to expel air. Breathing depends upon changes of pressure in the lungs and pleural space, functioning of the diaphragm and intercostal muscles (respiratory muscles – see Chapter 18), the elasticity of the lungs and airway resistance.

Person-centred Study Sid

Sid lives alone in an inner city neighbourhood where he rents a ground floor room. Before retiring he worked as a kitchen porter in a large hotel. A heavy smoker for the last 40 years, Sid has always had a cough with sputum but in the last year he has noticed that his cough is more persistent, the sputum is green on occasions and he gets very short of breath walking from the shops.

His GP refers him to the chest clinic at a nearby hospital where a diagnosis of COAD is confirmed.

NB Many people with chronic airways limitation have a combination of conditions, e.g. chronic bronchitis, emphysema, (see page 277) and asthma. The following regimen is planned with Sid in an effort to improve quality of life, minimize hospital admissions for acute exacerbations and postpone the development of cor pulmonale (see Chapter 10) or respiratory failure: (i) drugs – Salbutamol (see *Table 12.1*) is prescribed by nebulizer (see page 274) and antimicrobial drugs for the infection; he is asked to see his GP whenever his sputum looks infected (see below); (ii) he is advised to stop smoking (see *Special Focus*, page 277); (iii) the physiotherapist attached to the clinic teaches Sid how to maximize respiratory efforts to improve gaseous exchange and to expectorate sputum so that the risk of infection is reduced; (iv) a specialist nurse sees Sid at the clinic and again at home to ensure that he undertands his condition and regimen. Respiratory health workers have been found to be effective in improving a person's knowledge of their condition and treatments (Cockcroft *et al.*, 1986); (v) social services are asked to assess Sid's needs and to provide long-term support.

Nursing Practice Application Sputum

Sputum (phlegm) is excess mucus produced when respiratory disease affects the goblet cells and ciliary clearance. People with excess sputum should be encouraged to expectorate; they need privacy, suitable clean containers and a supply of tissues. Care plans should include oral hygiene and a fluid intake sufficient to prevent the sputum becoming thick and sticky. Physiotherapy which assists the person to cough and expectorate is reinforced by the nurse. Change of position and mobilization will help expectoration. If the person has pain on moving or coughing, e.g. following surgery, it is essential that they receive adequate pain relief. Various inhalations and mucolytic substances may help to loosen the sputum and ease discomfort. When the sputum is infected *(purulent)* an appropriate antimicrobial drug will be prescribed following microbiological examination of the sputum.

Sputum is an infection risk. Both sputum and used tissues should always be handled with care, disposed of correctly (suitably bagged prior to placing in the hazardous waste sack) and staff should protect themselves with disposable gloves, plastic aprons and masks as appropriate.

Observations of sputum include: the amount, odour, viscosity and colour, which may be white mucoid, yellow/green if purulent, frothy in left heart failure (see Chapter 10), 'rusty' with altered blood in pneumonia, or containing fresh blood with pulmonary embolus or tumour. Blood-stained sputum is known as *haemoptysis*. Nurses should also assess pain associated with expectoration, type of cough and how tired the person is becoming (see also breathing assessment, pages 280-282).

Intrathoracic pressures

Air pressure within the lungs/alveoli is always equalizing (to maintain equilibrium) with atmospheric pressure, which is 101 kPa (760 mmHg) at sea level. As the lung expands during inspiration its volume increases. This causes a pressure drop and air moves into the lungs to equalize the pressure. Movement of air into the lungs depends on **Boyle's law**, which states that at a constant temperature the pressure of a gas is inversely proportional to its volume.

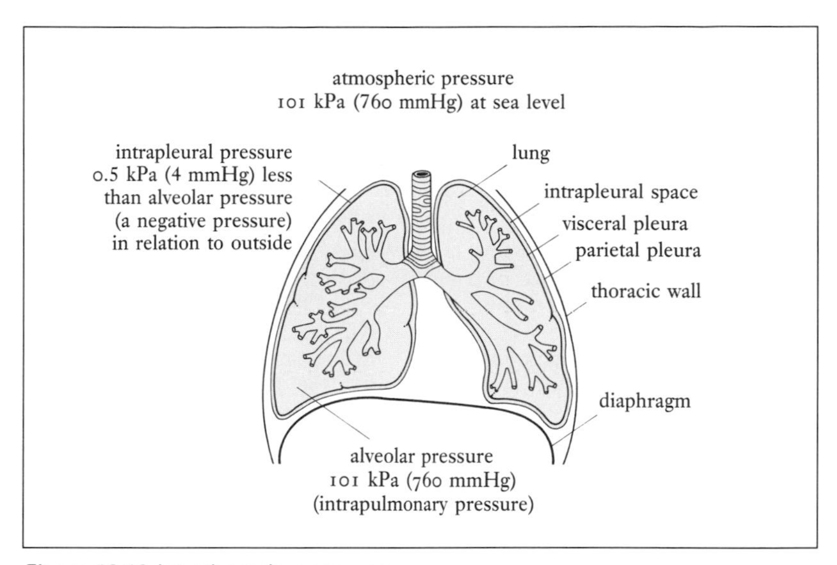

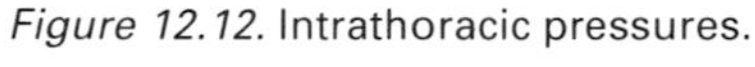
Figure 12.12. Intrathoracic pressures.

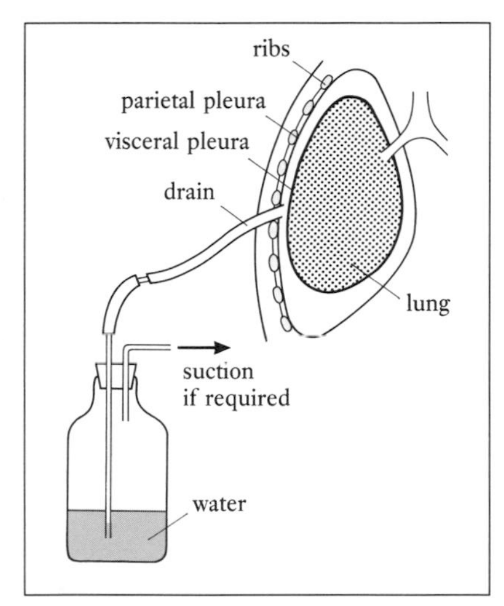

Figure 12.13. Underwater seal drainage.

Intrapleural pressure is 0.5 kPa (4 mmHg) less than alveolar pressure. This negative pressure produced by opposing forces ensures that the lungs adhere to the thoracic walls and do not collapse like deflated balloons (*see Figure 12.12*). The negative pressure also creates the 'respiratory' pump which assists venous return to the heart (see Chapter 10).

Loss of the 0.5 kPa difference between intrapleural and alveolar pressure, e.g. if air enters the chest through a wound, results in serious consequences which include lung collapse, impaired gas exchange and in some cases mediastinal shift and cardiovascular problems. This situation, where air is present in the thoracic cavity, is known as a *pneumothorax* or *haemothorax* if there is also blood. It occurs if the chest wall is damaged, e.g stab wounds, during chest surgery and with spontaneous rupture of diseased lung into the pleural space. It is usually treated by the insertion of a drain which allows air/fluid to escape from the thoracic cavity but prevents more air entering (see *Nursing Practice Application,* page 280).

Inspiration

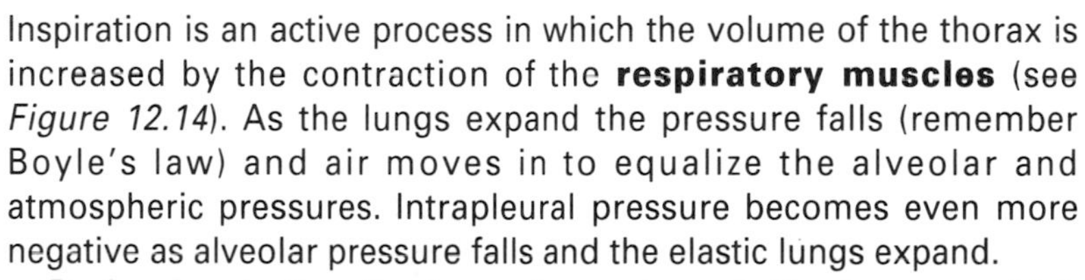
Inspiration is an active process in which the volume of the thorax is increased by the contraction of the **respiratory muscles** (see *Figure 12.14*). As the lungs expand the pressure falls (remember Boyle's law) and air moves in to equalize the alveolar and atmospheric pressures. Intrapleural pressure becomes even more negative as alveolar pressure falls and the elastic lungs expand.

During inspiration the large dome-shaped **diaphragm**, which separates the thorax from the abdomen, flattens and moves down. The diaphragm is the most important inspiratory muscle but the **external intercostal muscles** also contract to elevate the ribs. Contraction of these muscles increases chest volume laterally, vertically and from front to back.

Nursing Practice Application Pneumothorax/haemothorax

The development of a pneumothorax requires prompt diagnosis and the insertion of a drain (either with a one way valve or a water seal) into the pleural space. This is obviously planned for during chest surgery but in the case of trauma will be performed as an emergency measure.

Water seal drains – drainage tubes leading from the person's chest wall are placed under water to prevent air being sucked into the chest; they are sometimes attached to low-power suction (*Figure 12.13*). This arrangement allows air/blood, etc. to drain out and the lung to re-expand.

The level of fluid should fluctuate with breathing. If this does not occur it may indicate that the drain is no longer patent but could be due to lung re-expansion. Nurses should ensure that chest drain clamps are always available for use when moving the person, changing drainage bottles or if the closed system is breached, e.g. tubing becomes disconnected from drain.

It is also very important that the water seal bottle is kept below chest level to avoid fluid being 'sucked back' into the pleural space on inspiration.

Apart from these factors the nurse should be alert for changes in condition such as sudden dyspnoea or pain. Assessment will include: respiration, colour, pulse, blood pressure and pain. Frequency of observations will depend upon the reason for chest drainage and the condition of the person.

When inspiration is deep or forced such as during exertion we use accessory muscles to lift the ribs and maximize thoracic capacity. Use of accessory muscles, e.g. *sternocleidomastoid, scalenes* and *pectorals* (see Chapter 18) is seen in people with chronic respiratory disease as they strive to increase their air entry (see *Person-centred Study* – Sid, and page 278).

During chest expansion the lungs, which you will remember are attached to the chest wall by the pleurae, stretch by virtue of their elasticity.

Expiration

Normal quiet expiration is a passive process where the natural elastic recoil of the lungs causes air to be expelled (see *Figure 12.14*). As the diaphragm and external intercostal muscles relax, the thoracic capacity decreases, the alveolar pressure rises and air flows out to once again equilibrate with atmospheric pressure. As the alveolar pressure rises and the lungs recoil, the intrapleural pressure becomes less negative.

Forced expiration, on the other hand, is an active process involving the abdominal muscles. This may be observed during exercise or where airway obstruction exists.

Assessment of breathing

Normally, breathing is quiet, unlaboured and occurs 12–16 times/minute in adults. Each breathing cycle consists of inspiration, expiration and a short pause before the next cycle starts. It is important to remember that the rate is much faster in the newborn (40/min) but slows during childhood. Individual variations exist in healthy individuals and factors such as activity affect rate, as does the presence of abnormalities.

Apart from rate, other aspects noted during a complete nursing assessment are:

- The effort required to breathe, and the use of accessory muscles

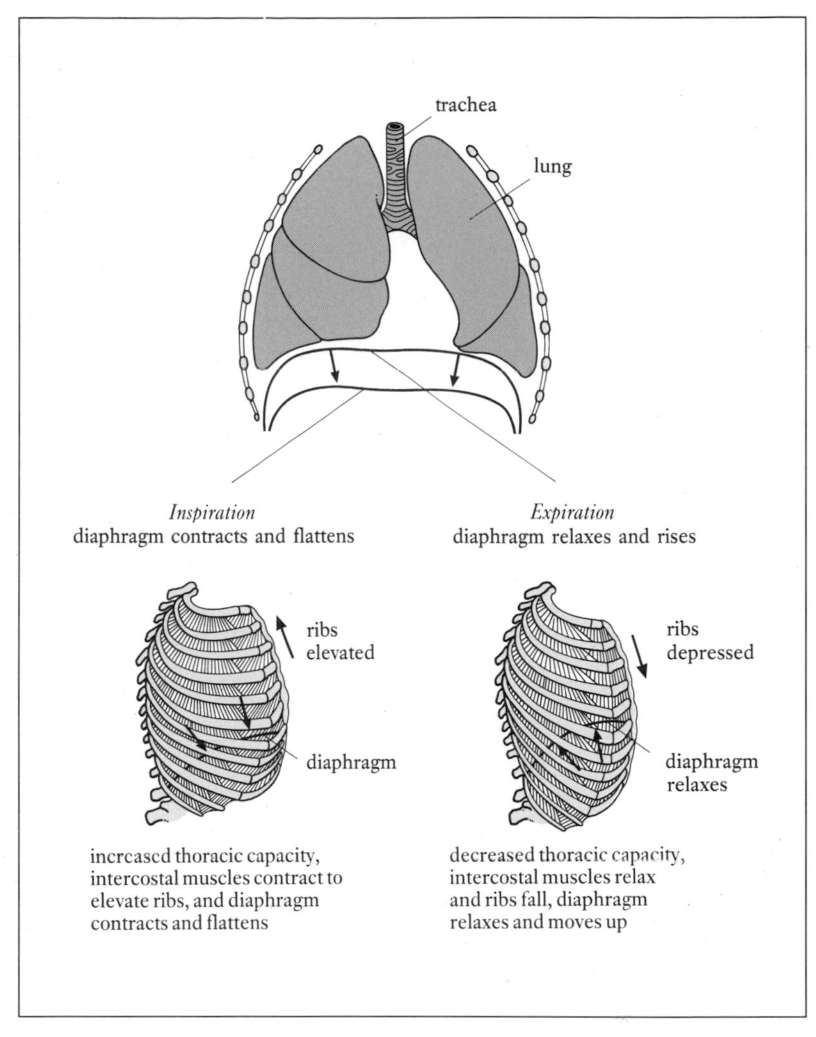

Figure 12.14. Inspiration and expiration.

(sternocleidomastoid, pectorals, scalenes), e.g. chronic respiratory disease.

- Depth, e.g. shallow breathing may indicate pleural pain (see page 276).
- Regularity: some abnormal patterns, e.g. *Cheyne–Stokes* where periods of *apnoea* (no breathing) are followed by *hyperventilation* (over-breathing; deep and rapid), may be present in severe respiratory malfunction.
- The colour of the skin: blue discolouration (*cyanosis*) is caused by the poor oxygenation associated with respiratory and other conditions. In Caucasians this is obvious but in Afro-Caribbean or Asian individuals it is easy to miss unless the nail beds and mucous membranes are checked for 'blueness' or dusky colouration.
- Noise, such as the wheezing associated with airway obstruction.

Considerable observational skills are required to make an objective assessment as people tend to change breathing rates when being watched. You can try this out on some friends: count their respiratory rate without them knowing, e.g. whilst they watch television, then repeat the count with their knowledge and compare the results.

Breathing: other factors

Earlier we looked at the role of pressure changes and inspiratory muscles in ventilation. A more complete coverage of the factors influencing ventilation includes:

Airway resistance

The **airways** produce **resistance** to air flow. If the bronchi are constricted there is more friction and it becomes more difficult for air to reach the alveoli, e.g. severe asthma. A greater pressure gradient must be overcome to move air when airway resistance increases.

Compliance and elasticity

Compliance is the ability of the lungs and thorax to stretch and distend. Healthy lungs need only a very small inflating pressure to move approximately 500 ml of air with each breath (*tidal volume*). Ventilation becomes more difficult if the lungs lose compliance and

Nursing Practice Application Oxygen administration

Many people with dyspnoea and poor oxygenation (*hypoxia* see page 290-292) will derive benefit from prescribed oxygen. Oxygen is usually given via face mask or nasal catheters but can be administered by endotracheal tube, tracheostomy (see page 271), oxygen tent and at higher than atmospheric pressure in a *hyperbaric* (high pressure) chamber, e.g. for carbon monoxide poisoning (see *Figure 12.15*).

Oxygen should be prescribed by the medical officer only after careful assessment because for some hypoxic states it is useless (cyanide poisoning) or harmful in high concentration (COAD). Where chronic respiratory diseases such as COAD have existed for many years the central chemoreceptors no longer respond to high carbon dioxide levels in the blood or CSF, and breathing is stimulated by hypoxia (see page 295 – chemical control). If too much oxygen is given it would cause respiratory depression and further carbon dioxide retention. It is vital that low-concentration oxygen, e.g. 24% administered via a *venturi* effect (the flow of oxygen is diluted by atmospheric air which is drawn into the mask) mask is prescribed for these people.

For all its life-saving uses oxygen therapy is a hazardous business and nurses need to be aware of other potential dangers including: (i) fire risk – oxygen supports combustion; (ii) high concentration oxygen can damage the lungs; (iii) newborn babies, given too much oxygen, may be blinded by fibrosis occurring behind the lens of the eye (*retrolental fibroplasia*).

Less dramatic but still important are the physical and psychological effects of oxygen therapy. Some people find a mask over their face claustrophobic and may be more comfortable with nasal catheters. For the already breathless person a face mask presents yet another barrier to effective communication which leads to fear, anger, frustration and withdrawal. The mask and attachments can cause pressure and soreness. Cheeks and nose should be checked regularly. Oxygen has a drying effect upon the mucous membranes and should be humidified before administration.

The care plan will include observations of vital signs and colour (skin and mucous membranes) and appropriate measures to minimize the effects of drying on the mouth and nose, e.g. adequate fluid intake.

Compliance will be improved if patients are educated about the benefits of oxygen therapy, encouraged to continue and helped to overcome discomforts (Foss, 1990). The patient and their visitors are asked not to smoke.

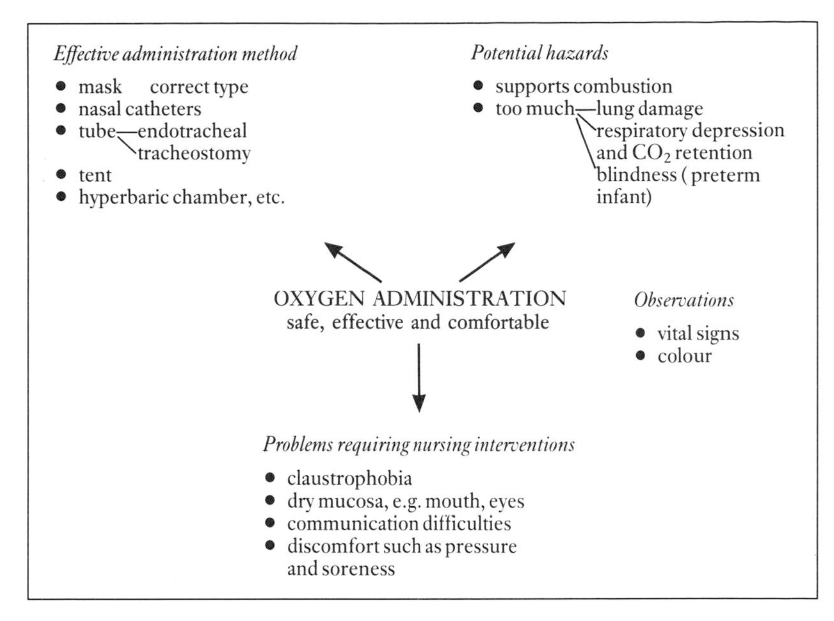

Figure 12.15. Oxygen administration.

become 'stiff' as in respiratory distress syndrome or pulmonary fibrosis.

Overstretching, with loss of elasticity, means that the lungs can no longer recoil to force air out (an analogy is when old elastic becomes overstretched and no longer recoils). In this situation, e.g. emphysema, the lungs are overinflated and the affected person works very hard to expire air from the lungs.

Surfactant and surface tension

The phospholipid **surfactant** reduces **surface tension** forces in the alveoli and allows lung expansion. Surfactant normally keeps the alveoli patent (open) between breaths: insufficient surfactant will result in alveolar collapse which requires considerable energy for reinflation on inspiration (see page 274 – respiratory distress syndrome).

Lung volumes and respiratory function tests

We can now consider the various volumes and capacities that exist in the lungs during inspiration and expiration (see *Figure 12.16*).

Changes in volume measured by *spirometry* (where an instrument called a *spirometer* is used to measure changes in lung volumes) can be used to assess respiratory function, aid diagnosis and give useful insights into progress and response to drug therapy (see *Table 12.2*). It is important to realize that they have limitations and, as with all investigations, they should form part of a holistic assessment. This would consist of a full nursing assessment which included breathing (see pages 280-282) and other appropriate investigations, e.g. chest radiography.

For most people the thought of having any investigation provokes anxiety, and respiratory function tests are no exception.

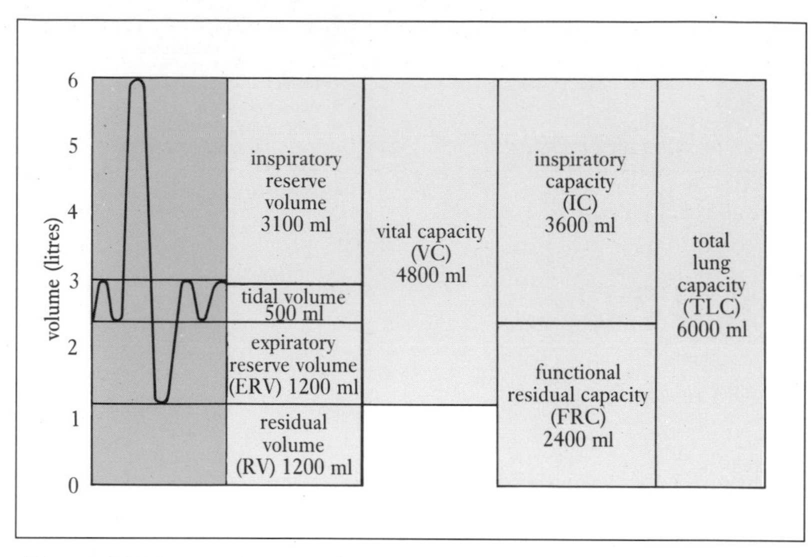

Figure 12.16. Lung volumes and capacities.

Nurses should ensure that adequate information is given and that the tests and their purpose are clearly understood. Careful preparation, which calms and relaxes, helps to ensure the best results (Twohig, 1984).

Pulmonary ventilation and tidal volume

The amount of air moved in and out of the lungs in one minute is termed the **pulmonary ventilation** or **minute volume**. This is about 6000 ml in an 'average' healthly adult at rest. The volume of air moved during one quiet breath is about 500 ml and is known as the **tidal volume**. Not all this air is involved in gaseous exchange: some 150 ml is always in the conducting part of the respiratory tract or dead space.

Detail of other respiratory volumes and capacities can be found in *Figure 12.16* and *Table 12.2* but readers who wish to explore this topic further are directed to the references, e.g. Twohig (1984) and further reading, e.g. Kendrick and Smith (1992).

Alveolar ventilation

This is the amount of inspired air reaching the alveoli in one minute available for gaseous exchange. It is the tidal volume minus the dead space (see page 272) multiplied by respiratory rate (Alveolar ventilation rate = [tidal volume – dead space] x respiratory rate), e.g:

[500 ml – 150 ml] x 12
= 4200 ml (alveolar ventilation rate/min)

These figures represent a healthy adult at rest but they vary with size, age, health and activity, e.g. alveolar ventilation rate increases greatly during exercise. Deep slow breathing produces an increased alveolar ventilation because the tidal volume increases and the constant dead

Table 12.2. Respiratory measurements and function tests. (Adapted from Marieb, E. *Human Anatomy and Physiology*. Copyright © 1989 by The Benjamin/Cummings Publishing Company, Inc., U.S.A.)

Test (volume/capacity)	***Definition/Description***	***Average values****
Vital capacity (VC)	Maximum volume expired after greatest inspiratory effort	4800 ml
Total lung capacity (TLC)	Volume of air in lungs following greatest inspiratory effort: TLC = FRC + IC	6000 ml
Inspiratory capacity (IC)	Maximum amount of air inspired after a normal expiratory effort	3600 ml
Expiratory reserve volume (ERV)	Maximum amount of air that can be forcefully expelled after a normal expiration	1200 ml
Residual volume (RV)	The amount of air remaining in the lungs after a maximal expiratory effort: RV = FRC – ERV	1200 ml
Functional residual capacity (FRC)	Volume of air in the lungs after a normal expiration: FRC = ERV + RV	2400 ml
Peak expiratory flow rate (PEFR) uses a peak flow meter with disposable mouthpiece	Measures the greatest flow rate of air during rapid exhalation. This test is commonly performed by nurses when assessing response to bronchodilator drugs. It is vital that everyone understands that this is a flow rate measurement not a volume	400–600 litres/min (depends on age, size and gender)
Forced vital capacity (FVC)	Volume of air expired forcefully soon after a maximal inspiratory effort	
Forced expiratory volume in one second (FEV_1)	The volume of air exhaled during the first second of FVC	normally 80% of the FVC
Forced expiratory ratio (FER)	The ratio between FEV_1 and FVC	normally 0.8

space is proportionally smaller. Rapid shallow breathing, on the other hand, decreases alveolar ventilation because the tidal volume is smaller and the dead space is a proportionally higher percentage. The effects of chronic respiratory disease also reduce alveolar ventilation rates, which leads to dyspnoea and hypoxia. It makes good sense to help an anxious, frightened or breathless person to change to a more efficient breathing pattern (see *Table 12.3*) and adopt an upright position which increases lung capacities.

Table 12.3. Breathing patterns and alveolar ventilation (hypothetical situations).

Breathing pattern	*Respiratory rate/min*	*Tidal volume (ml)*	*Pulmonary ventilation Minute volume (ml/min)*	*Alveolar Ventilation (ml/min)*	*% of Tidal volume as dead**
1. Normal rate and depth	16	500	8000	5600	30%
2. Shallow, rapid	32	250	8000	3200	60%
3. Deep, slow	8	1000	8000	6800	15%

N.B. Overall pulmonary ventilation is unchanged.

* Assumes the dead space to be a constant 150 ml.

GASEOUS EXCHANGE

Before we consider the process of gaseous exchange it is necessary to look at some fundamental principles of gas behaviour. These account for the bulk flow of gases (see page 278 – Boyle's law) and diffusion down pressure gradients in external and internal respiration.

Behaviour of gases and gas laws

Air is a mixture of nitrogen, oxygen, carbon dioxide, water vapour and an insignificant amount of inert gases, e.g. helium (see *Table 12.4*). Each of these gases individually exerts a **partial** (individual) **pressure**, measured in kPa (mmHg), proportional to its percentage concentration in the mixture. **Dalton's Law** (of partial pressures) states that the pressure of a gas mixture is a sum of the partial pressures which each gas would exert if it completely filled the space.

The partial pressure of nitrogen, which forms approximately 79% of air, is 79 kPa (590 mmHg) at sea level where atmospheric pressure is around 101 kPa (760 mmHg). Under the same conditions oxygen, at nearly 21%, has a partial pressure of 21.2 kPa.

We have looked at pressure at sea level, but as altitude increases the atmospheric pressure falls. The consequent fall in the partial pressure of the gases in air, although the proportions remain the same, will lead to problems with gas exchange in the lungs.

The opposite also causes difficulties, e.g. working deep under the sea where high pressure can cause decompression sickness (see page 296-297).

Gases diffuse down partial pressure gradients, i.e. from high to low pressure. These gradients are maintained in the body by the continual use of oxygen and production of carbon dioxide by the cells. The other gas law needed to explain diffusion is **Henry's Law** which states that the amount of a gas dissolving in a fluid is

Table 12.4. Composition (%) and gas partial pressures. (Adapted with permission from Hinchliff, S., Montague, S., eds. *Physiology for Nursing Practice.* Copyright © 1988 by Ballière Tindall, London).

	Atmospheric air	***Alveolar air***	***Expired air***
Oxygen	20.9% (21.2 kPa)	13.2% (13.3 kPa)	15.4% (15.6 kPa)
Nitrogen	78.6% (79.6 kPa)	75.3% (76.4 kPa)	74.6% (75.6 kPa)
Carbon dioxide	0.04% (0.04 kPa)	5.3% (5.3kPa)	3.8% (3.8 kPa)
Water vapour	0.46% (0.49 kPa)	6.2% (6.3 kPa)	6.2% (6.3 kPa)
Total	100% (101.3 kPa)	100% (101.3 kPa)	100% (101.3 kPa)

proportional to its pressure and solubility at a constant temperature. Diffusion rates will depend on the steepness of the pressure gradient, the solubility of the gas (carbon dioxide is the most soluble followed by oxygen and then nitrogen) and the state and surface area of any barrier, e.g. alveolar membrane.

Composition and partial pressures

Table 12.4 shows that there are differences in the composition and partial pressures of the gases in atmospheric, alveolar and expired air. These differences can be explained by:

- Humidification of atmospheric air in the airways.
- The continuous diffusion of gases across the alveolar membrane – oxygen into the blood and carbon dioxide to the alveoli.
- Expired air being a mixture of alveolar air and the air contained in the dead space.

Generally there is little change in alveolar air composition because air remaining after expiration (the functional residual capacity, FRC) is continually mixing with tidal air. The stability of composition prevents marked changes in the blood gases. It is possible, however, to improve oxygen content and remove carbon dioxide more rapidly by increasing alveolar ventilation (see *Table 12.3*).

External respiration

The diffusion of gases between alveolar air and blood in the pulmonary capillaries depends upon partial pressure gradients and solubilities, healthy functioning of the alveolar membrane (see *Figure 12.8b*) and the correct **ventilation/perfusion ratio.**

Partial pressures

The partial pressure of oxygen is 13.3 kPa in the alveoli and 5.3 kPa in the pulmonary capillaries. This means that oxygen passes from the alveolar air into the blood until equilibrium is reached (about 0.25s). It is then transported around the body, mainly by the erythrocytes. Carbon dioxide moves in the opposite direction because its partial pressure is 6.1 kPa in the capillary and only 5.3

kPa in the alveoli. The pressure gradient for carbon dioxide is small but because of its high solubility (20 times that of oxygen) it is still able to diffuse into the alveoli for removal during expiration (see *Figure 12.17*).

The relatively slow blood flow through the pulmonary capillaries allows time for gas exchange.

Alveolar membrane

Providing a huge surface area for gas exchange (see page 273) the thin, moist alveolar membrane is ideal for the rapid diffusion of gases. Any thickening or severe loss of area due to disease will impede diffusion and lead to hypoxia.

Ventilation/perfusion relationship

Homeostatic processes operate to maintain an efficient ratio between gases in the alveoli and the blood flow in the capillaries. These include the response of smooth muscle in the walls of the arterioles and the small airways to changes in the partial pressure of CO_2 (**PCO_2**) and the partial pressure of oxygen (**PO_2**). If a mismatch occurs, gaseous exchange will be seriously impaired with the development of hypoxia and possibly *hypercapnia* (high CO_2 levels

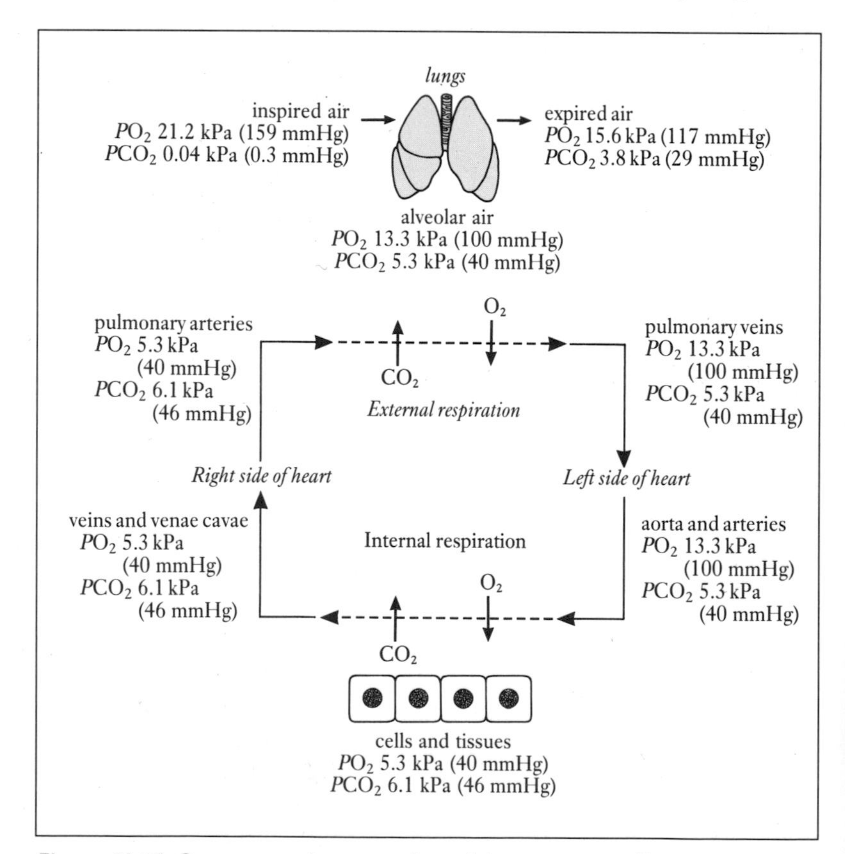

Figure 12.17. Gaseous exchange and partial pressure gradients – lungs and tissues.

Nursing Practice Application **Blood gases**

The measurement of gases and pH in arterial blood is a commonly performed investigation in critically ill patients, e.g. those with severe asthma, those needing artificial ventilation for respiratory failure and to differentiate between type I and type II respiratory failure.

A sample of arterial blood is obtained from a suitable vessel, e.g. the femoral or radial artery, and the results used in the diagnosis of hypoxia and in decisions regarding treatment, e.g. oxygen therapy.

The nursing role includes: (i) patient preparation with explanation; (ii) ensuring the sample, which is heparinized and kept cool by packing in an iced container, arrives at the laboratary quickly and in good condition; (iii) prevention of haematoma formation by applying pressure over an arterial puncture for at least 5 min. Blood gases may also be measured by electrodes placed on the skin or inserted into an artery.

Normal range of values
(local variations exist between laboratories)

PaO_2	11–13.3kPa (82.5–100 mmHg)
pH	7.35–7.45.
$PaCO_2$	4.7–6.0 kPa (35–45 mmHg)
Plasma hydrogen carbonate	22–28 mmol/l

See also – Respiratory role pH regulation, pages 297-299 and Chapter 2.

in the blood). This may be due to alveolar underventilation, e.g. with COAD (see page 277) or insufficient blood flowing in the pulmonary capillaries, e.g. after severe haemorrhage and shock (see Chapter 10) or following pulmonary embolus (see Chapter 9).

Internal respiration

It seems logical to consider internal respiration at this point, it being the reverse of external respiration in the lungs. Internal respiration depends upon the same factors but this time diffusion occurs between the capillaries, interstitial fluid and the cells (*Figure 12.17*).

Arterial blood arriving at the tissues has a higher partial pressure of oxygen (13.3 kPa) than the cells (5.3 kPa) which obviously leads to a movement of oxygen into the cells. Cellular carbon dioxide (6.1 kPa) meanwhile moves into the blood where the partial pressure is only 5.3 kPa. Venous blood drains from the tissues and returns to the lungs to recommence the cycle of events with external respiration. It is important that we remember that gases must be transported between lungs and tissues and vice versa before any exchange can occur.

Transport of Gases

Transport of oxygen

Most oxygen is carried to the tissues combined with haemoglobin in erythrocytes and a small amount (1%) is dissolved in the plasma (oxygen has a low solubility). The oxygen in the plasma, although not important for actual tissue oxygenation, is vital in maintaining oxygen pressure gradients between plasma and tissues.

Haemoglobin

Haemoglobin (see Chapter 9) is a specialized pigment–protein complex and is found within the erythrocytes. A molecule of haemoglobin (Hb) consists of four haem groups each containing an atom of ferrous iron, and four globin protein chains. Each molecule of Hb can combine reversibly with four molecules of oxygen. These join with the haem groups to form the bright red **oxyhaemoglobin**:

$$Hb + O_2 \rightleftharpoons HbO_2$$

Reduced (not saturated with oxygen) haemoglobin (HHb) is a dark purple colour prior to its **oxygenation**. One gram of haemoglobin combines with 1.34 ml of oxygen which means that given a haemoglobin level of 15 g per decilitre (dl) of blood (normal range 11.5–18 g/dl), each decilitre of blood, when fully saturated, can carry about 20 ml of oxygen. The actual amount depends on the PO_2 of the blood, influenced by the oxygen carried in simple solution (dissolved in the plasma) as well as the Hb content.

Oxygen dissociation curve

In Chapter 9 we discussed how oxygen molecules combine sequentially with haemoglobin. Each of the four haem groups has a different affinity for oxygen, which is illustrated by the sigmoid-shaped dissociation curve (*Figure 12.18*). The first combines with some difficulty, the next two much more easily but the last has the most difficulty. As the haemoglobin molecule becomes saturated with oxygen its folded globin chains change shape to facilitate the binding of the last oxygen molecule. The interaction between different parts of the haemoglobin molecule during oxygenation increases efficiency.

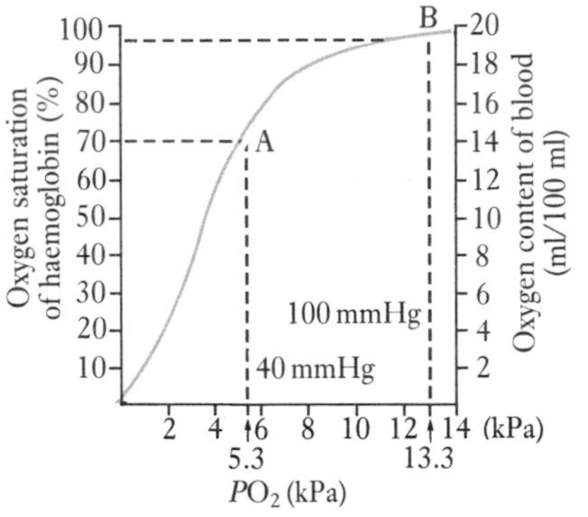

Figure 12.18. Oxygen dissociation curve (at 37°C, pH 7.4, Hb 15 g/dl).

Once blood reaches the tissues the oxygen molecules are unloaded sequentially with the release of one molecule assisting the unloading of the next molecule and so on. Under normal conditions only about 5 ml of oxygen is given up at the tissues which means the haemoglobin molecule is still 75% saturated when it returns to the lungs. However, if the PO_2 within an actively metabolizing tissue falls very low there is the facility for more oxygen to be released.

Various factors affect the rate at which haemoglobin combines with and unloads oxygen at a given PO_2. These include: temperature, PCO_2, pH and levels of **2,3-diphosphoglycerate (2,3-DPG)** a metabolite in erythrocytes (see *Figure 12.19*). An increase in temperature will decrease the affinity of haemoglobin for oxygen which means that more oxygen is unloaded in the most metabolically active tissues. A similar 'shift to the right' of the curve occurs if the PCO_2 or hydrogen ion concentration are raised. Again this ensures that oxygen release is facilitated in metabolically active tissue. The 'right shift' of the oxygen dissociation curve occurring as a result of reduced blood pH (becoming more acid as hydrogen ion concentration increases) is also known as the *Bohr effect*. The substance 2,3-DPG decreases haemoglobin affinity for oxygen and enhances oxygen release to the tissues. Its production by erythrocytes increases where oxygen partial pressures are reduced, e.g. chronic respiratory disease, high altitude.

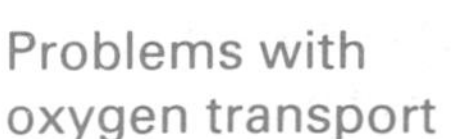

Problems with oxygen transport

Hypoxia

Hypoxia is the term used when there is insuffient oxygenation of the tissues or inability to utilize the available oxygen. Types include:

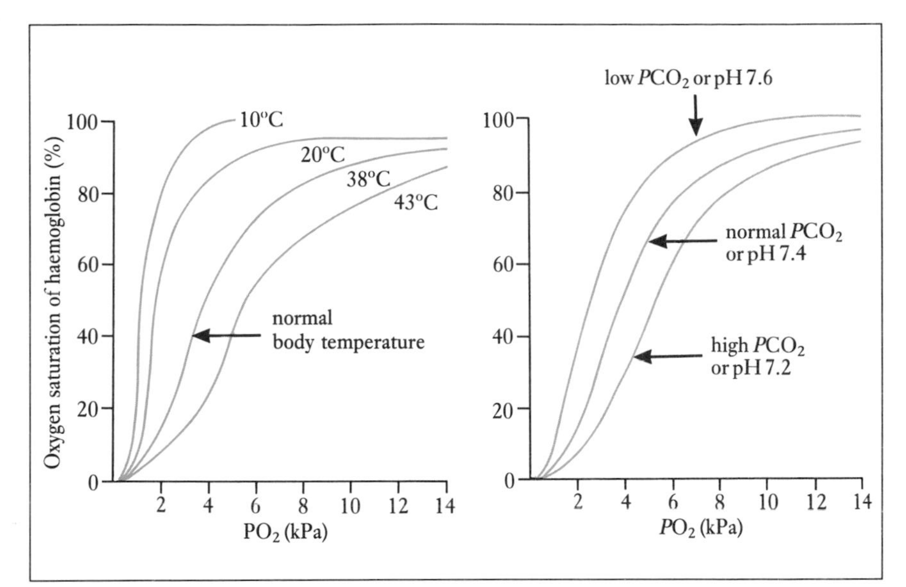

Figure 12.19. Effects of various factors on the oxygen dissociation curve.

- *Where PaO_2 is reduced*, e.g. respiratory disease which affects gaseous exchange.
- *Anaemic hypoxia*, where haemoglobin levels are reduced or not available for oxygen transport, e.g. iron deficiency anaemia, haemorrhage, carbon monoxide poisoning (carbon monoxide is produced in car exhaust gases and by the incomplete combustion of other hydrocarbons such as natural gas). At this point it is worth taking a closer look at carbon monoxide. This colourless, odourless gas causes problems because it combines more easily with haemoglobin (to form *carboxyhaemoglobin*) than does oxygen, with which it competes. The only antidote for carbon monoxide poisoning is oxygen (Compton, 1990).
- *Stagnant hypoxia*, where circulation of blood is impaired, e.g. cardiac failure.
- Cellular poisons such as cyanide cause *histotoxic hypoxia*, where cells are unable to utilize oxygen because their respiratory enzymes are impaired.

Hypoxia usually presents as hyperventilation, nausea, headache, tachycardia and changes in behaviour. If more than 5 g/dl of Hb is reduced (unsaturated with oxygen) in a person with a normal haemoglobin content, cyanosis will be evident (see page 281). If this was occurring in a person with anaemia, e.g. Hb 8 g/dl, the hypoxia would be severe with less than 40% of their Hb saturated before cyanosis was observed.

In many types of hypoxia the person will derive benefit from the administration of oxygen (see page 282) following careful assessment which may include the estimation of blood gases and pH. People who are chronically hypoxic respond physiologically by increasing erythropoiesis leading to polycythaemia (see Chapter 9) and producing more 2,3-DPG.

Transport of carbon dioxide

Carbon dioxide transport via the blood utilizes three methods: dissolved in plasma, combined with haemoglobin and as hydrogen carbonate ions (see *Figure 12.20*). At rest, cellular metabolism produces about 200 ml of carbon dioxide every minute which must be transported to the lungs for removal by gaseous exchange and expiration (see *Figure 12.17*):

- A small amount of carbon dioxide is transported in simple solution as *carbonic acid* (*H_2CO_3*)

$$CO_2 + H_2O \rightleftharpoons H_2CO_3$$

  The reaction occurs quite slowly without the help of an enzymic catalyst.
- Some carbon dioxide enters the erythrocyte to combine with amino acid groups in the globin chains of haemoglobin to form the neutral compound *carbaminohaemoglobin*.

$$Hb + CO_2 \rightleftharpoons HbCO_2$$

  The amount of carbon dioxide carried is determined by the PCO_2 and the degree of oxygen saturation. Most carbon dioxide is transported by reduced haemoglobin. Some carbon dioxide

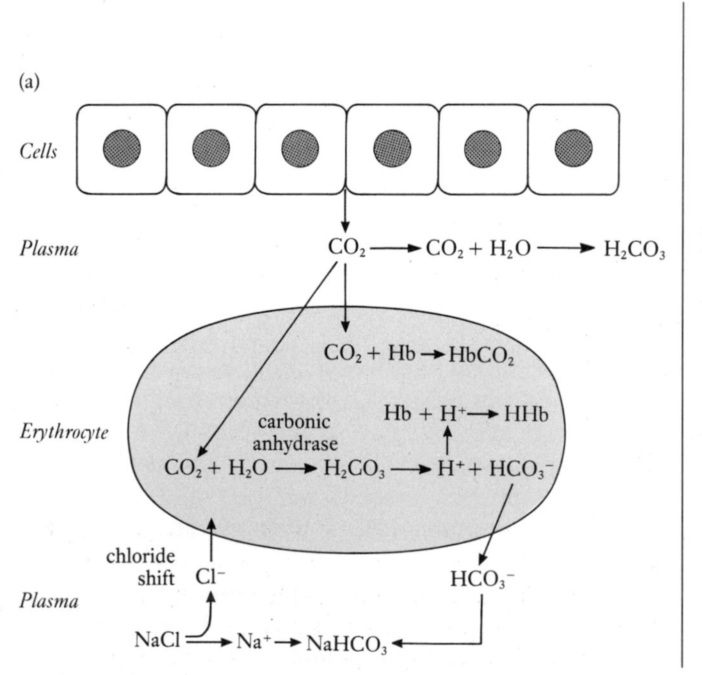

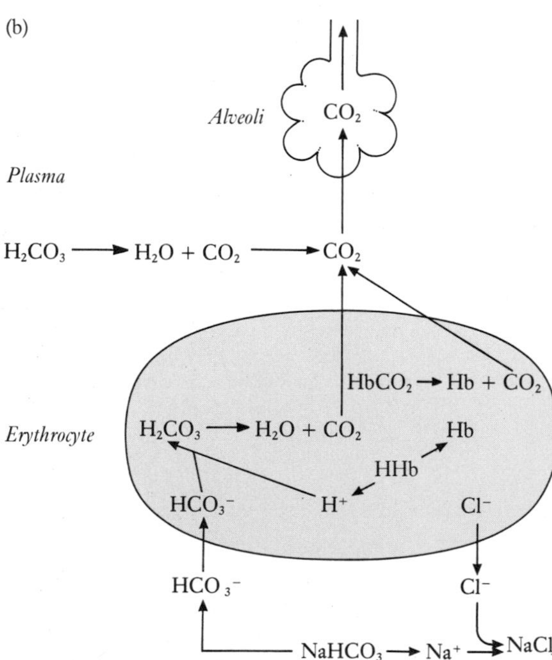

Figure 12.20. Carbon dioxide transport. (a) At the tissues; (b) in the lungs.

forms similar '*carbamino*' compounds with the plasma proteins. In the tissues, where the PCO_2 is higher than in the blood, the carbon dioxide combines easily with the haemoglobin. On arrival in the lungs it dissociates because alveolar air PCO_2 is lower than that in the blood.

- The most important method of transport involves a series of chemical reactions, mostly occurring in the erythrocyte, which facilitate the carriage of carbon dioxide as hydrogen carbonate. At the '*tissue end*' of the process the carbon dioxide diffuses into the blood and enters the erythrocyte where it combines with water to form carbonic acid. This time the reversible reaction proceeds rapidly because it is catalysed by the enzyme **carbonic anhydrase**. The carbonic acid formed within the erythrocyte dissociates into hydrogen carbonate and hydrogen ions

$$CO_2 + H_2O \rightleftharpoons H_2CO_3 \rightleftharpoons H^+ + HCO_3^-$$

The hydrogen ions combine with haemoglobin (remember the buffering role of haemoglobin), which limits their effect on pH, and most of the hydrogen carbonate diffuses out (because the increased concentration of hydrogen carbonate ions in the erythrocyte produces a concentration gradient) of the erythrocyte into the plasma. Hydrogen carbonate ions remaining in the erythrocyte are combined with potassium which, you will remember, is an important intracellular ion (see Chapter 2). As negatively charged hydrogen carbonate ions leave the erythrocyte the electrical balance is restored by the movement of negatively charged chloride ions from the plasma to the erythrocyte – a process known as the **chloride shift**. The hydrogen carbonate ions combine with sodium ions in the plasma to form sodium hydrogen carbonate.

When the blood arrives in the lungs the reactions are reversed and carbon dioxide is released into the plasma. From here it diffuses into the alveoli.

Carbon dioxide transport is dependent on the degree of blood oxygenation. Reduced haemoglobin is able to carry more carbon dioxide but haemoglobin can carry both gases at the same time because they utilize different sites on the molecule.

By this time you will have realized that the loading and unloading of both gases are closely related processes.

Problems with carbon dioxide levels

Hypercapnia

Hypercapnia is the term used when the arterial PCO_2 is abnormally high (normal range 4.7–6.0 kPa). It will happen when there is hypoventilation or where a ventilation/perfusion mismatch exists (see page 288). A high PCO_2 stimulates ventilation in an attempt to flush out excess carbon dioxide. Hypercapnia is always accompanied by hypoxia (see page 290-292); it can increase blood acidity (*acidosis*) and at very high levels will affect central nervous system (CNS) function. CNS effects include: raised intracranial pressure (see Chapter 4) leading to headaches, confusion, altered consciousness and eventual death if treatment is inadequate. Management is based on increasing ventilation and correcting hypoxia and acidosis.

Hypocapnia

The opposite situation to hypercapnia is *hypocapnia*, i.e. a reduction in the amount of carbon dioxide in arterial blood, which is usually caused by hyperventilation. As ventilation increases, excess carbon dioxide is flushed from the body. Some causes of hyperventilation include anxiety (may be observed in people admitted to hospital), hysteria, pain, high altitude, and when the body becomes too acidic, e.g. diabetic ketoacidosis (see Chapter 8). The high carbon dioxide respiratory drive will be lost and the blood becomes more alkaline as acid carbon dioxide is lost. This change in pH affects the amount of available calcium in the blood, resulting in *tetany* (see Chapter 8).

Control of Ventilation

As you can imagine, the vital rhythm, rate and depth of breathing/ventilation requires precise control mechanisms. This control is mostly involuntary but we have a voluntary override

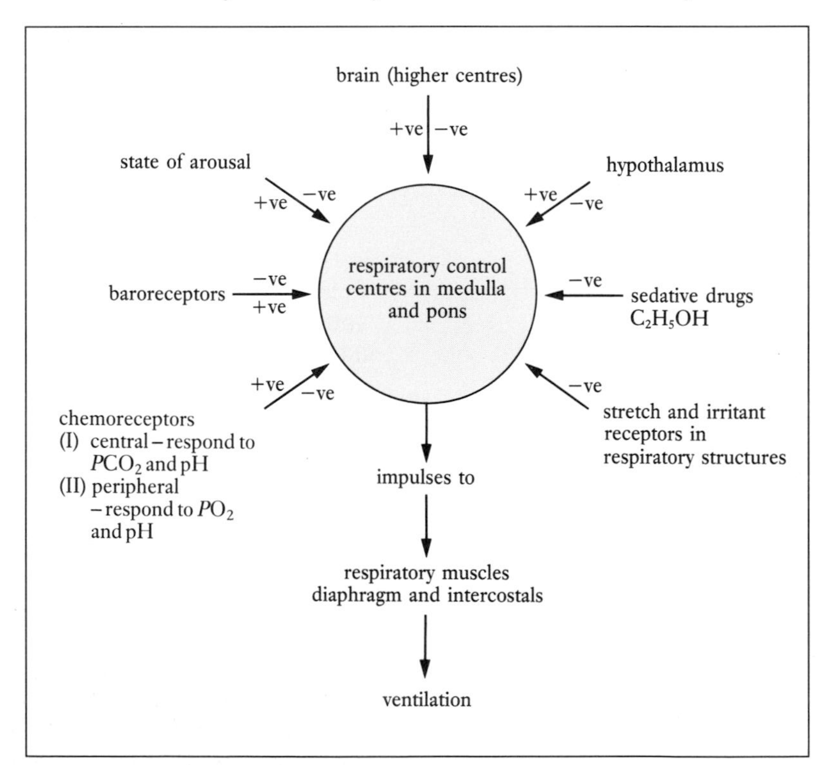

Figure 12.21. Control of ventilation (simplified); +ve, stimulatory; –ve, inhibitory.

ability in certain situations, e.g. singing.

The **respiratory control centres**, which are situated in the medulla and pons, receive inputs from chemoreceptors, stretch receptors (airways, pleura and lungs), higher centre voluntary overrides and the hypothalamus (*see Figure 12.21*). Just how the centres in the medulla actually work is not universally agreed but one hypothesis is described here.

The respiratory centre consists of **inspiratory** and **expiratory neurones** which are mutually inhibitory. Stimulation of the inspiratory neurones results in impulse transmission to the diaphragm and external intercostal muscles, which contract. Expiratory neurones are concerned only when active expiration is required; expiration is normally passive.

The frequency of impulse transmission and ventilation is geared to maintaining tissue oxygenation throughout a range of changing physiological needs, e.g. high altitude and exercise. The medullary respiratory centre is depressed by drugs such as morphine and by overdose of sedatives or alcohol (C_2H_5OH), which if severe will cause breathing to cease (*apnoea*).

Centres in the pons – the **pneumotaxic** and **apneustic centres** send inhibitory and stimulatory impulses to the inspiratory neurones which provides the precise tuning for a smooth respiratory rhythm.

Chemical control

Peripheral chemoreceptor cells situated in the aortic arch and carotid bodies, and some located centrally in the medulla, sample the PCO_2, PO_2 and pH of arterial blood. As we have already discussed, the major influence on the respiratory centre in healthy individuals is a raised PCO_2 and the drive for ventilation is high carbon dioxide level. This acts through the **central chemoreceptors**, which are particularly sensitive to chemical changes in the cerebrospinal fluid caused by increasing carbon dioxide and hydrogen ions.

The peripheral chemoreceptors transmit impulses to the medulla via the vagus and glossopharyngeal nerves. They are most sensitive to decreases in PO_2 and pH and less sensitive to increases in PCO_2. The hypoxic drive (low PO_2) is relatively insignificant in healthy people but a reduction in pH will rapidly stimulate ventilation even when blood gases are normal, as with ketoacidosis.

Hypoxia is an important control of ventilation where chemoreceptor sensitivity to PCO_2 has been lost, e.g. with longstanding COAD (see oxygen administration, page 282).

Stretch and irritant receptors

Stretch receptors within the airways and pleura send inhibitory impulses via the vagus nerve to the respiratory centres. This mechanism prevents overinflation of the lungs and helps to maintain the rhythm of breathing.

The **Hering–Breuer reflex**, which operates via the vagus nerve, causes a short period of apnoea if the lungs are overinflated. It may be an important reflex in the newborn, although it is probably more important as a protective measure than a control mechanism in adults.

Irritants such as dust stimulate receptors in the airways/lungs and initiate a protective reflex, e.g. coughing or sneezing. These impulses, which modify ventilation patterns, reach the respiratory centres via the vagus nerve.

Higher centre (voluntary) control

We can all think of situations where it is necessary to override involuntary control of ventilation. Swimming under water, singing, swallowing, laughing and 'having a good cry', all require some modification to the normal breathing pattern. These changes are for a limited time only: eventually the automatic chemical controls will take over.

The state of arousal/wakefulness also influences the respiratory centre, e.g. being 'wide awake' tends to stimulate the respiratory centres. During NREM sleep (stages 3 and 4) the respiratory rate slows but increases again with the onset of REM sleep when the oxygen demands of the brain are greatly increased (see Chapter 4).

Hypothalamic control

The gasp you give when jumping into cold water can be explained by the impulses sent to the respiratory centres from the hypothalamus. Intense emotions, pain and body temperature all modify the rate and depth of ventilation through hypothalamic/limbic system influences.

Effects of altitude, depth and exercise on ventilation

Altitude

Anyone who has visited a mountainous region will confirm that their breathing pattern was affected. At 18 000 feet (5486 m) above sea level the partial pressure of oxygen is only half that at sea level (see page 286). This change in PO_2 will initiate a series of physiological adaptations known as *acclimatization*. Some of these adaptions occur at once but others take some weeks, which is why climbers spend a period of time acclimatizing at increased altitude rather than rushing at the mountain on the day they arrive. Failure to acclimatize properly would lead to '*mountain sickness*' caused by hypoxia.

Changes with altitude include:

- Hyperventilation as the respiratory centre attempts to maintain adequate gaseous exchange. This is another example of the hypoxic drive in operation.
- Increased 2,3-DPG production allows easier oxygen unloading from Hb, which is not completely saturated at altitude.
- Changes in cardiac output.
- Increased erythropoiesis over some weeks as the kidneys produce more renal erythropoietic factor in response to hypoxia (see Chapter 9).

Depth

The increase in atmospheric pressure below sea level causes problems in the form of *decompression sickness*, which affects divers and people working deep underground.

At high atmospheric pressure more oxygen and nitrogen dissolve in the blood. During the return to normal pressure the nitrogen (poorly soluble - see page 287), which has been dissolved in the blood under increased pressure, can form bubbles in the blood. This is likely to occur if decompression time (return to atmospheric pressure) is too short. The gas bubbles form emboli which cause pain, abdominal distention and damage to the brain or other vital structures, possibly causing death. Decompression sickness or 'the bends' is a serious occupational hazard for divers. In the UK their work, including the use of decompression chambers, is controlled by stringent legislation.

Exercise

During strenuous exercise the hardworking skeletal muscles have a huge appetite for oxygen and produce much more carbon dioxide than usual. Obviously there needs to be considerable respiratory modification to cope with these changes and maintain homeostatic balance. In fit individuals ventilation can increase to 10–20 times the level at rest.

As you would expect, breathing becomes deeper but the reasons for the particular pattern are not yet fully understood. There is a sharp rise in ventilation when exercise starts, this rise slows to reach a plateau level which continues while the exercise is in progress. When the exercise ceases the ventilation rate falls sharply and then returns to the resting level more slowly. It may be that we consciously prepare for exercise and the cerebral cortex sends appropriate impulses to the respiratory centres which also receive excitatory impulses from muscle and joint proprioceptors.

Blood gases (PO_2 and PCO_2) change very little during exercise so the pattern cannot be explained solely by normal chemoreceptor mechanisms. While ventilation rates rise do not forget that cardiac output is also increasing to maintain the ventilation:perfusion ratio.

When contracting muscle fails to receive sufficient oxygen because contraction may interfere with blood supply it will start to utilize fuel molecules, e.g. glucose, without oxygen (*anaerobic respiration*). The muscles are really using credit: their *'oxygen debt'* must be repaid eventually, explaining the slow decline in ventilation rate. The slow decline is also linked to the lactic acid produced by anaerobic respiration. This waste product stimulates ventilation to prevent pH changes (see page 298). Lactic acid also accounts for the muscle ache and stiffness felt after unaccustomed or strenuous exercise.

Respiratory role in pH regulation

Respiration has a vital role in the regulation of pH (hydrogen ion concentration). The lungs and kidneys (see Chapter 15) provide the important homeostatic mechanisms by which the pH balance of the body is maintained. In the short term, however, it is the buffer systems that prevent any huge swings in pH. Before we continue it would be a good idea for readers to refresh their knowledge regarding the concepts of pH and buffers with a quick look at

Chapter 2. The *Henderson–Hasselbalch equation* (see below) explains why blood pH is dependent upon the amount of hydrogen carbonate (HCO_3^-) and dissolved carbon dioxide (H_2CO_3). Carbon dioxide levels (PCO_2) and blood pH as you already know, stimulate the respiratory centres and consequently ventilation.

Henderson–Hasselbalch equation

$$\text{Blood pH} = 6.1 + \log \frac{[HCO_3^-] \text{ hydrogen carbonate}}{[CO_2] \text{ carbon dioxide in solution}}$$

(N.B. the square brackets denote concentration.) The ratio of hydrogen carbonate (base) to carbonic acid (acid) must be 20:1 if blood pH is to remain at 7.4.

Example 1
If hydrogen carbonate levels fall (as it buffers acid) or the PCO_2 increases, the pH of the blood falls (*acidosis*) but an increase in ventilation expels more carbon dioxide to restore the ratio.

Example 2
If hydrogen carbonate levels rise or the PCO_2 decreases, an increase in blood pH (*alkalosis*) reduces ventilation, more carbon dioxide is retained and again the 20:1 ratio is restored.

These two examples illustrate the respiratory role in changing ventilation rates to restore and maintain the hydrogen carbonate/carbonic acid buffer system ratio.

NB Regulation of hydrogen carbonate in each case is part of kidney function (Chapter 15) and takes longer to regulate than PCO_2 by the respiratory tract, i.e. days rather than minutes.

Acidosis and alkalosis

From the examples above you can see that changes in blood pH may have a respiratory or metabolic cause although some changes have a mixed aetiology. When considering acid–base balance an important feature is the ability of the lungs and kidneys to compensate for disorders in each other, e.g. a high PCO_2 due to pulmonary disease is *compensated* by reabsorption of hydrogen carbonate ions in the kidney.

Acidosis (pH less than 7.35)
Respiratory acidosis is due to an increase in PCO_2, e.g. respiratory failure. **Metabolic acidosis** is caused by the production of acids which use up the buffering ability of the hydrogen carbonate, e.g. ketoacidosis in diabetes and lactic acid produced during exercise, or by the loss of alkali, e.g. severe diarrhoea (see blood gases, page 289).

Alkalosis (pH greater than 7.45)
Respiratory alkalosis results from a decrease in PCO_2 such as that caused by the hyperventilation of severe anxiety. **Metabolic alkalosis** can be due either to an increase in the amount of alkali, e.g. taking excessive amounts of alkalis for indigestion, or loss of acid, e.g. prolonged vomiting (see blood gases page 289).

SUMMARY/CHECK LIST

Introduction
Respiratory structures – Early development. Nose (sinuses). Pharynx. Larynx – swallowing, lining and vocal cords, Valsalva's manoeuvre, voice production, *Nursing Practice Application* – airway maintenance, Heimlich manoeuvre. Trachea – tracheostomy, cough and sneeze. Bronchial tree. Inhaled foreign bodies – factors affecting bronchial muscle tone, *Nursing Practice Application* – inhalers and nebulizers. Alveoli. Lungs. Pleura, *Nursing Practice Application* – pleural aspiration. *Special Focus* – smoking. Environmental/occupational hazards, *Person-centred study* – Sid, *Nursing Practice Application* – sputum.
Breathing/ventilation – Intrathoracic pressures, *Nursing Practice Application* – pneumothorax/ chest drains. Inspiration. Expiration. Assessment of breathing, *Nursing Practice Application* – oxygen administration. Airway resistance, compliance, elasticity, surfactant. Lung volumes/capacities, lung function tests, pulmonary ventilation and tidal volume, alveolar ventilation.
Gaseous exchange – Behaviour of gases. Composition/partial pressures. External respiration. Internal respiration. *Nursing Practice Application* – blood gases.
Transport of gases – Oxygen transport – haemoglobin, oxygen dissociation curve, hypoxia. Carbon dioxide transport – hypercapnia, hypocapnia.
Control of ventilation – respiratory centres. Controls. Effects of altitude/depth/exercise. Respiratory role in pH regulation – acidosis/ alkalosis.

SELF TEST

1 How is air modified on its journey from nose to alveoli?
2 What mechanisms exist to protect the respiratory tract during swallowing?
3 Which of the following statements are true?
(a) Intrapleural pressure is always greater than alveolar pressure.
(b) Inspiration is an active process involving the respiratory muscles.
(c) Expiration depends on the elastic recoil of the lungs.
(d) Surfactant prevents alveolar collapse.
4 Which of the following should be treated with low concentration oxygen?
(a) Cyanide poisoning.
(b) COAD.
(c) Carbon monoxide poisoning.

5 Explain the terms tidal volume and pulmonary ventilation. Why would you encourage an anxious or breathless person to breathe more deeply?
6 Complete the following:
(a) Atmospheric air has a PO_2 of ______.
(b) Alveolar air has a PCO_2 of ______.
(c) Expired air has a PCO_2 of ______.
7 The diffusion of gases between alveoli and blood depends on which factors?
8 Relate the shape of the oxygen dissociation curve to physiological events.
9 Which of the following transports most carbon dioxide:
(a) Hydrogen carbonate ions in the plasma.
(b) Dissolved in the plasma
(c) Carbaminohaemoglobin.
10. Explain the effects of PCO_2, PO_2 and pH on the respiratory centres.
11 What changes occur in blood pH and PCO_2 in the following situations:
(a) hyperventilation (b) hypoventilation?

Answers
1. See pages 266,287. 2. See pages 267-269. 3. b, c, d. 4. b. 5. See pages 284-286. 6 a. 21.2 kPa. b. 5.3 kPa. c. 3.8 kPa. 7. Gas partial pressures gradients, solubilities, state and area of alveolar membrane and ventilation/perfusion relationship. 8. See page 290. 9. a. 10. See page 295. 11. a. pH increases and PCO_2 falls. b. pH falls and PCO_2 increases.

References

Cockcroft, A., Bagnall, P., Heslop, A. *et al.* (1986) Controlled trial of a respiratory health worker visiting patients with chronic respiratory disability. *British Medical Journal*, **294**, 225–227.

Compton, L. (1990) Accidental poisoning. *Nursing Times*, **86** (15), 28–30.

Ellis, P. (1990) Asthma: meeting the demand for rapid relief. *Professional Nurse*, **6** (2), 76–81.

Foss, M. (1990) Oxygen therapy. *Professional Nurse*, **5** (4), 188–190.

Haverty, S., Macleod Clark, J. and Elliot, K. (1987) Helping people to stop smoking. *Nursing Times*, **83** (28), (occasional paper 87, **3**, 45–49).

Health Education Authority (1988) *Can You Avoid Cancer?* London: Health Education Authority.

Heimlich, H. J. (1975) A life-saving manoeuvre to prevent food-choking. *Journal of the American Medical Association*, **234**, 398–401.

Twohig, R. (1984) Respiratory function tests. *Nursing*, **2** (27), 807–810.

Further Reading

Brown, S.E. (1979) Respiratory physiotherapy and the nurse. *Nursing* (1st series), **6**, 257–259.

Bysshe, J. (1989) Politics in the air. *Nursing*, **3** (38), 15–18.

Ellis, P. (1990) A new direction for asthma relief? *Professional Nurse,* **6** (3), 145–147.

Erickson, R. (1989) Mastering the ins and outs of chest drainage: Part I. *Nursing (USA),* **19** (5), 36–43.

Erickson, R. (1989) Mastering the ins and outs of chest drainage: Part II. *Nursing (USA),* **19** (6), 46–49.

Green, J. H. (1976) *An introduction to Human Physiology,* 4th edn. Oxford: Oxford University Press.

Grenville-Mathers, R. (1983) *The Respiratory System,* 2nd edn. Edinburgh: Churchill Livingstone.

Kendrick, A.H. Smith, E.C. (1992) Simple Measurements of lung function. *Professional Nurse,* **7** (6), 395-404.

Kendrick, A.H. Smith, E.C. (1992) Respiratory Measurements 2 : interpreting simple measurements of lung function. *Professional Nurse,* **7** (11), 748-754.

Macleod Clark, J., Haverty, S. and Kendall, S. (1990) Helping people to stop smoking: a study of the nurse's role. *Journal of Advanced Nursing,* **15** (3), 357–363.

Smith, J. (1989) Editorial – Proposals by WHO for European tobacco policy. *Journal of Advanced Nursing,* **14** (6), 433–434.

Useful Addresses

Action on Smoking and Health (ASH)
5–11 Mortimer Street
London W1N 7RH

Health Education Authority
Hamilton House
Mabledon Place
London WC1H 9TX

National Asthma Campaign
Providence House
Providence Place
London N1 0NT

CHAPTER THIRTEEN

DIGESTIVE SYSTEM, METABOLISM AND NUTRITION

OVERVIEW

- Digestive tract structure and function including secretion, digestion and absorption.
- Metabolism and utilization of food.
- Nutrition.

LEARNING OUTCOMES

After studying Chapter 13 you should be able to:

- Describe the organs of the gastrointestinal tract and relate the adaptations in general structure to modified function.
- Discuss the role of the accessory digestive structures.
- Describe the location and functions of the peritoneum.
- Explain the processes of ingestion, mastication and swallowing.
- Describe the secretion, composition and functions of saliva.
- Explain how food is moved through the gastrointestinal tract.
- Outline the neural and hormonal controls of gastric secretion, motility and emptying.
- Describe the composition and functions of gastric juice.
- Discuss the digestive processes occurring in the small intestine including the role of bile and pancreatic juice.
- Name the enzymes involved in the digestion of carbohydrates, fats and protein, and the products of digestion.
- Describe absorption in the small intestine.
- Describe the functions of the large intestine including defaecation.
- Use physiological knowledge in the assessment of bowel function and identification of abnormalities.
- Describe the constituents of a healthy diet.
- Outline briefly the metabolism and utilization of glucose, fatty acids, glycerol and amino acids.
- Outline ways of assessing nutritional status.
- Discuss the nutritional requirements of certain special groups including methods of nutritional support.

Key Words

Absorption – passage of products of digestion, via the gastrointestinal tract epithelium, into the blood or lymph.
Alimentary canal/gastrointestinal tract – the whole digestive tract, a tube extending from the mouth to the anus.
Aerobic – requiring oxygen or in the presence of oxygen.
Anaerobic – without oxygen.
Colonic – relating to the colon.
Digestion – the catabolism (breakdown) of food into a form that can be absorbed.
Enteric – relating to the intestine.
Enzymes – specific protein catalysts which facilitate the breakdown of large food molecules into their chemical subunits, facilitating absorption.
Gastric – relating to the stomach.
Gut – the intestine or bowel.
Ingestion – taking in food or other substance into the body.
Peptic – pertaining to digestion.
Peristalsis – the rhythmic, wave-like contractions which convey food through the gastrointestinal tract. Peristalsis also occurs in other hollow structures.
Villi – projections from the mucosa of the small intestine which greatly increase the surface area available for absorption.

Introduction

The body requires raw materials for growth and repair and to produce the energy needed for cellular processes. These are obtained from the food and fluids we consume. Most of this food needs modification before it can be absorbed and utilized. The cheese sandwich for lunch will need some changes before it can provide energy for the afternoon and raw materials such as amino acids for the synthesis of new proteins.

The complex modification process involves: **ingestion** (taking in food or other substance into the body), **mastication** (chewing), propulsion, **digestion** (the catabolism (breakdown) of food into a form that can be absorbed), **absorption** (passage of products of digestion, via the gastrointestinal tract epithelium, into the blood or lymph), **utilization** and **elimination** (by *defecation*) of waste residue. The **alimentary canal** consisting of the *mouth, oesophagus, stomach, small intestine, large intestine* and *anus* is adapted to perform all these functions (see *Figure 13.1*), aided by accessory structures which include the *teeth, salivary glands, pancreas* and *liver* (see Chapter 14). The alimentary tract measures 8–9 m in a cadaver, but is shorter during life because of muscle tone.

Early development

The gastrointestinal tract develops from a simple tube or primitive gut which is open at both ends (mouth and anus) by the sixth week of fetal development. This primitive gut dilates and rotates during the following weeks to form the gastrointestinal structures in their final positions. The upper part, or *foregut*, develops to form the pharynx (development is closely linked with that of the trachea – see Chapter 12). The stomach forms from a dilation in the primitive gut and the

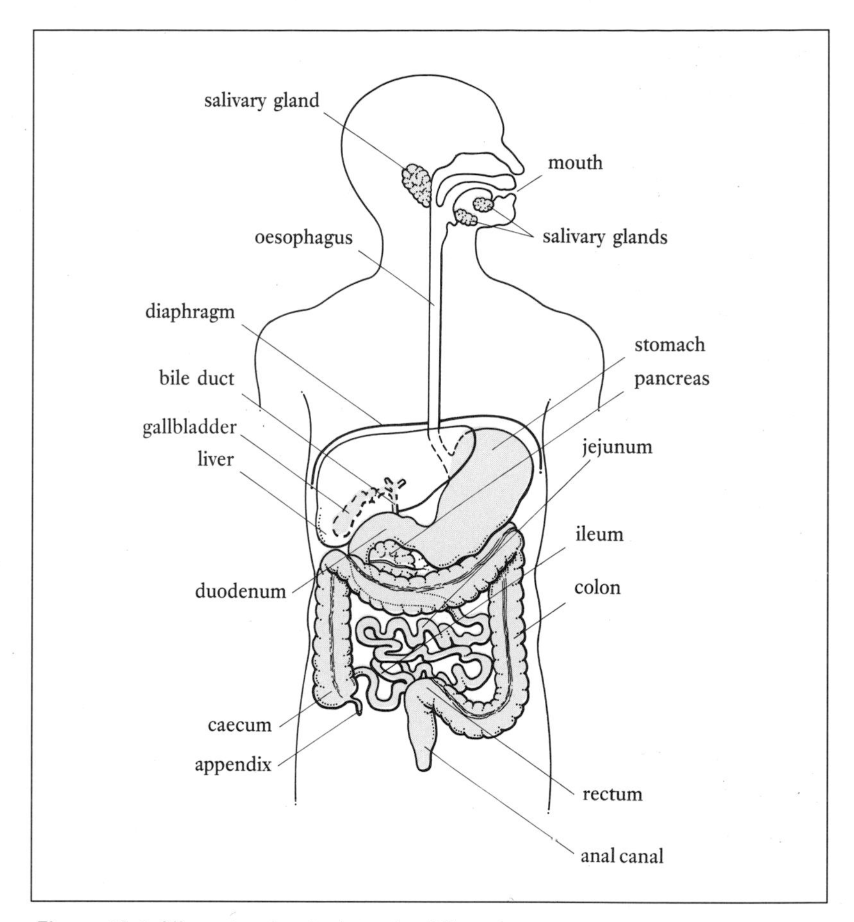

Figure 13.1. Alimentary/gastrointestinal/digestive tract.

lower part of the gut grows down to form the small and large intestine and the anal canal. The large intestine eventually rotates to 'frame' the small intestine.

During early development the intestines protrude into the umbilical cord and are outside the abdominal cavity, but by about the tenth week they are enclosed by the abdominal wall. Problems at this stage may result in the congenital abnormality *exomphalos* in which the intestines protrude through a gap in the abdominal wall, and are covered only by a thin membrane. Surgical repair of the defect is usually undertaken as soon after birth as possible. Other developmental problems include *atresia,* in which a part of the gastrointestinal tract fails to canalize, i.e. it comes to a full stop and ends as a blind tube. This may happen in the oesophagus, duodenum or the rectum/anal canal. The result, whichever site is involved, will be an obstruction in the gastrointestinal tract, which requires surgical intervention soon after birth.

General structure – digestive tract

Before we consider the individual digestive organs it is helpful to look at the basic structure common throughout the tract. Along its entire length can be found the same four layers (the *mucosa,*

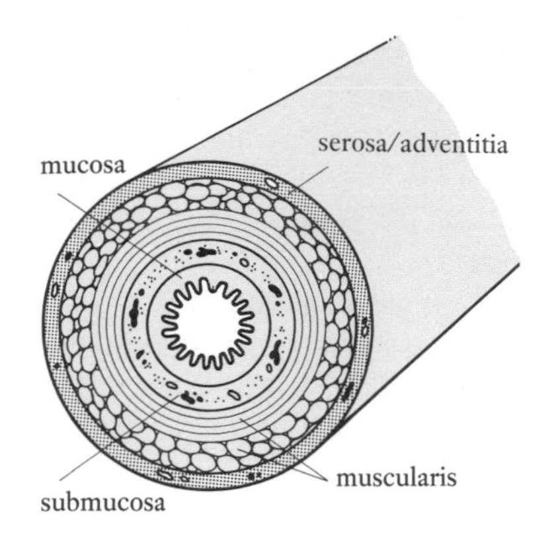

Figure 13.2. Basic structure of gut wall.

submucosa, muscularis and *serosa/adventitia*) with local adaptations related to specific function (*Figure 13.2*).

Mucosa: this is the inner mucous membrane lining and is stratified or columnar epithelium containing goblet cells and glands. Depending on location it is adapted to protect, secrete mucus and digestive juices or absorb products of digestion. Underlying the epithelium is the **lamina propria**, a layer of connective tissue which contains the lymphoid tissue (see Chapter 11) which helps to protect us from the various *pathogens* (disease producing organisms) entering the tract. In certain areas a layer of smooth muscle, the **muscularis mucosae**, produces folding of the mucosa which allows for distension and greatly increases the surface area available for secretion or absorption.

Submucosa: this layer, below the mucosa, consists of areolar connective tissue containing many blood vessels, lymphatics and nerves. The nerves which supply autonomic fibres to the mucosa form the **submucosal (Meissner's) plexus**.

Muscularis: this layer is smooth involuntary muscle. There are usually two layers, longitudinal and circular fibres, but the stomach has three layers. Running between the layers are blood vessels, lymphatics and the **myenteric** (**Auerbach's**) **plexus** of autonomic nerve fibres. The muscular layer contracts in response to nerve impulses and certain hormones to propel food along the tract (**peristalsis** see pages 317-318).

Adventitia/serosa: this is the outer protective layer and is either fibrous connective tissue (in the thorax) or a serous membrane, the *visceral peritoneum* (see page 307), in the abdominal cavity. The serosa prevents friction during gut motility.

Nerve supply

The gastrointestinal tract is innervated by the autonomic nervous system (see Chapter 6). Secretion and motility are generally increased by parasympathetic activity and decreased by sympathetic activity.

The most important parasympathetic nerve is the vagus nerve which innervates the oesophagus, stomach, pancreas, bile ducts and intestine. The salivary glands receive parasympathetic fibres via the facial and glossopharyngeal nerves and the last part of the colon is supplied via the sacral outflow (parasympathetic nerves which leave the sacral part of the spinal cord) (*nervi erigentes*).

Sympathetic fibres from the cervical, coeliac and mesenteric ganglia innervate the digestive structures, which have alpha or beta-2 receptors and both in the case of the stomach and intestine (see Chapter 6).

Apart from these autonomic nerve fibres the digestive organs are supplied by autonomic nerve plexuses (Meissner's and Auerbach's) which connect different parts of the tract rather like an internal telephone system. This arrangement allows the gastrointestinal tract to function as an integrated unit.

Peritoneum

The **peritoneum** is a double serous membrane forming a closed sac within the abdomen. The outer, **parietal**, layer lines the abdominal cavity and the inner, visceral, layer covers some of the abdominal and pelvic organs (*viscera*). Serous fluid is secreted into the potential space (normally the two layers are separated by a thin film of serous fluid, there is no actual space) between the two layers.

During embryonic development the organs invaginate into the peritoneal membrane from above, behind and below, resulting in a rather complex arrangement (*Figure 13.3*). Some organs are nearly enclosed, e.g. the stomach, whereas others, e.g the pelvic organs, are only partially covered by peritoneum; to get some idea of this try pushing objects into a partially inflated balloon.

In the male pelvis the peritoneum forms a closed sac, but in females it is open via the uterine tubes (see Chapter 20).

Functions of the peritoneum

The functions of the peritoneum are:

- Allowing the digestive organs to move easily without friction.
- Acting as a fat store.
- Anchoring and stabilizing some structures, e.g. the **mesocolon**, a double fold of peritoneum, joins the transverse colon to the posterior abdominal wall.
- Limiting the spread of infection within the cavity – the peritoneum

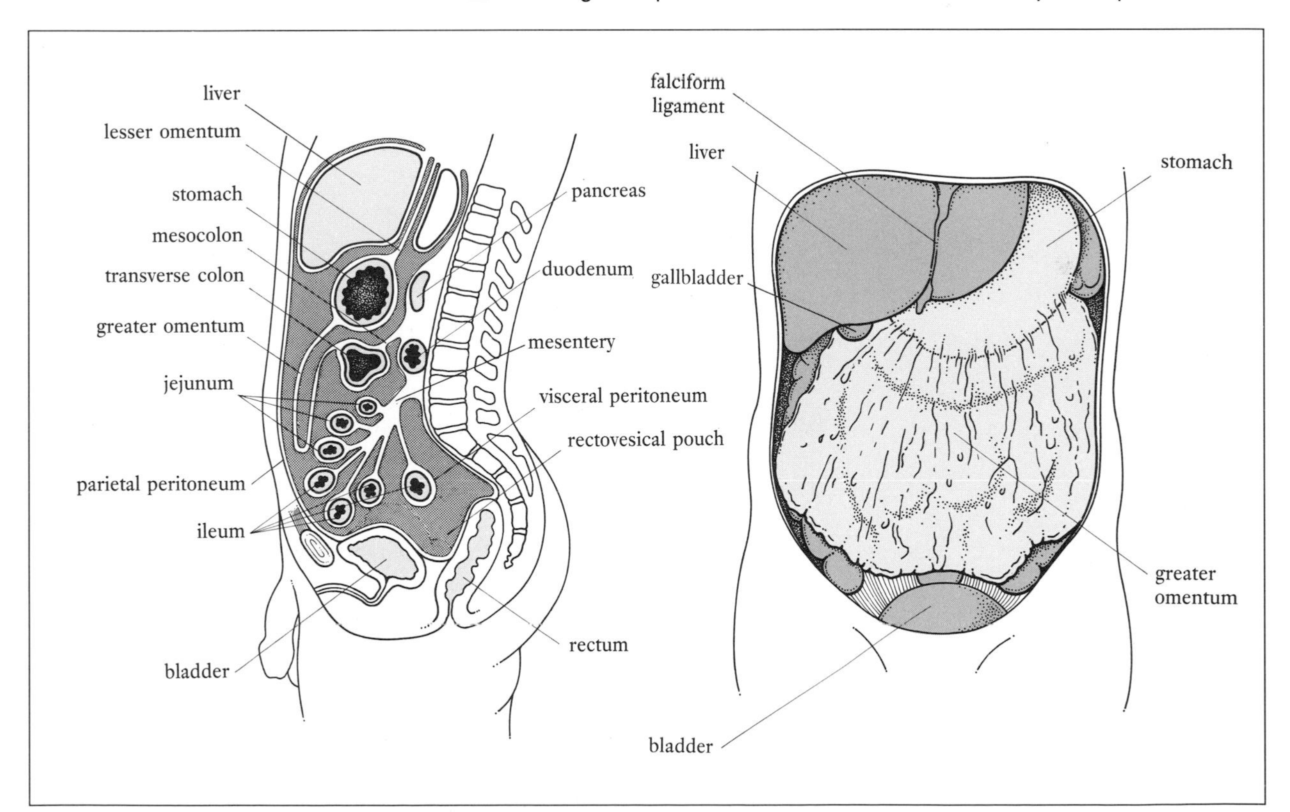

Figure 13.3. (a) The peritoneum (male); (b) greater omentum.

contains many lymph nodes (see Chapter 11). For example, the **greater omentum**, (see *Figure 13.3*), which hangs apron-like in front of the small intestine, can move to 'wall off' localized areas of infection (see *peritonitis*).

- Supporting blood vessels, lymphatics and nerves, e.g. the **mesentery**, which enfolds part of the small intestine, carries the vessels and nerves supplying the jejunum and ileum.
- Forming the protective outer coat of some organs.

Unfortunately, in common with other serous membranes the peritoneum can act as a route by which malignant cells spread across a cavity (see Chapter 1).

Peritonitis

Inflammation of the peritoneum (*peritonitis*) may be due to bacterial infection or chemical irritation. Initially defence mechanisms attempt to localize the area of inflammation but failure to do this results in spread and generalized peritonitis. Some causes include: *appendicitis*, **peptic** (pertaining to digestion) *ulcer perforation* and *bowel rupture*, e.g. *inflammatory bowel disease* and *biliary perforation* where the release of bile causes chemical irritation.

Generalized peritonitis is a serious condition requiring immediate treatment, which includes pain relief, antimicrobial drugs, intravenous fluids for dehydration and electrolyte imbalance (see Chapter 2), **gastric** (relating to the stomach) emptying via a nasogastric tube and surgery where appropriate. Even with prompt intervention the inflammation produces *exudate* (fluid) which sticks the inflamed structures together, these *adhesions* may later obstruct the gastrointestinal tract.

Basic Structure of the Organic Compounds – Proteins, Carbohydrates and Fats

Before we consider the gastrointestinal tract and the chemical processes involved in digestion, a quick look at the structure of the main nutrient compounds would be helpful.

Proteins

Proteins (see also Chapter 2 and pages 328, 329, 343 and 344) are large complex substances containing carbon, hydrogen, oxygen, nitrogen and sometimes sulphur and phosphorus. All proteins are long chains formed from the combination of the 20 amino acids (see pages 343-344) joined together in different sequences. A protein may consist of many hundreds of amino acids linked together by **peptide bonds** (see below). Each amino acid has an *amine end* (NH_2) and an acidic *carboxyl end* ($COOH$). To form amino acid chains the amine end of one amino acid reacts with the carboxyl end of another in a *dehydration synthesis* reaction i.e. a reaction in which a water molecule is released (condensation) to form a peptide bond.

NB During chemical digestion the peptide bonds of the protein chain are broken by the addition of water *(hydrolysis).*

An amino acid chain with peptide bonds:

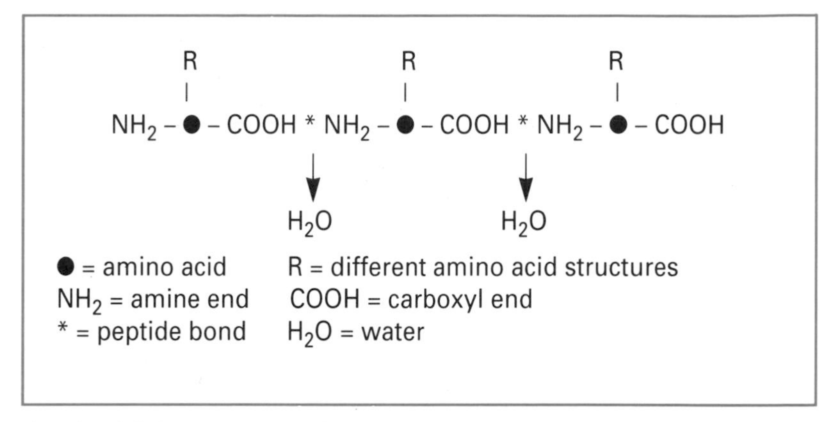

A pair of linked amino acids is called a **dipeptide** and a chain of three or more amino acids is called a **polypeptide**. Once the number of amino acids in the chain exceeds 50 or so the structure is known as a protein. Apart from the peptide bonds linking individual amino acids the more complex twisted or folded proteins are maintained by *hydrogen* and *covalent bonds* (see Chapter 2).

Carbohydrates

Carbohydrates (see pages 328-329 and 342-343) consist of carbon with hydrogen and oxygen in the same proportions as water. Carbohydrates include the fairly simple sugars and the more complex starches and cellulose (see page 310). The basic unit that forms the carbohydrates is the **monosaccharide** or single simple sugar molecule which usually has a closed ring structure. Examples of monosaccharides are the sugars; **glucose**, **galactose** and **fructose**. When two monosaccharides link together by dehydration synthesis (see above) they form a double sugar molecule or **disaccharide**, e.g. **sucrose**, **lactose** and **maltose**, joined by a **glycosidic linkage**. During chemical digestion the bond between the two sugar units is broken down by hydrolysis reactions (see above).

A monosaccharide unit, e.g. glucose:

```
            CH2OH
              |
  H           C ——— O       H
  |         /  |      \     |
  C            H    H   \   C
    \        OH     |   /
  |    \      |     |  /    |
  HO          C ——— C       OH
              |     |
              H     OH
```

A disaccharide (sucrose) and glycosidic linkage:

The most complex carbohydrates or **polysaccharides** consist of long chains and branches of monosaccharide units joined by glycosidic linkages. These large, non-sweet molecules include the storage carbohydrates – *starch* (plant) and *glycogen* (animal). Again, the processes of digestion eventually reduce most polysaccharides to their monosaccharide (glucose) subunits. However, the digestive process in humans is unable to break down some complex structural carbohydrates, e.g. cellulose, which provides dietary fibre (non-starch polysaccharide).

Fats

Fats (see pages 329, 331, 344, 346 and 347) consist of carbon, hydrogen and oxygen, but in different proportions to carbohydrates. Fats belong to a large group of **lipid** molecules which include **steroids**, e.g. *cholesterol*, **phospholipids** (which are important in cell membranes), and other lipids such as **lipoproteins** (see Chapter 10) and **prostaglandins** (see Chapters 8 and 20). Most of the fats consumed in our diet are **triglycerides**. Tryglycerides are formed from two types of subunit: *fatty acids* and *glycerol*. As its name suggests, a triglyceride consists of three fatty acids (hydrocarbon chains) linked to a glycerol molecule (sugar with alcohol groups) by dehydration synthesis reactions (see page 309).

The variations in triglycerides depend on the fatty acids, which have different carbon–carbon bonds (single or double) and hydrocarbon chains of differing lengths. Fats may be *saturated*, when all the carbon–carbon bonds (C–C) are single, these are usually solid

Structure of a triglyceride:

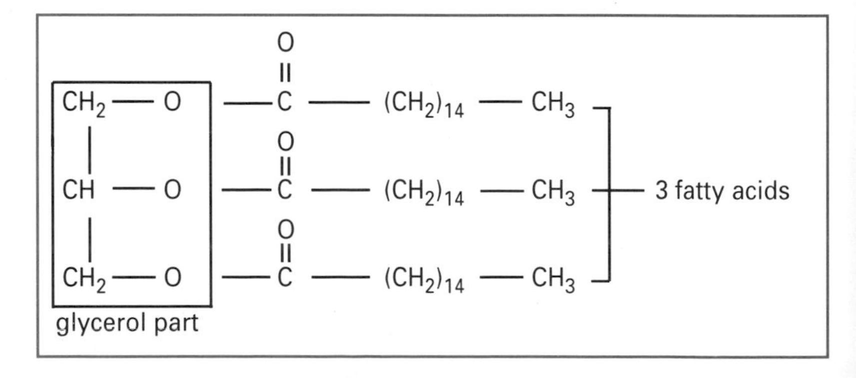

at room temperature, e.g. lard. *Unsaturated fats,* which are usually liquid at room temperature, may be *monounsaturated* where one C–C bond is double, e.g. olive oil, or *polyunsaturated* where two or more C–C bonds are double, e.g. sunflower oil.

ALIMENTARY TRACT - STRUCTURE AND FUNCTION

Mouth

The **mouth** (**buccal** or **oral cavity**) and accessory structures are concerned with the ingestion, mastication (chewing) and some chemical digestion of food. Apart from all this the mouth and accessory structures are important for taste (see Chapter 7) and are vital to our ability to communicate with the world. Without the proper functioning of the tongue and lips it is impossible to speak with clarity or to transmit non–verbal clues about our mood, e.g. smiling.

The mouth is divided into two parts: the **vestibule**, between the gums/teeth and the lips, and the part more properly called the **oral cavity** which lies behind the teeth. It is bounded by the lips anteriorly, the cheek muscles form its lateral walls and posteriorly it is continuous with the oropharynx (see Chapter 7). Inferiorly there is the tongue and superiorly the hard and soft palates (*Figure 13.4*).

Stratified squamous epithelium lines the mouth and covers the gums as protection against damage during eating. In areas liable to excessive wear and tear, e.g. gums and the external part of the lips, the epithelium is *keratinized* (see Chapter 1).

The lips and cheeks contain skeletal (voluntary) muscle which assists with ingestion, positioning food for chewing and swallowing and speech.

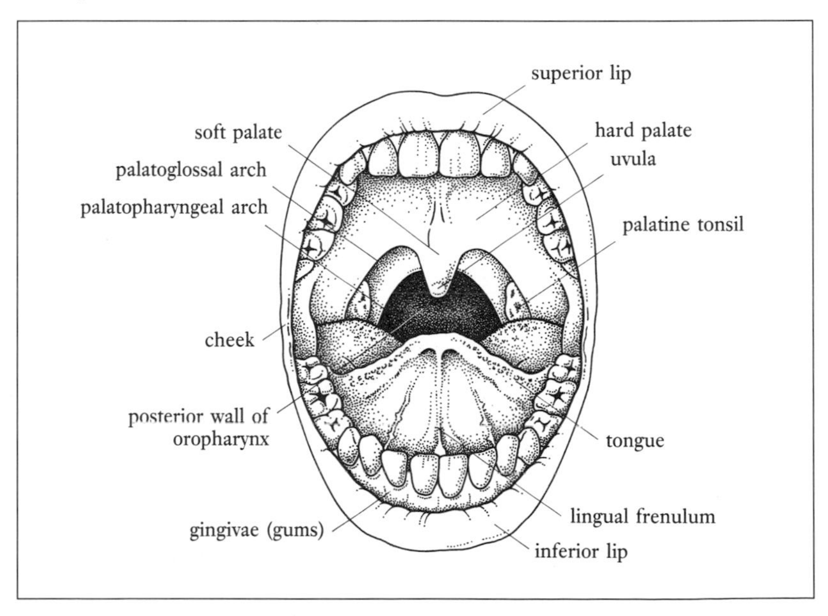

Figure 13.4. The mouth (not all the teeth are shown).

Tongue

The **tongue** (*Figure 13.4*) is a highly vascular structure consisting of voluntary muscle. It is attached to the hyoid bone (see *Figure 12.2*) and to the floor of the mouth by a fold of skin called the **frenulum**. The tongue is supplied with blood via the lingual artery and its venous return drains into the internal jugular vein. The tongue is extremely sensitive and its great mobility is important for speech, mixing food with saliva and the formation of a food *bolus* (ball of chewed food) for swallowing. Innervation to the tongue is through somatic sensory nerves and three cranial nerves – the facial, glossopharyngeal and hypoglossal.

The upper surface of the tongue, which should be pink and moist, is covered with epithelium which has projections or **papillae** containing the taste buds. Papillae are classified by their shape and may be *filiform* (filament-shaped), *fungiform* (mushroom-shaped) or *vallate* (surrounded by groove or trough); the last are the largest and form a 'V'-shaped pattern on the back of the tongue. Taste, which is discussed more fully in Chapter 7, is linked closely with olfaction to enhance appetite and ensure eating is a safe and enjoyable process not just the intake of nutrients. Different regions of the tongue are concerned with the detection of four basic tastes – bitter, sweet, salt and sour (see *Figure 7.23*). These stimulate the nerve impulses transmitted to the gustatory cortex for interpretation (see Chapters 4 and 7).

Hard and soft palates

The roof of the oral cavity consists of the **hard palate** anteriorly and the **soft palate** posteriorly (*Figure 13.4*). The hard palate is formed by the palatine bones and maxilla, which supply the hard surface needed during chewing.

The mobile soft palate is made of voluntary muscle covered with stratified squamous epithelium. From its free edge hangs a projection called the **uvula**, which can be seen by looking in a mirror and saying 'Ah'. Laterally it is secured to the tongue by the **palatoglossal arches** and to the oropharynx by the **palatopharyngeal arches**. It is between these arches that the palatine tonsils are located. During swallowing the soft palate moves upwards to protect the nasopharynx (see Chapter 12).

Salivary glands

Three major pairs of **salivary glands** – *parotid, submaxillary (submandibular)* and *sublingual* – secrete **saliva** (Figure 13.5). The

Nursing Practice Application Cleft palate and lip

These congenital abnormalities are due to developmental defects in the palate, which fails to fuse in the midline, and the top lip. Although a severe cleft lip is obvious it is essential that the examination of a newborn infant includes checking the palate. The immediate problems are those of feeding, because the infant cannot suck properly, and of the emotional distress caused by the infant's appearance. The family need a great deal of support while they come to accept that their baby will need special care including surgery to repair the lip and palate. A good repair is important to avoid later problems with teeth eruption, speech and to produce an acceptable cosmetic result.

secretion of saliva assists speech, mastication, taste and swallowing; we can all remember how difficult these are with a dry mouth associated with anxiety or fear. Saliva is vital for effective oral hygiene and people with dehydration and a dry mouth require nursing interventions to ensure oral well-being (see *Nursing Practice Application*, page 315). Although saliva contains an enzyme (*amylase*) it is of very little significance in digestion.

The **parotid glands** enclosed in fibrous capsules are situated just inferior and anterior to the ear. They are the largest salivary glands but produce only 25% of the total saliva. The saliva leaves via the *parotid ducts*, which open into the mouth at the level of the second upper molars (teeth). Parasympathetic fibres of the glossopharyngeal nerve stimulate the watery saliva secreted by the parotid glands.

The **sublingual glands** are situated at the base of the tongue in the floor of the mouth. They produce a small amount of thick mucus saliva, which enters the mouth via several small ducts. Their nerve supply is via the parasympathetic fibres of the facial nerve.

The **submaxillary glands** are situated below the maxilla (upper jaw) and have ducts entering the mouth either side of the frenulum. Most saliva is produced by these glands which are also innervated by the facial nerve.

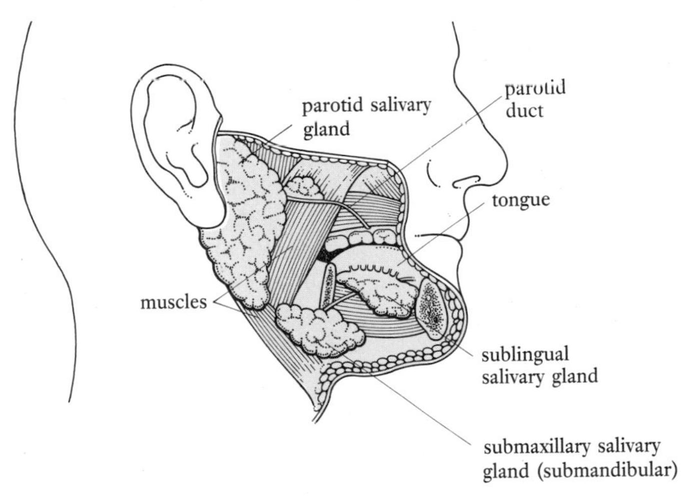

Figure 13.5. Salivary glands.

Mumps

Infectious parotitis (mumps) is a viral inflammation of the parotid glands. Apart from pain, swelling and difficulty opening the jaw it can also cause serious inflammation of the testes (*orchitis*), which may lead to male sterility, and more rarely of the ovaries (*oophoritis*). Mumps may also cause inflammation of the pancreas (*pancreatitis*). Routine immunization against mumps is available during childhood (administered as a combined Measles, Mumps and Rubella vaccine – MMR). Nurses should be able to give information regarding this and other infectious disease prophylaxis (see Chapter 19).

Saliva

Saliva is mostly water (99%) with electrolytes and varying amounts of mucus. It is generally slightly acid or neutral but may become alkaline when food is being chewed. The **enzyme** (specific protein catalysts which facilitate the breakdown of large food molecules into their chemical subunits, facilitating absorption) **alpha amylase** (previously called *ptyalin*) is present in saliva and commences the digestion of cooked starch, e.g. bread. Saliva also contains *lysozyme* and *immunoglobulin IgA,* which protect against micro-organisms (see Chapter 19).

Regulation of salivation

The daily production of saliva is between 1000 and 1500 ml depending on many factors, including hydration:

- A conditioned reflex linked to the anticipation, sight or smell of food will stimulate profuse secretion of watery saliva. Certain flavours or personal favourites can produce a more active response, e.g. the smell of a roast lunch cooking or the sharpness of a lemon drink.
- Secretion of saliva occurs in response to the mechanical stimulus of food in the mouth.

Receptors in the mouth send impulses to the salivary nuclei in the medulla and pons (sometimes via the cerebral cortex) which activate the parasympathetic fibres of the facial and glossopharyngeal nerves. These nerve impulses cause vasodilation in the salivary glands, which produce profuse watery secretions. Various drugs, e.g. atropine, block the action of the parasympathetic neurotransmitter acetylcholine (see Chapter 6) resulting in a reduction in salivary, gastric and bronchial secretions. Atropine and similar drugs are used preoperatively to dry secretions and so reduce the risk of inhalation while the cough reflex is absent.

Sympathetic activity causes vasoconstriction and the production of small amounts of viscous saliva.

Teeth

The teeth are primarily concerned with ingestion and mastication but they are also important to our appearance. We go to considerable trouble to keep our teeth white, bright and straight to ensure a dazzling smile.

There are four types of teeth (*Figure 13.6(c)*) embedded within the jaw – **incisor** (cutting), **canine** (tearing), **premolars** and **molars** (grinding). Each has the same basic structure (see *Figure 13.7*) but actual shape is adapted for its different action.

Dentition

We have two sets of teeth: the first dentition, of 20 deciduous teeth (milk teeth), which erupts between 6 months and 2 years and the second dentition, of 32 permanent teeth, which starts to replace the deciduous teeth at around the age of 6 (*Figure 13.6(a),(b)*). The process of replacement is nearly complete by about 12 years of age but the third molars (wisdom teeth), if they erupt, will appear

Nursing Practice Application **Oral hygiene**

An essential part of any nursing assessment is a thorough examination of the person's mouth. The information is obviously needed in planning care but it can also help in medical diagnosis. The following points should form part of the assessment.

Odour. The mouth should not smell and the presence of *halitosis* (bad breath) may indicate an abnormality. This may be due to a reduction in saliva and natural oral cleansing but will also depend on items ingested, e.g. alcoholic drinks or peppermints. Much more serious are the odours associated with various disease processes, e.g. the new-mown hay smell of diabetic ketoacidosis.

Lips. Dryness with cracking may indicate dehydration or exposure to wind and cold. This problem is overcome by the use of suitable lubricating applications and ensuring an adequate fluid intake. The presence of 'cold sores' should be noted. These viral (*Herpes simplex*) sores are often seen in debilitated individuals whose resistance is reduced. This allows the dormant virus to become activated.

Tongue. As already discussed the tongue should be pink and moist, but may become dry and coated in dehydration. Ulcers developing on the tongue feel huge and are extremely painful. They make eating or speaking a trial. Excessive smoothness of the tongue may indicate vitamin B_{12} deficiency anaemia (see Chapter 9).

Teeth and gums. Assessment should include the condition of the teeth (including dentures) and gums as inflammation and dental problems can cause eating difficulties. The presence of crowns should be noted preoperatively to avoid problems connected with anaesthetic administration.

Candidiasis (thrush). A fungal infection of the mouth and gums which commonly affects the very young or old, debilitated individuals, those taking antibiotics and people who are immunosuppressed, e.g. cytotoxic therapy for malignancy.

Nursing interventions which maintain good oral hygiene are based on stimulating the natural cleansing action of saliva, teeth/gum/mouth brushing and removal of debris or crusts. A vital part of effective 'mouth care' is good general hydration which ensures comfort and moistens the oral mucosa through normal salivation (Howarth, 1977).

between 18 and 25 years. The wisdom teeth are often impacted in the jaw and their failure to erupt causes problems, such as pain, usually remedied by removing the offending teeth.

Structure of a tooth

All teeth consist of a **root** embedded in the jaw and an exposed part or **crown** which extends above the gum (**gingiva**). The constricted region where the crown becomes the root is called the **neck**. The innermost part of the tooth is the **pulp cavity**, containing nerves and blood vessels. This is surrounded by a layer of bone-like **dentine**. Exposed areas, such as the crown, have a further protective layer of

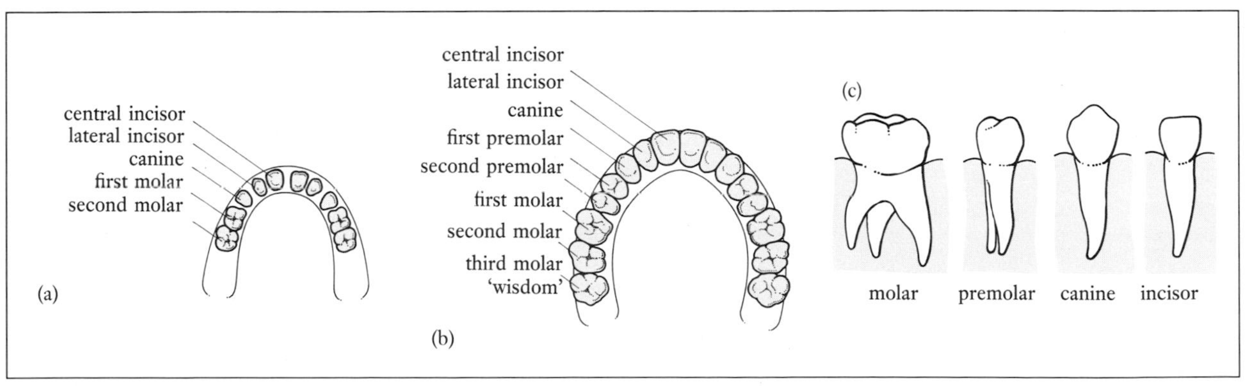

Figure 13.6. (a) First dentition; (b) second dentition; (c) tooth types.

extremely hard **enamel**; if the enamel layer is damaged or breached by decay the inner dentine and eventually the pulp cavity are destroyed by acids. The root is firmly fixed into its socket in the jaw by a substance called **cement**, which attaches it to the **periodontal membrane** or **ligament**.

The teeth are innervated by branches of the trigeminal nerve, which can be injected with local anaesthetic to block pain transmission during a dental procedure such as drilling out decay. The abundant arterial blood supply is via the maxillary artery and profuse bleeding can occur after dental extraction.

Dental caries and periodontal disease

Dental caries or cavities are one of the most common diseases of affluent societies. The problem starts with the formation of *plaque*, a mixture of bacteria, sugars and other food debris, which collects on the teeth and gums. Bacterial breakdown of the sugar produces acids which damage the enamel and eventually destroy the tooth.

Although dental caries are a serious problem most teeth are actually lost because of gum inflammation (*gingivitis*) and disease. Periodontal disease, which affects most adults over the age of 35, is caused by the calcification of plaque to form tartar. This disturbs the seal between gum and tooth, allowing bacteria to enter the socket with infection and bone destruction.

Prevention of dental caries/gum disease is based on a regimen of dental hygiene which includes brushing and flossing, regular dental examination with removal of tartar (every 6 months for adults) and avoiding food/fluids that encourage plaque formation, e.g. sugary foods and 'fizzy drinks' which are acidic and contain sugar.

The use of fluoride supplements or the fluoridation of the water supply (DHSS, 1969) have been found to reduce the amount of dental caries. This is especially important during early childhood when fluoride helps to build decay resistant teeth.

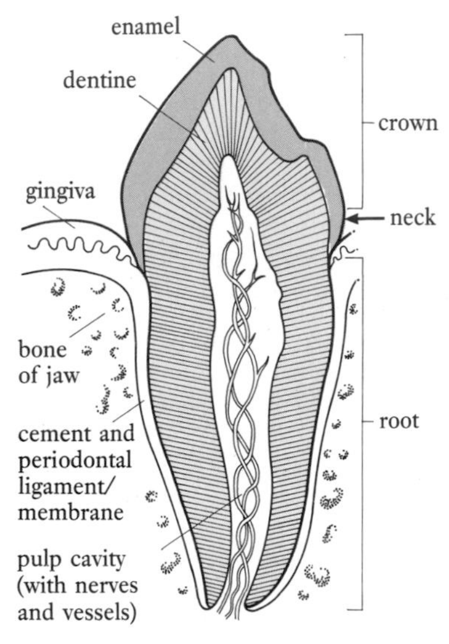

Figure 13.7. Tooth structure.

Pharynx and swallowing

In Chapter 12 we discussed the structure of the pharynx and swallowing in terms of the respiratory tract. You will remember that the oropharynx and laryngopharynx are common passages for food and air, and that various mechanisms keep food or fluids out of the larynx.

Ingested food is mixed with saliva, chewed and formed into a bolus by the tongue, which moves it to the back of the mouth ready for **deglutition** (swallowing). As the bolus arrives at the posterior pharyngeal wall it is enclosed by the muscular pharyngeal walls and swallowing ceases to have voluntary control (until this point is reached you decide whether to swallow or not). The involuntary phase now occurs as receptors in the pharynx stimulate the medulla which in turn initiates the *swallowing reflex*. Impulses via the vagus nerve cause contraction of the pharyngeal muscles and the food bolus is propelled into the oesophagus. During the involuntary phase the bolus can only go in one direction because all other routes are blocked: the soft palate blocks the nasopharynx, the larynx rises and the epiglottis covers the trachea, respiration stops and the mouth is closed (try swallowing with your mouth open). If all this fails, however, the cough reflex will operate in a conscious person to clear the airways if food does enter.

Oesophagus

The muscular **oesophagus** only conveys food from the pharynx to the stomach. It plays no part in digestion or absorption. The oesophagus is around 25 cm in length and runs from the laryngopharynx through the mediastinum to penetrate the diaphragm and enter the stomach (*Figure 13.8*).

The oesophagus has the four basic layers (see page 306) with certain modifications:

- The outer layer is fibrous adventia.
- The upper part has voluntary muscle which changes to smooth.
- The mucosa is stratified squamous epithelium containing numerous mucus glands which lubricate the bolus. The mucosa changes to columnar at the junction of the oesophagus and stomach.

When the oesophagus is empty its walls are collapsed into folds which distend as food enters.

The food bolus is propelled by a wave of muscular contraction and relaxation, known as **peristalsis** (*Figure 13.9*), towards the **gastro-oesophageal (cardiac) sphincter**. This is a physiological sphincter (the area is anatomically similar to the rest of the oesophagus) where oesophageal muscle contraction keeps the oesophagus closed until a food bolus is swallowed. It acts as a valve and with the action of the diaphragm the mechanism usually prevents the reflux of acidic gastric contents into the oesophagus. Sometimes this mechanism fails to operate properly if the pressure in the stomach increases, e.g. during pregnancy, with gross obesity or lying down after an enormous meal. Repeated reflux irritates the mucosa which leads to *oesophagitis* and ulcer formation. Reflux may be due to a structural defect known as *hiatus hernia* where the top of the stomach protrudes through the diaphragm into the chest.

Peristalsis occurs as stretching stimulates the parasympathetic fibres of the vagus nerve, which innervates the oesophagus and the nerve plexuses (myenteric) in the muscle layer.

The time taken for a bolus to reach the stomach depends on its consistency (fluids are faster), position (assisted by gravity) and the coordination of the muscular contraction and relaxation (*Figure 13.9*). If you try to eat lying down or too quickly the process is much more difficult.

The oesophagus receives arterial blood from the oesophageal and coeliac arteries but the venous drainage is more complex. The upper part of the oesophagus drains via the azygos system but the lower end drains into the hepatic portal vein and so into the liver (see

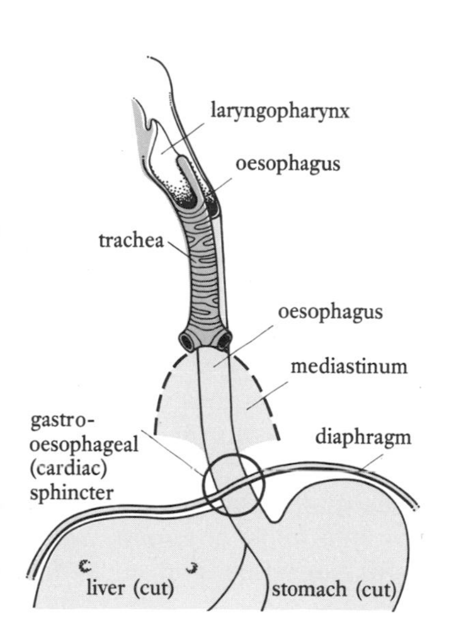

Figure 13.8. Oesophagus (diagrammatic). *NB* Lungs, heart and vessels are not shown.

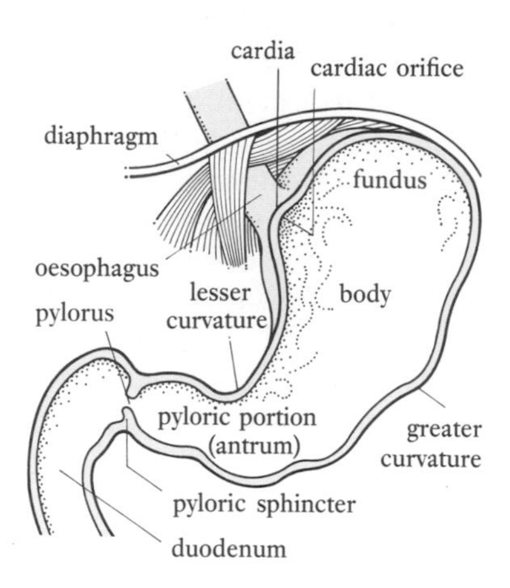

Figure 13.10. The stomach.

Figure 10.44). There is an anastomosis between the general and hepatic portal systems in the gastro-oesophageal region. *Varices* (dilations) may develop at this site if hepatic portal vein pressure is increased (see Chapters 10 and 14).

Stomach

The **stomach** is a roughly 'J'-shaped dilatation which can comfortably hold 1.5 litres of food and fluids, which it churns and mixes with gastric juices to form the semifluid **chyme**. The stomach is situated in the left upper abdomen, usually in the left hypochondriac and epigastric abdominal regions (see *Figure 1.23*).

The stomach can be divided into the **cardia** *(*which surrounds the **cardiac orifice***),* **fundus, body** and **pylorus**. A large lateral convex surface – the **greater curvature** – and a smaller medial concave surface – **lesser curvature** – provide attachments for folds of peritoneum (see *Figure 13.10* and *Figure 13.3*). The pyloric portion (*antrum*), which is continuous with the small intestine, is guarded by the **pyloric sphincter**, ensuring the coordinated release of gastric contents into the *duodenum* (first part of small intestine).

The basic four-layer structure, described on page 306, is present in the stomach with modifications to the muscle and the mucosa. Apart from longitudinal and circular muscle fibres the stomach also has oblique fibres which make it a highly mobile food mixer. This muscle arrangement allows very efficient mechanical mixing and breakdown of food. The **gastric** (relating to the stomach) mucosa is columnar epithelium with deep **gastric pits** containing mucus glands and **gastric glands** which secrete **gastric juice**. There is some variation within the stomach, e.g. more mucus glands in the cardia and pylorus. The gastric mucosa is thrown into folds, or **rugae**, which greatly increase the surface area for secretion and allow for considerable distention following a meal.

The stomach receives its arterial blood from several branches of the coeliac artery (see *Figure 10.42(a)*) and the venous return drains via the gastric and splenic veins, which empty into the hepatic portal

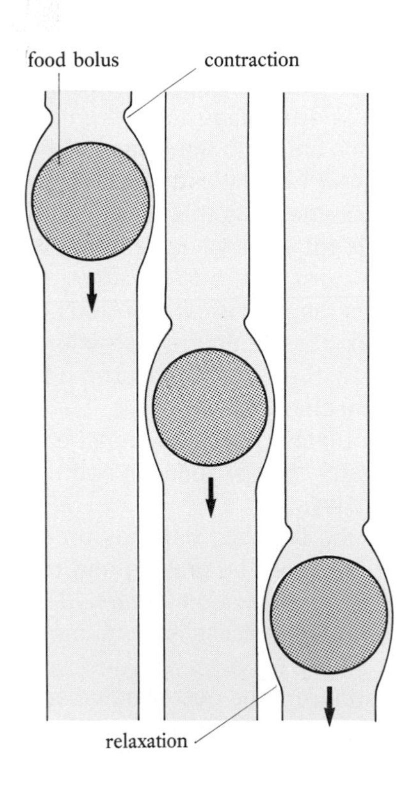

Figure 13.9. Peristalsis.

circulation (see *Figure 10.44*).

Parasympathetic fibres from the vagus nerves stimulate gastric secretion and motility (see below). The opposite effects are achieved by the stimulation of sympathetic fibres from the coeliac ganglion.

Gastric secretions and functions

Parietal or **oxyntic cells** situated in the gastric pits produce **hydrochloric acid** (*Figure 13.11*) and **intrinsic factor**, a glycoprotein needed for the absorption of vitamin B_{12} in the small intestine. Where a large part of the stomach has been removed (*gastrectomy*) it is vital that the person receives regular injections of vitamin B_{12}. Lack of vitamin B_{12} can cause a megaloblastic anaemia (see Chapter 9) and problems with nerve function (see Chapter 3).

Hydrochloric acid (HCl) is a strong acid which produces a pH of between 1.5 and 3 in the stomach. The acid is produced by a series of reactions (see *Figure 13.11*) involving chloride ions and carbon dioxide from the blood, water and the enzyme *carbonic anhydrase* (see Chapter 12). Hydrogen and chloride ions are actively secreted (against a gradient) into the stomach where they combine to form the acid. Meanwhile hydrogen carbonate ions move from the parietal cell into the blood as the *alkaline tide*.

The secretion of hydrochloric acid is stimulated by the hormone **gastrin**, which is released by the gastric mucosa and **histamine**; from the basophils and mast cells (see Chapter 9).

A synthetic gastrin (pentagastrin) or histamine is administered during a *test meal* when acid secretion is measured using samples obtained by aspirating gastric juice via a nasogastric tube. Test meals are used sometimes in the diagnosis of pernicious (megaloblastic) anaemia or lack of intrinsic factor (see Chapter 9).

Histamine stimulates acid secretion by binding to specific sites (H_2 receptors) in the cell. This has an important clinical application because drugs which block histamine are used to reduce acid production. These H_2 receptor blocker drugs, e.g. cimetidine, are

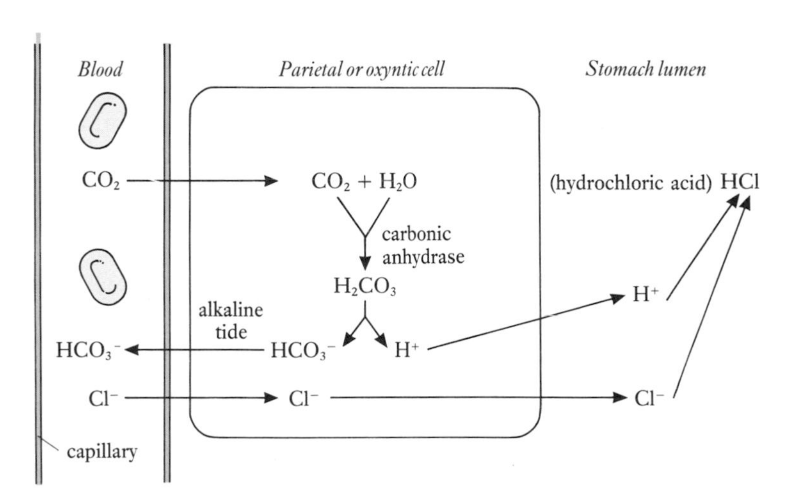

Figure 13.11. Hydrochloric acid production (basic mechanisms).

used to treat *peptic ulcers*.

The functions of gastric acid include destruction of many micro-organisms, inactivation of amylase (from the saliva), chemical breakdown of food, conversion of iron into an absorbable form, conversion of pepsinogen into pepsin and provision of optimum pH for pepsin activity.

Zymogen (chief) cells, also situated within the gastric pits, secrete the proteolytic enzyme precursor (forerunner) **pepsinogen**. This is converted by gastric acid to its active form **pepsin**, which in turn activates more pepsinogen. The pepsin starts the digestion of proteins by breaking them down into polypeptides (long chains of amino acids).

Mucus-secreting cells in the neck of the gastric pits produce thick alkaline mucus which adheres to the mucosa of the stomach wall. This protects the stomach wall from digestion by its own proteolytic (protein digesting) enzyme and damage from hydrochloric acid. Impairment of mucus production leaves the stomach vulnerable to damage and ulcer formation.

Hormone-producing cells secrete several hormones, including gastrin; a hormone concerned with gastric juice secretion (see below and page 321).

Control of gastric secretion and activity

Control of gastric secretions and activity has both neural and hormonal components and can be divided into three phases – *cephalic, gastric* and *intestinal* (*Figure 13.12*). Between 2 and 3 litres of gastric juice are produced by the mucosa each day.

The **cephalic phase** involves a conditioned reflex where the vagus nerve is stimulated by the thought, sight, smell or taste of food. The dramatic increase in gastric secretion of hydrochloric acid and enzymes is in anticipation of food arriving in the stomach. This reflex ceases to function following *vagotomy* (surgical division of the branches of the vagus nerve supplying the stomach, which may be performed for peptic ulceration – see page 323).

The **gastric phase** of secretion is both neural and hormonal. Food entering the stomach stimulates the local nerve plexus (submucosal) via stretch receptors activated by stomach distention. The neural activity (parasympathetic) leads to the release of acetylcholine, which in turn causes the gastric glands to secrete.

During the gastric phase the hormonal influence is also important. Certain foods (protein foods, caffeine-containing drinks: (tea, coffee, colas, and alcohol) arriving in the stomach cause the release of the peptide hormones known as gastrins from the gastric glands of the antrum (see Chapter 8). Gastrin travels in the circulation to the gastric glands where it stimulates the production of more gastric juice. The production of gastrin is also linked to the neural influences already described and to the release of other peptides.

Gastric secretion is inhibited by fear and anxiety (sympathetic nerves), and various regulatory peptides secreted by the intestine, e.g. **secretin, gastric inhibitory peptide (GIP), cholecystokinin**

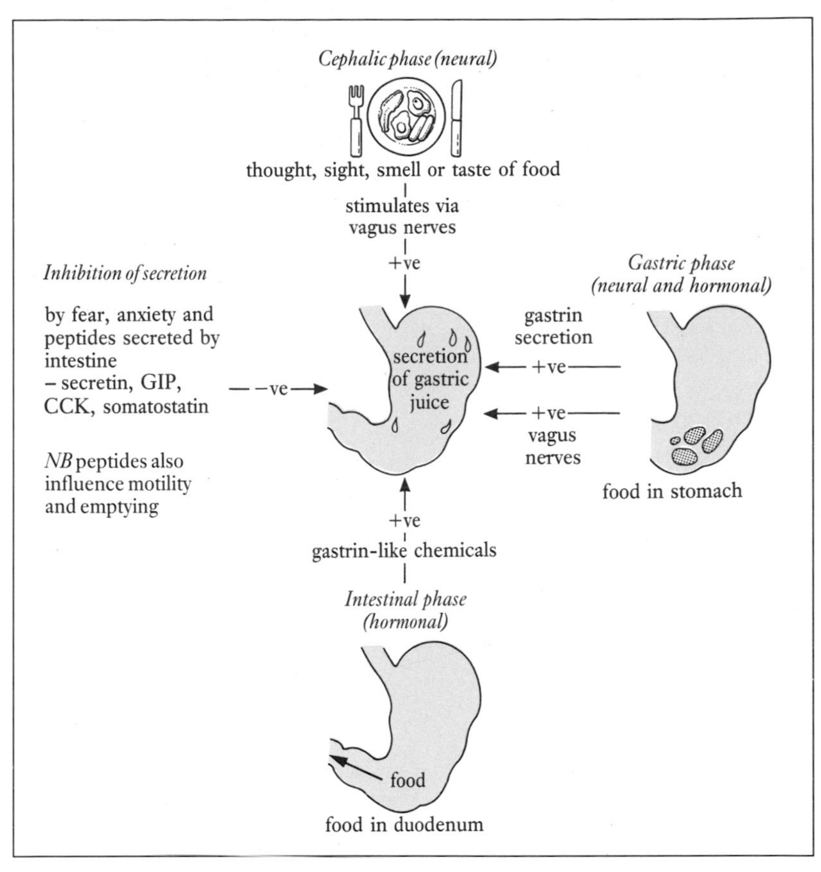

Figure 13.12. Control of gastric secretion (+ve, stimulatory; –ve, inhibitory).

(CCK) and **somatostatin** (see also GHRIH, Chapter 8).

During the **intestinal phase** small amounts of a gastrin–like chemical are produced by the duodenal mucosa as partially digested food starts to arrive in the small intestine.

Gastric motility and emptying

The length of time food stays in the stomach depends on its composition and consistency but emptying is usually complete within 4 hours of eating. As you would expect, fluids soon enter the duodenum but solid food is first mixed thoroughly with gastric juice by peristaltic contractions. A meal high in carbohydrates passes rapidly into the duodenum, followed by protein and lastly fats, which may stay in the stomach for 5–6 hours. This has important practice applications when deciding how long to fast a person before surgery (see Further Reading, Hamilton-Smith, 1972).

As acid chyme is squirted through the pyloric sphincter into the duodenum it initiates the **enterogastric reflex**. This involves the release of regulatory peptides, e.g secretin, and the stimulation of stretch receptors which inhibit gastric secretion and emptying. This ensures that the duodenum is not deluged and that there is time for intestinal mixing and digestion before the next batch of chyme arrives. This controlled release of hypertonic (see Chapter 2) gastric contents also prevents the osmotic movement of water from the blood to intestinal lumen resulting in the development of hypovolaemia (see Chapter 10). Chyme containing fat or glucose causes the release of GIP and **enterogastrone** from the duodenal mucosa, which also inhibit gastric sectretion, motility and emptying.

Summary – digestion in the stomach

- *Mechanical* – mixing and churning reduces solid food to liquid chyme.
- *Chemical* – hydrochloric acid alters the structure of proteins and the enzyme pepsin commences protein digestion.

Gastric absorption

Very little absorption is known to occur through the gastric mucosa, although water, alcohol and some drugs can be absorbed here. Anyone who has had an alcoholic drink on an empty stomach can vouch for the rapidity of this process: it is always a good idea to have something to eat and to drink alcohol within the recommended limits (see Chapter 14). Aspirin is one of the drugs absorbed by the stomach but in large doses over long periods it can cause *gastritis* (irritation of the stomach lining) with bleeding and possible *haematemesis* (vomiting blood) or *melaena* (passing altered blood in the stools). For this reason aspirin should not be taken by people with a history of ulcers, by those taking anticoagulants or by those with clotting problems (see Chapter 9).

Small intestine

The **small intestine** is a coiled tube 5–6 m in length which runs from the pyloric sphincter to the **ileocaecal valve**. It is divided into three parts: the **duodenum** *(25 cm)*, **jejunum** *(2 m)* and **ileum** *(3 m)* which lie in the abdomen surrounded by large intestine (see *Figure 13.1*). The duodenum (*Figure 13.13*), which loops around the head of the pancreas, receives the duct carrying bile and pancreatic juice. The **bile** and **pancreatic ducts** join at a point called the **hepato-pancreatic ampulla** (previously called *ampulla of Vater*) to empty via the **duodenal papilla** which is controlled by the **sphincter of Oddi**.

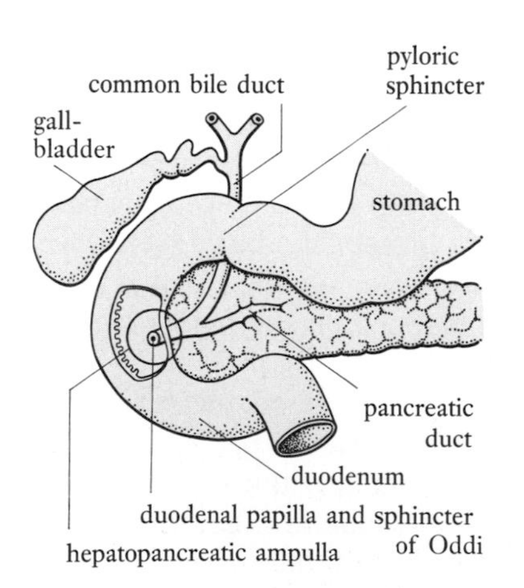

Figure 13.13. Duodenum and related structures.

The jejunum is the middle portion of small intestine between the duodenum and ileum. The ileum is the longest part of the small intestine and joins the large intestine at the ileocaecal valve. Both the jejunum and ileum are secured to the posterior abdominal wall by folds of peritoneum (*mesentery*) which also support its blood vessels, lymphatics and nerves.

The small intestine finishes the digestive processes started in the mouth and stomach, using secretions from the intestinal mucosa, pancreas and liver, and is responsible for nearly all the absorption of nutrients and water that occurs. Chyme takes up to 6 hours to move through the small intestine by **segmentation** which involves alternate contraction and relaxation of the muscle layer and peristalsis (see *Figure 13.9*). Segmentation also ensures mixing of chyme with intestinal juice.

Nursing Practice Application **Vomiting**

Vomiting is an extremely disagreeable experience occuring when gastric contents are reflexly expelled through the mouth. It is initiated by the vomiting centre in the medulla which sends impulses to the diaphragm and abdominal wall, which contract. As the pylorus is closed the increasing gastric pressure forces stomach contents out into the oesophagus and pharynx. The nasopharynx and larynx are normally closed during vomiting to prevent inhalation of vomitus, but if these mechanisms fail to operate, e.g. in an unconscious person vomitus may enter and block the airway (see Chapter 12).

Prior to vomiting the person often experiences *nausea*; he or she may be pale and sweaty and his or her mouth fills with watery saliva. Vomiting may be provoked by gastric/ intestinal distention or irritants e.g. bacterial toxins, alcohol and blood. Other causes of vomiting include motion sickness, certain drugs, e.g. cytotoxic agents, unpleasant sights or odours, fear, pain, anxiety and raised intracranial pressure (see Chapter 4). In some situations, e.g. following ingestion of 'bad' food, vomiting may be considered a protective mechanism.

A person who is vomiting generally needs a care plan which includes the following: (i) privacy and support; (ii) a suitable receptacle and a denture container if appropriate; (iii) facilities for teeth cleaning, a mouthwash and a wash afterwards. Postoperatively people should be encouraged to support wounds with their hands (see Chapter 19 for a fuller discussion of wound care). Apart from these simple measures nurses should ensure that the most appropriate *antiemetic* drugs (drugs that stop vomiting), e.g. metoclopramide, are administered as prescribed, they should observe the vomit and keep accurate records of fluid balance. Eburn (1989) considered the following to be important factors in choosing an antiemetic: (i) whether the drug is for prophylaxis or relief; (ii) which route is appropriate; (iii) which site of drug action is appropriate.

The problems of prolonged/ excessive vomiting include: (i) dehydration and electrolyte imbalance (see Chapter 2); (ii) pain and soreness felt in the abdominal muscles, especially following surgery; (iii) metabolic alkalosis (see Chapter 12) caused by the loss of hydrochloric acid. The parietal cells attempt to replace the acid and the hydrogen carbonate ions entering the blood makes it more alkaline (see *Figure 13.11*). Where vomiting is chronic the person will utilize body fat for energy which results in the production of ketoacids and metabolic acidosis rather than alkalosis (see Chapter 12)); (iv) weight loss.

Special Focus **Peptic Ulcer**

Peptic ulceration occurs in any area of mucosa exposed to pepsin/ gastric acid where the mucosal resistance is impaired. Ulcers are most common in the stomach and duodenum but can develop in the lower oesophagus, *Meckel's diverticulum* or in the jejunum following its surgical anastomosis to the stomach.

Ulcers may be acute or chronic, and some, but not all, are associated with excess production of gastric acid. Stress appears to be an important factor in their development; they occur as a result of severe physiological stress such as burns and are linked with stressful situations or occupations.

Other predisposing factors implicated in the aetiology include: heredity, bile reflux, excess alcohol, drugs such as aspirin, corticosteroids and other anti-inflammatory drugs, smoking and possibly the presence of the bacteria *Helicobacter pyloridis*.

The presence of a peptic ulcer usually causes a 'gnawing' pain in the upper abdomen which may be associated with food intake or an empty stomach, vomiting and chronic health problems such as anaemia caused by slight but persistent bleeding. For most people there is time lost from work and a general reduction in the quality of life. The dangers, however, are massive haemorrhage (haematemesis/melaena), perforation with peritonitis or gastric outlet obstruction due to scar tissue.

A very rare cause of peptic ulcer is *Zollinger–Ellison syndrome* where there is an *ectopic* (outside the normal site) source of gastrin production, usually a pancreatic new growth. As you can imagine this stimulates excessive gastric acid production, peptic ulcer formation and inactivation of intestinal enzymes which have an optimum pH of around 7.

Person-centred Study **Ken**

Ken, who is 39, has a very hectic lifestyle: he is a presenter on a local independent radio station, is renovating his Edwardian house and has a wide circle of friends. A few months ago he started taking 'indigestion' tablets which helped the gnawing pain which seemed to come when he missed meals or was particularly stressed at work.

Eventually the pain, which was waking him at night, sent Ken to his GP who organized some investigations (endoscopy). Ken received the news that he had a peptic ulcer with mixed feelings; relieved that it was not cancer but worried about the effects on a busy work schedule.

Ken is prescribed the following drugs: (i) antacids, e.g. aluminium hydroxide or magnesium trisilicate; (ii) H_2 receptor blocker, e.g. cimetidine. The doctor also discusses related health education aspects and the changes to lifestyle that Ken might consider. He is advised to stop smoking which should speed up the healing process. Previously his alcohol intake was between 25–30 units per week, Ken is given information regarding 'safe limits' (Chapter 14) and it is suggested that he restrict alcohol intake to the occassional glass of wine or beer after this exacerbation is over. The intake of alcohol during an exacerbation would tend to aggravate symptoms and increase pain. He is advised to review his eating habits with particular reference to having regular meals/snacks and obviously avoiding foods which give him indigestion. Strict 'ulcer' diets are of very doubtful value in symptom relief or healing (Macleod *et al.*, 1987) and are not generally used. Ken is advised to avoid drugs such as aspirin which are known to irritate the gastric mucosa.

The same four layers (see page 306) are present but the submucosa and mucosa (columnar epithelium) are highly modified for their role in digestion and absorption (see *Figure 13.14*). There are three adaptations which greatly increase the surface area of the small intestine:

- Circular folding which is permanent even when the intestine is distended.
- The mucosa has **villi** – finger-like projections about 0.5–1 mm long (see *Figure 13.16*) which contain capillaries and a central lymphatic *lacteal* (see Chapter 11).
- **Microvilli (brush border)** (see *Figure 13.16*) which project from the free edge of the villi. These microvilli increase the area available for absorption and contain some digestive enzymes.

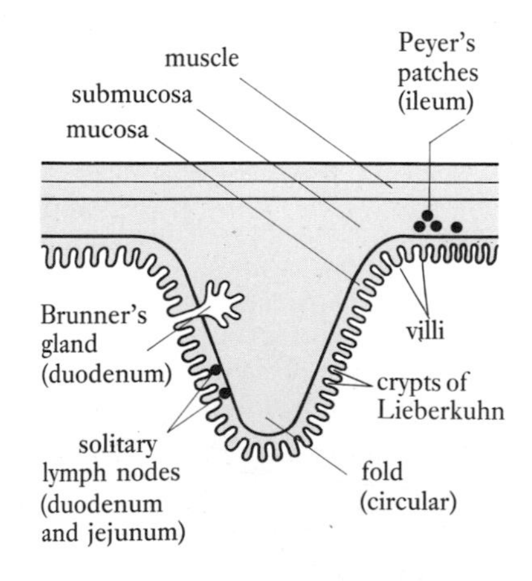

Figure 13.14. Structure of the small intestine (general) – diagrammatic.

Lymphoid tissue occurs as solitary mucosal nodes in the duodenum and jejunum and as submucosal aggregates known as **Peyer's patches** in the ileum. These help prevent bacteria from entering the blood.

Alkaline mucus secreted by **Brunner's glands** situated within the duodenal submucosa helps to neutralize the acidity of the chyme, protects the duodenal mucosa from damage and provides the optimum pH for intestinal enzyme activity. Deep folds or intestinal crypts (**crypts of Lieberkuhn**) between the villi form tubular glands containing several cell types, including mucus-secreting goblet cells (see Chapter 12) and columnar cells (**enterocytes**) which produce intestinal enzymes or are concerned with absorption (*Figure 13.14*).

Cells from the depths of the crypts are continually migrating to the surface of the villi where they replace cells lost through general wear and tear. The surface can be renewed in little more than a day.

The small intestine receives arterial blood from the superior mesenteric artery and drains its venous blood via veins which empty into the hepatic portal vein (see *Figures 10.42(b)* and *10.44*).

Role of the pancreas as an accessory organ of digestion

In an earlier section we discussed the dual endocrine and exocrine roles of the pancreas (see Chapter 8). The pancreas is divided into three: the head lying within the curve of the duodenum and the body and tail which lie behind the stomach and are *retroperitoneal* (behind the peritoneum *(Figure 13.15)*).

Exocrine *– a gland which has a duct through which its secretions are discharged. Compare with endocrine (Chapter 8).*

The pancreas, which has a vital role in digestion, produces approximately 1500 ml of alkaline (pH 8.0) pancreatic juice every day. Its exocrine *acinar* cells (grape-like secretory cells arranged in clusters around a tiny duct system) produce the juice, containing digestive enzymes, hydrogen carbonate ions, which further neutralize acid chyme, electrolytes and mucus. The juice leaves the pancreas in ducts, which join to form the main pancreatic duct, which in turn joins with the bile duct to enter the duodenum. Secretion of pancreatic juice is stimulated by:

- Regulatory peptides (hormones) – **cholecystokinin (CCK)** and **secretin**, which was the first hormone discovered (by Bayliss and Starling in 1902). These hormones are produced as chyme enters the duodenum.

  NB At one time it was thought that there were two separate hormones, cholecystokinin and **pancreazymin** *(PZ)*, but they are now known to be one substance (sometimes still called **CCK-PZ**) which affects the activity of both pancreas and gallbladder.
- Neural stimulation via the vagus nerve stimulates secretion but is of much less importance than the hormones.

Pancreatic enzymes

The pancreas produces enzymes concerned with the digestion of proteins, carbohydrates, fats and nucleic acids.

The three proteolytic (protein-digesting) enzymes are secreted as their inactive precursors – **trypsinogen**, **chymotrypsinogen** and **procarboxypeptidase**. These are not activated until they reach the duodenum, to avoid damage to the pancreas by *autodigestion*. Trypsinogen is coverted to the active enzyme **trypsin** by another enzyme **enterokinase** (sometimes called *enteropeptidase)* produced by the duodenal mucosa. The trypsin, once formed, acts on further trypsinogen and the other precursors, which it activates to form **chymotrypsin** and **carboxypeptidase**.

Trypsin and chymotrypsin act upon proteins and polypeptides to form peptides, which are short chains of amino acids. The peptides formed in this way are broken down into individual amino acids by the carboxypeptidase.

Amylase (see page 314) produced by the pancreas converts starch, which is a polysaccharide (composed of many linked monosaccharide units) to the disaccharide (two monosaccharide

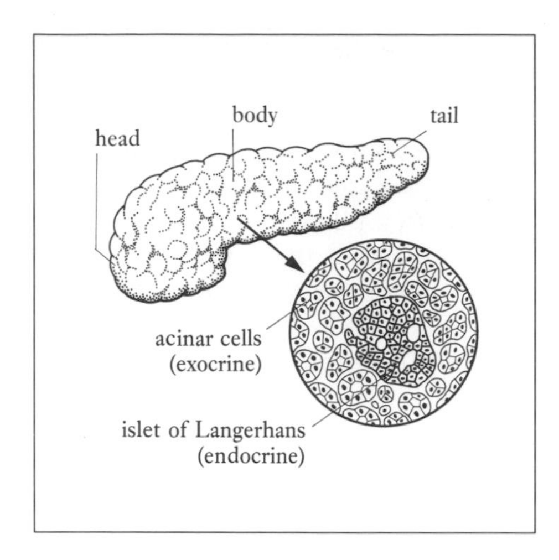

Figure 13.15. Pancreas.

units) maltose. (See pages 309,310 for an explanation of the terms mono-, di- and polysaccharide.)

Pancreatic lipase is the enzyme which digests fat. It acts on triglycerides converting them to glycerol and fatty acids. Lipase can only operate after the clumps of fat globules have been emulsified by *bile salts* (see below).

Pancreatic juice contains **nucleases** (**ribonuclease**, **deoxyribonuclease**) which convert RNA and DNA to nucleotides (see Chapter 1).

Bile and digestion

We have briefly discussed the formation of bile pigments from haem during the breakdown of erythrocytes (see Chapter 9) and the details of bile production and excretion by the liver can be found in Chapter 14.

Between 500 and 1000 ml of alkaline (pH 8.0) **bile** is produced each day by the liver. This is a continuous process and bile not immediately required is stored and concentrated by the gallbladder. Bile is a green–yellow fluid which contains water, **bile acids** (**salts**), **bile pigments** mainly **bilirubin** and some **biliverdin**, cholesterol, phospholipids and electrolytes. It also provides an excretory route for the breakdown products of steroid and other hormones (see Chapter 8) and some drugs which leave the body in the faeces. Bile is both a secretory (secreted by liver cells) and excretory (carries substances for excretion) product.

Bile secretion is stimulated by the intestinal hormone secretin and a high level of bile salts in the blood. Bile salts are reabsorbed in the ileum and returned to the liver by the hepatic portal vein by a recycling process known as the **enterohepatic circulation of bile salts**. The actual release of bile into the duodenum via the *common* *bile duct*, which unites with the pancreatic duct, depends on the simultaneous contraction of the gallbladder and relaxation of the sphincter of Oddi. The main stimulus for this is CCK, already mentioned in connection with pancreatic secretion, which is released in response to fatty and protein-rich chyme entering the duodenum. There is also a neural component via the vagus nerve. Following gall-bladder emptying the bile flows directly from the liver to duodenum via the bile ducts.

Functions of bile

- Bile acts like a detergent to emulsify fat globules in the partially digested food. Bile salts separate the fat into smaller droplets which can mix with the watery intestinal contents. The smaller droplets have a greater surface area which allows pancreatic lipases to start the chemical breakdown of fat into fatty acids and glycerol.
- Bile salts are also concerned with the absorption of digested fat, cholesterol and fat-soluble vitamins (see pages 330, 331 and 332). The bile salts form **micelles** (small 'clumps') with lipids (fatty substances) which enhances their 'transport' into the intestinal cells.
- Bile salts deodorize faeces.

NB Bilirubin is oxidized and converted by the gut bacteria to the substance **stercobilinogen**. Some of this is reabsorbed into the blood but the remainder becomes **stercobilin** which colours the faeces brown. When bile is not present in the **gut** (intestine or bowel), e.g. a gallstone is obstructing the common bile duct, the faeces will be pale (*clay coloured*) and the faeces will have an offensive odour due to the presence of undigested fat.

Intestinal juice

The secreting cells of the intestinal glands produce 2–3 litres of alkaline (pH 8.0) intestinal juice every day.

We might digress here, to remind ourselves just how much fluid is produced by the gastrointestinal tract and accessory organs:

saliva	1000–1500 ml
gastric juice	2000–3000 ml
pancreatic juice	1500 ml
bile	500–1000 ml
intestinal juice	2000–3000 ml
TOTAL	7000–10 000 ml

Taking the maximum figures we end up with the staggering total of 10 litres of fluid produced per day (enough to fill nearly 18 milk bottles), the majority of which is reabsorbed. From this it is easy to see why dehydration occurs with severe vomiting and diarrhoea (see Person-centred Study – 'Natalie', Chapter 2) and where intestinal obstruction prevents fluid reabsorption. It is important to grasp the concept that fluid in the gut is really 'outside the body'; it is not contributing to cell metabolism and collection of fluid in the large intestine due to obstruction causes dehydration although the fluid is not ejected from the gut.

Intestinal fluid contains water, mucus, electrolytes and enzymes, most of which are present in the brush border of the microvilli (see page 324). Secretion is stimulated by the mechanical and chemical effects of food on the cells of the jejunum and ileum and the release of regulatory peptides.

Intestinal enzymes

A whole series of intestinal enzymes complete protein and carbohydrate digestion. The duodenal mucosa secretes enterokinase, which activates pancreatic trypsinogen. The peptides formed by the pancreatic proteolytic enzymes are reduced to their individual amino acids by the action of **aminopeptidases** and **dipeptidases** present in the microvilli (brush border). Aminopeptidases act on the *amine* *(NH_2)* end of the peptide chain and dipeptidases cleave the bond between a pair of amino acids (dipeptide).

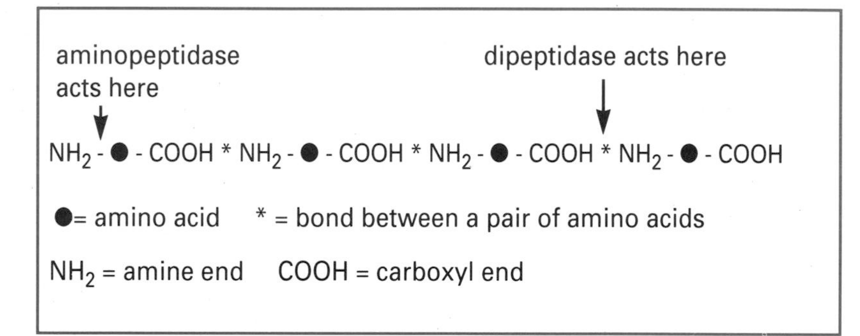

Carbohydrate digestion is completed by three enzymes which convert disaccharides into monosaccharides, which can then be absorbed: **maltase** converts maltose to glucose, **lactase** converts lactose (milk sugar) to glucose and galactose and **sucrase** converts sucrose to fructose and glucose.

Enzymes convert:

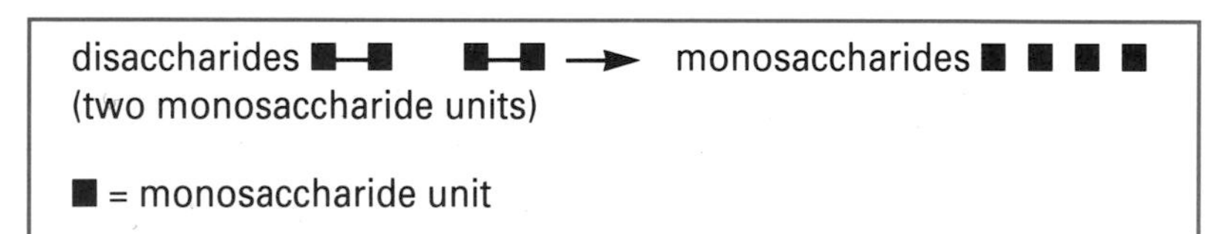

Fat digestion is completed by lipases in the micovilli but you will remember that this is mostly the responsibility of pancreatic lipases and takes place in the duodenum.

Nucleotides produced by the action of pancreatic nucleases on nucleic acids are further broken down. Intestinal enzymes cleave the nucleotides apart to release their component *pentose sugar, bases* and *phosphate*.

The digestive processes are now complete. The main products of digestion are the monosaccharides glucose, fructose and galactose, amino acids and fatty acids and glycerol. The various nutrients are now in a form which can be absorbed and utilized (see charts below).

Summary of chemical digestion

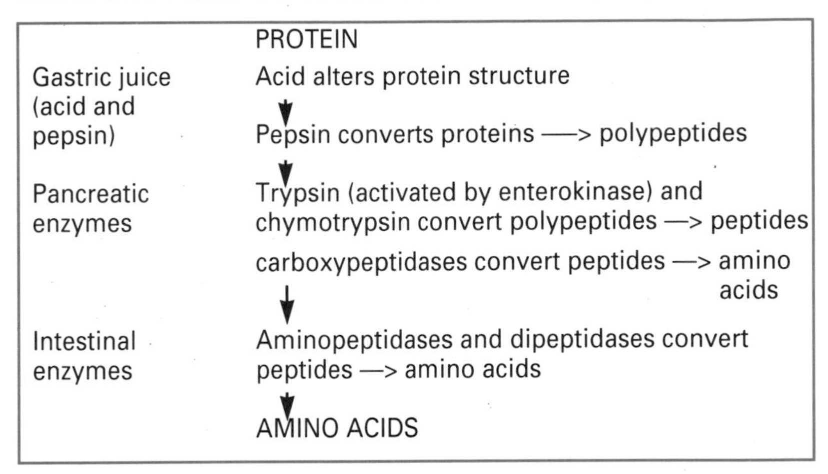

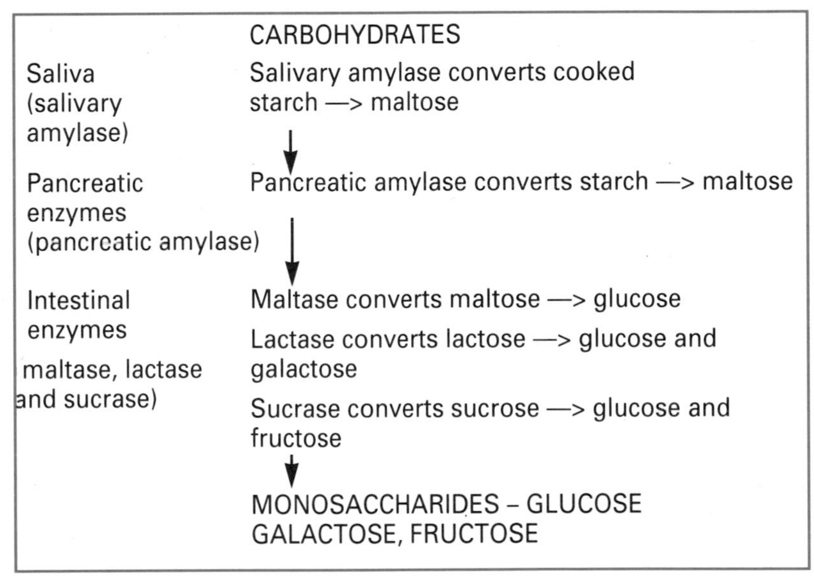

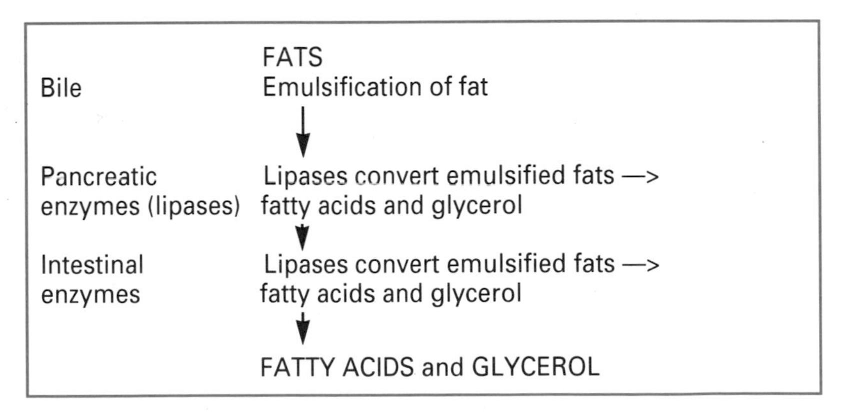

Absorption in the small intestine

As already stated the small intestine is the major site for the absorption of nutrients and water. To get some idea of the importance of the small intestine in absorption we have only to consider the fate of the 7–10 litres of fluid produced by the gastrointestinal tract and the assorted food and drinks consumed per day. Only about 1 litre of fluid and undigested food reaches the large intestine. This obviously means that an amazing feat of absorption is occurring in the small intestine.

Before we start it will be useful for you to have a quick look at Chapters 1 (pages 7 and 8) and 2 (pages 40-42). You should check your understanding of; active transport, carrier molecules, passive transport, facilitated diffusion and osmosis.

Some nutrients pass through the cell membrane of the intestinal cells by **active transport**, (see page 8) which requires metabolic energy (**ATP**). Active transport occurs against a concentration gradient and requires the presence of **carrier molecules** (see pages 7 and 8). Other nutrients move by **passive transport** (see page 7) down gradients, which does not use energy. When passive transport requires a carrier molecule it is known as **facilitated diffusion** (see pages 8 and 41). Water moves passively by **osmosis** (see pages 7, 41-42), which is linked to the active and passive movement of water-soluble nutrients.

The structure of the intestinal mucosa is ideally adapted for absorption – the villi and microvilli provide a huge surface area. Once inside the epithelial cells of the villi most nutrients enter the capillaries which eventually lead to the hepatic portal vein, but some fatty acids and glycerol are absorbed into the central lymphatic lacteal (*Figure 13.16*).

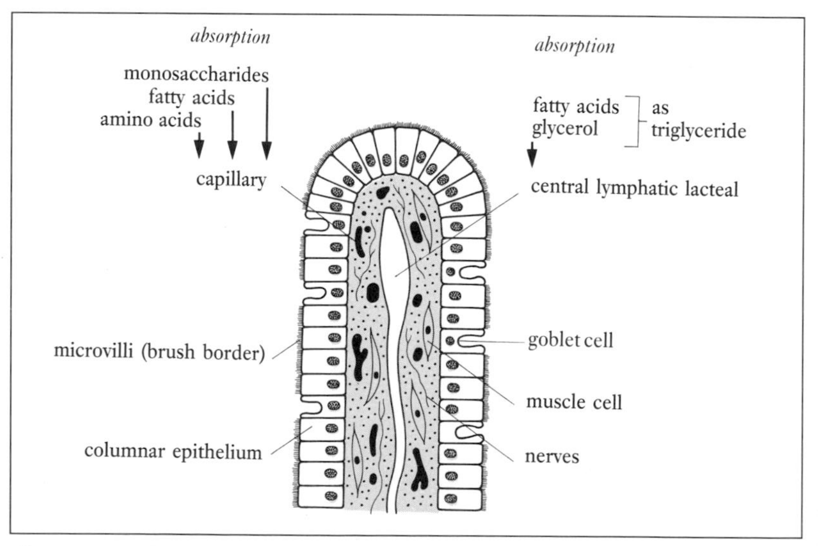

Figure 13.16. Detail of villus and absorption (diagrammatic).

Details of absorption

Amino acids

Amino acid absorption is an active process which involves different carrier molecules. Some of the carrier molecules are coupled with sodium transport which appears to increase their activity. Most amino acids are absorbed in the ileum and enter the capillary network of the villus.

Monosaccharides

Monosaccharides are actively absorbed. Glucose and galactose use the same carrier molecule and their transport is coupled with the active absorption of sodium. Fructose absorption uses a different carrier and is not linked with sodium transport. The monosaccharides enter the capillary network for transport to the liver.

Fats

Fats have an unusual mechanism for absorption. They are the only nutrient molecules which enter the lymph as well as the blood. Fatty acids, glycerol and other lipids, such as cholesterol, combine with bile salts (see pages 326-327) to form tiny clusters or **micelles**. The micelles transport the fats to the enterocytes (see page 324) and once there the fats move passively into the cells. The bile salts are left in the lumen of the intestine for re-use. Most fat absorption occurs in the duodenum and is completed in the ileum. Once inside the cell most of the fatty acids and glycerol re-form as triglycerides (three fatty acids and glycerol):

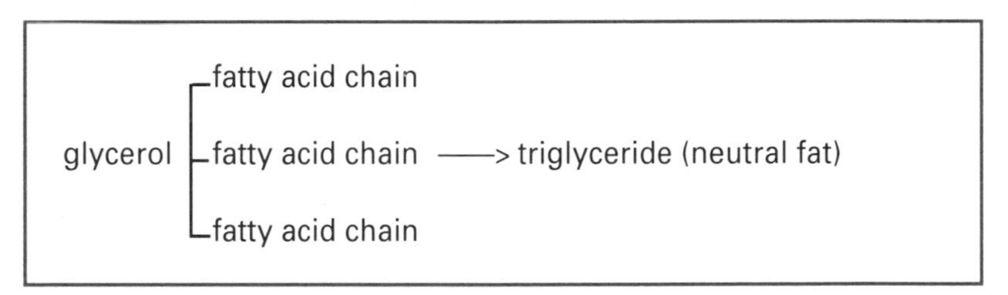

The reconstituted fat is formed into **chylomicrons** after being coated with other molecules, including lipoproteins and cholesterol. The chylomicrons enter the central lacteal to form the creamy **chyle**, which is transported via the lymph to the blood (see Chapter 11). The remaining fatty acids enter the capillary network. Both routes ensure that eventually the products of fat digestion arrive at the liver for utilization.

Cholesterol is transported in the blood combined with protein, a combination known as **lipoprotein**. There are various types of lipoprotein, which include **very low density lipoproteins (VLDL), low density lipoproteins (LDL)** and **high density lipoproteins (HD)**. The ratio of HDL:VLDL/LDL is very important for health (see Chapter 10). A decreased ratio of HDL:VLDL/LDL is associated with arterial disease, whereas an increased ratio of HDL:VLDL/LDL appears to have a protective function.

Vitamins

Vitamins are divided into two groups, the *water-soluble* (B complex and C) and the *fat-soluble* (A, D, E and K).

Water-soluble vitamins are mostly absorbed passively with water, except vitamin B_{12} which is absorbed in the terminal part of the ileum and requires the intrinsic factor (see page 319). Vitamin B_{12}, a large molecule, needs to form a complex with intrinsic factor before it can be absorbed. The vitamin B_{12}–intrinsic factor complex binds to receptor sites in the mucosa and the vitamin is absorbed into the capillaries.

Fat-soluble vitamins are absorbed with fats as part of the micelles. It follows that their efficient absorption is dependent upon the presence of bile in the small intestine and fat in the diet. In situations where the flow of bile is obstructed, such as when a gallstone (see Chapter 14) blocks the duct, there may be inadequate absorption of vitamin K. As you will remember, vitamin K is required for the production of several clotting factors (see Chapter 9). A person having surgery to remove the obstructing gallstone will be given vitamin K by intramuscular injection to ensure that haemostasis operates normally during and after surgery.

Minerals

Minerals are absorbed actively and passively throughout the small intestine and colon (see page 336). Sodium is absorbed actively in the small intestine, where it is linked with the transport of amino acids and monosaccharides, and in the colon. Chloride moves passively through the gut wall into the blood. Potassium ions are absorbed by passive transport from the ileum and colon. When a person loses water from the bowel he or she will also lose potassium in the watery stool. Loss of potassium leads to *hypokalaemia* (abnormally low levels of potassium in the blood) and may be a feature of severe diarrhoea or *ileostomy* (a *stoma* (opening) where the ileum discharges onto the abdominal wall).

A detailed account of mineral absorption is beyond the scope of this book but we should consider the absorption of iron and calcium. Their absorption rates are linked to demand, helping to ensure that homeostasis for these minerals is maintained.

Iron is required for erythropoiesis and details of its metabolism can be found in Chapter 9. Iron is most easily absorbed in the ferrous state. Gastric acid and vitamin C facilitate the conversion of ferric iron to ferrous iron, which is absorbed actively in the upper small intestine. Iron can be stored as a protein complex within the enterocytes until it is required in the blood. Iron absorption is inhibited by the *phytic acid* and *phosphates* present in a high fibre cereal diet. *Iron deficiency anaemia* may be a problem for some vegetarians because absorption is inhibited in this way and because of their avoidance of foods such as red meat which contains high levels of available iron. (*NB* Vegetarians can obtain iron from legumes, egg yolk (if eaten), dried fruit and wholegrain cereals. Remember that absorption of the iron present in cereals will be, to

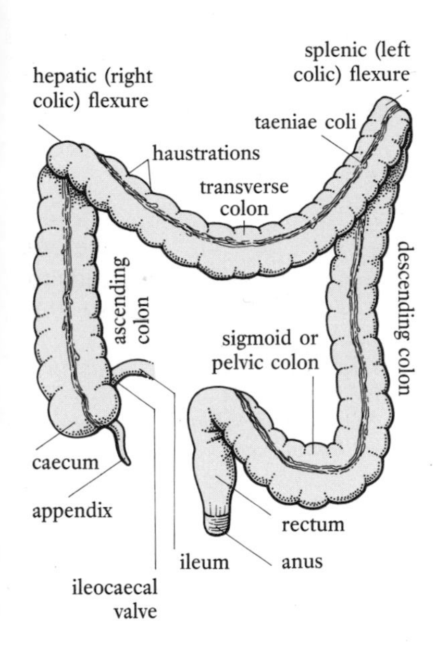

Figure 13.17. Large intestine.

some degree, inhibited by the phytic acid also present.)

Active calcium absorption from the upper small intestine is linked to the amount of calcium present in the blood. The rate at which calcium is absorbed is controlled by parathyroid hormone and calcitonin (see Chapter 8). The active form of vitamin D (1,25-dihydroxycholecalciferol) facilitates calcium absorption by the intestinal mucosa. In common with iron, the absorption of calcium is inhibited by phytic acid and phosphates.

A more comprehensive list of minerals required by the body can be found in *Table 2.1*.

Water

Water is mostly absorbed in the small intestine (80–90%) with the remainder by the colon. This explains why the discharge from an ileostomy is much more fluid than that from a *colostomy* (a stoma where the colon discharges onto the abdominal wall). Between 200 and 400 ml of water is absorbed every hour, and of the 9–11 litres of fluid (secretions + drinks) passing through the gastrointestinal tract, less than 200 ml/24 hours are lost in the faeces.

The actual amount of water in the faeces depends on the time food residues remain in the intestine. In constipation the food residues spend extra time in the colon and more water is absorbed, which leads to the characteristic small, hard *stools*. The watery stools of diarrhoea are due to lack of time for water absorption as the food residue 'rushes' through the intestine. Constipation and diarrhoea are discussed in greater depth on page 339. Water also moves in the opposite direction – from cells to intestinal lumen. This occurs if the intestinal contents become hypertonic (see also page 322).

(*NB* The metabolism and utilization of the main nutrients – amino acids, glucose and fats – is considered later.)

Large intestine

The large intestine runs from the ileocaecal valve to end at the anus and measures about 1.5 m in length. It can be divided into: **caecum**, **vermiform** (worm-like) **appendix**, **colon** (**ascending**, **transverse**, **descending** and **sigmoid**) and **rectum** which opens into the **anal canal** (*Figure 13.17*). More detail of the different parts can be found on page 335.

Major functions of the large intestine (see also pages 336-337)

- Absorb some water, minerals, some vitamins and drugs.
- Vitamin (K and some B) synthesis by bacteria present in colon.
- Stores food residues until eliminated by defaecation.
- Eliminates waste through defaecation.

The **colon** forms a frame around the small intestine by making two 90° turns – one under the liver and the other near the spleen, known respectively as the **hepatic (right colic)** and **splenic (left colic) flexures**.

Nursing Practice Application **Malabsorption**

Malabsorption will obviously result from any process which affects the ability of the intestinal mucosa to absorb water and nutrients. This may occur as part of an acute illness but may also be chronic. It is important that nurses are alert to the types of condition where malabsorption is a possibility (see nutritional assessment, page 351).

We have already mentioned chemicals such as phytic acid which inhibit mineral absorption. Iron or calcium absorption may be inadequate where a high fibre diet is consumed, e.g. vegetarians.

Infections affecting the intestine which cause diarrhoea will lead to dehydration and electrolyte imbalance, particularly potassium and sodium. These infections include *cholera* and food poisoning due to bacterial contamination,e.g. *Campylobacter, Salmonella.*

Inflammatory disease of the intestine, e.g *Crohn's disease* (*regional ileitis*) will cause malabsorption of nutrients and water.

Some people have an intolerance to a specific nutrient which causes malabsorption; commonly the intolerance is to lactose, gluten (a protein found in some cereals e.g. wheat) or fats.

Problems also arise if a large area of small intestine is resected (removed) or bypassed by the formation of an ileostomy. These are measures which may be required to alleviate the effects of inflammatory bowel disease.

Arterial blood reaches the colon from the superior mesenteric and inferior mesenteric arteries, which also supply the proximal part of the rectum (see *Figure 10.42(b)*). The distal part of the rectum and anus are supplied by branches of the internal iliac arteries. Venous return from the colon and proximal part of the rectum is via mesenteric veins which drain into the hepatic portal vein (see *Figure 10.44*). The rectum is another area where the hepatic portal and systemic circulations anastomose. Varicosities may develop in the rectum when pressure in the hepatic portal vein is raised such as with cirrhosis. The venous blood from the distal rectum and anus drains into the iliac veins.

The autonomic nervous system innervates the large intestine – the vagus and nervi erigentes are parasympathetic and sympathetic fibres from the coeliac and mesenteric ganglia (see page 306, Chapter 6). The external anal sphincter is, however, under voluntary control and is innervated by the pudendal nerve from the sacral plexus (see Chapter 5).

Structure of the large intestine

Structurally the large intestine has the basic layers (see page 306) but there are differences in the mucosa and muscle layer.

The mucosa is simple columnar epithelium without villi. There are numerous goblet cells within the mucosa of the colon and rectum which produce the mucus required for lubricating the movement of faeces. In the anal canal the mucosa becomes stratified squamous epithelium, which hangs in vertical folds called the **anal columns**. This region, which merges with the outside skin, is subjected to considerable friction (see defaecation pages 337 and 338).

Throughout the colon the submucosal layer contains abundant lymphoid tissue which generally prevents the numerous **colonic** (relating to the colon) bacteria from entering the blood.

There are several modifications to the muscle layer. Two horizontal folds of the circular muscle layer form the ileocaecal valve which controls the passage of food residues from small to large intestine. The longitudinal muscle layer in the colon consists of three bands called the **taeniae coli**. As these bands are shorter than the colon they produce puckering or **haustrations**. The circular muscle layer also forms the **anal sphincters**: the internal which is smooth muscle (involuntary) and the **external** which is skeletal muscle (voluntary).

Detail – large intestine

Caecum and appendix

The first part of the large intestine or **caecum** is a blind pouch projecting below the ileocaecal valve, which separates it from the ileum (*Figure 13.18*). The ileocaecal valve opens when peristalsis brings small bowel contents to the ileum and as a reflex in response to food entering the stomach. This **gastro-colic reflex** operates through the vagus nerve to initiate peristalsis in the colon, opening of the valve and the urge to defaecate. In grass-eating animals the caecum is a much larger structure which plays a significant part in their digestive processes.

The worm-like **vermiform appendix** hangs from the lower part of the caecum (*Figure 13.18*). The appendix is a blind tube which varies greatly in length but is usually around 5 cm. It is a common site of inflammation (*appendicitis*) although it contains a great deal of lymphoid tissue.

Appendicitis is a common condition occurring when the opening into the appendix is obstructed. This provides ideal conditions for bacterial growth within the appendix, which becomes inflamed and swollen. A person with acute appendicitis usually has abdominal pain which commences in the umbilical region and moves to the right iliac fossa (see *Figure 1.23*), anorexia, nausea and vomiting.

Treatment involves the immediate surgical removal of the appendix (*appendicectomy*). Complications of appendicitis include abscess formation or rupture of the appendix which causes *peritonitis* (see page 308) and later *adhesions*.

Colon

The **ascending colon** runs up the right side of the abdominal cavity to the level of the liver where it turns at the **hepatic flexure** to form the **transverse colon** (see *Figure 13.17*). The transverse colon runs across the abdomen where it passes in front of the stomach. It turns at the **splenic flexure** and runs down the left side of the abdomen as the **descending colon**. As the colon enters the pelvis it becomes the **pelvic** or **sigmoid colon**, so called because of its S-shaped curve.

Level with the third sacral vertebra, the sigmoid colon forms the slightly dilated **rectum**, which runs in front of the sacrum and coccyx. The rectum, which has no taeniae coli, is completely invested (covered, 'clothed') in longitudinal muscle; it also does not have any mesocolon (see page 307). The rectum terminates at the **anal canal** with its two sphincters and stratified squamous epithelium (*Figure 13.19*). The anal canal opens externally at the **anus**.

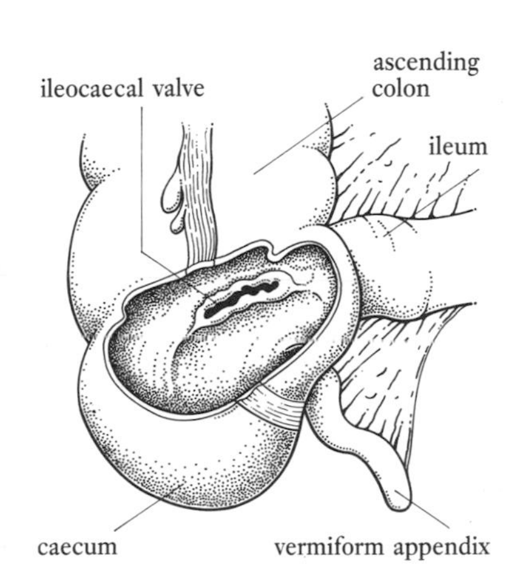

Figure 13.18. Caecum and appendix.

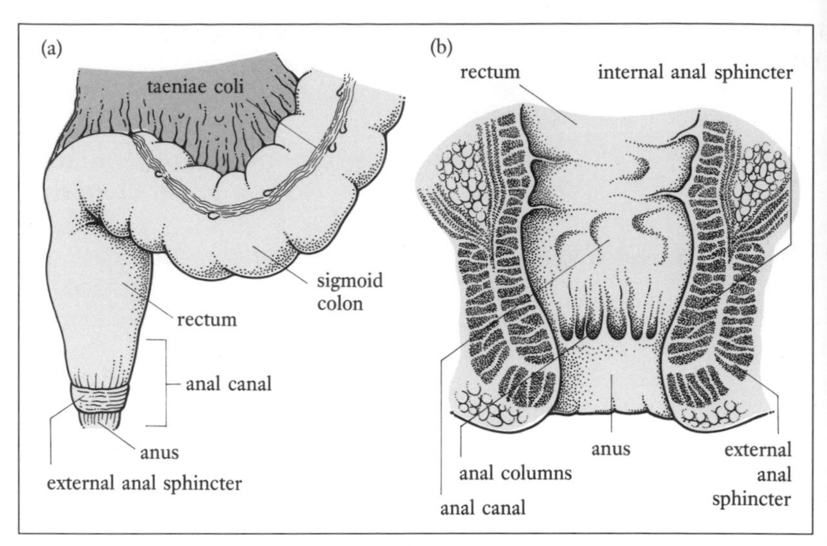

Figure 13.19. (a) Rectum; (b) anal canal.

Functions of the large intestine

Absorption

The colon absorbs water from food residues. The amount absorbed is related to the time that residues stay in the colon, and increases with time. The ability of the colon/rectum to absorb water and electrolytes can be utilized in fluid replacement via a rarely used rectal infusion (see Chapter 2).

Sodium ions are absorbed actively with chloride ions and water moving passively into the blood.

The colon is able to absorb a small amount of the vitamins (K and B complex) produced by its *commensal* bacteria. The amounts of vitamins involved are probably of little significance where nutrition is normal.

Mixing of the colonic contents by segmentation (see pages 322 and 337) aids absorption.

Synthesis

The large intestine is colonized by many *commensal* (meaning 'to eat at the same table') bacteria. This is a mutually beneficial relationship where the bacteria are nourished by the food residues and in return they synthesize vitamins K and some of the B complex (*thiamine, riboflavin, folate* and B_{12}). The bacteria involved include *Escherichia coli (E. coli), Enterobacter aerogenes, Streptococcus faecalis* and *Clostridium perfringens (C. perfringens)*. These bacteria are commensal in the intestine but can become *pathogenic* (disease-producing) if they find their way to some other site, e.g. *E. coli* contaminating the bladder or *C. perfringens* in a wound.

The intestinal bacteria also cause the fermentation of undigested food residues with the production of gases known as *flatus*. The gases include methane, carbon dioxide, hydrogen sulphide and hydrogen, some of which contribute to faecal odour. The amount produced depends on the type of food eaten; individuals vary but flatus-producing foods include beans, lentils, cabbage and onions. In large amounts flatus can cause considerable distension and discomfort, especially following abdominal surgery.

Storage

Most food residues pass through the large intestine within 12 to 72 hours but transit times depend upon many factors including the amount of indigestible fibre present. The intestine is able to store large amounts of waste which may take a week or more to travel to the rectum. After eating foods such as sweetcorn it is easy to establish how long it takes to appear in faeces.

Elimination/defaecation

The large intestine propels the waste towards the rectum by strong peristaltic movements called **mass movements**, these are linked to the gastro–colic reflex and occur after meals. Mass movement causes the rectum to fill with faeces which in turn initiates an urge to defaecate. Ordinary peristalsis also helps to move faeces toward the rectum and segmentation ensures that mixing occurs.

As the rectum is stretched by faeces (normally it is empty), the *defaecation reflex* is started through the sacral part of the spinal cord. Impulses also travel to the cerebral cortex, which can inhibit the reflex until defaecation is convenient. As well as this inhibition we have voluntary control through the pudendal nerve and can keep the external anal sphincter closed. When defaecation is convenient the spinal reflex can proceed; parasympathetic nerve activity causes the sigmoid colon and rectum to contract and relaxes the internal sphincter, and there is voluntary relaxation of the external sphincter.

Defaecation is aided by voluntary straining which involves *Valsalva's manoeuvre* (see Chapter 12). The amount of straining required depends on faecal consistency – it is much easier to pass a soft bulky stool than one that is hard and constipated.

In infants defaecation occurs as a reflex response to faeces in the rectum – voluntary control of the external anal sphincter is not achieved until the child is about 18 months. Reflex defaecation also occurs with sacral spinal cord lesions, after cerebrovascular accident or where the pudendal nerve is damaged.

Nursing Practice Application **Absorption of drugs**

The colon is able to absorb several types of drugs including corticosteroids, antimicrobials, analgesics and bronchodilators. The drugs may be administered as suppositories or a retention enema.

Nurses should ensure that the person and his or her family understand that the drug is absorbed through the colonic mucosa. Many people are anxious or confused about the rectal route. They often associate this with enemas used to evacuate the bowel contents and may find the concept that a drug given rectally can help their breathing rather difficult to understand. There is an obvious need for clear explanations and an education programme which deals with administration procedures, the time a preparation should be retained and drug storage.

Composition of faeces

Faeces contain mainly water with some epithelial cells, mucus, bacteria, non-starch polysaccharide (fibre) (Department of Health, 1991) such as undigested cellulose residue, electrolytes, stercobilin, which gives normal stools its brown colour, and various chemicals, which account for the characteristic odour.

Fibre and defaecation

Considerable evidence exists for increasing our daily intake of dietary fibre. A report from the National Advisory Committee on Nutrition Education (NACNE, 1983) recommends that fibre intake is built up to 30 g daily and that this increase comes mainly from wholegrain cereals (fibre may also be obtained from pulses, fruit and vegetables).

Faecal bulk and transit time through the large intestine depends on the amount of fibre present. Burkitt *et al.* (1972) found that stool weight varied greatly with fibre intake: in the UK the weight of stool per day was 39–223 g on a refined/processed diet and 71–488 g for vegetarians with a mixed diet; compared to this, rural Ugandans, having an unrefined diet, produced 178–980 g. The increased transit times associated with the low fibre Western diet leads to the formation of carcinogens (see Chapter 1) in the large intestine. There is a higher incidence of large *bowel* and *rectal carcinomas* in the West compared with developing countries. The risk of developing *haemorrhoids* may also be reduced by taking enough fibre to ensure that the stool is soft and passed without straining.

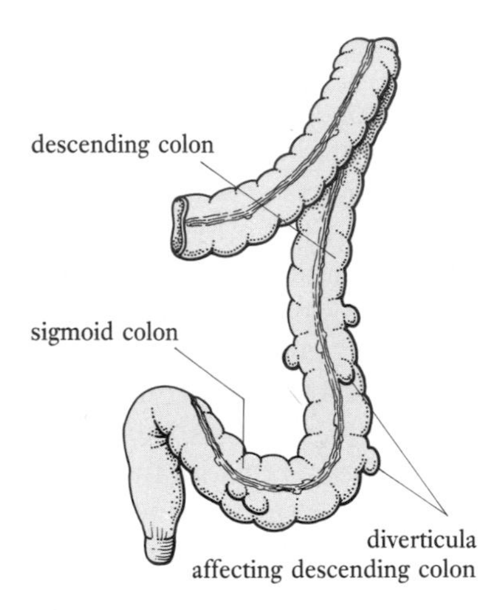

Figure 13.20. Diverticular disease.

Another common condition linked to fibre intake is *diverticular disease*. This occurs when *diverticula* or pouches (*Figure 13.20*) which form in the intestinal wall become inflamed (*diverticulitis*). The incidence of diverticular disease is higher in meat eaters than vegetarians, who generally have a high fibre intake. In Western countries the condition is more common than in developing countries where the diet contains more fibre and very little meat or processed food.

Faecal incontinence

The inability to control the anal sphincter, resulting in faecal soiling, is extremely distressing but less common than urinary incontinence (see Chapter 15).

Causes:

- Associated with faecal impaction/constipation. It is vital that this is diagnosed correctly and the bowel is evacuated.
- Drugs which cause diarrhoea, e.g. abuse of aperients, antibiotics.
- Rectal prolapse.
- Neurological cause, e.g. spinal cord lesion.
- Dietary indiscretions where anal sphincter control is not adequate to cope with the diarrhoea.

Management includes dealing with the cause if possible, discussion with continence advisor, establishment of 'good bowel habits', exercise and physiotherapy, adequate fluids and fibre and protective clothing/bedding only if essential.

Special Focus 'Normal bowel habit', diarrhoea and constipation

It is important to remember that 'normal bowel habit' will vary from person to person. The study of bowel function in hospital by Wright (1974) found that usual bowel habit prior to admission was 5–7 stools weekly. Earlier studies of bowel habit in two groups (factory employees and people visiting their family doctor) found that 99% of people fell within the range of three stools daily to three stools per week. Any departure from normal pattern should, however, be noted, as a change may be indicative of serious disease.

Diarrhoea is frequent loose stools which may, if severe or prolonged, lead to dehydration, hypokalaemia, metabolic acidosis and malabsorption of nutrients.

The causes of diarrhoea include: (i) dietary indiscretion such as eating too much fruit, change in diet; (ii) infection, e.g. food poisoning; (iii) inflammatory disease of the intestine, e.g. Crohn's disease; (iv) a food allergy or intolerance which causes malabsorption; (v) emotions, e.g. stress such as the morning of an examination; (vi) malignancy of the intestine, here the bowel habit may alternate between diarrhoea and constipation; (vii) drugs, e.g. antibiotics and iron; (viii) hyperthyroidism (see Chapter 8); (ix) following extensive small bowel resection.

NB It is essential to differentiate between true diarrhoea and the spurious leakage of liquid faeces associated with severe constipation/faecal impaction (see constipation).

Following a complete assessment the nursing care plan should include actions that provide privacy and commode/bedpans whenever required or accessibility to toilet facilities. Simple measures such as perianal hygiene, soft toilet tissue, soothing creams for perianal excoriation and clean linen do much to minimize discomfort. Embarrassment can be eased by providing a side room with toilet or, if this is not possible, ensuring good ventilation and air fresheners to dispel offensive odours. To avoid the possibility of cross-infection all faeces should be disposed of correctly and staff should wear disposable gloves and plastic aprons; hand hygiene for all concerned is essential for infection control. Monitoring the frequency of diarrhoea, fluid balance and skin condition are important components of what should be a continually updated assessment. In addition the colour, consistency and presence of blood or mucus in the stools should be noted. Nurses will also be involved with the collection of stool specimens for microscopy and culture, the management of oral or intravenous fluid replacement (see Chapter 2) and the administration of antidiarrhoeal drugs such as loperamide.

Constipation is the reverse of diarrhoea with hard, dry stools which are passed infrequently and with difficulty.

NB the longer the faeces stay in the colon the more water is reabsorbed (see page 336). Often, people feel that they are constipated if they do not defaecate daily. This is especially true in older people where 20% claim to be constipated but as many as 50% regularly take aperients (Gupta, 1980).

The causes of constipation include: (i) lack of fibre in the diet, change in diet; (ii) ignoring or the inability to respond to the urge to defaecate. This may be due to the problems of having to use a bedpan in hospital which include lack of privacy, physical difficulties and undue waiting. Wright (1974) found that 44% of people using bedpans /commodes became constipated compared with only 26% of people able to use the toilet. If a person in hospital postponed defaecation because toilet facilities were unavailable they may become constipated. Another disturbing finding of Wright's study was the lack of hand-washing facilities offered after defaecation; (iii) lack of mobility or pain on moving may also be a problem. The thought of having to climb stairs or go into a cold part of the house may cause the person to avoid defaecation; (iv) dehydration (see Chapter 2); (v) conditions which cause intense pain on defaecation, such as haemorrhoids or anal fissure, may lead to the person 'putting off' defaecation. This ill-judged action actually exacerbates the problem as the stool becomes even harder and more painful to pass; (vi) malignancy of the colon or rectum may present as alternating constipation and diarrhoea; (vii) drugs, e.g. iron, analgesics (such as codeine), anticholinergic drugs and some antidepressants; (viii) lack of exercise and immobility such as bed rest; (ix) depression; (x) hypothyroidism (see Chapter 8); (xi) hypokalaemia; (xii) hypercalcaemia; (xiii) hormones, e.g. progesterone during pregnancy.

At this point we might consider the problems experienced by 'Dorothy', page 340.

Person-centred Study Dorothy

Constipation is a common problem especially for older people. Since Dorothy, aged 88, moved in with her daughter she is not really 'going properly' and feels bloated. Dorothy mentions this during a routine visit to the practice nurse, who arranges for her to see the doctor. It is decided that the change in activity and diet are the most likely explanations but a rectal examination is performed to check for more sinister pathology.

Before leaving Dorothy has another chat with the nurse who suggests that she might consider the following actions: (i) increase fibre intake and visit her dentist if chewing is a problem; (ii) drink at least 2.5 litres of fluid daily; (iii) respond at once to the urge to defaecate; (iv) try to establish a routine for defaecation, e.g. after a hot drink on waking and wholewheat cereal for breakfast; (v) take some gentle exercise, e.g. walk the dog or potter in the garden.

If these measures are unsuccessful Dorothy is asked to make an appointment with the doctor, who may prescribe short-term aperients or decide that Dorothy's problem requires further investigation.

NB Dependence upon aperients may occur.

Nutrition - Metabolism and Utilization

Following our discussion of digestion and absorption it would seem sensible if we looked at some aspects of nutrition, metabolism and the uses the body makes of nutrients. By necessity this can only be a brief review and readers are directed to Further Reading, e.g. Dickerson and Booth (1985) and Marieb (1989). The emphasis here will be on carbohydrates, proteins and fats, with only brief mention of the other components of a '*healthy diet*'.

Introduction

The intake of the **essential nutrients** in the correct quantities is essential for health and, ultimately, life. We need a variety of nutrients which provide energy and produce the molecules needed for homeostasis and the materials for cell growth and repair. The nutrients required are: carbohydrates, proteins, fats, vitamins, minerals and water. As you already know we also need fibre (non-starch polysaccharide); although this is not absorbed as a nutrient it is important for health (see page 338).

A specific food usually contains several nutrients, e.g 'jacket' potatoes provide carbohydrate, protein, vitamins, water and fibre before you even consider the dressing. If we make a wide choice from the main food groups (wholegrain cereals and bread, fruit and vegetables, meat, fish and eggs or protein alternatives, dairy produce); it is very likely that dietary intake will contain all the essential nutrients. These essential nutrients are molecules which cannot be synthesized in the body – they must be provided by the diet. The body converts the essential nutrients into the many other molecules vital for health.

The metabolism and utilization of nutrients involves a multitude of interrelated, enzyme-facilitated biochemical reactions. These biochemical reactions may be:

- **Catabolic** – where fuel molecules are broken down within cells to produce chemical energy in the form of **adenosine triphosphate (ATP)** or heat energy. These reactions, in which glucose, fatty acids and amino acids are broken down, are together known as *cellular respiration.*
- **Anabolic** – where the ATP is used to drive the reactions which build up or synthesize other molecules.

Energy composition of food

Most of the food consumed is oxidized by the body to produce ATP. The energy value of a particular food is measured in the SI unit of kilojoules (kJ) or the widely used kilocalories (kcal) which are also called 'large Calories'.

4.2 kJ (approx) = 1 kcal or Calorie

NB 1 kcal is the amount of heat required to raise the temperature of 1 kg water by 1° C.

Table 13.1. Energy composition of nutrients.

Nutrient	*Energy yield from 1 g*
Carbohydrate	17 kJ or 4 kcal
Protein	17 kJ or 4 kcal
Fat	37 kJ or 9 kcal

NB Alcohol, which is not a nutrient, yields 29 kJ or 7 kcal/1g.

Our energy requirements vary according to gender, size, activity and health status, e.g. a raised body temperature increases the energy requirement. The **basal metabolic rate (BMR)** is the amount of energy needed by the body to drive the vital processes, e.g. respiration, when at complete rest but awake. BMR is calculated indirectly by measuring the amount of oxygen consumed in a given time (1 litre of oxygen used = 20 kJ of energy released) and is expressed as kJ or kcal per square metre body surface area per hour (kJ or kcal/m^2/h). Adult males require about 170 kJ or 40 kcal/m^2/h and females 155 kJ or 37kcal/m^2/h.

Example: A man weighing 67 kg, who is 172 cm tall, has a surface area of 1.8 m (calculated from a normogram). He will need 170 kJ x 1.8 = 306 kJ/h or 73 kcal/h. In 24 hours he needs approximately 7344 kJ or 1750 kcal to meet basal metabolic requirements.

If your energy consumption is greater than required the result is weight gain. Obesity is the most common nutritional disorder of developed countries, where it contributes significantly to ill health and premature death.

Carbohydrates

Carbohydrate molecules consist of carbon, hydrogen and oxygen. They are obtained mainly from plant sources with the exception of lactose (milk sugar) and glycogen. Carbohydrates (see pages 309-310 and 328) are divided into sugars, which may be monosaccharides (glucose, galactose and fructose) or disaccharides (maltose, lactose and sucrose) and the more complex polysaccharides (starch, cellulose and glycogen). Monosaccharides are formed from a single sugar unit, disaccharides consist of two linked monosaccharide units and polysaccharides are comprised of many monosaccharide units.

■	■–■	■–■–■–■–■–■–■–■–■–■
monosaccharide	disaccharide	polysaccharide

Sugars are obtained from fruit, jam, honey, sugar and milk. Polysaccharides are found in pulses, root vegetables, e.g. potato, cereals and products made from flour, e.g. bread, pasta, chapatis. During digestion the various carbohydrates are reduced to simple monosaccharides prior to absorption, but the polysaccharide cellulose remains undigested in humans to provide dietary fibre.

Monosaccharide metabolism and utilization

The monosaccharides absorbed in the intestine are delivered to the liver where the fructose and galactose are transformed to glucose. It is as glucose that the cells use carbohydrate. Although some cells can use fat as fuel, the erythrocytes and brain cells can only use glucose in the short term. It follows then that a fall in blood glucose levels can damage brain cells

Glucose in excess of immediate energy requirements is converted in the liver to the storage carbohydrate **glycogen**. This process, known as **glycogenesis**, requires the presence of insulin and provides around 100 g of carbohydrate which is stored in the liver. Skeletal muscles also store glycogen for muscle contraction during exercise. Once energy requirements have been met and glycogen stores replenished the glucose remaining is converted to fat. This fat is stored in the adipose tissue (see Chapter one).

Most glucose, however, is taken up by the cells under the influence of insulin (see Chapter 8) and oxidized to provide energy ATP (*Figure 13.21*). Initially glucose is broken down (**glycolysis**) by a series of mitochondrial enzymes to form **pyruvic acid**. The pyruvic acid is broken down **aerobically** (in the presence of oxygen) to produce **acetyl co-enzyme A (acetyl CoA)** which enters the **Krebs' citric acid cycle** (see *Figure 13.24*). The Krebs' cycle is a series of enzyme-controlled reactions where the acetyl CoA undergoes changes which produce ATP, water and carbon dioxide. If the pyruvic acid is broken down **anaerobically** (without oxygen), the waste product **lactic acid** is produced (see Chapter 12). This occurs in strenuous exercise but can also result from any hypoxic state.

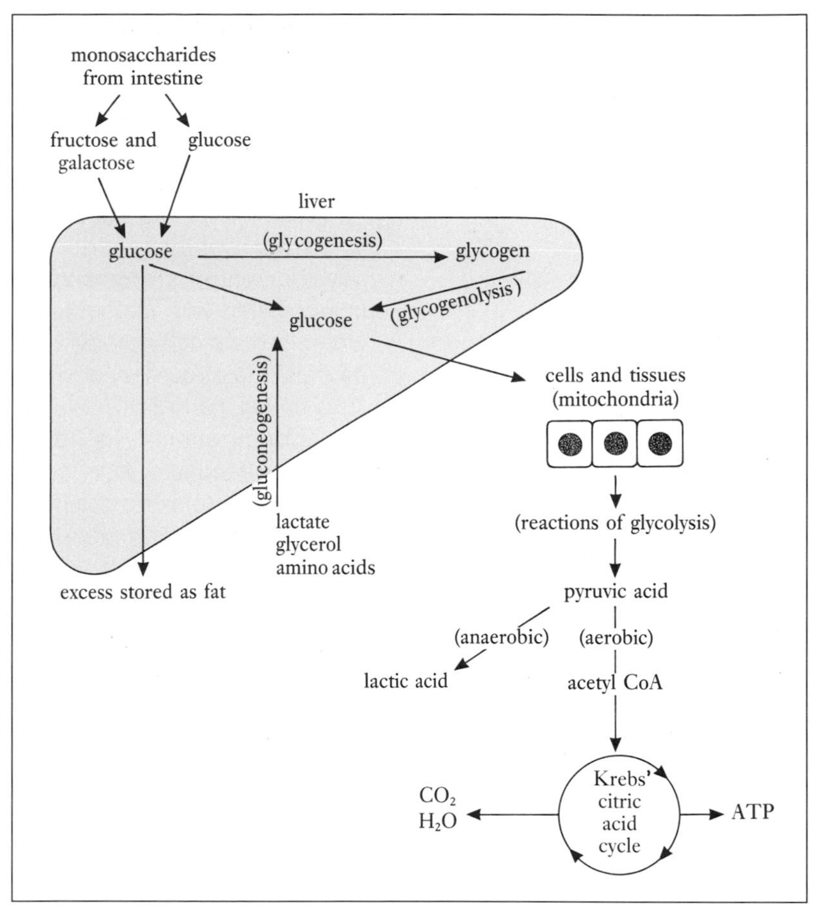

Figure 13.21. Metabolism of monosaccharides.

When the amount of glucose in the blood is reduced, the body must correct the deficit and maintain homeostasis. The stored liver glycogen is reconverted to glucose by **glycogenolysis**, a process regulated by glucagon and adrenaline (see Chapter 8). Where the glycogen stores are low the body is able to produce glucose from non-carbohydrate sources (*amino acids, glycerol* and *lactate*) by a process known as **gluconeogenesis**.

NB Gluconeogenesis also occurs when glucose, although in plentiful supply, is unavailable for use, e.g. in diabetes mellitus where a lack of insulin prevents the transfer of glucose into cells.

Protein

Proteins (see pages 308-309 and 328) contain carbon, hydrogen, oxygen, nitrogen and sometimes sulphur and phosphorous. During digestion proteins are broken down into the 20 amino acids, which, in different combinations form every protein.

Some amino acids are *essential* and/or *indispensible* (must be included in the diet because they cannot be synthesized)(Department of Health, 1991): *isoleucine, leucine, lysine, methionine, phenylalanine, threonine, tryptophan* and *valine*; during childhood *arginine* and *histidine* are also considered essential. Others are *non-essential*: *alanine, asparagine, aspartate (aspartic acid), cysteine, glutamate (glutamic acid), glutamine, glycine, proline, serine* and *tyrosine,* and can be synthesized by the body.

About 50 g (45–55 g) of protein is required each day by adults, to maintain a **positive nitrogen balance** where protein is being used for repair rather than to produce energy. The reverse, where protein is primarily being used to produce energy, will result in the development of a **negative nitrogen balance**. A negative balance is associated with starvation or following severe physiological stress where protein catabolism (breakdown) increases, e.g. burns, multiple injuries, major surgery and systemic infections (see *Nursing Practice Application*, page 345).

Although quantity is important it is vital that the quality of the protein is adequate. Most animal source protein, e.g. meat, eggs, fish and milk contain the essential/indispensible amino acids but plant proteins are not complete. Legumes such as beans are deficient in methionine and cereals lack lysine. This can be compensated for by mixing plant proteins, e.g baked beans on toast, lentil curry with rice. Protein is also obtained from nuts and vegetables such as potatoes.

Amino acid metabolism and utilization

Amino acids are not stored by the body. They circulate in a general pool (free amino acids available for utilization) from where they are 'selected' when required for cell division, growth or repair and to synthesize the functional proteins, e.g. enzymes, some hormones, plasma proteins and haemoglobin (*Figure 13.22*). As already mentioned, proteins can be oxidized to provide energy but this is rather wasteful and does not utilize their full potential.

Following absorption amino acids are transported to the liver where the non-essential ones are synthesized by a process called **transamination**. Various enzymes called *aminotransferases* (also called *transaminases*) move amine groups (NH_2) to form different amino acids.

When the body is deficient of stored energy, or amino acids are in excess of anabolic needs, they can be oxidized for energy, converted to glucose by gluconeogenesis or converted to fat.

Excess amino acids are **deaminated** in the liver, the process of deamination removes the nitrogenous amine group (NH_2) which is converted to ammonia (NH_3) and combined with carbon dioxide to form **urea** by the reactions of **Krebs' urea cycle**. These reactions convert highly toxic ammonia to less toxic urea, which travels in the blood to the kidneys to be excreted in the urine (see Chapters 14 and 15). Deamination reactions may be coupled chemically with those involving transamination.

Fats

Fats (see pages 310 and 331) consist of carbon, hydrogen and oxygen, but in different proportions to carbohydrates. Our consumption of fat is mostly as triglycerides formed from glycerol and three fatty acids but we also ingest other fatty substances, e.g.

Nursing Practice Application Meals in hospital and protein energy malnutrition (PEM)

Situations where the hospital patient receives inadequate protein and energy are unfortunately common and malnutrition may affect up to 50% of adults (Goodinson, 1987a). PEM may contribute to delayed recovery and complications such as poor wound healing, infection and depression.

Obviously any physical, psychological or social factor that affected intake or absorption of food prior to admission must be considered, e.g. anorexia, immobility or low income, but the problem may well start within the hospital setting. In recent years the serving and supervision of meals have been regarded by many as a non-nursing duty. This has led to considerable confusion regarding who is responsible for ensuring that diet is eaten. As with most aspects of care the team approach is essential with the patient/family, nurse, doctor, catering staff, dietitian and biochemist all involved in preventing and recognizing malnutrition (see nutritional assessment, page 351). The speech therapist may advise about swallowing and airway maintenance in certain neurological conditions. Where they exist the nutrition team, which includes a specialist nurse, should be consulted.

Over a century ago Florence Nightingale considered that nurses had a central role: 'If the nurse is an intelligent being, and not a mere carrier of diets to and from the patient, let her exercise her intelligence in these things' (Nightingale, 1859). It is important to ensure that at mealtimes the environment is quiet and free from activities such as treatments or ward rounds. Nurses should help the patient to select appropriate items from the menu, adopt suitable positions, provide any special utensils or skilled help with feeding if required and obviously assess what has been eaten.

The nutritional education of nursing and medical staff is a much neglected area with a low priority in basic courses. Dickerson (1986) states 'Nurses are not impressed with the importance of nutrition as part of total care'.

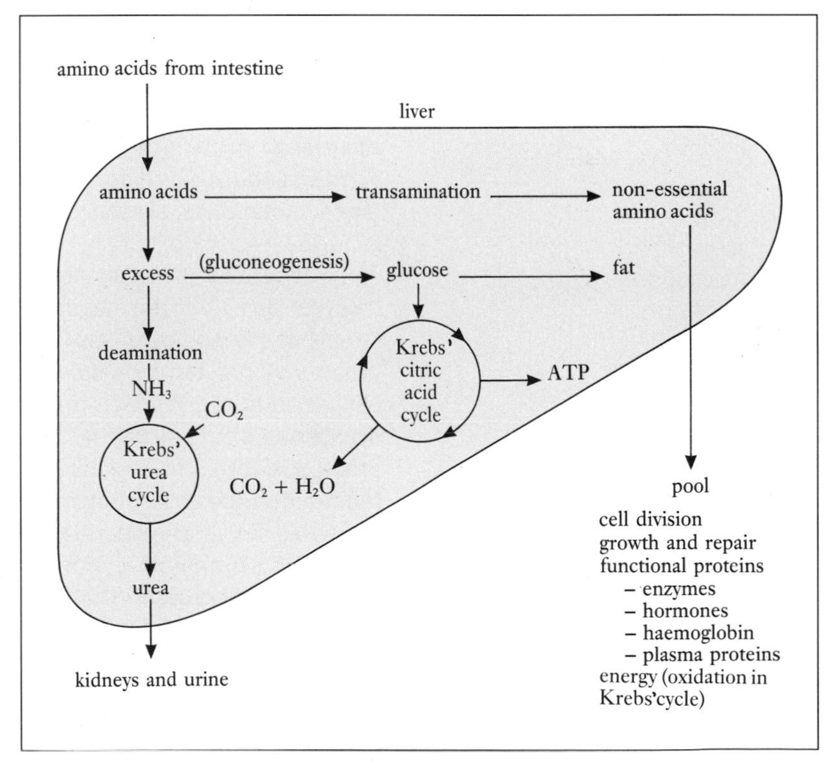

Figure 13.22. Metabolism of amino acids.

cholesterol found in offal and egg yolk.

When fat is oxidized it yields over twice the amount of energy as carbohydrate or protein. It is therefore a valuable fuel when energy requirements are high, but in sedentary individuals excess intake results in obesity.

Nursing Practice Application Phenylketonuria (PKU)

Phenylketonuria is an inborn error of metabolism where the enzyme (*phenylalanine hydoxylase*) required to convert phenylalanine to tyrosine is absent. Untreated this leads to an accumulation of phenylalanine and its toxic metabolites which cause brain damage and learning difficulties.

Luckily a routine blood test performed on all babies during the first days of life can detect the condition, which is managed with a diet low in phenylalanine. As phenylalanine is an essential amino acid it must be included in the diet but amounts given are carefully controlled by regular blood tests to measure the level of toxic metabolites. The phenylalanine-restricted diet is continued throughout childhood with its growth periods and, in the case of female sufferers, is recommenced if they decide to have a family, to protect the fetus. Affected individuals may have unusually pale colouring with fair hair and blue eyes because the tyrosine required for pigment production (melanin) is not synthesized.

A great deal is heard about whether a fat is saturated or not, this is simply a way of describing its particular chemical bond structure. Saturated fats (all carbon–carbon bonds are single) are generally found in animal products, e.g. fat around meat, lard, dairy products and are solid at room temperature. Unsaturated fats may be monounsaturated (with one double bond) or polyunsaturated (with two or more double bonds). These are found in nuts, seeds, some fish oils (omega-3 fatty acids) and most vegetable oils, e.g. sunflower oil, and are mostly liquid at room temperature. Three unsaturated fatty acids are essential for health – *linoleic acid, linolenic acid* and *arachidonic acid.* These **essential fatty acids (EFAs)** form the basis of structural lipids. Prostaglandins which form an important group of control molecules (see Chapter 8) are synthesized from arachidonic acid.

Fat metabolism and utilization

Fatty acids and glycerol absorbed in the intestine are reformed into triglycerides. Any not needed for immediate energy production are stored in the adipose tissue of the fat depots, e.g. under the skin. A process stimulated by insulin, by which glucose and amino acids are converted to triglycerides prior to storage in adipose tissue, is termed **lipogenesis**. The stored fat provides insulation, an energy source and protection for some organs, e.g. kidneys. Fat taken in the diet is also required for the absorption of fat-soluble vitamins (see page 332).

Stored fat as triglyceride is released from the adipose tissue by various hormones, e.g. cortisol, in a process called **lipolysis**. The triglycerides travel in the blood to the liver where they are again converted into fatty acids and glycerol which can be used by the cells for energy.

Glycerol is either transformed into pyruvic acid and hence to acetyl CoA which enters the Krebs' citric acid cycle to produce ATP, or it can be converted into glycogen and stored.

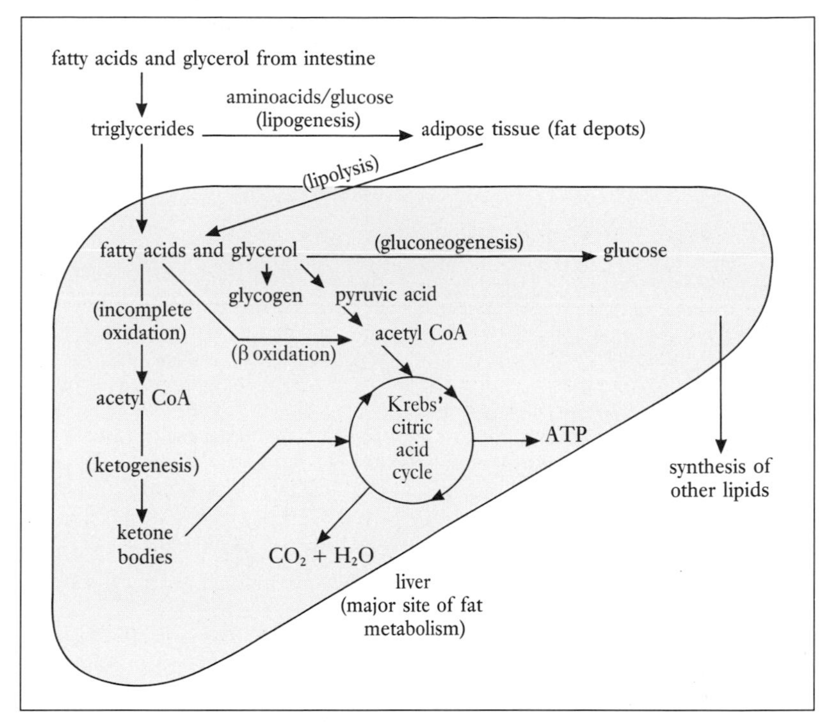

Figure 13.23. Metabolism of fats.

Fatty acids are transformed to acetyl CoA by *β-oxidation*, a process which requires oxygen and glucose. The resultant acetyl CoA is then oxidized in the Krebs' citric acid cycle with the production of ATP, carbon dioxide and water. When fatty acids are broken down in the absence of glucose the incomplete oxidation produces too much acetyl CoA which forms ketone bodies *(acetoacetic acid, β-hydroxybutyric acid* and *acetone*), a process called **ketogenesis**.

Small amounts of ketone bodies can be used as a fuel molecule in Krebs' citric acid cycle but an excess cause a life-threatening condition known as **ketoacidosis**. This occurs when glucose is either in short supply during starvation or unavailable in uncontrolled diabetes mellitus (see Chapter 8). High levels of the acidic ketone bodies leads to metabolic acidosis and dehydration (see *Figure 8.20*).

Fats, being very versatile, are also used as a material for gluconeogenesis; the liver cells use glycerol to produce glucose. *Figure 13.23* illustrates the metabolism of fats.

Summary - the final common pathway

The metabolites of the energy-producing nutrients – glucose, fatty acids, glycerol and amino acids – are all eventually oxidized to produce energy (ATP). The ATP is generated by the reactions of Krebs' citric acid cycle and **oxidative phosphorylation** (**electron transfer**) which operate within the mitochondria (*Figure 13.24*).

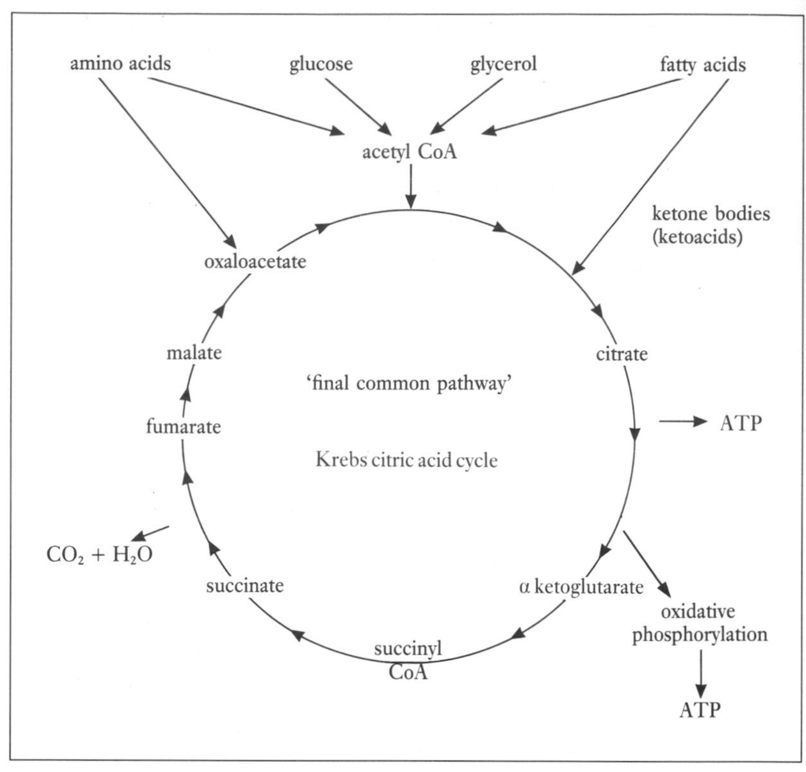

Figure 13.24. Final common pathway (simplified). *NB* Only some of the intermediates of the citric acid cycle are included.

Vitamins

Vitamins are organic molecules and are required in minute quantities for all metabolic processes. They do not produce energy but are vital to its release from the fuel molecules and the regulation of anabolic processes. Many vitamins act as **coenzymes** where they function with enzymes to faciltate reactions involving the metabolism and utilization of carbohydrates, proteins and fats. An example of this is vitamin B_1 thiamin, which is required for the oxidation of pyruvic acid to acetyl CoA in carbohydrate metabolism. Without the intake and absorption (see page 332) of the correct vitamins the body is unable to use the other nutrients.

Most vitamins must be ingested, but vitamins D, K and some of the B complex can be synthesized in the body. Vitamin D is produced by the action of ultraviolet light upon *7-dehydrocholesterol*, a cholesterol-based substance, found in the skin. Vitamin D may be deficient where an inadequate intake is combined with limited exposure to sunlight. This situation may lead to the development of *rickets* or *osteomalacia* (see Chapter 16). People at risk usually belong to those ethnic groups whose culture limits exposure to sunlight, or people in lower socioeconomic groups.

The commensal bacteria of the intestine produce vitamin K which, in common with the other fat-soluble vitamins, requires the presence of bile for its absorption (see page 332). B-complex vitamins are also produced by the gut bacteria but in fairly insignificant amounts.

The body is also able to convert certain **provitamins** such as the yellow colouring *carotene* found in carrots and dark green vegetables into active vitamin A.

Vitamins can be classified into two main groups:

- Water-soluble vitamins:
  Vitamin C – *ascorbic acid.*
  Vitamin B complex – B_1 *thiamin,* B_2 *riboflavin, niacin (nicotinic acid* and *nicotinamide),* B_6 *pyridoxine, biotin, folate (folic acid),* B_{12} *cobalamins* and *pantothentic acid.*
- Fat-soluble vitamins:
  Vitamin A – *retinol* which is found in animal products but is also synthesized from the provitamin carotene.
  Vitamin D – *cholecalciferol*
  Vitamin E – *tocopherols*
  Vitamin K – *phylloquinones*

Probably the most efficient way to take vitamins is as part of a 'healthy' balanced diet which includes items selected from all the food groups. It is important to remember that vitamin content of some foods may vary seasonally, e.g. old potatoes have less vitamin C than new, and that cooking and exposure to light can decrease vitamin levels. The practice of adding sodium hydrogen carbonate to greens which are then boiled destroys much of the vitamin C. Milk left on the doorstep in the sunlight will lose riboflavin as well as turn sour; it makes good sense to store it in a cool place until needed.

The habit of supplementing vitamin intake with 'pills and potions' is not necessary for healthy individuals taking a balanced diet. Water-soluble vitamins cannot be stored and are wastfully excreted in the urine. If fat-soluble vitamins are consumed in excess they accumulate in the body to cause *hypervitaminosis* with toxic effects. For certain special groups, which are discussed in *Table 13.2* it may be beneficial to consume vitamin supplements.

The Department of Health (1991) provides information regarding dietary reference values for the UK and further details of vitamins, their sources, uses, excesses and deficiencies can be found in Further Reading, Dickerson and Booth (1985) and Marieb (1989).

Minerals

Minerals (inorganic compounds) are vital to homeostasis and have a role in all metabolic processes. The major elements required by the body are: calcium, chlorine, iodine, iron, magnesium, phosphorus, potassium, sodium and sulphur. Other minerals, needed only in minute quantities (*trace elements*) include: chromium, cobalt, copper, fluorine, manganese, molybdenum, selenium, and zinc. Minerals are needed for:

- Utilization of nutrients.
- Enzymes, e.g. zinc required for carboxypeptidase (see page 325).
- Hormones, e.g. thyroid hormones need iodine (see Chapter 8).

Table 13.2 . Nutritional needs of selected special groups.

Special group	***Additional nutritional needs***
Infants and children	The requirement for energy and other nutrients, e.g. protein, calcium, is high in relation to body area. Extra vitamin D[a] until the child reaches 2 years of age. *NB* Children, adolescents and women living in communities where cases of rickets are likely to occur should have vitamin D supplements.
Pregnancy	Energy, protein, iron[a], folate[b], vitamins B, C, A and D, calcium, zinc and iodine. *NB* High levels of retinol may be associated with birth defects. Excessive vitamin A intake, in the form of supplements or liver, should be avoided as a precautionary measure in pregnant women and those intending to become pregnant.
Lactation	Energy, protein, calcium, folate, vitamins C, D, and A and extra fluid needed.
Older people	Reduction in energy requirements if they become less active. Important that diet provides adequate iron, calcium, and vitamins, especially B-complex, C and D. With inactivity may need more fibre.

[a] A supplement may be recommended.
[b] Results from a recent study Wald et al. (MRC, 1991) confirm that folate supplements should be taken by women planning to conceive and continued during early pregnancy where a previous pregnancy was affected by a neural tube defect, e.g. spina bifida. The same report also states that all women of childbearing age should have a diet containing adequate folic acid.

- Formation of other functional proteins, e.g. iron and haemoglobin (see Chapter 9).
- Maintenance of fluid compartments, e.g. sodium, chlorine and potassium (see Chapter 2).
- Nerve and muscle function, e.g. sodium, potassium and calcium (see Chapters 3 and 17).
- Structural strength, e.g. calcium, magnesium and phosphorus in bone (see Chapter 16).

Further information is given in Chapter 2 (*Table 2.1*). Again the dietary reference values are found in Department of Health (1991) and a detailed account in Marieb (1989).

Water

In health, water forms about 60% of the body mass (see Chapter 2). It is required for every metabolic process and is vital to all homeostatic mechanisms. Our requirement for water intake varies according to activity, climate and losses, e.g. in urine and faeces, but generally an adult needs to consume around 2.5 litres daily (remember that most foods also contain water).

Fibre/non-starch polysaccharide

Fibre, which forms an important component of a healthy diet, consists of indigestible plant polysaccharides such as cellulose. It adds bulk to the faeces, decreases transit times for food residues and gives a feeling of fullness without excessive energy consumption (see page 338). The recommended daily intake of 30 g (NACNE, 1983) is easily obtained from wholegrain cereals, pulses, fruit and unpeeled vegetables. There are some drawbacks of a high-fibre diet; it can cause flatulence and, more seriously, inhibit the absorption of minerals such as calcium (see pages 332-333).

Nursing Practice Application Nutritional assessment & support

Nutritional assessment

Nutritional status is important to many aspects of practice, e.g. wound care and healing, mobility and preventing pressure sores. It is essential to recognize those individuals who may be malnourished and assess nutritional status prior to identifying needs and planning care. Nutritional assessment is rather more than recording weight or asking about food preferences and involves the whole care team (see PEM page 345) plus the biochemist. In a series of articles, (Goodinson 1987 a,b,c,d) describes nutritional assessment and identifies the following methods: (i) biochemical tests, e.g. measuring serum proteins, muscle breakdown products; (ii) anthropometric, e.g. weight, skin fold thickness; (iii) subjective methods, e.g. dietary history, physical examination.

NB To ensure accuracy no method should be used in isolation as factors other than nutrition may influence the data obtained. Serial tests over a period of time are required to monitor changes which denote improvement or deterioration.

Nutritional support

When the intake of nutrients is inadequate or impossible the individual will need some form of nutritional support. This may take the form perhaps of extra energy and protein or it may be necessary to give all the nutrients as *total parenteral nutrition (TPN)*.

As already mentioned the people involved in maintaining nutrition include the patient/family, nurses, doctors, dietitian and catering staff, but the provision of extra support will also involve the pharmacist and biochemist.

Oral supplements/sip feeding should be used where possible because the oral route has many advantages. It maintains independence and normality as the person is still taking nourishment via the 'normal' route. The physiological mechanisms for oral cleansing still operate although their effectiveness may be reduced by other factors (see mouth care, page 315).

Enteral tube feeding involves the use of the gastrointestinal tract. It is usually achieved via a nasogastric tube, but tubes passed directly through the abdominal wall into the stomach (*gastrostomy*) or jejunum (*jejunostomy*) are sometimes indicated. The use of a fine-bore enteral tube is usually well tolerated and does not preclude oral nourishment so the two methods may be used together.

Nasogastric feeding is used in situations where oral intake is inadequate but the gut is functioning, e.g. altered consciousness, facial trauma/burns, *hypercatabolic* states. The complete nutritional needs including that for water can be met via this route. Nasogastric feeding is cheaper and involves fewer hazards than parenteral feeding (Taylor, 1989). The use of commercially prepared feeds reduces the risk of bacterial contamination.

Parenteral feeding via a central vein may be used to provide total parenteral nutrition (TPN) or to supplement oral and nasogastric feeding. Use of a central vein is essential for prolonged access and when hypertonic solutions are administered. Phlebitis (inflammation of a vein) results if hypertonic solutions are administered into a peripheral vein.

The intravenous route should only be chosen in the absence of gastrointestinal function or when it is impossible to provide sufficient nourishment by other methods. The indications for TPN include: after gastrointestinal surgery when peristalsis is absent (*paralytic ileus*), extreme hypercatabolic states, severe malabsorption or loss through gastrointestinal *fistulae*. TPN has serious disadvantages that include: problems with the insertion of the line, e.g. pneumothorax, infection, metabolic disturbances such as hyperglycaemia, and high costs.

Summary/Check List

Introduction – Early development. General structure of digestive tract. Nerve supply. Peritoneum – peritonitis.
Basic structure of organic compounds – Proteins. Carbohydrates. Fats.
Alimentary tract – Mouth. Tongue. Palate, *Nursing Practice Application* – cleft palate/lip. Salivary glands – mumps, composition of saliva, regulation of salivation, *Nursing Practice Application* – oral hygiene. Teeth – dentition, dental caries/periodontal disease. Pharynx, swallowing. Oesophagus – peristalsis. Stomach structure. Gastric secretions/functions – control of secretion and activity. Gastric motility and emptying – gastric absorption, *Nursing Practice Application* – vomiting, *Special Focus* – peptic ulceration, *Person-centred Study* – Ken. Small intestine – structure. Pancreas and pancreatic enzymes. Bile and Digestion – functions of bile. Intestinal juice and enzymes. Summary – chemical digestion. Absorption in the small intestine – amino acids, monosaccharides, fats, vitamins, minerals, water, *Nursing Practice Application* – malabsorption. Large intestine – caecum/appendix, appendicitis, colon, rectum and anal canal. Functions of large intestine – absorption, *Nursing Practice Application* – absorption of drugs, synthesis, storage, elimination (defaecation). Composition of faeces. Fibre and defaecation, faecal incontinence, *Special Focus* – normal bowel habit, diarrhoea, constipation, *Person-centred Study* – Dorothy.
Nutrition, metabolism and utilization – Introduction. Energy composition of food. Carbohydrates – metabolism and utilization of monosaccharides. Protein, *Nursing Practice Application* – PEM and meals in hospital, metabolism and utilization of amino acids, *Nursing Practice Application* – phenylketonuria. Fats – metabolism and utilization of fatty acids/glycerol. Final common pathway. Briefly – vitamins, minerals, water, fibre. *Nursing Practice Application* – nutritional assessment and support. Needs of some special groups.

?

Self Test

1 Put the following in their correct order:
(a) stomach
(b) jejunum
(c) oesophagus
(d) pharynx
(e) ileum
(f) rectum
(g) duodenum
(h) colon
(i) mouth
(j) anus.

2 Which of these statements are true?
(a) Enzymes in saliva are important in the digestion of fat.
(b) 1000 – 1500 ml of saliva is produced daily.
(c) The first dentition consists of 20 teeth.
(d) Periodontal disease is common in adults over 35.

3 What does gastric juice contain and how is its secretion controlled?

4 Explain how vomiting may cause:
(a) metabolic alkalosis
(b) metabolic acidosis.

5 Complete the following:
(a) Pancreatic secretion is stimulated by the regulatory peptides ________________ and ________.
(b) Trypsinogen is activated by _____________.
(c) Bile ___________ fat globules prior to their chemical breakdown by _______.

6 What are the products of protein, fat and carbohydrate digestion, and how are they absorbed?

7 Briefly describe the following:
(a) gastro-colic reflex, (b) mass movements, (c) importance of dietary fibre, (d) defaecation reflex.

8 Match the pairs correctly:

(a) calcium	(b) monosaccharide
(c) essential fatty acid	(d) carotene
(e) phenylalanine	(f) mineral
(g) fructose	(h) linoleic acid
(i) vitamin A	(j) amino acid

9 What is produced by the oxidation of fuel molecules?

10 Which of the following statements are true:
(a) Amino acids are a major energy source.
(b) Pyruvic acid is converted aerobically into acetyl Co A.
(c) Ketone bodies are produced by the incomplete oxidation of fatty acids.
(d) Fat is stored as triglycerides.

Answers
1. i, d, c, a, g, b, e, h, f, j. 2. b, c, d. 3. see pages 319-321. 4. See page 323. 5. a. cholecystokinin, secretin. b. enterokinase. c. emulsifies, lipases. 6. See pages 329-331. 7. See pages 335,337,338. 8. a – f, b – g, c – h, d – i, e – j. 9. ATP, carbon dioxide and water. 10. b, c, d.

References

Burkitt, D.P., Walker, A.R.P. and Painter, N.S. (1972) Effect of dietary fibre on stools and transit times and its role in the causation of disease. *Lancet*, **ii**, 1408–1412.

Department of Health (1991) *Dietary Reference Values for Food Energy and Nutrients for the United Kingdom*. Report on Health and Social Subjects, no. 41. Report of the Panel on Dietary Reference Values of the Committee on Medical Aspects of Food Policy. London: HMSO.

DHSS (1969) *The Fluoridation Studies in the United Kingdom and Results Achieved After Eleven Years*. Reports on Public Health and Medical Subjects, no.122. London: HMSO.

Dickerson, J.W.T. (1986) Hospital induced malnutrition: a cause for concern. *The Professional Nurse*, **1** (11), 293–296.

Eburn, E. (1989) Choosing the right antiemetic. *Nursing Times*, **85** (24), 36–38.

Goodinson, S.M. (1987a) Assessment of nutritional status. *The Professional Nurse*, **2** (11), 367–369.

Goodinson, S.M. (1987b) Anthropometric assessment of nutritional status. *The Professional Nurse,* **2** (12), 388–393.

Goodinson, S.M. (1987c) Biochemical assessment of nutritional status. *The Professional Nurse,* **3** (1), 8–12.

Goodinson, S.M. (1987d) Assessing nutritional status: Subjective methods. *The Professional Nurse,* **3** (2), 48–51.

Gupta, K. (1980) Constipation. *Geriatric Medicine,* **10** (12), 45.

Howarth, H. (1977) Mouth care procedures for the very ill. *Nursing Times,* **73** (10), 354–355.

Macleod, J., Edwards, C., and Bouchier, I., Eds. (1987) *Davidson's Principles and Practice of Medicine,* 15th edn. Edinburgh: Churchill Livingstone.

MRC Vitamin Study Research Group, Wald, N. et al. (1991) Prevention of neural tube defects: results of the Medical Research Council Vitamin Study. *Lancet,* **ii**, 131–137.

National Advisory Committee on Nutrition Education (NACNE) (1983) *Proposals for Nutritional Guidelines for Health Education in Britain.* London: Health Education Council.

Nightingale, F. (1859) *Notes on Nursing.* London: Duckworth 1970.

Taylor, S.J. (1989) A guide to enteral feeding. *The Professional Nurse,* **4** (4), 195–200.

Wright, L. (1974) *Bowel Function in Hospital Patients.* Royal College of Nursing Research Project, Series 1, Number 4. London: Royal College of Nursing.

FURTHER READING

Dickerson, J.W.T and Booth, E.M. (1985) *Clinical Nutrition for Nurses, Dietitians and Other Health Care Professionals.* London: Faber and Faber.

Dyer, S. (1988) The development of stoma care. *Professional Nurse,* **3** (7), 226–230.

Finnegan, S. and Oldfield, K. (1989) When eating is impossible: TPN in maintaining nutritional status. *Professional Nurse,* **4** (6), 271–275.

Hamilton–Smith, S. (1972) *Nil by Mouth?* London: Royal College of Nursing.

Heading, C. (1987) Nursing assessment and management of constipation. *Nursing,* **3** (21), 778–780.

Marieb, E.N. (1989) *Human Anatomy and Physiology.* Redwood City California: Benjamin/Cummings, pp.811–817.

Shearman, D.J.C. and Finlayson, N.D.C. (1987) *Diseases of the Gastro-intestinal Tract and Liver,* 2nd edn. Edinburgh: Churchill Livingstone.

Truswell, A.S. (1986) *ABC of Nutrition.* London: British Medical Association.

Wilson-Barnett, J. and Batehup, L. (1988) *Patient Problems: A Research Base for Nursing Care* (Chapter 7). London: Scutari Press.

CHAPTER FOURTEEN

LIVER AND BILIARY TRACT

OVERVIEW

- Structure and functions of the liver and biliary tract.

LEARNING OUTCOMES

After studying Chapter 14, you should be able to:

- Describe the gross structure and position of the liver.
- Describe blood flow through the liver.
- Describe the histology of the liver lobules.
- Discuss the formation and secretion of bile.
- Describe the biliary tract and explain how bile reaches the duodenum.
- Discuss the types and causes of jaundice.
- Outline the synthetic functions of the liver.
- Discuss the importance of hepatic detoxification processes.
- Discuss the effects of excess alcohol on healthy liver function.
- Outline the role of the liver as a storage organ.
- Briefly describe the metabolic role of the liver.
- Outline ways in which liver function is investigated for pathological processes.
- Discuss the consequences of failing liver function.

KEY WORDS

Biliary tract – series of ducts which transport bile from liver to gallbladder and duodenum.
Bilirubin/biliverdin – the bile pigments derived from haem.
Cholelithiasis – formation of gallstones.
Cholestasis – an obstruction to the flow of bile. Unrelieved it leads to **jaundice.**
Conjugation – joining together. **Bilirubin** is conjugated with glucuronic acid within the liver cells.
Glycogenesis – a process by which glucose, in excess of immediate needs, is converted to glycogen in the liver. Glycogen is a storage carbohydrate.
Glycogenolysis – the process by which liver glycogen is converted to glucose to meet body needs and restore homeostasis.
Glycolysis – series of reactions where glucose is broken down to form pyruvic acid.
Gluconeogenesis – the formation of glucose from non-carbohydrate sources, e.g. lactate, glycerol and amino acids. Used when glycogen stores are low.
Hepatic – pertaining to the liver.
Hepatocytes – the parenchymal or functional cells of the liver.
Hepatomegaly – enlarged liver.
Hyperbilirubinaemia – elevated levels of **bilirubin** in the blood.
Jaundice – yellow discoloration of the skin, sclera and mucous membranes which is due to an increased serum **bilirubin.**
Ketogenesis – the formation of ketone bodies from the acetyl CoA formed from the incomplete oxidation of fats.
Lipogenesis – the deposition of excess triglycerides (formed from non-lipid sources) in adipose tissue.
Lipolysis – the release of stored triglycerides from adipose tissue for use as energy.

INTRODUCTION

The liver, with its many essential functions, is vital to life. Some of these functions have already been mentioned – erythrocyte breakdown (Chapter 9), role of bile in the digestion and absorption of fats (Chapter 13) and the metabolism of nutrients (Chapter 13). The liver also synthesizes proteins, stores vitamins and minerals, and is involved in the detoxification of many of the molecules produced by or taken into the body. With all these functions occurring, the high metabolic rate of the liver is responsible for the production of a considerable amount of heat.

Luckily for us the liver has huge functional reserves, which is illustrated by the fact that people with serious liver disease may exhibit no ill effects until considerable damage has occurred. Liver cells can regenerate if damage is not too severe, e.g. when part of the liver is removed after trauma or in transplant surgery of the liver when liver tissue from a living donor is used (extremely unusual). What the liver cannot cope with is continuous damage from a toxic agent, e.g. alcohol. In this situation liver cells are destroyed and fibrosis occurs.

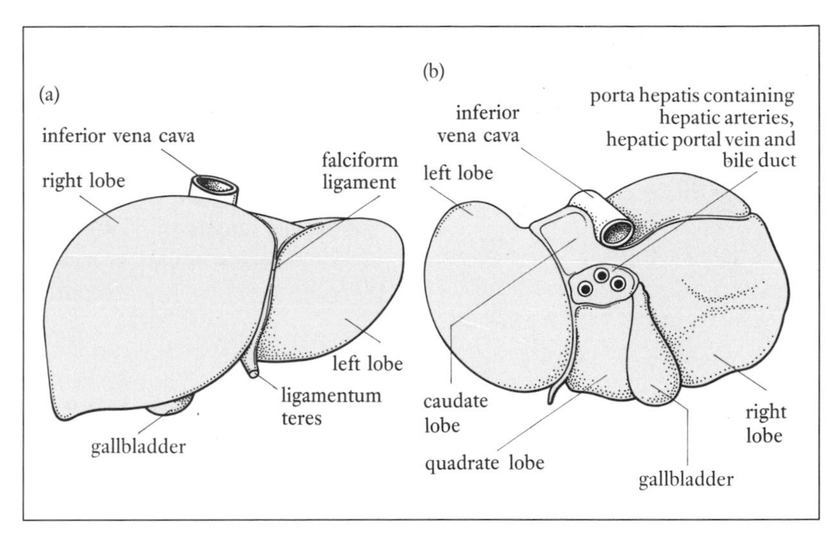

Figure 14.1. The liver. (a) Anterior surface; (b) posterior surface.

The Liver and Biliary Tract

Early development

The liver, gallbladder and bile ducts, in common with the other accessory digestive structures, develop as 'offshoots' from the primitive embryonic gut. The liver, which develops very rapidly during fetal life, is proportionally very large in the newborn infant.

Gross structure of the liver

The liver, which weighs 1.2–1.5 kg, is the largest gland in the body. Located in the right upper abdomen, it fills most of the right hypochondriac region and extends into the epigastric and left hypochondriac regions (see *Figure 1.23*). The smooth superior and anterior surfaces of the liver are situated under the diaphragm and are well protected by the rib cage. The liver cannot normally be felt below the costal margin. The liver displaces the right kidney downwards.

The liver consists of four lobes: *right*, *left* and two small lobes, the *caudate* and *quadrate*, which are situated on its irregular posterior surface (*Figure 14.1*). The gallbladder can be found in a depression on this posterior surface. The liver is enclosed within a thin connective tissue capsule and partially covered by peritoneum. It is attached to the diaphragm and abdominal wall by ligaments formed from folds of peritoneum, e.g. **falciform ligament**, which also divides the right and left lobes. Within the falciform ligament is the fibrous remnant of the fetal umbilical vein (see Chapter 20) known as the **ligamentum teres**.

The **porta hepatis (portal fissure)** is the area on the posterior surface where blood vessels (hepatic artery and hepatic portal vein), nerves, lymphatic vessels and bile ducts enter and leave the liver.

Blood supply of the liver

As you will remember, the liver has a special arrangement of blood vessels but actual blood flow through the liver will be covered in more detail when we consider the liver lobule (see page 358)

To meet the needs of its metabolic role the liver receives an abundant supply of oxygenated blood from the hepatic artery. In addition, venous blood from the digestive organs is conveyed to the liver via the hepatic portal vein (see Chapters 10 and 13). The advantages of this arrangement are obvious when you consider the role of the liver in the metabolism and utilization of nutrients. All the venous blood eventually leaves the liver via three main hepatic veins.

Hepatic portal vein → in → LIVER

Hepatic artery → in → LIVER → out → Hepatic veins

Liver blood flow is autoregulated by vascular sphincters which control the amount of blood entering via the hepatic artery. This allows for variations in flow via the hepatic portal vein and keeps total **hepatic** (pertaining to the liver) blood flow fairly constant.

Nerve supply

The liver is innervated by the parasympathetic fibres of the vagus nerve and sympathetic fibres from the coeliac ganglia (see Chapter 6).

Microscopic structure of the liver

The liver consists of many hexagonal-shaped functional units called **lobules** (*Figure 14.2*). A lobule, which measures 1–2 mm in diameter, contains **hepatocytes** (*parenchymal* or functional cells of the liver), blood vessels, bile ducts and phagocytic cells of the immune system (**Kupffer cells,** see below).

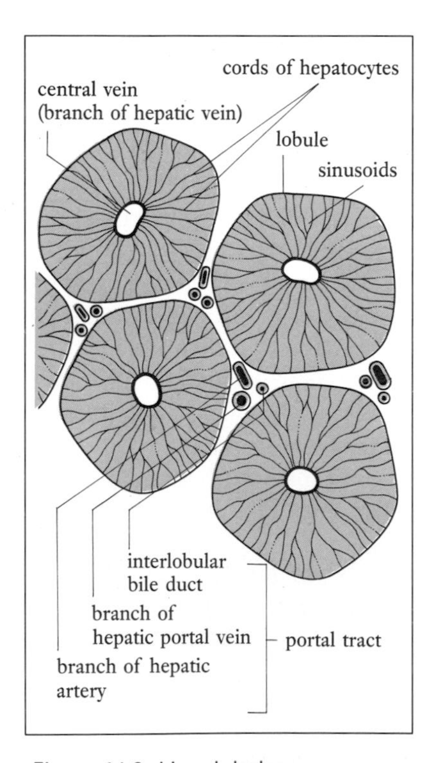

Figure 14.2. Liver lobules.

Each lobule has a **central vein** (branch of hepatic vein) and cords of hepatocytes which radiate like the spokes of a wheel (*Figure 14.2*). Blood-filled channels called *sinusoids* (see Chapter 10) permeate the cords of hepatocytes, which allows for slow blood flow and very close contact between cells and blood.

At each corner of the hexagonal lobule is a **portal tract**, which contains a branch of both the hepatic artery and hepatic portal vein and a bile duct. Blood from the hepatic artery and hepatic portal vein flows through the sinusoids to drain via the central vein and eventually via the hepatic veins and inferior vena cava. The direction of blood flow is therefore away from the portal tract whereas bile secreted by the hepatocytes flows in the opposite direction (*Figure 14.3*).

The bile secreted is collected in a network of tiny tubes called **bile canaliculi**, which are situated between the hepatocytes. The canaliculi unite to form the **interlobular bile ducts** of the portal tracts, which eventually become the **right** and **left hepatic ducts** carrying bile from the liver.

The special Kupffer cells lining the sinusoids are phagocytic macrophages (see Chapters 1 and 9) which are part of the more widespread *mononuclear phagocytic system*. Kupffer cells are concerned with the phagocytosis of bacteria and 'spent' erythrocytes.

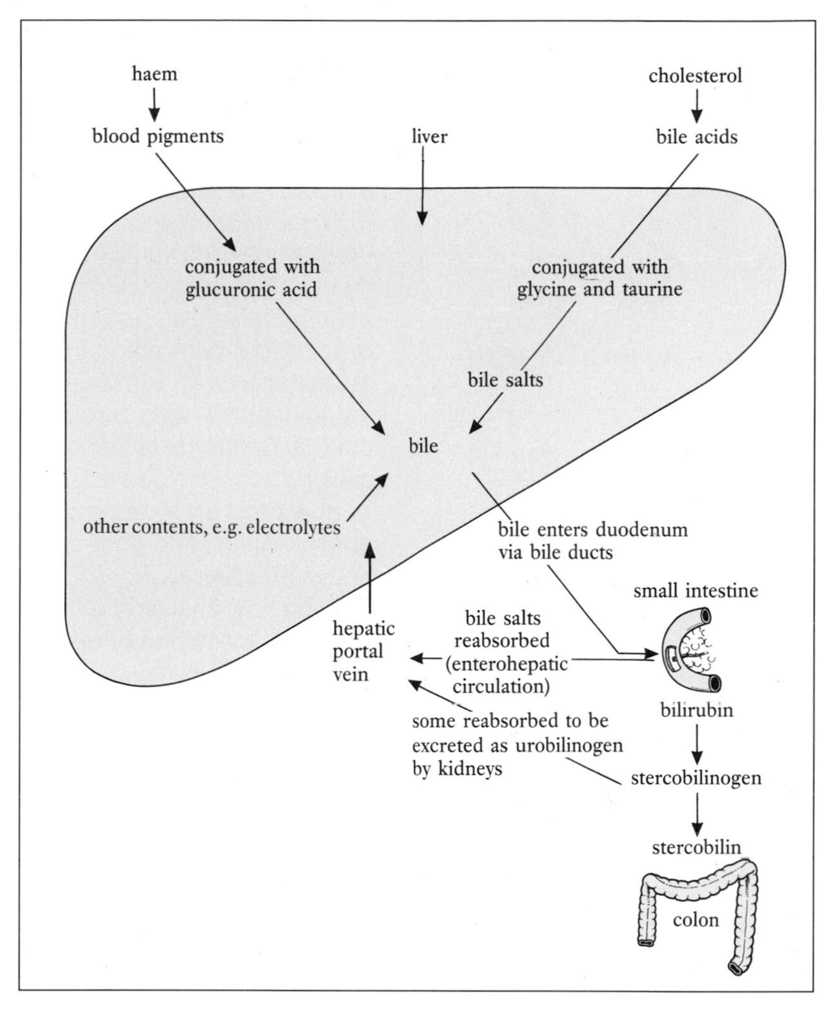

Figure 14.4. Production of bile and reabsorption of bile salts.

acids with *glycine* and *taurine* occurs in the liver and the conjugated acids form **bile salts** with sodium – **sodium glycocholate** and **sodium taurocholate**. The bile salts are secreted into the bile ducts and enter the intestine where most are reabsorbed. In a recycling process known as **enterohepatic circulation** the bile salts return to the liver via the hepatic portal vein. Returning bile salts stimulate further bile and bile acid production (*Figure 14.4*).

The **bile pigments**, **bilirubin** and **biliverdin**, are formed from the *haem* molecule produced mainly from the breakdown of erythrocytes. The unconjugated fat-soluble bilirubin, which is toxic to cells, is transported in the plasma bound to albumin. It is conjugated with **glucuronic acid** within the liver cells and the less toxic water-soluble product passes with the bile into the duodenum.

Microbial activity in the bowel converts bilirubin to **stercobilinogen**. Some is reabsorbed and passes through the liver into the general circulation and is excreted as **urobilinogen** by the kidney (see Chapter

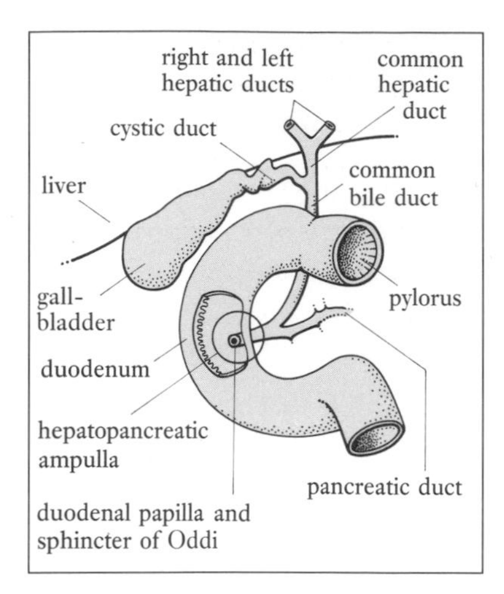

Figure 14.5. Bile ducts and gallbladder (diagrammatic).

15). The remaining stercobilinogen is converted in the bowel to **stercobilin**, which is excreted in the faeces.

Many drugs interfere with the complex, enzyme-catalysed reactions involved with bilirubin transport and conjugation. Drugs such as salicylates interfere with the transport of unconjugated bilirubin and should not be given late in pregnancy or to neonates (infants under 4 weeks of age) suffering from **jaundice** (yellow discoloration of the skin, sclera and mucous membranes which is due to an increased serum bilirubin). The increase in serum bilirubin may cause *kernicterus* (staining of brain cells especially those of the basal nuclei with bilirubin) and brain damage. Those drugs which stimulate liver enzymes (called *enzyme inducers*) include antiepileptic drugs, barbiturates and rifampicin. Enzyme inducers may affect the metabolism of other drugs, e.g. oral contraceptives, causing a lower blood level and reduced effectiveness.

Bile also provides a route for the excretion of cholesterol, phospholipids, and drug and hormone metabolites, which leave the body in the faeces.

Secretion of bile is stimulated by bile acid production and the intestinal hormone **secretin** (see Chapter 13). The actual release of bile into the duodenum depends on the hormone **cholecystokinin** and to a lesser extent the activity of the vagus nerve. *NB* The role of bile in digestion is discussed in Chapter 13.

Biliary tract, gallbladder and the excretion of bile

Bile drains from the main lobes of the liver via the **right** and **left hepatic ducts**, which drain their respective lobes. The two ducts unite to form the **common hepatic duct**, which leaves the liver at the porta hepatis (see *Figure 14.1(b)*). The **cystic duct** leads from the common hepatic duct to the **gallbladder**, where bile not needed immediately is stored and concentrated.

The gallbladder is a pear-shaped sac, about 10 cm long, which lies in a fossa on the posterior surface of the liver. It has a muscular layer and a mucosal lining of columnar epithelium thrown up into *rugae* very similar to those in the stomach. The rugae allow the gallbladder to distend when storing bile. When bile is required for digestion the muscle layer contracts and bile is expelled into the cystic duct. The cystic duct forms the **common bile duct** which joins with the pancreatic duct to enter the duodenum at the **hepatopancreatic ampulla** (see Chapter 13). The opening into the duodenum is guarded by the **sphincter of Oddi**, which regulates the flow of bile. Sphincter opening is controlled by the release of cholecystokinin, which simultaneously contracts the gallbladder (see Figure 14.5).

Gallstones (cholelithiasis)

Gallstones (cholelithiasis – formation of gallstones) are extremely common in developed countries. In the UK they are known to affect around 15% of adults, with many more being asymptomatic. The incidence is higher in women and is becoming increasingly common in younger adults (under 40).

In the UK the most common type of gallstones are those containing mostly cholesterol. They are formed by the crystallization of cholesterol in the gallbladder, which occurs when the cholesterol level is high or the level of bile salts is reduced. The factors which predispose to the formation of cholesterol gallstones include multiparity, obesity, oral contraceptives, drugs which reduce serum lipid levels and problems with the reabsorption of bile salts.

The other type of gallstone, which consists of bile pigments, is much less common and is associated with conditions where excess bilirubin is produced, e.g. haemolysis (see Chapter 10).

Although many people with gallstones are asymptomatic the stones can cause various problems, which include *cholecystitis* (gallbladder inflammation) which may be acute (see *Person-centred Study* – Daphne); or chronic, *biliary colic* characterized by extremely intense pain caused by a stone moving along the ducts, obstruction of the common bile duct with jaundice, and the remote possibility of perforation of the gallbladder with peritonitis (see Chapter 13).

Person-centred Study **Daphne**

Daphne, who had previously enjoyed good health, became quite unwell whilst her daughter and grandchildren were visiting. She complained of severe abdominal pain (epigastrium and right hypochondrium) which seemed to bore right through to her back. Thinking the pain was 'indigestion' caused by a rather late supper she took some 'stomach tablets' and retired to bed. By the next morning Daphne felt very hot, had been sick several times and the pain was still severe. Her daughter called out the family doctor, who thought that she might have acute cholecystitis and arranged for Daphne to be admitted to hospital.

Once in hospital, where the doctors agreed with the provisional diagnosis of acute cholecystitis, Daphne is given intramuscular analgesia (pethidine), an antiemetic (metoclopramide) and because infection is present an antimicrobial (a cephalosporin). The management regimen consists of bed rest, intravenous fluid replacement and careful monitoring of pain, vital signs and fluid balance.

When the acute episode has settled Daphne will have investigations to confirm the presence of gallstones which are found obstructing the cystic duct in 96% of people with acute cholecystitis (Sherlock, 1989). The possible investigations include *ultrasonography, computerized axial tomography (CAT scan), radionuclide scan* and possibly *cholecystography.* If the diagnosis of gallstones/inflammation is confirmed it is likely that Daphne will be offered surgical treatment to remove the gallbladder (*cholecystectomy*). This is usually performed as soon as possible to reduce the chance of further attacks.

Other treatments for gallstones include: (i) chemical dispersal of the gallstones with bile acids, e.g. chenodeoxycholic acid taken orally for many months; (ii) removal of a stone located within the bile ducts by *endoscopic retrograde cholangiopancreatography (ERCP)*; and (iii) the *lithotripter,* a machine which produces shock waves to disintegrate the gallstones. Use of the lithotripter, which is noninvasive, has obvious advantages – reduced discomfort, shorter stay in hospital and less risk of complications.

Nursing Practice Application **Biliary surgery**

After cholecystectomy (removal of the gallbladder) it is important that the person understands that his or her digestion can function without the gallbladder, as bile will trickle continuously into the duodenum.

Where the common bile duct has been explored it is usual for the bile to be drained externally via a T tube until the swelling subsides and the duct is again patent (open). Considerable fluid and electrolyte losses occur and nurses should ensure that accurate records of drainage are kept and that replacement fluid/electrolyte is administered as prescribed.

Prehepatic or haemolytic	*Hepatocellular*	*Obstructive (cholestatic)*
 haemolysis and release of haem	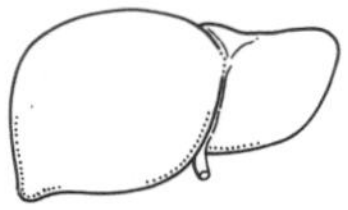	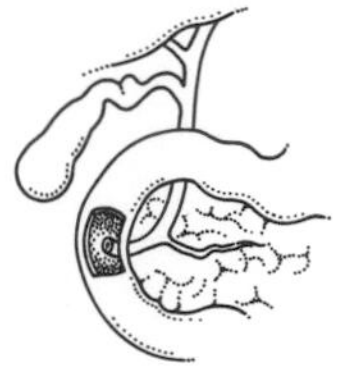
causes: overactive spleen rhesus incompatibility ABO mismatch abnormal RBCs drugs infections	causes: hepatitis drugs alcohol viruses conjugation problems in the newborn Gilbert's syndrome	causes: *a* intrahepatic cirrhosis drugs metastatic cancer *b* extrahepatic gallstone in CBD cancer in head of pancreas other tumours, etc.
results: mild jaundice (lemon) urine contains increased urobilinogen but no bilirubin (acholuric) stools – dark due to increased stercobilin	results: jaundice variable or urine may have bilirubin and urobilinogen varies stools – normal or paler	results: jaundice severe (green) urine contains bilirubin but no urobilinogen stools – pale (clay coloured)

Figure 14.6. Types and causes of jaundice.

Jaundice

Jaundice (see page 356) is not a disease but a sign of some malfunction of bile production, transport or excretion. Probably the most simple classification of jaundice considers the three types: (i) **prehepatic**; (ii) **hepatocellular**; and (iii) **obstructive** (*Figure 14.6*).

Apart from the yellow discoloration of skin, mucous membranes and sclera which becomes apparent when the serum bilirubin reaches 35–50 μmol/l (normal 2–17 μmol/l), the affected person may also complain of intense skin irritation (*pruritus*).

Prehepatic or haemolytic jaundice

The **hyperbilirubinaemia** (elevated levels of bilirubin in the blood) is usually due to excessive breakdown of erythrocytes. This occurs in haemolytic disease of the newborn (see Chapter 9) where the unconjugated bilirubin can cross the blood–brain barrier to cause kernicterus (see page 362).

Other causes of prehepatic jaundice include excessive haemolysis by the spleen, drugs (such as methyldopa), incompatible ABO blood transfusion, severe infections and erythrocyte abnormality, e.g. spherocytosis, thalassaemia.

Prehepatic jaundice is usually mild and is accompanied by an increase in urobilinogen excretion in the urine and excess stercobilin in the faeces, which may be very dark. Levels of unconjugated bilirubin in the serum are increased and this type of jaundice is sometimes termed *acholuric* because unconjugated bilirubin does not pass through the kidney and is consequently absent from the urine. The affected person usually has some degree of anaemia (see Chapter 9).

Hepatocellular jaundice

This type of jaundice is due to some defect in the transport of bilirubin within the liver. It may be due to the hepatitis caused by agents such as viruses, alcohol or drugs. Hepatocellular jaundice also occurs in newborn infants because their immature livers lack the enzymes required for the conjugation of bilirubin. As with haemolytic disease, the unconjugated bilirubin crosses into the brain tissue, especially the basal nuclei, to cause kernicterus and damage which may lead to learning difficulties. A more unusual cause of jaundice is the inherited *Gilbert's syndrome* where a deficiency of conjugation enzymes leads to a mild but fluctuating jaundice.

Hepatocellular jaundice varies in severity, and associated features depend upon the cause, e.g. bilirubin may appear in the urine in hepatitis but not with Gilbert's syndrome. The hyperbilirubinaemia is unconjugated if the required enzymes are deficient, but where hepatocyte damage is present it is a conjugated hyperbilirubinaemia.

Obstructive or cholestatic jaundice
(**cholestasis** – *an obstruction to the flow of bile)*

Obstructive or cholestatic jaundice may be due to intrahepatic obstruction to the flow of bile or the extrahepatic blockage of a large bile duct.

Intrahepatic causes, where the tiny bile ducts are affected, include: cirrhosis, hepatitis, drugs, e.g. oral contraceptives, and widespread metastatic malignacy.

Extrahepatic obstruction is commonly due to a gallstone occluding the common bile duct but other causes include carcinoma of the head of pancreas, other tumours, bile duct strictures (narrowing) and even parasitic worms which migrate from the gut.

When bile flow to the duodenum is obstructed the affected person will have dark urine, which contains bilirubin but no urobilinogen, and pale faeces, due to the absence of stercobilin. The jaundice is usually severe, and may produce a greenish tinge to the skin and a metallic taste in the mouth. There is an increase in both unconjugated and conjugated bilirubin in the serum, but in practice usually only the total bilirubin is measured. Unrelieved it will lead to hepatocyte damage, malabsorption and coagulation problems due to lack of vitamin K and prothrombin (see Chapters 9 and 13).

Synthesis in the liver

The liver can synthesize many different molecules, ranging from clotting factors to active vitamins.

The plasma proteins (see Chapter 9) albumin, α- and β-globulins and fibrinogen are made in the liver from amino acids. The γ-globulins are made elsewhere in the mononuclear phagocytic system. A useful test of liver function is to measure plasma proteins and the albumin:globulin ratio; a fall in albumin may indicate liver malfunction or protein malnutrition.

Apart from fibrinogen, the liver also manufactures other clotting factors, which include prothrombin and factors V, VII, VIII, IX, X, XI, XII and XIII (see *Table 9.3*). The liver is responsible for the production of heparin, a natural anticoagulant, which is involved with other substances in the prevention of excess coagulation.

During embryonic life the liver is a site of haemopoiesis and this function can be revived if the demand for blood cells exceeds supply.

Transport proteins such as transferrin (which carries iron) are made in the liver as are the proteins of the '*complement*' defence system (see Chapter 19).

The liver is concerned with the production and activation of certain vitamins; it can convert the provitamin *β-carotene* to *vitamin A (retinol)* and the amino acid *tryptophan* to *nicotinic acid* (see niacin). The inactive *vitamin D (cholecalciferol)* is converted by the liver to a storage form *(25-hydroxycholecalciferol)* which is converted to the active vitamin by the kidney.

Detoxification

The liver is an important organ in the modification of alcohol, drugs (see enzyme inducers, page 362) and some hormones prior to their elimination from the body. The metabolites produced by the liver are excreted via the kidneys or leave in the bile.

Alcohol (see Special Focus – alcohol and the liver)

The alcohol in alcoholic drinks is *ethanol*. This is oxidized in the hepatocytes, mainly by an enzyme called **alcohol dehydrogenase**, which produces a toxic intermediate called *acetaldehyde*. The acetaldehyde is further oxidized to acetyl CoA, which is used to produce energy (7 kcal/g of alcohol), carbon dioxide and water.

Ethanol ——> Acetaldehyde ——> Carbon dioxide + Water + Energy

Other enzyme systems also concerned with the oxidation of ethanol are actually stimulated by the intake of alcohol, which may explain why heavy drinkers develop some tolerance to alcohol.

Drugs (see Nursing Practice Application – Liver function in old age)

The liver metabolizes drugs in several ways, which either modify

Special Focus **Alcohol and the liver**

Throughout this book we have stressed how excess alcohol affects normal function and health. Here, only the hepatic effects will be considered.

It is the production of acetaldehyde during the oxidation of ethanol which directly damages the hepatocytes. The hepatocytes are infiltrated with fat and the whole liver becomes 'fatty'. Acute hepatitis (see page 359) occurs and may, if the abuse continues, become chronic with the eventual development of fibrosis, scarring and cirrhosis. Other liver problems associated with alcohol abuse include iron overload and primary hepatic carcinoma, which is uncommon in the United Kingdom.

Individual response to alcohol varies for reasons not fully understood, but it is known that women suffer liver damage with smaller amounts of alcohol than men. This is due in part to the amount of water present in the body, women have a lower percentage of water and consequently the alcohol they consume is less diluted. There are also factors concerned with the actual metabolism of alcohol within the liver that make women more susceptible to damage.

The recommended safe intake per week is up to 13 standard units of alcohol for non-pregnant women and up to 20 standard units of alcohol for men (Health Education Council, 1983). There should be alcohol-free days and in fact the Health Education Authority (previously Health Education Council) defines sensible drinking for women as 2–3 units on 2–3 occasions per week.

Advice to pregnant women or those planning to conceive is generally that they limit their alcohol intake to 1–2 units once or twice a week. Some authorities go further than this and suggest that only total abstinence from alcohol can be considered safe.

Although alcohol consumption has increased, generally in the UK it is in women and young people that the rise is most marked. In their study of at-risk drinkers in a health visitor's case load Robinson and Gaskell (1989) identified various factors associated with alcohol use, e.g. marital status, number and age of children and whether employed. They found that the group with the highest alcohol intake were single, unemployed mothers with more than two children.

Questioning of young people aged 16–18 revealed that 97% had tried alcohol and that 71% felt that they had associated physical or social problems (Smith and Collins, 1989). There would appear to be a great need for information about the results of excess alcohol intake if there is to be a reduction in the effects of abuse. This is especially so in the case of alcohol-induced cirrhosis, which is occurring more frequently – in the United States it is the fourth most common cause of death (Sherlock, 1989).

Support and information for those with an alcohol-associated problem (the drinker or his or her family/friends) is available from various organizations, e.g. Alcoholics Anonymous – a selection of addresses can be found on page 374.

Nursing Practice Application **Liver function in old age**

Liver size and function declines with normal ageing and in people over 70 there may be a significant loss of hepatocytes. As we have discussed, the liver plays a vital part in the detoxification of drugs prior to their excretion. These functions may be seriously impaired in older people and nurses should be particularly vigilant in observing for side-effects which may indicate that drug metabolism is abnormal, e.g. confusion and ataxia with benzodiazepines. Nurses can also help by checking that only essential drugs are prescribed or self-administered and by providing effective drug education for the person and his or her family or friends.

their activity or render them water-soluble prior to excretion in the urine or bile. Various drugs are altered chemically prior to conjugation with substances such as *glucuronic acid, glutathione* (see *Person-centred Study* – Steve) and *amino acids.*

Hormones

Several hormones are broken down in the liver prior to their excretion in the bile. These include those from the thyroid gland, glucocorticoids (see Chapter 8) and other steroid hormones such as oestrogens. In health the liver is able to prevent an accumulation of hormones in the blood and tissues. The failure of the liver to metabolize oestrogen may account for the *gynaecomastia* (male breast development) seen when hepatic function is seriously impaired.

Chronic parenchymal disease (cirrhosis and chronic hepatitis)

A detailed classification and diagnosis of chronic liver disease does not really concern us here and readers are directed to the reading list for specific information.

We shall concentrate on *cirrhosis,* where the liver shows areas of fibrosis with hepatocyte destruction. In Western countries the most common cause of cirrhosis is alcohol abuse, but other factors, including those listed for acute hepatitis (see page 359), are possible causes. The fibrosis and necrosis within the liver can lead to *portal hypertension* with the development of *oesophageal varices* (see Chapters 10 and 13 and page 371), *ascites*, jaundice (see pages 364-366) and liver failure (see pages 370-372).

Storage

The liver stores iron, vitamin B_{12}, folic acid and the fat-soluble vitamins A, D, E and K.

Iron derived from erythrocyte destruction is stored as *ferritin* and *haemosiderin* (see Chapter 9) until needed for the synthesis of new haem. Deposition of excess iron in the liver is known as *haemochromatosis*. This may be primary, due to an inability to control iron absorption, or may result from excess iron intake, e.g. dietary overload, repeated blood transfusion.

Large reserves of vitamin B_{12} are held in the liver and conditions associated with its deficiency, e.g. megaloblastic anaemia, may take several years to develop. There are much smaller stocks of folic acid which must soon be replenished by dietary intake.

The liver stores most of the vitamin A and D reserves and small amounts of E and K. It is worth remembering that hypervitaminosis A or D, which results from an excessive intake (see Chapter 13) can cause serious problems, e.g. **hepatomegaly** (enlarged liver) with vitamin A.

Metabolism

In Chapter 13 we discussed in some detail the role of the liver in the metabolism of carbohydrates, protein and fats, so here we need only restate the main points.

Carbohydrates

The liver has a central role in blood glucose homeostasis (see *Figure 8.19*):

- Converts fructose and galactose to glucose.
- Stores glucose as glycogen **(glycogenesis)** which can be broken down when required to provide glucose **(glycogenolysis).**
- Utilizes glucose to produce the energy required for its considerable metabolic activity **(glycolysis** and **glucose oxidation).**

- Converts excess glucose to fats for storage.
- Produces glucose from amino acids, glycerol and lactate by **gluconeogenesis.**

Protein

The liver is important in metabolism of amino acids, their conversion to structural and functional proteins (see synthesis, page 366) and the safe disposal of the waste ammonia.

- Production of non-essential amino acids by the process of *transamination*, which is facilitated by enzymes known as **aminotransferases** (previously transaminases). The two main enzymes are **alanine aminotransferase (ALT)** and **aspartate aminotransferase (AST)** both of which are produced in the liver. Measurement of AST and ALT in the serum can be used in the diagnosis and assessment of liver disease (see below). The enzymes are also produced by other tissues, e.g. myocardium and skeletal muscle; estimation of AST levels can be useful in the diagnosis of myocardial damage (see Chapter 10).
  NB AST was previously called glutamic-oxaloacetic transaminase (GOT) and ALT was called glutamic-pyruvic transaminase (GPT).
- Deamination of amino acids, which occurs prior to their oxidation or conversion to glucose. The toxic ammonia formed in this process is combined with carbon dioxide in the liver cells to form urea. The high level of ammonia in the blood, resulting from serious hepatic malfunction, is responsible for *hepatic coma.*

Table 14.1. Liver function tests.

Blood test	***Normal range***
Bilirubin	2–17 μmol/l
Enzymes	
alkaline phosphatase	98–280 i.u./l
aminotransferases	
ALT	10–40 i.u./l
AST	10–35 i.u./l
gamma glutamyl transferase	10–55 i.u./l (males)
	5–35 i.u./l (females)
Plasma proteins	
albumin	35–50 g/l
globulins	23–35 g/l
Coagulation ability	
prothrombin time	11–15 s

Reference ranges vary between centres and depend upon the type of analytical equipment and temperatures used. i.u. = international unit.

Nursing Practice Application Liver biopsy

Liver biopsy (*Figure 14.7*) is very useful in the diagnosis of certain conditions, e.g. cirrhosis, but is contraindicated where coagulation is seriously impaired. As with all investigations the person concerned should have adequate information about what is going to happen and care plans must include measures which ensure safety and comfort. It is likely that the person already has serious hepatic malfunction which increases the risks associated with the procedure. Prior to the biopsy blood is taken to measure the platelet count and prothrombin time. Obviously the risk of bleeding is increased where these are abnormal and two units of blood should be available, for transfusion, if required. Where the prothrombin time is extended it might be necessary to administer vitamin K to minimize any tendency to bleed.

The biopsy is performed after local anaesthesia of the tissues, and in some cases sedation may also be given to relieve anxiety. A special needle is introduced via the intercostal route and a sample of liver tissue is obtained when the person has exhaled and is holding his or her breath to avoid lung damage.

Following the procedure, the person stays in bed for 24 hours and the nurse monitors blood pressure, pulse, temperature, respiration, leakage from puncture site and pain, so that bleeding, haematoma, biliary peritonitis, pleural problems and infection may be detected.

Fats

Many cells use fats but a considerable part of fat metabolism occurs within the liver, which:

- Is an important site for the β-*oxidation* (see Chapter 13) of fats for energy.
- Excess acetyl CoA is converted into ketone bodies, which can be used in small amounts by body cells **(ketogenesis).**
- Stores fat and forms triglycerides which are stored in the fat depots **(lipogenesis)** and releases stored fats for energy use **(lipolysis).**
- Produces other lipids, e.g. lipoproteins for transport, cholesterol (for steroid hormones and bile acids) and phospholipids.

Liver function tests

As you would expect, the investigation of liver function requires a whole battery of biochemical tests. Samples of venous blood are obtained and various constituents in the serum are measured (Table 14.1). The results assist in the diagnosis of parenchymal disease, biliary obstruction and the progress of established disease.

Hepatic failure

Failure of liver function may result from many of the conditions already discussed, e.g. hepatitis B, alcoholic hepatitis, cirrhosis and drug toxicity (see *Person-centred Study* – Steve).

There is considerable variation in clinical presentation, depending upon the causation. Linking commonly occurring problems with the failure of normal function emphasizes just how much we depend on a healthy liver:

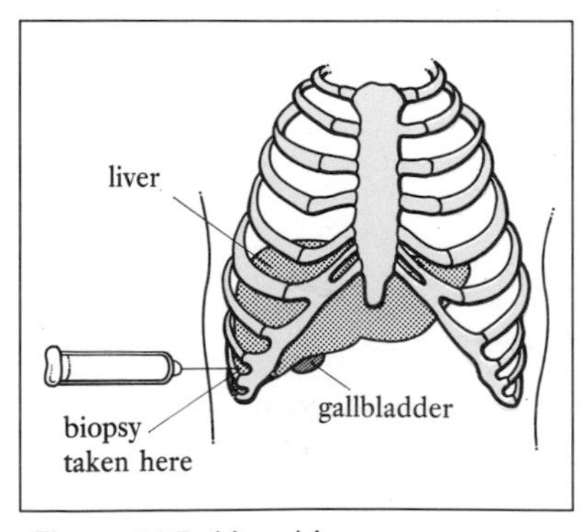

Figure 14.7. Liver biopsy.

- *Jaundice,* due to hepatocellular failure and intrahepatic obstruction.
- *Ascites,* caused by portal hypertension, failure of albumin production and water and sodium retention.
- *Cerebral oedema,* which also results from *hypoalbuminaemia.*
- *Coagulation problems* as the liver is unable to produce prothrombin and other clotting factors, which results in gastrointestinal bleeding,

bruising (intramuscular injections best avoided) and *purpura* (red/purple spots or patches caused by bleeding into the skin).

- *Abnormal reaction to drugs* – metabolism and conjugation fail to occur, this results in accumulation of the drug and side-effects (see also the elderly and drugs, page 367).
- *Gynaecomastia* occurs because hormones are not being degraded by the liver, which allows the build-up of oestrogens and other molecules. Other problems associated with lack of hormone metabolism include amenorrhoea (absent menstruation), testicular atrophy and loss of libido.
- *Hypoglycaemia* occurs because blood sugar homeostasis is seriously impaired.
- *Hepatic encephalopathy* and *coma* are due mainly to high levels of ammonia produced during amino acid metabolism as urea (levels normal or low) production is impaired. The nitrogenous metabolites produced by the gut bacteria and disturbances in electrolyte balance also contribute. Abnormal metabolites of amino acids excreted by the lungs possibly account for the characteristic musty, sweet odour of the breath known as *fetor hepaticus*.
  NB Energy requirements during hepatic failure are met with a high carbohydrate intake, protein must be restricted to minimize the toxic effects of the ammonia produced. Laxatives such as lactulose may be given to produce adequate evacuation which reduces the absorption of nitrogenous material from the gut.

Other manifestations of liver failure include:

- *Portal hypertension* and *varices* (see Chapter 10) caused by liver fibrosis and structural disorganization. This may result in haematemesis which if severe will cause hypovolaemia (see Chapter 10), and the blood (which, remember, contains protein) digested in the gastrointestinal tract adds further to the problems of protein metabolism and encephalopathy (brain disease caused as

Person-centred Study **Steve**

Steve, aged 17, has been very depressed about the difficulty in finding work; although he has been offered a training place at the local garage, he wanted a job with more money to start. One evening it was all too much and after drinking two cans of strong lager he took 20 (500 mg) paracetamol tablets. After an hour Steve became very frightened and told his parents who took him immediately to the nearest accident and emergency department.

His emergency management consisted of a gastric lavage (washout of stomach contents), estimation of plasma paracetamol levels, and intravenous acetylcysteine, which restores glutathione levels in the liver. Glutathione protects against liver/kidney damage by combining with paracetamol and so preventing the formation of toxic metabolites. Steve is admitted for observation and monitoring of liver function, which luckily remains unaffected. Whilst in hospital Steve is able to talk about his feelings and money worries with his parents and the social worker. Eventually he decides to take up the training vacancy and accept financial support from his parents for a few months.

NB Methionine might be given orally as a protective agent instead of acetylcysteine. If Steve had delayed (approximately 10–12 hours) getting help these antidotes would have been of no value and haemoperfusion could have been used in an attempt to remove the paracetamol.

toxins accumulate). The management includes:
- measures to control haematemesis, such as compression with a *Sengstaken–Blakemore* tube or operative procedures;
- blood transfusion;
- administration of *vasopressin* (see Chapter 8) which reduces portal pressure and also stimulates bowel emptying which evacuates blood that has entered the bowel;
- administration of neomycin (antimicrobial drug) to reduce bacterial action in the gut.

- Possibly *hepatomegaly* but the liver may in some cases become smaller.
- *Flapping tremor of the hands.*
- *Circulatory changes*, e.g. red palms and spider naevi.
- *Associated renal failure* (hepatorenal syndrome).
- *Systemic effects* such as nausea, abdominal pain and pruritus.

Paracetamol poisoning

Paracetamol is one of the most common drugs used in accidental or intentional self-poisoning. It is freely available without prescription and its accessibility in most homes makes it a favourite choice for both planned and impulse suicide attempts.

What is not generally known is that it damages the liver and kidneys in quite moderate overdoses. The person concerned recovers from the suicide attempt only to succumb to hepatic or renal failure 3–5 days later.

Paracetamol is the most common cause of hepatic necrosis in the UK and accounts for 200 deaths/year (Henry and Volans, 1984). It is particularly dangerous when combined with dextropropoxyphene as co-proxamol, and when taken with alcohol. As with all drugs, paracetamol should be kept away from children. The recommended dose should not be exceeded and alcohol should be avoided.

Summary/Check List

Introduction

Liver structure – Early development – Gross structure. Blood supply, *Nursing Practice Application* – liver injuries. Nerve supply. Microscopic structure. Acute hepatitis.

Liver functions – Bile production. Biliary tract, gallbladder and bile excretion. Gallstones, *Person-centred Study* – Daphne, *Nursing Practice Application* – biliary surgery. Jaundice. Synthesis. Detoxification, *Special Focus* – alcohol and the liver, *Nursing Practice Application* - liver function in old age and drugs, chronic parenchymal diseases. Storage. Metabolism. Liver function tests, *Nursing Practice Application* – liver biopsy. Hepatic failure, *Person-centred Study* – Steve. Poisoning, paracetamol.

?

SELF TEST

1 Which of the following statements about the liver are true?
(a) The adult liver weighs about 1.5 lbs.
(b) Vessels and ducts leave at the porta hepatis.
(c) The caudate and quadrate are the smallest lobes.
(d) All its venous blood leaves via the hepatic veins.

2 Draw and label a liver lobule.

3 Complete the following:
(a) The daily production of bile is ____ ml.
(b) Bile has a pH of __ .
(c) Bilirubin is the major bile ______ .

4 Which of the following would cause jaundice and what type would it be?
(a) Gallstone in the cystic duct.
(b) Transfusion of group A blood to a group B recipient.
(c) Acute hepatitis.
(d) Carcinoma in the head of the pancreas.

5 Why is measuring the albumin:globulin ratio a good test of liver function?

6 Explain the following:
(a) Development of alcohol tolerance in heavy drinkers.
(b) Increased drug toxicity seen in the elderly.

7 Name six molecules stored in the liver.

8 Put the following in their logical pairs:

(a) urea	(f) prothrombin
(b) ferritin	(g) carotene
(c) ethanol	(h) acetaldehyde
(d) retinol	(i) iron
(e) vitamin K	(j) ammonia.

9 Describe the role of the liver in blood glucose homeostasis.

10 Explain the following features of hepatic failure: ascites, bruising, encephalopathy and jaundice.

Answers
1. b, c, d. 2. See pages 358-359 3. a. 500–1000 ml. b. 8.0 c. Pigment. 4. b. prehepatic c. hepatocellular d. obstructive. 5. See page 366. 6. See pages 366-367. 7. iron, vitamin B_{12}, folic acid, vitamins A, D, E, K and glucose as glycogen. 8. a–j, b–i, c–h, d–g, e–f. 9. See pages 368-369. 10. See pages 370-371.

REFERENCES

Health Education Council (1983) *That's the Limit: A Guide to Sensible Drinking*. London: Health Education Council.

Henry, J. and Volans, G. (1984) *ABC of Poisoning – Part I.* London: British Medical Association.

Robinson, B. G. and Gaskell, K. (1989) Prevalence and characteristics of at-risk drinkers among a health visitor's case load. *Health Visitor*, **62**, 242–243.

Sherlock, S. (1989) *Diseases of the Liver and Biliary System*, 8th edn. Oxford: Blackwell Scientific Publications.

Smith, I. and Collins, F. (1989) Patterns of drug abuse. *Nursing Times*, **85** (10), 55.

Further Reading

Plant, M. (1990) Advising on alcohol. *Nursing Times* (Midwives Journal), **86** (12), 64–65.

Sherlock, S. (1989) *Diseases of the Liver and Biliary System,* 8th edn. Oxford, Blackwell Scientific Publications.

Vale, J. A. and Meredith, T. J. (1985) *A Concise Guide to the Management of Poisoning*, 3rd edn. Edinburgh: Churchill Livingstone.

Useful Addresses

Health Education Authority
Hamilton House
Mabledon Place
London WC1H 9TX

Al Anon Family Groups
61 Great Dover Street
London SE1 4YF

Alcoholics Anonymous
11 Redcliffe Gardens
London SW10

National Council on Alcoholism
3 Grosvenor Street
London SW1

CHAPTER FIFTEEN

URINARY SYSTEM

OVERVIEW

- Urinary system structure and function.
- The role of the kidneys in maintaining homeostasis.
- Micturition.

LEARNING OUTCOMES

After studying Chapter 15 you should be able to:

- Describe the gross structure and position of the kidneys.
- Describe renal blood flow.
- Describe the structure of a nephron.
- Discuss the homeostatic role of the kidneys.
- List the normal constituents of urine.
- Describe the formation of urine.
- Outline mechanisms by which the kidney varies the volume and concentration of urine.
- List some abnormal constituents of urine and describe their cause.
- Discuss the causes and consequences of renal failure.
- Describe the structure of the ureters, bladder and urethra.
- Describe the process of normal micturition.
- Use physiological knowledge in the assessment of micturition and the identification of abnormalites.

KEY WORDS

Anuria – cessation of urine production by the kidneys.
Dysuria – difficult or painful micturition.
Glomerular filtration rate (GFR) – the amount of plasma filtered by the kidneys in one minute.
Micturition – passing or voiding urine. Urination.
Nephron – the functional unit of the kidney.
Oliguria – diminished volume of urine produced by the kidneys.
Polyuria – increased volume of urine produced by the kidneys.
Renal – relating to the kidney.
Renin – proteolytic enzyme produced by the kidney which activates angiotensin, in turn causing the release of aldosterone.

Introduction

The kidneys are central to homeostatic regulation. They excrete soluble waste, help to maintain the water and electrolyte composition of body fluids and regulate pH in conjunction with the lungs and other buffer systems. Without the complex **renal** (relating to the kidney) processes the body's 'waste' builds up, pollution occurs and the delicate balance of the internal environment is lost (see Chapter 2).

Other functions of the kidney include the secretion of *erythropoietin*, which stimulates erythropoiesis (see Chapter 9) and **renin** (proteolytic enzyme produced by the kidney which activates angiotensin), which raises blood pressure. The kidney also converts vitamin D to *1,25-dihydroxycholecalciferol,* its most active form (see Chapter 8).

The Urinary System

Apart from two *kidneys* which produce urine, the structures of the **urinary system** are two *ureters* which transport urine, the *bladder* acting as a temporary reservoir for urine and the *urethra* which conveys urine to the exterior during **micturition** (passing or voiding urine) (*Figure 15.1*).

Early development

During the fifth week of fetal life the primitive kidneys start to develop within the pelvis. Later, after many changes, they migrate upwards to the adult position in the abdomen. One part of the early lobulated kidney will become the *renal pelvis, calyces, collecting ducts* and a ureter, which by the eighth week is connected to a bladder that develops from the *cloaca* (embryonic gut structure) and *urogenital sinus.* The other part of the developing kidney is destined to form the *nephrons.* At this early stage of development there is a close link between the urinary tract, external genitalia and the internal reproductive structures which develop from the *Mullerian* and *Wolffian ducts* (see Chapter 20).

The fetal kidneys are structurally complete but remain functionally immature (although they do produce urine as early as the twelfth week of gestation) until some weeks after birth, when they become capable of producing concentrated urine.

Problems occurring during the early weeks of development may result in:

- The formation of only one kidney.
- *Polycystic disease*, where normal fusion of the two kidney components fails to occur.
- A single *horseshoe kidney* which results from an abnormal fusion between both kidneys.

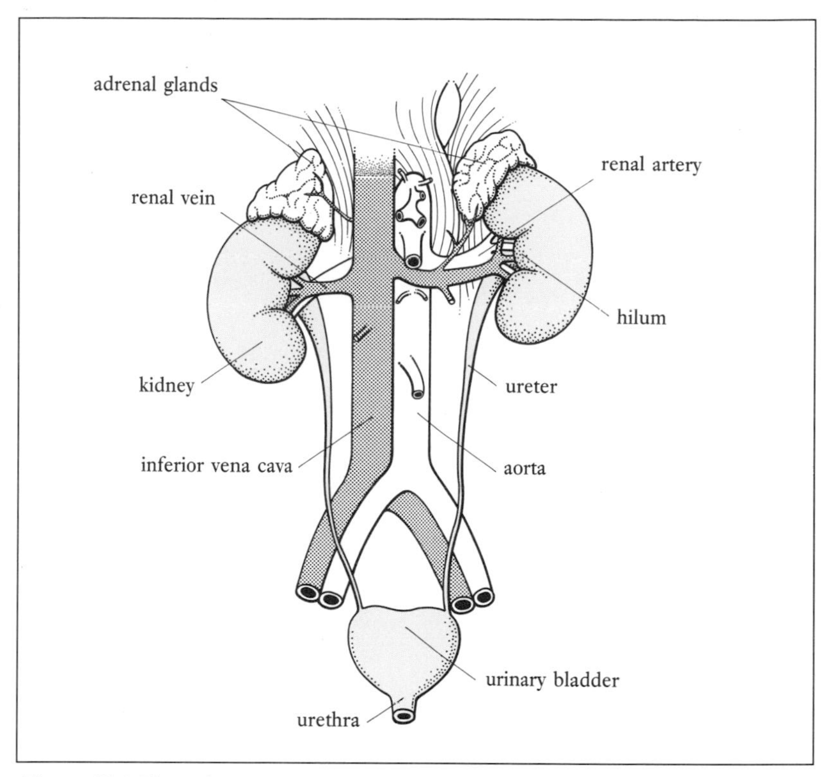

Figure 15.1. The urinary system.

The Kidney

Gross structure of the kidney

The paired **kidneys** are situated behind the peritoneum (*retroperitoneal*) and attached to the posterior abdominal wall by adipose and fibrous tissue. The kidneys lie, one either side of the vertebral column, at a level extending from T12 to L3, with the right kidney slightly lower to accommodate the liver (see Chapter 14). Each bean-shaped kidney carries on its superior surface an adrenal (suprarenal) gland, which we discussed at some length in Chapter 8.

In adults each kidney weighs around 150 g and is about 12 cm long, 7 cm wide and 3 cm thick. On its concave medial surface there is a depression known as the **hilum** where blood vessels, lymphatics, nerves and a ureter enter or leave the kidney.

The kidney is enclosed within several layers of supporting tissue: an outer fibrous connective layer, a middle adipose layer which helps to cushion the kidney and, around the kidney itself, a transparent fibrous **renal capsule**.

If we look at a cut kidney (*coronal section*) several separate regions can be recognized (*Figure 15.2*). There is an outer **cortex** which forms a pale red–brown layer under the capsule. Below the cortex is the darker **medulla** containing the cone-shaped striations called **renal pyramids**. At the apex of each pyramid a **papilla** opens into a **minor calyx** which communicates with **major calyces** and

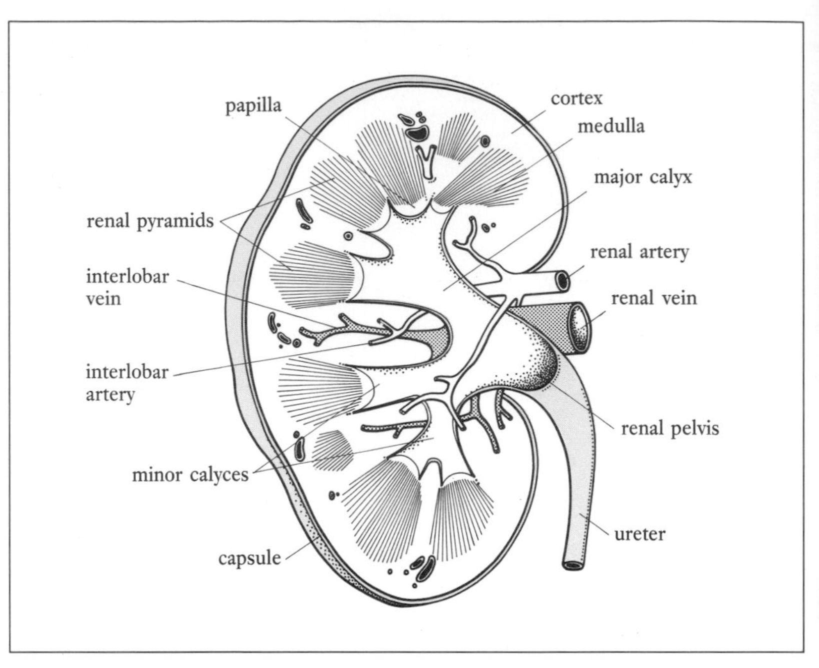

Figure 15.2. Section through a kidney.

the funnel-shaped **renal pelvis**, which distends to receive urine. Contraction of smooth muscle in the walls of the calyces and renal pelvis conveys the urine into the ureter.

Blood and nerve supply of the kidney

The kidneys receive around 1200 ml of blood per minute, 25% of the resting cardiac output, via renal arteries which branch directly from the abdominal aorta (see Chapter 10). The continuity of this supply is vital to effective renal function and later we will discuss how the kidney autoregulates blood flow (see page 385). Venous blood leaves the kidney via renal veins which empty into the inferior vena cava. The smaller vessels and the two-capillary system of the kidney are covered in more detail when we consider the nephron (see below).

The kidney is well supplied with sympathetic nerve fibres (see Chapter 6) from the coeliac plexus, which when stimulated cause renal vasoconstriction, secretion of renin and decreased urine production. A few parasympathetic fibres derived from the vagus are present but their role is unknown. Sensory fibres also leave the kidney and are responsible for the transmission of pain.

Nephron

One kidney contains about one million microscopic **nephrons** (functional units of the kidney) of two types: *cortical* (most abundant; they are located mainly in the cortex but some may extend into the medulla) and *juxtamedullary* (located close to the junction between cortex and medulla, they extend well into the medulla). The number of nephrons is more than we actually need but this forms a useful reserve. It is entirely possible to function with only one kidney (see page 376), e.g. congenital absence of a kidney which occurs in about

Nursing Practice Application Renal calculi (kidney stones)

Renal stones (calculi) are often associated with recurrent infections of the urinary tract (see pages 403-404). Other causes include disturbances of calcium and phosphate metabolism, such as may occur in hyperparathyroidism (see Chapter 8) where the excess calcium salts excreted in the urine (*hypercalciuria*) are deposited as stones.

Renal calculi may be large enough to fill the renal pelvis and calyces (*staghorn calculus*); they may obstruct the outflow of urine, which leads to *hydronephrosis* (collection of water in the kidney), chronic infection and renal failure (see page 395). If the stone moves into and obstructs the narrow lumen of the ureter it causes *renal colic,* which is characterized by extremely severe loin pain, nausea and *haematuria* (blood in the urine). Nursing care plans must make provision for adequate pain relief, a fluid intake of at least 2–3 litres/24 hours, testing urine for blood (see page 394) and straining all urine passed for small stones/debris.

Treatment of renal calculi has undergone great changes in the last decade. Non-invasive techniques such as *lithotripsy* (shock waves) may be used to disintegrate the stone, but where a stone is impacted in the ureter with distal obstruction a *percutaneous nephrolithotomy* (a procedure where a *nephroscope* – an instrument with a light source used to view inside the kidney – passed through a small skin incision, is used to locate the stone in the renal pelvis. The stone can then be removed via the nephroscope without the need for a large incision) is the best method (Wright, 1987). When available these developments can replace major surgical intervention, which reduces pain and inconvenience, stay in hospital and complications associated with traditional operations such as *pyelolithotomy.* Obviously the underlying cause of the calculus formation is treated, e.g. antimicrobial drugs for infection.

1 in 500 infants (Macleod *et al.,* 1987) or after *nephrectomy* (removal of a kidney).

A nephron comprises a **renal tubule** and the **glomerulus**, a knot of capillaries, which lies within the invaginated blind end of the tubule. The tubule, which is lined with cuboidal epithelium, is divided into: **Bowman's capsule** which encloses the glomerulus, the **proximal convoluted tubule (PCT)**, the **loop of Henle**, the **distal convoluted tubule (DCT)** and the **collecting ducts/tubules** which drain urine from several nephrons (*Figure 15.3*).

Within the cortex of the kidney the glomerulus is formed from a wide **afferent arteriole**, which supplies blood at a rate that ensures that the pressure of blood in the capillaries is sufficient to force fluid out into the tubule. The epithelium of the capillaries, which have **fenestrations** (pores), and Bowman's capsule, are highly permeable and adapted for filtration. These two epithelial layers, divided by a basement membrane (see Chapter 1), form an efficient filtration membrane which allows the passage of water and molecules with a molecular weight of 69 000 or less. Selectivity based on size is important as it means that blood cells and plasma proteins do not normally cross into the **filtrate** (fluid within the tubule derived from blood).

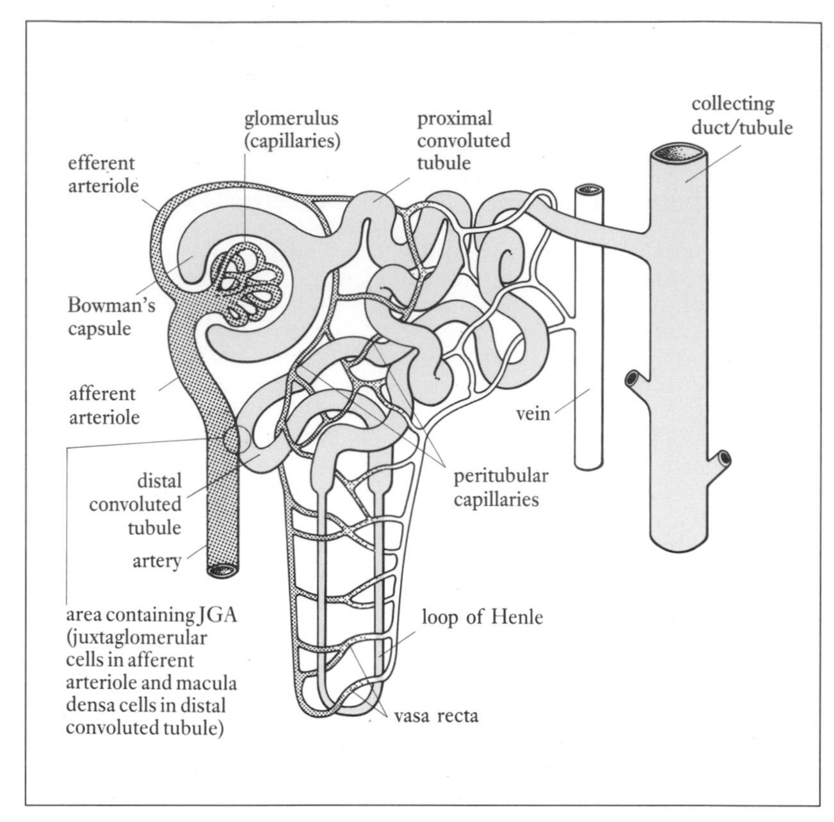

Figure 15.3. A nephron.

Most unusually, the glomerulus is drained via a smaller **efferent arteriole**, which forms a second, low-pressure **peritubular capillary network**. The lower pressure in these capillaries is a feature which allows reabsorption of filtrate from the tubule.

Leading from Bowman's capsule is the much coiled proximal convoluted tubule. With its specialized epithelial lining containing *microvilli*, and thus a greatly increased surface area, the proximal convoluted tubule is well adapted for reabsorption.

The loop of Henle, which consists of a descending and ascending limb, is a simple loop. The epithelium lining in the loop of Henle consists of flattened cells containing few microvilli. In certain nephrons (the *juxtamedullary nephrons*) the loop of Henle is surrounded by looped capillaries known as the **vasa recta**, in addition to the peritubular capillary network. The juxtamedullary nephrons have a long loop of Henle which extends into the medulla (in contrast to the shorter cortical nephrons which are mostly confined to the cortex). The advantage of having two types of nephron is that the kidney is able to adjust the concentration of urine and the amount of sodium excreted to maintain homeostasis under differing conditions.

From the ascending limb of the loop of Henle the distal convoluted tubule twists and coils so as to have contact with the afferent arteriole. The area in contact with the arteriole is known as the

macula densa; it contains special cells which form part of the **juxtaglomerular apparatus (JGA)**. The distal convoluted tubule leads into a collecting tubule. These unite to form larger ducts that eventually empty into the minor calyces.

Juxtaglomerular apparatus (JGA)

The JGA is a region formed from modified cells of the distal convoluted tubule and the afferent arteriole. The special cells of the macula densa monitor the levels of sodium and chloride in the filtrate. Modified muscle cells in the arteriole (juxtaglomerular cells) which contain renin granules are sensitive to blood pressure in the arteriole. The JGA plays an important part in renal autoregulation and will be discussed further (see page 385).

Renal blood flow

As the renal arteries enter at the hilum they divide to form *interlobar arteries*, (see *Figure 15.2*) which pass between the pyramids. Further subdivision gives rise to smaller *arcuate* and *cortical interlobular arteries*, which eventually become the afferent arterioles supplying the capillary bed of the glomerulus. The efferent arteriole draining the glomerulus forms a second capillary bed around the tubule (peritubular). Most of the filtrate, which is reabsorbed in the tubule, returns to the circulation in the peritubular capillaries (see *Figure 15.3*). The peritubular capillary bed drains via *venous plexi*, which become larger veins corresponding to the arteries. Eventually the venous blood leaves the kidney via the renal vein to empty into the inferior vena cava.

Functions of the Kidney

- Helps to maintain water and electrolyte balance.
- Helps to regulate acid–base balance.
- Excretes nitrogenous waste.
- Excretes the metabolites of many drugs and some hormones.
- Haemopoietic role, with the secretion of erythropoietin which stimulates the red bone marrow (see Chapter 9).
- Secretes renin, which raises blood pressure (see pages 385, 386 and 388).
- Produces an active form of vitamin D (see Chapters 8 and 16).

Figure 15.4 illustrates the functions of the kidney. The kidney accomplishes the first four functions above through the formation of urine. This involves *filtration, reabsorption* and *secretion,* which occur exclusively within the nephrons.

Composition and characteristics of urine

Normal physical characteristics

In adults between 1 and 1.5 litres of urine is produced each day. The volume and composition depend upon many factors which include fluid intake, diet, climate, activity and health.

Urine is normally clear straw-yellow to amber in colour, the colour is produced by the presence of *urochrome,* a pigment derived from haemoglobin breakdown. Concentrated urine is darker in colour than

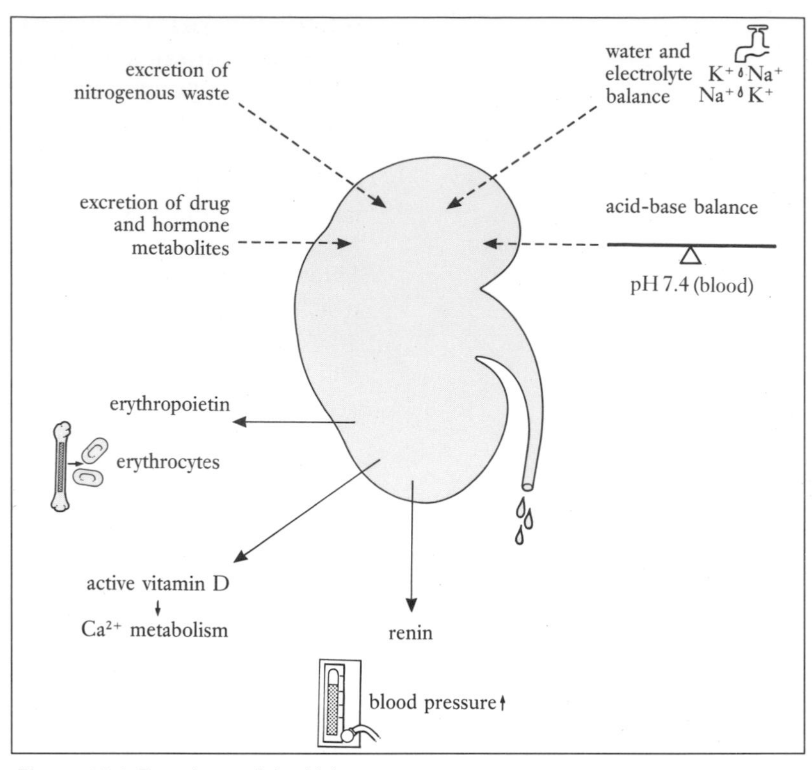

Figure 15.4. Functions of the kidney.

dilute urine, which may resemble water. Drugs such as rifampicin colour the urine orange and eating beetroot causes red urine, which may lead to some alarm until you remember what you have eaten.

The normal pH of urine is around 6.0 but a range of 4.5–8.0 is possible; a vegetarian diet tends to produce alkaline urine and a diet containing high levels of animal protein will make the urine more acidic.

The specific gravity (SG) is usually in the range 1.010–1.030 and depends upon the amount of solid material or *solute* present; the SG of urine measures its density compared with that of distilled water, which has a SG of 1.000.

Urine has a characteristic aromatic odour when fresh; the ammonia odour only develops on standing or when infection is present. Other odours are associated with certain foods, e.g. asparagus, or conditions where abnormal substances are excreted, such as *diabetes mellitus* (see Chapter 8) and *maple syrup disease*, when amino acids passed in the urine cause it to smell of maple syrup.

Composition of urine

Urine is 96% water and 4% solids. The solids comprise:

- Nitrogenous waste, which includes **urea** (see Chapter 14), **uric acid** (a product of nucleic acid metabolism) and **creatinine** derived from skeletal muscle.

- Sodium, potassium, calcium, magnesium, phosphates, sulphates, nitrates, chlorides and hydrogen carbonate.
- Organic acids, ammonia as ammonium salts.
- Drug and hormone metabolites.

NB The presence of *casts* (cellular debris) is usually indicative of renal damage.

Urine formation

To produce the 1–1.5 litres of urine daily the kidneys **filter** about 180 litres of plasma, which means that **reabsorption** in the tubule must be a very efficient process or you would soon become dehydrated. The cells of the tubule put the finishing touches to the filtrate by **secreting** electrolytes, drugs and hydrogen ions into the lumen of the tubule, much as you might add seasoning to a meal (*Figure 15.5*).

When you consider this mighty undertaking it becomes obvious why the kidneys need 25% of the resting cardiac output to provide the necessary oxygenated blood. We shall now consider the three stages of filtration, reabsorption and tubular secretion in more detail.

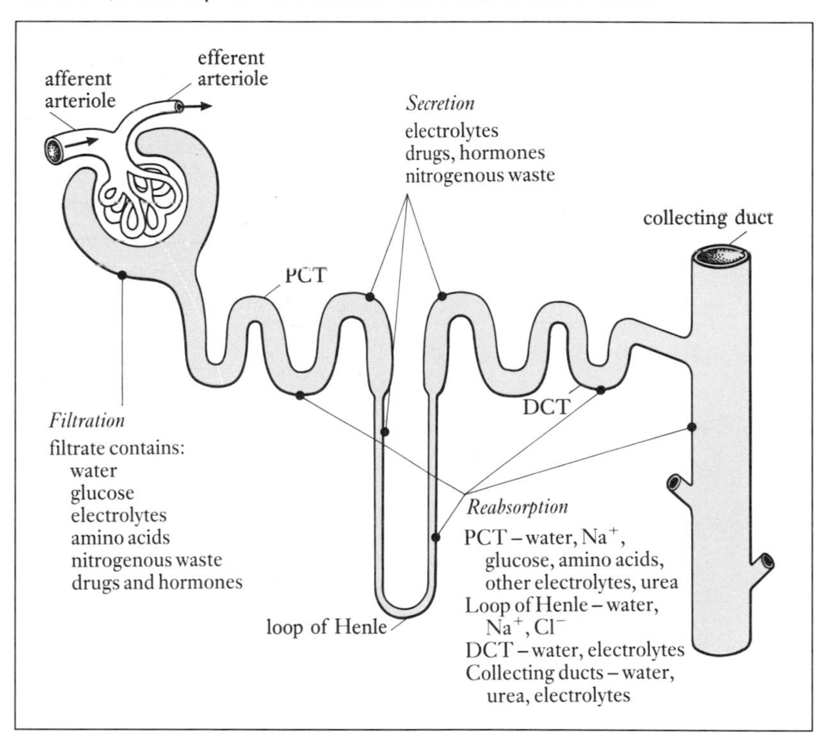

Figure 15.5. Formation of urine – filtration, reabsorption and secretion. PCT = proximal convoluted tubule; DCT = distal convoluted tubule.

Filtration

Every minute about 125 ml of plasma is forced through the glomerular membrane into the tubule by the hydrostatic pressure within the glomerulus. This is known as the **glomerular filtration rate (GFR).** This rate is much less at the extremes of the normal life-span and accounts for the inability of small infants to cope with extra fluid and problems associated with drug excretion in elderly people. Generally, in people over 70 the GFR has fallen by half, which may lead to the accumulation of drugs within the body and the development of toxic side-effects, e.g. anorexia and nausea with digoxin (see Chapter 10). Others with impaired renal function are similarly affected and great care is required in the prescibing and subsequent management of drug therapy.

Water and small molecules pass passively down the pressure gradient as they do in other capillary beds, but this time, because of the specialized structure of the filtration barrier, this occurs more efficiently. Normally the filtration barrier prevents loss of larger proteins and blood cells (see page 379), but when the membrane is damaged, e.g. by inflammation in *glomerulonephritis,* the pores enlarge, making the membrane 'leaky', and albumin (molecular weight 69 000) appears in the urine (see page 394).

The hydrostatic pressure (see Chapter 10) within the glomerulus, at around 7.3 kPa, (55 mmHg) is higher than in other capillaries because the afferent arteriole has a larger calibre than the efferent arteriole. However, we are left with a net filtration pressure of around 1.3 kPa (10 mmHg) after the pressures which oppose the movement of fluid from the blood to the tubule, the osmotic pressure of the plasma proteins in the capillaries (see Chapter 10) and the hydrostatic pressure exerted by fluid in Bowman's capsule have been taken into account.

Glomerular hydrostatic pressure – (osmotic pressure + Bowman's capsule pressure) = net filtration pressure

7.3 kPa – (4 kPa + 2 kPa) = 1.3 kPa
or
55 mmHg – (30 mmHg + 15 mmHg) = 10 mmHg

The small molecules which pass with the water into the filtrate in the tubule include glucose, amino acids, electrolytes, drug and hormone metabolites, urea, uric acid and creatinine (see *Figure 15.5*). Many of these, which are needed by the body, are later reabsorbed, e.g. glucose, which does not normally appear in the urine.

By the time blood reaches the end of the glomerulus, filtration has all but stopped. This is because the movement of fluid from blood to filtrate increases the osmotic pressure to a point where it equals the glomerular hydrostatic pressure.

Glomerular filtration rate and regulation

The glomerular filtration rate of about 125 ml/min in healthy adults is dependent upon the pressure gradients produced by sufficient blood reaching the glomerulus. Although the kidneys already receive 25% of the cardiac output there are intrinsic (*autoregulatory*) mechanisms which operate to ensure that enough blood at the correct pressure reaches the nephrons. There are also extrinsic neural controls which modify blood flow in stress situations. In Chapter 10 we discussed shock due to hypovolaemia, which leads to marked or sustained fall in systemic blood pressure. This reduces renal perfusion and filtration, which explains the **oliguria** (reduced volume of urine) associated with shock.

Autoregulation

Autoregulatory mechanisms which operate under everyday conditions include the ability of smooth muscle in the afferent arterioles to constrict when systemic blood pressure is high and dilate when pressure is low. In this way the GFR remains stable whilst systemic blood pressure fluctuates within a wide range (80–200 mmHg). Autoregulatory mechanisms are not, however, able to maintain GFR during shock if systolic blood pressure falls below 80 mmHg. The changes in arteriolar diameter may be linked to the release of prostaglandins (see Chapter 8) by renal cells.

The other autoregulatory mechanism involves the cells of the **juxtaglomerular apparatus** (see page 381). Juxtaglomerular cells in the afferent arteriole, which are sensitive to changes in pressure, release the enzyme **renin** into the blood if pressure falls. Renin acts upon the plasma protein **angiotensinogen** which it converts to **angiotensin I**. Other enzymes, situated mainly in the lungs, transform angiotensin I to **angiotensin II**, which causes vasoconstriction and a rise in systemic blood pressure (*Figure 15.6*). You will remember that chemoreceptor cells of the **macula densa** monitor sodium content of the filtrate; if sodium levels are low or flow is slow there will also be release of renin.

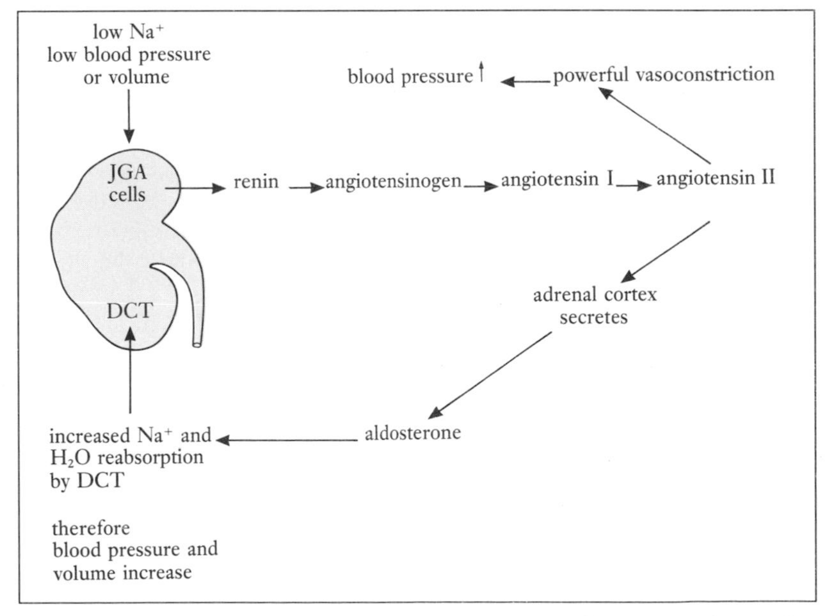

Figure 15.6. Renin–angiotensin–aldosterone system. JGA, = juxtaglomerular apparatus; DCT = distal convoluted tubule.

Apart from being a powerful vasoconstrictor the angiotensin II formed stimulates the adrenal cortex to release **aldosterone** (see Chapter 8) which causes more sodium to be reabsorbed in the distal convoluted tubule (*Figure 15.6*). As the sodium ions move back into the blood they are accompanied by water, which increases blood volume and hence pressure. We shall discuss renin– angiotensin–aldosterone again in the consideration of water/electrolyte homeostasis.

Extrinsic mechanisms

Whilst autoregulation deals with everyday needs, the extrinsic mechanisms come into play during stress, exercise, pain and abnormal situations, such as severe hypovolaemic shock. Sympathetic nerve stimulation and the release of adrenaline cause renal vasoconstriction as blood is diverted to the heart and brain, which reduces GFR and urinary output. Renin release is also stimulated, which helps by causing general vasoconstriction and reaborption of fluid in an attempt to maintain blood volume and pressure.

Nursing Practice Application Renal clearance

Renal clearance is a way of measuring the GFR and assessing renal function. This is done by calculating how well the kidneys can clear a particular substance from the blood in a given time, usually a minute. For a marker substance, which is filtered but not significantly reabsorbed or secreted, such as creatinine or the polysaccharide inulin (see Chapter 13), the amount of blood cleared in one minute will equal the GFR. It is calculated using the formula below, used here to illustrate creatinine clearance:

$$\text{GFR ml/min} = \frac{U \times V}{P}$$

where U is the concentration of creatinine (mg/ml) in the urine, P its concentration (mg/ml) in plasma and V the amount of urine produced per minute. Normally the creatinine clearance is between 120 and 125 ml/minute.

For the laboratory to ascertain the values it is necessary to have an accurate 24-hour urine collection and a sample of venous blood taken during this time. Collecting urine is usually the responsibility of the nurse who should ensure that: (i) the collection is timed to finish during the morning of a weekday; (ii) the person concerned understands that all urine passed must be saved (after first emptying the bladder); (iii) the correct receptacle is available; (iv) adequate privacy is provided; (v) all staff are aware of the collection; (vi) the sample, properly labelled, arrives in the laboratory as soon after completion as possible on a weekday morning so that it can be processed.

Creatinine clearance estimation is used in the diagnosis of renal impairment and to monitor progress in existing disease. It is straightforward to execute and causes minimal inconvenience and discomfort but, as with all investigations, the nurse should be able to offer a clear explanation as part of the care plan.

Reabsorption

Reabsorption of water and molecules from the filtrate is a vital activity in the production of 1–1.5 litres of urine from 180 litres of filtrate. Of the 125 ml of filtrate formed per minute, only 1 ml becomes urine. Filtrate, which is very similar to plasma, but without the proteins, needs considerable modification in the tubule before it conforms with the composition of urine (see pages 381-383).

The renal tubules provide a large surface area for the reabsorption of water and other molecules from the filtrate. Reabsorbed

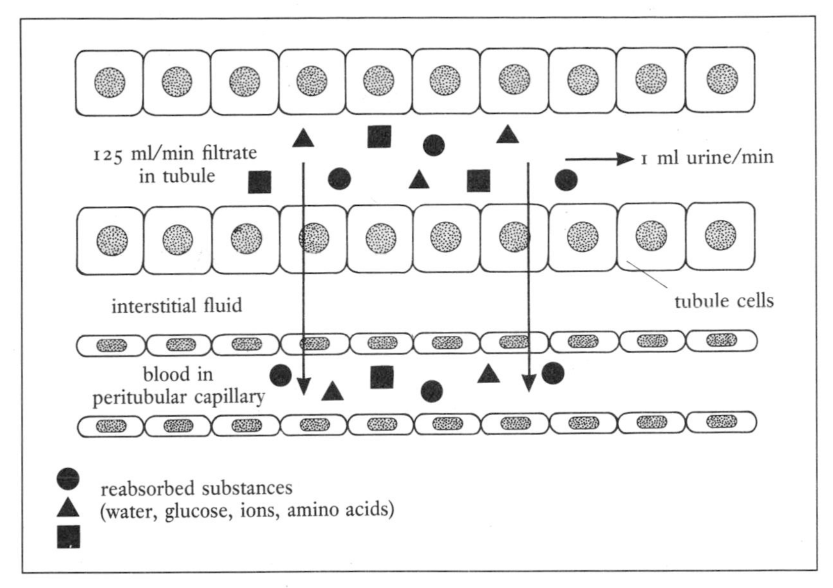

Figure 15.7. Reabsorption.

substances cross the epithelial layer of the tubule cells and return to the blood via the peritubular capillaries (*Figure 15.7*).

The cells of the proximal convoluted tubule, with their microvilli, are particularly well adapted for reabsorption and this part of the tubule is responsible for reclaiming over 65% of the filtered water and sodium ions. The proximal convoluted tubule reabsorbs the nutrients, amino acids and glucose along with water and electrolytes, which are also reabsorbed in the loop of Henle, distal convoluted tubule and collecting ducts (see *Figure 15.5*).

In Chapters 1 and 2 we looked at the transport of water and other molecules across cytoplasmic membranes, and you might care to revise the main points, which are highly relevant to our consideration of tubular reabsorption.

Passive transport mechanisms, which include osmosis, diffusion and facilitated diffusion with a carrier molecule are all used for reabsorption. Water moves passively down osmotic gradients and certain solutes diffuse down chemical concentration or electrical gradients (potential difference – see Chapter 3) without the use of energy. Water molecules, sodium ions, chloride ions, hydrogen carbonate ions and some urea are absorbed in this way.

The other form of transport is active and requires ATP to move molecules against electrochemical gradients. Actively transported molecules include glucose, amino acids and sodium, calcium, potassium, phosphate and chloride ions. This type of reabsorption is often interdependent, e.g. glucose depends on the active reabsorption of sodium in a process called *cotransfer*. Active transport mechanisms only work up to a maximum level, known as the **transport maximum (Tm)**, for any substance. When all the carrier molecules are occupied, it is not possible for any more molecules of that substance to be reabsorbed and it will appear in the urine. This occurs in diabetic hyperglycaemia, when the amount of glucose filtered exceeds its Tm value, which results in glycosuria (see Chapter 8). The amount of a substance, such as glucose, that can be reabsorbed is also termed the **renal threshold**. If the level of a substance rises beyond its threshold it will appear in the urine. It is worth remembering that glycosuria does not always mean that the person has diabetes mellitus.

Further considerations of reabsorption

Details of molecules reabsorbed can be found in *Figure 15.5*, but we should note certain other important aspects.

Ions such as sodium and chloride are reabsorbed both passively and actively depending upon the location. Active transport produces the gradients required for some passive transport and osmosis, e.g. active sodium reabsorption is accompanied by the osmotic movement of water.

As we have already said, most reabsorption takes place in the proximal convoluted tubule. The reabsorption which occurs in the remainder of the tubule depends on homeostatic needs and is often regulated by hormones:

- The permeability of the distal tubule and collecting ducts to water is influenced by **antidiuretic hormone (ADH)** from the posterior pituitary gland (see Chapter 8). This change in permeability, which allows the kidney to produce dilute and concentrated urine, is discussed more fully on pages 389-391.
- Calcium and phosphate reabsorption is regulated by active vitamin D, parathyroid hormone and calcitonin (see Chapter 8).
- From our discussion of renal autoregulation you already know that sodium reabsorption in the distal tubule is influenced by secretion of aldosterone (see *Figure 15.6*). The adrenal cortex secretes aldosterone in response to renin–angiotensin release stimulated by low plasma sodium levels (*hyponatraemia*), a reduction in blood volume and or blood pressure, and sympathetic nerve stimulation. Direct aldosterone release occurs when potassium is high (*hyperkalaemia*). Here potassium ions are secreted in exchange for the reabsorption of sodium in the distal tubule. **Natriuretic peptide hormones** (see Chapter 10) have the reverse effect of reducing sodium and water reabsorption by their inhibition of renin and ADH secretion.

Reabsorption is also important in regulating the pH of the blood, further discussion can be found on pages 391-394.

Secretion

Tubular secretion is the passage of molecules from the blood in the peritubular capillaries, through the tubule cell, and into the filtrate. As with reabsorption the molecules are moved either actively or passively and secretion occurs throughout the tubule.

The molecules secreted may be those completely unwanted by the body, e.g. drug and hormone metabolites, or those present in excess amounts, e.g. potassium. When potassium ions are secreted into the filtrate they are exchanged for sodium, which we discussed earlier. Other molecules which find their way into the filtrate by tubular secretion include hydrogen ions in acidosis, urea, creatinine and ammonia.

The much modified filtrate is now urine, which travels in the collecting ducts (see *Figure 15.3*) and through the calyces to the renal pelvis and ureter (see *Figures 15.1* and *15.2*).

Further consideration of urine concentration and volume

The kidneys can respond to changes in body fluid solute concentration or *osmolarity* (see Chapter 2) by producing either dilute or concentrated urine (see *Figure 15.8*). The mechanisms by which this response is achieved involve osmoreceptors, the secretion of antidiuretic hormone (ADH), permeability of the tubule to water and **countercurrent multiplication theory**, a hypothesis used to explain the production of osmolarity gradients within the medullary interstitial fluid.

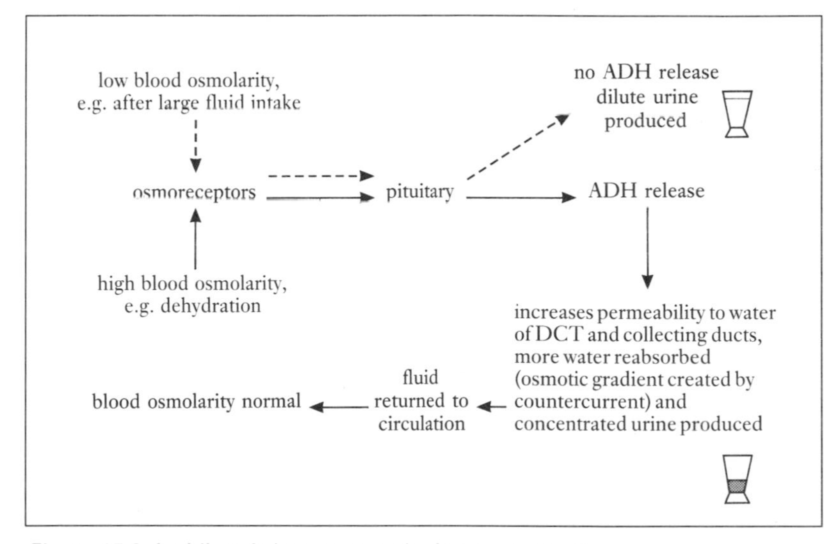

Figure 15.8. Antidiuretic hormone and urine concentration.

For our purposes it might be easier to consider the production of concentrated and dilute urine separately.

Concentrated urine

Osmoreceptors in the hypothalamus monitor the osmolarity of the blood and, when this is too high (in dehydration), stored ADH is released by the posterior pituitary gland (see Chapter 8). ADH affects the distal tubule and collecting duct by making the cells more permeable to water. However, this is only part of the story because the countercurrent movement of water and sodium and chloride ions between the loop of Henle, medullary interstitial fluid and the vasa recta (see Further Reading, e.g. Marieb, 1989) produces osmolarity gradients in the medulla, which ensure that dilute filtrate reaches the distal tubule.

Water moves passively from the descending limb of the loop of Henle to increase filtrate concentration. The ascending limb is impermeable to water but sodium and chloride ions move out into the medulla. The sodium and chloride ions, with the urea absorbed in the collecting ducts, diffuse into the descending limb to maximize filtrate concentration at the bottom of the loop. It is this movement of solutes without water that sets up the *multiplier system* which further increases the osmostic gradients.

The vasa recta, associated with juxtamedullary nephrons, act with the loop of Henle as part of the countercurrent exchange to maintain the osmotic gradients. Water from deep in the medullary region is reclaimed by the vasa recta and returned to the circulation.

Thus the filtrate entering the distal tubule, having lost both water and solutes, is dilute. The difference in osmolarity between the filtrate in the distal tubule and the medullary region produces the osmotic movement of water through the highly permeable tubule, which results in concentrated urine. The water returns to the circulation via the peritubular capillaries, which results in fluid conservation, the production of a smaller volume of concentrated urine and a fall in blood osmolarity – thus restoring homeostasis.

Nursing Practice Application **Diuretics**

Drugs which increase urine production are known as diuretics. They act in a variety of ways but all rid the body of excess fluid. Many everyday substances, including tea, coffee and alcohol, have a similar diuretic effect. One of the more unpleasant effects of a 'hangover' is the dehydration which results from the alcohol-induced diuresis.

Different types of diuretic drugs include:

(i) *Thiazides,* e.g. bendrofluazide, used to control the oedema of cardiac failure (see Chapter 10). They act on the distal tubule increasing excretion of water, sodium, potassium and chloride.

(ii) *Loop diuretics,* e.g. bumetanide, are more powerful drugs used for oedema and oliguria caused by renal failure. These drugs act by preventing reabsorption in the loop of Henle.

(iii) *Potassium sparing,* e.g. spironolactone, used for the oedema of hepatic and cardiac disease. Spironolactone acts by competing with aldosterone at its tubular receptor sites, which results in retention of potassium and increased excretion of water and sodium. It also potentiates the action of thiazide and loop diuretics.

(iv) *Osmotic diuretics,* e.g. mannitol, a sugar, given intravenously to reduce cerebral oedema or produce a diuresis after drug overdose. The diuresis is achieved by the osmotic 'pull' created by the sugar as it is excreted.

During diuretic therapy drug effectiveness can be assessed by monitoring fluid balance, extent of oedema, weight, blood pressure and serum electrolyte levels (see Chapter 2).

Nurses should observe for side-effects, which include hypokalaemia (which may present as cardiac arrhythmias and muscle weakness), dehydration and hypotension. It is important that diuretics are given early enough in the day to avoid *nocturia* (passing urine at night) and that potassium supplements are given as prescribed. The person is advised to maintain an adequate (2–3l) fluid intake unless this is specifically contraindicated, e.g. renal failure and for those affected by postural hypotension changing position slowly will help to avoid dizziness or fainting. Compliance can be increased by ensuring that the person and his or her family have adequate information about the drug: this should include details of side-effects and the need to report any adverse reactions.

Dilute urine

If the blood reaching the osmoreceptors has a low osmolarity, such as following a large fluid intake, no ADH is released and the distal tubule and collecting ducts remain impermeable to water. Consequently the dilute filtrate produced by the countercurrent mechanisms is excreted as dilute urine, which ensures the loss of excess water and a return to homeostatic balance.

Role of the kidneys in the regulation of pH

The kidneys provide the long-term means of maintaining the normal blood pH in the range 7.35–7.45. They function in conjunction with the more immediate effects of the blood buffer systems and lungs (see Chapters 2 and 12). Every day we ingest acids and the body produces acidic substances during metabolism: carbon dioxide, lactic acid, ketone bodies and acids, such as sulphuric and phosphoric acids, produced from the breakdown of proteins and lipids. Sulphuric and phosphoric acids are termed the *metabolic* or *fixed acids* (i.e. they cannot be converted to CO_2). The buffer systems 'mop up' or counteract the effects of the acids as a temporary solution but it is the lungs and kidneys which operate to remove or retain the acids and prevent pH changes.

Depending upon body chemistry the kidneys are able to produce urine with a pH 4.5–8.0 to maintain homeostasis. Renal mechanisms for preventing pH changes, which take place over several hours, are responsible for excreting fixed acids (see pages 393-394) as required and for conserving or eliminating hydrogen carbonate ions.

If either one of the main (renal or respiratory) pH regulation systems is impaired the other will attempt to *compensate*, e.g. if ventilation decreases, causing PCO_2 to rise, the kidneys conserve more hydrogen carbonate ions to restore blood pH to normal.

The kidneys help to maintain acid–base balance with a variety of

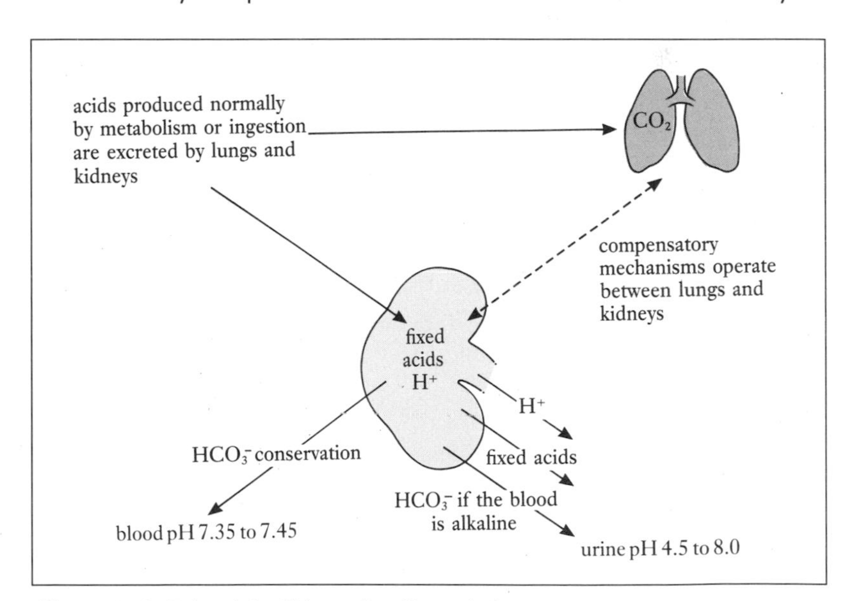

Figure 15.9. Role of the kidneys in pH regulation.

complex mechanisms which we outline only (*Figure 15.9* and *Flowcharts 15.1–15.3*). Readers requiring more information are directed to Further Reading (e.g. Robinson, 1975).

Secretion of hydrogen ions into the filtrate

Hydrogen ions are released into the filtrate by the dissociation of carbonic acid formed in the tubule cells. Carbonic acid, you will remember, is produced from water and carbon dioxide in the presence of the enzyme *carbonic anhydrase*. Hydrogen ion secretion is closely linked to sodium and potassium balance. For every hydrogen or potassium ion excreted a sodium ion is reabsorbed to maintain the electrochemical balance.

Problems arise in hyperkalaemia where the increased potassium excretion results in the retention of hydrogen ions and metabolic acidosis.

H^+ or K^+ secreted ⟶ | ⟵ Na^+ reabsorbed

Regulation and replenishment of alkali reserve in the blood

The alkali reserve in the blood is replenished by the conservation and reabsorption of hydrogen carbonate ions, which are produced as hydrogen ions are secreted or buffered. The return of hydrogen carbonate ions to the blood is dependent upon the presence of hydrogen ions in the filtrate and, with this mechanism, the kidneys provide a second means of reducing acidity. In a situation where the alkali reserve is high and few hydrogen ions are being secreted, e.g. respiratory alkalosis, the tubule cells will produce an alkaline urine containing hydrogen carbonate ions.

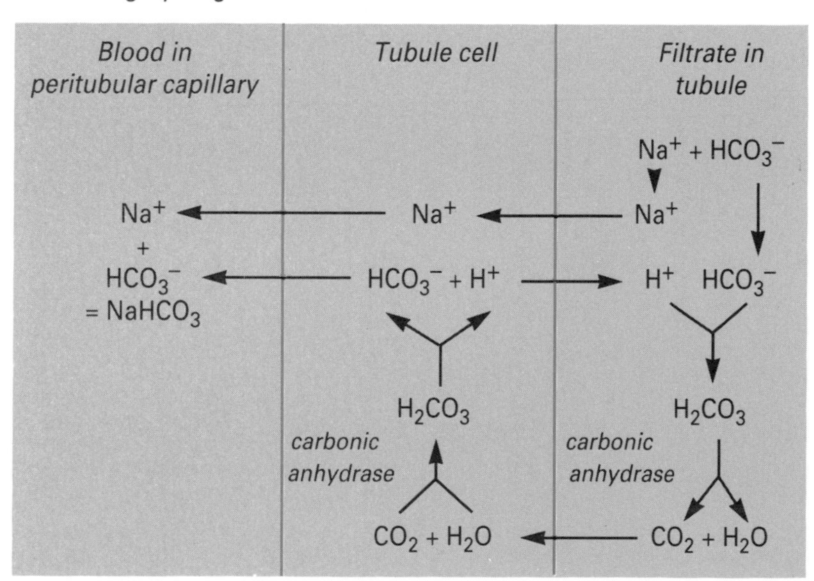

Flowchart 15.1

Excretion of the hydrogen ions and anions

Excretion of hydrogen ions and anions is achieved through the two buffer systems, the *ammonia–ammonium* and *hydrogen phosphate* systems, present in the kidney. Hydrogen ions secreted into the filtrate are buffered to prevent the urine becoming too acidic, and

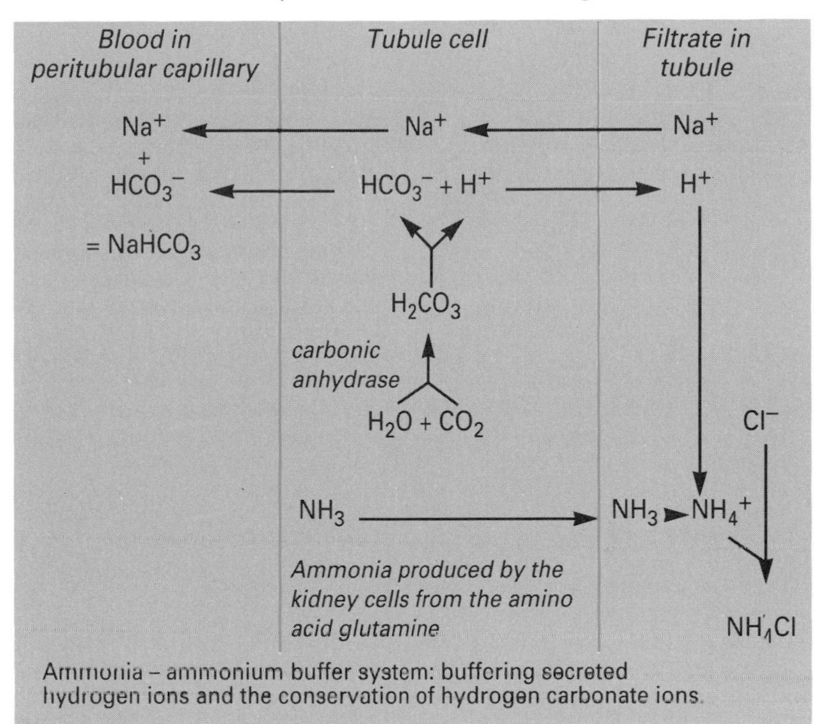

Flowchart 15.2.

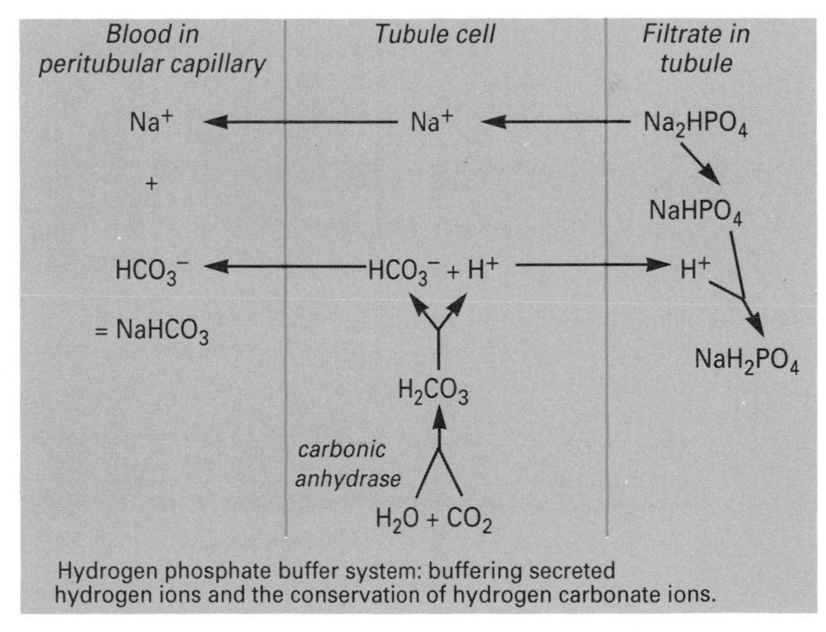

Flowchart 15.3.

Nursing Practice Application Observation and testing of urine

Routine screening of urine for abnormalities is important in: (i) early identification of potential problems, e.g. *albuminuria* (albumin in the urine) during pregnancy may indicate the development of *eclampsia* (a condition characterized by oedema, albuminuria and high blood pressure which may progress to fits); (ii) diagnosis of disease, e.g. the presence or not of bilirubin in the urine is helpful in differentiating between types of jaundice (see Chapter 14); (iii) monitoring of existing conditions, e.g. the amount of *glycosuria* (glucose in the urine) gives some guide to diabetic control where blood glucose monitoring is not possible.

The volume of urine passed will obviously reflect the fluid intake but abnormally large or small amounts may be associated with disease processes. **Polyuria** (increased volume of urine)is caused by an inability to form concentrated urine or the presence of large amounts of solutes in the urine. It may be a feature of early renal failure, diabetes insipidus due to a lack of ADH (see page 388 and Chapter 8) and diabetes mellitus where glycosuria causes an osmotic diuresis (see Chapter 8). Oliguria or **anuria** (cessation of urine production) indicate a decrease in GFR such as may occur in renal failure or shock. Problems occur when the urinary volume falls below 500 ml/24 hours; this is the minimum amount required to excrete the solutes, e.g. urea.

Before testing a fresh sample of urine the nurse should first note the colour, clarity and odour. The colour gives clues about concentration (urine is usually pale when dilute and dark when concentrated) and abnormal contents such as blood (*haematuria*) give the urine a smoky appearance, or it may be frankly blood-stained. Urine is normally clear with occasional slight turbidity caused by mucus; however, excessive cloudiness may indicate the presence of pus and infection. Fresh urine should not have an offensive odour but if left to stand the urea undergoes bacterial conversion to ammonia. A 'fishy' odour usually means the urine is infected and in diabetes mellitus the urine may have a sweet odour.

Following this important initial examination the specific gravity (SG) can be measured with a *urinometer.* SG obviously depends on the amount of solutes present and is a guide to other abnormalities, e.g. pale (apparently dilute) urine with a high SG may contain glucose.

Only now are we ready to 'test' the urine with one of the variety of commercially produced reagent sticks/tablets. The reagent sticks generally measure or detect the following: pH, protein, glucose, ketones, blood, nitrites, bilirubin and urobilinogen. For possible significance of abnormal substances in urine see *Figure 15.10.*

anions, such as sulphates and chlorides, combine with ammonia produced in the tubule cells to form ammonium salts, e.g. ammonium sulphate and ammonium chloride, which are excreted in the urine.

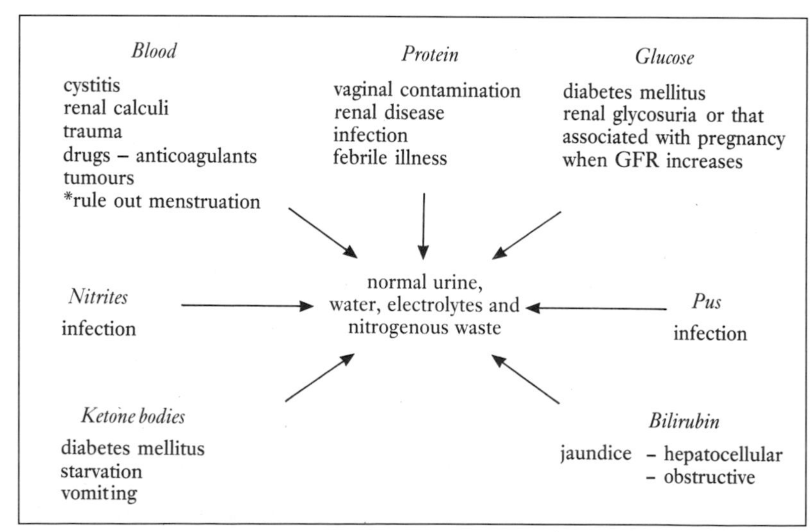

Figure 15.10. Abnormalities in urine.

Special Focus Renal failure

The failure of any major organ is always very serious, but support for the failing kidneys is available through *dialysis* and renal transplant, both of which are now commonly performed and successful procedures.

Renal failure may occur as an acute event where renal function is often reversibly impaired, or as a chronic illness where deterioration with the loss of nephrons progresses irreversibly over months or years.

We will look at some common causes of acute and chronic failure separately but consider the homeostatic effects of impaired renal function as a whole.

Acute renal failure

Usually the causes are divided into prerenal, renal and postrenal: (i) prerenal causes include: situations where the blood volume or cardiac output is reduced, e.g. haemorrhage, dehydration and cardiac failure, and haemolysis, e.g. caused by a mismatched blood transfusion, which reduces renal perfusion; (ii) renal causes include drugs, toxins and ischaemia, which directly affect renal function by causing acute tubular necrosis, e.g. heavy metal poisoning, certain types of glomerulonephritis may also cause the kidneys to fail acutely; (iii) postrenal causes include obstruction of the urinary tract, e.g. prostatic hypertrophy, bilateral calculi or external pressure, e.g. from advanced tumours of the cervix.

Chronic renal failure

The causes of this gradual reduction in renal function leading to failure include: congenital abnormalities of the kidney, e.g. polycystic disease, hypertension, chronic infection, glomerulonephritis and diabetic nephropathy.

Effects on homeostatic balance and management

The effects of failing renal function do not become apparent until GFR has been reduced by 75%. What happens to homeostatic balance, when the kidneys fail, can be compared with a situation where your household rubbish is not collected and the waste still being produced simply adds to the pile.

When little or no urine is produced the body cannot excrete waste and a toxic state known as *uraemia* develops. Nitrogenous waste accumulates and the blood urea level (normal 2.5–6.0 mmol/l) rises.

The electrolyte and acid–base balance is impaired and potassium and hydrogen ions are retained. Excess potassium results in hyperkalaemia, which leads to life-threatening cardiac arrhythmias and the inability to excrete hydrogen ions produces a metabolic acidosis. Water and sodium are retained which results in oedema and cardiac failure.

Apart from these serious effects the person concerned, who is as you can imagine desperately ill, also suffers from: (i) nausea and vomiting – toxins, which include urea, irritate the gastrointestinal tract; (ii) hiccups – toxins irritate the phrenic nerve (see Chapter 5); (ii) pruritus – the accumulation of toxic waste causes intense skin itching; (iv) dyspnoea – due to anaemia associated with renal failure, fluid retention will lead to heart failure and pulmonary oedema (see Chapter 10), acidosis will also contribute to the dyspnoea; (v) confusion, fits and eventual coma – due to accumulation of toxic waste (uraemia) and possibly severe hypertension, both of which affect brain function (see Chapter 4).

The management of renal failure depends on its cause plus the age and general condition of the individual. Any regimen would include measures to treat the cause if possible and relieve distressing symptoms such as hiccups.

When oliguria/anuria is present the fluid intake is restricted to an amount which reflects urinary and insensible losses and takes account of the water produced during metabolism. Dietary modifications in the form of restricted sodium, potassium and protein are usual, these are combined with a high energy carbohydrate intake which prevents the use of protein or fat and catabolism of body tissue. Sometimes it is possible to treat hyperkalaemia by the oral or rectal administration of an ion exchange resin.

Where these measures are inadequate to control the problems it will be necessary to commence either peritoneal dialysis (see *Figure 15.11*) or haemodialysis. Both forms of dialysis depend upon the basic principles of water movement by osmosis and solute movement by diffusion across a selectively permeable membrane (see Chapter 2) to produce a concentration equilibrium either side of the membrane.

Peritoneal dialysis utilizes the peritoneum as the selectively permeable membrane and in haemodialysis a synthetic membrane is part of the 'artificial kidney'. Whichever type of dialysis is used the person's blood is separated from the dialysis fluid only by the selectively permeable membrane; the dialysis fluid contains no nitrogenous waste or potassium, which allows the movement of these molecules from the blood to the dialysis fluid.

Hypertonic dialysis fluid can be used to remove excess water and various additions to the fluid can pass in the opposite direction, e.g. to counteract the acidosis. Dialysis may be a temporary measure for a person with acute failure but in chronic situations some form of intermittent dialysis is required unless a renal transplant is possible.

Peritoneal dialysis may be undertaken as a continuous process (continuous ambulatory peritoneal dialysis, CAPD). Here the homeostatic balance is maintained by the use of 3–6 exchanges of dialysis fluid each and every day. CAPD is considered to provide a more constant internal environment than does haemodialysis (Jerrum, 1991). Unfortunately not everyone is suitable for transplant, this and the shortage of donor kidneys means that many people must depend for long periods upon dialysis with its attendant physical, psychological and social problems.

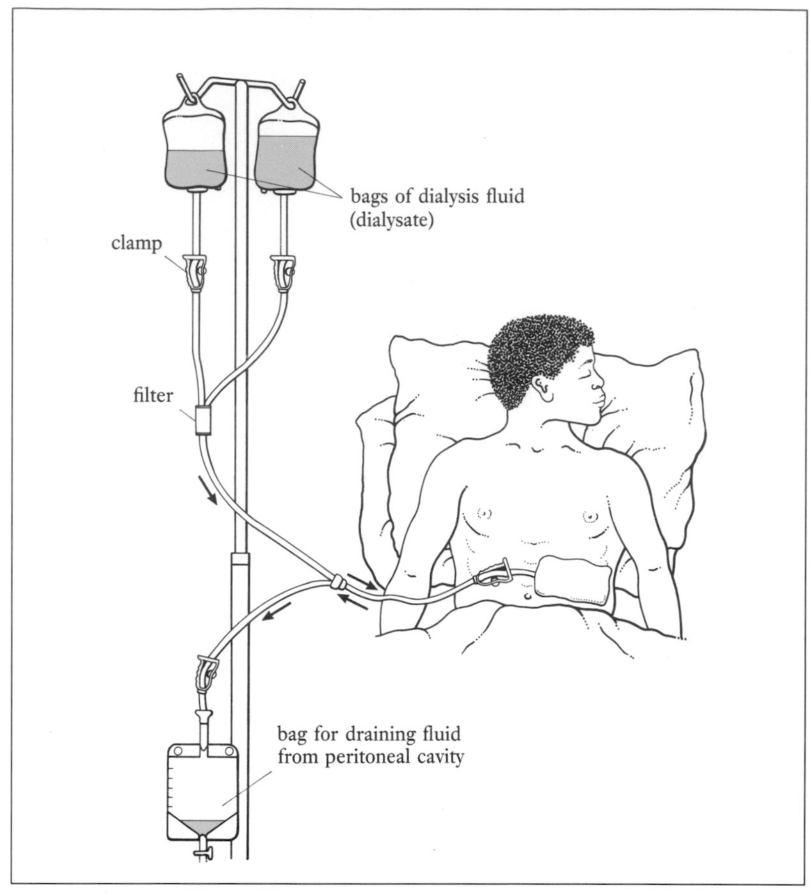

Figure 15.11. Peritoneal dialysis.

Lower Urinary Tract

Ureters

The two **ureters** convey urine from the renal pelvis to the bladder (see *Figure 15.1*). They are 25–30 cm in length with a very small diameter of about 3–6 mm (see *Nursing Practice Application*, page 379). The ureters run behind the peritoneum close to the psoas muscle to enter the bladder through its posterior wall. The portion of ureter in the bladder wall runs obliquely (see *Figure 15.12*), a feature which prevents the reflux of urine by compressing the ureter as the bladder fills and empties.

The ureters have three basic layers: inner transitional epithelial mucosa (see Chapter 1) continuous with that of the renal pelvis and bladder, two smooth muscle layers and an outer fibrous connective coat.

Urine is moved down the ureters by peristaltic contractions (see Chapter 13). These waves of muscular contraction occur as urine distends the renal pelvis and ureter. A calculus entering the ureter will cause severe *colic* (see *Nursing Practice Application*, page 379)

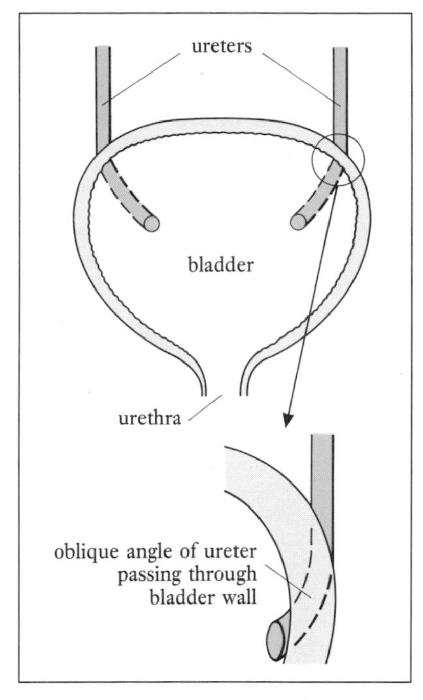

Figure 15.12. Ureters and bladder.

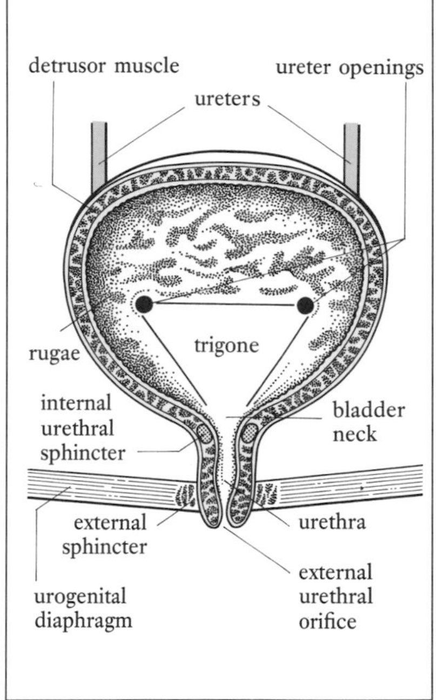

Figure 15.13. Bladder interior.

as intense muscular contractions attempt to dislodge the calculus and move it into the bladder.

Bladder

The **urinary bladder**, which lies in the pelvic cavity, acts as temporary storage for urine. It is a collapsible sac which, when empty, is behind the symphysis pubis, but rises into the abdominal cavity on filling. In females the bladder is close to the uterus and vagina; it is close to the prostate gland and rectum in males (see *Figure 1.24*). The bladder has four layers:

- Transitional epithelial mucosa able to withstand the normal variations in urinary pH and composition. This lining forms folds, or *rugae*, which allow for considerable distention as the bladder fills. We become aware of rising pressure when the bladder contains around 250–300 ml of urine and usually void when it contains 400–500 ml. The bladder can hold more than a litre, but would be very distended, painful and easily palpated (felt) abdominally.
- A submucosa of connective tissue, vessels and nerves.
- A smooth muscle layer formed from the **detrusor** muscle which thickens at the bladder–urethral junction to form the **internal urethral sphincter**. During micturition (see pages 399-400) involuntary contraction of the detrusor and opening of the sphincter empties the bladder and allows urine into the urethra. Micturition is further aided by relaxation of the pelvic floor muscles (*levator ani,* Chapter 18) on which the bladder sits.

- The outer coat is part peritoneum (superior surface) and part fibrous tissue.

Inside the bladder is a smooth triangular area, formed from the openings of the ureters and bladder neck, known as the **trigone** (*Figure 15.13*).

Person-centred Study Harry

Harry has been a postman for 32 years; he is due to retire next year and plans to spend his extra leisure time in the garden and with his racing pigeons. He and his wife Pat feel relieved and rather fortunate that they can now look forward to their retirement. Only six months ago Harry was diagnosed as having early bladder cancer and they thought it was the end of all their plans.

Harry had passed some blood in his urine but as he had no pain did not really think it much to worry about, but Pat happened to mention 'the trouble' to their neighbour, a charge nurse at the local hospital. He suggested that Harry see his doctor despite having no pain.

Luckily for Harry his painless haematuria was investigated at once and a *cystoscopy* (endoscopic examination of the bladder) and biopsy confirmed an early bladder papilloma which was destroyed by diathermy via the cystoscope. He will have regular check cystoscopies with further diathermy as required.

Often painless haematuria is the only sign of bladder tumours and it is essential that all such occurrences are investigated. Other treatment modalities include radiotherapy, cytotoxic drugs and surgery if the tumour is more widespread.

The development of *photodynamic therapy* (photosensitization followed by laser therapy) is seen as a possibility for the treatment of early bladder tumours (Tootla and Easterling, 1989).

Urethra

The **urethra**, which conveys urine from the bladder to the outside, extends from the bladder neck to the **external urethral orifice** (*Figure 15.14*). As the urethra leaves the bladder it is lined with transitional epithelium which becomes stratified squamous epithelium. In females this tissue is similar to that lining the vagina and is also influenced by the hormone oestrogen. The involuntary internal sphincter (see page 397) keeps the urethra closed except during micturition. A further sphincter, the **external urethral sphincter** is formed from skeletal muscle at the level of the pelvic floor. This sphincter is under voluntary control.

The female urethra, which is around 4 cm in length, is attached to the anterior vaginal wall, and opens externally just anterior to the vaginal orifice (*Figure 15.14(a)*). The short urethra and its proximity to the anus make women particularly susceptible to urinary tract infection. When oestrogen secretion declines during the climacteric (see Chapter 20) it can lead to urethral discomfort (Gould, 1990).

In males the urethra is around 20 cm in length and has dual excretory and reproductive functions (see Chapter 20). It has three parts – **prostatic**, where it passes through the *prostate gland*, the **membranous** urethra, surrounded by the external urethral sphincter, and the **penile** urethra, which passes through the penis to open at the external urethral orifice. The *ejaculatory* and *prostatic ducts* enter the prostatic urethra and those of the *bulbourethral glands* empty their secretions lower down in the penile part (*Figure 15.14(b)*). The secretions and their role is examined in Chapter 20.

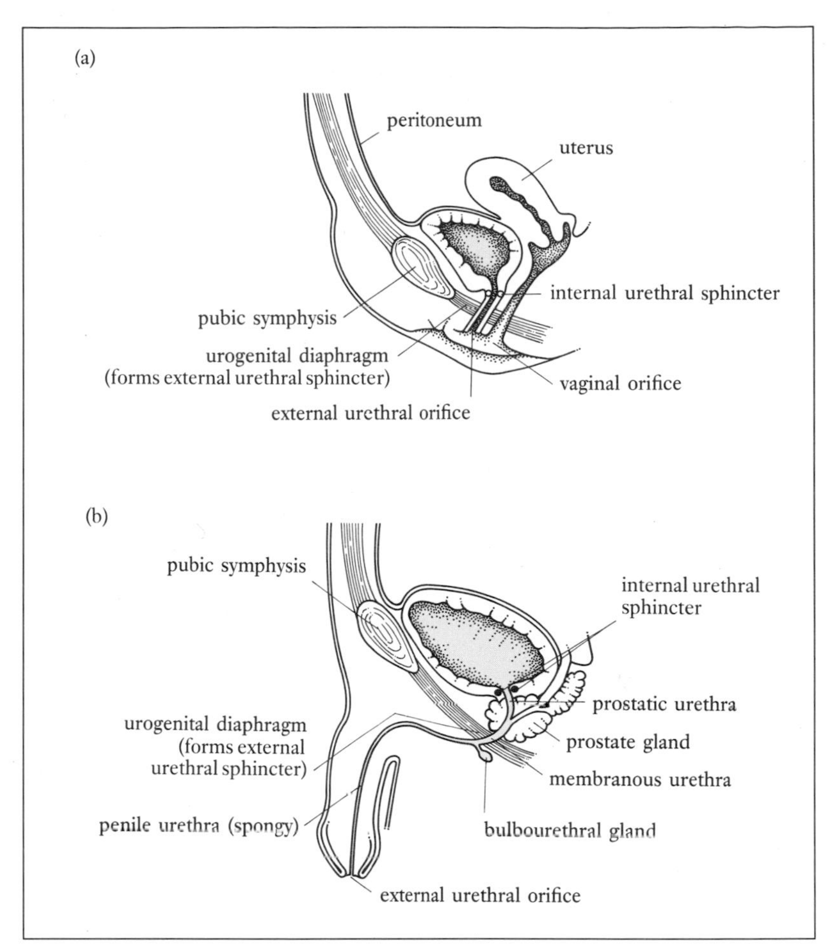

Figure 15.14. Urethra. (a) Female; (b) male.

Micturition

Micturition is controlled through autonomic and voluntary nerves (*Figure 15.15*). In infants it operates entirely through a spinal reflex, but by the age of 2 years nervous system development is sufficient to give awareness of bladder filling and the ability to inhibit the reflex and control the external urethral sphincter.

As the bladder fills with urine (250–300 ml), stretch receptors in the wall transmit sensory (afferent) impulses in parasympathetic fibres to the sacral region of the spinal cord. Impulses travelling up the cord to higher brain centres make us aware of the need to void urine.

When sufficient urine is present (400–500 ml), parasympathetic motor (efferent) impulses from the spinal cord cause detrusor contraction, bladder neck relaxation with *'funnelling'* and internal sphincter relaxation which allows urine into the urethra. Actual voiding occurs when the somatic motor nerves (*pudendal*), which normally keep the external sphincter closed, are reflexly inhibited. This relaxes the external sphincter and *urogenital diaphragm* (pelvic floor) and

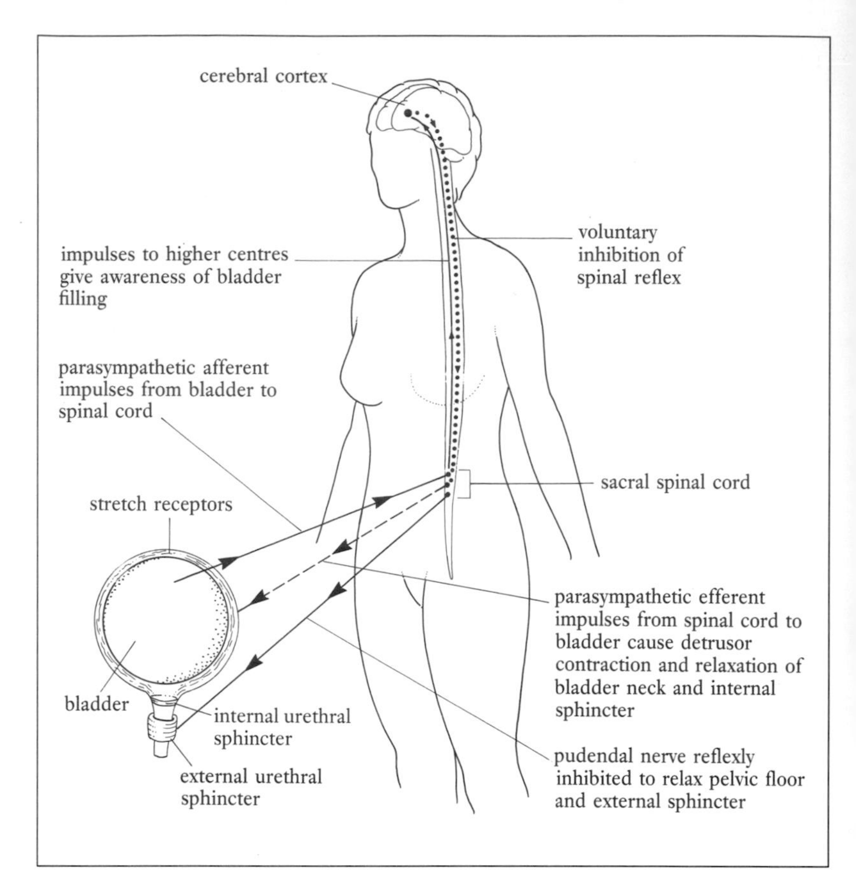

Figure 15.15. Micturition.

allows urine to flow out. Flow is aided by increased intra-abdominal pressure (Valsalva's manoeuvre) and contraction of the abdominal muscles.

Once we have learnt control, the micturition reflex can be voluntarily inhibited by impulses from higher centres and contraction of the external sphincter. This gives us some choice about where and when voiding occurs but eventually this option is removed, and when urine volume reaches a certain level the bladder will empty whether convenient or not.

Sympathetic fibres innervating the bladder may be involved with detrusor relaxation but are thought to have only a minor role in micturition (Hinchliff and Montague, 1988).

The process of micturition is not fully understood and to date much of the current knowledge has been obtained through *urodynamic studies*. Readers are directed to Further Reading (Abbott, 1992; Blandy, 1982; Smith, 1984) and References (Gould, 1990) for information regarding urodynamic studies.

Abnormalities of micturition

It is not possible to define *'normal'* micturition as there are large variations in healthy individuals. What is important is the recognition of departures from normal or factors which predispose to abnormality for the individual. At this stage it is appropriate to include a brief outline of some problems associated with the elimination of urine but readers are also directed to the Further Reading and References. Lowthian (1977) states that most micturition problems involve increased frequency and social inconvenience especially when control is lost.

Frequency

Passing small amounts of urine frequently is often a feature of *cystitis* (see pages 403-404). This occurs because bladder stretch receptors are stimulated constantly, resulting in detrusor contraction before the bladder is really full. It should not be confused with increased voiding associated with a large fluid intake, e.g. getting up many times to void after an evening drinking beer. Frequency may also occur with anxiety, such as that associated with hospital admission, examinations or fear of an 'accident'. It also accompanies pregnancy due to pressure on the bladder and more seriously with bladder tumours.

Retention

Retention is the inability to pass urine. It is commonly due to *prostatic hypertrophy* in older men (see Chapter 20) but may be associated with other obstructions of the bladder or urethra. Retention with overflow, frequency or incontinence is easily misdiagnosed and it is essential to discover whether the bladder is distended by abdominal examination.

The immediate management of retention involves the introduction of a urinary catheter (see page 402) to drain urine. The catheter is usually introduced via the urethra but the *suprapubic* route may be used where the urethra is obstructed.

Incontinence

Inability to control the voiding of urine is extremely common. Norton (1986) states that 2–3 million people in the United Kingdom are incontinent of urine, although for many this is very minor and is not seen as a problem. Whatever the exact numbers they are much greater than those known to health and social service agencies. The results of a postal survey by Thomas *et al.* (1980) confirm this for all age groups and both genders. The accepted rate for females aged 15–64 was 0.2%, but the study revealed that 8.5% had continence problems. Contrary to popular opinion, incontinence is not confined to the elderly population but affects all age groups and both genders. Mandelstam (1989) gives examples which include a young man who wets the bed, a new mother, a middle-aged woman and an isolated elderly person. Research undertaken by Armstrong-Ester and Hawkins (1982), based on the hypothesis that the disturbance of circadian rhythms would manifest as incontinence and nocturnal disturbances, found that many elderly people had desynchronized renal circadian rhythms in which the cycle was 28 hours, not 24 hours. Their small sample showed that renal circadian rhythms were affected in 15 people – renal and temperature – and 8 people – renal. Obviously the scale of the problem represents an enormous challenge for practice, and Norton (1986) sees the promotion of continence as the business of every nurse, midwife and health visitor.

Incontinence can be divided into several types which include:

- *Stress incontinence,* which is characterized by a leakage of urine when intra-abdominal pressure rises, e.g. coughing, laughing. This type of incontinence, which affects women, is due to bladder neck displacement and loss of the urethral–vesicular angle caused by weakening of the pelvic floor which follows childbirth, uterine prolapse and the climacteric.
- *Detrusor instability,* which leads to *urge incontinence.*
- *Overflow,* where obstruction to the flow of urine, such as an enlarged prostate gland, leads to retention with overflow.
- *Neurological incontinence* where conditions such as spinal injury or multiple sclerosis (see Chapters 3 and 4) damage the nervous pathways involved in bladder control. The resultant problems depend on where the damage has occurred and include: atonic bladder with overflow, reflex emptying of the full bladder and loss of higher centre inhibition with urge incontinence.

Catheterization

We have already mentioned the use of urinary catheters in the management of retention. Catheterization is a common procedure and 10–12% of people in hospital will have a catheter at some stage. This may be only short-term, such as following gynaecological repair surgery, but could be longer term, e.g. management of incontinence where other strategies have been ineffective. Many studies have shown the presence of a catheter to increase greatly the risk of infection (see page 403) and nursing interventions which prevent contamination are required. Nursing care plans should include maintaining the integrity of the closed drainage system, adequate cleaning, hand washing for staff, a fluid intake of at least 2.5 litres and appropriate teaching for the patient. In a study of 294 catheterized patients in general hospitals Crow *et al.* (1988) found that the closed system was broken in 42%, that only 48% had the bag positioned correctly and that cleaning and hand washing by nurses was poor.

Roe (1989a) found that the information given to patients/their carers by nurses (60 hospital, 46 community) about catheter care was neither consistent with the literature nor comprehensive. She identified a need for more catheter care education for nurses, teaching for patients/carers, and for research which considers the effects of teaching on the ability of people to care for their own urinary drainage. Another study by Roe (1989b) identified the number of catheterized people being looked after by the community nurses of one district health authority; for the period of the study this was 61 (4%). Roe concludes that this knowledge is important in assessing education needs and costs.

Urinary tract infection (UTI)

Complete and regular bladder emptying and the secretion of immunoglobulins by the bladder mucosa affords some protection against infection, but urinary tract infection is still a common occurrence. UTI represented 30% of hospital-acquired infections in a study by Meers *et al.* (1981). The infection usually enters the urinary tract via the urethra but it can be blood-borne or spread directly from other structures, e.g. diverticular disease of the colon (see Chapter 13).

The vast majority of urinary infections are caused by intestinal bacteria, e.g. *Escherichia coli* and *Streptococcus faecalis,* which spread from the anal/perineal area, but other culprits include *Proteus spp., Klebsiella spp., Pseudomonas aeruginosa* and *Staphylococcus epidermidis.*

Infection is an unusual event in males with a normal urinary tract. However, for reasons we have already discussed UTI is very common in females, and up to 50% will be affected at some stage (Asscher, 1982). The frequency of UTI increases in the population over 65.

Predisposing factors

The predisposing factors include:

- Trauma to the urethra during intercourse and childbirth.
- Catheterization, bladder instrumentation, e.g. cystoscopy.
- Urinary stasis, e.g. urinary obstruction, immobility.
- Poorly functioning immune system, e.g. people having immunosuppression therapy, the elderly.
- Pregnancy.
- Diabetes mellitus.
- Any severe debilitating condition.

Types of infection

- Asymptomatic or covert *bacteriuria* (bacteria in the urine) is characterized by at least 10^5 colony-forming organisms/ml in freshly voided urine or a specimen obtained suprapubically (MRC, 1979). With these numbers, and remembering that bladder urine is normally sterile, it is unlikely that the presence of bacteria is due to contamination. Urine testing, which should form part of every nursing assessment, is important in the detection of asymptomatic UTI (Pitt, 1989).
- Infection causing inflammation of the bladder (*bacterial cystitis*) and urethra (*urethritis*), leading to **dysuria** (painful or difficult micturition), haematuria, offensive cloudy urine which contains leucocytes, frequency, urgency, suprapubic pain and possibly

pyrexia. The diagnosis is confirmed by culturing a midstream specimen of urine (MSSU). Usually treatment with an appropriate antimicrobial drug, e.g. trimethoprim and a fluid intake of at least 2.5 litres is effective. Analgesia such as paracetamol and a mild alkalis, e.g. potassium citrate, to reduce urine acidity will minimize the scalding pain on micturition.

- The infection may travel up the ureters to involve the kidneys (*pyelonephritis*). In addition to the problems of cystitis the person has loin pain and pyrexia. Pyelonephritis can cause permanent kidney damage which if severe can lead to renal failure (see page 395).

Person-centred Study **Sue**

It may seem to you that UTI is an inevitable consequence of being female but there are several simple preventive measures that any woman can be taught.

Sue, who works in a food processing factory, has lost considerable time from work due to several attacks of cystitis treated with antimicrobial drugs. Following Sue's return the occupational health sister suggests a visit to the health centre, to which she agrees.

Sue is able to discuss her problem with one of the nurses who gives her a leaflet (HEA, 1987), invites her to contact the centre if she is worried and makes the following suggestions: (i) take 2–3 litres of fluid a day avoiding tea, coffee and alcohol which are diuretic (to avoid dehydration and keep bladder 'flushed out'); (ii) attention to hygiene especially after using the toilet, always wipe 'front to back' (to avoid bringing bowel organisms towards the urethra); (iii) empty bladder before and after intercourse (to 'flush' organisms from the bladder and urethra); (iv) avoid intercourse if symptoms of cystitis are present (to avoid urethral trauma); (v) if dysuria/frequency occurs increase intake of water or fruit juice, take a proprietary 'cystitis mixture' (to reduce urine acidity and hence discomfort), start simple measures which give relief such as analgesia or a warm bath and arrange to see GP.

If Sue continues to experience attacks of cystitis her GP may decide to investigate more fully with an intravenous urogram (radiographic examination of the urinary tract, the dye injected intravenously is excreted by the kidneys) and cystoscopy.

Summary/Check List

Introduction – early development.
Kidney – Gross structure. Blood and nerve supply, *Nursing Practice Application* – renal calculus. Nephron, juxtaglomerular apparatus. Blood flow.
Functions of the kidney – Urine, physical characteristics, composition. Formation of urine. Filtration. GFR and regulation, *Nursing Practice Application* – renal clearance. Reabsorption. Secretion. Urine concentration and volume, *Nursing Practice Application* – diuretics. Kidney and pH regulation, *Nursing Practice Application* – observation and testing of urine. *Special Focus* – renal failure.
Lower urinary tract – Ureters. Bladder, *Person-centred Study* – Harry. Urethra.
Micturition – normal. Abnormal. Catheterization. Urinary tract infection. Person-centred study – Sue.

Self Test

1 Draw and label the urinary system.

2 Put the following in the correct order:
(a) distal convoluted tubule
(b) Bowman's capsule
(c) collecting ducts
(d) loop of Henle
(e) proximal convoluted tubule.

3 Which of the following statements are true?
(a) The kidneys receive around 25% of the cardiac output.
(b) The afferent arteriole of the glomerulus is larger than the efferent arteriole.
(c) All nephrons have a vasa recta.
(d) The nephron has two sets of capillaries.

4 Briefly describe the three processes involved in the formation of urine.

5 Put the following in their correct pairs:

(a) GFR/min	(b) 1200 ml
(c) daily urine volume	(d) 180 litres
(e) renal blood flow/min	(f) 125 ml
(g) filtrate formed daily	(h) 1000–1500 ml

6 Explain briefly how the kidney produces concentrated urine to conserve water.

7 What information can be gained from noting the colour, clarity and odour of a fresh sample of urine.

8 Which of the following statements are true?
(a) The detrusor muscle is involuntary.
(b) The ureteric and urethral openings are found in the bladder area known as the trigone.
(c) The urethra is closed except during micturition.
(d) Sympathetic nerves control bladder contraction and sphincter relaxation.

9 What type of incontinence is present in the following people?
(a) Farida, who has three small children, is incontinent when she laughs or coughs.
(b) Maggie finds she has little warning of voiding and often cannot get to the toilet in time.
(c) Declan, aged 65, finds that he 'dribbles' small amounts of urine without warning.

10 What factors predispose to UTI and how might micturition be affected by bacterial cystitis?

Answers
1. See page 376, 377. 2. b, e, d, a, c. 3. a, b, d. 4. See page 383. 5. a – f, c – h, e – b, g – d. 6. See pages 389-390. 7. See page 394. 8. a, b, c. 9. a, stress; b, detrusor instability (urge incontinence); c, overflow with retention. 10. See page 403.

References

Armstrong-Ester, C.A. and Hawkins, L.A. (1982) Day for night circadian rhythms in the elderly. *Nursing Times*, **78** (30), 1263–1265.

Asscher, A.W. (1982) *Urinary Tract Infection*. London: Update Publications.

Crow, R., Mulhall, A. and Chapman, R. (1988) Indwelling catheterization and related nursing practice. *Journal of Advanced Nursing*, **13** (4), 489–495.

Gould, D. (1990) *Nursing Care of Women*. London: Prentice Hall.

Health Education Authority (1987) *Cystitis and What to Do About it!* London: Health Education Authority.

Hinchliff, S.M. and Montague, S.E. (1988) *Physiology for Nursing Practice*. London: Bailliere Tindall.

Jerrum, C. (1991) CAPD: The state of the art. *Nursing*, **4** (30) 28–30.

Lowthian, P. (1977) Frequent micturition and its significance. *Nursing Times*, **73** (46), 1809–1813.

Macleod, J., Edwards, C. and Bouchier, I. (1987) *Davidson's Principles and Practice of Medicine*, 15th edn. Edinburgh: Churchill Livingstone.

Mandelstam, D. (1989) *Understanding Incontinence: A Guide to the Nature and Management of a Very Common Complaint*. London: Chapman and Hall (Published for the Disabled Living Foundation).

Medical Research Council (MRC) Bacteriuria Committee (1979) Recommended terminology of urinary tract infection. *British Medical Journal*, **2**, 717–719.

Meers, P.D. *et al.* (1981) Report on the National Survey of Infection in Hospitals 1980. *Journal of Hospital Infection*, **2**, (suppl.), UTI, 23–28.

Norton, C. (1986) *Nursing for Continence*. Beaconsfield: Beaconsfield Publishers.

Pitt, M. (1989) Fluid intake and urinary tract infection. *Nursing Times*, **85** (1), 36–38.

Roe, B.H. (1989a) A study of information given by nurses for catheter care to patients and their carers. *Journal of Advanced Nursing*, **14** (3), 203–210.

Roe, B. H. (1989b) Long–term catheter care in the community. *Nursing Times*, **85** (36), 43–44.

Thomas, T.M., Plymat, K.R., Blannin, J. and Meade, T.W. (1980) Prevalence of urinary incontinence. *British Medical Journal*, **281**, 1243–1245.

Tootla, J. and Easterling, A. (1989) PDT: Destroying malignant cells with laser beams. *Nursing 89* (US), **19** (11), 48–49.

Wright, E. (1987) Percutaneous nephrolithotomy. *The Professional Nurse*, **3** (3), 76–79.

Further Reading

Abbott, D. (1992) Objective assessment ensures improved diagnosis. Principles and techniques of urodynamics. *Professional Nurse*, **7** (11), 738–742.

Baer, C. L. (1990) Acute renal failure. *Nursing 90* (US), **20** (6), 34–39.

Blandy, J. (1982) *Lecture Notes on Urology*, 3rd edn. Oxford: Blackwell Scientific Publications.
de Wardener, H. E. (1985) *The Kidney: An Outline of Normal and Abnormal Function*, 5th edn. Edinburgh: Churchill Livingstone.
Evans, D. B. and Henderson, R. G. (1985) *Lecture Notes on Nephrology*. Oxford: Blackwell Scientific Publications.
Mandelstam, D. (Ed.) (1986) *Incontinence and its Management*, 2nd edn. London: Croom Helm.
Marieb. E.N. (1989) *Human Anatomy and Physiology*. Redwood City, California: Benjamin/Cummings.
Meers, P. D. and Stronge, J. L. (1980) Hospitals – should do the sick no harm. *Nursing Times*, **76** (30) Supplement.
Robinson, J. R. (1975) *Fundamentals of Acid–Base Regulation*, 5th edn. Oxford: Blackwell Scientific Publications.
Smith, D.R. (Ed.) (1984) *General Urology*, 11th edn. Los Altos, California: Lange Medical Publications.
Wheeler, V. (1990) A new kind of loving? The effect of continence problems on sexuality. *Professional Nurse*, **5** (9), 492–496.
Wilson–Barnett, J. and Batehup, L. (1988) *Patient Problems: A Research Base for Nursing Care* (Chapter 7). London: Scutari Press.
Wright, E. (1988) Catheter care: the risk of infection. *The Professional Nurse*, **3** (12), 487–490.
Wright, E. (1988) Minimising the risks of UTI. *The Professional Nurse*, **4** (2), 63–67.
Wright, E. (1989) Teaching patients to cope with catheters at home. *The Professional Nurse*, **4** (4), 191–194.

Useful Addresses

Incontinence Advisory Service
c/o Disabled Living Foundation
380–384 Harrow Road
London W9 2HU

SPOD (Association to promote the sexual
and personal relationships of people with disabilities)
286 Camden Road
London N7 0BJ

CHAPTER SIXTEEN

BONE TISSUE

OVERVIEW

- Structure and functions of bone tissue.
- Osteogenesis.
- Bone homeostasis.

LEARNING OUTCOMES

After studying Chapter 16 you should be able to:

- Discuss the functions of bone.
- Describe the two forms of lamellar bone tissue – compact and cancellous.
- Explain the classification of bones by shape.
- Describe the gross structure of a long bone.
- Describe the microscopic structure and chemical compostion of bone.
- Outline the formation of bone from membrane and cartilage.
- Explain how bones grow.
- Outline normal bone homeostasis and remodelling.
- Discuss abnormalities of bone homeostasis.
- Describe the different types of fracture.
- Explain how fractures heal.

KEY WORDS

Cancellous bone - light, spongy bone tissue.
Compact bone – dense, hard bone tissue.
Diaphysis – the shaft or middle part of a long bone.
Endosteum – thin membrane lining the medullary cavity.
Epiphysis – the ends of a long bone. In growing bones the epiphyses and diaphysis are separated by a plate of cartilage.
Medullary cavity – the central cavity of a long bone which contains marrow.
Ossification – formation of bone tissue, either from cartilage and membrane in the fetus or from bone growth after birth.
Osteo – a prefix meaning bone, e.g. *osteology*: the study of bones.
Osteoblasts – cells involved in bone formation and deposition.
Osteoclasts – cells involved in bone reabsorption and remodelling.
Osteocytes – bone cells derived from osteoblasts.
Osteogenesis – formation of bone (see ossification).
Periosteum – fibrous connective tissue covering a bone.

INTRODUCTION

Try to imagine for a moment what we would be like without bone; no movement would be possible and we would be forever constrained within a mass of flesh. Bone, the hardest body tissue, gives us form, protection, stability and movement. Sometimes mistakenly considered rather dull and static, bones are in fact formed from extremely dynamic tissue which undergoes constant structural changes.

In this chapter we will consider bone tissue and homeostasis prior to looking at muscle tissue in Chapter 17. Chapter 18 brings together the skeleton, joints and major muscles, and considers the integration required for movement and stability.

Functions of bone

Movement and stability
Bones form a supporting body framework and provide attachments for skeletal muscles, ligaments and tendons. They act as a system of levers to allow movement (see Chapter 18).

Protective
The bony framework protects internal structures, e.g. the skull encloses the brain and the rib cage protects the thoracic organs.

Storage
Bone is involved in mineral homeostasis. It is a repository for minerals such as calcium, phosphorus and magnesium.

Haemopoietic
Red marrow found in the marrow cavity of some bones produces blood cells (see Chapter 9).

BONE TISSUE

Types of bone tissue

Bone is a connective tissue (see Chapter 1) which has been *mineralized* to produce an extremely hard substance. Mineralization of connective tissue by the addition of mainly calcium salts gives it physical properties that include great tensile (capable of stretching) strength and the ability to withstand considerable compressive forces.

Bone may be the *woven* type, which is a fragile tissue produced during some stages of rapid **ossification** (formation of bone tissue from cartilage or membrane in the fetus, or bone growth after birth) as in fetal development (see pages 416-417) or fracture healing (see page 421). The other type is *lamellar* bone, the tough tissue which forms healthy adult bones.

The two basic types of lamellar bone tissue are **compact** (dense, hard bone tissue) (*cortical*) and **cancellous** (light, spongy bone tissue) (*trabecular*) which differ in structure, function and in location (*Figure 16.1*).

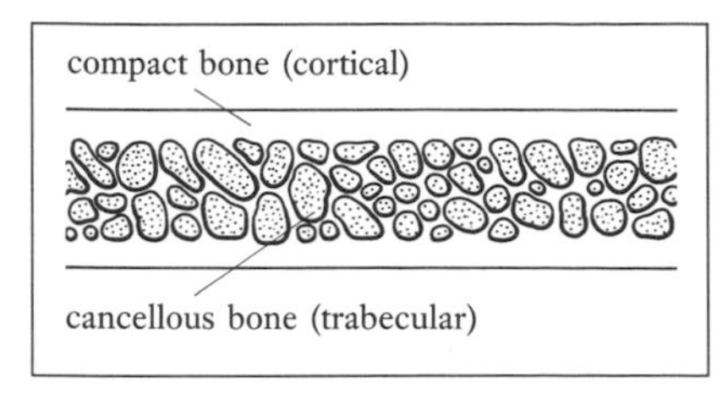

Figure 16.1. Bone tissue – compact and cancellous.

On naked-eye examination, compact bone looks hard and dense, although it actually consists of many structural units called **Haversian systems** or **osteons** (see pages 414 and 415 and *Figure 16.4*). Compact bone is found in flat bones, in the shaft **(diaphysis)** of long bones and as the covering of all bones.

Cancellous bone is spongy in appearance; it has fewer Haversian systems and is arranged in a network of **trabeculae** (tiny, beam-like pieces of bone) which surround marrow-filled cavities. The trabecular arrangement ensures that cancellous bone is well adapted to withstand physical stresses whilst being comparatively light. It is found in the ends (**epiphyses**) of long bones, in short bones, flat bones and irregular bones.

Types of bones

The bones which form the skeleton can be classified by their shape as long, short, flat or irregular (*Figure 16.2*).

Long bones

The **long bones** form the limb bones, e.g. the femur. As already mentioned they have a diaphysis and two epiphyses. The diaphysis, which is formed from compact bone, encloses the central **medullary cavity**. In young children the medullary cavity is filled with haemopoietic red marrow which is replaced by yellow fatty marrow in adults. Cancellous bone, containing red marrow, forms the epiphyses and the whole bone is covered with a layer of compact bone. Long bones act as levers which, with muscles, facilitate movement.

Short bones

The **short bones** are small, light, but very strong. They form, for example, the bones which contribute to the wrist and ankle joints, the carpals and tarsals respectively. Short bones are constructed from an inner layer of cancellous bone with a thin outer covering of compact bone.

Flat bones

Flat bones consist of a sandwich of cancellous bone between two layers of compact bone. Examples of flat bones include sternum,

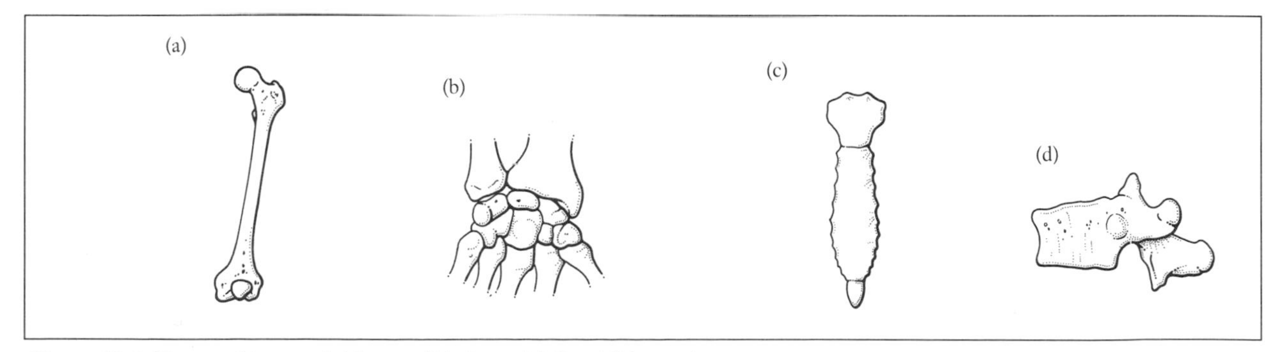

Figure 16.2. Types of bones. (a) Long; (b) short; (c) flat; (d) irregular.

ribs, scapulae and skull. They provide protection, attachment for muscles and, in adults, most contain haemopoietic red marrow.

Irregular bones

Irregular bones are those that cannot be included in the previous groups, including the vertebrae and some facial bones. They consist of cancellous bone covered with a thin layer of compact bone. In adults the red marrow of irregular bones is an important site of haemopoiesis.

In addition to these groups are the **sesamoid** bones, a type of short bone forming within tendons. They develop close to joints and include the patella found at the front of the knee joint.

Structure of Bone

Gross structure of bone

At this stage it is useful to discuss the gross structural features of bones before looking at the microscopic features and composition in more detail. For our purposes this is best achieved by considering the features of a typical long bone such as the humerus or femur (*Figure 16.3*).

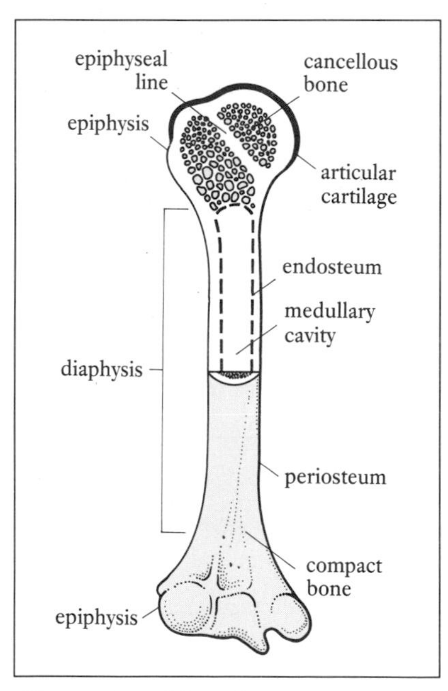

Figure 16.3. A long bone.

Long bones consist of a diaphysis and two epiphyses; during childhood, cartilaginous **epiphyseal plates** separate the epiphyses from the diaphysis. Growth in bone length occurs at these epiphyseal plates from the growing region of the diaphysis called the **metaphysis**.

The **periosteum**, a tough fibrous connective tissue, closely covers the bone except for the articular surfaces of synovial joints (see Chapter 18) which are covered with hyaline cartilage (see Chapter 1). An outer fibrous layer of periosteum contains the nutrient arteries and nerves which supply the bone, and under it lies an inner layer of **osteoblasts** (cells involved in bone formation) concerned with growth in bone girth. The periosteum is attached to the bone by collagen fibres known as **Sharpey's fibres**. The periosteum also protects the underlying bone and provides attachments for muscles and ligaments.

As already mentioned there is a central **medullary cavity** (central cavity containing marrow) running within the diaphysis. The marrow-filled cavity is lined with a thin membrane called the **endosteum** which contains osteoblasts and **osteoclasts** (cells involved in bone reabsorption).

The various physical features which form the bony landmarks, are discussed in Chapter 18.

Microscopic features

The basic structural unit of bone tissue is the **Haversian system** or **osteon**. In compact bone tissue there are numerous osteons of regular structure but in cancellous bone the osteons are far fewer in number.

Compact bone

An osteon comprises a central Haversian canal which runs along the long axis of the bone surrounded by concentric rings of bone known as **lamellae**. The Haversian canals carry the blood vessels, a few lymph vessels and nerve fibres which supply each osteon; these are also carried by smaller channels called **Volkmann's canals** which run at 90° to the Haversian canals.

Between the lamellae are spaces or **lacunae** which contain **osteocytes** (bone cells derived from osteoblasts) and tissue fluid from which the bone cells obtain nutrients and dispose of waste

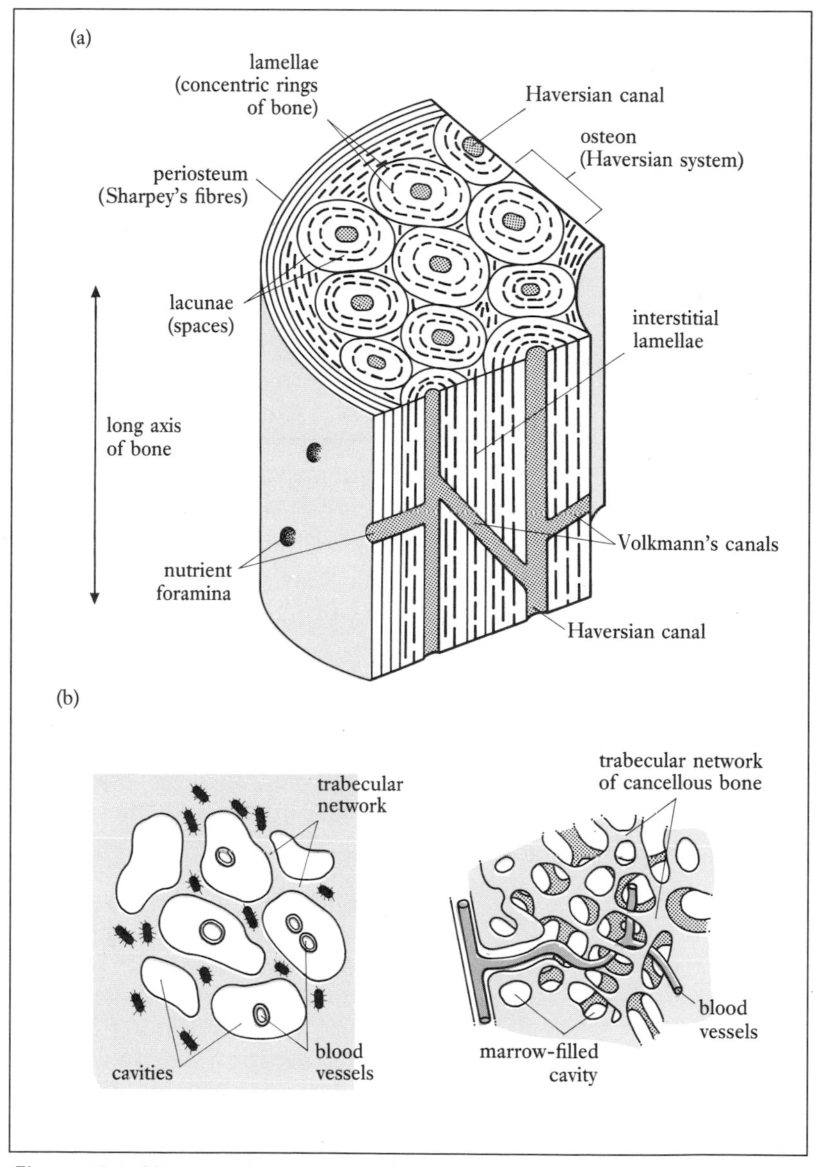

Figure 16.4. Microscopic structure of bone (canaliculi not shown). (a) Compact; (b) cancellous.

(*Figure 16.4(a)*). The lacunae connect with each other and the Haversian canal by channels called **canaliculi** which are formed by osteoblasts during bone formation. This arrangement of bone tissue gives strength without weight and is sometimes referred to as lamellar bone (see page 412).

Cancellous bone

Cancellous bone is also formed from lamellar bone but with fewer and larger osteons. In contrast to the regular compact structure it is formed from a beam-like or trabecular network interspaced with marrow cavities, an arrangement well able to resist stresses placed upon the bone (*Figure 16.4(b)*)

Composition

Bone contains water, organic molecules such as collagen, bone cells (osteoblasts, osteocytes and osteoclasts) and inorganic mineral salts. The organic matrix or **osteoid**, which consists of collagen and other organic molecules, is produced by the osteoblasts. It forms around a third of bone weight and gives bone its incredible tensile strength.

Inorganic salts deposited within the organic matrix during mineralization form the remaining two-thirds of the matrix and confer the extreme hardness characteristic of bone. The inorganic salts are *hydroxyapatites*, mainly calcium phosphate with some calcium carbonate and calcium hydroxide. Other minerals, present in smaller quantities, include magnesium and fluoride. Mineral crystals align themselves along the collagen fibres to produce bone tissue able to cope with compression forces of over 1400 kg/cm^2. Operating together, the collagen and minerals provide us with a material that is immensely strong and stretchy without being brittle, which is very reassuring when you consider the forces to which our bones are subjected during most activities.

Ossification (Osteogenesis) and Bone Growth

The formation and growth of bone occurs from around the 6th week of fetal life (embryonic mesenchymal cells have been converted into cartilage) until the late teens or early 20s. *Figure 16.5* illustrates the development of bone in a simple long bone.

Ossification

In the fetus, bone develops from the **ossification (osteogenesis**) of either hyaline cartilage (**endochondral**) or fibrous membranes (**membranous**) which act as frameworks or 'patterns' for the developing bone.

Endochondral ossification of cartilage results in the formation of the long and short bones. The process starts as a **primary ossification centre** in the shaft of the 'bone', and the development of a bone collar around the shaft. Primary ossification commences in the cartilage rods destined to become the long bones by the 12th week of fetal life. Accompanying these structural alterations are changes in blood and nutrient supply, the destruction of

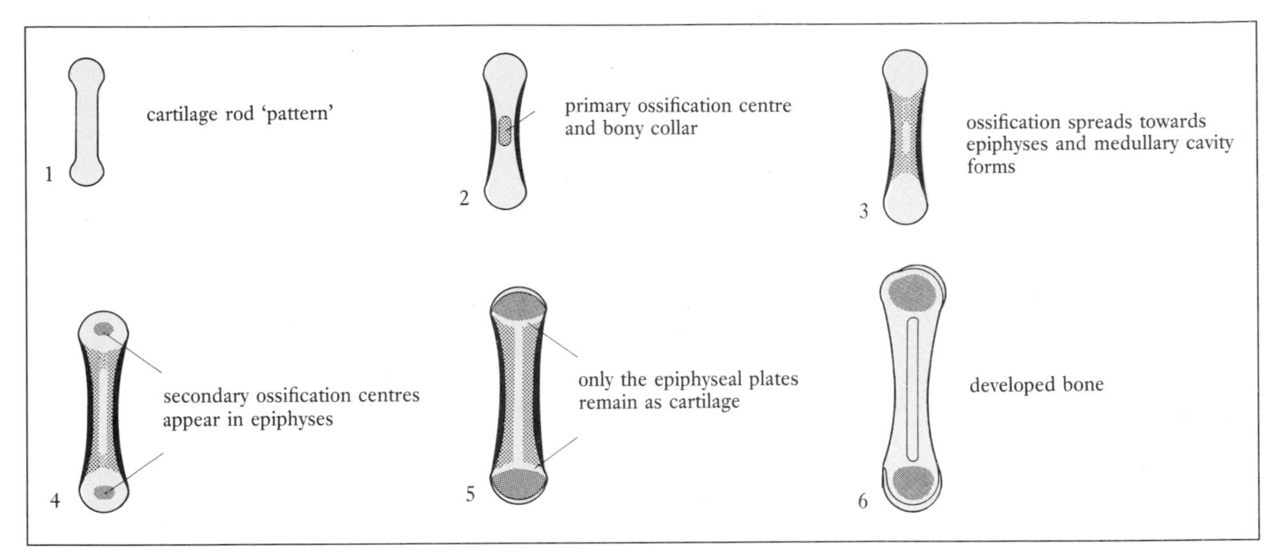

Figure 16.5. Bone development.

chondrocytes (cartilage-forming cells) and the development of osteoblasts, osteoclasts (see remodelling, pages 418-419), and red marrow cells.

Osteoblasts secrete collagen bony matrix (osteoid) into the cartilaginous framework and promote mineralization by the production of the enzyme *alkaline phosphatase* (see also Chapter 14). As bone is produced the osteoblasts ensnared within the lacunae differentiate to form osteocytes, the exact role of which is not yet fully understood. Ossification spreads out towards the epiphyses, which are still mainly cartilaginous at birth, and the medullary cavity is formed.

Secondary ossification centres form in the epiphyses during early childhood, resulting in further bone formation until only the epiphyseal growth plates remain as cartilage.

Membranous ossification is a less complex process where bony replacement of fibrous membranes is responsible for flat bones and some irregular bones. A single ossification centre usually occurs in the membrane as *mesenchymal* cells (embryonic connective tissue) differentiate into osteoblasts. These cells secrete osteoid matrix which is soon mineralized to form trabeculae. As before, the osteoblasts are trapped within the lacunae, where they differentiate to osteocytes, and the woven bone is replaced by either compact or cancellous lamellar bone.

Bone growth

Bone growth after birth, which continues throughout childhood and adolescence, is usually complete by the late teens or early 20s. Bone growth takes slightly longer in males and certain bones such as the clavicle do not stop growing until as late as 25 years of age. Even when true growth has ceased there is considerable structural change in adult bones through remodelling (see pages 418-419).

Growth in length

Growth in bone length continues by endochondral ossification which occurs at the epiphyseal plates. The process involves the production of new hyaline cartilage by chondrocytes followed by ossification as previously described. Growth ceases when the epiphyseal plate is completely replaced by bone and the epiphyses fuse with the diaphysis. A remnant of the original cartilaginous pattern remains as the articular cartilage covering the ends of long bones.

Growth in girth or thickness

Growth in bone girth occurs by appositional growth, where osteoblasts situated in the *osteogenic layer* of the periosteum produce bone matrix which is then ossified.

Hormones and bone growth

The major hormone which stimulates bone growth at the epiphyseal plate is *growth hormone* (*GH*). The problems caused by disordered GH secretion prior to epiphyseal plate ossification are *gigantism* (hypersecretion) and *dwarfism* (hyposecretion). When hypersecretion occurs after epiphyseal fusion it results in a condition called *acromegaly* which is characterized by the growth of the facial bones and those of the hands and feet. GH and its effects are discussed in Chapter 8.

The thyroid hormones (T_3 and T_4) also stimulate bone growth. A deficiency of these hormones in infancy (*cretinism*) leads to stunted growth (see Chapter 8).

During puberty (see Chapter 20) the sex hormones *oestrogens* and *testosterone* influence bone growth and cause the adolescent 'growth spurts'. In girls the secretion of oestrogens leads to the particular female bony changes, e.g. the wide pelvis adapted for child bearing, and in boys testosterone secretion accounts for the masculinization of the skeleton.

The many other hormones concerned with bone homeostasis and remodelling, which include parathyroid hormone, calcitonin and cortisol, are discussed on page 419.

Nutritional requirements for ossification and bone growth

Healthy bone formation, both before and after birth, is dependent upon the availability of the correct nutrients in sufficient amounts. Nutrients of particular importance in the formation of bone include protein, calcium and vitamins A, D and C (see Chapter 13). The intake of these, and the energy intake, will need to be increased during periods of bone formation such as pregnancy, infancy, childhood, adolescence and to facilitate the healing of fractures (see page 421).

HOMEOSTASIS OF BONE AND REMODELLING

Earlier we mentioned that bone is subject to constant change, with areas of new bone deposition by osteoblasts and bone reabsorption by enzymes produced by osteoclasts. Deposition and reabsorption occur at both the periosteal and endosteal surfaces. In young adults around 5% of bone tissue is usually involved in the two remodelling processes but this decreases later in life. The ability to remodel means that our bones can respond to mechanical forces in a way that adds bone tissue and consequently strength to an area subjected to particular stresses.

Bone remodelling is controlled by various hormones, especially those concerned with calcium and phosphate homeostasis, vitamin D and mechanical forces acting upon the skeleton. These controls maintain the precise balance between osteoblastic and osteoclastic activity which ensures that the overall bone mass changes very little in healthy adults (*peak bone mass* is reached at 35–40 years), although a decline in bone mass occurs with normal ageing.

Parathyroid hormone (PTH) secreted by the parathyroid glands and *calcitonin* secreted by the thyroid gland control calcium homeostasis and hence bone remodelling (see Chapter 8). When blood calcium levels are low PTH is secreted, which acts to raise calcium levels in several ways. Osteoclasts are stimulated to reabsorb bone matrix, which releases calcium and phosphate into the blood. Meanwhile the PTH increases renal reabsorption of calcium and phosphate excretion. The kidneys also produce 1,25-dihydroxycholecalciferol, the active form of vitamin D, which increases calcium absorption in the gut. A daily calcium requirement of around 700–800 mg in adults must be absorbed by the gut each day.

When blood calcium levels rise, the secretion of PTH is inhibited and calcitonin is secreted. Calcitonin inhibits bone reabsorption and favours the activity of osteoblasts, which produce new bone matrix; as calcium is deposited within the newly formed matrix its level in the blood will fall. At the same time renal and intestinal calcium absorption declines. Although the action of PTH and calcitonin do influence the activity of osteoclasts and osteoblasts, it must be stated that their primary role is that of calcium homeostasis.

Cortisol from the adrenal cortex increases bone reabsorption, which explains why loss of bone matrix density (*osteoporosis*) occurs in *Cushing's syndrome* (see Chapter 8) and is a side-effect of corticosteroid therapy.

Oestrogens and testosterone, so important in bone growth, continue to have an influence on bone even after growth has ceased. Oestrogens stimulate osteoblastic activity and their decline during the climacteric (the menopause is a single event (last period) during the climacteric) is associated with *postmenopausal osteoporosis* in older women (see below and Chapter 20).

Mechanical forces and gravity have a profound effect upon the skeleton and bone remodelling. These forces, unlike hormones, are primarily concerned with bone dynamics rather than mineral homeostasis. The old adage of 'use it or lose it' is definitely true when applied to bone. Periods of prolonged immobility lead to loss of bone mass, such as that seen when people become housebound or confined to bed (see *Nursing Practice Application* – osteoporosis, page 420). Bone mass decrease resulting from immobility and lack of gravity which occurred during early space exploration programmes is now prevented by planned physical activity. That bone remodelling and deposition occurs in response to stresses is accepted, but the actual mechanisms involved are more controversial. One possibility is that electrical activity produced by forces deforming the bone stimulate the cells required for remodelling.

Rickets and osteomalacia

To associate *rickets* and its adult equivalent, *osteomalacia*, exclusively with Victorian inner-city life or the developing countries would be incorrect. Unfortunately they both occur in people living within developed economies such as the UK, where they are especially common in British Asians.

Rickets is caused by lack of vitamin D during childhood, which results from a low intake of the vitamin or lack of exposure to sunlight. (The production of vitamin D from the action of ultraviolet light on sterols present in the skin is discussed in Chapters 13 and 19.) Insufficient vitamin D prevents adequate calcium absorption and normal ossification processes cannot proceed. Development is delayed and, if untreated, bony changes occur which lead to permanent deformity, e.g. bow legs.

In adults the lack of vitamin D may also be due to malabsorption (see Chapter 13) and serious kidney or liver disease. Osteomalacia is a softening of the bones caused by a failure to mineralize the osteoid produced. The affected person has bone pain, deformity and an increased risk of fractures, which heal poorly.

Susceptible groups need information about foods rich in vitamin D and available calcium, and in certain situations the advisability of supplementing vitamin D intake, e.g. children living in areas where rickets is endemic. The benefits of exposure to ultraviolet light should be explained and where appropriate encouragement given to get out in the sunshine. However, this will not always be possible, as certain cultures find it unacceptable to uncover the limbs and body.

Osteoporosis

Osteoporosis is the loss of bone mass due to excess reabsorption without the usual balance of bone deposition. The bones, which retain their normal composition, are lighter and weaker. These structural changes lead to bones that deform and fracture more easily.

Osteoporotic changes are associated with ageing in both sexes, but are especially common in females when oestrogen declines during the climacteric. Other factors involved in the aetiology include: level of calcium intake, which may be important in achieving peak bone mass (Francis, 1989) and phosphate intake, immobility, corticosteroid therapy and Cushing's syndrome.

Nursing Practice Application **Immobility and osteoporosis**

Prolonged immobility at any age can lead to osteoporotic changes. Nurses should encourage exercise and physiotherapy for those who are housebound or confined to bed. It is particularly important that older people continue to exercise, as this appears to slow or prevent age-related bone loss (Chow *et al.*, 1987; Hamdy,1990).

A diet containing adequate calcium, phosphate, protein and vitamin D should be encouraged, but a study by (Stevenson *et al.*, 1988) suggests that dietary calcium supplements in menopausal women have little influence on bone loss. For some older women the use of oestrogens in the form of *hormone replacement therapy (HRT)* may be appropriate, and here the nursing responsibilities include the planning of an education programme which meets the person's information needs with regard to this type of therapy and its effects.

NB HRT is mentioned again within the general discussion of changes occurring during the climacteric in Chapter 20.

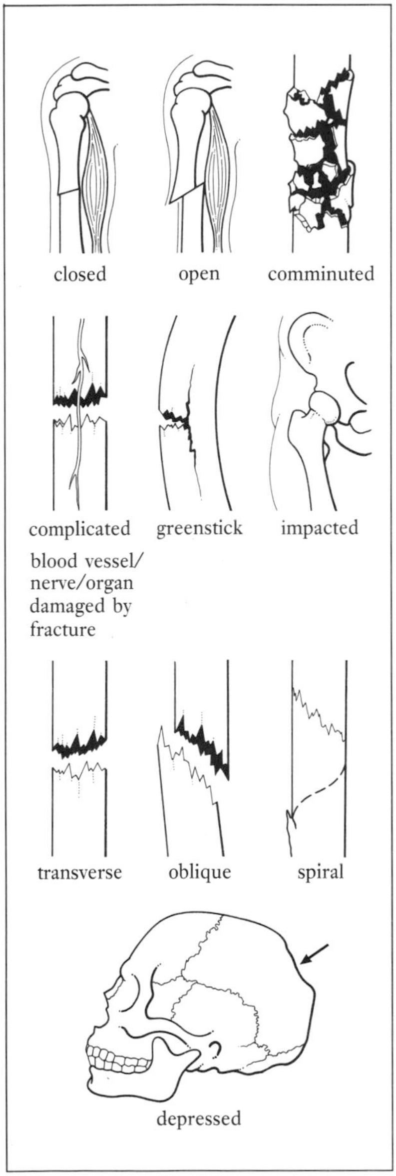

closed open comminuted

complicated greenstick impacted

blood vessel/ nerve/organ damaged by fracture

transverse oblique spiral

depressed

Figure 16.6. Types of fracture.

Special Focus **Fractures**

A fracture or break in a bone is not an uncommon occurrence, despite the incredible strength of bone tissue. Most fractures are caused by trauma resulting from road traffic accidents, sports injuries and falls. Older people with their weaker osteoporotic bones are particularly vulnerable to fractures as a result of falls, e.g. fractures of the neck of femur in elderly women. Fractures associated with existing bone disorders are said to be *pathological*. These fractures, which occur spontaneously or after minimal trauma, may be due to conditions such as osteomalacia, *osteogenesis imperfecta* (brittle bone disease) and metastatic malignancy from a primary growth in the breast, prostate or bronchus.

Types of fracture (*Figure 16.6*)
Fractures can be classified in a variety of ways, which include severity, direction of the break and location on the bone, e.g. supracondylar fracture of the elbow, see Chapter 17, *Person-centred Study* – Dave).

Healing of fractures (see *Figure 16.7*)
The healing of fractures involves the following stages: haematoma formation and organization (see Chapter 19), *callus* formation and remodelling.

Haematoma and organization
The intial events of fracture healing are those of the *inflammatory process*. The bleeding which occurs at the fracture site produces a haematoma and, within 24 hours, *organization* has commenced within the haematoma. Organization involves the phagocytosis of debris and growth of capillaries and *fibroblasts* (young cells of fibrous connective tissue) within the haematoma to form *granulation tissue*. Fibroblasts eventually become the chondroblasts and osteoblasts which produce the materials for the bony repair.

Callus formation
Fibrous connective tissue (*callus*) appears at the fracture site after about a week. The cartilage and *osteoid* (woven material) produced form a network which is later mineralized and converted to lamellar cancellous bone. The early callus, formed in about 3 weeks, is irregular in structure and relatively fragile. Later, as further ossification occurs, the bony callus becomes stronger and exhibits the usual orderly bone structure of Haversian systems.

Remodelling
Over many months the activity of osteoblasts and osteoclasts return the bone to its original form. The osteoblasts deposit new compact bone as required and osteoclasts restore the medullary cavity and reabsorb excess callus.

NB Readers may care to compare the process of fracture healing with wound healing (Chapter 19).

The time taken for fracture healing to occur obviously depends on the size of the bone involved, e.g. the shaft of the femur will take much longer to heal than the clavicle. Apart from differences between bones the factors which affect healing include blood supply, age, adequate immobilization of the fracture and apposition of the bone ends, nutritional status and freedom from sepsis (infection).

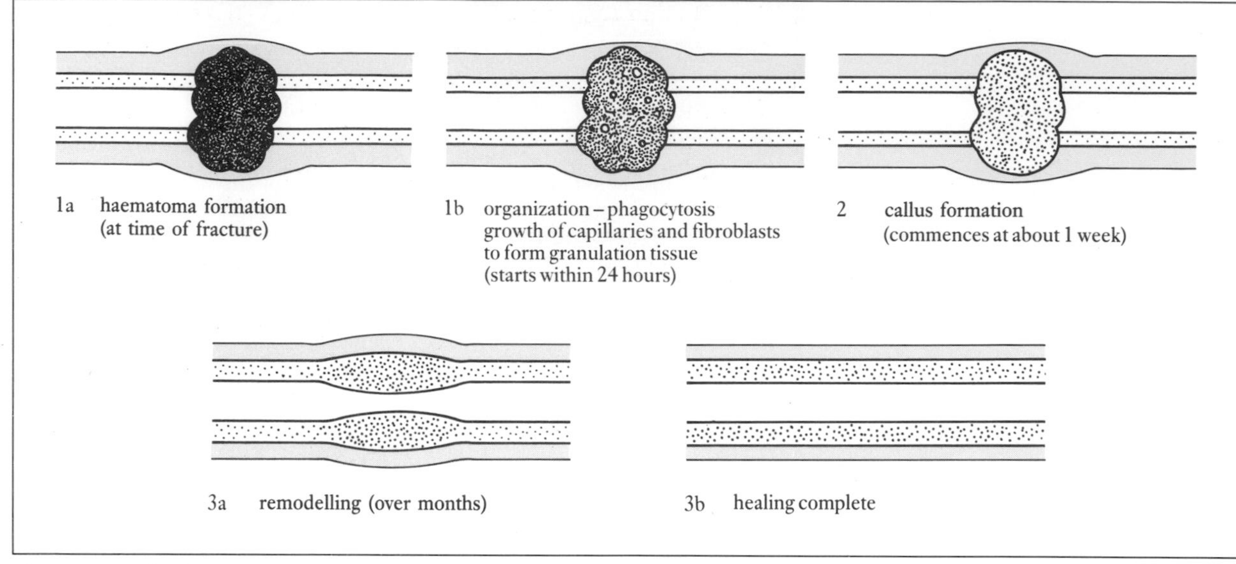

Figure 16.7. Healing of fractures.

Summary/Check List

Introduction – Functions of bone.
Bone tissue – woven, lamellar, compact, cancellous. Bone types – long, short, flat, irregular, sesamoid.
Bone structure – Gross structure (long bone). Microscopic features. Bone composition.
Ossification – endochondral, membranous. Bone growth – length, girth, hormonal influences. Nutritional requirements for ossification and growth.
Bone homeostasis and remodelling – hormones, mechanical forces. Problems with homeostasis – rickets and osteomalacia, osteoporosis, *Nursing Practice Application* – immobility and osteoporosis. *Special Focus* – fractures (types and healing).

?

Self Test

1 Briefly describe the functions of bone.
2 Complete the following:
(a) The two types of lamellar bone are _______ and __________.
(b) _______ bone which forms the outer layer of all bones is also known as ________.
(c) The inner part of the epiphysis is formed from __________ or __________ bone.
3 Give one example of each type of bone:
(a) long (b) short (c) flat (d) irregular (e) sesamoid.
4 Draw a diagram of a typical long bone to illustrate the location of the following:
(a) epiphyses (b) epiphyseal line (c) diaphysis (d) medullary cavity (e) periosteum (f) endosteum.

Answers
1. See page 412. 2. a. compact, cancellous. b. compact, cortical. c. cancellous, trabecular. 3. Examples include: a. femur, tibia, humerus, radius; b. tarsals, carpals; c. ribs, sternum, skull; d. vertebrae, facial; e. patella. 4. See page 414 and *Figure 16.3.* 5. See pages 414-416. 6. a–g, b–h, c–e, d–f. 7. b, c. 8. See page 419. 9. See page 420. 10. See page 421 and *Figure 16.7.*

5 Describe the basic structural unit (osteon/Haversian system) of compact bone.

6 Put the following in their correct pairs:

(a) osteoclast	(e) woven bone
(b) hydroxyapatite	(f) osteoid
(c) lamellar bone	(g) osteoblast
(d) collagen	(h) calcium salts

7 Which of the following statements are true?
(a) All fetal bone formation is endochondral ossification.
(b) Bone growth may continue until the mid 20s.
(c) Growth hormone is a major influence on bone growth.
(d) Increase in bone girth arises from the epiphyseal plate.

8 Briefly describe the control of bone remodelling in a young adult.

9 What information would you give to Tom, aged 65, who asks how he might avoid or delay the bony changes related to ageing.

10 Briefly describe the stages of fracture healing.

References

Chow, R., Harrison, J.E. and Notarius, C. (1987) Effects of two randomized exercise programmes on bone mass of healthy postmenopausal women. *British Medical Journal,* **295,** 1441–1444.

Francis, R.M. (1989) Calcium's role in preventing and treating osteoporosis. *Geriatric Medicine,* **19** (7), 24–26.

Hamdy, R.C. (1990) Identifying risk factors for osteoporosis. *Geriatric Medicine,* **20** (4), 49–51.

Stevenson, J.C., Whitehead, M.J., Padwick, M., Endacott, J.A., Sutton, C., Banks, L.M., Freemantle, C., Spinks, T.J., and Hesp, R. (1988) Dietary intake of calcium and postmenopausal bone loss. *British Medical Journal,* **297**, 15–17.

Further Reading

Tortora, G.J. and Anagnostakos, N.P. (1990) *Principles of Anatomy and Physiology,* 6th edn. New York: Harper and Row.

Vaughan, J.M. (1981) *The Physiology of Bone,* 3rd edn. Oxford: Oxford University Press.

CHAPTER SEVENTEEN

MUSCLE TISSUE

OVERVIEW

- Functions and structure of muscle tissue.
- Contraction of skeletal muscle tissue.
- Muscle metabolism.

LEARNING OUTCOMES

After studying Chapter 17 you should be able to:

- Discuss the functions of muscle.
- Outline the characteristics of muscle.
- Describe briefly the three types of muscle.
- Compare and contrast skeletal muscle with smooth muscle.
- Describe the gross structure of a skeletal muscle.
- Discuss the effects of immobility and disuse on skeletal muscle.
- Describe the microscopic structure of skeletal muscle.
- Outline the events occurring when skeletal muscle contracts.
- Describe briefly the types of muscle contraction and the phases of a twitch contraction.
- Explain how metabolism provides the energy required for muscle contraction.

KEY WORDS

Actin - one of the contractile proteins forming the thin filaments of the myofibril.
Insertion – the point at which a muscle is attached to the bone that it moves.
Myofibril – longitudinal bundles of contractile proteins (filaments) within the muscle cell/fibre.
Myoglobin – a haem–protein molecule found in skeletal muscle. It acts as a temporary oxygen store for use when muscle is contracting.
Myopathy – a disease affecting muscle.
Myosin – the contractile protein forming the thick filaments of the myofibril.
Origin – the fixed or immovable point at which a muscle is attached to the bone.
Sarcomere – a segment of myofibril which forms the smallest contractile unit of skeletal muscle.
Tendon – band of white fibrous connective tissue which attaches muscle to bone.
Tropomyosin – one of the proteins found in the thin filaments of the myofibril.
Troponin – one of the proteins found in the thin filaments of the myofibril.

Introduction

Muscle is formed from excitable cells (muscle cells and neurones which are able to produce an action potential by reversibly altering potential difference across their membrane) which, in common with nervous tissue, are highly specialized and responsive to stimuli. These specially adapted elongated muscle cells (fibres) produce the powerful contractions required for body movement and the movement occurring within body structures, e.g. peristalsis in the gut. Muscle cells take chemical energy as ATP and convert it to mechanical energy and work.

This chapter is concerned with movement and stability, and we will concentrate on skeletal muscle, which forms over 600 individual muscles and accounts for around 40% of adult body weight. Smooth muscle is discussed in sufficient depth for you to compare it with skeletal muscle, but further coverage of smooth muscle is included in more appropriate chapters (12,13,15 and 20), as is cardiac muscle (see Chapter 10).

Muscle Tissue

Functions of muscle tissue

Movement

Almost every movement involving the body requires the contraction of muscle. Skeletal muscles provide the diverse abilities of gross and very precise movements which allow us to move within our environment, and to investigate and respond as appropriate. Contraction of skeletal muscle is also responsible for: eyeball movement (see Chapter 7), control of certain sphincters, e.g. external urinary sphincter (see Chapter 15), mastication and facial expression.

The smooth muscle of the alimentary tract moves food and waste (Chapter 13), the bladder empties when its smooth muscle contracts (Chapter 15), smooth muscle controls the diameter of the bronchi (Chapter 12) and uterine smooth muscle contracts during the birth of a baby (Chapter 20).

The vital circulation of blood around the vascular system is totally dependent upon the rhythmic contraction of the cardiac muscle tissue which forms the myocardium (Chapter 10).

Posture and stability

The state of partial contraction (**muscle tone**) which exists in skeletal muscles helps us to maintain a stable posture. The ability to stand or sit for considerable periods depends on the activity of *proprioceptors* (see page 431 and Chapter 3) in muscles, tendons and joints. These signal minute changes to the nervous system, which can if necessary modify the partial muscle contraction required to maintain posture.

Heat production and temperature regulation

When you drive a car the engine gets hot whilst producing the mechanical energy required to move the wheels. When muscle is contracting it also produces heat energy in addition to mechanical energy required for movement. This considerable amount of heat energy is used by the body to maintain temperature homeostasis (see Chapter 19). For instance, on a cold day the contraction of skeletal muscle produces shivering, which acts to increase body temperature.

General characteristics of muscle tissue

All types of muscle tissue possess (to some degree) the following physical properties:

- *Excitability (irritability)* – the ability of the muscle fibres to respond to a stimulus, which may be electrical, chemical or mechanical. Skeletal muscle responds to a nerve impulse which crosses the chemical synapse to produce an action potential in the muscle (see pages 434-435 and Chapter 3). Muscle tissue is also stimulated by hormones, changes in the local chemical environment and stretching.
- *Contractility and extensibility* – when sufficiently stimulated, muscle contracts (shortens). As muscle relaxes it extends and, in the case of smooth muscle, is capable of considerable stretching. This property allows for distension in hollow organs, e.g. the stomach after a meal.
- *Elasticity* – this property ensures that muscle length returns to resting proportions following stretching or contraction.

Types of muscle tissue

Skeletal muscle (striated, voluntary) tissue
The **skeletal muscles** cover the bony skeleton and facilitate movement. The individual cells or fibres are **multinucleate** and form extended cylinders (see *Figure 17.1(a)*). A microscopic examination reveals banding or *striations* on the muscle fibre, which produces its striped appearance.

Skeletal muscle is the only type of muscle tissue which may be controlled consciously, which explains the term **voluntary** muscle. However, in many cases this control operates through reflexes which we do not consciously control (see Chapter 5).

Skeletal muscle is stimulated by the voluntary motor division (somatic) of the PNS to produce rapid, forceful contraction of short duration. Although the fibres of skeletal muscle contract in series (one after the other – as the fibres of one motor unit contract those comprising other motor units are relaxed), any muscle will eventually tire if required to contract for any length of time without rest, a fact of which you need no reminding of after a busy day of intense physical activity.

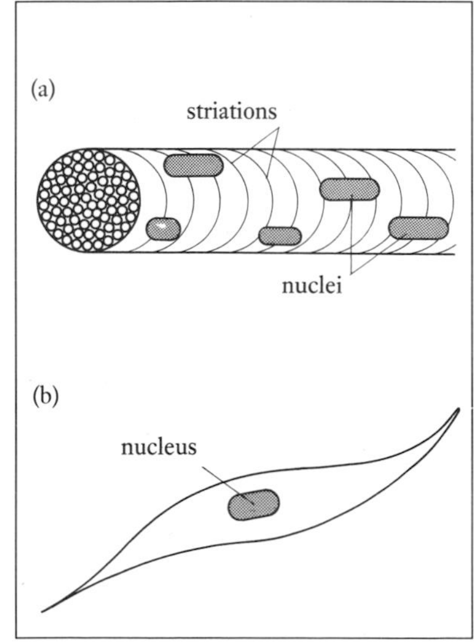

Figure 17.1. (a) Striated muscle fibre; (b) smooth muscle fibre.

Smooth muscle (unstriated, visceral or involuntary) tissue
Smooth muscle forms part of the structure of the stomach, gut, bronchi, ureter, bladder, uterus and blood vessels. The spindle-shaped smooth muscle fibres, which are **uninucleate**, form flat

sheets of contractile units (see *Figure 17.1(b)*).

All smooth muscle is controlled involuntarily and there is no conscious regulation of its contraction. Stimulation of smooth muscle occurs through the action of the ANS and various hormones or regulatory peptides, e.g. gastric motility is affected by both vagal activity and locally produced peptides (see Chapter 13).

The contraction of smooth muscle tissue differs from that of skeletal muscle in several respects. Smooth muscle responds slowly, and the contraction, which is less intense than the contraction of skeletal muscle, tends to be more sustained. Contractions of smooth muscle are more widespread, with all the fibres within the sheet able to contract in unison. In addition some smooth muscle fibres act as '*pacemakers*' which are capable of initiating inherent, rhythmic contractions that may be modified by neural or hormonal influences.

Cardiac muscle tissue (see Chapter 10)

Cardiac muscle is found only in the myocardium and, although it forms a distinct muscle type, it has common features with both skeletal and smooth muscle. Cardiac muscle has short branching fibres which may be **uni-** or **binucleate**. The fibres, like skeletal muscle, are striated, but contraction is involuntary. The presence of inherent control via a 'pacemaker' produces steady rhythmic contractions which are modified by the ANS and hormones as physiological needs alter. Boundaries between individual fibres, which are not well defined, are formed by **intercalated discs** (see *Figure 10.3*). This feature allows the wave of contraction to pass easily across the myocardium, which behaves like a **syncitium** (i.e. cell boundaries do not exist or are poorly defined).

Skeletal Muscle Tissue

Gross structure

The 600 or so skeletal muscles are separate organs, which, in addition to muscle fibres arranged in bundles called **fasciculi**, contain blood vessels, nerves, lymphatics and connective tissue.

Earlier we mentioned that skeletal muscle forms around 40% of the body weight. However, there are gender and age differences. In males muscle forms a greater percentage of body weight, and this, with their generally larger body size, accounts for comparatively greater strength on average. Individual muscle fibres are of the same strength in both genders but androgens are responsible for the increased muscle development in the male.

In normal ageing there is a gradual loss of muscle mass as fibres are replaced with connective tissue. These changes account for the decline in strength and weight loss experienced by elderly people, especially those over 80.

At every level a skeletal muscle is enclosed by fibrous connective tissue. The whole muscle is covered by the **epimysium**, bundles of fibres (fasciculi) are enclosed within the **perimysium** and a thin layer

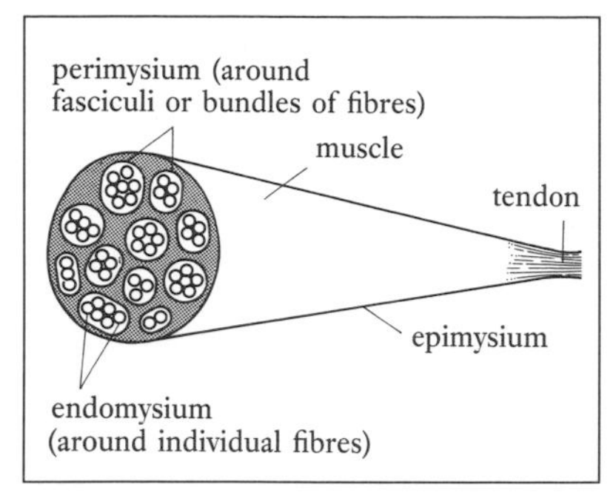

Figure 17.2. A muscle showing connective tissue.

of **endomysium** is found around individual fibres. These layers of connective tissue support the muscle as a whole and protect the fragile fibres (*Figure 17.2*).

Muscle attachments to the skeleton may be direct, with the outer connective tissue joining with the periosteum of the bone (see Chapter 16), or indirect by means of **tendons** (band of white fibrous connective tissue which attaches muscle to bone). The epimysium merges into the tendons, formed of white fibrous connective tissue and collagen, which attach muscles to bones and other structures. Tendons are either rope-like structures, e.g. those of the *biceps brachii* in the arm, or flat sheets known as **aponeuroses** which form the attachments of the abdominal muscles. The tendons, which also transfer the force of muscle contraction to the bone, and their collagen fibres are well adapted to withstand the stresses involved.

Muscles are generally attached to the skeleton at two or more points: the **origin** (fixed or immovable point at which a muscle is attached to the bone), which is usually fixed, and the movable **insertion** (the point at which a muscle is attached to the bone that it moves). When a muscle contracts, the two points move closer together.

Muscle fibre arrangement

Skeletal muscles can be classified by the arrangement and direction of their fasciculi. The arrangement of the fasciculi will determine a muscle's movement range and power. Two basic patterns exist – the fasciculi may lie parallel or oblique to the direction of the muscle pull. Generally muscles with long parallel fasciculi tend to have the greatest degree of movement, whereas the obliquely arranged muscles, with more fibres, have increased power. Variations on the two basic patterns include: *strap, fusiform, unipennate, bipennate, multipennate, circular, spiral* and *triangular (convergent)* some of which are illustrated in *Figure 17.3. NB* Pennate means arranged like feathers.

Muscle blood supply

Skeletal muscles require an abundant supply of oxygenated blood and receive about 1 litre of blood/min, which represents around 20% of the cardiac output at rest. During strenuous exercise the amount of arterial blood required to supply oxygen to the actively contracting muscles may increase to 15 litres/min or more as precapillary sphincters (see Chapter 10) in the muscles open. The increase in blood flow, which also depends

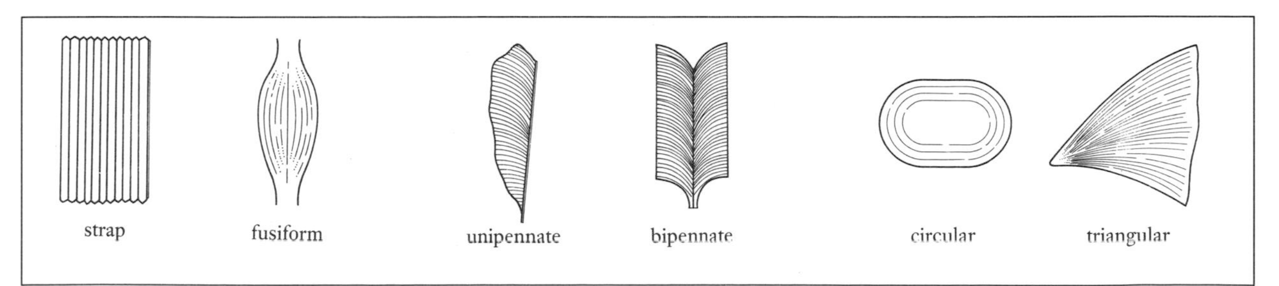

Figure 17.3. Arrangement of muscle fasciculi.

Person-centred Study **Dave**

Dave, aged 11, is in the accident and emergency department after a fall from his bicycle. He and his mother have just been told that his left arm (humerus) is fractured near the elbow and that he needs to be admitted for observation after fracture reduction.

Fractures of the humerus occurring near the elbow (*supracondylar fracture*) may cause compression of the artery which supplies the muscles of the forearm (see Chapter 16). Any prolonged interference to the flow of blood would result in the development of *Volkmann's ischaemic contracture* and deformity of the forearm. Following this type of fracture a vital part of the nursing assessment is to observe for pain and monitor forearm circulation by taking the radial pulse and observing skin colour and temperature on the affected side. In this way abnormalities are detected early and measures which prevent permanent damage can be initiated.

Dave and his mother are naturally concerned about the injury, but cannot understand all the fuss for 'just a broken arm'. Prior to Dave's treatment a nurse gives them a full explanation about the fracture and why these observations are needed. This helps to reduce their anxiety and ensures their co-operation, a process which is continued on the ward by providing information in terms Dave and his mother can understand.

on the ability of the heart to maintain supply, is termed *exercise* or *active hyperaemia*.

The individual artery which supplies a muscle and its tendons branches to form an extensive capillary network around individual muscle fibres. Eventually blood drains into the one or more veins required to remove the considerable metabolic waste produced by actively contracting muscle.

If the circulation to or from muscle tissue is impeded the muscle will, like cardiac muscle, undergo ischaemic changes. The results of ischaemia include necrosis, fibrosis, shortening, formation of deforming *contractures*, (see *Person-centred Study* – Dave) and muscle atrophy (see Chapter 1).

Muscle nerve supply

Skeletal muscles are innervated by nerves which contain voluntary and autonomic motor fibres and sensory fibres.

Voluntary motor fibres

Muscle fibres are innervated by axons of **lower motor neurones**. Each axon supplies a group of fibres known as a **motor unit** (*Figure 17.4*). The motor units vary in size. Those which cause precise movements contain few fibres, e.g. finger movements, but where movement is less precise the number of fibres in the motor unit increases, e.g. large muscles of the thigh.

You may wish to refresh your memory with a quick look at Chapter 4 but, very briefly, the pyramidal motor pathway is a two-neurone system. An **upper motor neurone** arising in the motor cortex initiates voluntary muscle contraction on the opposite side of the body. This occurs because the neurone decussates in the medulla prior to continuing down the spinal cord to the appropriate level. Here it synapses, in the anterior horn, with the lower motor neurone which transmits between spinal cord and neuromuscular junction (see muscle contraction, page 433). Extrapyramidal motor pathways

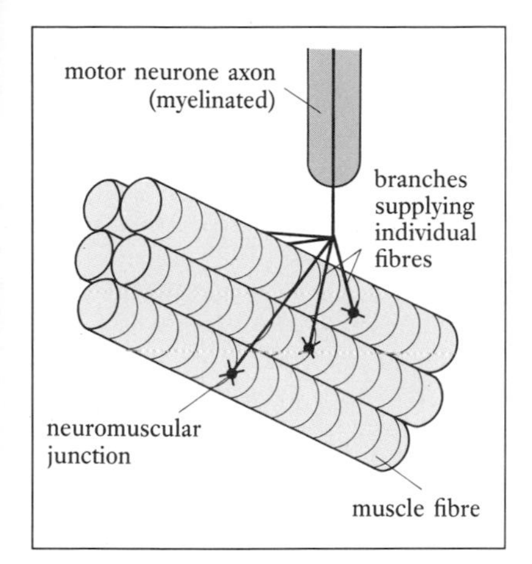

Figure 17.4. Muscle nerve supply – a motor unit.

provide innervation for the synergistic muscle contraction necessary for smooth co-ordinated movement and posture.

Autonomic motor nerves

Autonomic motor nerves are concerned with the regulation of blood flow through the muscle by control of the arteriolar smooth muscle and precapillary sphincters (see Chapters 6 and 10). Apart from neural controls the metabolites produced within contracting muscle are responsible for the vasodilation and increased blood flow of exercise.

Sensory nerves

Sensory nerves relay information from the muscle **proprioceptors** found in **muscle spindles**, which are formed from modified non-contractile muscle fibres. This information is required by the CNS for maintaining posture through changes in muscle tone. Sensory fibres also transmit pain impulses such as those you experience the day after unaccustomed exercise.

Microscopic structure of muscle

The extended, cylindrical skeletal muscle fibres have a diameter of 10–100 μm but there is considerable difference in length, which may range from a few millimetres to several centimetres.

The muscle fibre has a **sarcolemma** (plasma membrane) surrounding the **sarcoplasm** (called cytoplasm in other cells) which contains several peripheral nuclei (see *Figure 17.5(a)*). Within the sarcoplasm large quantities of stored glycogen and **myoglobin** (a haem-protein molecule found in skeletal muscle, which acts as a temporary oxygen store for use when muscle is contracting) provide the metabolic needs of contracting muscle.

The sarcoplasm consists mainly of bundles of longitudinal **myofibrils** (see *Figure 17.5(b)*) which are formed from filaments of contractile proteins. In each myofibril there are thick filaments

Nursing Practice Application Muscle changes in immobility

Muscle bulk decreases due to the atrophy resulting from immobility (disuse atrophy), denervation and certain **myopathies** (disease affecting muscle), e.g. muscular dystrophy (see page 432-433). Problems of immobility may affect isolated muscles which are under used when a fractured limb is immobilized, or the changes may be more general if someone is confined to bed. These changes, apart from atrophy where muscle bulk is replaced with fibrous tissue, also include obvious wasting, muscle weakness and a reduction in mobility which further exacerbates the problem. Elderly people who may be at particular risk, should remain active, which according to (Wickham, 1989) will maximize their muscle strength, maintain co-ordination and reduce the risk of falls.

Care plans must anticipate these known effects of disuse with the inclusion of counter-measures which include a full range of passive or active physiotherapy exercises and, where possible, early mobilization.

The risk of other problems of immobility may also be minimized by a regimen of planned exercise/movement, e.g. deep vein thrombosis (see Chapter 9), urinary stasis (see Chapter 15), loss of bone (see Chapter 16), joint stiffness (see Chapter 18) and pressure sores (Chapter 19).

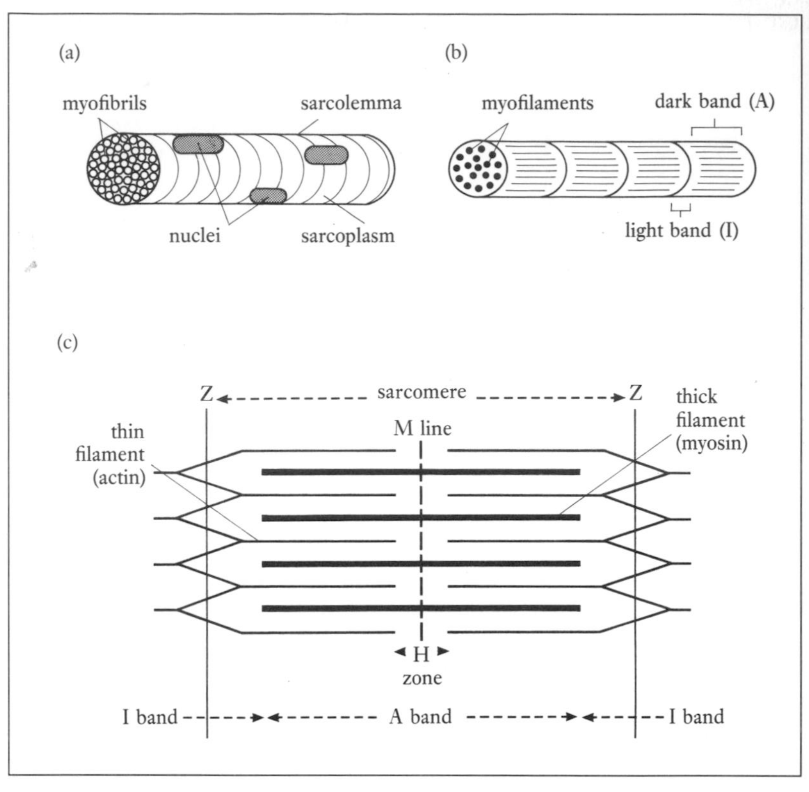

Figure 17.5. (a) Muscle fibre; (b) myofibril; (c) myofilaments.

formed from **myosin** and thin filaments comprising **actin, tropomyosin** and **troponin**. The thick and thin filaments are organized alternately in a repeating pattern (see *Figure 17.5c)*, which because of overlap results in a banded appearance with dark **A bands** each containing an **H zone** and **M line**, and light **I bands** each containing a **Z line**. It is these bands in the myofibrils, which match across the muscle fibre, that give skeletal muscle its characteristic striations.

Each myofibril is divided into several **sarcomeres** which form the smallest contractile unit of skeletal muscle, the region between two *Z* lines (see *Figure 17.5(c)*) constitutes one sarcomere.

In each muscle fibre/cell there are two sets of tubules – one, formed from tubules of the **sarcoplasmic reticulum** (type of endoplasmic reticulum), surrounds the myofibrils in a loose network, and the other, a system of **transverse** or **T tubules** lined with sarcolemma, runs from the extracellular space to penetrate deep into the sarcoplasm. These tubular systems maintain the correct concentration of calcium and transmit the action potential to the contractile units.

Degenerative myopathies – muscular dystrophies

Several myopathies can cause progressive structural changes within muscle fibres, e.g. *Duchenne-type muscular dystrophy*. This is inherited as a sex-linked recessive condition (see colour blindness,

Chapter 7 and haemophilia, Chapter 9), where the female carrier transmits the abnormal gene which is expressed in her male children (theoretically 50% of males are affected, i.e. a1 in 2 chance). The condition is characterized by progressive muscle weakness and wasting which leads ultimately to death from infection or respiratory failure during early adulthood. Muscle fibre degeneration, atrophy and necrosis occur but replacement with fat and connective tissue may appear to increase the muscle size.

Muscle Contraction

The ability of the filaments in the myofibrils to shorten results in muscle contraction. This can result in shortening of the whole muscle by as much as 50%. Muscle contraction can be explained by the **sliding filament hypothesis** (see *Figure 17.7*), which proposes that filaments slide against each other to cause shortening. The molecular structure of the contractile proteins facilitates the movement of the thin filament (actin, tropomyosin and troponin), by the thick filament (myosin), towards the centre of the sarcomere. Protrusions or '*heads*' on the myosin filaments form **cross-bridges** which attach to and pull the actin filament (see **excitation–contraction coupling**, page 434).

Events occurring before, during and after contraction

At this point it might be helpful for you to look again at the transmission of the nerve impulse and synapse in Chapter 3, as skeletal muscle contraction depends on nerve stimulation.

Nerve stimulation and transmission of the action potential

The axons of the motor nerves transmit action potentials to the muscle. Within the muscle the axon forms unmyelinated terminal branches, which conduct the impulse to the **neuromuscular junction** or **motor end plate** (see *Figure 17.6(a)*), which forms the communication between axon terminal and the muscle fibre. This, however, is not journey's end, as the action potential must still cross the **synaptic cleft** (see *Figure 17.6(b)*).

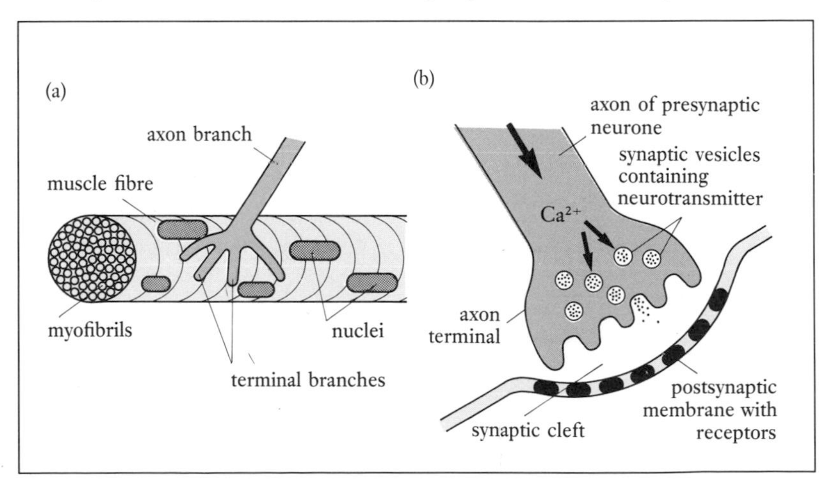

Figure 17.6. (a) Neuromuscular junction; (b) synapse.

This is facilitated by the the influx of calcium ions and the release of the neurotransmitter *acetylcholine* from synaptic vesicles. Following synaptic transmission the acetylcholine is broken down by the enzyme *acetylcholinesterase.*

The sarcolemma, which has a resting potential of –90 millivolts (mV), is depolarized as sodium ions move into the fibre. This movement of sodium produces an action potential of +30 mV prior to repolarization as potassium ions move out. The action potential spreads into the myofibrils via the T tubules.

The conduction of the action potential in a muscle fibre obeys the same rules as a nerve impulse; the **all-or-none law** operates, a **threshold** level of stimulation must first be reached and **absolute** and **relative refractory periods** occur during and after stimulation (see Chapter 3). The actual stimulation to the muscle fibre lasts only a few milliseconds (ms) but its effects in terms of muscle contraction last for several hundred milliseconds.

Problems with neuromuscular conduction may be caused by electrolyte imbalance, especially calcium (see tetany, Chapter 8) and *myasthenia gravis,* a condition characterized by muscle weakness and tiredness that occur because acetylcholine receptors are blocked by autoantibodies produced by the thymus which circulate in the blood. Most people with myasthenia gravis have an abnormality of the thymus gland (see Chapters 8, 11 and 19). People with this condition should structure their daily activities to provide sufficient rest to reduce the tiredness which follows excessive muscle activity. The management of myasthenia gravis may include drugs which prevent the breakdown of acetylcholine (e.g. neostigmine), thymectomy, measures which oppose the thymic antibodies (such as immunosuppressive drugs) and plasmapheresis where the antibodies are removed from the blood during plasma exchange.

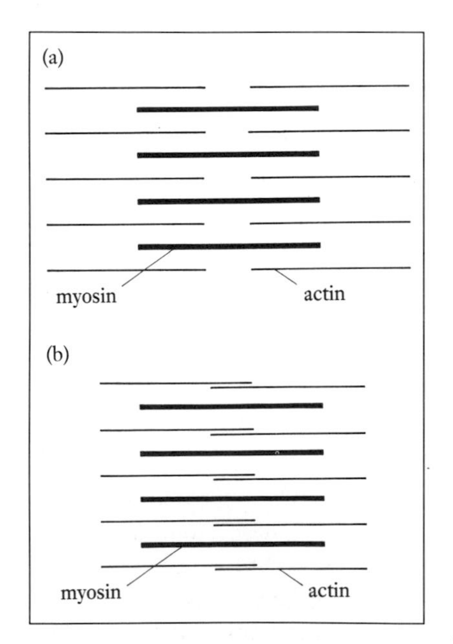

Figure 17.7. Muscle contraction – 'sliding filaments'. (a) Relaxed; (b) contracted.

Excitation–contraction coupling

The contraction of the myofibrils occurs in response to an increase in intracellular calcium levels which is initiated by the arrival of an action potential via the T tubules. As calcium combines with troponin the position of tropomyosin is altered to expose binding sites on the actin filament. In a process requiring ATP the myosin filament heads form **cross-bridges** which attach to the actin filaments. The myosin heads attach and detach several times to slide the actin filament towards the centre of the sarcomere, which causes contraction of the fibre.

Relaxation

Relaxation occurs when action potentials cease and calcium moves back into the sarcoplasmic reticulum. The cross-bridges break as tropomyosin again blocks the binding sites and the filaments return to their resting length.

Types of contraction in a skeletal muscle

Muscles contract to produce the tension which, when transmitted to the skeleton, produces movement or stabilizes the body. However, the tension or force of contraction must be sufficient to overcome the resistance of the load before the muscle will shorten, e.g. your arm muscles must produce enough tension to overcome that exerted by heavy shopping before you can lift the bag.

Twitch contraction occurs in response to an isolated action potential; there is a latent period (a period following muscle fibre stimulation and before contraction commences) prior to the contraction, which is followed by relaxation (*Figure 17.8*).

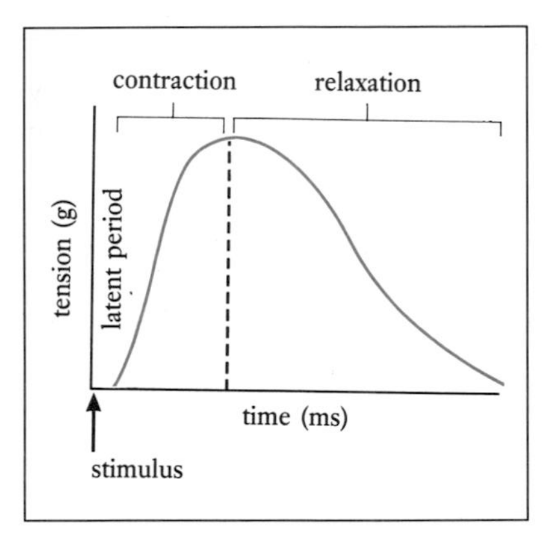

Figure 17.8. Twitch muscle contraction.

In reality the single twitch contraction is unusual; what usually happens is a **graded response** where **summation** of contractions occurs. Summation produces smooth, sustained contraction where tension is increased as required. Summation which increases force is achieved by increasing the frequency of stimulation to a maximum level known as **tetanus** (tetanus is also the name of a bacterial disease where bacterial toxins affect motor nerves causing muscle spasm, rigidity and convulsions), or by **recruitment**, where extra motor units become involved to increase the force of contraction.

A contraction where the tension is constant is termed **isotonic**. Here the muscle shortens to produce movement as the load tension is overcome, e.g. walking or lifting your shopping.

The other type of contraction – **isometric** contraction – is where tension in the muscle increases without shortening. This occurs when you push against a fixed object or hold a posture. The tension is insufficient to overcome that of the load and no movement occurs. Most of our muscle contraction is a combination of isotonic and isometric.

Metabolism in Muscle Tissue

Active muscle requires adenosine triphosphate (ATP) which provides the energy to power contraction. The ATP is produced in three ways:

- At the start of contraction from a coupled reaction where **creatine phosphate**, a high-energy molecule stored in muscle, reacts with **adenosine diphosphate** *(ADP)* to form ATP:

Creatine phosphate + ADP ———> ATP + creatine

The reaction is facilitated by the enzyme **creatine kinase**, if muscle is damaged the enzyme is released into the blood where levels can be measured to assist in diagnosis (see Chapter 10).

- The *aerobic* use of glucose, glycogen and fatty acids (see Chapter 13) when oxygen is available from the blood and myoglobin. The ATP is produced during the complex processes of **oxidative phosphorylation**.

- After prolonged muscle contraction the oxygen supply and utilization is reduced and ATP is produced by the *anaerobic* metabolism of glucose. This process results in an '**oxygen debt**' (see Chapters 12 and 13) and the formation of **lactic acid**, the waste product which causes muscle ache felt after exercise.

There is a limit to these processes and when insufficient ATP is forthcoming a condition known as **muscle fatigue** develops. No amount of 'will power' can overcome the fatigue as myofibrils cannot contract until ATP is replenished. However much you want to complete the last mile of a sponsored run, the muscle fatigue, if it occurs, will dash your hopes. The development of fatigue may include factors such as a fall in pH in the muscle caused by the accumulating lactic acid and local electrolyte imbalance.

Exercise and muscle

Obviously exercise requires more ATP, oxygen and fuel than is needed at rest, and regular exercise can produce beneficial physiological effects. The lungs and cardiovascular system function more efficiently and the muscle fibres use oxygen more effectively, tire less easily and eventually strengthen as individual muscles increase in size as fibres *hypertrophy* (increase in size; see Chapter 1). Other benefits of exercise include the maintenance of bone (see Chapter 16), avoidance of obesity and prevention of constipation (see Chapter 13).

However, it is important to remember some basic rules before embarking upon an exercise programme. You should build up slowly as your fitness increases: with intense unaccustomed exercise there is a real risk of *cramp* (prolonged tetanic contraction), muscle injury and exhaustion. There should always be a 'warm-up' period prior to the activity as muscle functions better when warm, and equally important is a gradual slowing down at the end of a session. Some guidelines for exercise in adults are covered in a booklet produced by the Health Education Authority (1987).

Muscle damage

Highly specialized skeletal muscle has very little regenerative ability, but relatively minor injuries where nerve and blood supply remain intact will 'heal' with no adverse functional effects. This healing probably occurs by the hypertrophy of existing fibres, not through the production of new fibres. However, more major damage is repaired by fibrosis, which in some situations leads to impaired function.

Very serious *crush injuries (crush syndrome)*, such as being trapped under fallen masonry, may be complicated by myoglobin release from the crushed muscles into the circulation (myoglobin 'leaks out' when muscle cells are disrupted). This type of crush injury results in severe shock with circulatory failure and possibly renal failure (see Chapter 15).

SUMMARY/CHECK LIST

Introduction – Muscle – functions. General characteristics. Types.
Skeletal muscle – Gross structure, attachments, fibre arrangements, blood supply and problems, *Person-centred Study* – Dave, muscle innervation, *Nursing Practice Application* – muscle changes in immobility. Microscopic structure. Muscular dystrophy.
Muscle contraction – 'sliding filament'. Contraction events – neuromuscular transmission and problems, e.g. myasthenia gravis, excitation–contraction coupling, relaxation. Types of contraction.
Muscle metabolism – ATP production, creatine phosphate, aerobic, anaerobic, oxygen debt, fatigue. Exercise and muscle. Muscle damage/crush injuries.

?

SELF TEST

1 Complete the following:
Muscle functions include body ---------, maintaining ------- and --------- and ---- production.
2 Explain the terms excitability, contractility, extensibility and elasticity.
3 Compare and contrast skeletal and smooth muscle.
4 Which of the following statements are true?
(a) Individual muscle fibres are of the same strength in both males and females.
(b) The endomysium encloses the whole muscle.
(c) Most muscles have at least two attachments, the origin and insertion.
(d) Muscle fasciculi arrangement is either parallel or oblique to the direction of pull.
5 Outline the effects of ischaemia on skeletal muscle.
6 Put the following in decreasing size order:
(a) myofibril
(b) muscle
(c) muscle fibre
(d) myofilament.
7 Briefly describe 'sliding filament' hypothesis.
8 Draw a diagram to illustrate the stages of a twitch contraction.
9 (a) How does muscle tissue produce the ATP required for contraction.
(b) What happens when insufficient ATP is available?

Answers
1. Movement, posture, stability, heat. 2. See page 427. 3. See pages 427-428. 4. a, c, d. 5. See page 430. 6. b, c, a, d. 7. See pages 433-434. 8. See *Figure 17.8*. 9. a. Creatine phosphate + ADP, aerobic use of fuel molecules, anaerobic use of glucose. b. muscle fatigue

References

Wickham, C. (1989) Falls in the elderly. *Nursing Times*, **85** (40), 50–51.

Further Reading

Health Education Authority (1987) *Exercise. Why Bother?* London: Health Education Authority.

Huxley, H.E. (1965) The mechanism of muscular contraction. *Scientific American*, **213** (6), 18–27.

Keele, C.A., Neil, E. and Joels, N. (1982 reprinted with corrections 1984) *Samson Wright's Applied Physiology (part V)*, 13th edn. Oxford: Oxford Univerisity Press.

Tortora, G.J. and Anagnostakos, N.P. (1990) *Principles of Anatomy and Physiology*, 6th edn. New York: Harper and Row.

CHAPTER EIGHTEEN

SKELETON, JOINTS AND MUSCULAR SYSTEM

OVERVIEW

- The axial and appendicular skeleton.
- Levers.
- Joints – types and movement.
- Muscles – naming and major groups.

LEARNING OUTCOMES

After studying Chapter 18 you should be able to:

- Name the bones forming the axial and appendicular skeleton.
- Name and describe the bones of the skull and face.
- Describe how the infant skull differs from that of an adult.
- Describe th location and functions of the paranasal air sinuses.
- Describe the structure of the vertebral column, including the intervertebral discs and ligaments.
- Describe the normal spinal curvatures.
- Describe the common structural features of all vertebrae and the characteristic adaptations found in each type which relate to function.
- Outline the causes of prolapsed intervertebral disc and discuss ways in which it may be prevented.
- Describe the sternum and ribs.
- Name and describe the bones of the pectoral girdle, arm and hand.
- Name and describe the bones of the pelvic girdle, leg and foot.
- Discuss the structural differences between the female and male pelvis relating these to function.
- Describe the arches of the foot.
- Outline the principles of levers and relate them to body movement.
- Classify joints by structure and function.
- Describe the general features of joints and give examples of fibrous, cartilaginous and synovial joints.

Learning Outcomes (continued)

- Name the different types of synovial joint and the movements possible at each.
- Describe some common joint disorders.
- Outline the structure of the shoulder, elbow, wrist, hip, knee and ankle joints; describe the movements possible at each and the major muscles involved.
- Explain how muscles are named.
- Explain muscle movement relationships and describe the function of prime movers, antagonists and synergists.
- Name and briefly describe the location of some major muscle groups with an overview of their action.

Key Words

Amphiarthrosis – a slightly movable joint.
Antagonist – a muscle which opposes or limits the action of a prime mover.
Appendicular skeleton – the limbs and limb girdles.
Articulation – a joint formed between two or more bones.
Axial skeleton – the bones which form the longitudinal axis – skull, sternum, ribs and spine (vertebral column).
Bursa – a small sac lined with synovial membrane found between bones and tendons.
Diarthrosis – a freely moveable or synovial joint.
Ligament – fibrous connective tissue which joins bones and stabilizes joints.
Pectoral (shoulder) girdle – the clavicle and scapula, which join the arms to the axial skeleton.
Pelvic (hip) girdle – the two innominate bones, which join with the sacrum posteriorly and attach the legs to the axial skeleton.
Prime mover – a muscle primarily responsible for a particular movement.
Synarthrosis – an immovable joint.
Synergist – a muscle which cancels out unwanted movement or contracts to stabilize joints not involved in a particular movement.
Vertebral column (spine) – consists of 33 vertebrae, of which 24 are unfused (individual bones) and 9 or 10 fused (joined).

Introduction

This section completes our study of movement and stability with coverage of the skeleton, joints and muscles. It is important to understand the high level of co-ordination existing between bones, joints and muscles, and your knowledge from Chapters 16 and 17 should be integrated with the material in this chapter.

Your understanding will be greatly enhanced if you look at, and feel, the structures in question. Where possible you should use the 'working model' provided by yourself, or specimen bones and models.

It is not the purpose of this book to give every detail of every bone and muscle, and readers requiring more information are directed to Further Reading.

SKELETON

The bony skeleton or framework of the body consists of some 206 bones (*Figure 18.1*). Its incredibly well designed structure, which acts as levers for movement, supports and protects organs (see also Chapter 16) can be divided into two parts:

- The **axial skeleton** – skull, spine (vertebral column), sternum and ribs and hyoid bone (see Chapters 12 and 13).
- The **appendicular skeleton** – pectoral girdle, upper limb, pelvic girdle and lower limb.

Bony landmarks

Before considering specific bones it might be useful to discuss some bony landmarks. The bones comprising the skeleton have various 'holes, hollows and bumps', none are smooth or completely regular. These landmarks are useful in describing the surface geography of the bone just as we use hills and dales to describe a landscape. Their real purpose, however, is to allow the passage of blood vessels and nerves, provide muscle attachments and form joints (see *Table 18.1*).

Table 18.1. Common bony landmarks.

Landmark and description	*Examples of particular clinical importance*
Openings and indentations	
Fissure – narrow cleft	Orbital fissure of sphenoid bone
Foramen – an opening	Foramen magnum in occipital bone
Fossa – a shallow cavity	Pituitary fossa in sphenoid bone
Meatus – a passageway	External auditory meatus in temporal bone
Sinus – air-filled cavity within a bone	Maxillary sinuses in the maxillae
Prominences and projections	
a. *Providing muscle/ligament attachment*	
Crest – narrow ridge	Crest of the tibia
Epicondyle – bony eminence upon or above condyle	Medial and lateral epicondyles of the humerus
Process – bony eminence	Mastoid process of temporal bone
Spine – sharpter bony eminence	Vertebral spines (spinous processes)
Trochanter – large, irregular prominence	Greater and lesser trochanters of the femur
Tuberosity – rough prominence	Tibial tuberosity
Tubercle – small round prominence	Tubercle at lower end of femur
b. *Forming joints*	
Condyles – rounded articular surface	Condyles at the upper end of the tibia
Facet – flat articular surface	Facets on thoracic vertebrae for ribs
Head – bony projection on a neck	Head of humerus which articulates with the scapula
Ramus – thin bony projection	Rami of the lower jaw

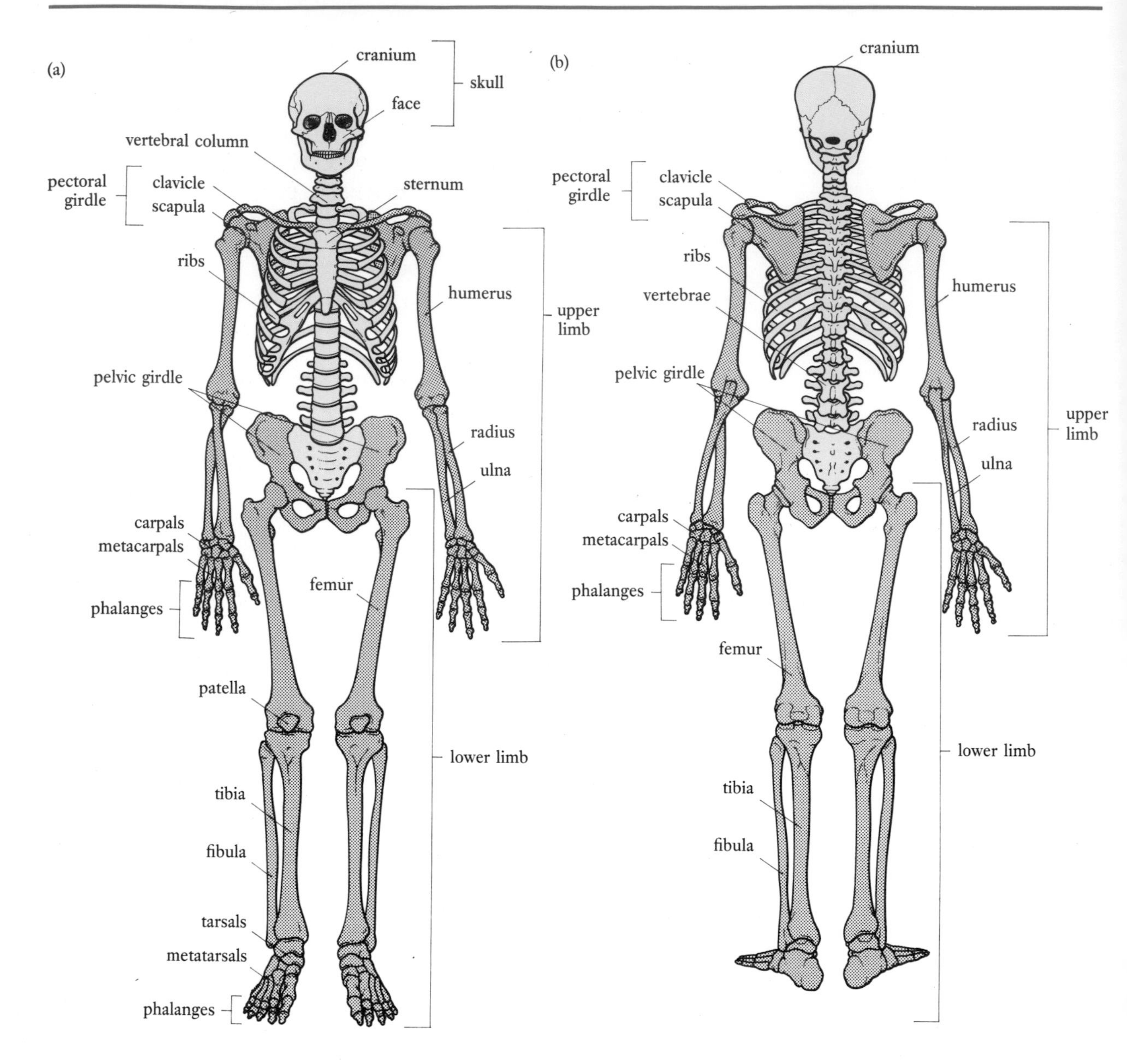

Figure 18.1. Skeleton (axial = unstippled, appendicular = stippled). (a) Anterior view; (b) posterior view.

Axial Skeleton

The skull

The **skull** is formed from flat or irregular bones (see Chapter 16), eight in the **cranium** (*calvaria*) and fourteen **facial** bones (*Figure 18.2*). The infant skull is large in relation to the face and the joints (**sutures**) are not fully ossified (see page 446), which allows for some overlap of skull bones as the infant's head passes through the birth canal.

The cranium

The **cranium** forms a bony box which surrounds and protects the brain. It has an upper part which forms the **vault** and a lower part called the **base**. The bones which form the cranium – the **frontal**, two **parietal**, two **temporal**, **occipital**, **sphenoid** and **ethmoid** – are smooth on the outside but grooved and ridged internally to accommodate the brain and vessels. Ridges divide the base of the skull into three **fossae** – anterior, middle and posterior – each containing a different part of the brain (*Figure 18.3*). Numerous *foramina* and other openings in the cranial bones, especially those of the base, allow the passage of nerves and blood vessels.

Frontal bone

The **frontal bone** forms the forehead, part of the orbit and the ridges (**supraorbital margins**) above the eyes. The supraorbital margin contains a notch through which vessels and nerves pass. The two halves of the frontal bone are joined by the **frontal suture** and the **coronal suture** unites the frontal and parietal bones. It also joins with several other skull bones (*Figure 18.2*). Two paranasal sinuses are situated in the frontal bone.

Parietal bones

The two **parietal bones** form the lateral walls and the roof of the skull. They are joined to the frontal bone, to the temporal bone by the **squamous suture**, to the occipital bone by the **lambdoidal suture** and to each other by the **sagittal suture**.

Temporal bones

The **temporal bones** form the lower lateral walls and part of the base of skull (*Figure 18.4*). Each bone is divided into four parts:

- **Squamous part**, which projects upwards to join with the parietal bone.
- **Zygomatic process**, which joins the zygomatic bone to form the zygomatic arch, the external auditory meatus is situated just below this area.
- **Mastoid process**, which is the prominent part felt behind the ear; it contains tiny air cells which communicate with the middle ear. Infection spreading from the middle ear may cause serious infection of the air cells (*mastoiditis*) which can spread to the brain.
- **Petrous portion**, which forms part of the middle cranial fossa of the base of skull and contains the organ of Corti (see Chapter 7).

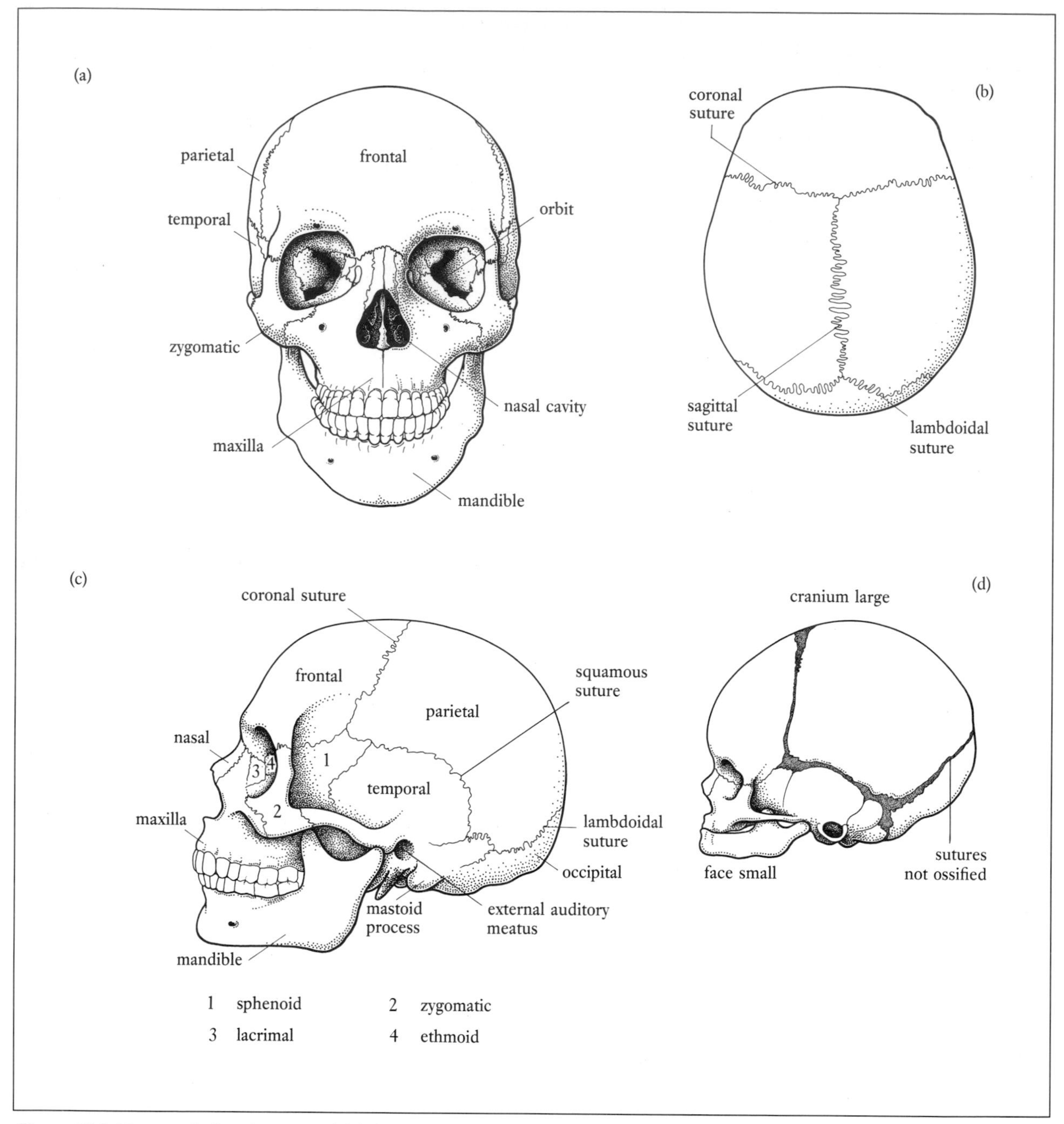

Figure 18.2. Human skull and sutures. (a) Adult skull – anterior view; (b) adult skull – superior view; (c) adult skull – lateral view; (d) infant skull (see also *Figure 18.8*).

The temporal bones provide attachment for many muscles and articulate with the mandible at the *temporomandibular* joints.

Occipital bone

The **occipital bone** forms the back of the cranium and part of the base of the skull (*Figure 18.5*). It joins with the temporal, parietal and

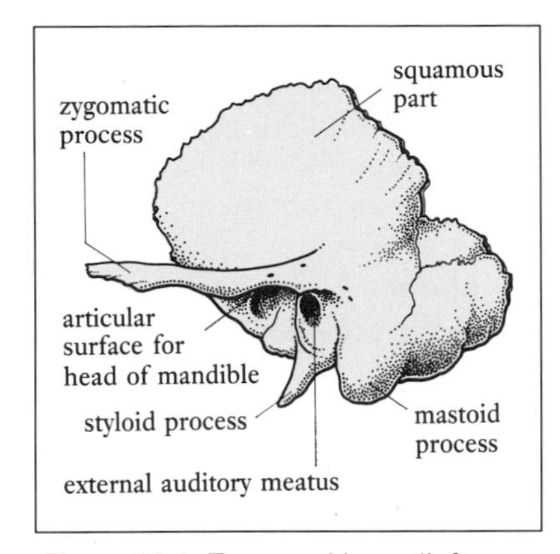

Figure 18.4. Temporal bone (left, viewed from the side).

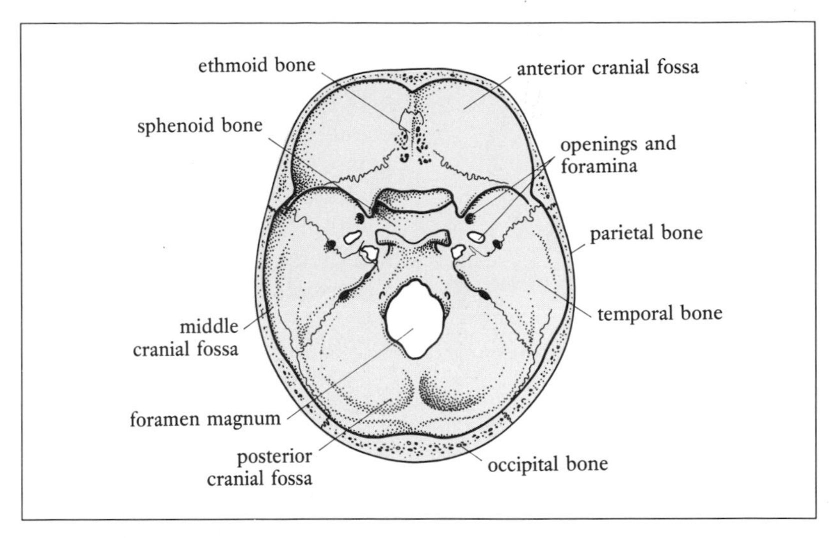

Figure 18.3. Interior of the base of skull.

sphenoid bones. It contains an opening, the **foramen magnum**, through which the spinal cord passes. On each side of this foramen are *condyles* (see *Table 18.1*) which form articulating surfaces for the first cervical vertebra (atlas). The base of the occipital bone forms the posterior cranial fossa which contains the cerebellum and part of the cerebrum.

Sphenoid bone

The bat-shaped **sphenoid bone** forms part of the middle cranial fossa of the base of the skull (*Figure 18.6*). A depression in its body, called the *sella turcica (pituitary fossa)*, accommodates the pituitary gland (see Chapter 8). The sphenoid, which joins with all the other cranial bones, contains paired paranasal sinuses. *Optic foramina* near the sella turcica give passage to the optic nerves and arteries en route to the eyes (see Chapter 7).

Ethmoid bone

The complex **ethmoid** bone, located between the sphenoid and nasal bones, forms part of the orbital and nasal cavities (*Figure 18.7*). The **cribriform plate**, perforated by numerous foramina for the olfactory nerves, forms the roof of the nose. A **perpendicular plate** forms the upper part of the nasal septum. On each side of the perpendicular plate the **lateral masses** of the ethmoid extend to form the **superior** and **middle nasal conchae** or *turbinates*, which project into the nasal cavity (see Chapter 12). The lateral masses contain many paranasal sinuses which make the bone spongy and light.

Skull sutures and fontanelles

In adults the serrated joints or **sutures** between the skull bones are fibrous and immovable, which means that any pressure rise within the closed cavity will compress the brain (see Chapter 4). The exceptions are the freely movable joints between the mandible and temporal bones, and the joints between the middle ear ossicles.

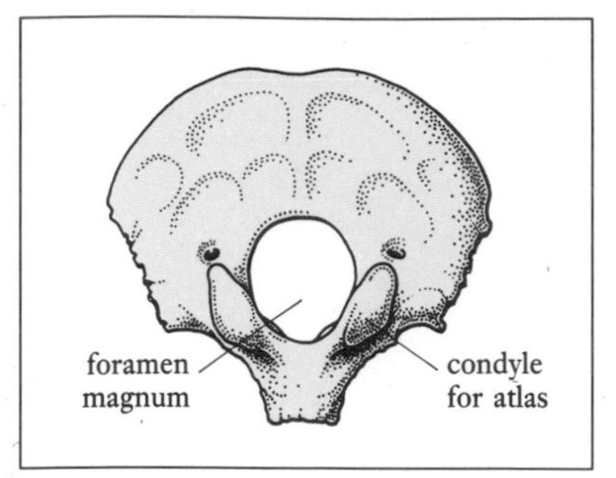

Figure 18.5. Occipital bone (viewed from below).

However, at birth the sutures are not fully ossified and several membranous spaces known as **fontanelles** are present (*Figure 18.8*). The **anterior** fontanelle forms a diamond-shaped opening where the coronal, frontal and sagittal sutures meet; it has usually ossified by 18 months. The smaller **posterior** fontanelle, situated where the lambdoidal and sagittal sutures meet, is usually closed within 3 months (see *Figure 18.2*). Abnormally wide sutures and large fontanelles may be observed where the skull has expanded in a baby with *hydrocephaly* (see Chapter 4).

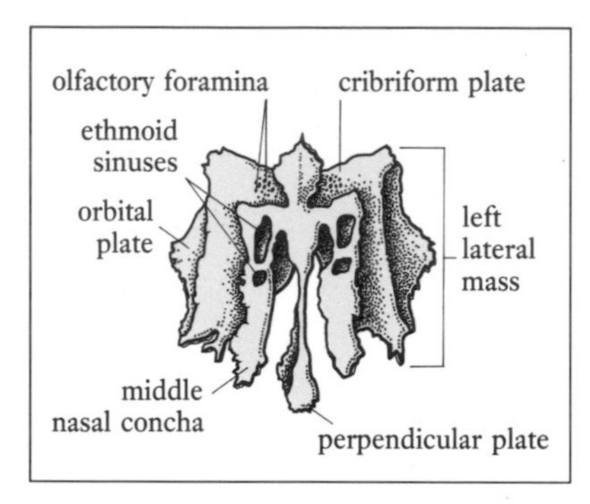

Figure 18.7. Ethmoid bone (anterior view).

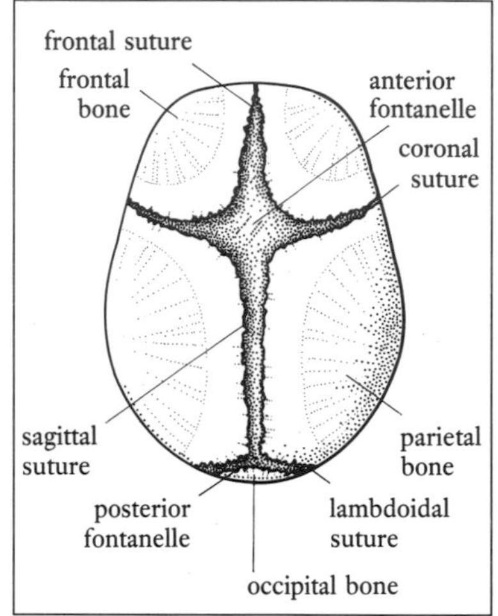

Figure 18.8. Infant skull showing fontanelles.

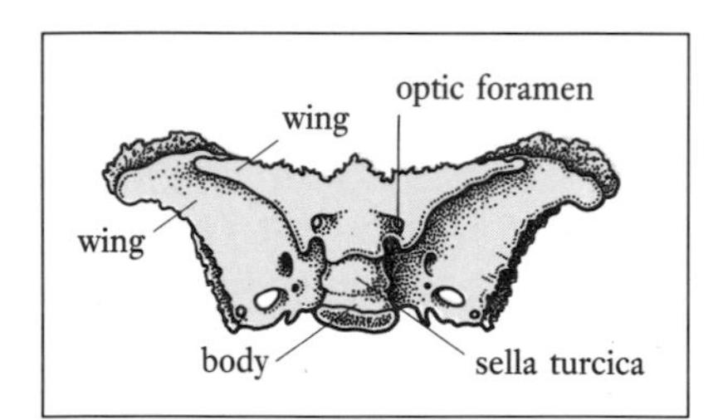

Figure 18.6. Sphenoid bone (superior view).

The facial bones

The 14 bones of the face are, except the mandible, joined by immovable sutures (*Figure 18.9*). There are six paired bones – **nasal, palatine, inferior conchae, zygomatic, lacrimal, maxillae** – and a single **vomer** and **mandible**.

Nasal bones

Two small **nasal bones** join medially to form the bridge of the nose.

Palatine bones

The horizontal parts of the two L-shaped **palatine bones** form the posterior hard palate (see Chapter 13). The vertical portions project upwards to form part of the nose and orbit.

Inferior conchae

Two curved **inferior conchae** form part of the lateral walls of the nasal cavity. They protrude into the nasal cavity below the middle conchae (ethmoid) and are the largest pair of nasal conchae.

Zygomatic (malar) bones

Paired **zygomatic bones** form the prominences of the cheeks and part of the orbit. They join with the zygomatic processes of the temporal bone to form the zygomatic arch.

Lacrimal bones

Two small **lacrimal** bones form part of the medial wall of the orbit.

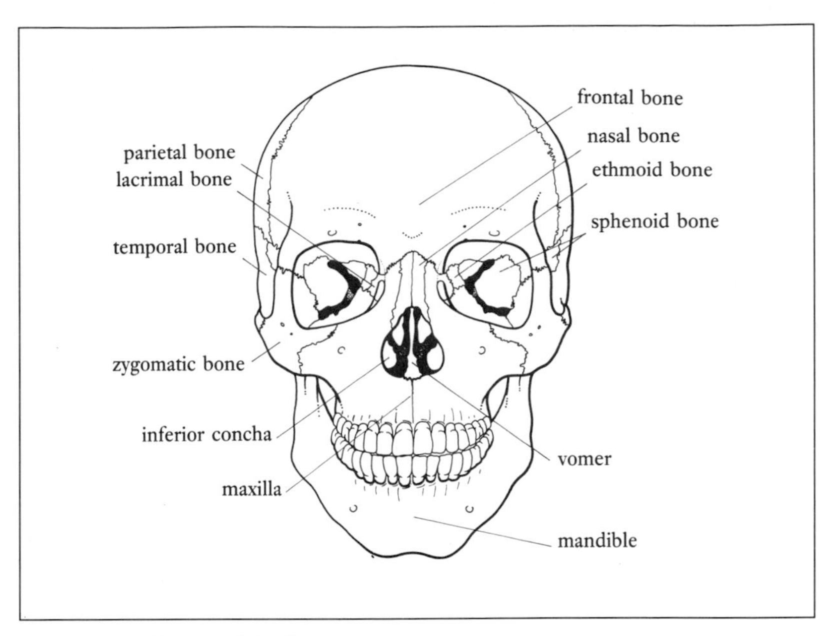

Figure 18.9. Bones of the face.

The nasolacrimal ducts carrying tears to the nasal cavity pass through foramina in the lacrimal bones (see Chapter 7).

Maxilla

Forms the upper jaw, anterior hard palate and part of the orbit and nasal cavity. The **maxillae** develop in two halves which normally fuse, in the midline, before birth (see cleft palate, Chapter 13). An **alveolar ridge** contains the upper teeth. A large paranasal sinus or **maxillary antrum**, running from the orbit to the upper teeth, is present on each side.

Vomer

A single slender **vomer** forms the lower part of the bony **nasal septum**. Details of the structure of the orbit and nasal cavity are illustrated in *Figure 18.10.*

Mandible

The **mandible** or lower jaw forms the chin and contains the lower teeth in an **alveolar ridge** (*Figure 18.11)*. It has a curved horizontal body, and on each side a **ramus** extends upwards from the body from a point known as the angle of the jaw. The ramus terminates in two processes; the **coronoid process**, which provides attachment for a muscle (*temporalis*) concerned with mastication, and the **mandibular condyle,** which articulates with the temporal bone at the temporomandibular joint. This joint allows you to open and close your mouth, retract and protract the jaw and move it slightly from side to side.

Paranasal sinuses

The **paranasal sinuses** are membrane-lined cavities or air sinuses which communicate with the nose and are located within the frontal, maxillary, ethmoid and sphenoid bones (*Figure 18.12*). These air–filled

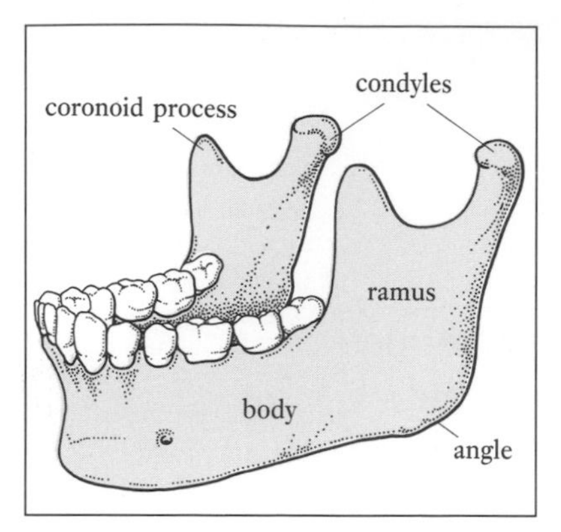

Figure 18.11. Mandible (oblique lateral view).

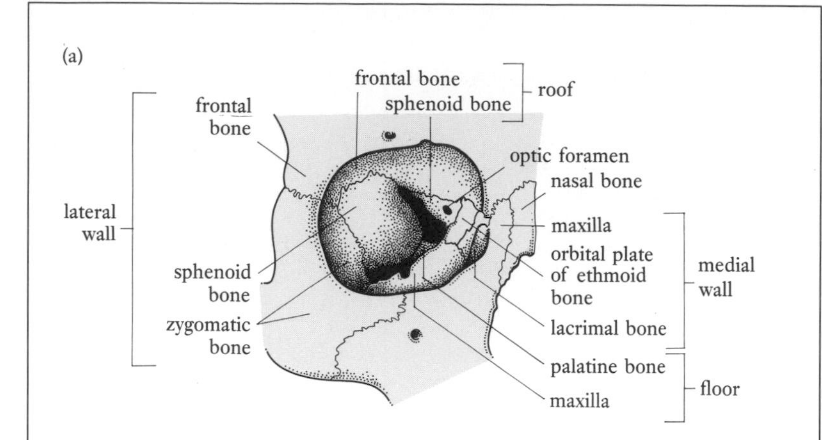

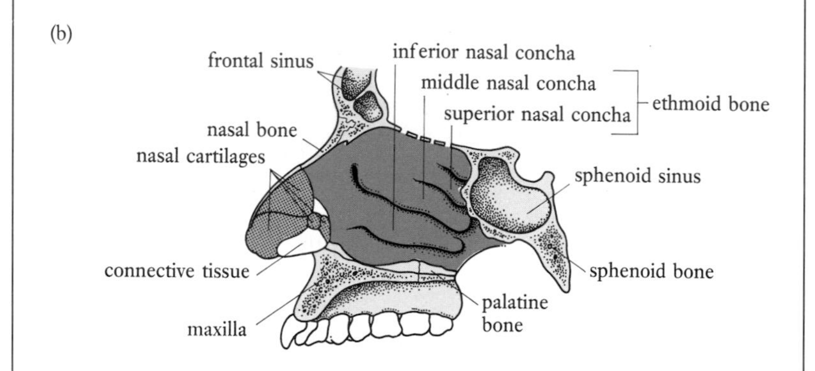

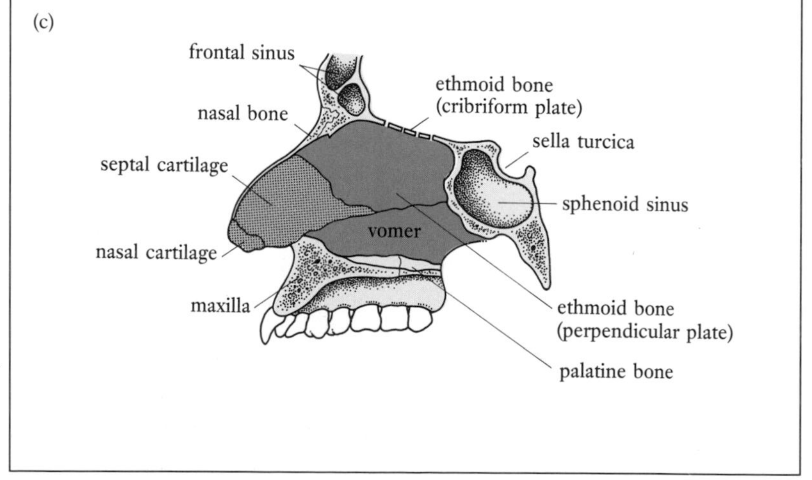

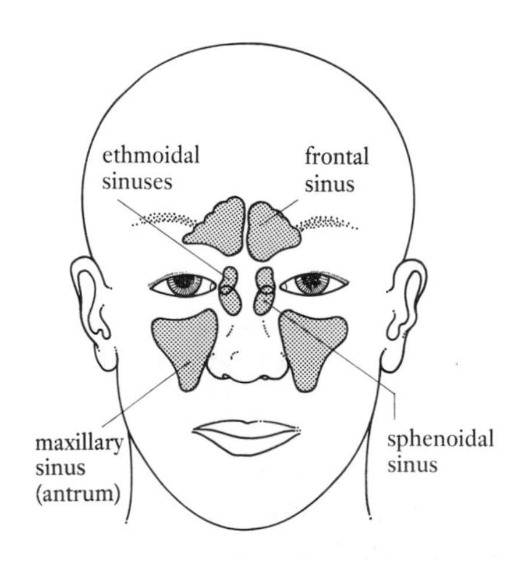

Figure 18.12. Paranasal sinuses.

Figure 18.10. (a) Orbit; (b) nasal cavity (lateral wall); (c) nasal septum.

Nursing Practice Application Mandibular fracture

Severe jaw injuries, such as those sustained in a car accident, may be stabilized with wires or bands. Care plans should include: (i) Maintaining an airway – whilst vomiting is a possibility, it is vital that equipment for releasing the wires and/or bands and for clearing the airway is to hand. Remember that when the mandibular fracture is stabilized the teeth are secured shut. There is a real risk of inhalation of vomit if vomiting occurs in a person who cannot open his or her mouth. Intially, a nasogastric tube (passed into the stomach via the nose) is used, to aspirate gastric contents, and anti-emetic drugs are prescribed; both of these reduce the possibility of vomiting with inhalation. (ii) Setting up an effective system for communication until the person is able to talk. The considerable swelling and pain associated with facial injuries and the inability to open the mouth makes speech very difficult. Nurses should be alert to the person's non-verbal communications, administer prescribed pain relief, anticipate the need for clear information and provide resources for writing messages. (iii) Providing adequate oral hygiene (see Chapter 13) is an essential part of care. The normal cleansing mechanisms are inhibited and the person has actual tissue damage. Keeping the mouth clean, e.g. by gentle irrigation, will minimize discomfort and the risk of complications, such as infection. (iv) Providing nutritional and fluid requirements in an acceptable and effective way (see Chapter 13). Initially, the person will probably have intravenous fluid replacement, but if his or her gut is functioning a quick return to oral or nasoenteric feeding is desirable. Imaginative food presentation, which takes account of what the person likes, needs and can manage to eat, is most likely to be successful. It is worth remembering that this person will require extra calories (they may well have other injuries) to facilitate healing. The nurse, after consultation with the nutrition team, can provide these calories in the form of high-energy fluid supplements, which can be sipped, or, if necessary, administered via the nasoenteric route. Attention to oral hygeine after eating or drinking is of particular importance. (v) Helping the person and/or family to cope with the altered body image and long-term treatment that will be required. (vi) An individual education programme will be required if the person is discharged with the wires and/or bands still in place.

cavities lighten the skull and give resonance to the voice, but their close proximity to the nose renders them easy prey to infection (see Chapter 12).

Spine

The **spine** or **vertebral column** (*Figure 18.13*) supports the head, allows us to adopt the upright position by providing axial support, protects the all-important spinal cord (see Chapter 4) and acts as a 'shock absorber', e.g. when you jump down the last few stairs. It also provides attachment for muscles and other parts of the skeleton. The importance of the spine is well illustrated by expressions such as 'spineless' which implies weakness, and describing a person of strong character and courage as having 'backbone'.

The spine, which measures some 60–70 cm in an adult, consists of twenty four individual vertebrae (*7 cervical, 12 thoracic, 5 lumbar*) and nine or ten fused bones (*5 sacral, 4–5 coccygeal*). The bony arrangement with its supporting muscles and ligaments gives the spine great strength and the number of joints present confers considerable collective flexibility.

Spinal curves

At birth the spine shows a C-shaped curve, caused by two **primary curves** where the **thoracic** and **sacral** regions are *convex posteriorly (fetal position)*. The two **secondary curves**, which are *convex anteriorly*, occur in the **cervical** region when the child lifts its head and the second in the **lumbar** region when the child stands. All four curves together give the normal adult spinal curvature (*Figure 18.14*).

Various abnormal curvatures exist; these may be congenital, due to disease or result from age changes. A lateral curvature known as *scoliosis* may occur in the thoracic region. An exaggerated lumbar curve which gives rise to *lordosis* may be due to disease, e.g. rickets (see Chapter 16) or an alteration in the centre of gravity such as late pregnancy. *Kyphosis* is an abnormally pronounced thoracic curve, which may result from osteoporotic age changes (see Chapter 16) and intervertebral disc degeneration.

General features of vertebrae

Vertebrae all have certain features in common (*Figure 18.15*). There is an anterior part called the **body** and a **vertebral (neural) arch**, which together surround the **vertebral foramen**. The foramina of the unfused vertebrae form the **vertebral (neural) canal**, in which the spinal cord is situated and protected. The vertebral arch consists of two **pedicles**, forming the sides, and two **laminae** posteriorly. Protruding from the arch are three processes and four articular surfaces. There are two **lateral transverse processes** and a **spinous process (spine)**, which provide attachment for muscles and ligaments. The paired **articular surfaces** articulate with the vertebra above and below.

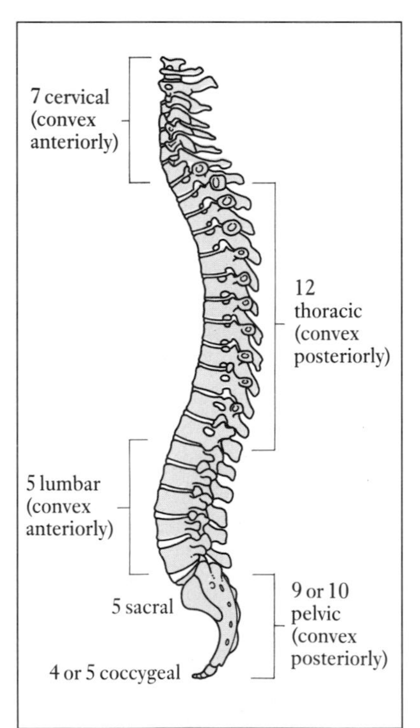

Figure 18.13. Vertebral column.

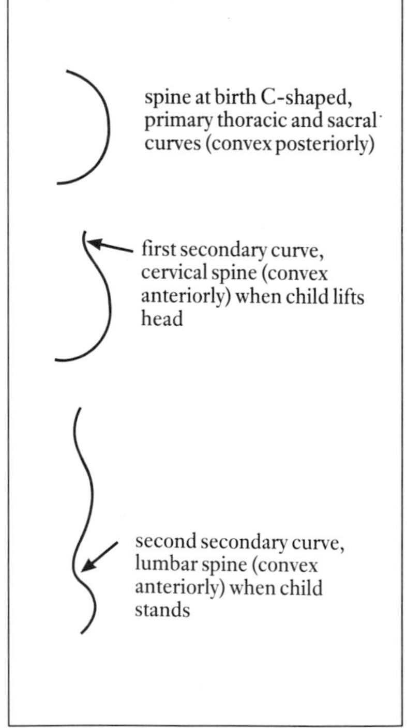

Figure 18.14. Curves of the spine.

Individual vertebrae

Cervical (seven vertebrae)

The first two vertebrae, the **atlas** and **axis**, are atypical and will be described separately. Generally the **cervical vertebrae** have a body which is wider side to side than front to back, and all except the seventh have a *bifid* spinous process (*Figure 18.16(a)*). The spinous process of C7 (the seventh vertebra) can be felt as the prominence at the base of the neck. Their transverse processes contain foramina for the vertebral arteries which supply the brain and the vertebral foramen is enlarged to accommodate the spinal cord.

The atlas (C1), a ring of bone without body or spinous process, supports the skull (*Figure 18.16(b)*). It articulates with the occipital condyles to allow you to nod your head. It might be easier to remember the position of the atlas if you think of Atlas who, according to mythology, supported the universe.

Below the atlas is the axis (C2) which has a large process called

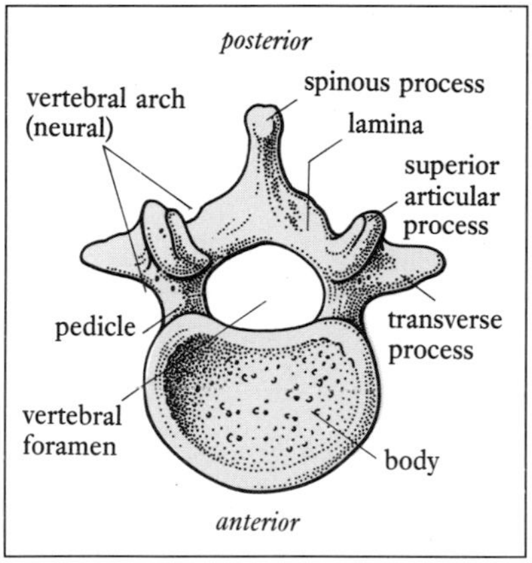

Figure 18.15. Vertebra, showing general features.

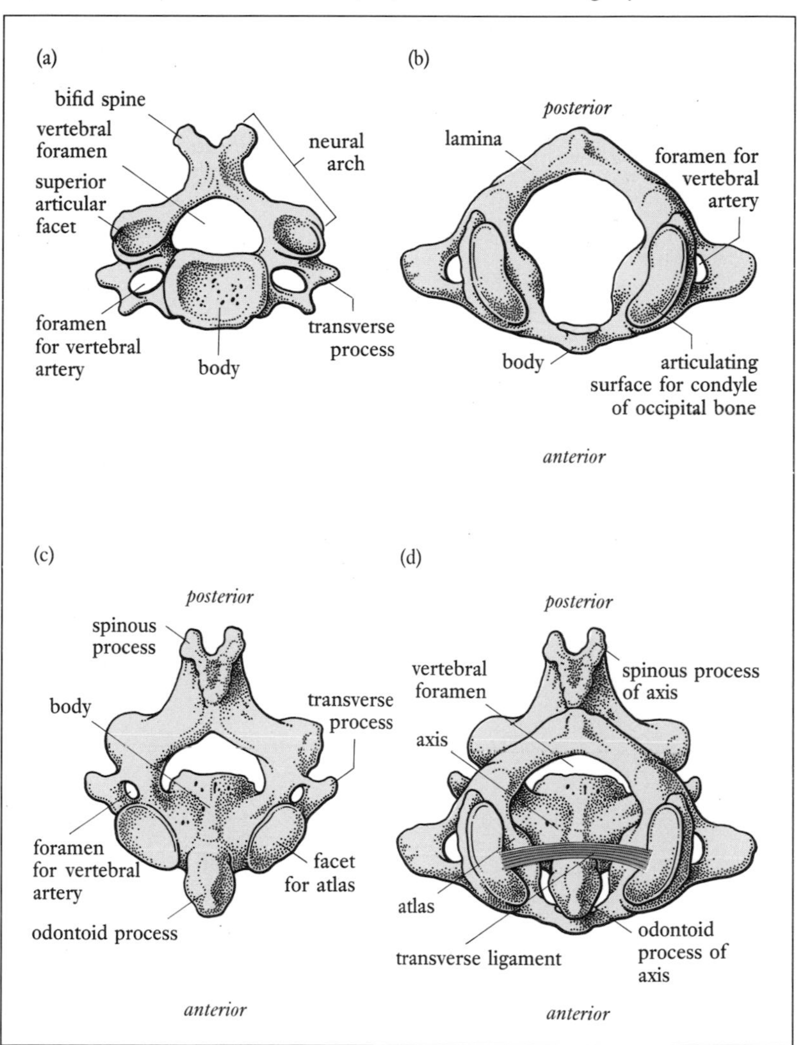

Figure 18.16. (a) Cervical vertebra; (b) atlas; (c) axis; (d) atlas and axis. All viewed from above.

the **odontoid process (dens)** protruding superiorly from its body (*Figure 18.16(c)*). The odontoid process, which represents the body of the atlas, fits into the vertebral foramen of the atlas. The atlas rotates around the odontoid process which acts as a pivot for the movement of shaking your head (*Figure 18.16(d)*).

Thoracic (12 vertebrae)

The **thoracic vertebrae**, all larger than the cervical vertebrae, show an increase in size from T1 to T12. The heart-shaped body has facets on both sides which articulate with the ribs (heads) (*Figure 18.17*). Facets on the transverse processes also articulate with the rib tubercles. Their spinous process is long and projects downwards.

Lumbar (five vertebrae)

The strong **lumbar vertebrae**, with kidney-shaped bodies, are the largest of the vertebrae, a fact that reflects their weight-bearing role. The transverse processes are long and thin. The spinous process, which provides attachment for back muscles, is short, wide and hatchet-shaped (*Figure 18.18*). The positioning of the articular surfaces prevents rotation and increases stability in this region.

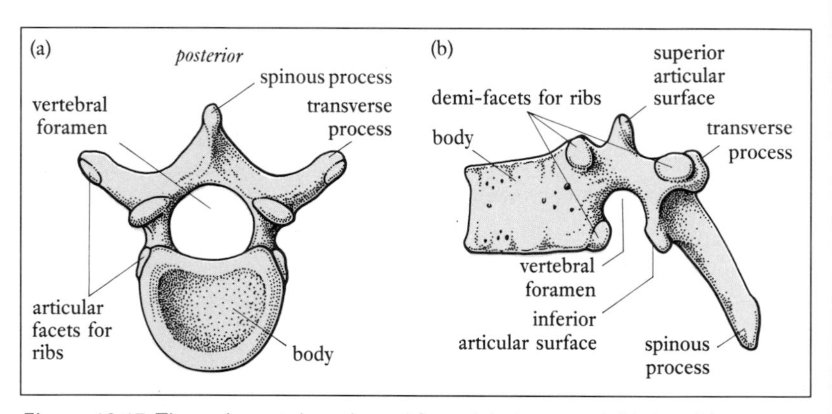

Figure 18.17. Thoracic vertebra viewed from (a) above and (b) the side.

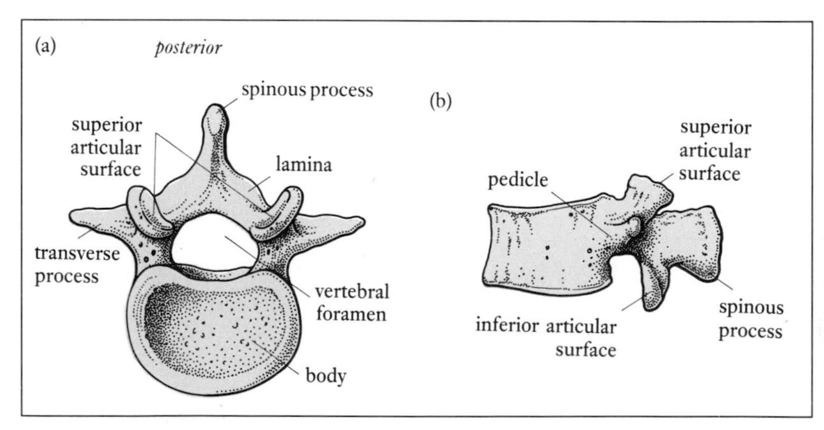

Figure 18.18. Lumbar vertebra viewed from (a) above and (b) the side.

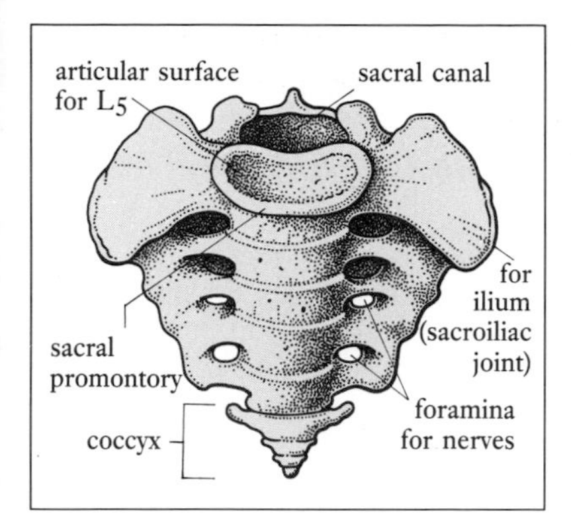

Figure 18.19. Sacrum and coccyx (anterior surface).

Sacrum

The wedge-shaped, triangular **sacrum** consists of five fused bones in the adult (*Figure 18.19*). The base, which is uppermost, articulates with the fifth lumbar vertebra. The anterior edge of the base forms the **sacral promontory**, which bulges forwards into the pelvic cavity. The sacrum, which forms the posterior part of the pelvis, also articulates with the innominate bones at the **sacroiliac joints**. The anterior surface of the sacrum is concave with foramina for the passage of nerves and vessels. Running within the sacrum is the **sacral canal**, which is a continuation of the vertebral canal. The sacral apex articulates with the coccyx below.

Coccyx

Four or five fused vestigial tail bones form the triangular **coccyx** in adults (*Figure 18.19*). It articulates with the sacrum above.

We only really become aware of its existence when it is damaged. Injuries to the coccyx commonly follow a heavy fall, e.g. slipping on a step, and result in pain and problems with sitting. Recovery may take many months and the unfortunate person is forced to rely heavily upon analgesics and several cushions. Those people who continue to experience pain may obtain relief from local injections of anaesthetic and corticosteroids or surgery.

Spinal ligaments

The vertebral column is connected and stabilized by various **ligaments** (fibrous connective tissue which joins bones and stabilizes joints) which include (*Figure 18.20(a)*):

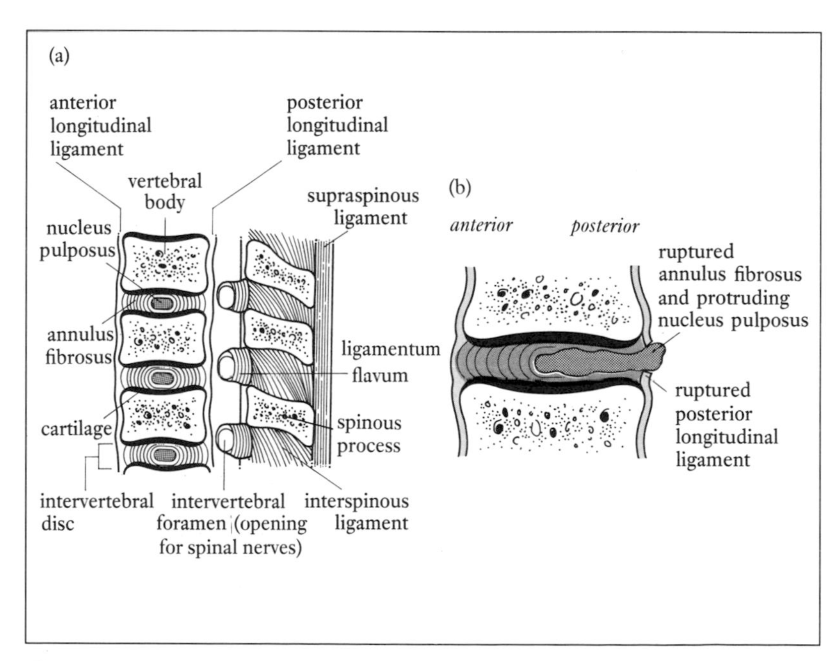

Figure 18.20. (a) Section through the vertebral column showing intervertebral joints, discs and ligaments; (b) prolapsed intervertebral disc.

- **Anterior** and **posterior longitudinal ligaments**, which run in front and behind the vertebral bodies for the entire length of the column.
- **Transverse ligament** (see *Figure 18.16(d)*) which secures the position of the odontoid process.
- Several ligaments which connect adjacent vertebrae, e.g. **supraspinous**, **ligamenta flava** and **interspinous ligaments**.

The muscles of the trunk are also involved in supporting the vertebral column.

Nursing Practice Application Back pain

Members of the caring professions, especially nurses, are at risk from back problems resulting from their work. In a survey of nurses in a teaching hospital, Griffiths (1988) found that nearly 70 per cent of the sample reported 'back problems' which occurred after they became nurses. Obviously the severity of the problem varies considerably; an aching back which soon recovers, right up to a disc prolapse with cord or nerve compression (see *Figure 18.20(b)* and Chapter 5).

At worse the injury results in severe pain, immobility, neurological effects, considerable time lost from work/study and the possibility of permanent disability and discontinuation of nursing career. Damage occurs when the spine is subjected to sudden severe trauma or prolonged abuse which occurs during the bending, twisting and lifting when handling patients. A *prolapsed intervertebral disc (PID)* occurs when trauma ruptures the annulus fibrosus and allows the nucleus pulposus to protrude (see *Figure 18.20(b)*).

The management involves rest, analgesics, anti-inflammatory drugs, physiotherapy and in some cases surgery. All this causes endless pain and inconvenience before you even consider the financial implications of treatment costs and sick leave. The prevention of back injuries is much the preferred option, but where problems do occur they should be recognized and treated early.

Some areas to consider in the prevention of back injuries include: (i) consideration of the work environment and uniform, both of which may militate against safe practices; (ii) adequate staffing levels; (iii) provision of lifting equipment; (iv) effective staff education programmes with regular updating which include correct lifting techniques, use of equipment, postural and back care aspects; (v) formulating a proper lifting policy and reporting procedures for accidents.

Intervertebral foramina and discs

Intervertebral foramina formed by the pedicles of adjacent vertebrae allow the passage of spinal nerves (see Chapter 5) and blood and lymph vessels. The foramina are present on both sides of the vertebral column along its entire length.

The cartilaginous joints (see page 465) between the vertebrae, except C1 and C2, are formed from an **intervertebral disc** of fibrocartilage (**annulus fibrosus**) surrounding a semisolid core (**nucleus pulposus**). The discs, which contribute to spinal flexibility and act as shock absorbers, are held in place by the posterior longitudinal ligament. There is very little movement between individual vertebral joints, but collective flexibility produces considerable movement. This flexibility allows bending forwards (*flexion*), bending backwards (*extension*), bending from side to side (*lateral flexion*) and *rotation*. Back pain and *prolapsed intervertebral disc* (slipped disc) are occupational hazards for carers.

Thoracic (thorax) cage

The structures forming the thorax or chest are the sternum, costal cartilages, 12 pairs of ribs and the thoracic vertebrae (*Figure 18.22*) (see also page 452). The **thoracic cage** protects the organs and great vessels of the chest, provides attachment for back and chest muscles and supports the pectoral girdle. The *intercostal muscles* are situated between the ribs (see Chapter 12).

The sternum

The **sternum** (breastbone) is the flat dagger-shaped bone situated in the midline of the anterior part of the chest (*Figure 18.21*). It is around 15 cm in length, and consists of an upper part, the **manubrium**, connected to the **body** by a cartilaginous joint which normally remains unossified. Projecting downwards from the body is the **xiphisternum (xiphoid cartilage)** which provides attachment for some abdominal muscles and the diaphragm. The xiphisternum usually becomes ossified later in life.

The manubrium articulates with the clavicles at its **clavicular notches** and with the first two pairs of ribs. The body articulates with the costal cartilages of ribs 3 to 7.

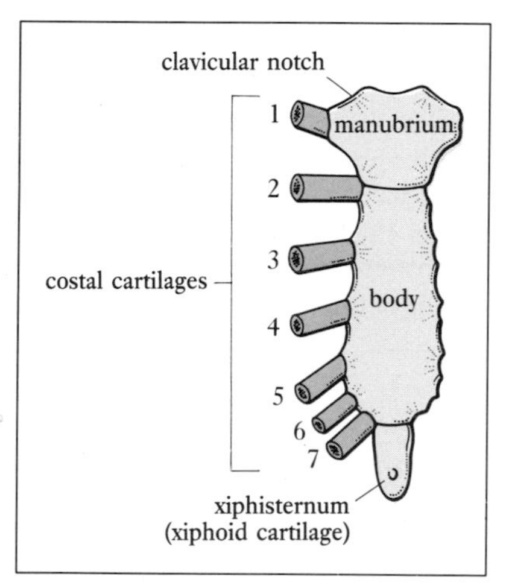

Figure 18.21. Sternum (anterior view).

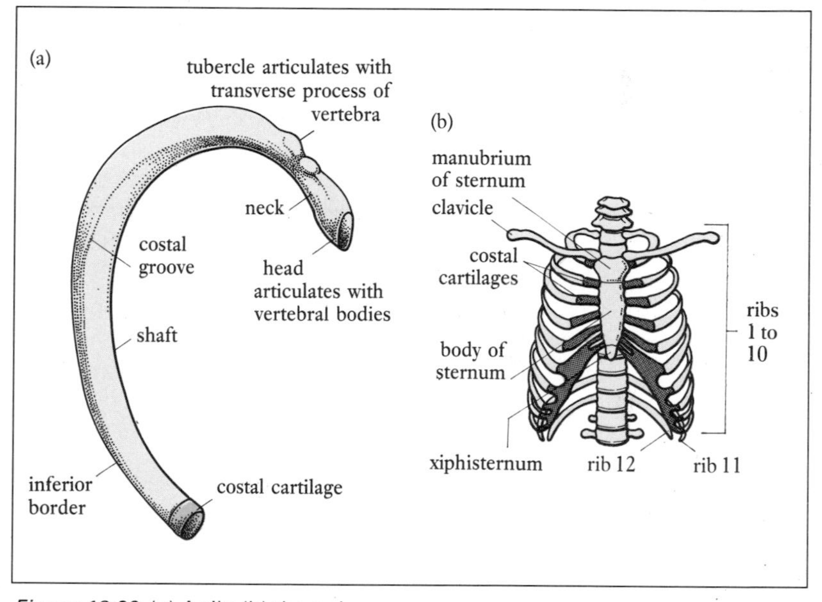

Figure 18.22. (a) A rib; (b) thoracic cage.

The ribs

The lateral walls of the thorax are formed from12 pairs of **ribs**. They are curved bones attached posteriorly to the thoracic vertebrae (see page 452).

The ribs can be divided into seven pairs of **true ribs** which articulate with the sternum via **costal cartilages** and five pairs of **false ribs**. The false rib pairs 8 to 10 are attached indirectly to the sternum by their costal cartilage joining to the costal cartilage above, but pairs 11 and 12, which have no sternal attachment, are known as **floating ribs**.

Ribs generally have a curved **shaft**, **tubercle**, **neck** and **head**. The inferior border is grooved for the passage of blood vessels and nerves. The head of the rib has two facets which articulate with facets on adjacent vertebral bodies and the tubercle articulates with the transverse process of the lower vertebral body.

The first, eleventh and twelth pairs are atypical ribs: the first pair is short and broad and the three pairs (1, 11 and 12) articulate with only one vertebra.

Anteriorly the ribs join with the costal cartilages, made of hyaline cartilage, which allow the flexibility required for chest movements during respiration. The chest movements which change thoracic volumes (see Chapter 12) are facilitated by the *external* and *internal intercostal muscles* (see page 477).

APPENDICULAR SKELETON

Shoulder (pectoral) girdle and arm

A shoulder (**pectoral**) girdle, which joins the arm to the axial skeleton, consists of a **clavicle** (collar bone) and **scapula** (shoulder blade) (see *Figure 18.1*). The upper limb is formed from a **humerus**, **radius**, **ulna**, **carpals** (8), **metacarpals** (5) and **phalanges** (14).

Clavicle

The two clavicles are long bones with a double curvature (*Figure 18.23*). They articulate medially with the sternum and laterally with the scapulae at the **acromioclavicular joints**. The clavicles act as an anterior brace (support) for the shoulder, and provide muscle attachments. *Clavicular fractures* are common in contact sports such as rugby and following a fall on the outstretched arm.

Scapula

The **scapulae** are large, flat, triangular bones situated on the posterior thoracic wall. A layer of muscle separates the scapulae and ribs (*Figure 18.24*). Each has two surfaces, three borders and three

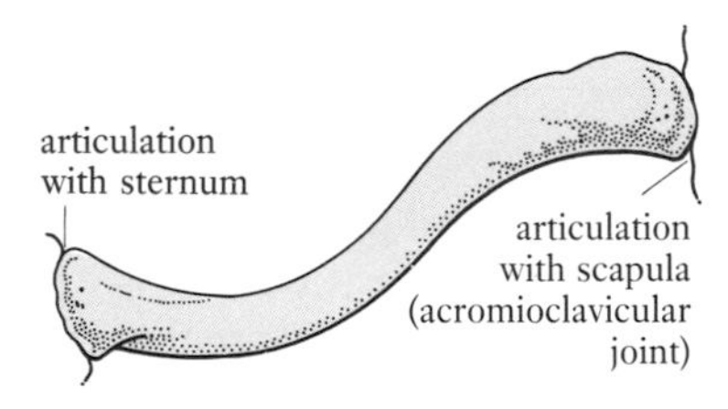

Figure 18.23. Clavicle (left – upper surface).

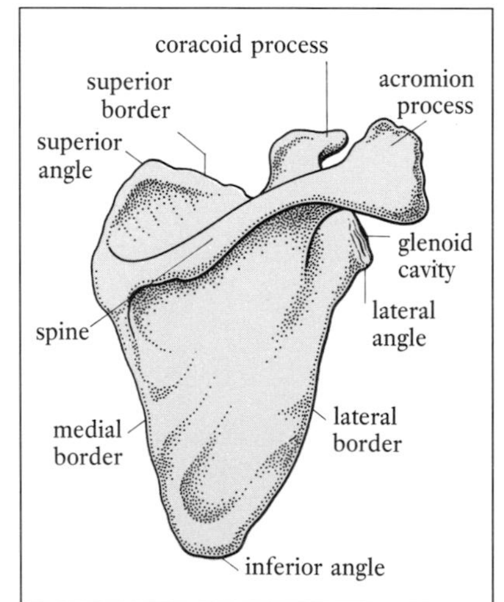

Figure 18.24. Scapula (right – posterior surface).

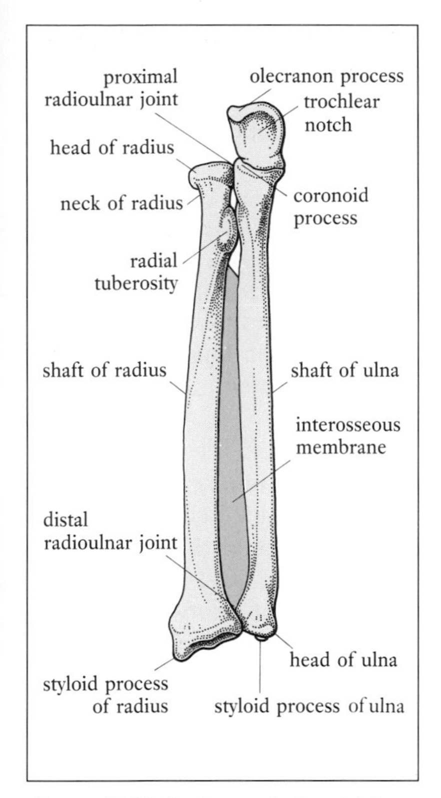

Figure 18.26. Radius and ulna (right – anterior view).

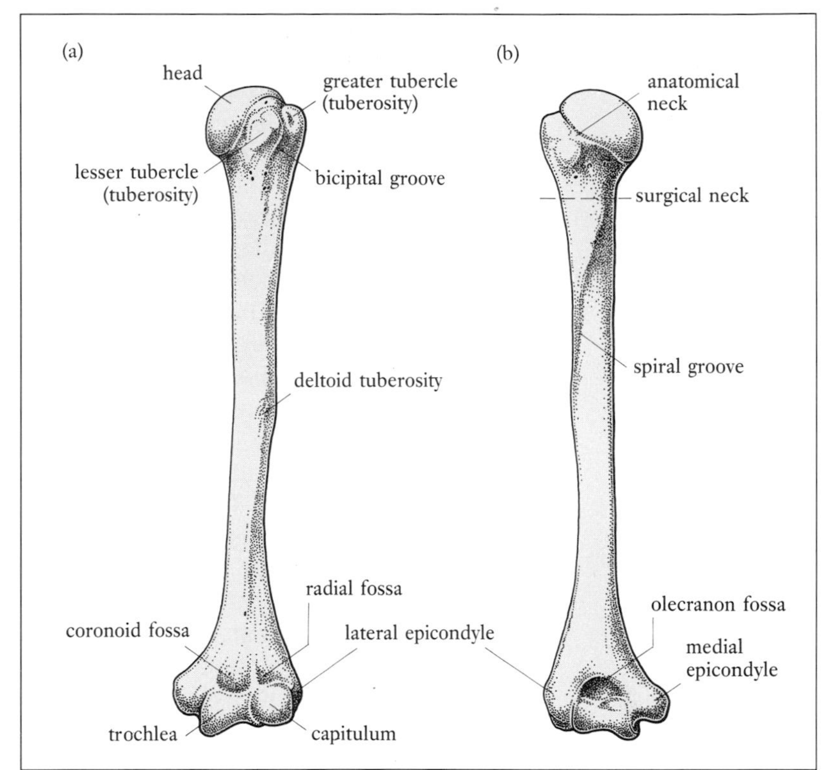

Figure 18.25. Humerus left – (a) anterior and (b) posterior views.

angles. The posterior surface is marked by a prominent ridge or **spine** which extends to the **acromion process**. The scapula articulates with the clavicle at the acromioclavicular joint. A shallow fossa known as the **glenoid cavity**, situated at the lateral angle, accommodates the head of the humerus to form the shoulder joint. The **coracoid process** provides attachment for the *biceps brachii muscle.*

Humerus

The humerus is the long bone forming the upper arm (*Figure 18.25*). It articulates with the scapula at the shoulder joint and the radius and ulna at the elbow joint.

The head of the humerus fits into the glenoid cavity to give a good range of movements at the shoulder. Below the head is the **anatomical neck** and two projections called the **greater** and **lesser tubercles** *(tuberosities* by some authorities), which are separated by a groove which accommodates the *biceps tendon.* The region where the extremity becomes the shaft is called the **surgical neck**. This is a frequent fracture site. The shaft, which is triangular in section, has a tuberosity which provides attachment for the *deltoid muscle* and a groove for the *radial nerve.* At its distal end the humerus is broad and flattened. It has a **lateral** and **medial epicondyle** and two articulating condyles – the **capitulum** for the radius and the **trochlea** for the ulna. Above the articulating surfaces are fossae (which receive the processes of the ulna and radius when the elbow flexes or extends), the **coronoid** and **olecranon fossae** for the ulna and the **radial fossa** for the radius. The ulnar nerve, situated behind the medial epicondyle, accounts for the excruciating pain felt when you knock your elbow ('funny bone').

Radius and ulna

The two long bones of the forearm are the radius (lateral side) and ulna (medial side) (*Figure 18.26*). With the arm in the anatomical position (see *Figure 1.25*) they lie parallel, with the radius lateral to the ulna. The two bones articulate with the humerus at the elbow, with the carpals at the wrist and with each other at the **proximal** and **distal radioulnar joints**. They are also joined along their complete length by the **interosseous membrane**.

The ulna is the longer of the two bones. It stabilizes the forearm and contributes to the elbow joint with the **coronoid** and **olecranon processes**. The olecranon prevents elbow hyperextension by 'fixing' the arm when the elbow is straight. The shaft of the ulna gives attachment to many muscles of the forearm and, at its lower extremity or head, a **styloid process** provides attachment for some wrist muscles.

The radius is less important at the elbow than the ulna, but is more significant at the wrist joint where it 'carries the hand'. The small radial head articulates with the capitulum of the humerus at the elbow. Below the head the **radial tuberosity** provides attachment for the *biceps brachii muscle*. The shaft, which provides attachment for many muscles, is narrow and rounded but widens at its lower end. The lower extremity has a lateral **styloid process** and a concave surface which articulates with the carpal bones.

Wrist and hand

The eight **carpals** (singular **carpus**), which form the wrist joint, are short bones connected by ligaments (*Figure 18.27*). They are arranged roughly in two rows – a proximal row (lateral to medial) of **scaphoid**, **lunate**, **triquetral** and **pisiform**, and a distal row (lateral to medial) of **trapezium**, **trapezoid**, **capitate** and **hamate**. The scaphoid and lunate articulate with the radius at the wrist, and with some distal carpals.

Five small long bones, the **metacarpals** (**metacarpus**) (numbered from thumb inwards) form the palm of the hand. Their proximal ends articulate with the carpals (**carpo-metacarpal joints**) and the distal ends (heads) articulate with the proximal phalanges.

The digits are formed from 14 small long bones called **phalanges** – two in the thumb and three in each finger. Articulation occurs between the proximal phalanges and the metacarpals and with each other.

Pelvic (hip) girdle and leg

The **pelvic (hip) girdle**, which joins the leg to the axial skeleton, consists of two **innominate bones** and the **sacrum** (see page 453 and *Figure 18.1*). The lower limb is formed from a **femur**, **tibia**,

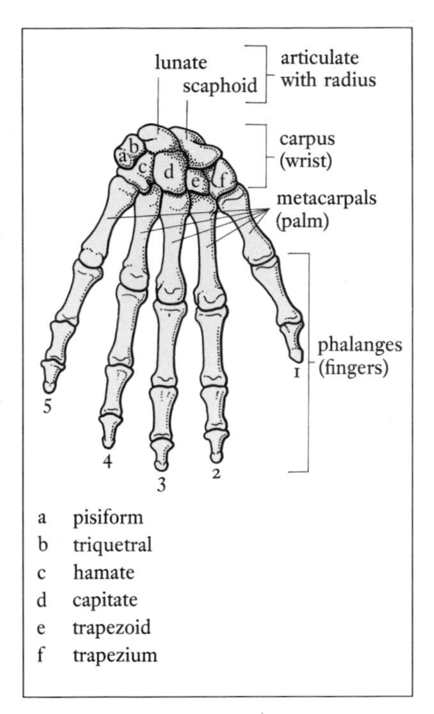

Figure 18.27. Bones of the wrist and hand (left – anterior aspect).

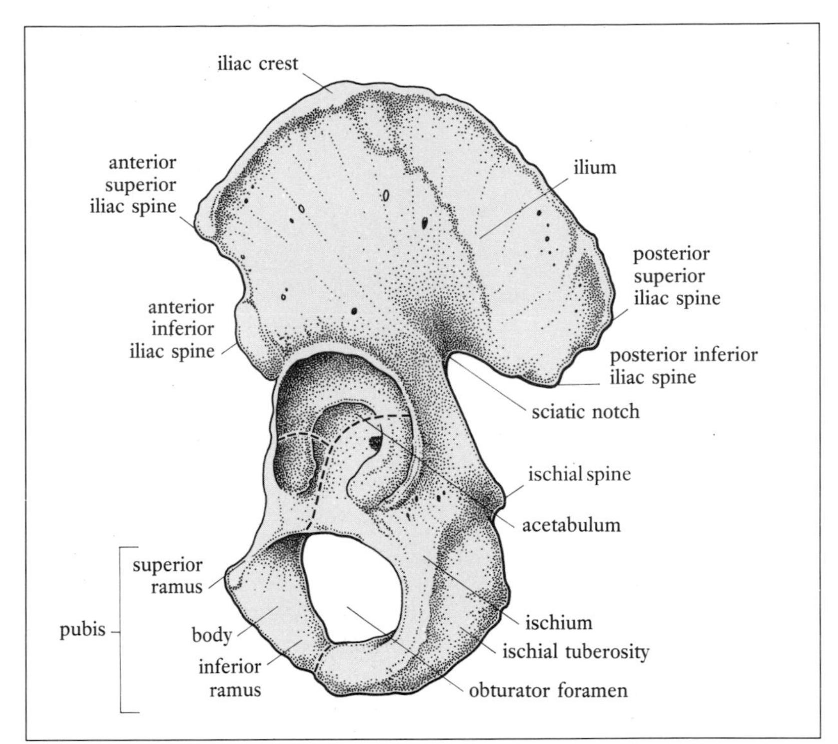

Figure 18.28. Innominate bone (left – external surface).

fibula, **patella**, **tarsals** (7), **metatarsals** (5) and **phalanges** (14).

Innominate (hip) bones

The **innominate bone** is formed from the fusion of the **ilium, ischium** and **pubis**. These unite at the deep depression (**acetabulum**) which accommodates the femoral head to form the hip joint. The innominate bones articulate with the sacrum to form the **sacroiliac joints**, and with each other at the **symphysis pubis**, a slightly movable joint between the pubic bones (*Figure 18.28*).

The ilium is the large upper bone and has a **crest** bounded by **anterior** and **posterior superior iliac spines**; it is at the anterior spines that you measure your hips. Many large muscles of the back, buttock and abdomen are attached to the ilium.

The ischium forms the strong posterior and inferior parts of the innominate bone. It has an **ischial spine** and, at its lowest point, an **ischial tuberosity** which supports your weight when sitting.

The pubis, which forms the anterior part of the innominate, has a **body** and two **rami**. The two pubic bones unite in the midline at the symphysis pubis where they form the **pubic arch**.

The ischium and pubis together surround the **obturator foramen**, which although being nearly filled with membrane, allows the passage of vessels and nerves from pelvis to thigh.

Pelvis

The pelvis, as already mentioned, consists of two innominate bones and the sacrum, which form the bony and cartilaginous ring

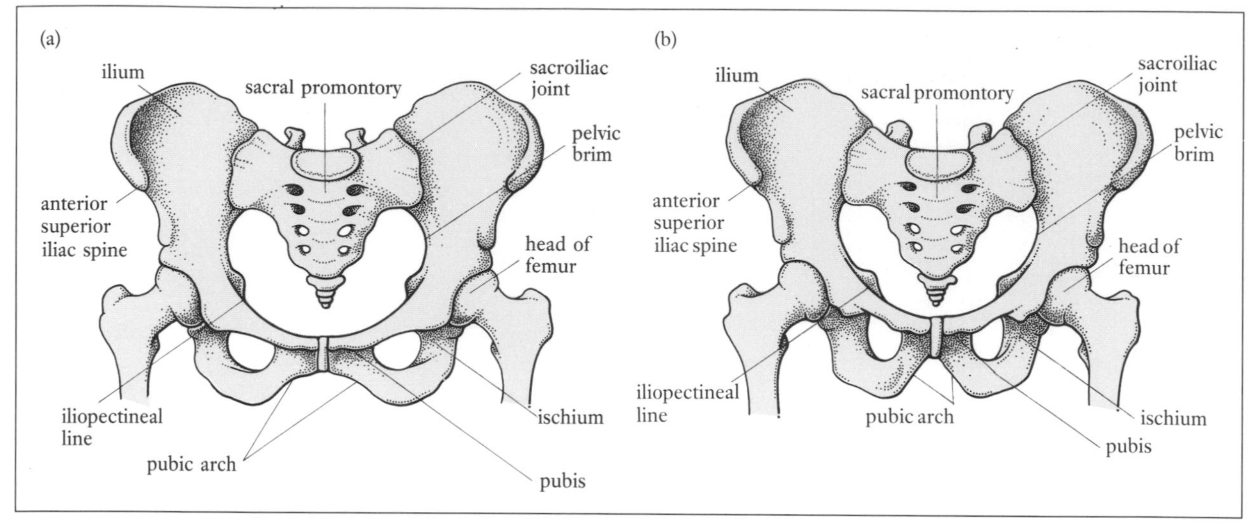

Figure 18.29. (a) Female pelvis; (b) male pelvis.

surrounding the lower part of the trunk (*Figure 18.29*). Apart from functions already stated the pelvis protects the organs of the pelvic cavity e.g. bladder, and in the case of the female provides space for a developing fetus and its exit into the world at the end of pregnancy (see Chapter 20).

The pelvis is divided into two by the **pelvic brim** formed from the **iliopectineal line** and the **sacral promontory**; the part above is the **false pelvis** and that below the **true pelvis**.

Structural differences between the female and male pelvis

Most babies are born by the vaginal route, a process for which the female pelvis is generally well adapted.

The female pelvis is lighter than the male and the true pelvis is wider and more shallow. Females have a wider pelvic arch than males and their acetabula are further apart. The pelvic brim in females is rounded, whereas in males it is heart–shaped (*Figure 18.29(a)*).

Occasionally the fetus cannot fit into and through the pelvis. This is termed *disproportion*. If this situation exists it is necessary to deliver the infant by *caesarian section,* which with early diagnosis can be performed as an elective (planned) operation.

Femur

The **femur** is the long bone forming the thigh, and is the largest and strongest bone of the body. It articulates with the innominate bone at the hip joint and the tibia at the knee joint (*Figure 18.30*).

The large femoral head fits into the acetabulum to give a stable joint which retains a good range of movements. A small pit on the head provides attachment for the **ligamentum teres**, which holds the femur in the acteabulum. Below the head is the **femoral neck**, which is a frequent site for the 'hip' fractures which occur so

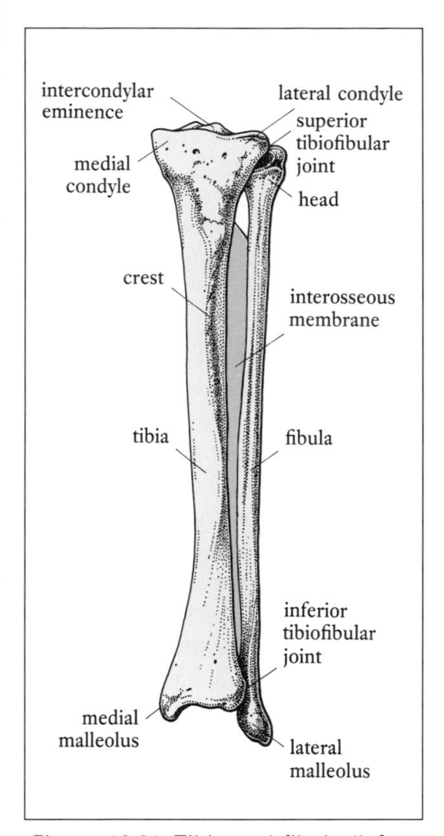

Figure 18.31. Tibia and fibula (left – anterior view).

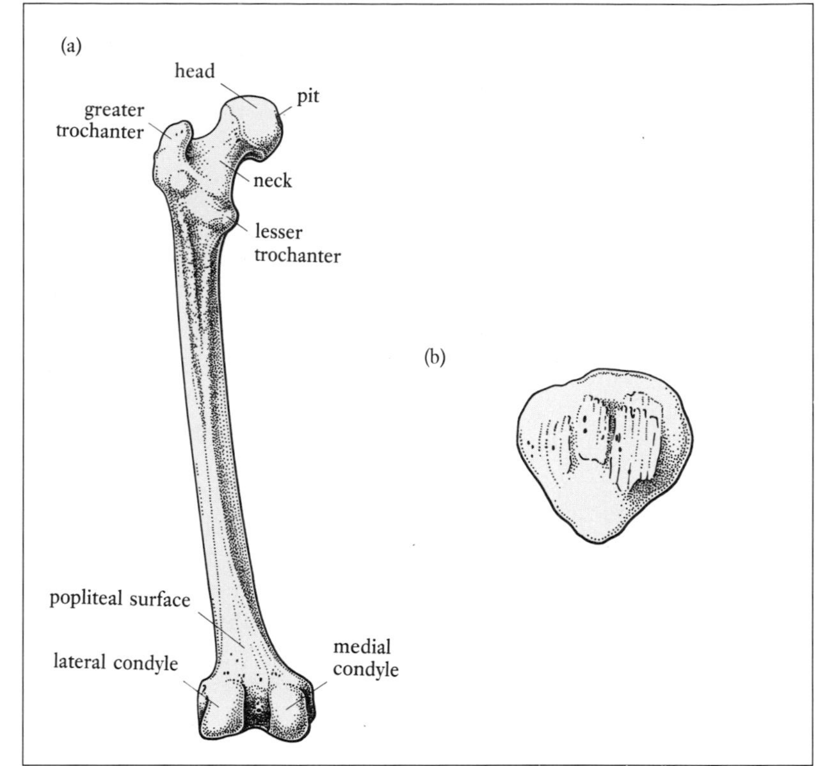

Figure 18.30. (a) Femur (left – posterior view); (b) patella (anterior view).

commonly in elderly women as a result of falls (see Chapter 16).

The **greater** and **lesser trochanters** mark the boundary between the head and the shaft. The shaft is cylindrical and provides attachment for thigh muscles. At its lower extremity the femur forms **medial** and **lateral condyles**, which articulate with the patella and *tibial condyles*. On the posterior aspect, above the condyles, is the **popliteal surface** for vessels and nerves.

The **patella** (kneecap) (*Figure 18.30(b)*) is a *sesamoid* bone which forms within the *patellar tendon* of the *quadriceps femoris* muscle. It is positioned (see *Figure 18.1*) over the femoral condyles and lies anterior to the knee joint which it protects.

Tibia and fibula

The two long bones of the lower leg are the **tibia** and **fibula** (*Figure 18.31*). The tibia forms the shin bone and is positioned medial to the fibula. The two bones articulate with each other at the **superior** and **inferior tibiofibular joints**; the tibia articulates with the femur and both articulate with the talus at the ankle. They are also joined by an interosseous membrane.

The tibia is the more important bone, at its upper extremity **lateral** and **medial condyles** present flat articulating surfaces for the femur. The *semilunar cartilages* of the knee joint are situated on the tibial condyles. The tibial shaft is triangular with a very prominent

crest, which can be felt just under the skin. Poorly protected, this area is easily damaged, e.g. walking into furniture. At its lower extremity the tibia has an articulating surface for the talus and a projection, the **medial malleolus** (felt as lump on inner aspect of the ankle).

The fibula is the lateral non-weight-bearing bone. At its upper extremity the head articulates with the tibia, the shaft provides muscle attachment and its lower extremity projects to form the **lateral malleolus** which articulates with the talus.

Ankle and foot

The seven short **tarsals** (**tarsus**) which form the ankle and heel and support the body weight are the **talus**, which forms part of the ankle joint, the **calcaneus** or heel bone which provides attachment for the massive *Achilles tendon* (see *Figure 18.50(b)*) of the calf muscle, and the **navicular**, **cuboid** and **three cuneiforms** (*Figure 18.32 (a)*).

Five **metatarsals (metatarsus)** (numbered from medial to lateral aspects), which are small long bones, form the **dorsum** or 'instep' of the foot. The proximal extremities articulate with the tarsals, and the distal extremities with the proximal phalanges.

The 14 **phalanges** which form the toes are much shorter than the fingers, but are arranged in the same way. There are three in each toe, apart from the great toe, which has two.

Arches of the foot

Look at a wet footprint and you can see that the foot is not flat on the ground, it is arched (*Figure 18.32 (b)*). The four arches of the foot

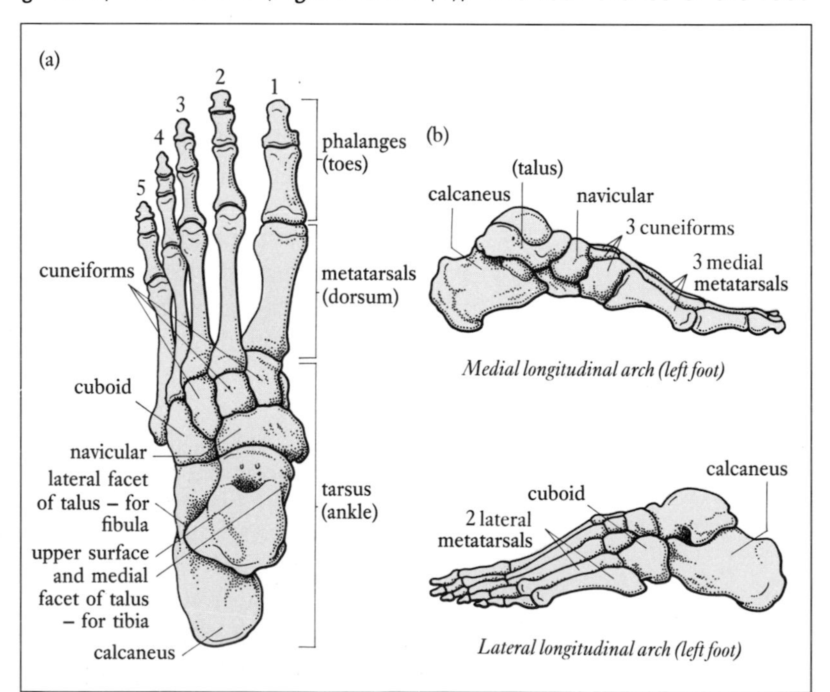

Figure 18.32. (a) Bones of the ankle and foot (left – dorsal aspect); (b) arches of the foot.

produce a bridge-like structure capable of supporting weight whilst remaining flexible enough to give 'spring' on walking. The arches are joined by ligaments and supported by muscles.

The **medial longitudinal arch** is the highest and is formed from the calcaneus, navicular, the cuneiforms and the three inner metatarsals. A **lateral longitudinal arch** is formed by the calcaneus, cuboid and two outer metatarsals. Two **transverse arches** are present, one in the tarsals and the other in the metatarsals (see *Figure 18.32 (b)*).

The problem of 'flat feet' is due to loss of the foot arches. It may be caused by the prolonged standing associated with certain occupations or damage caused by running on hard surfaces without proper footwear.

Levers and Movement

Bones, joints and muscles operating together produce movement through a system of levers (leverage). A brief look at the principles of levers is appropriate before the consideration of body movement.

A **lever** is a rigid structure, with one fixed point or **fulcrum** (plural, *fulcra)*, which moves if force is applied to it. This **force (effort)** is used to shift a **load (resistance)**. The distance between the effort and the fulcrum is termed the **effort arm** and that between the load and the fulcrum is the **load** or **resistance arm**.

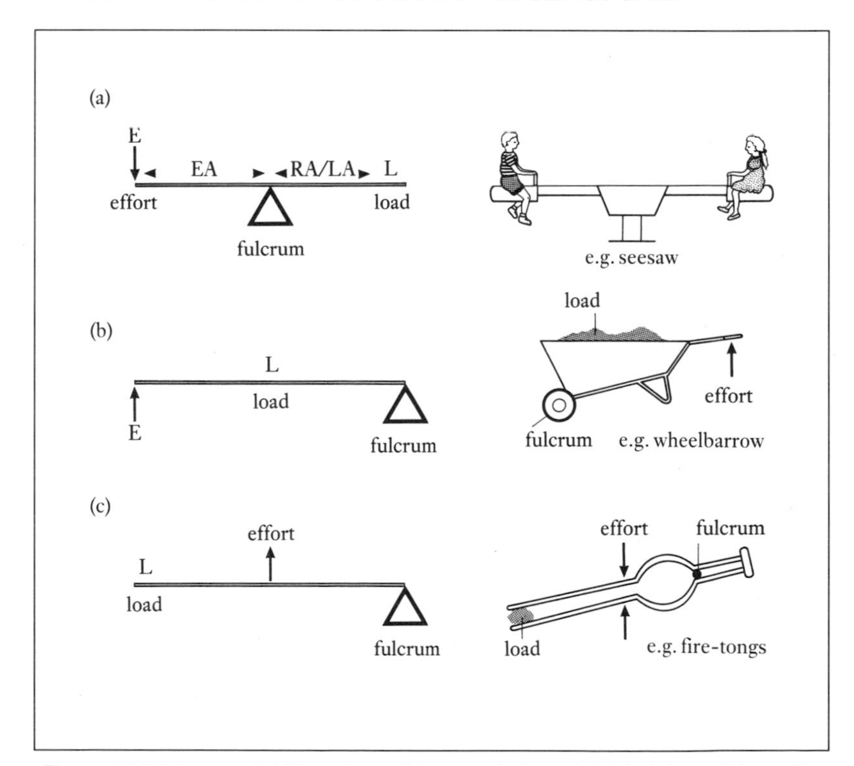

Figure 18.33. Levers. (a) First class; (b) second class; (c) third class. EA = effort arm; RA/LA = resistance or load arm.

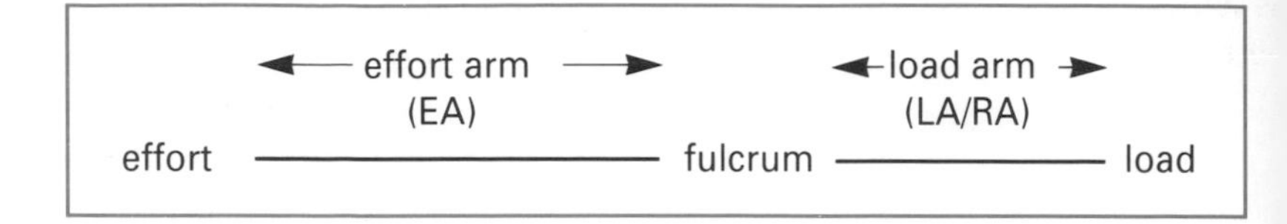

In the body the bones are the levers and the joints are the fulcra. The effort is supplied by the muscles at their insertions on bones, and load consists of the bone, the relevant part of the body and whatever you are lifting or carrying.

Levers may be classified as first, second or third class according to the positions of the load, effort, and fulcrum relative to each other (*Figure 18.33*). A **first class lever** has the fulcrum between the load and effort, e.g. a seesaw. This type of lever may require an effort less than the load which makes it very efficient, but in the body, e.g. elbow extension, the effort required is greater than the load. **Second class levers**, such as a wheelbarrow, have the load between the effort and the fulcrum; these are generally efficient levers and a body example would be standing on tip toes. **Third class levers**, e.g. fire tongs, have the effort between the fulcrum and the load, they are not efficient and always require an effort greater than the load. The levers operating in the body are generally of this type, e.g. elbow flexion and adduction of the thigh.

Different lever types in the body reflect their particular role: rapid movement, stability, moving large loads.

The equation below shows the equilibrium situation with the lever balanced:

> Effort x EA (effort arm length) = Load x RA (resistance arm length)
> e.g. 5 kg x 10 cm = 25 kg x 2 cm
> (50 = 50)

Lever efficiency is dependent upon the distances EA and RA, and when these are altered the effort and load relationship changes:

> 1. If EA is increased then effort required will be less:
>   2.5 kg x 20 cm = 25 kg x 2 cm
>   (50 = 50)
>
> 2. If EA is decreased then effort required will be more:
>   10 kg x 5 cm = 25 kg x 2 cm
>   (50 = 50)

In the body, where EA and fulcrum are fixed, it is the RA which changes to maintain the equilibrium. When the RA is lengthened a greater effort is required than if the RA is shortened. With this in mind it becomes clear that carrying heavy objects close to the body requires less effort than holding the same weight away from the body.

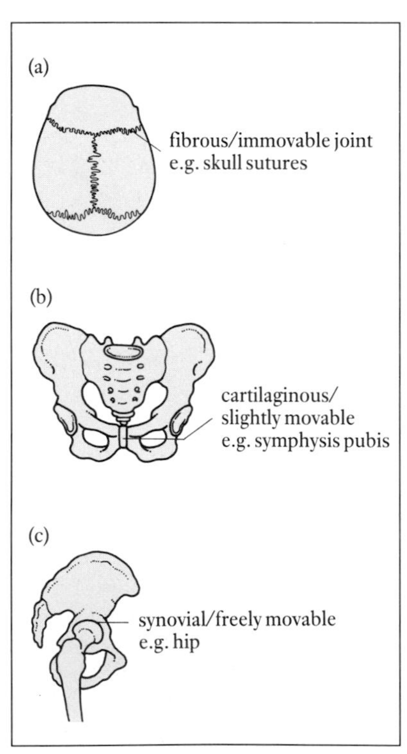

Figure 18.34. (a) Fibrous – skull sutures; (b) cartilaginous – symphysis pubis; (c) synovial – hip.

Joints

A joint or **articulation**, which is the point at which two or more bones meet, can be classified by its structure and the degree of movement it allows.

Types of joint

Joints may be *fibrous* (immovable), *cartilaginous* (slightly movable) or *synovial* (freely movable). The three main types are all described, but synovial joints and movement are discussed in more detail (*Figure 18.34*).

Fibrous joints (synarthroses) – immovable joints

The bones in a **fibrous joint** are united by fibrous tissue. For example, **sutures**, e.g. skull bones (page 445) where the joint is fixed and allows no movement; **gomphoses**, e.g. the teeth in their sockets, where a tiny amount of movement occurs; and **syndesmoses** where a membrane connecting the bones allows some 'give', e.g. the inferior tibiofibular joint.

Cartilaginous joints (amphiarthroses) – slightly movable joints

In **cartilaginous joints** the bones are joined by cartilage and slight movement is possible. There are two types: (i) **synchondroses** formed from the epiphyseal plates present in long bones during growth, these are ossified in adults and no movement is possible; and (ii) a **symphysis** where a fibrocartilage pad separates the hyaline cartilage articular surfaces, which provides strength and some movement. Examples of symphyses include the intervertebral joints (page 454), the manubrium and body of the sternum (page 455) and the symphysis pubis (page 459).

Synovial joints (diarthroses – freely movable joints)

Synovial joints form the freely movable joints of the skeleton which allow a variety of movements, e.g. hip, elbow. Basically the articulating bones are separated by a membrane-lined joint cavity.

Synovial Joints

General characteristics

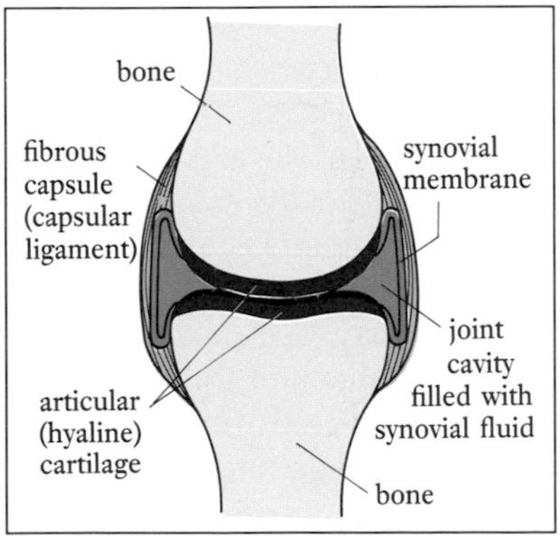

All synovial joints have the following common characteristics (*Figure 18.35*):

Figure 18.35. A synovial joint.

- **Joint capsule** – a capsule of fibrous tissue encloses the **joint cavity**, which may be further reinforced by **intracapsular** and **extracapsular ligaments**. The capsule is well supplied with blood vessels, motor nerves and specialized sensory nerves or *proprioceptors* which convey data regarding position to the brain.
- **Articular cartilage** – the articular surfaces of the bones are covered with smooth hyaline cartilage (see Chapter 1).
- **Synovial membrane and fluid** – the joint cavity, except areas covered with hyaline cartilage, is lined with synovial membrane. The membrane secretes a small amount of viscous synovial fluid which occupies the spaces in the joint cavity. The synovial fluid contains protein and *hyaluronic acid*, which confer the

characteristic viscosity. Synovial fluid, which thins on warming during joint movement, lubricates joint activity and reduces friction and erosion in much the same way as the oil in a car engine. The fluid has a role in joint nourishment, and contains phagocytic cells which remove debris from the joint.

Associated with some joints, e.g. the knee, are small sacs or **bursae** which are lined with synovial membrane. They act to prevent friction between structures such as bone and tendon. *Bursitis* (inflammation of a bursa) is an extremely painful condition caused by excessive or prolonged use, e.g. *tennis elbow* or *housemaid's knee.* Rest, anti-inflammatory drugs and analgesics usually relieve the condition but in some cases local corticosteroid injections are required.

The muscles and tendons which cross a joint are involved with its movement and increasing stability.

The special features present in individual joints are included in the coverage of specific joints (pages 469-474).

Movements occurring at synovial joints

(*Figure 18.36*)

- *Flexion* – bending. Reducing the angle of a joint and bringing the bones closer, e.g. bending the elbow.
- *Extension* – straightening. The opposite to flexion, e.g. straightening the elbow.
- *Abduction* – movement away from the midline, e.g. moving the straight arm away from the body.
- *Adduction* – movement towards the midline, e.g. moving the arm back to the body.
- *Circumduction* – a movement which combines flexion, extension, abduction, and adduction where the limb traces a cone in the air, e.g. circular movement of the straight arm from the shoulder.
- *Rotation* – the turning of a bone as it moves around its axis, e.g. the movement between the atlas and axis when you turn your head.
- *Inversion* – turning the sole of the foot medially (inwards).
- *Eversion* – the opposite movement where the sole is turned laterally (outwards).
- *Pronation* – movement of the radius across the ulna which turns the palm downwards.
- *Supination* – the radius and ulna are parallel and the palm is turned upwards.
- *Elevation* – movement upwards, e.g. the scapulae when the shoulders are lifted.
- *Depression* – downward movement the reverse of elevation.
- *Protraction* – forward movement, e.g. thrusting out the mandible.
- *Retraction* – the opposite backward movement returning the jaw to its normal position.

Movement at a specific joint is limited by the shape of the bones, the ligaments and soft tissue such as muscle. The normal range of movement will also be modified by increasing age and pathological processes.

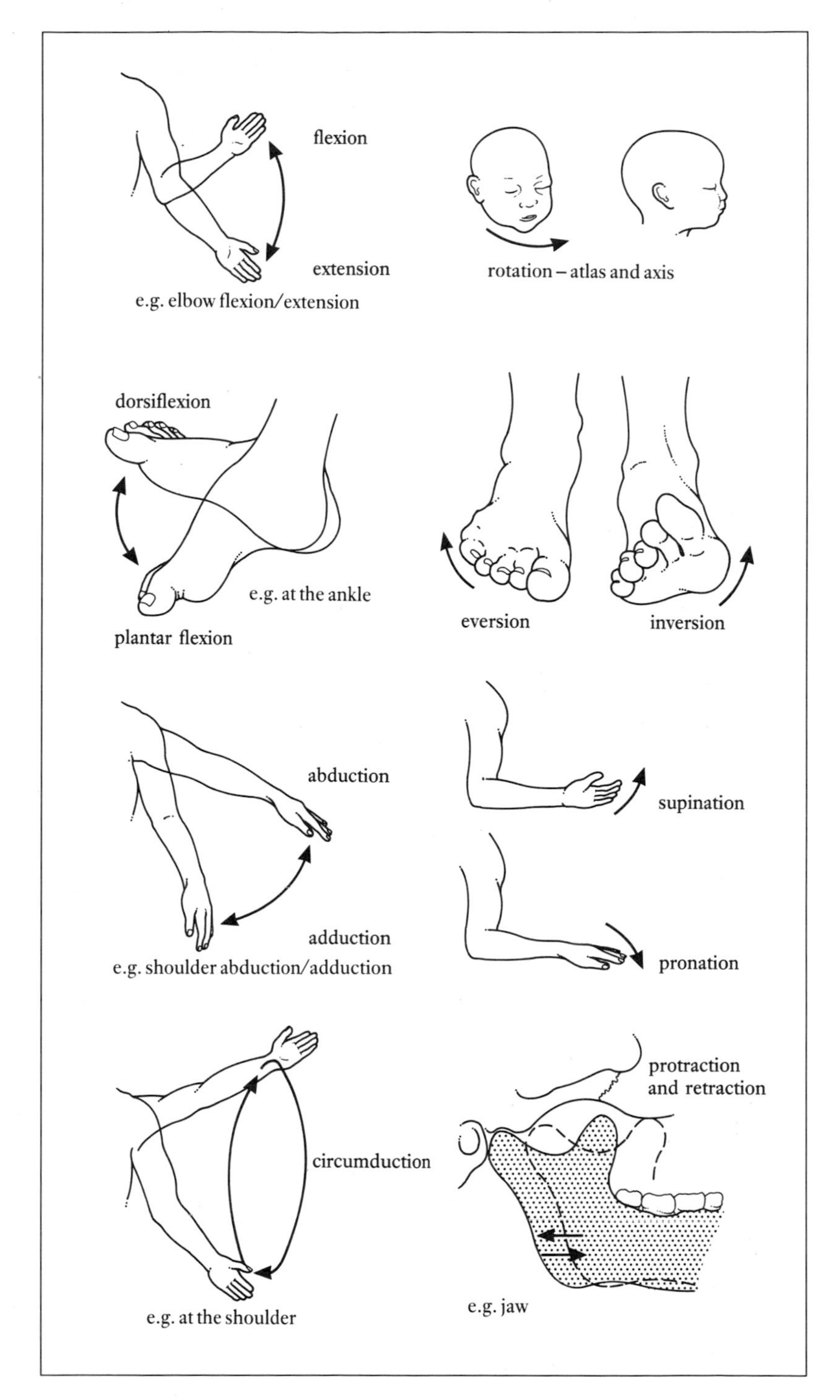

Figure 18.36. Some movements at synovial joints.

Varieties of synovial joints and movements possible

Ball and socket joint
A **ball and socket joint** has a rounded extremity which articulates within a cavity in another bone, e.g. hip and shoulder joints. This type of arrangement allows extensive movement – flexion, extension, adduction, abduction, circumduction and rotation.

Hinge joint
A **hinge joint** has a bony projection which articulates with a concave surface on another bone to give movement in one plane, e.g. elbow, knee, ankle and interphalangeal joints. Only allows flexion and extension.

Pivot joint
A **pivot joint** allows rotation only, e.g. the ring of the atlas rotates around the odontoid peg of the axis. Another example is the proximal radioulnar joint, which allows supination and pronation.

Saddle joint
A **saddle joint** consists of two bones with both convex and concave surfaces which fit together. The only example is the joint at the base of the thumb (carpometacarpal). Movement in two planes is possible: extension, flexion, adduction, abduction and circumduction. The thumb can also be opposed to the fingers.

Condyloid joint
Condyloid joints are similar to a hinge joint, but movement in two planes is possible. Examples include the wrist and metacarpophalangeal joints where extension, flexion, adduction, abduction and circumduction are possible.

Gliding (plane) joint
A **gliding joint** comprises two flat surfaces which glide on each other in all directions, e.g. intercarpal, intertarsal and sternoclavicular joints. Movement is generally very limited because of the ligaments present.

Injuries to joints

Injuries occur when the joint structures are stretched, stressed or torn; they range from sprains to the major disruption of dislocation, which may be associated with a fracture.

Nursing Practice Application Joint stiffness and immobility

Your knowledge of the movements occurring at joints can be put to good use in preventing one of the problems of immobility.

Joints suffer from lack of use just as do bones and muscles (Chapters 16 and 17). Disuse causes joint stiffness which can seriously affect mobility and stability. Activity should be encouraged, but where this is not possible the nurse can work with the physiotherapist and occupational therapist to prevent joint stiffness. This can be achieved by careful positioning and use of aids such as bed cradles, encouraging active movements or by moving the person's joints through a full range of passive movements.

Sprains
Joint sprain occurs when the supporting ligaments are stretched or torn. A common example is the sprain which results when the ankle is twisted.

Dislocation
The damage caused by dislocation is more severe than that caused by a sprain, with joint disruption and displacement of articular surfaces. Dislocations, e.g. shoulder and finger joints, are commonly sustained during contact sports or result from a fall. A partial dislocation is termed a *subluxation*.
NB See also congenital dislocation of the hips (page 473).

Joint injuries can reduce stability with risk of repeated injury, and an already damaged joint is more likely to develop degenerative changes of *osteoarthritis* (see *Special Focus* – Joint conditions, page 475).

Specific synovial joints – structure, movements and muscles

A description of some joints is given here, along with their range of movements and some of the muscles involved with the movement. The major muscle groups are mentioned again later in this chapter.

Shoulder joint (humero-scapular) – ball and socket

The shoulder joint consists of the head of the humerus articulating within the glenoid cavity of the scapula. The glenoid cavity, which is relatively shallow, is deepened by a rim of cartilage known as the **glenoidal labrum** (*Figure 18.37(a)*). Extensive movement at the shoulder (*Table 18.2*) is possible because the cavity is shallow and because of the looseness of the ligaments around the joint. The price, however, of increased mobility is reduced stability and dislocation (see above). The major ligaments reinforcing the shoulder are the **coracohumeral ligament**, running from coracoid process to humerus, the three **glenohumeral ligaments** and a **transverse humeral ligament** (*Figure 18.37(b)*). Some shoulder muscles also help to stabilize the joint.

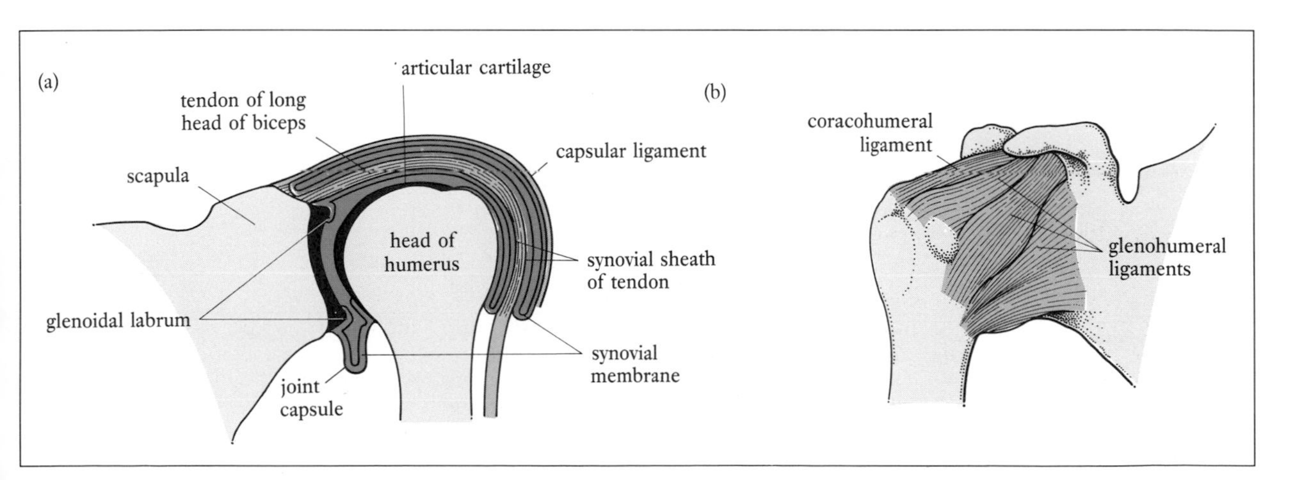

Figure 18.37. (a) Shoulder joint; (b) ligaments reinforcing the shoulder (anterior view).

Table 18.2. Movements and muscles at the shoulder (see also *Figure 18.45*).

Movement	Muscles involved
Extension	Latissimus dorsi, teres major, deltoid (posterior fibres)
Flexion	Pectoralis major, deltoid (anterior fibres), coracobrachialis
Adduction	Pectoralis major, latissimus dorsi, teres major, coracobrachialis
Abduction	Supraspinatus, deltoid
Rotatation – medial	Pectoralis major, teres major, deltoid (anterior fibres), latissimus dorsi, subscapularis
– lateral	Deltoid (posterior fibres), infraspinatus
Circumduction	Shoulder muscles acting in sequence

Elbow joint – hinge

The elbow joint is formed by the trochlear surface of the humerus and the ulnar notch, and the capitulum of the humerus and the radius (*Figure 18.38(a)*). The joint is stabilized by **medial** and **lateral collateral ligaments** and the muscles crossing the elbow (*Figure 18.38(b)*).

Table 18.3. Movements and muscles at the elbow (see also *Figure 18.45*).

Movement	Muscles involved
Extension	Triceps brachii, anconeus
Flexion	Biceps brachii, brachialis

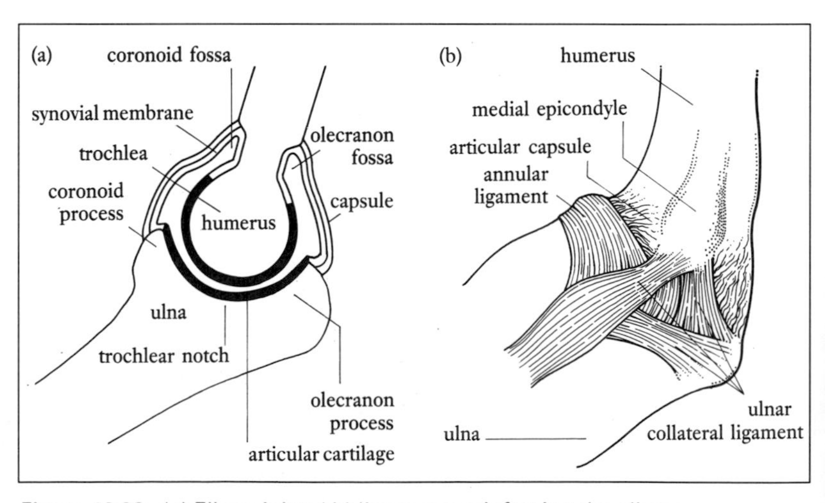

Figure 18.38. (a) Elbow joint; (b) ligaments reinforcing the elbow.

Radioulnar joints

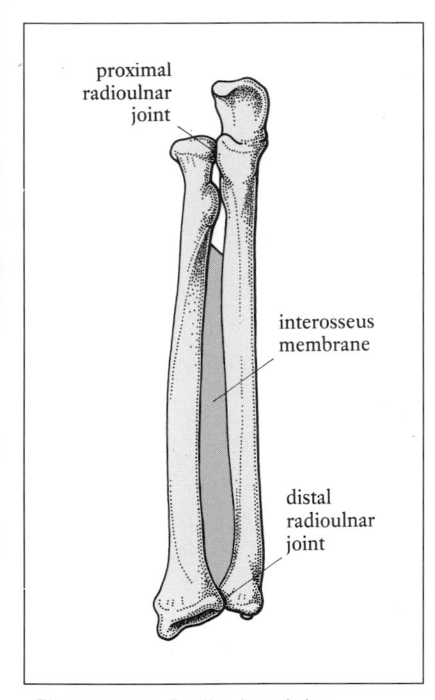

Figure 18.39. Radioulnar joint.

The proximal radioulnar joint (*Figure 18.39*) is formed by the radial head which rotates in a notch on the ulna. This joint is stabilized by the **annular ligament** (*Figure 18.38(b)*).

The lower end of the radius rotates on the head of the ulna to form the distal radioulnar joint.

These joints produce the movements – pronation and supination of the hand – needed when using a screwdriver or turning a door knob (*Table 18.4*).

Table 18.4. Movements and muscles at the radioulnar joints (see also *Figure 18.45*).

Movement	Muscles involved
Pronation	Pronator teres, pronator quadratus
Supination	Supinator, biceps brachii

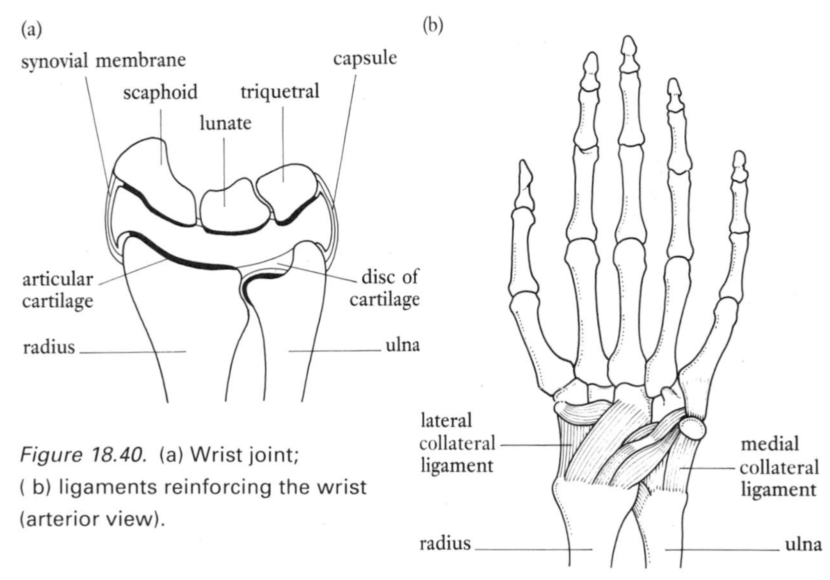

Figure 18.40. (a) Wrist joint; (b) ligaments reinforcing the wrist (arterior view).

Wrist joint – condyloid

The wrist or radiocarpal joint is formed by the distal radius and a disc of cartilage below the ulna which articulate with the scaphoid, lunate and triquetral carpal bones (*Figure 18.40(a)*). **Lateral** and **medial collateral ligaments** and **transverse ligaments** stabilize the joint (*Figure 18.40(b)* and Table 18.5, page 472).

Hand and finger joints

The joints between the carpals are gliding, as are the carpometacarpal joints, except the joint at the base of the thumb, which is a saddle joint. The metacarpophalangeal joints are condyloid type which allow extension, flexion, adduction and abduction. The interphalangeal joints are hinge type, which allow flexion and extension. The thumb can be opposed to the fingers which provides our ability to grasp objects.

The powerful and fine movements of the hand and digits are produced by various forearm muscles (extensors and flexors) and

Table 18.5. Movements and muscles at the wrist (see also *Figure 18.45*).

Movement	***Muscles involved***
Extension	Extensor carpi radialis (longus, brevis), extensor carpi ulnaris, extensor digitorum
Flexion	Flexor carpi radialis, flexor carpi ulnaris, flexor digitorum
Adduction	Flexor and extensor carpi ulnaris
Abduction	Flexor and extensor carpi radialis, abductor pollicis longus
Circumduction	Wrist muscles acting in sequence

NB Usual movements at the wrist are extension/abduction and flexion/adduction.

many small intrinsic hand muscles, e.g. *thenar, lumbricals*. The extensor and flexor tendons run to the hand within synovial sheaths and are secured by strong ligaments – the **extensor** and **flexor retinacula**.

Hip joint – ball and socket

The hip joint is formed by the femoral head and the acetabulum of the innominate bone. A rim of cartilage called the **acetabular labrum** deepens the acetabulum (*Figure 18.41(a)*). The hip, which is less mobile than the shoulder, is very much more stable. This stability is provided by the **ligamentum teres** and a strong capsular ligament, which are reinforced by the **iliofemoral**, **pubofemoral** (*Figure 18.41(b)*)and **ischiofemoral ligaments**. Some of the strongest muscles in the body also contribute to hip stability.

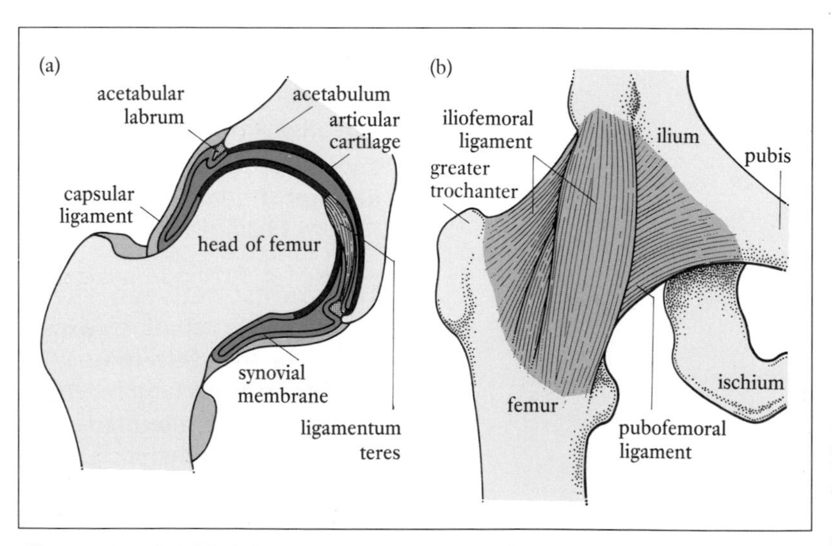

Figure 18.41. (a) Hip joint; (b) ligaments reinforcing the hip.

Nursing Practice Application Congenital dislocation of the hip (CDH)

The hip is the most common site for congenital dislocation, which occurs when the acetabula are abnormally shallow. The examination of all infants, soon after birth by the midwife/medical officer, must include tests to exclude congenital dislocation of the hip. Various tests exist, e.g. Ortolani's – with the knees and hips flexed, the hips are abducted, externally rotated and extended. If the hip is dislocated there is limitation of movement and a 'click' may be heard as the femur slips back into the acetabulum. It is important that a complete explanation is given, by the midwife or doctor, to the parents regarding the examination of their new baby.

Early diagnosis and treatment is essential in preventing problems when the child starts to walk and later degenerative joint changes. Management depends on the severity of the problem but usually involves splints or plaster, physiotherapy, frequent monitoring and possibly corrective surgery.

Table 18.6. Movements and muscles at the hip (see also *Figure 18.50)*.

Movement	***Muscles involved***
Extension	Gluteus maximus, hamstrings
Flexion	Iliopsoas (iliacus and psoas), rectus femoris, sartorius
Adduction	Adductors
Abduction	Gluteus medius and minimus
Rotation – medial	Gluteus medius and minimus
– lateral	Gluteal muscles (lateral rotators)
Circumduction	Hip muscles acting in sequence

Knee joint – hinge

The knee is a modified hinge joint formed by the femoral condyles, tibial condyles and the patella, which glides over the patellar surface of the femur (*Figure 18.42(a)*). The knee is stabilized by two intracapsular **cruciate ligaments**. The tibial articular surfaces are deepened by the **semilunar cartilages** (**menisci**) which also stabilize the knee by preventing lateral movement, and act as shock absorbers (*Figure 18.42(b)*). The joint is further strengthened by a capsular ligament, various extracapsular ligaments and muscles and tendons.

Table 18.7. Movements and muscles at the knee (see also *Figure 18.50)*.

Movement	***Muscles involved***
Extension	Quadriceps femoris (rectus femoris and three vasti)
Flexion	Hamstrings, gastrocnemius
Medial rotation (slight)	Popliteus

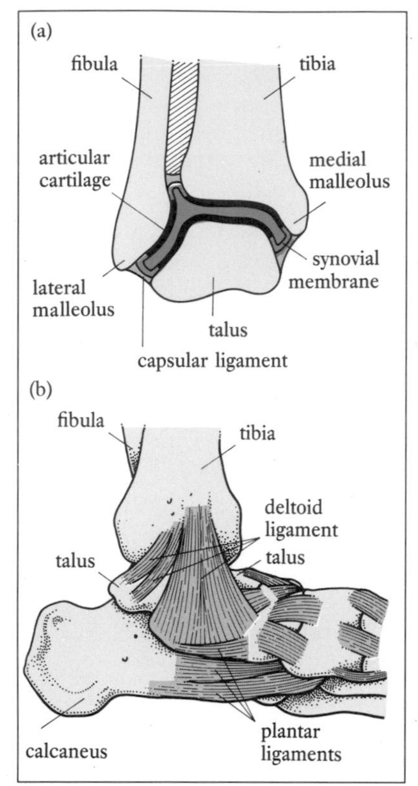

Figure 18.43. (a) Ankle; (b) ligaments reinforcing the ankle.

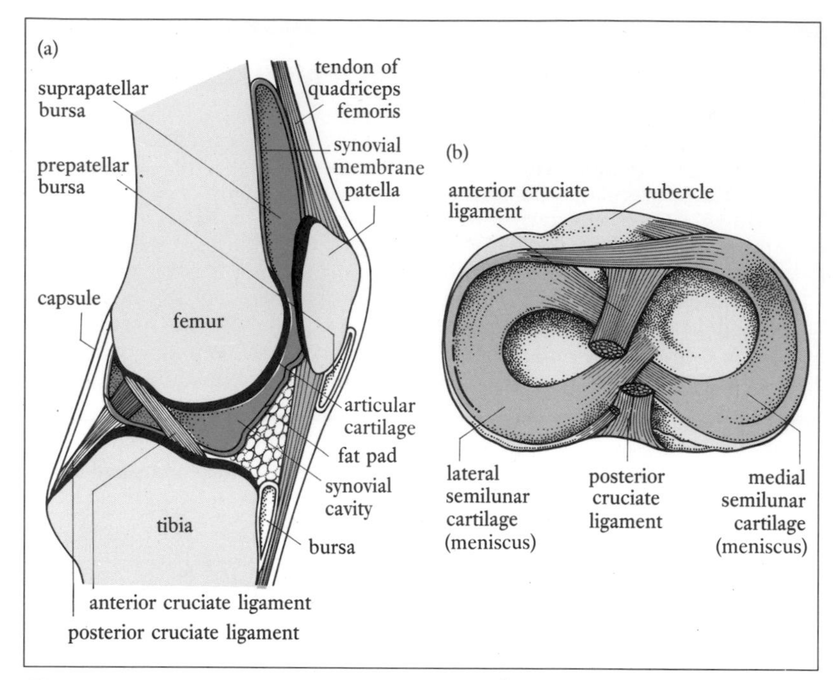

Figure 18.42. (a) Knee (side view); (b) semilunar cartilages.

The knee joint has the largest synovial membrane in the body. It lines the joint cavity, extends upwards beneath the patellar tendon to form the **suprapatellar bursa** and lines other bursae associated with the joint (see page 466).

Ankle joint – hinge

The ankle joint is formed by the tibia and its medial malleolus and the lateral malleolus of the fibula, which both articulate with the talus (*Figure 18.43(a)*). The joint is strengthened by ligaments which include a strong **medial ligament** (**deltoid**) (*Figure 18.43(b)*).

Table 18.8. Movements and muscles at the ankle (see also *Figure 18.50)*.

Movement	***Muscles involved***
Extension (plantarflexion)	Gastrocnemius, tibialis posterior, soleus, long flexors of the toes
Flexion (dorsiflexion)	Tibialis anterior, long extensors of the toes

Foot and toe joints

The joints between the tarsals are gliding, as are the tarsometatarsal joints. In addition to the gliding movements there is some slight adduction and abduction possible between the talus and calcaneus. Movement between the talus and navicular, and calcaneus and cuboid, facilitates eversion and inversion of the foot.

The metatarsophalangeal joints are condyloid type, which allow flexion, extension, adduction and abduction. Interphalangeal joints are hinge type and allow only extension and flexion.

Special Focus Joint conditions - arthritis and gout

Arthritis is not a single condition – it comes in many forms, affects all age groups and may vary in severity from a mild ache to severe joint pathology with pain, deformity and disability. The word arthritis, which means joint inflammation tends to be an over-simplification, because, arthritis may be infective, inflammatory or degenerative.

Classifications of arthritis are complex with dozens of different types, but for our purposes a brief consideration of *rheumatoid arthritis, osteoarthritis (osteoarthrosis)* and *gout* will give you an outline of the common joint damaging conditions.

Rheumatoid arthritis

Rheumatoid arthritis (RA) is a generalized chronic inflammatory condition of connective tissue, mostly affecting joints. It is characterized by polyarthritis (fingers, hand, wrist, hip, knee and ankle) and systemic effects, which include fatigue, high temperature, weight loss and anaemia. Serious side-effects which may accompany RA include heart valve disease.

The aetiology of RA is auto-immune (Chapter 19) but, as yet, the exact initiating mechanisms are unclear. The abnormal immunological situation causes *synovitis* with proliferation of the synovial membrane and eventual loss of articular cartilage and joint destruction. These joint changes cause pain, swelling, severe deformity and progressive loss of mobility with increasing dependence and disability.

Management includes:

- General rest.
- Rest for joints with splints to reduce deformity.
- Measures to deal with systemic effects, e.g. treatment of anaemia.
- Local warmth, e.g. wax baths, gloves and socks, warm baths or cold applications.
- Physiotherapy and mobility aids.
- Occupational therapy and aids to maintain independence, e.g. large handles on cutlery.
- Drugs: analgesics, e.g. aspirin (primarily prescribed for its anti-inflammatory effects); anti-inflammatory e.g. ibuprofen; systemic corticosteroids, e.g. prednisolone prescribed for its anti-inflammatory effects (*NB* aspirin, ibuprofen and prednisolone may cause peptic ulceration (see Chapter 13); chloroquine; penicillamine; gold salts; immunosuppressives, e.g. azathioprine.
- Joint surgery, e.g. synovectomy, arthrodesis.
- Joint replacement surgery (*replacement arthroplasty*), introduced some 30 years ago, has revolutionized the management of arthritis. Commonly used to replace hip and knee joints, arthroplasty is a highly effective and efficient means of improving quality of life over long periods, e.g. a hip replacement should function maintenance-free for 10–15 years (Green, 1989).

Osteoarthritis

Osteoarthritis is characterized by the destruction of the articular cartilage, which usually affects the large weight-bearing joints, e.g. knee and hip, and the intervertebral and phalangeal joints. It occurs commonly in older people (60+) as joints 'wear out', but can be secondary to existing joint problems, e.g. congenital dislocation of the hips (see page 473) or trauma. The loss of cartilage, development of bony spurs (*osteophytes*) and thickening of the synovial membrane results in pain, crepitus, swelling, stiffness and loss of function with reduced mobility.

Management includes:

- Minimizing damage, e.g. weight reduction, as obesity will contribute to damage, or reducing activities to rest the joint.
- Analgesia, anti-inflammatory drugs, intra-articular corticosteroids.
- Physiotherapy and mobility aids.
- Arthroplasty (see RA).

Gout

Gout is a metabolic condition caused by excess uric acid in the blood. Uric acid is a normal waste product of purine-containing nucleic acid metabolism (Chapter 1) which is eliminated in the urine (see Chapter 15).

However, problems occur if production of uric acid increases or its elimination via the urine decreases. The excess uric acid, in the form of crystals, is deposited in joints, e.g. first metatarsophalangeal, where it causes swelling and severe pain.

Anti-inflammatory drugs are used to relieve an acute attack and drugs such as allopurinol are used to control gout after the acute episode. Advice regarding resting the joint, reducing intake of high purine foods, e.g. offal, and restricting alcohol intake which inhibits uric acid elimination is also useful.

Person-centred Study Brian

Knee injuries involving the medial meniscus or anterior cruciate ligament commonly occur when the knee is twisted in sports such as soccer or squash. Brian, who plays with a local soccer team, has had pain and increasing problems with his knee, which has 'locked up' completely on a couple of occassions. Brian's GP referred him to the orthopaedic consultant who diagnosed damage to the medial meniscus. It was possible to rectify the damage via an *arthroscope* (instrument for viewing the inside of the joint cavity) which meant that Brian was able to go home the same day without the need for more extensive surgical treatment (menisectomy – removal of the meniscus). Following this procedure Brian's knee will function as before, apart for some possible loss of stability (remember the menisci help to stabilize the knee joint by preventing lateral movement).

The movements of the foot and digits are produced by various leg muscles (extensors and flexors) and many intrinsic foot muscles, e.g. *interossei, lumbricals*. The intrinsic muscles and the very long tendons of the leg muscles are also important in supporting the foot arches (see page 462-463). The long extensor and flexor tendons which run to the foot are enclosed within synovial sheaths and are secured by strong ligaments the **retinacula**.

Muscular System

It is appropriate here to give a brief outline of the muscular system, a process that includes: naming muscles, movement relationships and the major muscle groups. Some muscles associated with joints have already been discussed in terms of their action at a specific joint and you should refer to the appropriate tables.

Naming muscles

The naming of skeletal muscles, which may at first appear rather complex, is often based on characteristics which include size, position, function, shape, number of origins, direction of fibres and a combination of these:

- *Size* – e.g. gluteus maximus (largest) and minimus (smallest).
- *Position* – e.g. intercostal (between the ribs).
- *Function* – e.g. pronator teres (pronates).
- *Shape* – e.g. deltoid (triangular).
- *Number of origins* – e.g. triceps brachii has three.
- *Direction of fibres* – e.g. rectus (straight) femoris.
- *Combination* – e.g. extensor digitorum longus (function, location and size).

Movement relationships

The great variety of movements initiated by skeletal muscles are interdependent and co-ordinated just like the steps of a dance. Imagine for a moment the chaos which would ensue if muscles or dancers operated in isolation. Muscles around a joint most often function in pairs: if a muscle initiates a certain movement there must be another muscle that has the opposing effect. Muscles which have the main responsibility for an action are known as **prime movers**, e.g. the biceps brachii flexes the elbow. The muscles which limit and counter their action are **antagonists**, e.g. the triceps brachii extends the elbow.

Opposing muscle groups also act as **synergists**: they contract together to stabilize a joint or to cancel out unwanted movements produced by prime movers where the muscles cross more than one joint.

Major muscle groups

It is unnecessary for you to attempt to memorize the details of every muscle. However, the ability to apply the principles of movement and posture to your practice is of paramount importance. The following overview is not intended to provide a complete coverage of skeletal muscles and readers requiring more detail are referred to Further Reading (Williams *et al.*, 1989). It is worth noting that attempts to classify muscles into rigid groups can cause difficulties, e.g. latissimus dorsi is a back muscle which acts at the shoulder.

Muscles of the head and neck

The paired and unpaired muscles of the head and neck (*Figure 18.44*) are concerned with movement, support, facial expression, e.g. **zygomaticus** (smiling), speaking, chewing and swallowing.

Damage to the **sternocleidomastoid** muscle of the neck, which may occur during birth, can result in *torticollis* (wry neck), where the infant's head is tilted and rotated to one side.

Muscles of the shoulder and arm

The muscles of the shoulder and arm (*Figures 18.45* and *18.46*) are concerned with the movements of the shoulder, elbow, wrist, hand and fingers (see also *Tables 18.2–18.5*).

Muscles of the trunk

The superficial and deep muscles of the trunk (*Figures 18.46–18.49*) are collectively concerned with movement of the trunk, maintaining erect posture, support, respiration and increasing intra-abdominal pressure to assist defaecation, micturition and parturition.

Back

The large back muscles found either side of the spinal column include the **trapezius**, **latissimus dorsi**, **erector spinae group**, e.g. **sacrospinalis** (see *Figure 18.46*), and the **quadratus lumborum** and **psoas** which also form part of the posterior abdominal wall (*Figure 18.48(b)*). These powerful muscles support, maintain posture and allow movement at the trunk.

Thoracic wall

The thoracic wall has a superficial layer, which includes muscles involved with movements of the upper limb, e.g. **pectoralis major** (*Figure 18.45(a)*) and **minor**, those which move the scapula such as the **serratus anterior** and **rhomboids** (*Figure 18.46*), and those confined to the chest wall, e.g. **serratus posterior** (*Figure 18.46*). A middle layer consists of the **external** and **internal intercostal** muscles (*Figure 18.47(a)*) which lift the rib cage to enlarge the thorax during inspiration (see Chapter 12). The inner layer of muscle is known as the **transversus thoracis**. During forced inspiration other muscles are used, e.g. **scalenes** and sternocleidomastoids, and the back and abdominal muscles assist with forced expiration.

Logically the **diaphragm(***Figure 18.47(b)***)**, which divides thoracic and abdominal cavities, is included here as it is concerned with respiration. During inspiration it contracts and flattens to enlarge the thorax, and during expiration it relaxes to decrease thoracic size. The diaphragm also increases intra-abdominal pressure to facilitate venous return (see Chapter 10), defaecation (see Chapter 13), micturition (see Chapter 15) and parturition (see Chapter 20).

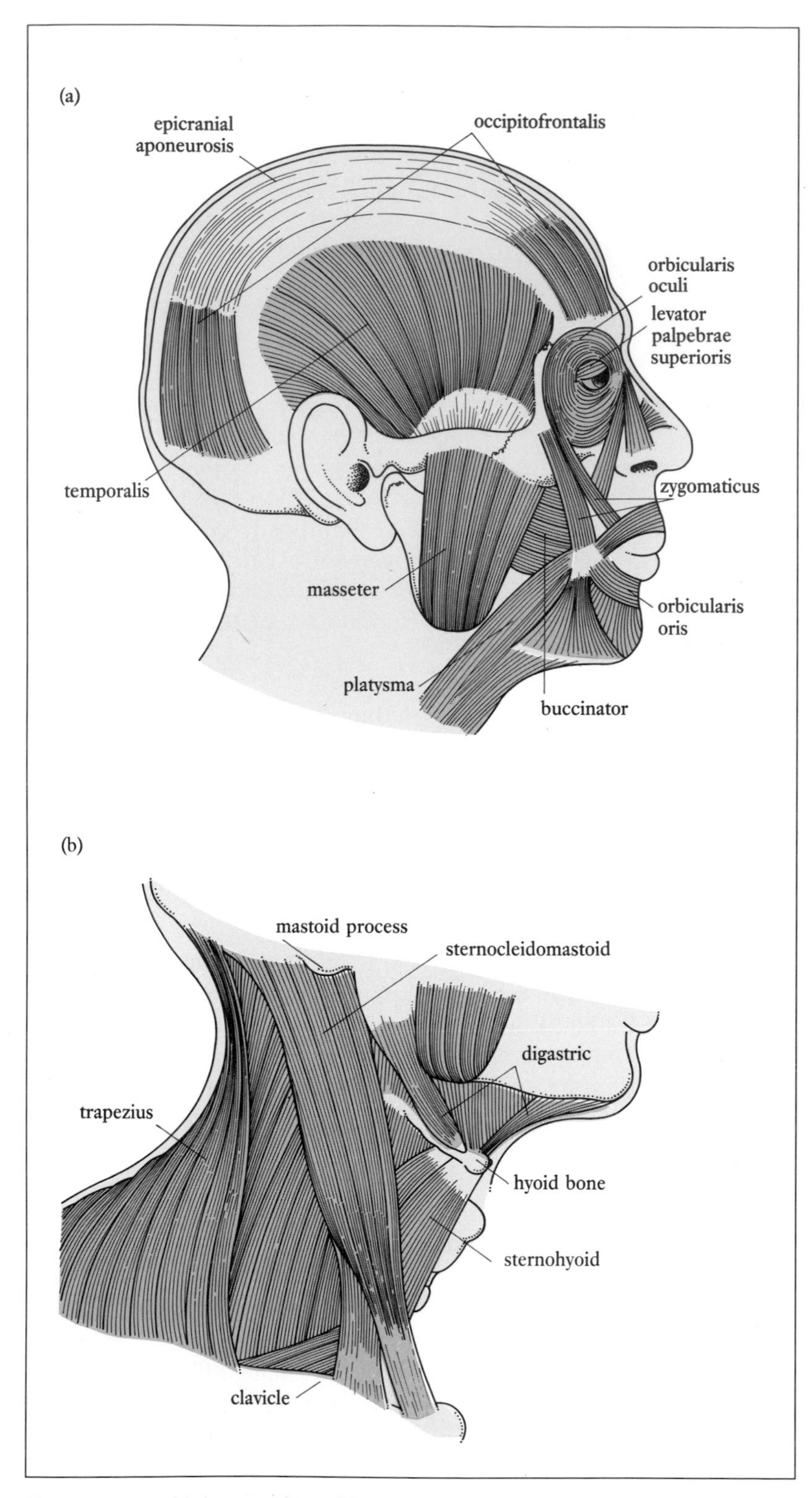

Figure 18.44. Main muscles of (a) head and face; (b) neck (platysma not shown in latter).

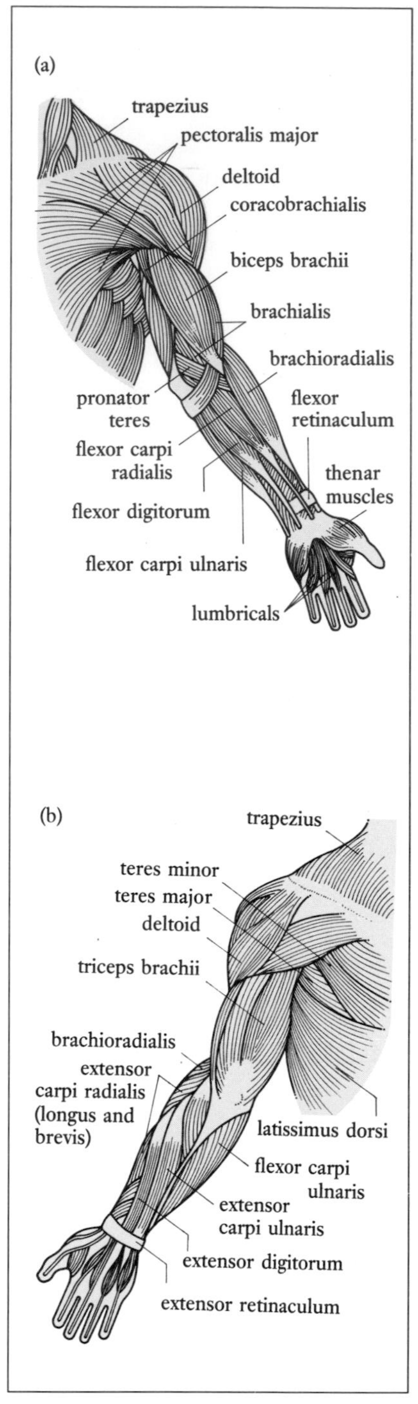

Figure 18.45. Main muscles of the shoulder and arm. (a) Anterior view; (b) posterior view. *NB* Some deep muscles are not shown.

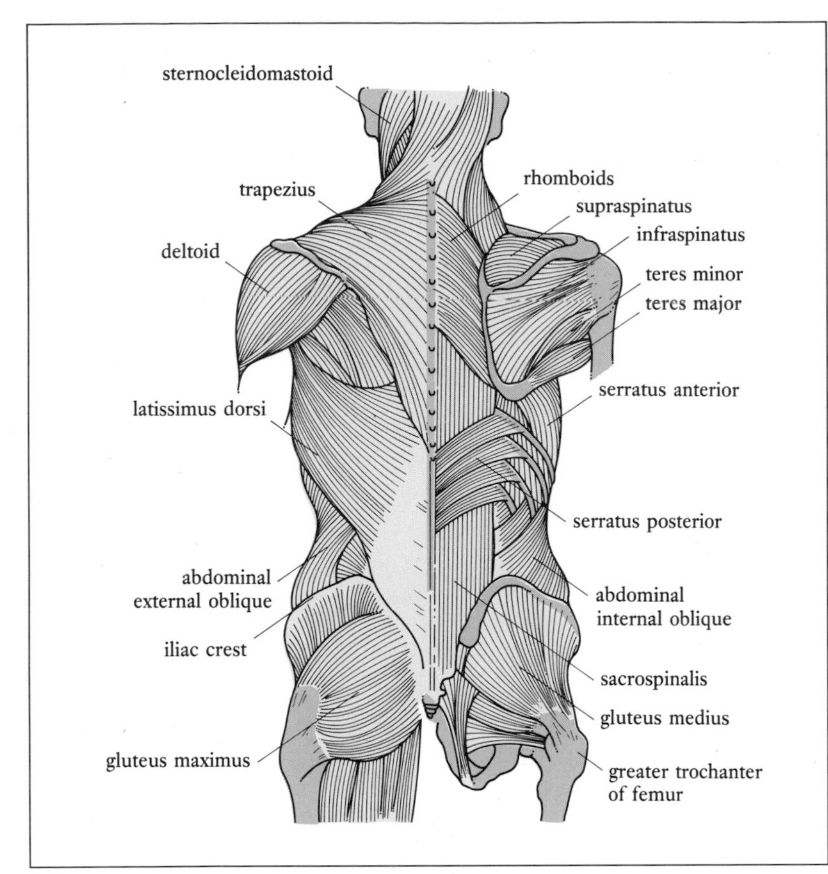

Figure 18.46. Main muscles of the back (superficial muscles removed on the right).

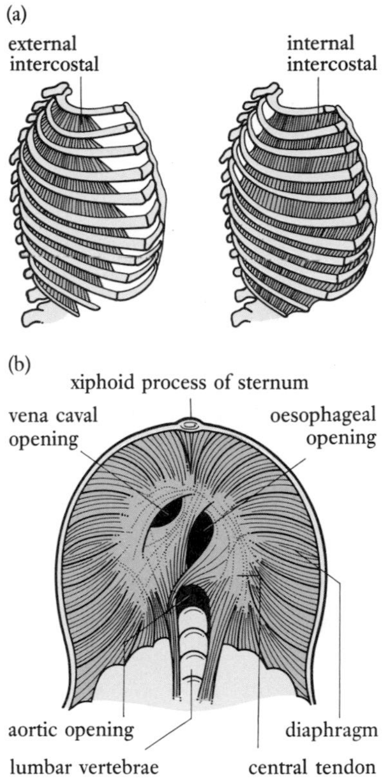

(a)
external intercostal
internal intercostal
(b)
xiphoid process of sternum
vena caval opening
oesophageal opening
aortic opening
diaphragm
lumbar vertebrae
central tendon

Figure 18.47. (a) Deep muscles of the thorax; (b) diaphragm (inferior view).

Abdominal wall

Three layers of muscle form the anterolateral walls of the abdomen (*Figure 18.48(a)*). They allow trunk movement, support viscera, compress the abdomen during evacuation and lifting and assist with expiration. Working with the diaphragm they increase abdominal pressure during the Valsalva manoeuvre (see Chapters 12 and 13). The different directions of the fibres of the three layers produces a very strong wall.

The layers, from the outside, are the **external oblique**, **internal oblique** and **transversus abdominus**. The aponeuroses of these three muscles merge with the strap-like **rectus abdominus**, the medial muscle running from sternum to pubis. The **linea alba**, a tendinous cord formed from the aponeuroses of the abdominal muscles, divides the rectus abdominus in the midline.

The **inguinal canal**, which carries the round ligament in females and the spermatic cord in males (Chapter 20), is an intermuscular canal running obliquely through the abdominal muscles at the level of the inguinal ligament formed from the aponeurosis of the external oblique muscle.

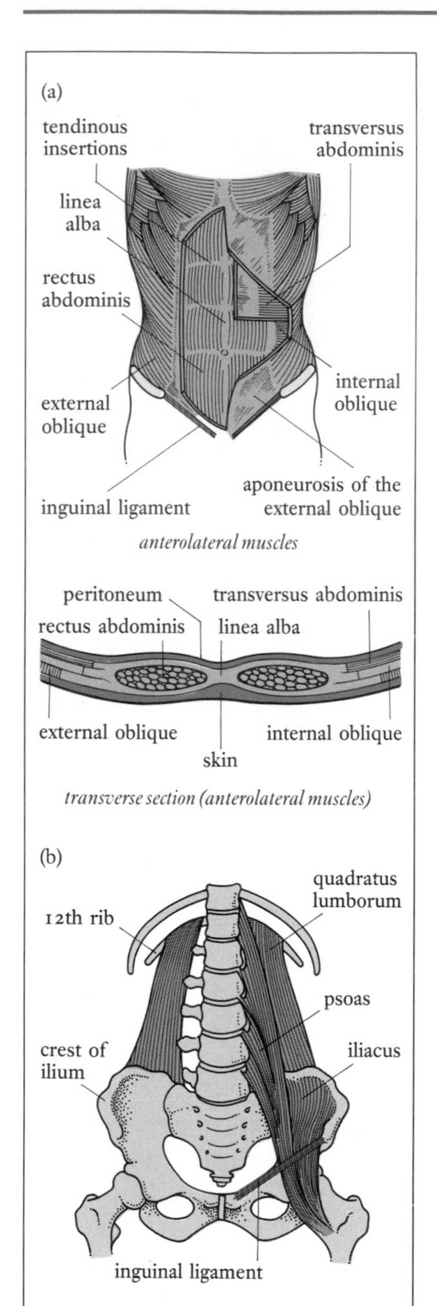

Figure 18.48. Abdominal wall. (a) Anterolateral; (b) posterior.

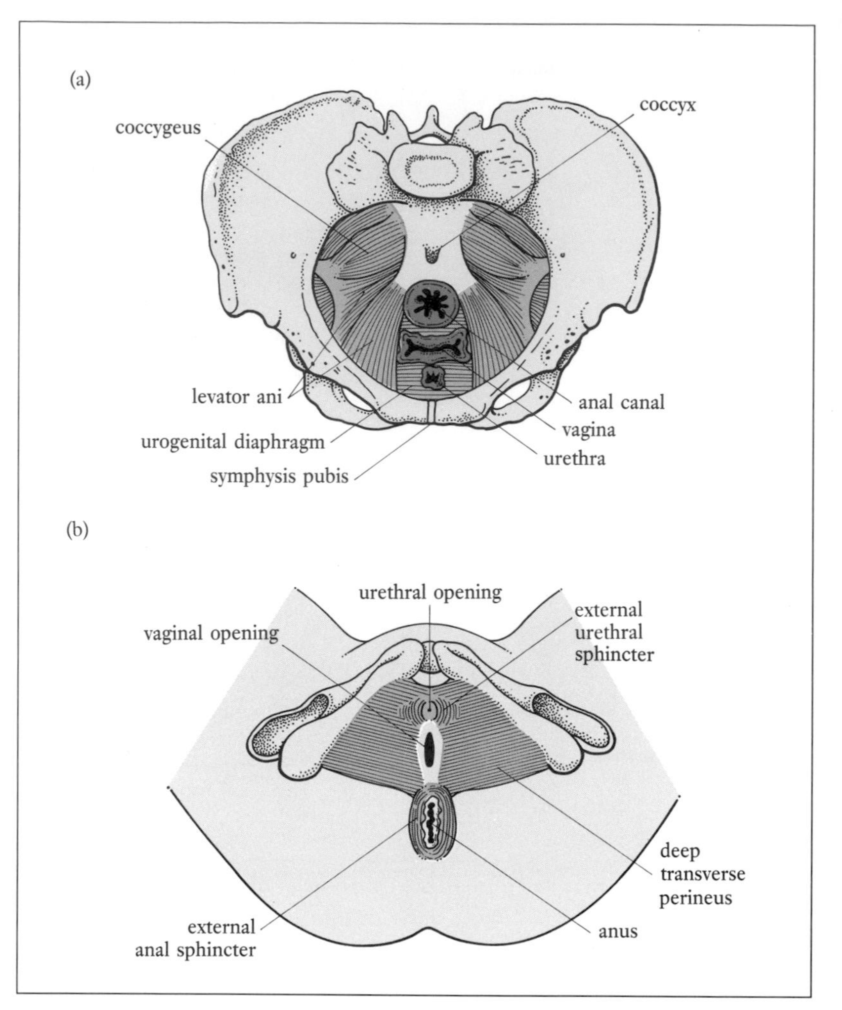

Figure 18.49. Pelvic floor (female). (a) Main muscles; (b) urogenital diaphragm.

The inguinal canal is a 'weak spot' where *herniation* occurs (*inguinal hernia*); other sites include the femoral canal *(femoral hernia),* around the umbilicus (*umbilical hernia*), oesophageal opening in the diaphragm (see *hiatus hernia*, Chapter 13) and at the site of an incision (*incisional hernia*).

The posterior abdominal wall is formed by the quadratus lumborum, psoas and iliacus muscles with their fascia (*Figure 18.48(b)*).

Pelvic floor

The **pelvic floor** consists of the muscles and their ligaments which support the pelvic organs and form the **urogenital diaphragm** and the **perineum**. Two main muscles, the **levator ani** and **coccygeus**, form a funnel-shaped structure consisting of identical halves which join in the midline (*Figure 18.49(a)*). Structures forming the urogenital diaphragm and perineum include the **transverse**

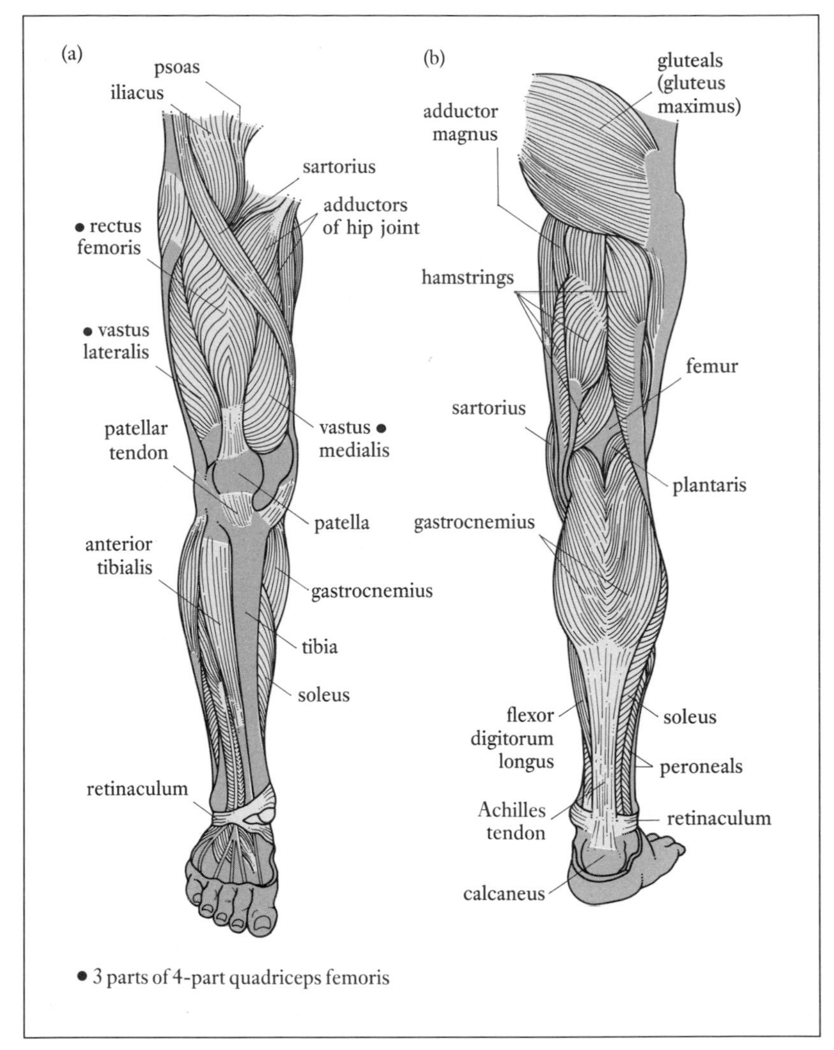

Figure 18.50. Main muscles of the hip and leg. (a) Anterior; (b) posterior. NB Some deep muscles not shown.

perineus muscles and the **urinary** and **anal sphincters** (*Figure 18.49(b)*). The pelvic floor is perforated by the urethra and anus, and the vagina in the female. The role of the pelvic floor is to control voiding and defaecation, maintain continence (see Chapters 13 and 15) and support the pelvic structures in position.

Problems which occur when the pelvic floor malfunctions, e.g. *genital prolapse*, are discussed in Chapter 20.

Muscles of the hip and leg

The muscles of the hip and leg (*Figure 18.50*) are concerned with the movements of the hip, knee, ankle, foot and toes (*Tables 18.6–18.8*), and maintaining stable postures.

Summary/Check List

Introduction.
Skeleton – Bony landmarks.
Axial skeleton – Skull, cranium, sutures, fontanelles, facial bones, *Nursing Practice Application* – mandibular fracture, paranasal sinuses. Spine, curves, general vertebral features, individual vertebrae, spinal ligaments, intervertebral foramina and discs, spinal movements, *Nursing Practice Application* – back pain. Thoracic cage, sternum, ribs.
Appendicular skeleton – Pectoral girdle and arm, clavicle, scapula, humerus, radius, ulna, wrist and hand. Pelvic girdle and leg, innominate bones, pelvis (female/male), femur, patella, tibia, fibula, ankle and foot, arches.
Levers and movement.
Joints – Types, fibrous, cartilaginous, synovial.
Synovial joints, general characteristics, movements, varieties, *Nursing Practice Application* – joint stiffness/immobility, injuries to joints. Specific synovial joints, movements and muscles, shoulder, elbow, radioulnar, wrist, hands/fingers, hip, *Nursing Practice Application* – CDH, knee joint, *Person-centred Study* – Brian, ankle, foot/toes. *Special Focus* – joint conditions.
Muscular system – Naming muscles. Movement relationships. Major muscle groups.

?

Self Test

1 Name the bones of the axial and appendicular skeletons.
2 Put the following in their correct pairs:
(a) occipital bone (b) odontoid process
(c) axis (d) nucleus pulposus
(e) ribs (f) foramen magnum
(g) annulus fibrosus (h) costal cartilage
3 How does the female pelvis differ from the male?
4 Explain why lifting is more efficient when loads are carried close to the body.
5 Give an example of:
(a) fibrous (b) cartilaginous (c) synovial joints.
6 Draw a diagram of a synovial joint to illustrate its general characteristics.
7 You are asked to teach Maud a full range of active arm exercises. What movements would you teach for the shoulder joint?
8 Outline the function of the following: bursae, semilunar cartilages, foot arches, glenoidal labrum, ligamentum teres.
9 What can the name tell you about the characteristics of a muscle?
10 Complete the following:
(a) The _________ separates the thoracic and abdominal cavities.
(b) The levator ani and coccygeus are muscles of the ___________.
(c) the Achilles tendon is the insertion of the _____________ muscle into the __________.

Answers
1. See page 441. 2. a–f, b–c, d–g, e–h. 3. See page 460. 4. See page 464. 5. a, skull sutures, teeth, tibio-fibular; b, symphysis pubis, intervertebral, epiphyseal plate, sternum; c, hip, shoulder. 6. See page 465. 7. Extension, flexion, adduction, abduction, medial and lateral rotation and circumduction. 8. See pages 466, 473, 463, 469, 472. 9. Shape, position, size, function, number of origins and fibre direction. 10. a, diaphragm; b, pelvic floor; c, gastrocnemius, calcaneus.

References

Green, A. (1989) What's happening to hips? *Nursing Times*, **85** (46), 32–33.

Griffiths, B.L. (1988) Have you ever had a pain in your back? *The Professional Nurse*, **3** (4), 125–130.

Further Reading

Bird, H.A (1988) *Update Postgraduate Centre Series – Rheumatoid Arthritis*, Guildford: Update–Siebert Publications, pp. 12– 23.

Burton, S. (1989) Drugs to treat rheumatic disorders. *Nursing*, **3** (34), 27–29.

Dieppe, P.A., Doherty, M., Macfarlane, D.G. and Maddison, P.J. (1985) *Rheumatological Medicine*, Edinburgh: Churchill Livingstone.

Griffiths, B.L. (1987) Back pain: why bother? *The Professional Nurse*, **3** (3), 97–101.

Huskisson, E.C. (1985) *Update Postgraduate Centre Series – Osteoarthritis: Pathogenesis and Management*, London: Update Publications, pp. 5–10.

Molitor, P. (1989) Arthroplasty: a joint venture. *Nursing*, **3** (34), 9–12.

Pugh, J. (1989) Nurses' perceptions of lifting techniques. *Nursing Times*, **85** (46), 55.

Royal College of Nursing (1979/1985) *Avoiding Low Back Injury among Nurses*. Report of RCN working party. London: RCN.

Tortora, G.J. and Anagnostakos, N.P. (1990) *Principles of Anatomy and Physiology* (chs 7, 8, 9, 11), 6th edn. New York: Harper and Row Publishers.

Williams, P.L., Warwick, R., Dyson, M. and Bannister, L.H. (1989) *Gray's Anatomy*, 37th edn. Edinburgh: Churchill Livingstone.

Useful Addresses

Arthritis and Rheumatism Council
41 Eagle Street
London WC1R 4AR

section seven

defence and survival strategies

CHAPTER NINETEEN

BODY DEFENCES AND THE SKIN

OVERVIEW

- Non-specific defences: skin and membrane barriers, cells and chemicals.
- Specific defences: humoral and cell-mediated.

LEARNING OUTCOMES

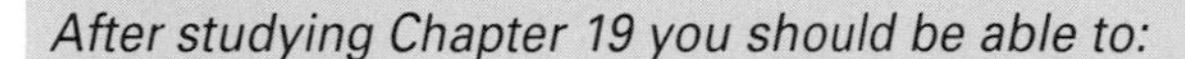

After studying Chapter 19 you should be able to:

- Describe the structure of the skin and its appendages.
- Discuss normal skin pigmentation and colour changes which may indicate departures from normal.
- Describe the functions of the skin and relate these to structure.
- Explain the mechanisms by which the skin contributes to temperature homeostasis.
- Use physiological knowledge to explain the rationale for the nursing practices involved in monitoring body temperature, and to explain failures of temperature homeostasis.
- Outline the protective role of other innate (membrane) barriers.
- Describe phagocytosis and the role of non-specific cellular defences.
- Describe the events of the inflammatory response and discuss its defensive role.
- List important inflammatory and antimicrobial chemicals.
- Describe the events of wound healing and use this knowledge in the care of wounds.
- Discuss the factors which predispose to pressure sores, and measures used in their prevention.
- Outline the characteristics of a specific defence.
- Describe humoral immunity.
- List the classes of antibodies and briefly outline their functions.
- Explain active and passive immunity.
- Outline the immunization programme available to children in the UK.
- Describe cell-mediated immunity.
- Outline the role of cell-mediated immunity in cancer surveillance.

Learning Outcomes (continued)

- Discuss cell-mediated immunity and its relevance to organ transplants.
- Briefly describe abnormal immune responses: hypersensitivity reactions, autoimmunity and immunodeficiency.
- Discuss the aetiology and effects of acquired immune deficiency syndrome (AIDS).
- Discuss aspects of health education aimed at reducing the spread of AIDS.

Key Words

Antibody – a protein produced by plasma cells which binds to a specific antigen.
Antigen – a molecule or part of a molecule which is recognized as foreign by body defences and which therefore stimulates the immune response, e.g. a virus.
Antiserum – serum, obtained from animals, containing antibodies to a specific antigen.
Autoimmune – abnormal immune response where the body defences fail to recognize 'self'. It results in tissue damage.
Cell-mediated immunity – part of the immune response involving the action of specific lymphocytes which destroy abnormal/ foreign cells and release regulatory chemicals.
Commensals – micro-organisms which live in close association with a host. They benefit from the relationship and generally cause no harm.
Complement – a collection of proteins involved in both non-specific and specific defences.
Dermis (true skin) – layer below the epidermis.
Epidermis – superficial layer of the skin.
Humoral – part of the immune response involving the production of antibodies by plasma cells.
Hyperaemia – excess blood in a particular area.
Hypothermia – abnormally low body temperature where core temperature is below 35°C.
Immune response – specific defences provided by B and T lymphocytes.
Immunity – resistance to disease.
Immunoglobulin Ig (see antibody) – proteins present in blood and other body fluids.
Inflammatory response – tissue reponse to injury. Local tissue changes are initiated by components which may be chemical, cellular, vascular and exudate.
Integument – the skin and its appendages.
Keratin – a tough fibrous protein found in epidermis, hair and nails.
Lysis – breakdown/disruption of cells, e.g. *haemolysis* – breakdown of erythrocytes.
Lysozyme – bactericidal enzyme present in many body fluids, e.g. tears.
Melanin – dark pigment found in varying degrees within the skin, hair and other sites, e.g. iris.
Normal flora – the micro-organisms which normally live on or in the body, e.g. *Escherichia coli* in the gut.
Pathogen – a disease-producing micro-organism.
Phagocyte – a cell able to engulf micro-organisms and other debris by the process known as phagocytosis.
Pyrexia (fever, febrile) – elevated body temperature, usually between 37.2°C and 41°C. Temperatures above 41°C are termed 'hyperpyrexia'.

Introduction

Body defence systems, which are vital to our survival, consist of two main functional components. The first line of defence is provided by the non-specific innate defences such as intact skin which prevents the entry of **pathogens** (disease-producing micro-organism) into the tissues.

Our defence strategies are completed by the specific adaptive functions of the **immune response** (specific defences provided by B and T lymphocytes), e.g. **antibody** (protein produced by plasma cells which binds to a specific antigen) production. The components of the immune response can deal with pathogens that slip past the first-line barriers and also recognize and destroy abnormal or foreign cells, e.g. malignant cells.

Non-specific and specific defence systems work closely in their protective role, e.g. **phagocytosis** (process by which phagocytes engulf micro-organisms and other debris) is enhanced by the presence of **immunoglobulin IgG** (see antibody).

Defence, which is part of most body systems or structures, cannot be attributed to a particular organ or system. For a wider understanding of surveillance and defence, it will be necessary for you to make frequent reference to other relevant chapters (Chapters 9 and 11).

Throughout this book great emphasis has been placed upon the degree of integration required to maintain homeostasis. There is a necessary interdependence between all body systems. The innate defences and immune response, which protect most body systems, have important links with both endocrine and nervous control systems (Section Three).

Examples of these links include:

- Stress hormones – corticosteroids – are known to suppress both inflammation and immune response (see wound healing delay, page 505, and Chapters 6 and 8).
- Thymic hormones are involved in T lymphocyte processing (see Chapter 8 and page 512).
- Some neurotransmitter chemicals may have a role in regulating the immune response, as may our emotional state.

Non-specific Components – Skin and Membrane Barriers

Skin

The skin (*Figure 19.1*), which we often take for granted, covers and protects the body, and is continuous with the mucous membranes lining body cavities and orifices opening at the surface. Amazingly your skin or **integument** (skin and appendages) weighs around 4 kg, covers an area of some 1.4 m^2 and normally varies in thickness between 0.5 mm on the eyelid to 3–4 mm on the soles of the feet. The skin can become much thicker in certain situations, e.g. badly fitting shoes cause 'hard skin' to form.

Functions of the skin include: control of water loss, protection, sensation, storage, temperature regulation, excretion and vitamin D synthesis. The skin is divided into two layers, the **epidermis** (superficial layer of the skin) and **dermis** (hide) (true skin – layer below the epidermis), which are secured to the underlying structures by the **subcutaneous** tissue (**hypodermis**) and adipose.

Structure of the skin

Epidermis

The **epidermis** is formed from several layers of stratified epithelium (see Chapter 1) which have no blood vessels or nerves (*Figure 19.2*). It has an outer **horny zone** consisting of three layers: **stratum corneum** (**horny cells**), **stratum lucidum** (**clear cells**) and **stratum granulosum** (**granular cells**) – which overlay the **germinative zone**. The germinative zone has two layers: **stratum spinosum** (**prickle cells**) and **stratum basale** (**basal cells**) which contains the **melanocytes** responsible for the production of **melanin** (dark pigment found in varying degrees within the skin, hair and other sites e.g. iris) and skin pigmentation.

The epidermis undergoes renewal as the stratum corneum is continually shed (*exfoliation*) to be replaced by new cells formed in the stratum basale which move upwards through the layers over a period of 6–8 weeks. Various changes occurring as the cells migrate upwards are nuclear disintegration, *keratinization* (addition of

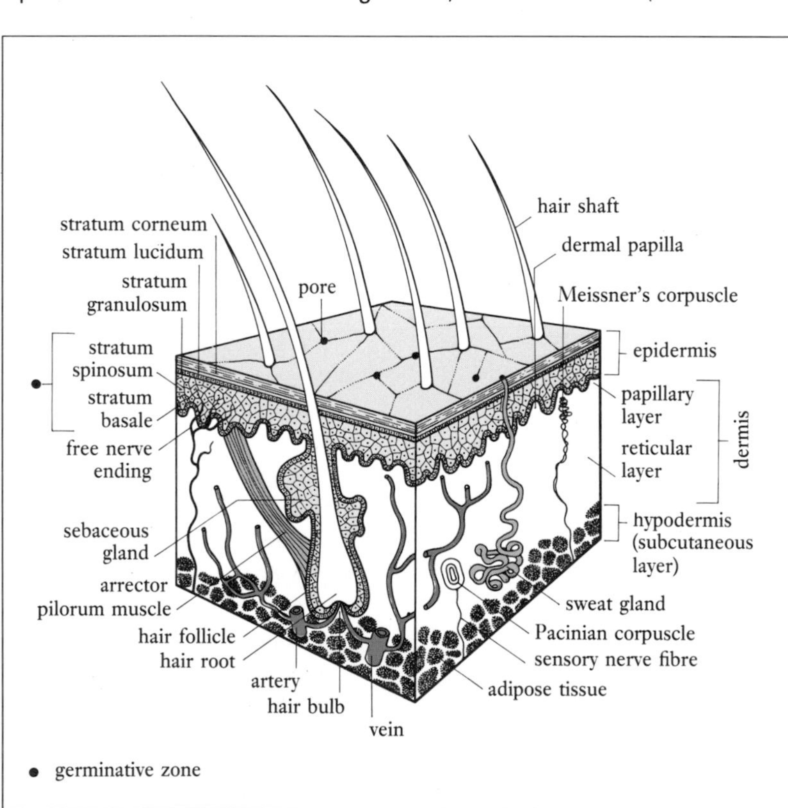

Figure 19.1. The skin.

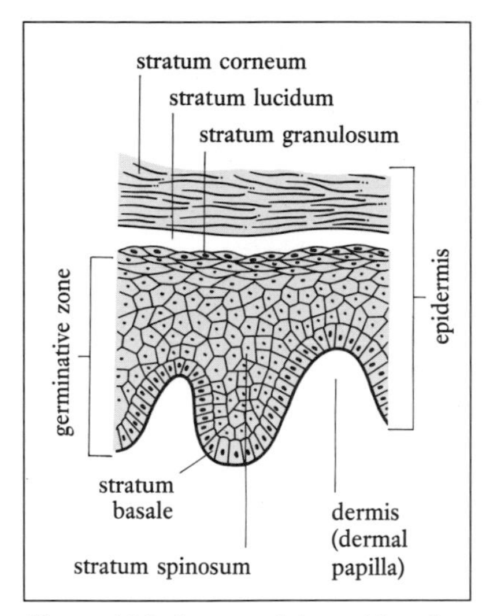

Figure 19.2. Layers of the epidermis.

keratin – a fibrous protein) and flattening. As a result the stratum corneum comprises dead cells containing **keratin** (tough fibrous protein found in epidermis, hair and nails).

This renewal process, which is most rapid during childhood, stabilizes during adult life to decline as ageing occurs, a point which has relevance when you consider the time required for wounds to heal in different age groups.

Next time you undress or make your bed take note of the dust, which is in fact your exfoliated epidermal cells. The dust is only a nuisance at home but in a surgical ward it can present a real infection risk as micro-organisms which are shed with the epidermis can be transferred to wounds when the dust is disturbed, e.g. drawing curtains.

The epidermis is attached to the dermis by an undulating line consisting of **dermal papillae** which are reproduced on the epidermis as the whorls and ridges of your 'fingerprints'. The interlocking layers of epidermis and dermis may be disrupted when subjected to the excessive shearing forces which predispose to pressure sores (page 506) and heel blisters.

Dermis

The **dermis**, or true skin, is connective tissue containing collagen, elastin and reticular fibres, cells such as fibroblasts and macrophages, blood and lymph vessels and nerve endings. The dermis is divided into **papillary** and **reticular layers** (*Figure 19.1*). The arrangement of fibres gives rise to a pattern of **cleavage lines** which were first described by Langer in 1861. These cleavage lines are of particular interest to surgeons because wounds following the lines heal with less scarring than do those which cross cleavage lines.

The wrinkles of normal ageing are caused by loss of dermal fibres, a process much accelerated by exposure to wind and sunlight. If the skin is stretched, e.g. during pregnancy or severe obesity, the rupture of elastin fibres leads to the formation of 'stretch marks' or **striae**.

Structures derived from the skin

Structures derived from the skin include: *sweat glands, ceruminous glands, sebaceous glands, hair follicles* and *nails*.

Sweat glands

Sweat glands are widely distributed over the body and are of two types: **eccrine** and **apocrine**. Eccrine sweat glands are the most abundant. They are concentrated on the palms, soles of the feet and forehead. These coiled glands produce watery sweat which leaves via a duct to empty onto the surface through pores (see *Figure 19.1*). The major role of eccrine sweat glands is temperature homeostasis, where the evaporation of sweat from the body causes heat loss. This principle applies when tepid sponging is used to reduce elevated body temperature.

Sweat, which contains electrolytes (sodium and chloride), nitrogenous waste and other waste molecules, such as drugs and food residues, also has a minor excretory role. Sympathetic nerve stimulation and adrenaline (see Chapters 6 and 8) cause sweat production in response to changes in skin temperature and emotional states such as nervousness. Insensible water loss also occurs by diffusion, and this plus sweat may total between 800–1000 ml/day, under normal conditions in a temperate climate. This amount is obviously very variable and depends on factors such as environmental temperature and activity level. In situations where sweating is excessive, e.g. heavy work in hot conditions where 10 litres/day may be lost, there is risk of dehydration and electrolyte imbalance (see Chapter 2).

Apocrine glands are found in the axillae, areola, groin and external genitalia. They produce a thicker fluid which, when subjected to bacterial action, has a distinctive musky odour. It is possible that apocrine secretions contain chemicals known as **pheromones**; these may influence human sexual behaviour as they do in other species. This view is supported by the fact that apocrine activity does not commence until puberty and is influenced by sex hormones and the catecholamines produced during stressful situations.

Ceruminous glands are modified sweat glands present in the external auditory meatus. They produce **cerumen** (wax) which traps particles and prevents them entering the ear (see Chapter 7).

Sebaceous glands

Sebaceous glands are abundant on the face, neck and back. They secrete a fatty substance, containing cholesterol and other lipids, known as **sebum**. The glands may discharge their sebum into hair follicles or directly onto the surface of the skin (see *Figure 19.1*). Sebum is important in keeping hair and skin supple and resistant to cracking, in waterproofing and in providing a bactericidal/ fungicidal barrier.

Sebaceous gland activity increases at puberty and is especially stimulated by the action of androgens. Bacterial action on the sebum with associated blockage of the follicles, inflammation and bacterial infection may explain the development of 'spots' (*acne*) which cause so much distress during adolescence. The activity of sebaceous glands is low at the extremes of age, which renders the skin more prone to damage and reinforces the view that babies and old people need particularly gentle skin care.

Hair and follicles

Hair helps to protect the skin, assists minimally in temperature regulation and is to some extent used as a form of expression (a new hairstyle may do wonders for your self-image).

The **hair follicles** are formed from epidermal tissue which grows down into the dermis or subcutaneous layer (see *Figure 19.1*). Blood vessels and nerves supplying the follicle enter at its base via a papilla. Hair, which consists of keratinized cells, has a **bulb** or growing region at the base of the follicle, a **root** within the follicle and a **shaft** above the surface of the epidermis. Associated with the hair follicle is the involuntary **arrector pilorum** (plural, pili) muscle (see *Figure 19.1*), innervated by sympathetic fibres which erects the

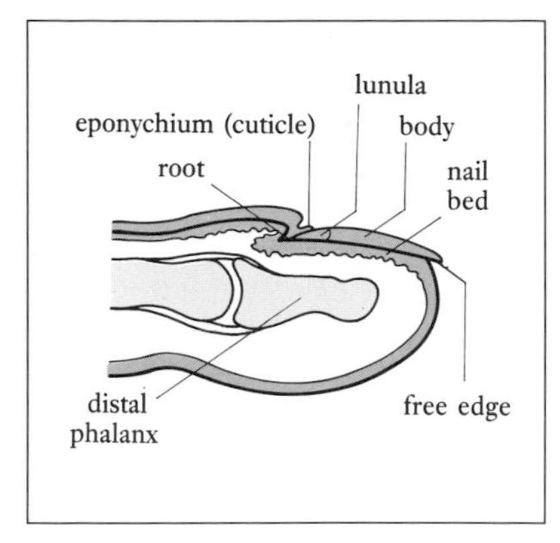

Figure 19.3. Section through distal phalanx and nail.

hair to give you 'goose flesh' or 'pimples' when cold or frightened. Animals such as cats can make their fur 'stand on end', making themselves appear larger and more ferocious. This mechanism is also important in heat conservation in such animals, as a layer of warm air is trapped next to the skin. Humans can in no way match this, but you can still feel the neck hairs become erect whilst watching a really scarey film.

Hair colour is genetically determined and depends on the amount of melanin present. When the amount of melanin declines, which is also genetically controlled, the hair turns grey or white. Natural loss of head hair or *male pattern baldness*, which has an inherited basis, only affects males and is linked with testosterone secretion which stimulates hair growth on the face (not the head). The hair thins with normal ageing in both sexes. Causes of unnatural hair loss or thinning include: cytotoxic drugs (see Chapter 9), hypothyroidism (Chapter 8) and protein deficiency (Chapter 13).

Nails

The **nails** are keratinized sheets which protect the distal ends of the digits (*Figure 19.3*). They are derived from epidermal tissue and are similar to the hooves and claws of other species. Each nail grows on a **nail bed** from a matrix of germinative cells. It has a **root**, **body** and **free edge**. At its proximal edge the nail is thickened to form the white **lunula** which is covered by **eponychium** (cuticle). In health the nail appears pink because of numerous capillaries in the dermis.

Skin pigmentation

Skin pigmentation or colour is due to the presence of haemoglobin, melanin and **carotene**. The pink colour of Caucasian skin is caused by haemoglobin. Melanin is a brown–black pigment responsible for racial differences in skin colour, freckles and the changes occurring when light skin tans. Its presence gives the skin some protection from ultraviolet light. Melanin is derived from the amino acid tyrosine (see phenylketonuria, Chapter 13) and is produced by the epidermal melanocytes when they are stimulated by the pituitary hormone *melanocyte stimulating hormone (MSH)* (see Chapter 8). Yellow–orange pigments are produced by the deposition of carotene and the presence of melanin in the skin. People in oriental racial groups have increased carotene.

Functions of the skin

Normally functioning skin plays a vital part in maintaining homeostasis and provides protection against many physical and chemical hazards.

Waterproofing

The skin forms an effective waterproof layer which prevents excess water loss or entry – just as well if you like to soak in the bath. The outer layers and sebum work as well as any waxed jacket. If this barrier is breached, as with severe burns, the consequent fluid loss results in hypovolaemic shock (Chapter 10). Indeed, fluid replacement, based on the percentage of body area ('rule of nine'), is an initial priority in the management of burns.

Nursing Practice Application **Skin colour and condition**

During a nursing assessment considerable information about general health status can be gained from careful observation of the colour and condition of the skin. Some examples include: (i) *pallor* – cold conditions, shock, anaemia; (ii) *redness (erythema)* – hot conditions, emotions, e.g. nervousness or anger, or due to pyrexia, allergy, inflammation, polycythaemia (Chapter 9), 'hot flushes' of the climacteric (Chapter 20) and after exercise; (iii) *bruising/haematoma/petechiae* (escape of blood into the tissue) – bleeding disorder (Chapter 9) or abuse; (iv) *cyanosis* (Chapter 12) – blue discoloration of the skin caused by poor oxygenation of the haemoglobin. Occurs in chronic respiratory conditions and heart failure. Cyanosis, which may be peripheral or central, is easily observed in Caucasians, but is much more difficult to detect in an Afro-Caribbean person in whom a careful examination of the nail beds and mucous membranes is required; (v) *yellow* – jaundice (Chapter 14), fading bruises (due to the breakdown of haemoglobin in the tissues), suntan; (vi) *normal pigmentation changes* – increased pigmentation, e.g. of the face during pregnancy, dark 'age spots' on the hands of elderly people; (vii) *abnormal pigmentation* – the bronze discoloration seen in Addison's disease (Chapter 8) caused by increased secretion of adrenocorticotrophic hormone (ACTH), which is very similar chemically to MSH. Patchy areas of depigmentation known as *vitiligo*; (viii) *rashes/local swellings or lesions/ dryness/oedema* – may indicate a variety of abnormal states, e.g. dry skin in dehydration.

Temperature regulation

The skin is a major contributor to temperature homeostasis. Humans are *homeothermic* ('warm-blooded') and must maintain a constant core body temperature of around 37°C by balancing heat gained with heat lost (*Table 19.1*). The control mechanism or thermostat is situated within the heat regulating centres of the hypothalamus (see Chapter 4).

A constant body temperature is critical for the proper functioning of the enzymes which facilitate the biochemical reactions required for physiological processes. A rise in temperature of 1°C will cause the metabolic rate to increase by some 10%, a fact which explains why **pyrexial** (having an elevated body temperature – usually between 37.2°C and 41°C) people need more calories.

Temperature shows a diurnal variation – it is lowest in the night and rises during the day to reach a peak late afternoon/early evening. This pattern will be reversed in people who do shift work involving long periods of night work.

Earlier we mentioned *core temperature* – the centres of the skull, chest and abdomen are much warmer than the skin surface. When recording body temperature it is important to remember that rectal recordings are nearest to core temperature, followed by the oral and then axillary/groin recordings.

Temperature control

The cells of the heat regulating centre receive impulses from skin *thermoreceptors* and monitor the temperature of the blood passing through the hypothalamus.

When temperature rises above 37°C the hypothalamus initiates sweat gland activity (sympathetic stimulation, Chapter 6) and peripheral vasodilatation (sympathetic inhibition) (*Figure 19.4*).

Table 19.1. Heat gain and loss.

Heat gain	***Heat loss***
Metabolism of fuel molecules.	Through the skin by:
Muscular activity/shivering	evaporation
Environmental temperatures	conduction
Hormones increasing	convection
metabolism e.g. thyroxine	radiation
Female sex hormones	Via the lungs
Body fat	Urine and faeces
Hot food/fluids (minimal)	

Sweating causes heat loss as moisture *evaporates* from the surface whilst environmental humidity is low, which is why we find humid conditions so much more uncomfortable than dry heat. Vasodilation of the arterioles in the dermis, which diverts blood to the skin capillaries and accounts for the red colour of hot skin, causes heat to be lost.

This loss occurs by:

- *Radiation* to the surrounding air.
- *Conduction* to objects such as clothes or a chair.
- *Convection* as warm air currents move away from the skin to be replaced by cooler air.

Most heat is lost from the body by radiation, conduction and convection and all can be utilized to reduce pyrexia, e.g. fanning, reducing clothing, cool bath, change of bed linen.

Conversely when the temperature falls below 37°C the heat regulating centre initiates heat conserving mechanisms via the sympathetic pathways. Sympathetic stimulation causes vasoconstriction and the skin becomes pale and cold as blood is diverted from the surface to prevent heat loss. Shivering, which is intense muscular contraction, produces heat and a small amount of warm air is trapped by body hair. The subcutaneous adipose layer serves as insulation, which we enhance by wearing several layers of clothes in cold weather. An increase in catecholamines (short term) and thyroxine (long term) increases metabolic rate and more heat is produced (see Chapter 8), catecholamines also increase sympathetic vasoconstriction.

Certain adaptative features operate in very cold weather: *cold-induced vasodilation* prevents hypoxic changes if skin blood flow is shut down for too long. In response to local hypoxia (see Chapter 12) the vessels dilate to restore blood flow, which accounts for the red hands and nose on a cold morning. Another feature is an arrangement by which venous blood regains heat before returning to the body core. The veins (**venae comitantes**) run close to the main arteries of the limbs, and use them as a countercurrent heat

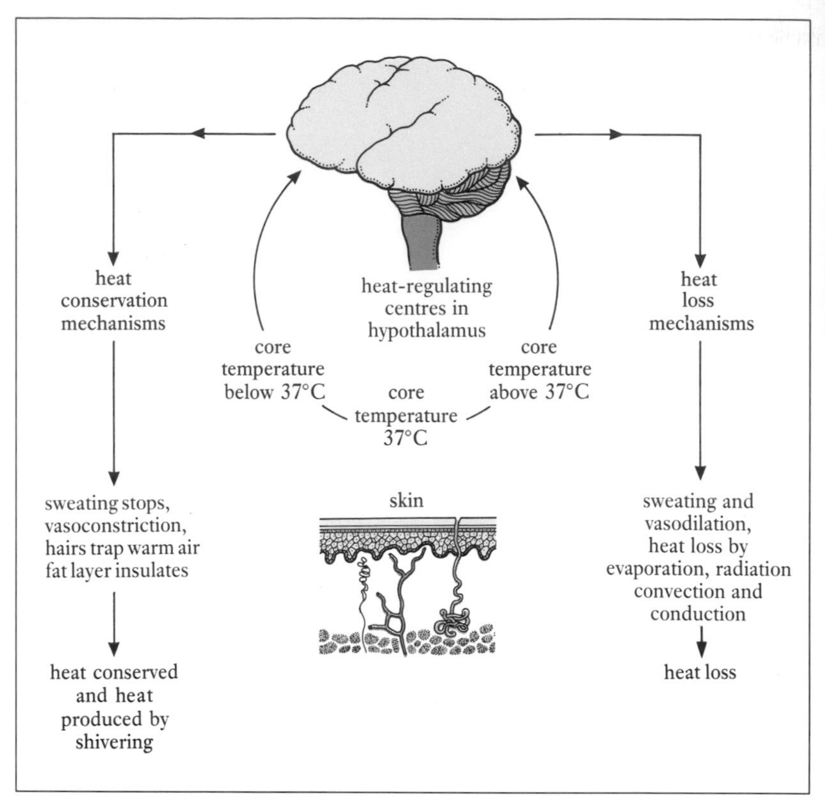

Figure 19.4. The skin and temperature regulation.

exchanger in much the same way that a cold water supply is warmed by running close to hot water pipes. This prevents cold blood reducing core temperature.

Blood flow through the skin, which is central to temperature control, is regulated by arteriovenous shunts and precapillary sphincters (see Chapter 10) which divert blood to or from the skin capillaries as needs change.

Problems with temperature homeostasis

Fever

In this situation the thermostat in the hypothalamus is reset to a higher temperature by the presence of **pyrogens** (chemicals producing fever) released by micro-organisms, blood cells and injured tissue. Physical problems within the heat-regulating centre, e.g. severe head injury (see Chapter 4) have similar effects on the thermostat. The hypothalamus continues to initiate heat conserving/producing mechanisms which explains the vaso-constriction, feeling cold, shivering and *rigors* as your core temperature rises to the new setting.

Once treatment with antibiotics or antipyretics, or natural defence mechanisms, have dealt with the cause of the fever the thermostat is reset to normal and the whole system goes into reverse to initiate heat loss. There is vasodilation, profuse sweating and the person feels very hot.

Nursing Practice Application **Recording temperature**

Accurate recording of temperature, as part of a nursing assessment, requires consideration of a number of factors:

Choice of site – depends on age and condition. For instance during cardiac surgery the requirement for precise core temperature necessitates the use of sensitive internal (rectal) probes, but rectal temperature using a glass thermometer would suffice in most situations where great accuracy is needed, e.g. **hypothermia** (abnormally low body temperature where core temperature is below 35°C) in the elderly. There are problems with all sites and Closs (1987) reminds us that the oral site, which is the most convenient and commonly used site in adults, has great scope for error and inaccuracy. Activities and diurnal variation – exercise, eating, smoking and drinking hot or cold liquids prior to the recording can change results. We have already discussed diurnal variation which you should also consider when interpreting results. If you are recording temperature to detect an elevation which may indicate infection, it is sensible to plan your assessment to coincide with the natural peak.

Another factor, important for accuracy, is the length of time that the thermometer remains *in situ*. The duration depends upon factors which include environmental temperatures; Nichols and Kucha (1972) found that accurate readings were obtained after 8 minutes for men and 9 minutes for women when room temperature was between 18 – 24°C. In a review of the literature Sims-Williams (1976) states that many patients' temperatures were under-recorded because insufficient time was allowed.

For accuracy it is also necessary to use the correct type/range thermometer, e.g. where hypothermia is suspected a 'low reading' rectal thermometer is required. However, the use of specialized equipment is rarely needed and Closs (1987) concludes that for most patients good estimates of core temperature can be obtained from glass/mercury and electronic thermometers.

High body temperature is one of the defence mechanisms of the body, but on the negative side an excessive or prolonged fever will have profound effects upon physiological processes. The metabolic rate increases, with an increase in pulse and respiration, and with it the demand for energy and protein. With **hyperpyrexia** (body temperature above 41°C) there will be serious loss of enzyme efficiency and brain damage (see below).

Heat stroke

If body temperature reaches 41–42°C the hypothalamus ceases to function properly and heat loss mechanisms fail to operate. The skin is hot and dry, and sweating and vasodilation do not occur. The rise in core temperature increases metabolic rate which in turn produces more heat. If this dangerous state is not reversed the person will suffer convulsions and brain damage or death, when the body proteins are denatured at around 43°C. Heat stroke may result from unaccustomed or prolonged exposure to high environmental temperatures. It is especially likely to follow strenuous exercise in very hot conditions.

Hypothermia

At the opposite end of the scale is accidental hypothermia, which affects young babies, the elderly (see *Person-centred Study* – Mavis) and those whose occupation or lifestyle may expose them to harsh environmental conditions, e.g. people 'living rough' or those who go ill-prepared for climbing and walking expeditions.

Small babies are at risk because they may become uncovered, they do not shiver and their heat-regulating centre is underdeveloped. The elderly, who also have reduced adipose tissue and poor homeostatic control (see Chapter 2), may simply not feel cold or shiver as their temperature falls.

When body core temperature falls below 34°C the normal heat conservation mechanisms, such as shivering, fail to work. All body systems become sluggish, the person becomes drowsy and eventually unconscious. Unresolved hypothermia leads to cardiac arrhythmias (see Chapter 10) and death when the temperature falls to around 20°C.

Various factors which increase the risk of accidental hypothermia include:

- Socio-economic situation of elderly people, who may have poor housing, reduced income for heating or food and less motivation.
- Physiological causes, e.g. falls, hypothyroidism, alcohol abuse, sedating drugs, confusion and reduced mobility in the elderly population.

NB Hypothermia is used during certain types of cardiac surgery to reduce metabolic rate and hence the oxygen requirements of the cells.

Sensation

The skin is a major part of the environmental sampling kit vital to survival. It contains different sensory receptors which are able to differentiate between the touch of a feather to extremes of temperature. Receptors which respond to temperature, touch, pressure and pain relay the sensory data to the CNS for integration and, where appropriate, the initiation of motor activity (see Chapter 3).

Skin receptors include: **free dendritic nerve endings** (pain, touch and temperature), **Meissner's corpuscles** (light pressure), **Pacinian corpuscles** (deep pressure), **Krause's bulb** (cold) and the

Person-centred Study Mavis

Mavis, aged 81, is widowed and lives alone in a small terraced house in the town centre. Her neighbour Val works from home and visits Mavis most afternoons for a chat and cup of tea.

When Val got no answer at the door she used the key that Mavis had given her for just such an eventuality. Val feared the worst, and when she saw Mavis slumped on the sofa thought that she had died. On closer examination Val found that Mavis was drowsy but felt very cold.

While Val waited for the ambulance to arrive she put Mavis into the recovery position (Chapter 12), covered her with two blankets and switched on the gas fire. The paramedical staff wrapped Mavis in an aluminium foil 'space blanket' prior to the journey to hospital. The underlying principle here is to prevent further heat loss by radiating the heat back on to the body. Mavis was admitted to hospital for gradual rewarming (temperature 30°C) and rehydration. Investigations after her recovery revealed that Mavis had fallen on her way to bed and just managed to get on the sofa but felt too poorly to light the fire or summon help.

Ruffini organ (heat). There is, however, some overlap of function between some receptors. Receptors in the skin protect the body by alerting us to external factors which are harmful, e.g. you move your hand rapidly (by reflex) when you touch a hot cooker ring. They also allow you the pleasurable sensations such as the feel of fresh bed linen.

Protection

The skin protects in a variety of ways – prevention of water loss, temperature homeostasis and the recognition of harmful stimuli by the sensory receptors have already been discussed. The epidermis is also able to withstand some chemicals, e.g. mild acids; melanin blocks some ultraviolet light; and some types of ionizing radiation do not penetrate the skin, e.g. alpha radiation. The fat layer provides varying amounts of 'padding' which gives some protection against trauma.

Intact skin, which is usually acid (due to its secretions being slightly acidic), keeps out micro-organisms by providing a physical barrier, and sebum and sweat both have bactericidal properties. Another way in which the skin protects against infection is through its own **normal flora** (micro-organisms which normally live on or in the body) of **commensal** (micro-organisms which live in close association with a host. They benefit from the relationship and generally cause no harm) organisms resident on the surface or within the deeper layers. These commensals, which are well adapted to their environment, tend to 'crowd out' other, more harmful organisms, which find it difficult to become established. The normal flora changes after admission to hospital where the skin may be colonized by a different set of organisms. Although we have talked in terms of 'normal' flora it should be remembered that these organisms may become pathogenic if they find their way into a different site, e.g. a wound.

Synthesis of vitamin D

Vitamin D (see Chapter 13) is produced by the action of ultraviolet light on the sterol **7-dehydrocholesterol**, which is present in the skin. Although some exposure to sunlight is beneficial, it is important to stress the dangers inherent in excessive 'sunbathing' which include: 'sunburn', sun stroke, dehydation, premature ageing and, most serious of all, malignant skin changes. The incidence of skin cancers including *malignant melanoma* is increasing (in the Caucasian population) in areas which enjoy good weather, e.g. Australia and the southern United States.

Information regarding safe exposure to the sun and possible skin changes includes: identifying skin type, using a sunscreen, limiting the time spent in full sun especially around midday, covering the head and wearing dark glasses, ensuring adequate fluid intake especially for small children and remembering that working or playing in the sun can result in over-exposure and that certain drugs, e.g. phenothiazines, increase photosensitivity, as do perfumes. It is especially important that individuals note and report any skin changes, e.g. an enlarging or bleeding mole.

Nursing Practice Application **Skin hygiene**

Before starting we should perhaps reflect upon what Armstrong-Esther (1981) had to say about the subject of daily bed baths, 'the patient is not in contact with the water long enough to suffer effects other than mild exposure, hypothermia and discomfort'. Put this way, the daily bed bath does seem to fail in terms of providing hygiene and comfort without harm.

The concept of skin as a physical, chemical and biological barrier to bacteria has important implications for practice. It is imperative that nursing interventions aimed at assisting with personal cleansing do not in fact compromise the integrity of these barriers. Moisture should not be allowed to accumulate in areas such as axilla, groin and where skin surfaces meet, as this encourages the growth of pathogens. Another aspect to consider is the frequent and often ritualistic use of alkaline soap and water which alters the pH of the skin and has a drying effect which leads to cracking. This is particularly serious in elderly people who already have a thinner drier skin with reduced sebaceous gland activity. Pembroke (1983) suggests that skin dryness in the elderly can be avoided by reducing the use of soap and water and by using emollients instead.

Excretion

Compared with the lungs and kidneys, the skin excretes only small amounts of waste. Water, sodium, chloride, urea and a minute amount of carbon dioxide leave via the skin.

Storage

The skin acts as a store for water, which can augment blood volume as required. Subcutaneous adipose tissue forms a substantial part of our fat reserves. These can be utilized as a source of energy when the body is otherwise deprived. The amount and type of fat stored depends on many factors which include:

- *Gender* – females have a greater percentage of adipose tissue.
- *Inheritance* – one tends to inherit a lean or plump build.
- *Age* – new babies have special *brown fat cells* which more easily yield their energy and some individuals may retain some brown fat into adult life. Older people have less subcutaneous fat to act as a reserve.

Other innate barriers

Most of these are covered in the relevant chapters, but during our discussion of non-specific defences a summary of the innate protective barriers and their particular features is helpful.

Gastrointestinal tract (Chapter 13)

- Saliva contains **lysozyme** (bactericidal enzyme present in many body fluids) and immunoglobulin IgA.
- Hydrochloric acid (pH 1.5–3) in the stomach kills most organisms. It cannot deal with large numbers of food poisoning organisms.
- Gastric mucus protects lining from acid attack.
- Reflex vomiting, e.g. after ingesting 'bad' food.
- Normal flora of the gastrointestinal tract helps to prevent colonization by pathogens. Fungal infections, e.g. *Candida albicans* which causes 'thrush' can be a problem following

antibiotic therapy which may kill off the normal flora.
- Lymphoid tissue and immunoglobulins present in the intestinal wall.

Genitourinary tract (Chapters 15 and 20)
- Frequent voiding flushes bacteria from the urethra/bladder.
- Length of the male urethra.
- Normal flora of the vagina produces an acid environment (pH 4.5) which, during the reproductive years, keeps most pathogens out.
- Vagina has a 'tough' lining of stratified squamous epithelium (Chapter 1).
- Regular shedding of the endometrium during menstruation may militate against chronic uterine infections.

Eyes (Chapter 7)
- Lashes.
- Blinking reflex.
- Tears contain lysozyme.

Respiratory tract (Chapter 12)
- Shape of upper respiratory tract, hairs, cilia and mucus prevents the entry of debris and many pathogens.
- Nasal secretions contain lysozyme.
- Sneeze reflex prevents the entry of unwanted material.
- Mechanisms which protect the larynx during swallowing.
- Cilia and mucus in trachea and bronchi trap micro-organisms and move them towards the pharynx.
- Cough reflex helps to expel unwanted material from the lungs.
- Phagocytic cells and lymphoid tissue in the bronchial walls.

Non-specific Components – Chemicals and Cells

There are many cellular and chemical processes which operate as part of the non-specific body defences. The major components are: **phagocytes, natural killer cells**, the **inflammatory response** (tissue response to injury. Local tissue changes are initiated by components which may be chemical, cellular, vascular and exudate) and various antimicrobial chemicals, e.g. **complement** (collection of proteins involved in both non-specific and specific defences) and **interferons**.

Phagocytes and phagocytosis

The process of **phagocytosis**, in which phagocytic cells are able to engulf and destroy foreign particles or dead cells, is a vital part of non-specific defences. **Phagocytes** are found in most tissues and include some types of leucocyte, e.g. neutrophils, eosinophils and monocytes (Chapter 9) and macrophages. Some phagocytic cells are fixed, e.g. Kupffer cells of the liver (Chapter 14) whilst others, such as neutrophils, are very mobile and can congregate in areas of inflammation.

The phagocytes are aided by immunoglobulins and complement, which both prepare the particle for destruction. After dealing with the foreign particle leucocytes are destroyed and form part of the pus resulting from the encounter. The macrophages are not destroyed and can perform their role many times. Phagocytosis occurs as part of the inflammatory and healing processes (see healing of fractures, Chapter 16, and wound healing, (pages 503-505) and performs a vital role in both parts of the immune response (see page 489).

Natural killer cells

Natural killer (NK) cells, found in the blood and tissues, are part of the non-specific defences. These leucocytes can recognize and destroy, by **lysis** (breakdown/disruption of cells), a cell infected by a virus or one showing malignant changes. In the case of viral infection the natural killer cells are attracted by the release of chemicals called **interferons** (see page 503).

Inflammatory response (inflammation)

Inflammation is the response of tissues to physical or chemical injury. The tissue injury may be due to micro-organisms, trauma, extremes of temperature (burns, scalds and frostbite), ultraviolet light, extremes of pH and ionizing radiation.

The non-specific inflammatory response operates locally to limit tissue damage, remove the cause and 'clear away' dead cells and other debris so that healing may occur. Its components are chemical, vascular, cellular and exudate, which interact to produce a very effective defence mechanism.

Acute inflammation is characterized by: *redness (erythema)* and *heat* as **hyperaemia** (excess blood in a particular area) occurs, *swelling* due to fluid exudation, *pain* caused by inflammatory chemicals and pressure on nerve endings and *loss of function* especially if a joint is involved. The degree of loss of function is directly related to the amount of swelling and pain present as well as the injury site.

Events of the inflammatory response

- Immediately following the tissue injury there is short-lived vasoconstriction.
- Vasodilation occurs in response to the release of inflammatory chemicals, e.g. *histamine*, by damaged tissue, mast cells and basophils. This leads to local hyperaemia as blood flow increases to the area.
- The release of inflammatory chemicals, e.g. *5-hydoxytryptamine (5-HT), kinins, complement* and *prostaglandins*, increases capillary permeability. Some of these chemicals also help to attract extra leucocytes to the area by **chemotaxis** (see Chapter 9).
- The increase in capillary permeability allows exudate of fluid and plasma proteins to leak from the blood into the tissues. This movement of plasma proteins results in a slowing of blood flow which allows the neutrophils to **marginate** (move to the sides of

the capillaries). The exudate within the tissues dilutes toxic substances, carries extra oxygen, nutrients and leucocytes to the area and commences the 'walling off' process with fibrin, which limits the spread of damage.

- Margination of neutrophils occurs prior to their movement through the vessel wall by a process called **diapedesis** (see Chapter 9). The influx of neutrophils at the site of injury is, as already discussed, facilitated by increased blood flow, inflammatory chemicals and chemicals produced by micro-organisms.
- Once the neutrophils, and later the macrophages, reach the area of injury they start to remove micro-organisms and damaged tissue by phagocytosis. Where inflammation is chronic the macrophage is the predominant cell. The phagocytic cells engulf the particles, which are then destroyed by lytic enzymes present in the leucocyte. Sometimes the phagocyte is itself destroyed during the process. Pus, if formed, is a mixture of dead leucocytes, tissues debris, micro-organisms and exudate. Phagocytosis is enhanced by the presence of immunoglobulins and complement.
- The last stage of the inflammatory response is to clear the debris from the battle zone so that the processes of healing can proceed (see wound healing, below).

NB Before healing can occur it may be necessary for a collection of pus (*abscess*) to be drained from the wound surgically.

Interferons

Cells infected by viruses produce antiviral proteins called **interferons**. These proteins help to prevent the destruction of surrounding cells by viruses, and stimulate macrophages and natural killer cells which also destroy malignant cells. This has led to considerable interest in their development as antiviral agents and as a treatment for malignancy, but as yet the results have been very variable. Interferons are also produced by T lymphocytes.

Complement

Complement is a complex of about 20 plasma proteins which are concerned with both non-specific and specific defences. Complement is activated in a variety of ways and acts through cascades where the first protein activates the second and so on. The activated complement is concerned with enhancing phagocytosis, bacterial lysis, inflammatory chemical release, vasodilation and chemotaxis.

Wound healing

Wound healing and the inflammatory response are closely related processes. Healing, which aims to restore tissue integrity and normal function, may be by rapid **primary/first intention** (*Figure 19.5*) or the more protracted **secondary intention**.

Obviously the desired method is primary intention healing which occurs in clean surgical incisions where the wound edges are in apposition. Where there is tissue loss, such as pressure sores or a gaping wound, the healing occurs slowly by secondary intention as the cavity heals from the bottom.

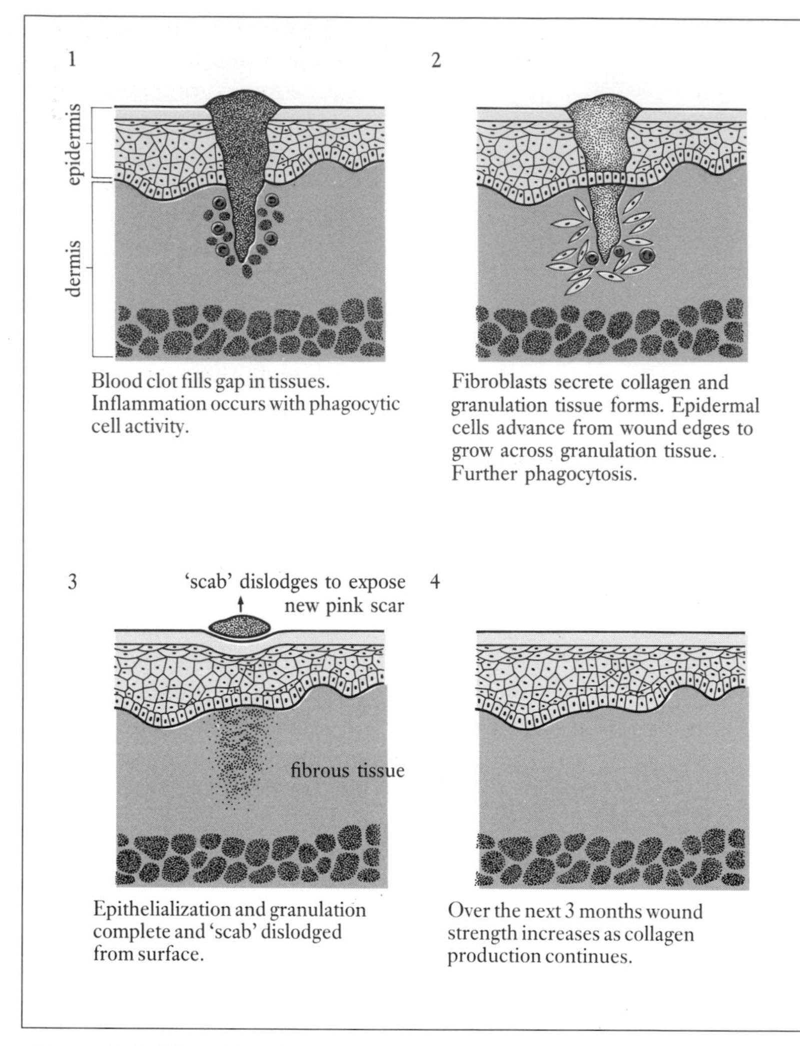

Figure 19.5. Wound healing – primary intention.

Stages of wound healing

The stages of healing are: *acute/traumatic inflammation* and *destructive phase* (days 0–6), *proliferative phase* (days 3–4) and *maturation* which can take months (Torrance, 1986; Westaby, 1985). The inflammatory and proliferative stages do, in fact, occur concurrently.

The major events of healing by first intention are:

- Blood clot fills the wound and the inflammatory response is initiated.
- Neutrophils and macrophages commence clearing clot and cellular debris.
- Macrophages stimulate the development of fibroblasts, which secrete collagen.
- **Granulation tissue**, formed from the collagen and newly formed

capillary network, fills the gap.

- Epidermal cells regenerate from the wound edges and grow across the granulation tissue by a process termed *epithelialization*. *NB* at this stage the wound is still very fragile and requires gentle handling.
- When epithelialization is complete the 'scab' is dislodged to expose the new pink scar tissue which gradually fades and becomes white. The wound repair continues to be strengthened by further synthesis of collagen and usually reaches maximum strength by about 3 months (Johnson, 1988).

Healing by secondary intention or granulation is basically the same but requires some modification:

- Infection must first be overcome by the inflammatory response.
- Necrotic tissue must be removed by the macrophages.
- Foreign bodies, e.g. splinter of wood, must be removed by natural mechanisms. When this proves ineffective it may be necessary to intervene surgically.
- The process of filling the cavity with granulation tissue, which takes much longer, commences on the floor of the cavity.
- Epithelialization will take longer.
- The resultant fibrous scar will be larger because wound edges were not in apposition.

Factors which delay healing

The following list includes some factors known to influence wound healing:

- Infection.
- Poor apposition of wound edges/tissue loss.
- Foreign bodies in the wound and the presence of necrosis.
- Repeated wound trauma and inappropriate dressings.
- Stress (physiological and psychological) and other reasons for high corticosteroid levels, e.g. Cushing's disease, 'steroid' therapy (see Chapter 8).
- Poor blood supply, e.g. in cardiovascular disease (see Chapter 10).
- Diabetes mellitus (see Chapter 8).
- Nutritional status (protein depletion, lack of vitamin C and zinc).
- Widespread malignancy and immune problems.

For many of these factors competent nursing practices can prevent or minimize the negative effects. Care strategies should also take account of those factors which fall within the responsibilities of the medical officer, such as diabetes. Holistic practice which includes high standards of aseptic technique, choice of appropriate dressings (see *Nursing Practice Application*) use of research-based wound care protocols, consideration of nutritional status with a multidisciplinary team (see Chapter 13) and minimizing stress with information and support, can contribute considerably to wound healing.

Nursing Practice Application Wound healing and dressings

Earlier we discussed the importance of holistic practice, where all requisites for wound healing are ensured, e.g. adequate nutrition and prevention of infection. Another major nursing responsibility is the provision of the correct microenvironment for wound healing, by the choice of appropriate dressings. The characteristics of an 'ideal' dressing, which provides such a microenvironment, have been identified (Turner, 1983, 1985): (i) thermal insulation; (ii) non-adherent, causing no trauma on removal; (iii) removes excess exudate and toxic substances; (iv) maintains high humidity; (v) allows gaseous exchange; (vi) impermeable to micro-organisms; (viii) leaves no contaminants in the wound.

NB any dressing must also be comfortable and acceptable to the patient.

Special Focus Pressure sores (decubitus ulcer)

Pressure sores cause endless suffering and use huge amounts of scarce resources. The development of a pressure sore means that the person will require expensive nursing time, dressings and drugs, and where the sore is extensive surgical *debridement* and skin grafting may be necessary. Pressure sores increase morbidity and in some situations may contribute to death, e.g. from infection.

A complete coverage of pressure sores is not appropriate here but a brief discussion is included.

Aetiology

Pressure sores develop when part of the body is subjected to unrelieved pressure which is sufficient to collapse the capillaries and disrupt the microcirculation. This happens when the skin and underlying tissues are squeezed between bone and a hard surface, e.g. operating table. Unrelieved pressure leads to tissue hypoxia, ischaemia and necrosis with inflammation and ulcer formation.

Shearing forces also disrupt the microcirculation when they cause the skin layers to move against one another. Shearing occurs when a person slips down the bed or is dragged instead of lifted properly. This type of tissue injury damages the deeper layers and results in a very extensive pressure sore.

Friction caused by continual rubbing of the skin leads to blisters, abrasions and superficial pressure sores. This type of tissue injury is exacerbated by the presence of moisture such as sweat or urine.

Assessment of risk

Initial risk assessment should be completed as soon as possible, and again at regular intervals whenever circumstances change. Financial considerations make this essential as prevention costs money and nursing interventions should be targeted at those people at risk. Prevention is obviously better in every respect and is a great deal cheaper than the management of an existing sore.

Risk scales are in common use, e.g. Norton *et al.* (1962), Waterlow (1985, 1988, 1991), but all may have limitations and should be adapted to individual circumstances. Spenceley (1988) thought that although risk calculators were valuable there was nothing to replace good nursing observations.

Factors which predispose to pressure sore formation

(i) Age over 65–70; (ii) immobility, e.g. due to pain; (iii) incontinence (see Chapters 13, 15); (iv) systemic infections, circulatory disorders, e.g. shock, heart failure, diabetes and malignancy; (v) altered consciousness, e.g. sedating drugs; (vi) sensory loss, e.g. paraplegia after spinal injury (see Chapter 4); (vii) dehydration (see Chapter 2) and or malnutrition, e.g. protein energy malnutrition (see Chapter 13), vitamin C, iron and zinc; (viii) damage sustained from wrinkled sheets or clothes, nurses' rings and watches, plasters and splints; (ix) oedema (see Chapter 10).

NB For some 'high risk' people there may be many factors operating together, e.g. they may be unconscious, immobile and incontinent following a stroke (see Chapter 4).

Factors important in prevention

(i) Relief of pressure, e.g. special mattresses and beds, use of 'turning clocks' to record regular turning (Lowthian, 1979); (ii) frequent assessment of high-risk areas, e.g. heels; (iii) avoiding skin trauma and shearing, e.g. correct lifting techniques (which has other benefits, see back injuries, Chapter 18); (iv) assessing nutritional status (for a full account of nutritional assessment methods see Goodinson, 1987a–d) and providing adequate hydration and nutrition; (v) planning care that minimizes the effects of incontinence, insomnia, confusion; (vi) skin care and hygiene (see page 500).

A considerable body of reseach-based knowledge is available regarding all aspects of wound healing and pressure sores, and readers are urged to make full use of this in their practice. Gould (1986) uses the prevention and treatment of pressure sores as an example of where nurses fail to implement research findings. She explored the reasons (proposed by Hunt, 1981) why nurses do not apply research findings: they do not know about them, do not understand them, do not believe them, do not know how to apply them or are not allowed to use them.

The further reading list includes just a few of the very wide selection of articles and books available, and reference is made to others within the text.

Specific Components - The Immune Response

Specific adaptive defences of the immune system form our last line of defence against foreign molecules such as pathogens which slip past the non-specific defences, as well as transplanted cells and abnormal cells arising in the body.

The components of the specific defences (*Figure 19.6*), which produce the immune response, are two lymphocyte populations – **B lymphocytes** (**humoral immunity** – part of the immune response involving the production of antibodies by plasma cells) and **T lymphocytes** (**cell-mediated immunity** – part of the immune response involving the action of specific lymphocytes which destroy abnormal foreign cells and release regulatory chemicals), **antibodies** and various chemicals. The ubiquitous macrophages which, although part of the innate non-specific defences, are also concerned with the activities of B and T lymphocytes.

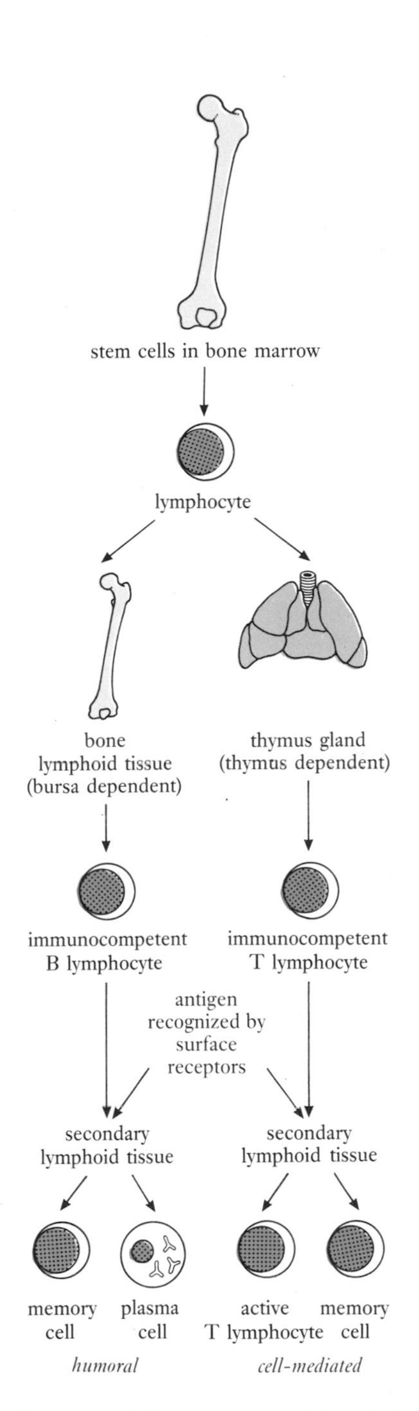

Figure 19.6. Development of B and T lymphocytes.

Co-operation between the specific defences confers **immunity** (resistance to disease) against a wide range of **antigens** (a molecule or part of a molecule which is recognized as foreign by the body defences and which therefore stimulates the immune response), by the recognition and subsequent neutralization or destruction of the foreign molecule. The correct functioning of the immune system, where the T and B lymphocytes have become **immunocompetent**, provides the body with considerable protection. There are numerous examples of what can happen when the immune system 'gets it wrong': **autoimmune** (abnormal immune response where body defences fail to recognize 'self', results in damage) disease, allergies, tumours and *immunodeficiency*, e.g. *acquired immune deficiency syndrome (AIDS)*.

The immune system is not fully developed at birth and neonates derive some passive protection from maternal antibodies, received via the placenta, until their own antibodies are produced and T and B lymphocyte immunocompetence is established. Antibodies conveyed in *colostrum* (see Chapter 20) and breast milk to the infant may have a role in preventing pathogens becoming established in the gastrointestinal tract.

At the other extreme the immune system in elderly people tends to degenerate which may help to explain why malignant and autoimmune diseases occur more commonly as we age.

A specific defence has characteristics which include:

- Antigen specificity, where the defence is mobilized by a specific antigen, e.g. a bacterium. However, it is possible for antibodies to act against similar antigens (see page 509).
- 'Memory' cells remember how to destroy a particular antigen so that when it is confronted on subsequent occassions the defence response is more vigorous.
- Specific defences operate throughout the body and in this respect are quite different to the rather parochial inflammatory response.

It is worth reminding ourselves just how much the non-specific and specific defences co-operate to enhance each other's activity, e.g. phagocytosis is enhanced by the presence of antibodies.

Humoral immunity

Humoral immunity is conferred by the presence of antibodies within the body fluids. Some lymphocytes, derived from the bone marrow stem cells, become the B lymphocytes (**bursa dependent**) which eventually produce antibodies (*Figure 19.6*). The B lymphocytes are so called because in birds they are processed by gut lymphoid tissue known as the *bursa of Fabricius*; in humans the equivalent processing probably occurs in the lymphoid tissue of the bone (see Chapters 9 and 11). When the B lymphocytes in the blood are exposed to an antigen which their surface receptors recognize, they move into the secondary lymphoid tissue where further changes occur. Now two distinct cell types emerge: **plasma cells** and **memory cells** (*Figure 19.7*). The transient plasma cells secrete the antibody which disarms the antigen, and the clone of memory cells, which 'remember' how to produce that particular antibody, circulate in the blood until required. The memory cells, which may persist for many years, retain the ability rapidly to produce the specific antibody if they are again challenged by the antigen. This explains the increased vigor of response to second and subsequent antigen exposures, and forms the basis of immunization (see page 510-512).

Antibodies (immunoglobins)

Antibodies are proteins secreted by the activated B lymphocytes in response to a specific antigen (also usually a protein), e.g. bacteria, bacterial toxin or virus particle. When the antigen is a bacterial toxin the antibody which neutralizes its effects is termed an **antitoxin**. Antibodies which circulate in the gamma globulin part of the plasma proteins (see Chapter 9) are also known as **immunoglobulins (Ig)** of which there are five classes: IgG, IgM, IgA, IgD and IgE.

Although they have the same basic structure – four polypeptide chains (2 heavy chains, 2 light chains) – they perform many different defensive roles in a variety of body sites (*Table 19.2*).

Table 19.2. Antibody/immunoglobulin classes.

Ig Class	*Site and function*
IgG (gamma,γ)	Commonest antibody found in plasma, also in other body fluids. Passes across the placenta to confer passive immunity on the fetus. Binds to antigens, enhances phagocytosis and fixes complement.
IgM (mu,μ)	Largest antibody, found only in the plasma. First Ig released in response to a new antigen. Concerned with agglutination and fixes complement. Does not cross the placenta.
IgA (alpha,α)	Present in the plasma, respiratory, and gastrointestinal tracts, and other secretions, e.g. milk and tears. Acts to protect the mucosal surfaces.
IgD (delta,δ)	Large Ig present only in the plasma. Role not entirely clear but may be involved with B lymphocyte activity.
IgE (epsilon,ε)	Small amounts present in the plasma and bound to basophils and tissue mast cells (see Chapter 9). Involved with histamine release from these cells in allergic reactions such as asthma.

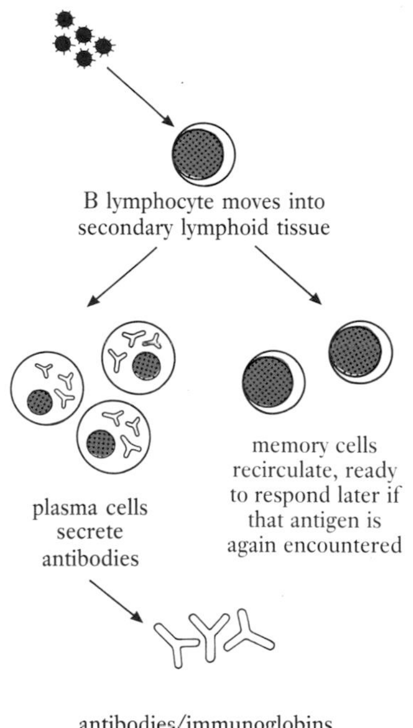

Figure 19.7. Summary of humoral immunity.

Antibodies function in different ways to inactivate antigens but all involve the formation of an **antibody–antigen complex** (**immune complex**) where the specific (or very similar) antibody and antigen interact together (*Figure 19.7*). A helpful analogy is that of a jigsaw puzzle, in which each piece is specifically shaped to fit a space; a very similar piece may also fit, but a differently shaped piece will never fit the space in the puzzle.

Once the complex has been formed it can be destroyed by *neutralization* of toxins, *complement fixation, precipitation* and *agglutination*. These immune-complex-driven processes enhance inflammation, phagocytosis and cell lysis which follow on to destroy and 'clear up'. It is worth noting that some of the reactions, such as agglutination, may adversely affect the body, e.g. agglutination of donor red cells by IgM during a mismatched blood transfusion (see Chapter 9).

Many of these immunoglobulin reactions are used in diagnostic tests, and include: blood grouping, testing for certain infections and pregnancy. A recent development is the use of identical **monoclonal antibodies**, which are specific to a single antigen, for research and diagnosis of diverse conditions including some early cancers, but most interesting of all is their use in cancer treatment. Monoclonal antibodies can be joined to cytotoxic drugs which, when introduced into the body, 'target' the specific antigen (cancer cells) and facilitate their destruction by the cytotoxic drug. This area of work is at an early stage but the possibilities for effective cancer treatment without side-effects are very promising; this is especially so where the monoclonal antibody can also be made to stimulate body defences.

Immunity and immunization

As we have already mentioned, the presence of antibodies confers humoral immunity against many microbial diseases. During the first exposure to an antigen there is a lag phase of 2–3 weeks before antibody production (**primary response**) reaches a protective level. If we survive this first exposure the B lymphocyte memory cells can produce the antibody much more quickly (**secondary response**) with a very much reduced lag phase, if that antigen is again encountered. This exposure to the antigen, e.g. having the disease is a form of **active immunity** in which the immune system is naturally stimulated to produce antibodies.

Active immunity, which is generally of a long duration, can also be produced artificially by injecting *vaccines*. The vaccines contain inactivated organisms or *attenuated* (weakened/changed strain) organisms or toxins which retain sufficient *antigenicity* to stimulate antibody production without the *pathogenicity* which would cause the disease. Many vaccines provide protection by active immunity, including those against: tetanus, diphtheria, pertussis (whooping cough), poliomyelitis, mumps, morbilli (measles) and rubella (German measles).

The type of vaccine and its level of antigenicity will determine the number of doses required: vaccines containing inactivated organisms, such as pertussis, are administered on three occassions, whereas those vaccines with live attenuated organisms need only be given once, e.g. measles.

The other type of immunity is **passive immunity**, which results from the transfer of antibodies from one person or animal to another. This type of immunity is short-lived because the antibodies 'on loan' are eventually destroyed and, without antigenic stimulation, the immune system does not 'learn' to produce that particular antibody. Passive immunity occurs naturally, e.g. when maternal immunoglobulins are transfered to the fetus via the placenta.

Artificial passive immunity can be provided by injecting various immunoglobulins obtained from immune humans or **antisera** (serum, obtained from animals, containing antibodies to a specific antigen) from animals (but see note below). There are several situations where the provision of temporary but immediate immunity is appropriate: for non-immunized individuals during an epidemic, protection during short trips abroad, following contact with the organism and where an individual is immunosuppressed. The injection of gamma globulin, specific immunoglobulins and antitoxins give protection against diseases, which include hepatitis A and B, rabies, tetanus and botulism.

NB Antiserum, which causes allergic reactions, has been replaced with immunoglobulins.

The prevention of rhesus (Rh) blood group incompatibility and haemolytic disease of the newborn is another area which utilizes passive immunity. Anti-D immunoglobulin (containing antibodies [AntiD] to rhesus Rh-positive blood) administered to rhesus Rh-negative women within 72 hours of delivery, miscarriage or termination stops natural antibody production by destroying any fetal Rh-positive red cells which have entered the maternal circulation. Without this injection of anti-D, the woman will produce her own anti-D and, next time she is pregnant with an Rh-postive fetus, the memory cells produce anti-D antibodies which cross the placenta and harm the fetus (see Chapter 9).

NB There is no danger to future pregnancies because the injected anti-D is short-lived.

Immunization information

Nurses may be required to answer basic questions and give information regarding immunization programmes during their work. Varied questions from new parents, people worried about contact with an infectious disease or someone planning a trip abroad, may all be encountered.

Table 19.3. Basic immunization programme: birth to leaving school.

Immunization	***Timing***	***Remarks***
Diphtheria Tetanus Pertussis (Whooping cough)	2, 3 and 4 months	Injection (DPT) Pertussis can be excluded if parents wish or medical contraindications exist
Haemophilus influenzae type b (Hib) (causes a type of meningitis and other serious conditions)	2, 3 and 4 month	Injection (introduced Oct 1992)
Polio	2, 3 and 4 months	Oral
Measles Mumps Rubella	12–18 months	Injection (MMR introduced 1988)
Preschool booster: Diphtheria Tetanus Polio	3–5 years	Injection + oral
Rubella	10–14 years	All girls; injection
BCG (Bacillus of Calmette–Guerin) tuberculosis protection	13 years	For tuberculin-negative children
Boosters: Tetanus Polio	school leaving 15-19 years	Injection Oral

Slight variations may exist where individual needs differ. Hib vaccine which is offered at 2, 3 and 4 months will also be available for children aged 5 months to 4 years during 1992–93.

A basic immunization programme is outlined in Table 19.3, but sources of detailed information include: health visitors, child health and school health clinics, health centres/family doctors and health promotion units. Where travel is involved, occupational health departments in companies where employees are required to travel, Department of Health leaflets obtained direct or from travel agents and the appropriate embassy can all help.

Many special groups such as health care workers and people travelling abroad require an individualized programme, e.g. health care workers should be offered hepatitis B protection.

Cell-mediated immunity

Cell-mediated immunity is facilitated by the activation of T lymphocytes (*Figure 19.6* and *Figure 19.8*). Some lymphocytes derived from the bone marrow stem cells become the T lymphocytes, which eventually form specific *killer cells*. The T lymphocyte population is processed by the *thymus gland* in the mediastinum (see Chapters 8 and 9) and is said to be **thymus-dependent**. The thymus does most of the T lymphocyte processing before birth and during early childhood, which explains the gradual reduction in thymus size throughout adult life.

When the circulating T lymphocytes are exposed to an antigen which their surface receptors recognize, they move into the secondary lymphoid tissue where further changes occur. The distinct cell types which emerge are: **killer cells**, which are short lived, **delayed hypersensitivity cells**, **helper** and **suppressor T cells** and **memory cells**, which continue to divide.

The killer cells attach themselves to the specific antigen and destroy it. Here the entire T lymphocyte acts much like an antibody by 'locking on' to the antigen. The memory cells continue to circulate between the blood and lymphatics until called upon to 'remember' that particular antigen, should it again threaten the body.

In normal circumstances the T lymphocytes 'learn' to differentiate between 'self' and 'non-self': they do not react to the surface antigens they recognize as 'self'. However, if this ability is impaired the T lymphocytes fail to recognize parts of the body as 'self' and their attack on body cells results in a form of autoimmune disease (see pages 517 and 518), e.g. *autoimmune thyroiditis*. Cell surface antigens (see transplants pages 513 and 514) are coded for by genes of the **major histocompatibility complex** (**MHC**), also called **human leucocyte antigen** (**HLA**) because they were first described on leucocytes.

T lymphocytes

The several types of T lymphocyte can be classified functionally as:

- **Cytotoxic** or **killer cells**, which directly destroy cells carrying antigens to which they are sensitized. These T lymphocytes also provide a mobile immune surveillance as they continually search for malignant cells.

- **Delayed hypersensitivity cells**, involved in the response to chronic inflammation and cell-mediated delayed hypersensitivities (see pages 516 and 517).
- *Regulatory cells*, **T suppressor cells** which, as their name suggests, stop or slow T and B lymphocyte activity once immune responses have dealt successfully with the antigen. In this way they act to limit the immune response to an appropriate level and are thought to have a role in preventing autoimmunity. **T helper cells** stimulate the proliferation of killer cells and B lymphocytes. The helper cells are vital to the immune response.

Loss of balance between suppressor/helper cells may result in abnormally intense immune responses or an immunodeficiency such as that caused when T helper cells are destroyed by HIV infection (see *Special Focus, page 519*). Over-activity of T suppressor cells may be linked to this and other types of *immunodeficiency* (see page 518).

T lymphocyte cell-mediated immunity protects the body against antigens, which include virus-infected cells, malignant cells and transplanted foreign cells. The destruction of these antigens may be by direct attack and lysis of its cell membrane, or the release of chemicals (**lymphokines**) which stimulate macrophages, other lymphocytes and the inflammatory response.

The lymphokines produced by activated T lymphocytes are a group of non-specific chemicals which include interferons, which are also made by virus-infected cells, *interleukins, macrophage activating factor (MAF), migration inhibitory factor (MIF), lymphotoxin (LT), chemotactic factors* and various helper and suppressor chemicals. In addition the macrophages also produce certain lymphokines which enhance the immune response.

Organ transplants and rejection

Organ transplant is now a common life-saving measure for people with serious kidney, heart, lung or liver disease. The degree of success, however, depends upon finding a 'good match' between recipient and donor and, where necessary, T lymphocyte immune response suppression to prevent *graft* rejection. Body cells have genetically determined surface antigens, e.g. ABO blood groups on erythrocytes (see Chapter 9) and other cells such as leucocytes have antigens coded by the MHC/HLA gene cluster on chromosome 6. The closer the match between MHC antigens the greater is the chance of a successful transplant. In a study of 26 patients following heart transplant, 18 were alive and well at 10–33 months, with an overall success rate of 69% (Mulcahy *et al.*, 1988). First-time kidney transplants from cadaveric donation are even more successful, with 80% functioning for at least a year (Lancet, 1990).

The very best 'match' possible is an **autograft**, where tissue is transplanted from one site to another in the same person, e.g. a skin graft. Here the MHC antigens are identical and, in the absence of

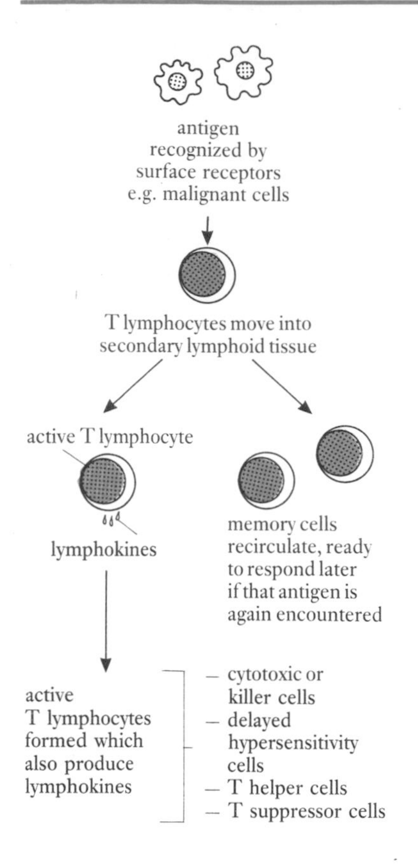

Figure 19.8. Summary of cell-mediated immunity.

other problems, e.g. infection, the graft is likely to be successful. This will also be the case where an **isograft** between identical twins is used, again the genetic make-up and hence the MHC antigens are identical.

Most of us, however, don't have an identical twin, and the best MHC match between non-identical individuals, called an **allograft** (**homograft**), offers the best chance. Allografts are usually cadaver donations (see brain death criteria, Chapter 4) and, more rarely, when a living person donates a kidney to a close relative (genetically similar) suffering from end-stage renal failure (Chapter 15). A survey by Donnelly *et al.* (1989) found that the numbers of living donors varied between 0 and 25% depending on the centre, they concluded that these numbers could be increased to provide much-needed kidneys.

To reduce the risk of allograft rejection, a close MHC match, determined by tissue typing, is highly desirable, and obviously there must be ABO compatibility. However, even when tissue typing reveals a close match between donor and recipient MHC antigens it will be necessary to suppress the immune response, especially the activity of T lymphocytes and macrophages, which would recognize and destroy the 'foreign' cells of the graft. Organ transplants may also fail because antibody-mediated reactions cause thrombus production and other vascular problems in the donated organ. Various measures are taken to prevent rejection:

- *Drugs*
  - Azathioprine (cytotoxic immunosuppressant).
  - Corticosteroids, which suppress the immune response.
  - Antilymphocyte serum (immunoglobulin from horses).
  - Cyclosporin, which prevents rejection by the T lymphocytes, is also used to prevent *graft versus host (GVH) disease* which follows bone marrow transplant where the donor T lymphocytes become sensitized to the host, which they regard as foreign.
- *Radiation* may be used to produce immunosuppression.

An important side-effect of immunosuppression is bone marrow depression with a reduction in the leucocytes which would normally defend against infection. This is overcome by giving 'just enough' immunosuppressive therapy to prevent rejection, administration of antimicrobial drugs and very careful monitoring for signs of infection.

The last type of graft, a **xenograft** (**heterograft**) between different species, is fraught with immunological problems and is not used at present. However there is research being undertaken in this area.

MHC (HLA) antigens and disease

The presence of certain MHC antigens appears to make that individual more susceptible to diseases as diverse as insulin-dependent diabetes, autoimmune Addison's disease (see Chapter 8), some types of rheumatoid arthritis (see Chapter 18), multiple sclerosis (see Chapter 3) and myasthenia gravis (see Chapter 17). Some of these conditions have an autoimmune or allergic basis but considerable work remains to be done in this field.

Abnormal Immune Responses

We have given considerable thought to the protective role of the immune response, but sometimes the reaction to an antigen can cause harm. The same highly versatile immune system that carries out cancer surveillance and produces immunoglobulins is also responsible for the 'over the top' anaphylactic response to bee stings or the production of autoantibodies which destroy your own erythrocytes.

At the other extreme there may be an immunodeficient state, where a much-needed immune response does not occur, leaving the individual exposed and vulnerable to antigen attack.

A brief mention of some abnormal responses is appropriate here but readers requiring more details should consult Further Reading, e.g. Roitt *et al.*, 1985.

Hypersensitivity (allergy)

A hypersensitivity reaction to a particular **allergen** (antigen, which produces an allergic reaction) may be classified by its timing and whether it involves antibodies or is cell mediated.

Immediate (within minutes or hours of exposure)

Antibody-mediated type I reaction (*anaphylaxis*)

An **antibody-mediated type I** reaction occurs when IgE binds to an allergen such as dust or pollen. This causes mast cells and basophils to release chemicals, which include histamine, 5-hydroxytryptamine, kinins, prostaglandins and other lipids (see also inflammatory response, pages 502). The chemicals produce vasodilation, vessel permeability and smooth muscle contraction, which may be local or systemic. An **anaphylatic reaction** causes oedema and other problems specific to the site – skin redness, diarrhoea, excess mucus with nasal discharge and bronchoconstriction. Anaphylactic reactions, in common with normal immune responses, are more intense following a second exposure to the allergen.

Abnormalities caused by an anaphylactic reaction include: drug, food and bee sting sensitivities, hay fever and allergic asthma. Treatment involves avoiding the allergen where possible, antihistamine drugs and desensitization programmes, where increasing doses of the allergen are injected over a period of time to stimulate an immune response thought to prevent IgE–allergen binding. Some families, where the individuals are described as *atopic,* show a definite propensity to anaphylactic reactions such as eczema, hay fever and asthma.

Rarely, a very severe systemic reaction known as *anaphylactic shock* occurs. This is due to widespread chemical release following the second exposure of the susceptible individual to the allergen.

Often the allergen enters the circulation by injection, such as an insect sting or drug administration. Anaphylactic shock may follow exposure to bee stings or to drugs, e.g. penicillin. It is characterized by bronchospasm, laryngeal oedema, hypovolaemia and shock. The immediate life-saving management includes adrenaline by injection, provision of an airway, e.g. tracheostomy (see Chapter 12) and antihistamines, e.g. chlorpheniramine.

Antibody-mediated type II reaction (cytotoxic)

An **antibody-mediated type II** reaction involves IgG or IgM, the activation of complement, lysis and subsequent phagocytosis of the cell-bound antigen. An example of a cytotoxic allergic reaction is seen following a mismatched (ABO) blood transfusion, where donor red cells are agglutinated (see Chapter 9). A similar reaction occurs in rhesus incompatibility as Rh-positive fetal red cells are destroyed by maternal anti-D immunoglobulin (see pages 510-511).

Antibody-mediated type III reaction (immune complex)

An **antibody-mediated type III** reaction involves antibodies and the activation of complement. Immune complexes are lodged within the tissues or vessel walls with effects that vary according to severity and site involved.

There are various forms of immune complex hypersensitivity which include localized reactions, such as farmer's lung (*extrinsic allergic alveolitis*), which is caused by inhaling an antigen in mouldy hay. Immune complexes may cause a more local *Arthus* reaction with swelling, inflammation and necrosis at the point of antigen entry by injection. In this case the antigen may be found within a drug, e.g. insulin.

A more widespread immune complex hypersensitivity called *serum sickness* may follow injection of a foreign antigen such as bacteria, e.g. streptococcal infections, viruses and drugs, e.g. penicillin. The immune complexes are lodged in various sites: heart (myocarditis or endocarditis), kidneys (glomerulonephritis) and joints (arthritis), in addition there may be local swelling, enlarged lymph nodes and *urticarial rashes* (nettle rash or hives).

Immune complex hypersensitivity may be linked to some autoimmune diseases in which the antibodies contributing to the immune complexes are autoantibodies which act against body cells.

Delayed (24–48 hours after exposure)

Cell-mediated type IV reaction

Cell-mediated type IV hypersensitivity, which involves T lymphocyte activity, lymphokine release and macrophages, occurs 24–48 hours after exposure to the antigen. There is chronic inflammation and a tissue reaction which may endure for weeks. Examples of commonly implicated antigens are the bacteria responsible for tuberculosis and syphilis, intracellular malarial parasites, metals (such as nickel), cosmetics and plants, e.g. poison ivy. Delayed hypersensitivity affecting the skin is called *allergic* *contact dermatitis*. The inflammatory effects of delayed hypersensitivity, which are due to lymphokines, may be treated with corticosteroids as anti-histamines are ineffective.

An example of type IV hypersensitivity is the local reaction seen after skin testing (e.g. with the Heaf or Mantoux tests) with tuberculin antigens to determine the need for BCG immunization.

Mixed antibody/cell-mediated type V reaction

Mixed reactions involving antibodies, T lymphocytes and phagocytes are implicated in responses to some viruses and a type of autoimmune hyperthyroidism (Graves disease, see Chapter 8), where autoantibodies stimulate hormone production.

Autoimmune disorders

A large number of autoimmune conditions exist which range from organ specific conditions, e.g. autoimmune Addison's disease (see Chapter 8) to the more generalized such as rheumatoid arthritis (see Chapter 18) and *systemic lupus erythematosus.*

Autoimmune conditions result from changes in the normal immunological tolerance which we have to our own cells. The changes may occur spontaneously or be due to some extrinsic factor such as drugs or micro-organisms.

Various mechanisms by which autoimmunity develops have been proposed and include:

- Exposure of body proteins not previously in contact with the immune system, e.g. spermatozoa, which are not produced until long after immunocompetence has been completed. The development of autoantibodies to spermatozoa, which may occur after inflammation of the testes, is a cause of male infertility. The proteins of the lens of the eye are also normally isolated from the immune system and if the two should meet, e.g. after an eye injury, the resultant autoantibodies can damage the uninjured eye.
- An extrinsic agent, e.g. micro-organism or drug, may alter the 'self' antigens in such a way that the immune system regards them as foreign. Numerous examples include exposure to viruses, which are linked with the development of type I diabetes and multiple sclerosis, and the antihypertensive drug methyldopa, which is associated with immune haemolytic anaemia. 'Self' antigens may also be altered by their combination with antibodies formed against bacterial infection e.g. with streptococci. This explains the glomerulonephritis (see Chapter 15) caused by immune complex deposition following a streptococcal sore throat.
- There may be changes in suppressor T lymphocytes, with age or again caused by micro-organisms and drugs, which prevents their recognition of 'self' and suppression of T and B lymphocytes.

The actual damage of autoimmunity is caused by autoantibodies, immune complexes or T lymphocytes sensitized against body cells, or a combined antibody/cell-mediated effort. By now you will have realized, even after our brief discussion, that many links exist between normal immune responses, hypersensitivities and autoimmune reactions.

Immunodeficiency

The large number of immunodeficiency states reflects the complexity of a body defence system which involves so many innate and adaptive components. Classification is difficult and readers should consult Further Reading for detailed descriptions.

For our purposes it will be sufficient to describe briefly the major immunodeficiencies which may be present at birth (congenital) or acquired after birth:

Congenital defects (which are genetic)

- Lack of complement.
- *Agammaglobulinaemia* where the individual is unable to produce immunoglobulins or has a deficiency of B lymphocytes.
- T lymphocyte deficiency and thymic defects.
- Absence of both B and T lymphocytes – *severe combined immunodeficiency* (*SCID*).
- Abnormal or decreased phagocytes.

Acquired defects

- Phagocyte deficiency results from bone marrow depression which may be caused by malignant disease or drugs.
- Inadequate complement when general protein synthesis is reduced.
- *Hypogammaglobulinaemia* (reduced immunoglobulins) due to protein deficiency or loss, e.g. malnutrition, kidney disease.
- Deficiency of lymphocytes caused by radiation, immunosuppressive or cytotoxic therapy, *Hodgkin's disease* (see Chapter 11), corticosteroid therapy, infections, e.g. AIDS (see page 519).
- Abnormal immunoglobulins produced at the expense of the normal range, e.g. multiple *myeloma*.

Whatever the cause of the immunodeficiency the major problem is infection due to an increased vulnerability to pathogens and organisms which may normally cause little harm. The management of immunodeficient states depends on individual circumstances but may include provision of a clean or even sterile environment, administration of immunoglobulins and antimicrobial drugs, possible bone marrow transplant, meticulous hygiene, careful monitoring for signs of infection and, where possible, measures to reduce the feelings of isolation and loneliness.

Special Focus Human immunodeficiency virus (HIV)/acquired immune deficiency syndrome (AIDS)

AIDS is one of the most serious and potentially catastrophic health problems to emerge this century. It was first diagnosed over a decade ago on the west coast of the US but is known to have existed in Africa for much longer. Infection with HIV, a retrovirus, depresses the T lymphocyte immune response by reducing T helper cell numbers. A reduction in interleukin production contributes to a reduced number of T helper cells, macrophages and natural killer cells. In addition there is an increase in T suppressor lymphocytes and B lymphocyte activity.

As you can imagine, this leaves the individual open to attack by opportunistic micro-organisms, e.g. fungal infections, meningitis and a protozoal pneumonia caused by *Pneumocystis carinii*, which usually occurs only after immuno-suppression. Another problem linked to the breakdown in immune processes is the development of the otherwise rare *Kaposi's tumour* which affects blood vessels.

There is a great deal yet to learn about HIV infection. It has a very variable incubation period which may be as long as 10 years and the virus is present in the blood, semen, vaginal secretions and to a limited extent in other body fluids such as saliva and tears. The state of having HIV antibodies in the blood (*seropositive*), which does not confer immunity, means that the person is capable of passing on the virus although they may still be asymptomatic. This idea has caused some confusion for both public and health professionals. Searle (1987) found that 38% of senior doctors, dentists and nurses at sister level did not know what a positive HIV serotest meant.

AIDS, which is characterized by weight loss, lymphadenopathy, diarrhoea, night sweats and repeated infection may be preceded by *AIDS-related complex (ARC)* which is similar to but less severe than fully developed AIDS.

Throughout the 1980s AIDS infection was most commonly associated with 'high risk' activities, such as sharing needles or syringes and some types of sexual behaviour, e.g. anal intercourse. However, individuals were also infected by contaminated blood or blood products, e.g. those suffering from haemophilia (see Chapter 9) or via the placenta, where the virus passes from mother to fetus. Rarely, infection has occurred, following 'needlestick' accidents, in health workers. The risk of HIV infection from 'needlestick' injury, where the patient is HIV positive, has been estimated to be 0.25% (McKeown, 1992).

There are differences in infection transmission worldwide, with marked regional variations within the UK. Current patterns of infection in the UK appear to indicate that more children are now infected with HIV and that people are being exposed to the virus while travelling abroad.

The considerable prejudice and ignorance concerning AIDS is unfortunate, because many people still think that the infection has 'nothing to do with them'. In a study of student attitudes, Cornish (1989) found that only 29% (sample 202) thought that heterosexuals were at risk from AIDS.

Another commonly held, but erroneous, view is that you can 'catch' AIDS from normal social contact, such as shaking hands. This is certainly untrue, as AIDS is spread by sharing needles or syringes, 'high-risk' sex (which includes vaginal sex with an infected person), exposure to infected body fluids, blood, or blood products, and via the placenta.

There is as yet no cure for AIDS, which at present is almost always fatal, but there are drugs which may be used to prolong and improve quality of life, e.g. antimicrobial drugs, zidovudine (which used to be known as *azidothymidine (AZT)*), *acyclovir, ganciclovir, foscarnet sodium* and the *interferons*.

Until effective treatment or immunization is available the most important aspects of AIDS management and containment are health staff education at all levels (Searle, 1987) and health education programmes which inform all groups about high risk activities. Other initiatives, such as heat- treated factor VIII, screening blood donors and needle exchange schemes for drug users, are all helpful, but the major task is to convince people that they are at risk (Cornish, 1989) and help them to take responsibility for and modify their behaviour. The important health message is that 'all sexually active individuals have the potential to be infected and to infect others' (Kuykendall, 1992).

Various studies have considered the risks of HIV infection in people who inject drugs. In 1989 Hart *et al.*, studied the risk behaviours in drug users in London; most were sexually active with one-third having experienced between 2–20 sexual partners and most had shared injection equipment. Skidmore *et al.* (1990) found that many of the drug users studied in Edinburgh had modified their intake of drugs, but the changing patterns of HIV spread suggests that heterosexual transmission will be a future problem.

Health education aimed at reducing the spread of AIDS includes: (i) practice 'safer sex', e.g. use of condoms, limiting sexual partners and consider alternatives to 'high risk' activities such as anal intercourse; (ii) injecting drug users should not share needles/syringes and used equipment should be disposed of safely; (iii) it is safest not to share razors, toothbrushes and other equipment which may be contaminated with blood, e.g. earpiercing equipment.

NB There is no risk of being infected with AIDS when donating blood for transfusion; only sterile, single-use equipment is used.

Developments in immunology and clinical applications

Many exciting developments can be expected as immunological research and clinical application continues to expand. Specific areas include: immunization, AIDS control, organ transplant/control of rejection, contraception, treatment of some types of infertility, cancer diagnosis and treatment, autoimmune mechanisms and MHC/HLA disease susceptibility.

SUMMARY/CHECK LIST

Introduction

Non-specific defences skin and membrane barriers – Skin structure, epidermis, dermis, sweat glands, sebaceous glands, hair and follicles, nails. Skin pigmentation, *Nursing Practice Application* – skin colour/condition. Skin functions, waterproofing, temperature regulation, *Nursing Practice Application* – recording temperature, problems with temperature homeostasis – fever, heat stroke, hypothermia, *Person-centred Study* – Mavis, sensation, protection, *Nursing Practice Application* - skin hygiene, synthesis of vitamin D–exposure to UV light, excretion, storage. Other innate membrane barriers – gastrointestinal tract, genitourinary tract, eyes, respiratory tract.

Non-specific components – chemicals and cells – Phagocytes, Phagocytosis, Natural killer cells, Inflammatory response, Interferons, Complement, Wound healing, *Nursing Practice Application* – wound healing and dressings, *Special Focus*– pressure sores.

Specific components – the immune response – Humoral immunity – B lymphocytes. Antibodies/ immunoglobulins. Immunity/ immunization. Cell-mediated immunity – T lymphocytes. Organ transplants and rejection. MHC and disease.

Abnormal immune responses - Hypersensitivities (allergies). Autoimmune conditions. Immunodeficiency, *Special Focus* – HIV and AIDS.

?

SELF TEST

1 Put the following in their correct pairs:
(a) sebum (b) eccrine gland
(c) ceruminous gland (d) sebaceous gland
(e) germinative layer (f) wax
(g) sweat (h) epidermis

2 Describe the heat loss mechanisms of the skin which operate to maintain body temperature homeostasis on a hot day.

3 Describe how the following features of the membrane barriers protect the body: lysozyme, cilia, normal flora, mucus.

4 Describe the events of acute inflammation and explain why it is characterized by: redness, heat, swelling, pain, possible loss of function.

5 Which of the following statements are true?
(a) Phagocytes, natural killer cells and complement are part of the non-specific innate defences.
(b) Clean surgical incisions usually heal by primary intention.
(c) A wound repair reaches maximum strength at about 14 days.
(d) circulatory disorders such as shock predispose to the formation of pressure sores.

6 What is meant by antigen specificity and 'memory' in relation to specific adaptive defences?

7 Complete the following statements:
(a) Humoral immunity is conferred by the presence of __________.
(b) B lymphocytes are said to be _____ dependent.
(c) Antibodies are also known as ________________.
(d) Antibodies function in different ways to destroy antigens but all involve the formation of an ________/_______ _______.

8 Explain active and passive immunity and give examples of how each may be achieved.

9 Complete the following statements:
(a) Activation of T lymphocytes confers ____ _________ immunity.
(b) T lymphocytes are ______ dependent.
(c) The four types of T lymphocyte are __________, ______, __________ and _______ ________________.
(d) Chemicals such as interleukin and lymphotoxin are known collectively as ___________.

10 What information would you give to the following people:
(a) Vicky, a health care assistant, asks how HIV causes immunodeficiency.
(b) During a home visit you and the health visitor are asked about the risks of working in an office with a person who is HIV positive.
(c) Your friend says that he will not donate any more blood because of the AIDS risk.
(d) Paul, who injected drugs for a short while 5 years ago and has never had an AIDS blood test, asks if he and his girlfriend should use condoms.

Answers
1. a – d, b – g, c – f, e – h. 2. See pages 494-496. 3. See pages 500, 501. 4. See pages 502, 503. 5. a, b, d. 6. See page 508. 7. a. antibodies, b. bursa, c. immunoglobulins, d. antibody– antigen complex 8. See pages 510-511. 9. a. cell-mediated, b. thymus, c. cytotoxic, helper, suppressor, delayed hypersensitivity cells, d. lymphokines. 10. See page 519.

References

Armstrong-Esther, C.A. (1981) Skin introduction. *Nursing,* (1st Series), 1115.

Closs, J. (1987) Oral temperature measurement. *Nursing Times*, **83** (1), 36–39.

Cornish, M. (1989) Students' attitudes towards AIDS. *Nursing Times*, **85** (25), 62–63.

Donnelly, P.K., Clayton, D.G. and Simpson, A. R. (1989) Transplants from living donors in the United Kingdom and Ireland: a centre study. *British Medical Journal*, **298**, 490–493.

Goodinson, S.M. (1987a) Assessment of nutritional status. *The Professional Nurse*, **2** (11), 367–369.

Goodinson, S.M. (1987b) Anthropometric assessment of nutritional status. *The Professional Nurse*, **2** (12), 388–393.

Goodinson, S.M. (1987c) Biochemical assessment of nutritional status. *The Professional Nurse*, **3** (1), 8–12.

Goodinson, S.M. (1987d) Assessing nutritional status: subjective methods. *The Professional Nurse,* **3** (2), 48–51.

Gould, D. (1986) Pressure sore prevention and treatment: an example of nurses' failure to implement research findings. *Journal of Advanced Nursing,* **11**, 389–394.

Hart, G.J., Sonnex, C., Petherick, A., Johnson, A.M., Feinmann, C., and Adler, M.W. (1989) Risk behaviours for HIV infection among injecting drug users attending a drug dependency clinic. *British Medical Journal,* **298**, 1081–1083.

Hunt, J. (1981) Indicators for nursing practice: the use of research findings. *Journal of Advanced Nursing,* **6** (3), 189–194.

Johnson, A. (1988) Natural healing processes: An essential update. *The Professional Nurse,* **3** (5), 149–152.

Kuykendall, J. (1992) Families and HIV – rebels at risk. *Nursing Times,* **88** (5), 26–28.

Lancet (1990) Organ donors in the United Kingdom – getting the numbers right. (Editorial). *Lancet,* **335** 80–82.

Lowthian, P. (1979) Turning clocks system to prevent pressure sores. *Nursing Mirror,* **148** (21), 30–31.

McKeown, M. (1992) Sharpening awareness. *Nursing Times,* **88** (14), 66–68.

Mulcahy, D., Wright, C., Mockus, L., Yacoub, M., and Fox, K. (1988) Cardiac transplantation in severely ill patients requiring intensive support in hospital. *British Medical Journal,* **296**, 817–819.

Nichols, G. A. and Kucha, D. H. (1972) Taking adult temperatures: oral measurements. American Journal of Nursing, **72**, 1091–1092.

Norton, D., McLaren, R., and Exton-Smith, A. (1962) *An Investigation of Geriatric Nursing Problems in Hospital.* Edinburgh: Churchill Livingstone.

Pembroke, A.C. (1983) Preventing skin problems. *Geriatric Medicine,* **13** (11), 797–781.

Searle, E.S. (1987) Knowledge, attitudes and behaviour of health professionals in relation to AIDS. *Lancet,* **i**, 26–28.

Sims-Williams, A.J. (1976) Temperature taking with glass thermometers: a review. *Journal of Advanced Nursing,* **1** (6), 481-493.

Spenceley, P. (1988) Norton v Waterlow. *Nursing Times,* **84** (32), 52–53.

Skidmore, C.A., Robertson, J.R., Robertson, A.A., and Elton, R.A. (1990) After the epidemic: a follow up study of HIV seroprevalance and changing patterns of drug use. *British Medical Journal,* **300**, 219–223.

Torrance, C. (1986) Physiology of wound healing. *Nursing,* **3** (5), 162–168

Turner, T.D. (1983) Absorbents and wound dressings. *Nursing,* **2**, 12 (suppl.), 1–7.

Turner, T.D. (1985) Semiocclusive and occlusive dressings. In: Ryan, T. J. (Ed.) *An environment for healing: the role of occlusion.* Royal Society of Medicine – International Congress and

Symposium Series No 88 (held 1984).
Waterlow, J. (1985) A risk assessment card. *Nursing Times*, **81** (48), 49–55.
Waterlow, J. (1988) The Waterlow card for the prevention and management of pressure sores: towards a pocket policy. *Care –Science and Practice*, **6** (1), 8–12.
Waterlow, J. (1991) A policy that protects: The Waterlow pressure sore prevention/treatment policy. *Professional Nurse*, **6** (5), 258–264.
Westaby, S. (1985) (Ed.) *Wound Care*. London: Heinemann Medical.

Further Reading

Adler, M. W. (Editor) (1987) *ABC of AIDS*. London: British Medical Journal Publications.
Barton, A. and Barton, M. (1981) *The Management and Prevention of Pressure Sores*. London: Faber and Faber.
Buckley, E.G., Donald, A.G., Drury, V.W.M., Pereira Gray, D.J., Hill, P., Murfin, D., and Robertson, J.R. (1988) Human immunodeficiency virus infection and the acquired immune deficiency syndrome in general practice. Discussion paper of the working party of the Royal College of General Practitioners. *Journal of the Royal College of General Practitioners*. **38** (310), 219–225.
Crow, R. David, J.A. Cooper, E.J. (1981) Pressure sores and their prevention. *Nursing* (1st Series), 1139–1142.
Crow, R. (1988) The challenge of pressure sores. *Nursing Times*, **84** (38), 68–78 (literature review).
David, J. A. (1986) *Wound Management: A Comprehensive Guide to Dressing and Healing*. London: Martin Dunitz.
Fry, L. (1984) *Dermatology: An Illustrated Guide*, 3rd edn. London: Butterworth–Update.
HEA (1992) *Immunization: The Safest Way to Protect Your Child.* London: Health Education Authority
HEA (1992) *Protect Your Child with the New Hib Immunization.* London: Health Education Authority.
Help the Aged (Winter of action on cold homes) (1988) *The Cold Facts – A Programme for Action.* London: Help the Aged.
Hillman, H. (1987) Hypothermia: The cold that kills. *Nursing Times*, **83** (4), 19–20.
Hutchinson, A. (1988) Rubella prevention – a new era? *Journal of the Royal College of General Practitioners*, **38** (310), 193 - 194.
Marson, S. (Ed.) (1989) *AIDS: Meeting the challenge*. London: English National Board for Nursing, Health Visiting and Midwifery.
Nursing 85 Books. Various contributors. *Nurses' Clinical Library – Immune Disorders*. Springhouse, PA: Springhouse Corporation.
Pratt, R.J. (1988) *AIDS: A Strategy for Nursing Care*, 2nd edn. London: Edward Arnold.
Roitt, I.M., Brostoff, J. and Male, D. (1985) *Immunology*. London: Gower Medical Publishing.

Royal College of Nursing (1986) *AIDS -- Nursing Guidelines*. Second report of the RCN AIDS working party. London: RCN.
Scott, P. (1990) *National Aids Manual, Vol I – Topics, Vol II – Directory*, 2nd edn (revised). London: NAM Publications. (An extremely good source of information and help.)
Solomons, B. (1983) *Lecture Notes on Dermatology*, 5th edn. Oxford: Blackwell Scientific Publications.
Thompson, J. (1989) Fact sheet – the cold that kills. *Nursing Times*, **85** (46), (Community Outlook) 21–25.
Torrance, C. (1983) *Pressure Sore: Aetiology, Treatment and Prevention*. Beckenham: Croom Helm.
Wilson-Barnett, J. and Batehup, L. (1988) *Patient problems: A research base for nursing care* (Ch 11 Problems with wound healing.) London: Scutari Press.

Useful Addreses

Terrence Higgins Trust
52–54 Grays Inn Rd
London
(Tel: 071 242 1010)

National AIDS Helpline
(Tel: 0800 567 123)
(Calls are free)

section eight

ensuring continuity

CHAPTER TWENTY

REPRODUCTION

OVERVIEW

- Structure and function of the male and female reproductive structures including the breast.
- Conception, pregnancy, parturition and lactation.

LEARNING OUTCOMES

After studying Chapter 20 you should be able to:

- Describe the structures of the male reproductive system, ducts and accessory glands.
- Describe meiosis and outline the process of spermatogenesis.
- Discuss the hormonal control of male reproductive function.
- Describe pubertal changes occurring in the male.
- Describe how male reproductive function changes with normal ageing.
- Describe the structures of the female reproductive system.
- Outline the structure of the mammary glands.
- Describe oogenesis and the ovarian cycle.
- Describe hormonal control of the ovarian cycle.
- Describe the menstrual (uterine) cycle.
- Describe pubertal changes occurring in the female.
- Describe the climacteric and discuss the use of hormone replacement therapy (HRT).
- Discuss the sexual response in both male and female.
- Describe the sexually transmitted diseases: syphilis, gonorrhoea, chlamydia, genital herpes, trichomoniasis and candidiasis. Discuss how they might affect the reproductive system.
- Discuss methods of contraception.
- Describe fertilization and the determination of genetic sex.
- Outline the events of early embryonic development, implantation and formation of the placenta.
- State the ways in which the fetal circulation is adapted for uterine life.

Learning Outcomes (continued)

- Describe important maternal changes occurring during pregnancy and discuss the major hormones involved with pregnancy.
- Outline the initiation of labour and describe its three stages.
- Describe the changes occurring at birth by which the infant adjusts to its extrauterine environment.
- Describe the process and control of lactation.

Key Words

Autosome – a chromosome which is not a sex chromosome. In human somatic (body) cells there are 44 autosomes arranged in 22 pairs. The gametes contain 22 autosomes.
Climacteric – the period of time during which changes in the female reproductive tract result in the decline and eventual cessation of reproductive function.
Diploid (2n) – a cell which has a full set of paired chromosomes (46 arranged in 23 pairs; in humans n = 23 therefore 2n = 46), seen in all cells except the gametes. There are 44 autosomes and 2 sex chromosomes.
Embryo – the early developmental stage which lasts from fertilization until the end of the eighth week of gestation.
Fertilization – union of the spermatozoon (male) and ovum (female) nuclei to form the diploid (2n) zygote or embryo.
Fetus – the developmental stage from the ninth week of gestation until birth.
Gamete – the haploid (n) reproductive cells, ova and spermatozoa.
Gametogenesis – the process by which the ova and spermatozoa are formed in the ovary (ova) or testis (spermatozoa).
Gonad – the primary reproductive structure, ovary (female) and testis (male).
Gonadotrophins – hormones produced by the pituitary gland which control gonad functioning, e.g. FSH, LH.
Haploid (n) – a cell which has a set of unpaired chromosomes (23 only), seen only in the gametes following meiosis. There are 22 autosomes and 1 sex chromosomes.
Meiosis – a complex reduction division by which an immature diploid reproductive cell eventually produces four haploid cells. The process also allows for genetic variability of the gametes.
Menarche – the commencement of menstruation. An event occurring as part of puberty.
Menopause – the cessation of menstruation. An event occurring during the climacteric.
Oogenesis – the formation of ova (female gamete) by the ovary.
Ovulation – the release of the secondary oocyte from the surface of the ovary.
Puberty – the period during which the reproductive structures become functional and the secondary sexual characteristics appear.
Semen – fluid produced at ejaculation. It contains spermatozoa and fluid from the male accessory glands, e.g. prostate gland.
Sex chromosomes – the chromosomes (X and Y) which determine genetic sex. One sex chromosome is inherited from each parent. An individual has a pair of sex chromosomes – XX in females and XY in males.
Spermatogenesis – the formation of spermatozoa (male gamete) by the seminiferous tubules in the testis.
Zygote – the fertilized ovum following fusion of the female and male nuclei.

INTRODUCTION

Each chapter so far has dealt with functions that are vital to homeostasis and life. However, in this chapter we will discuss reproductive function, which, while being essential to species continuity, has nothing to do individual survival. The reproductive organs do not even function until **puberty** (the period during which reproductive structures become functional and the secondary sexual characteristics appear) and in the female cease to function during mid-life.

Chapter 20 looks at male and female reproductive function and includes the incredible events of conception, development and birth, and some problems associated with reproduction. It is important to remember that reproductive functioning is much more than the workings of just another physiological system: the complexity of human sexual behaviour goes far beyond the basic need to 'continue the line'. The business of expressing sexuality, which includes reproduction, is subject to considerable environmental and sociocultural influence as well as physiological controls. For even the most superficial understanding it is essential to take a holistic view which includes a consideration of the ways in which an individual expresses sexuality. Another feature is that, for the most part, sexual behaviour is not intended by the participants to lead to reproduction. In this humans differ from other species.

MALE REPRODUCTIVE SYSTEM

The function of the male reproductive system is the production of male **gametes** (haploid (*n)* reproductive cells) and the transfer of these **spermatozoa** to the female during sexual intercourse. However, research during the latter part of the 20th century has led to developments which allow for fertilization without sexual intercourse, e.g. *in vitro fertilization (IVF).*

Structure of the male reproductive system

The male reproductive system (*Figure 20.1*) consists of: **testes** (two) in the **scrotum**, **epididymis** (two), **vas deferens** (two), **ejaculatory ducts** (two), **seminal vesicles** (two), **prostate gland**, **bulbo-urethral glands** (two) and the **urethra** as it passes through the *penis*.

NB The **scrotum** and **penis** are known as the **external genitalia**.

Testes

The male **gonads** (primary reproductive structures) or **testes** produce the spermatozoa and act as endocrine organs (see Chapter 8) by secreting the male hormone **testosterone** which is discussed further on page 539. Each testis is around 4.5 cm in length and 2.5 cm in diameter (*Figure 20.2*).

During fetal development the testes, which start off high in the abdomen, descend with their **spermatic cord** through the right and left inguinal canals (see Chapter 18) and are present in the scrotum by the eighth month of pregnancy. Occasionally the testes fail to

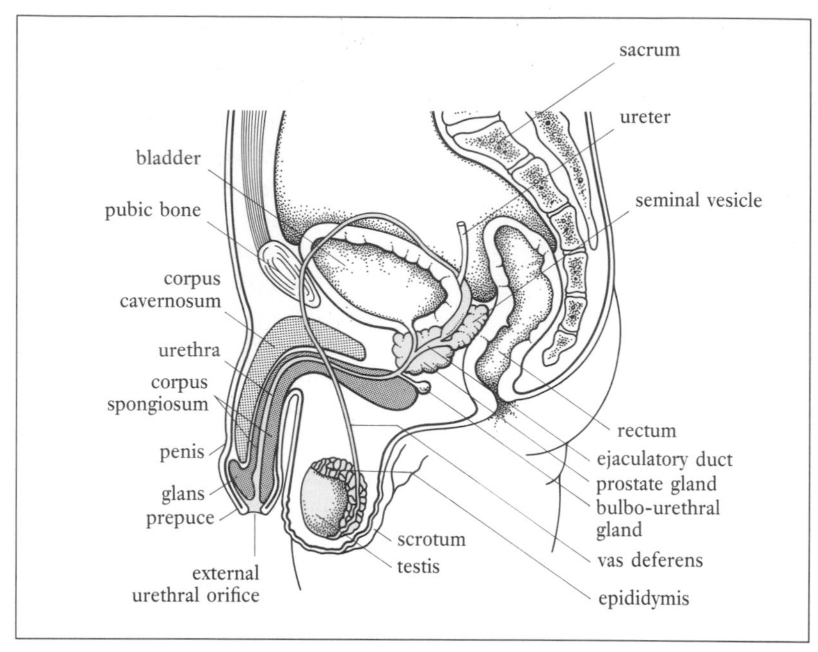

Figure 20.1. Male reproductive structures (midsagittal section).

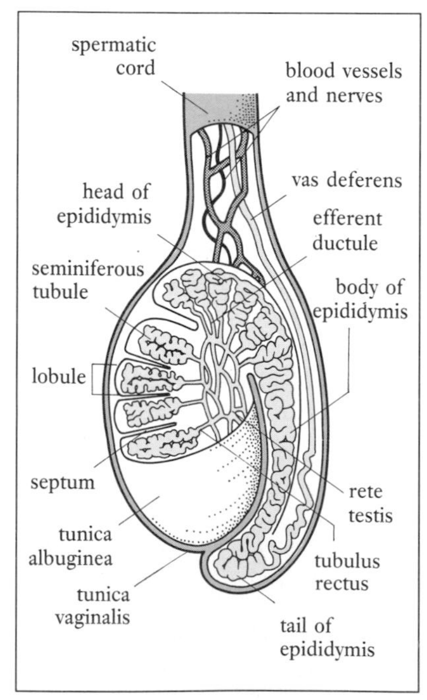

Figure 20.2. Testis.

descend. If both testes fail to descend and this is not corrected, it could lead to sterility because the testes need to be 2–3°C cooler than body temperature. Testes which remain in the abdomen are more likely to become malignant, another reason for surgical correction.

Normally the oval testes, suspended by their spermatic cords within the scrotum, are outside the body cavity. The scrotum is a bag of pigmented skin and fascia (fibrous connective tissue), which hangs between the thighs anterior to the rectum. It may appear to be somewhat risky having the structures which confer maleness in such a vulnerable position, but this adaptation keeps the testes at the slightly lower temperature (2–3°C below body temperature) required for the production of healthy spermatozoa. To maintain a constant temperature the position of the scrotum can be changed by muscle contraction. This occurs in response to temperature changes, e.g. heat causes the scrotum to hang loosely away from the body but it moves closer to the body when cold.

Each testis, which occupies a separate compartment (formed from fascia) within the scrotum, has two coverings – a double serous layer, the **tunica vaginalis**, and the inner fibrous **tunica albuginea**, which forms the septa dividing the testis into 200–300 wedge-shaped **lobules**. Individual lobules contain the convoluted **seminiferous tubules**, the site of **spermatogenesis** (formation of spermatozoa), and **Leydig (interstitial) cells** which secrete androgen hormones. The seminiferous tubules from the lobules form **tubuli recti** which converge at the **rete testis** from where the spermatozoa enter the **efferent ductules** and the duct system of the **epididymis** (see below).

The spermatic cord, which passes through the inguinal canal, encloses blood vessels, lymphatics and autonomic nerves of both divisions travelling to and from the testes and the **vas deferens** (**ductus deferens**), which carries the spermatozoa. Arterial blood supply to the testes is via the *testicular arteries,* which branch directly from the aorta (see Chapter 10) and venous blood leaves in the *pampiniform plexus* of veins. Occasionally the spermatic cord becomes twisted (*testicular torsion*), which results in damage to the testis if treatment is delayed.

Testicular tumours

Testicular tumours are comparatively rare and represent between 1 and 2% of male malignancies. Although this is a small number each year it is worth remembering that it commonly affects young men 20–40 years and early diagnosis increases the chance of successful treatment.

A study by Thornhill *et al.* (1986), involving 500 men aged 21–65, revealed that 32% did not know about testicular tumours, 92% did not know about self-examination and 90% wanted more information.

Regular self-examination of the testes can be helpful in detecting changes and nurses can encourage men to undertake this simple screening test. Any alterations such as a lump, general enlargement or a testis that 'feels' different, although they may not be due to malignancy, should be reported to their family doctor as soon as possible.

Some men may find the topic of screening for conditions which directly affect sexuality distressing or embarrassing, but nurses can be sensitive to individual needs when giving information about testicular tumours and self-examination. The leaflets published by the Health Education Authority (HEA, 1989) and other visual aids can be used to overcome these difficulties.

Duct system

The convoluted **epididymis**, a tube some 6 m in length, is coiled on the posterior part of each testes (see *Figure 20.2*). Immature spermatozoa leaving the testes are conveyed into the epididymis by the action of cilia and muscle contraction. The non-motile spermatozoa take about 3 weeks to travel through the epididymis, during which time they become fully mature and motile.

The epididymis is continuous with the **vas deferens**, which is around 45 cm in length and travels in the spermatic cord to enter the pelvic cavity. The vas deferens runs anterior to the pubic bone, over the ureter and behind the bladder where it dilates to form the **ampulla**. The vas deferens contains a muscle layer capable of the rapid contraction which conveys mature spermatozoa into the urethra during *ejaculation*.

Male sterilization or *vasectomy* involves removal of a small portion from each vas deferens, which ensures that they are no longer continuous. It should be considered to be permanent although reversal is possible in some cases. Vasectomy, which is usually performed under local anaesthetic as an outpatient, should have no effect other than sterility. It is important that couples are informed that sterility is not immediate as spermatozoa already in the duct system may remain viable for some weeks. An alternative method of contraception is required for about 2–3 months until all spermatozoa have been ejaculated. The actual time depends upon the frequency of ejaculation and usually samples of ejaculate are examined to confirm the absence of spermatozoa before contraception is abandoned. Spermatozoa which continue to be produced are broken down and absorbed within the seminiferous tubules.

A duct from a seminal vesicle (an accessory gland; there are two of these) opens into the ampulla of each vas deferens, which is then known as the **ejaculatory duct**. The two ejaculatory ducts merge with the urethra as it passes through the prostate gland.

The male **urethra**, which is some 20 cm in length, transports both urine (see Chapter 15) and semen. It is divided into three parts: *prostatic*, *membranous* and *penile* urethra, which opens at the *external urethral orifice* and forms around 75% of urethral length (see penis, page 533).

The Accessory glands

Seminal vesicles

The **seminal vesicles** are tubular glands with muscular walls (see *Figure 20.1*). They are lined with secretory epithelium which produces a viscous alkaline fluid. This secretion which contains fructose, amino acids, ascorbic acid and prostaglandins forms about 60% of the volume of **semen** (fluid produced at ejaculation).

Prostate gland

The first part of the male urethra is surrounded by the **prostate gland** (*Figure 20.3*). This walnut-sized structure consists of several glands, fibrous tissue and smooth muscle, all enclosed within a capsule. Prostatic secretion, which accounts for around 30% of the total volume, is added to semen during ejaculation. It is a thin milky fluid, containing chemicals such as calcium and enzymes, which may assist spermatozoa motility and modify vaginal acidity.

Enlargement of the prostate gland is extremely common from mid-life onwards. The enlargement, which may be caused by benign hypertrophy (see Chapter 1) or malignant changes, will obstruct the urethra and affect micturition in a variety of ways. The man may complain of hesitancy, dribbling, poor stream, nocturia and retention with overflow incontinence (see Chapter 15). Prostatic problems usually affect men over 60, but may be seen earlier. Other manifestations are urinary tract infection and renal damage due to urinary stasis or acute retention.

The management of benign hypertrophy involves relief of retention, if present, by catheterization prior to surgery. In many instances the prostatic enlargement is removed by *transurethral resection (TURP)*. Following transurethral resection the ability to achieve erection should not be affected but ejaculation may be altered (volume of ejaculate is less and ejaculation may be

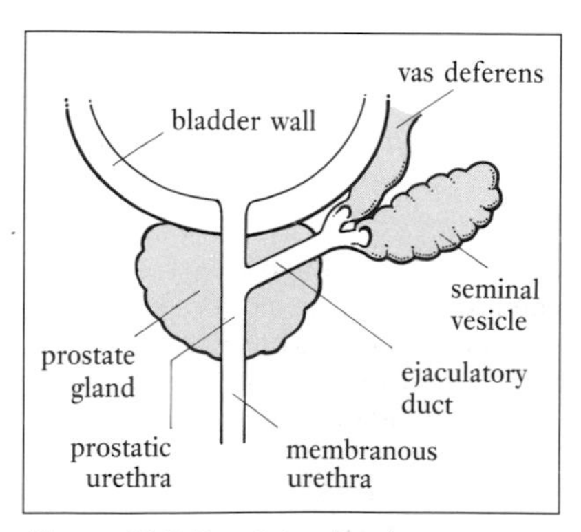

Figure 20.3. Prostate gland.

retrograde into the bladder).

Prostatic malignancy, which spreads locally in the pelvis and to nearby bone, e.g. pelvis and vertebrae, is increasingly common with age. These tumours are hormone-dependent and the management usually involves some modification to androgen levels: removal of the testes (*orchidectomy*) or the administration of synthetic oestrogens. These treatments, which may cause feminization, e.g. breast development and impotence with oestrogens, will cause considerable distress to many men and their families. Very careful explanation and support will be required for individuals needing this treatment. Unfortunately prostatic tumours are not usually amenable to surgery.

Bulbo-urethral glands (Cowper's glands)

The two tiny **bulbo-urethral glands** are located inferior to the prostate (see *Figures 20.1* and *20.4*). These pea-sized glands secrete a small amount of lubricating mucus into the penile urethra before ejaculation.

Penis

The **penis**, through which the spongy (penile) urethra runs, is the male organ of *copulation.* The penis has a *root* embedded in the perineum (see Chapter 18) and a *body/shaft* which terminates at the expanded **glans penis** (*Figure 20.4*). The glans is normally covered with a loose double fold of skin known as the **prepuce** (foreskin).

The penis consists of three columns of erectile tissue containing vascular spaces, connective tissue and involuntary muscle. There are two lateral columns, the **corpora cavernosa** and the ventral **corpus spongiosum** surrounding the urethra. Normally the penis is flaccid but during sexual excitement the vascular spaces fill with blood which causes *erection* as the penis becomes enlarged and rigid (see page 563). The penis is well supplied with arteries, veins and autonomic parasympathetic nerves which run in its dorsal aspect.

Functioning of the male reproductive structures

In order to understand **spermatogenesis** or **oogenesis** (formation of ova) it is necessary for us to first consider the process of meiosis. Readers will also find it helpful to revise the events of DNA replication and mitosis (Chapter 1) so that the two processes may be compared.

Nursing Practice Application Phimosis and paraphimosis

Sometimes an abnormally tight prepuce cannot be drawn back over the glans penis, a condition known as *phimosis*. This is treated by *circumcision* (excision of the prepuce) to prevent hygiene problems where secretions collecting under the prepuce cause inflammation of the glans penis (*balanitis*).

The opposite problem of *paraphimosis* occurs when the retracted prepuce cannot be drawn back over the glans penis, which swells. Paraphimosis may occur when the prepuce is not drawn over the glans after catheterization or where a tight prepuce has been forcibly retracted.

Meiosis

Meiosis is the term given to the complex nuclear divisions by which a **diploid (2*n*)** (cell with a full set of paired chromosomes – 46 arranged in 23 pairs) cell forms four **haploid (*n*)** (cell with a set of unpaired chromosomes – 23 in total) **gametes**. This process ensures that the resultant gametes have only half the normal chromosome number (23) and that random genetic variation occurs. When the male and female gametes fuse at **fertilization** the diploid chromosome number of 46 is again restored.

NB Meiosis, which is divided into stages, is preceded by **interphase** and replication of DNA which doubles the amount of **chromatin** in the nucleus.

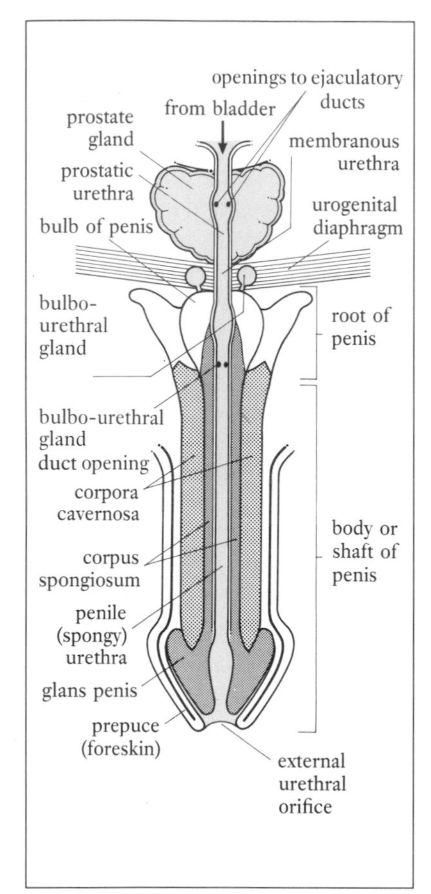

Figure 20.4. Penis.

Stages of meiosis I (Figure 20.5)

- *Prophase I.* In meiosis this stage takes longer than in mitosis. During prophase I the chromosomes coil, condense and become visible. In a process unique to meiosis there is pairing or **synapsis** of **homologous chromosomes** to form a **bivalent**, consisting of two chromosomes. The paired chromosomes coil around each other. Towards the end of prophase I the chromosome pairs undergo incomplete separation and each chromosome can be seen to have two **chromatids**; now each unit consists of four strands and is **quadrivalent** (two bivalents). At this stage several '*crossover*' points or **chiasmata** can be seen between the chromatids. These 'crossover' points are the means by which genetic material is exchanged between the chromosomes of a homologous pair to produce gametes of infinite genetic variability. Separation of the two bivalents continues and the nucleolus and nuclear membrane disappear ready for the next stage.

  NB The process of oogenesis, which commences in fetal life, is arrested during prophase I and does not resume until puberty (see pages 552-554).
- *Metaphase I.* This stage sees the appearance of the spindle apparatus consisting of contractile proteins. The bivalents position themselves on the spindle equator in such a way as to ensure that each new cell receives only one chromosome from a homologous pair.
- *Anaphase I.* One chromosome, consisting of two chromatids, from each homologous pair migrates to opposite poles of the cell. Here there is no separation of the chromatid strands at the **centromere** as there is during mitosis. Each pole of the cell will contain the haploid (*n*) number of 23 chromosomes.
- *Telophase I.* During this stage nuclear membranes form around each haploid set of chromosomes.

Cytokinesis, which follows telophase, forms two haploid 'daughter' cells.

To summarize: the events of meiosis I produces two haploid (*n*) cells which are genetically quite unlike the original cell.

NB There follows a second interphase, but in this instance no DNA replication occurs prior to the start of meiosis II.

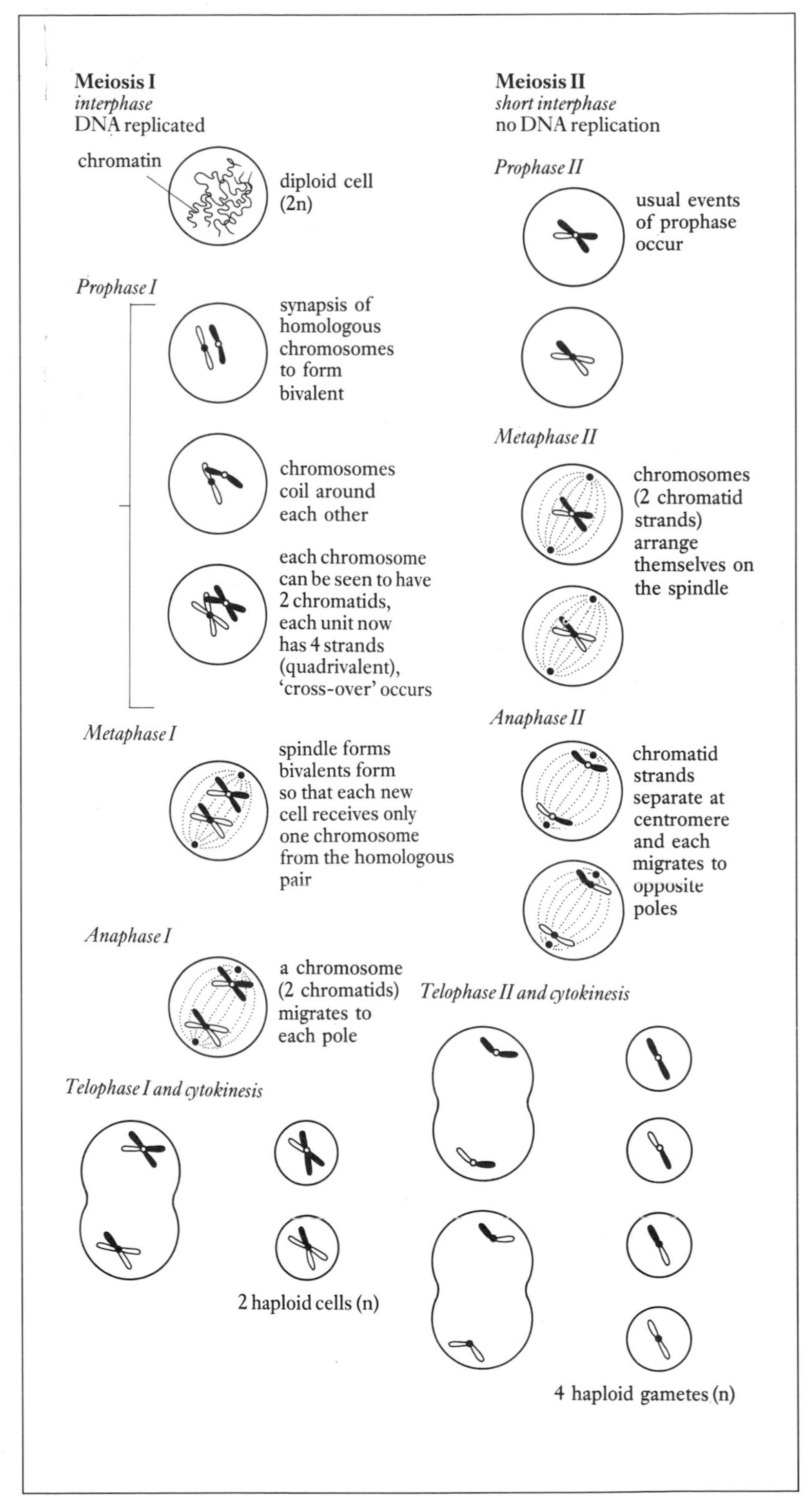

Figure 20.5. Stages of meiosis (events are shown in a hypothetical cell with only one pair of chromosomes).

Stages of meiosis II (Figure 20.5)
Meiosis II, which commences with two haploid cells, follows the same four stages as mitosis. During prophase II the events already described occur but with only 23 chromosomes, not homologous pairs. During metaphase II the chromosomes, which consist of two chromatid strands, arrange themselves on the spindle equator. This time, however, the chromatid strands separate at the centromere, as they do in mitosis, and during anaphase II the daughter chromatid strands from each chromosome migrate to opposite poles of the cell. The events of telophase II and cytokinesis result in each of the two haploid 'daughter' cells becoming two haploid cells. In this way four haploid gametes with a mix of genetic material are formed.
To summarize:

One diploid cell, $(2n)$ $\xrightarrow{\text{Meiosis I}}$ Two haploid cells, $(n) + (n)$

Two haploid cells, $(n) + (n)$ $\xrightarrow{\text{Meiosis II}}$ Four haploid gametes, $(n) + (n) + (n) + (n)$

Spermatogenesis

Spermatogenesis, which occurs within the seminiferous tubules of the testes, is a continuous process from puberty (see pages 539–540). Situated in the seminiferous tubules of an adult testis are two cell types: **germ cells** and **Sertoli cells**. The germ cells are all at different stages of development and will eventually become spermatozoa; the supporting Sertoli cells may have a role in nourishing the immature germ cells and provision of the *blood-testis barrier* (see autoimmunity, Chapter 19).

Diploid stem cells known as **spermatogonia** produce germ cells by continuous mitosis (*Figure 20.6*). Prior to puberty this mitotic division results in the production of identical spermatogonia, but following puberty some cells formed will differentiate into **primary spermatocytes**. The first meiotic division involves the primary spermatocytes which divide to form two haploid **secondary spermatocytes**. The secondary spermatocytes then undergo the second meiotic division which results in four haploid cells known as **spermatids**.

Each spermatid still needs considerable modification before it becomes a highly specialized motile **spermatozoa** with a head, midpiece and tail (*Figure 20.7*). These modifications, which occur in the seminiferous tubules and the duct system, include changes to the nucleus, loss of excess cytoplasm and the formation of a tail. The genetic material of the spermatid nucleus is condensed to form the head of the spermatozoon, which is topped by an **acrosome** formed from the Golgi region (see Chapter 1). Lytic enzymes present in the acrosome allow the spermatozoa to pass through the cervical mucus and ultimately to penetrate the ovum.

The midpiece contains many mitochondria (see Chapter 1),

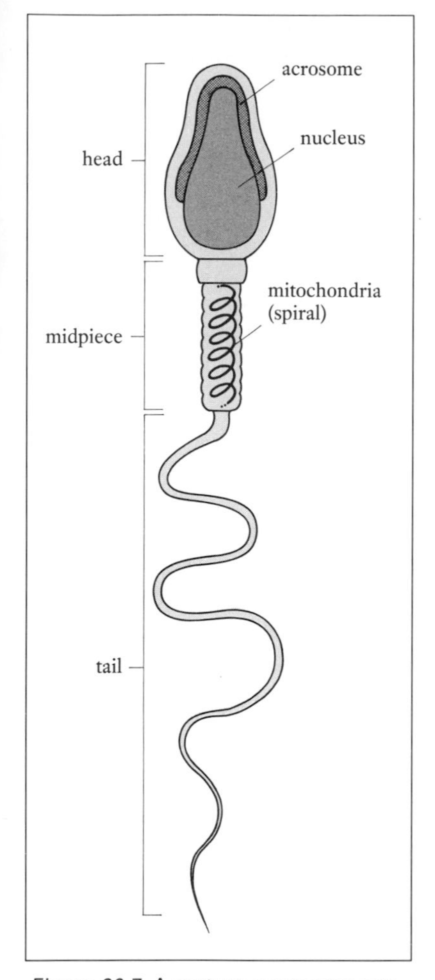

Figure 20.7. A mature spermatozoon.

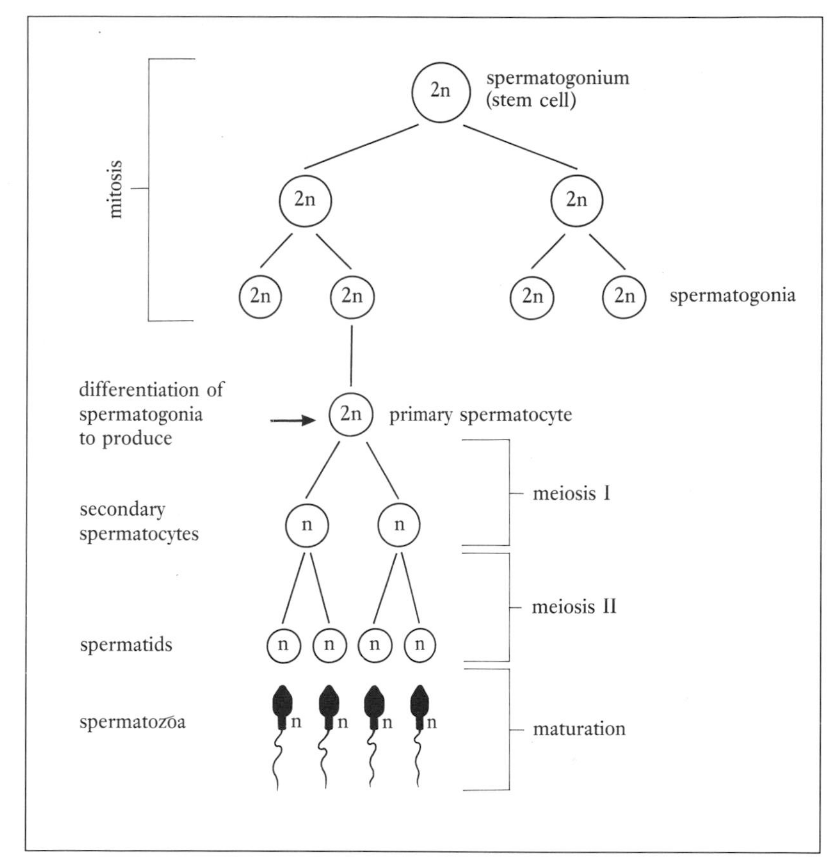

Figure 20.6. Spermatogenesis: $2n$ = diploid; n = haploid.

arranged in a spiral, which produce the ATP required to power the whip-like movements of the tail. The tail, which has a typical flagellum configuration of two central microtubules surrounded by a further nine pairs (2+9), is formed from one of the centrioles present in the spermatid (see Chapter 1).

The process of spermatogenesis takes about 70-75 days but further maturation occurs within the epididymis where the spermatozoa become motile and fully fertile. Many millions of spermatozoa are produced each day and those not ejaculated are liquefied and reabsorbed.

Semen

Semen is the viscous white fluid consisting of motile spermatozoa and the secretions of the accessory sex glands. The fluid part of semen provides a vehicle for the spermatozoa and contains nutrients and chemicals necessary for spermatozoa survival.

Semen is slightly alkaline (pH 7.2–7.6) which helps to counteract the vaginal acidity – an important property as spermatozoa need an alkaline environment for motility. Nutrients such as fructose are supplied by the secretions from the seminal vesicles. The presence of hormones and prostaglandins in semen facilitates the movement of spermatozoa through the female reproductive tract and enzymes enhance their ability to penetrate the cervical mucus.

The volume of semen produced at each ejaculation is normally 2–6 ml and contains 50–150 million spermatozoa/ml; this volume and number of spermatozoa decreases as the frequency of ejaculation increases. During investigations for infertility (see Further Reading, e.g. Winston, 1986, for details) the numbers (sperm count), motility and morphology (shape) of the spermatozoa are determined, as are the pH, chemical content and volume of the semen.

Hormones and control of male reproductive function

Spermatogenesis and the release of testicular androgens are controlled by hormones produced by the hypothalamus, anterior pituitary and by the testes (see Chapter 8 and *Figure 20.8*).

At puberty a **gonadotrophin-releasing hormone** (**GnRH**) released by the hypothalamus stimulates the anterior pituitary to produce two **gonadotrophins** (hormones controlling gonad function): **follicle stimulating hormone** (**FSH**) and **luteinizing hormone** (**LH**) which is also known as **interstitial cell stimulating hormone** (**ICSH**) in the male.

FSH acts upon the seminiferous tubule cells to stimulate spermatogenesis, a process which also requires some LH. The major role of LH, however, is to stimulate the Leydig cells surrounding the

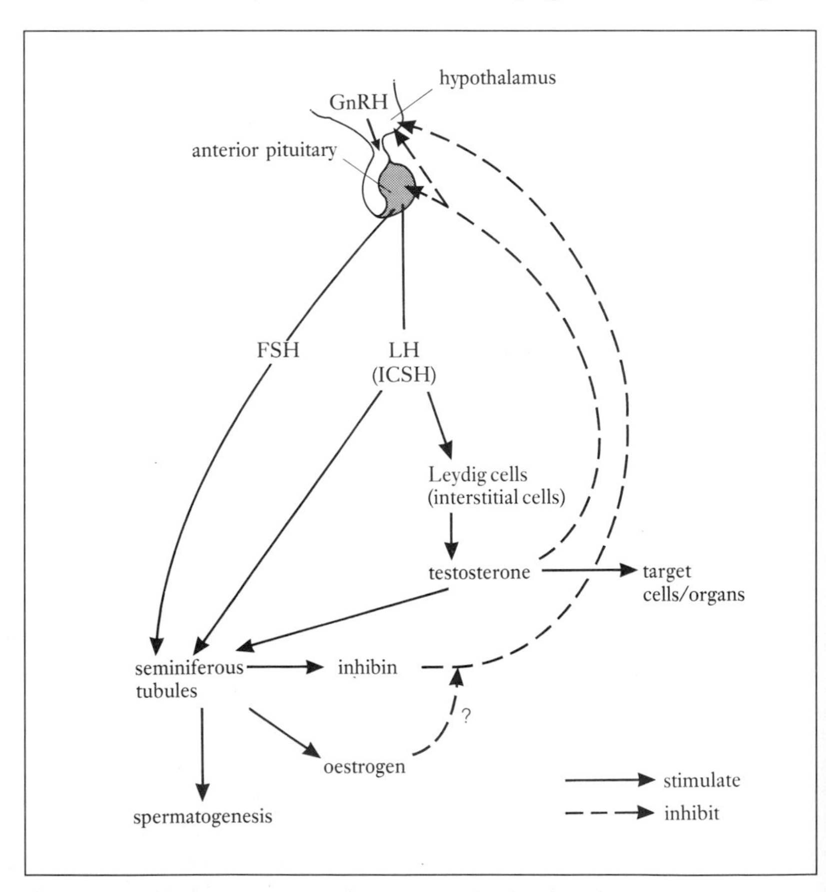

Figure 20.8. Hormone control of male reproductive function.

seminiferous tubules to produce testosterone, an androgen steroid hormone, which in turn stimulates spermatogenesis. Testosterone circulates in the blood and shows effects elsewhere in the body, which we will consider separately. Paradoxically LH also stimulates the testes to produce small amounts of **oestrogens** (female steroid hormone) which may have some regulatory role.

In common with many hormones the gonadotrophins are subject to *negative feedback/inhibition.* High levels of testosterone act either to inhibit the release of the hypothalamic and or pituitary hormones. Also involved in the inhibition of FSH and LH release is the protein hormone, **inhibin**, which is thought to be produced by the testicular Sertoli cells in response to the spermatozoa count.

Following puberty the adult hormonal patterns and level of spermatogenesis remain fairly constant. They do not, as in the adult female, follow cyclical patterns. Hormonal control of male reproductive function is dependent upon the relationships between hypothalamic releasing hormone, FSH and LH from the anterior pituitary, and testicular testosterone and inhibin.

Testosterone

Testosterone is the major androgen hormone, it has a *steroid structure* and is derived from cholesterol. Most testosterone originates from the testes but small amounts are produced by the adrenal glands in both sexes. Apart from its effects upon spermatogenesis, testosterone exerts widespread anabolic (see Chapter 13) effects by stimulating protein synthesis in both reproductive and somatic tissue. Testosterone circulates in the blood to reach its many target cells where it influences *libido* (sex drive) in both sexes, and is responsible for the development of the male secondary sexual characteristics at puberty and the continued correct functioning of the accessory reproductive structures. A drastic fall in testosterone production, e.g. bilateral orchidectomy or testicular insufficiency, causes spermatogenesis to decline and atrophy of the reproductive structures resulting in impotence and loss of fertility.

Sex differentiation and puberty in the male

The development of a male genitourinary tract from the primitive embryonic *Wolffian duct* (see Chapter 15) is stimulated by the presence of androgens secreted by the fetal testes. Obviously the type of gonad is determined by the **sex chromosomes** (pair of chromosomes which determine genetic sex) present (see *Figure 20.22*), the male Y chromosome influences the formation of testes, but androgens are required for the proper formation of a male tract and external genitalia.

During childhood small amounts of adrenal sex hormones inhibit GnRH release but, as **puberty** nears, the hypothalamus becomes less sensitive and GnRH is released, which in turn stimulates FSH, LH and gonadal testosterone.

At puberty the secretion of hormones (see page 538) activates the individual's reproductive potential, spermatogenesis is stimulated and the development of the secondary sexual characteristics occurs.

The timing of this great event varies; it usually starts between 10 and 14 years but some changes may continue into the 20s, e.g. growth of chest hair.

As already discussed it is testosterone which initiates the changes of puberty. These include:

- Enlargement of the testes, penis and accessory glands.
- Growth of the seminiferous tubules and spermatogenesis.
- Enlargement of the larynx (see Chapter 12) and deepening of the voice; it is this 'breaking' of the voice which rather limits the singing life of a 'boy soprano'.
- Growth of facial, pubic, axillary and chest hair.
- Increased sebaceous gland activity which may be linked to acne (see Chapter 19).
- Formation of the male physique, 'growth spurts' and increase in muscle mass (see Chapters 16–18). All of which are linked to the anabolic properties of testosterone.
- Development of libido, spontaneous erection and nocturnal emissions of semen.
- Possible influences on behaviour, but many other sociocultural factors contribute to this aspect of pubertal change.

Changes with ageing

Male reproductive function, unlike that in the female, continues well into old age. Men have no physiological equivalent of the climacteric, and, although there is a general decline in reproductive capacity, they may be capable of fathering children into their 70s and beyond. Testosterone levels decrease with ageing and the ability to achieve erection and ejaculation may be affected by other health factors, e.g. antihypertensive drugs. Reduced testosterone levels causes structures such as the seminal vesicles to become smaller and the seminiferous tubules to produce fewer germ cells.

Female Reproductive System

The female reproductive system has an intricate role which consists of the production of female gametes, providing the site for fertilization and the provision of an environment suitable for the nurture of a developing **embryo** (early developmental stage from fertilization to the end of the eighth week of gestation)/**fetus** (developmental stage from the ninth week of gestation until birth) should conception occur. It is more complex than that of the male and is completely separate from the urinary tract.

Structure of the female reproductive system

The female reproductive system (*Figure 20.9*) can be divided into two parts. The **internal genitalia** consists of the **ovaries** (two) in the pelvis, **uterine/Fallopian tubes** (two), **uterus** and **vagina**, and the **external genitalia** which consists of the structures comprising the **vulva**. The **mammary glands** (breasts) are usually included in a discussion of reproductive structures because of lactation.

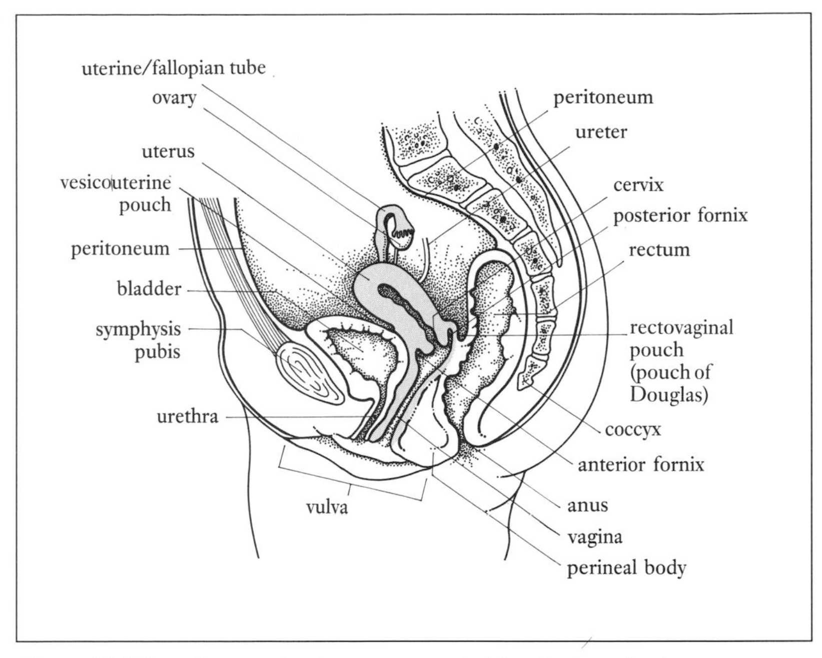

Figure 20.9. Female reproductive structures (midsagittal section).

Ovaries

The female gonads or **ovaries** produce the cells destined to become ova, and act as endocrine organs (see Chapter 8) by secreting female hormones – **oestrogen** and **progesterone** and small amounts of male androgens. During fetal life the ovaries, in common with the testes, develop high in the abdomen, but unlike the testes the ovaries descend only to the pelvic cavity, where they remain.

The almond-shaped ovaries lie in shallow fossae on the lateral pelvic walls. One ovary is situated on either side of the uterus, to which they are attached by **ovarian ligaments**. The ovaries are attached by the **mesovarium** to the posterior part of the **broad ligament**, the fold of peritoneum (see Chapter 13) which encloses and supports the internal genitalia. Blood vessels and autonomic nerves of both divisions run in the mesovarium to enter the ovary at its hilum. The arterial supply is via the *ovarian arteries*, which branch directly from the aorta and from a branch of the *uterine artery.* A plexus of veins drains venous blood from the ovaries.

The ovary is covered by a thin layer of *germinal epithelium,* which is so named because it was once thought, quite erroneously, to be the site of ova development. Under this layer is the fibrous **tunica albuginea** which encloses an outer *cortex* of connective tissue stroma and follicles containing the **oocytes** and an inner *medulla* containing blood vessels (*Figure 20.10*). Scarring on the outer surface, which is due to repeated follicle rupture at ovulation, increases with the age of the woman.

The adult ovary varies in size during the ovarian cycle (see pages 554 and 555) and contains follicles at different stages of maturation

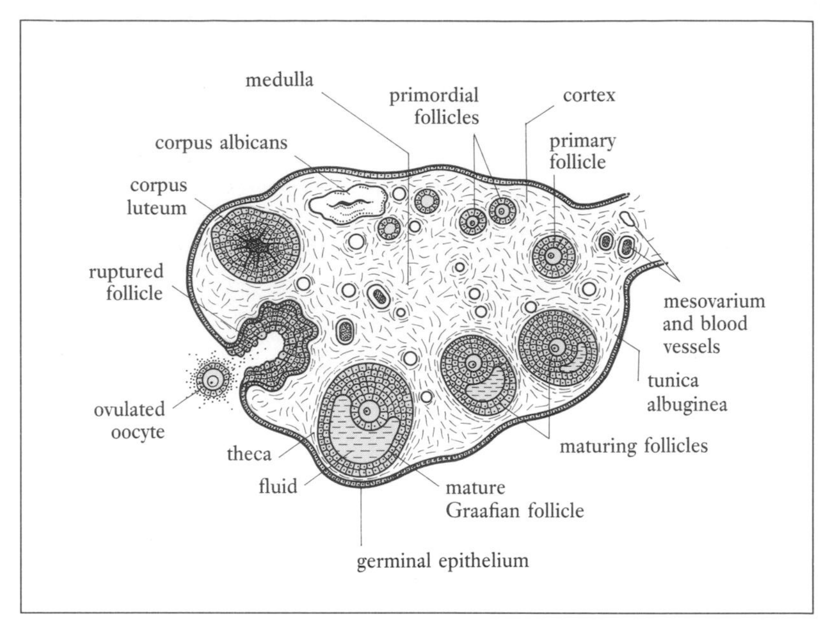

Figure 20.10. Ovary and follicles at all stages.

(*Figure 20.10*) – **primary**, **maturing**, **mature Graafian follicles** and a structure known as the **corpus luteum** which develops following follicle rupture (see page 555).

Uterine/fallopian tubes

The paired **uterine tubes** (**fallopian tubes**) extend laterally from the uterus to open into the peritoneal cavity; they are about 10 cm long with a diameter of about 1 cm, but the tube varies considerably in diameter along its length.

NB This arrangement of the uterine tubes opening into the peritoneal cavity means that infection from other parts of the

Person-centred Study Lucy

Lucy, aged 57, had been worrying about cancer since the death of a close friend, from ovarian cancer, the previous month. It was some years since she had thought much about her health but her friend's death made Lucy think about taking care of herself.

Ovarian tumours, which kill around 4500 women each year in the UK, are more commonly diagnosed as investigations improve and represent an important malignancy of the female reproductive structures.

Unfortunately few women are even aware of the possibility of ovarian cancer, probably because it has received little publicity compared with tumours of the cervix (see pages 546-547) and breast (see pages 551-552). Many ovarian tumours are diagnosed too late for an effective cure with surgery and cytotoxic drugs, because their growth deep in the pelvis gives rise to few symptoms in the early stages.

Lucy was able to discuss her fears about cancer and her doctor suggested that she have a pelvic examination which proved to be normal. Her doctor told Lucy that had there been any doubt she would have referred her for further tests which usually include *ultrasound scan (USS)* and a *computerized axial tomography (CAT or CT scan)* if indicated. What she did suggest, however, was that Lucy make an appointment for *Well-woman Screening* with the practice nurse (see page 551).

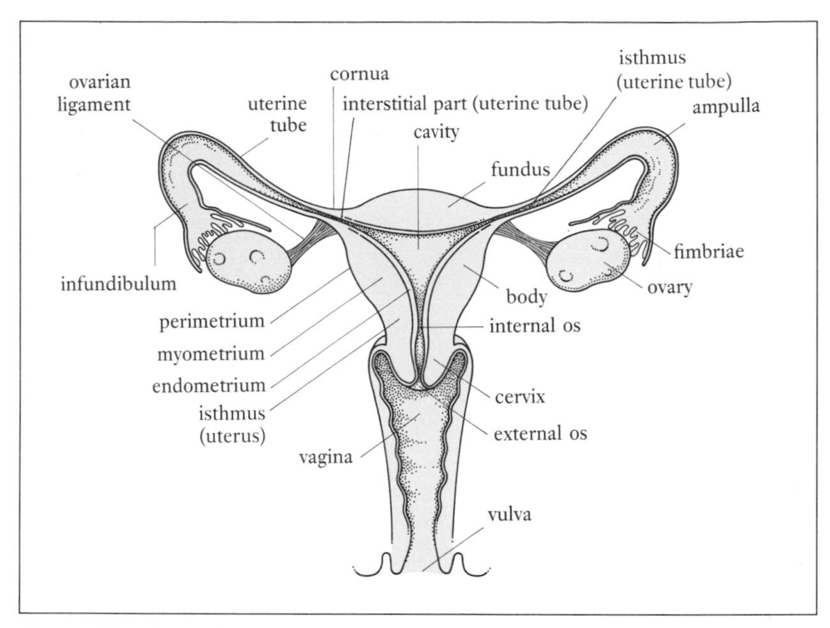

Figure 20.11. Female reproductive structures (anterior view).

reproductive tract can spread to the pelvic cavity and cause *pelvic inflammatory disease* (*PID*, see below and 565).

Each tube is divided into a funnel-like **infundibulum** ending in the **fimbriae** (finger-like projections) which help to waft the oocyte from the ovary into the tube, a dilated **ampulla** where fertilization usually occurs, an **isthmus** (1 cm) and an **interstitial part** within the uterine wall which is very narrow (1 mm) (*Figure 20.11*).

The uterine tubes are covered with and supported by part of the broad ligament known as the **mesosalpinx**. A middle layer of smooth muscle is responsible for the *peristalsis* which helps to convey the oocyte towards the uterus. The mucosal lining is highly specialized with ciliated cells and secreting cells with microvilli. The cilia beat rhythmically to assist the oocyte in its journey to the uterus and the secretions help to keep the oocyte and spermatozoa in a viable condition.

The uterine tubes have an extremely good arterial blood supply via the ovarian and uterine arteries. This vascularity leads to considerable blood loss if a tube were to rupture.

Any extrauterine pregnancy is termed *ectopic*. Most commonly this occurs within the uterine tube but rarely implantation may be in the pelvis or ovary. As you can imagine, a structure as delicate as the uterine tube is easily occluded by congenital defects, pelvic surgery, *pelvic inflammatory disease* (see page 565) or by blood clot from a previous ectopic pregnancy. A partial or unilateral blockage may lead to ectopic implantation where the early embryo, finding its route to the uterus barred, starts to develop within the uterine tube.

This abnormal event may have several outcomes, but all result in the loss of the pregnancy. Most dramatically the growing embryo ruptures the uterine tube resulting in pain, severe haemorrhage, hypovolaemia and shock (see Chapter 10). The management of tubal rupture includes blood transfusion followed by surgical removal of the affected tube (*salpingectomy)* as life-saving measures. Where the clinical presentation is subacute or atypical the diagnosis is confirmed by ultrasound and *laparoscopy* (endoscopic examination of the abdominal cavity).

Complete bilateral tubal blockage, for whatever reason, will cause infertility. This type of infertility may sometimes be overcome by specialized tubal surgery techniques or possibly by in vitro fertilization (IVF) (see page 571), which bypasses the blocked tube.

Uterus

The uterus is an amazing structure; it: provides the correct environment for embryo implantation, nurtures the developing fetus and expels the infant, usually at the end of pregnancy.

A healthy *nulliparous* (never been pregnant) uterus is pear-shaped and approximately 7.5 cm long, 5 cm wide and 2.5 cm thick, but may be slightly larger following pregnancy.

The uterus is a pelvic structure situated between the bladder and rectum. Basically, the thick-walled uterus is a hollow muscular organ which can be divided into a **fundus** *(top),* **corpus** *(body)* and **cervix** *(neck)* (see *Figure 20.11*); the body and cervix are separated by a narrow area (also called the **isthmus**), which becomes the *lower segment* in pregnancy. The two uterine tubes insert laterally into the uterus at the **cornuae** inferior to the fundus, and the cervix protrudes into the vagina.

An abundant arterial supply reaches the uterus via two *uterine arteries* which branch from the internal iliac arteries. The uterine arteries divide at the level of the cervix, the lower branch supplies the cervix and the vagina and the upper branch the body, fundus, tubes and ovaries. The branches which run up to the fundus are convoluted to allow for uterine enlargement during pregnancy. We will discuss the further modifications to arterial supply during our consideration of the endometrium. Venous blood leaves the uterus in corresponding veins.

The uterus is innervated by both divisions of the autonomic system, parasympathetic fibres from the sacral outflow and sympathetic fibres from the lumbar outflow. Lymphatic vessels drain via aortic and iliac lymph nodes.

A fold of peritoneum called the **perimetrium** (see *Figure* 20.11) is draped over the uterus to form an adherent outer covering, this is reflected laterally to form the broad ligament. Anteriorly the pelvic peritoneum forms the **vesicouterine pouch** between the uterus and bladder and posteriorly the **rectovaginal pouch** (also called **rectouterine pouch** or **pouch of Douglas**) between the vagina/uterus and rectum (see *Figure 20.9*). The latter is a common site for the collection of pus or blood. The fluid which results from pelvic disease drains to the lowest point by gravity.

Uterine muscle or **myometrium**, found in the fundus and body, is formed from a thick layer of interlocking smooth muscle fibres. The myometrium is influenced by both autonomic nerves and hormones when it contracts strongly during labour to expel the infant. Following delivery of the infant and placenta the interlocking fibres contract as a 'living ligature' and compress the uterine blood vessels to limit blood loss.

The mucosal lining of the body of the uterus, known as the **endometrium**, consists of highly vascular epithelium containing many tubular glands (*Figure 20.12*). From the **menarche** (commencement of menstruation) to the **menopause** (cessation of menstruation) this unique tissue undergoes cyclical changes in response to ovarian hormones (see page 555-557) and if fertilization occurs the embryo implants into a specially prepared endometrium (see page 571).

The endometrium consists of two layers – the **stratum basalis** which is permanent, and the **stratum functionalis** which is shed every 28 days or so during menstruation and regenerates under the influence of ovarian hormones.

It is appropriate here to consider the special features of the endometrial blood supply. Branches of the uterine arteries divide

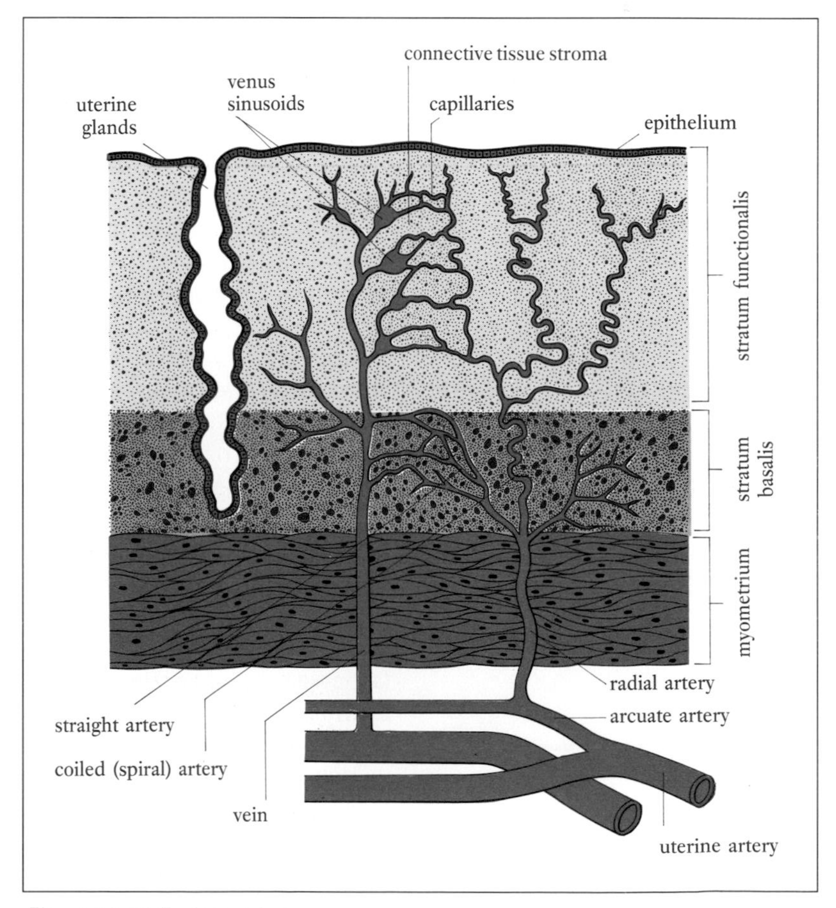

Figure 20.12. Endometrium.

within the myometrium to form the *arcuate arteries* which send radial branches to the endometrium. These *radial arteries* form *straight arteries* which supply the stratum basalis and the *coiled (spiral) arteries*, which supply the stratum functionalis and which degenerate with each menstrual flow.

The uterine cervix contains mostly fibrous tissue and its mucosa does not show the cyclical changes which affect the endometrium. What does change is the type and quantity of mucus produced by the glands in the cervical columnar epithelium during the menstrual cycle. Examination of this mucus forms the basis of a 'natural' method of family planning (see page 567). The cervical canal has two openings: the **internal os** which communicates with the uterine cavity and **external os** which opens into the vagina (see *Figure 20.11*). The columnar epithelium lining the cervix is replaced by stratified squamous epithelium in the part of the cervix which protrudes into the vagina.

Cervical carcinoma

Cervical carcinoma kills around 2000 women in the UK each year and many of these deaths could be avoided with regular and effective screening (see *Nursing Practice Application*, page 547). The disease develops through an early phase known as *cervical intraepithelial neoplasia (CIN)* where potentially malignant precancerous changes are present. CIN and the next stage of early non-invasive malignant change may be detected by 'smear test' and treated successfully with lasers and or cone biopsy. The abnormal cells are either destroyed by laser or removed by cone biopsy. The cellular changes occurring in the cervix appear to be initiated by viral agents, such as the *human papilloma virus (HPV)* or *herpes simplex virus (HSV2)*, which are passed to the woman during sexual intercourse (coitus).

Precancerous and malignant lesions of the cervix are associated with risk factors that include:

- Early first coitus.
- Multiple sexual partners (least common in women who have never been sexually active).
- Sexual activity of a partner who may have had multiple partners.
- Multiparity.
- Chronic cervical inflammation.
- Sexually transmitted diseases.
- Lower socio-economic groups.
- Smoking.
- Sexual partners not circumcised (cervical carcinoma is virtually unknown in Jewish women, whose husbands are circumcised). *NB* Carcinoma of the cervix may be present in some populations where the male is circumcised. This may be due to multiple sexual partners.
- Oral contraceptives (see page 567).

Nursing Practice Application Screening for cervical disease

Women are naturally anxious about the tests and should be given adequate information and the opportunity to express their fears. Although screening and diagnostic tests usually cause no pain and are performed without anaesthesia, the care plan should take account of the need for emotional support. A major problem can be the time waiting for appointments or results, both of which increase anxiety.

Exfoliative cytology – here cervical cells are obtained, via a *speculum* using a wooden spatula or brush, and examined microscopically for changes. A cervical smear (*Papanicolaou*) should cause no pain but speculum insertion may be uncomfortable for women who are anxious or embarrassed. False results do occur but if abnormal cells are found the diagnosis is confirmed by *colposcopy*.

Colposcopy is the examination of the cervix with a colposcope, an instrument which magnifies the cells. Again this involves speculum insertion, the cells are stained with iodine and any abnormal tissue seen can be biopsied and examined in the histology department.

Opinion differs regarding the frequency of cervical screening and which age groups should be included. Currently the Department of Health (UK) recommends that women aged 20–64 should have a smear at least once every 5 years but many experts suggest that tests be repeated more frequently (1–3 years). Hiscock and Reece (1988) recommended that there should be more pressure for 'high risk' women to attend triennial (at least) screening.

Uterine supports

The position of the uterus is normally *anteverted* (inclined forward) and *anteflexed* (bent forward) over the bladder (see *Figure 20.9*). The uterus is supported in this position by the muscles of the pelvic floor (see *Figure 18.49*), ligaments and, to a limited extent, by the pelvic peritoneum (broad ligament). Apart from support, these structures, when healthy, are able to maintain continence, assist with micturition and defaecation and stretch sufficiently to permit the birth of a baby. The supporting ligaments are:

- **Round ligaments** – one from each uterine cornua runs through the inguinal canal to the *labia majora* (see page 549) to offer limited support.
- **Transverse cervical ligaments** (**cardinal**) – two ligaments which run from the cervix to the lateral walls of the pelvis.
- **Uterosacral ligament** – extends backwards from the cervix to the sacrum, it divides on reaching the rectum but the two parts rejoin at the sacrum.
- **Pubocervical ligament** – extends from the cervix to the symphysis pubis, it supports the bladder and urethra.

Uterovaginal prolapse, which commonly occurs around the menopause, is due to damage incurred by the supporting structures over many years. The damage occurs primarily during childbirth, especially during a difficult or prolonged second stage, but other risk factors include constipation and straining (Chapter 13), poor lifting techniques (Chapter 18) which has particular significance for both professional and informal carers, poor posture, chronic cough and obesity. The problem occurs during the **climacteric** (period of time during which changes in the female reproductive tract result in decline and eventual cessation of reproductive function) because declining oestrogen levels leads to muscle/ligament atrophy with loss of function.

The problems experienced by individual women depends on the type and severity of the prolapse but the proximity of the bladder and urethra quite often results in frequency or stress incontinence. Some women complain of defaecation difficulties and, where the uterus is prolapsed, they may have discomfort and the feeling that 'something has dropped down'. Management may include: information regarding contributing factors such as constipation, a supporting pessary (rarely) or surgery, e.g. *anterior colporrhaphy* – an operation to repair the anterior vaginal wall.

NB Prevention is, however, much the preferred option, with careful management of labour and suturing of perineal lacerations and *episiotomies* (incision made in the perineum during the second stage of labour), pelvic floor exercises and avoidance of other risk factors.

Vagina

The **vagina**, (also called the 'birth canal'), is a canal extending from the cervix to the external genitalia (see *Figure 20.9*). Its anterior wall is about 7.5 cm long and lies close to the urethra and bladder, and the posterior wall, which is longer at around 9 cm, has contact with the rectum and the rectovaginal pouch. Normally the vaginal walls are in apposition (touching) but because of **rugae** (folds) the vagina is capable of the considerable stretching required to facilitate coitus and childbirth.

The vagina runs at an angle of 45° which is an important point to note prior to inserting instruments or medication, nurses should also remember to mention this fact when teaching women about pessaries or contraceptive diaphragms. The projection of the cervix into the vagina forms four **fornices** (deep folds or gutters); the posterior fornix which receives the semen is deeper than the lateral or anterior fornices. At its distal orifice the vagina is partially occluded by the **hymen**, a perforated membrane which is usually ruptured during the first coitus. It may also have been ruptured by tampon use or exercise and any remnants are completely destroyed by the birth of a woman's first child.

The vagina receives arterial blood from branches of the uterine arteries and a *vaginal artery* which may branch direct from the internal iliac artery or from the uterine artery. The sensitive lower vagina is innervated by the voluntary *pudendal nerve* and the upper part by autonomic sympathetic fibres.

Structurally the vagina consists of an outer layer of fibrous tissue, a layer of involuntary muscle, a loose areolar layer and the stratified squamous epithelial lining. The stratified squamous epithelium, which is adapted to withstand the trauma of coitus and childbirth, has no secretory glands. Some fluid may leak through the walls but the bulk of vaginal moisture originates from the cervical mucus and that produced during sexual excitement (see page 564).

During the reproductive years the vagina is acidic (pH 4.5) due to the production of **lactic acid** by bacteria known as *Döderlein's bacilli* (*Lactobacillus* species) which form part of the normal flora (see

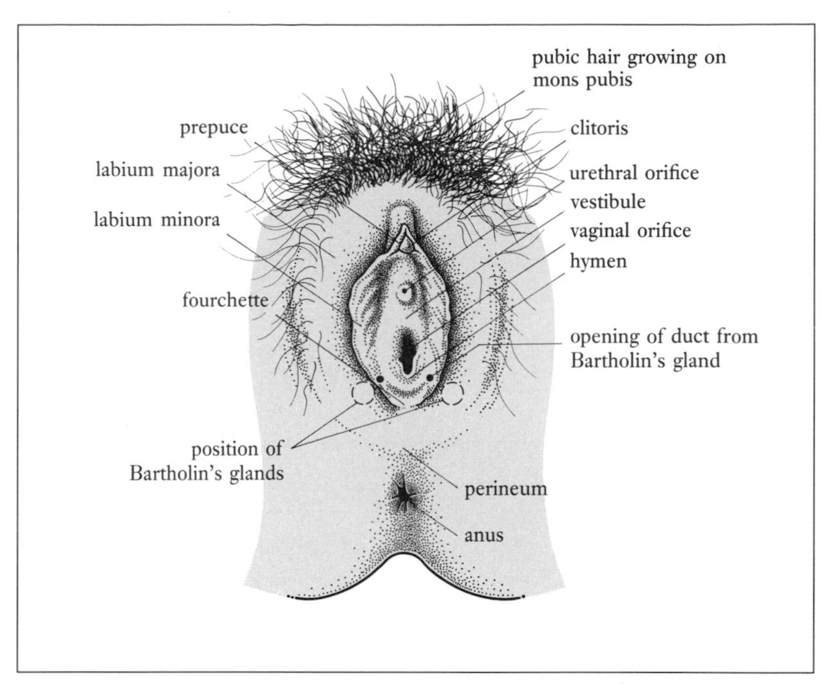

Figure 20.13. External female genitalia.

Chapter 19). The lactic acid produced from bacterial metabolism of glycogen, stored in the mucosal cells, helps protect the vagina from many pathogenic organisms. Vaginal acidity, is however, hostile to spermatozoa, which are normally protected by the alkaline semen and cervical mucus.

Vulva

The **vulva** consists of the structures known collectively as the **external genitalia** (*Figure 20.13*); these structures include the **mons pubis**, **labia majora**, **labia minora**, **clitoris**, structures within the **vestibule**, hymen (see page 548), **fourchette** and **perineum.**

The **mons pubis** is a pad of fatty tissue situated over the symphysis pubis, it is covered with pubic hair which first grows during puberty. Moving back from the mons pubis are the fatty **labia majora** which merge posteriorly with the perineal skin. The labia majora are skin-covered and contain numerous sebaceous glands; pubic hair grows on the outer part, whereas the inner surface is smooth. The labia majora are homologous with the male scrotum – they develop from similar embryonic tissue.

Inside the protective outer labia are two smaller folds called the **labia minora**. The labia minora formed from smooth skin enclose the area known as the **vestibule**. Anteriorly the labia minora fuse to form the **prepuce** which covers the **clitoris** and posteriorly they form the **fourchette**. The clitoris is homologous with the penis; it contains erectile tissue and is abundantly innervated. It is extremely sensitive and is involved in the female sexual response.

Apart from the clitoris the vestibule contains the openings of the urethra and vagina and the vestibular glands. The ducts of two tiny **Skene's glands** (**lesser vestibular glands**) open into the urethral meatus. There are two larger **Bartholin's glands** (**greater vestibular glands**) whose ducts open into the vaginal orifice. Bartholin's glands, which are homologous with the male bulbo-urethral glands, produce lubricating mucus which increases during sexual stimulation to facilitate coitus.

The **perineum** or **perineal body** is the wedge-shaped structure situated between the vagina and rectum. Extending from fourchette to anus it consists of muscle and connective tissue covered by skin. The perineum overlies the midline junction between the two halves of the pelvic floor muscles.

Nursing Practice Application Episiotomy and perineal laceration

Following episiotomy or perineal laceration a careful repair of the perineum is essential to promote healing (see Chapter 19) and prevent later vaginal prolapse. The perineum is extremely well supplied with nerve endings and the episiotomy wound or laceration is likely to be painful. This may make sitting uncomfortable and the establishment of breast feeding more difficult. Defaecation will be painful and constipation should be avoided, e.g. regular aperients and high fibre intake. Meticulous hygiene regimens should be followed to avoid infection in this damp, warm area and women should be encouraged to wash their hands before and after perineal contact, change pads frequently, wash and dry the area after using the toilet and report signs of infection, e.g. increasing pain and discharge.

Pelvic floor exercises, undertaken soon after delivery, will improve perineal blood flow which will aid healing, minimize infection risk and reduce congestion and with it discomfort (Gould, 1990).

Mammary glands

The **breasts** (**mammary glands**) are highly specialized glands adapted to produce milk (*lactation*), when stimulated by the correct hormone environment, following the birth of a baby. Although breasts are significant in expressing sexuality their only physiological function is that of lactation. At puberty in the female (see pages 560-561) the rudimentary breast, which is present in both sexes, responds to oestrogen stimulation by developing and enlarging. The breasts also show cyclical changes during the menstrual cycle. However, it is only during pregnancy that the breast enlarges and becomes physiologically active in readiness for lactation. When oestrogen levels decline, during the climacteric, there is breast atrophy and later loss of adipose tissue.

Each breast has an area of pigmented skin, the **areola**, into which the **nipple** opens. Smooth muscle fibres in the nipple and areola cause nipple erection in response to cold and sexual excitement. The many sebaceous glands (**Montgomery's tubules**) in the areola help to keep the skin in that area supple and lubricated. The areola darkens during a first pregnancy, when it becomes permanently brown in colour.

Situated anterior to the pectoral muscles (see Chapter 18), each breast consists of 15–20 **lobes** containing alveolar glands which

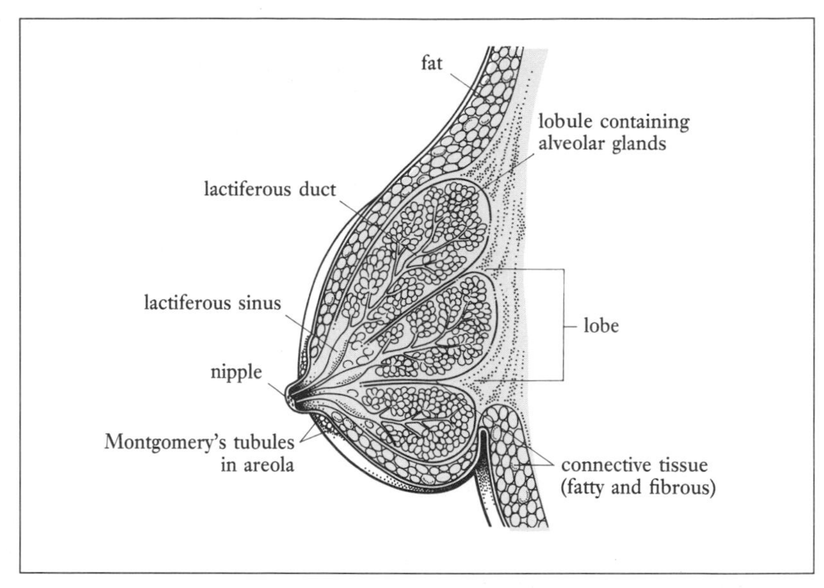

Figure 20.14. The breast.

radiate out from the nipple (*Figure 20.14*). The lobes are supported by fatty and fibrous tissue and separated by suspensory ligaments which also secure the breast to the chest wall. Normal breasts show a large variation in size and shape.

Within the breast lobes are the **small lobules** of alveolar glands whose role is milk secretion. A **lactiferous duct** draining each lobule dilates to form a **lactiferous sinus** *(ampulla)*, just below the areola, where milk is stored prior to discharge through the ducts opening on the nipple surface. Further coverage of breast function can be found in the discussion about lactation (pages 579-580).

Breast cancer

In the UK, breast cancer is the most common malignant condition affecting women, where it accounts for 20% of female malignancies and kills at least 15 000 women annually. Properly planned and supported screening programmes, which include breast awareness and reporting changes, can detect lesions at an early stage when cure may be possible and treatment options are more acceptable to the women concerned.

Person-centred Study Lucy (Continued)

Lucy attended an appointment at the Well-woman clinic where the following checks were carried out: weight, urine test, blood pressure, cervical smear (see page 547), pelvic examination and breast examination. She also had the opportunity to discuss lifestyle influences on health and worries with the staff. The practice nurse discussed breast awareness with Lucy and explained the importance of noticing changes such as lumps or nipple discharge and offered her a leaflet (Cancer Research Campaign, 1991). She also told her about the *mammography* (breast X-ray) screening programme which is offered, by the NHS in the United Kingdom, to women aged 50–64 every 3 years. This service is also available to women with suspicious breast lumps and those over 64 on request.

The management of breast cancer usually involves a combination of:

- Surgery – *lumpectomy* where only the lump is excised or some form of *mastectomy* (breast removal) with its implications for body image, sexuality, self-esteem and relationships.
- Radiotherapy.
- Some tumours are oestrogen-dependent and hormone modification, such as removal or destruction of the ovaries in premenopausal women or the administration of *tamoxifen* (oestrogen antagonist), may be indicated. After the menopause oestrogen production is adrenal and the adrenal glands may be removed or drugs such as *aminoglutethimide* or tamoxifen administered.
- Cytotoxic drugs may be used following excision.

Readers are directed to Gould (1990) (Chapter 13) who includes a very sensitive and clear discussion of breast tumours.

Functioning of the female reproductive structures

Oogenesis

The process of **oogenesis** occurring within the cortex of the ovary differs from spermatogenesis (page 536-537) in several ways. Notably the entire stock of female gametes are already present, in an immature state, within the ovaries of a female fetus before birth. The time spent in the immature state may be up to 50 years. Maturation occurs only on a cyclical basis which commences at puberty and continues until the cessation of reproductive function after the climacteric. During maturation there is a second resting stage and an unequal cytoplasmic division.

Readers may care to refresh their memories regarding meiosis (page 534-536) before our discussion of oogenesis (*Figure 20.15*). We will, however, first consider the processes occurring in the ovary prior to birth. Early during fetal life **primordial germ cells** differentiate to become diploid stem cells called **oogonia** ($2n$). The oogonia multiply by rapid mitotic division to form several hundred thousand **primary oocytes** ($2n$). Mitosis is followed by a period of growth and the development of **primordial follicles** which surround the primary oocytes. At this stage the first meiotic division commences in the primary oocyte but is arrested at prophase I. The primary oocyte commences a variable-length resting stage before meiosis I is completed, in some 450 cells, some time during the reproductive life of the woman.

At puberty the production of functional ova commences with the activation of several primary oocytes during each ovarian cycle. Ovarian follicular changes occurring as part of the ovarian cycle and oogenesis are discussed on pages 554 and 555.

The arrested meiosis I is recommenced in several primary oocytes but usually only one is 'chosen' to complete the process. During meiosis I the diploid primary oocyte produces two very dissimilar haploid cells: the **secondary oocyte** (n) and the **first polar body** (n). The secondary oocyte, which still contains the first polar body, is discharged from the surface of the ovary at **ovulation**.

The first polar body, which has very little cytoplasm to sustain its existence, usually undergoes meiosis II but eventually all polar bodies degenerate. Meiosis II starts in the secondary oocyte, only to be arrested, this time in metaphase II, and will only be completed if the secondary oocyte is penetrated by a spermatozoon.

If spermatozoon penetration does occur the secondary oocyte completes meiosis II with the production of two haploid cells, a

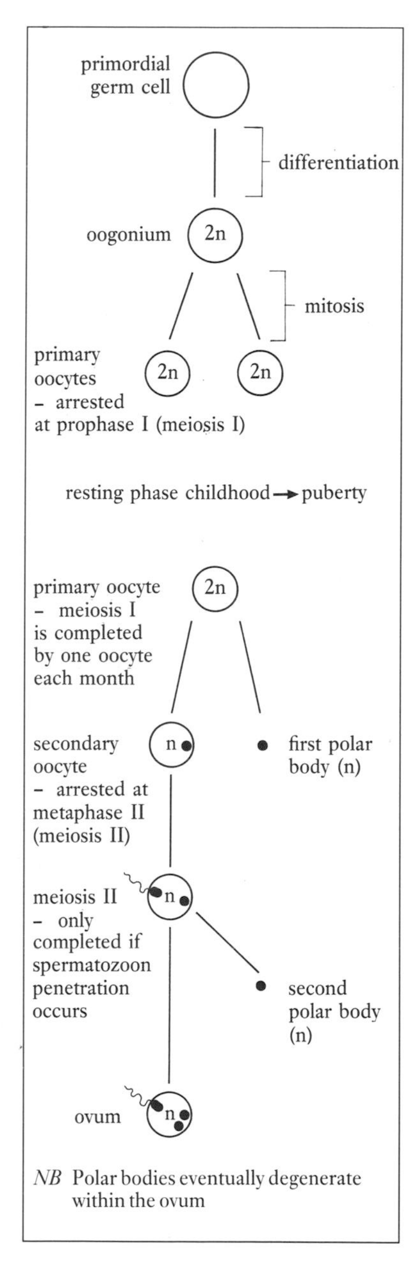

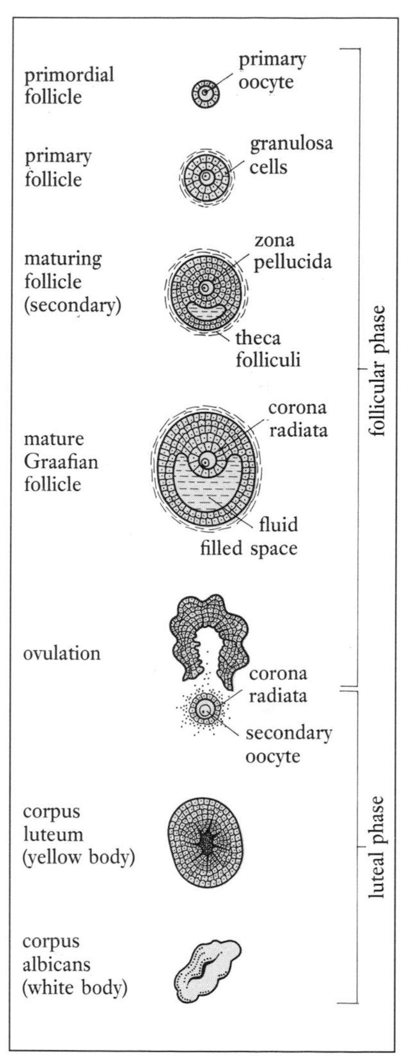

Figure 20.16. Ovarian cycle – follicular development.

Figure 20.15. Oogenesis. 2*n* = diploid; *n* = haploid.

viable ovum (*n*) and a **second polar body** (*n*). The actual events of fertilization, which take place in the uterine tube, are discussed on pages 566 and 568-569. Interestingly the events of oogenesis, which result in four haploid cells, yield only one functional ovum (and thus ensure that the one ovum has enough nutrients to survive prior to implantation), whereas the same meiotic divisions during spermatogenesis produces four viable spermatozoa (large numbers of spermatozoa are lost on their journey to the uterine tube, see page 566).

Ovarian cycle

As already mentioned the events of oogenesis are accompanied by changes in the ovary which can be divided into two phases: **follicular** (days 1–14), which includes ovulation and **luteal** (days 14–28) (*Figures 20.16 and 20.17*). The ovarian cycle lasts for around 28 days in most women but length may vary (21–35 days); any changes in cycle length are reflected in the follicular phase; the luteal phase remains fairly constant.

The hormonal influences which control oogenesis and the ovarian cycle are discussed separately, in more detail, on pages 555-556.

During the follicular phase, primordial follicles, enclosing the primary oocytes, develop layers of **granulosa cells** to become **primary follicles**. Maturation of the primary follicle into a **secondary follicle** involves the formation of a **zona pellucida** around the oocyte and a **theca folliculi** around the outside of the follicle. Cells of the theca produce oestrogens during maturation. Further development results in the formation of a fluid-filled space

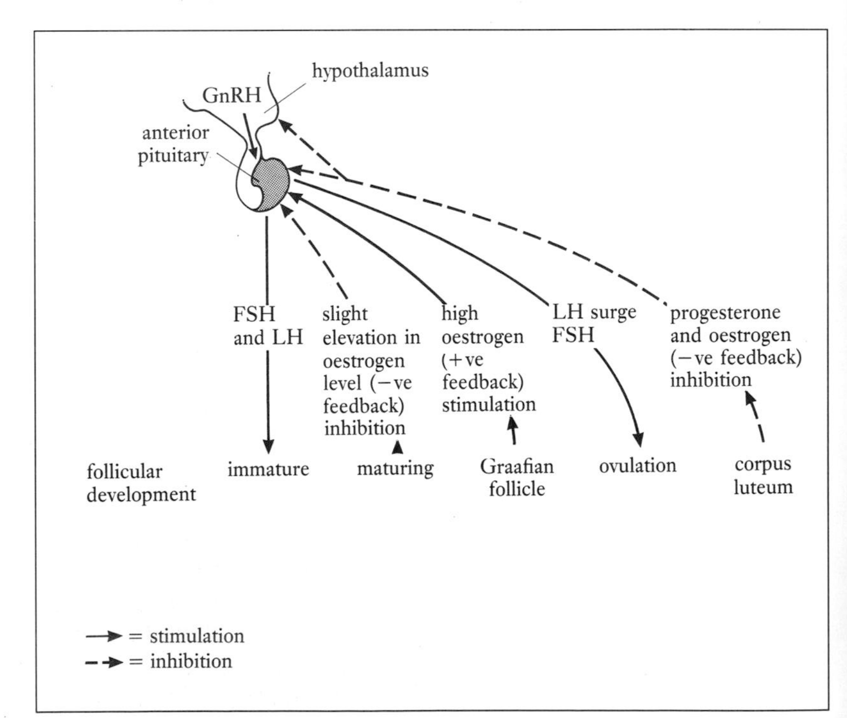

Figure 20.17. Hormonal control of ovarian cycle.

around the oocyte and some granulosa cells develop into a **corona radiata** which surrounds the oocyte. The follicle, by this stage, is a **mature Graafian follicle** ready to release a secondary oocyte at ovulation (remember the ovum is not formed until after ovulation).

Ovulation, which usually occurs in alternate ovaries each month, involves rupture of the follicle and discharge of the secondary oocyte into the abdominal cavity. From the abdomen it enters the fimbriated end of a uterine tube. Ovulation is caused by an increase in LH (see page 556) and is probably linked to pressure increases in the Graafian follicle and the secretion of prostaglandins. Some women (around 25%) experience 'cramping' pain in the lower abdomen at ovulation; this is known as *mittelschmerz* which is German for 'middle pain'.

Usually only one oocyte is released during the ovarian cycle, but the release of two or more oocytes can result in a multiple birth, which is not the norm for humans. In this situation the individuals are not identical because they result from the fertilization of different oocytes by different spermatozoa. They are no closer genetically than siblings born at different times.

The follicular phase ends with ovulation and is replaced by the luteal phase. The ruptured follicle collapses and fills with blood clot. Granulosa and theca cells increase to form an endocrine structure called the **corpus luteum** (yellow body). The corpus luteum secretes progesterone and oestrogen and is important in preparing for possible fertilization and maintaining pregnancy should it occur. The lifespan of the corpus luteum depends upon the fate of the oocyte released at ovulation. If fertilization does not occur, which is the usual course, the corpus luteum degenerates after about 12–14 days and hormone production stops. The area left after the corpus luteum has degenerated is filled with scar tissue and becomes the **corpus albicans** (white body). However, if fertilization does occur the corpus luteum persists and functions to maintain the pregnancy until placental and fetal hormone production is sufficiently developed (see page 573).

Hormones and control of the ovarian cycle

The ovarian cycle and the events of oogenesis are controlled by the gonadotrophins-follicle stimulating hormone (FSH) and luteinizing hormone (LH) (*Figure 20.17*). These two hormones are released cyclically by the anterior pituitary gland in response to hypothalamic gonadotrophin releasing hormone (GnRH). Although FSH and LH are the same in both sexes, their role in female reproductive functioning is more complex than in the male. FSH and LH levels fluctuate during the ovarian cycle and they control the release of two ovarian steroid hormones, oestrogen and progesterone. As you can imagine this is a finely balanced system requiring precise controls and considerable integration if it is to function.

During puberty the secretion of GnRH causes cyclical release of FSH and LH, which act upon the ovaries to cause the development of adult hormone patterns. Eventually the menarche occurs, but cycles remain largely **anovulatory** (without ovulation) for the first 2 years until hormonal cycles are stabilized.

In adult women the release of FSH and LH stimulate initial follicular development (see pages 554 and 555) and oocyte maturation (see pages 552-554). At this stage the small amount of oestrogen produced by the follicle inhibits pituitary secretion of FSH and LH by negative feedback. Oestrogen, however, actually increases the effects of FSH and LH locally on the follicle where development continues and oestrogen levels rise. When oestrogen plasma levels reach a critical point they stop inhibiting the hypothalamus/anterior pituitary and stimulate, by postive feedback, the release of further LH and FSH. Now, at around midcycle, there is a sudden surge of LH which stimulates ovulation and the formation of the corpus luteum, which it helps to maintain. FSH released at this time is involved with follicle growth, ovulation and corpus luteum formation.

Following ovulation the corpus luteum secretes progesterone and oestrogen, which prepare the body for possible pregnancy and together exert negative feedback/inhibition on further LH and FSH release.

If pregnancy does not occur the decline in LH causes the corpus luteum to degenerate, resulting in reduced levels of oestrogen and progesterone. Now the combined ovarian hormone inhibition is removed and the pituitary can again secrete FSH and LH to start a new ovarian cycle.

NB For events if pregnancy occurs see pages 571 and 572.

The ovarian cycle is controlled through the integration of hypothalamic releasing hormone, FSH and LH from the pituitary and ovarian oestrogen and progesterone, which exhibit both negative and positive feedback.

Ovarian female steroid hormones

Oestrogens

Oestrogens are steroid hormones, derived from cholesterol and include: **oestradiol**, **oestrone** and **oestriol**. Oestrogens are produced by the ovaries, placenta and, to a limited extent, by the adrenal glands. These hormones, which are vital for reproductive function, are involved with oogenesis and follicle maturation, development of female secondary sexual characteristics and pubertal growth spurt, and growth and maintenance of reproductive organs.

Oestrogens cause wider metabolic effects such as a lower serum cholesterol level, which helps to explain why *premenopausal* women have a lower incidence of coronary heart disease (see Chapter 10) compared with men of the same age. Another important metabolic influence is the effect of oestrogen levels upon calcium homeostasis, demonstrated by the increase in osteoporosis (Chapter 16) after the menopause.

Progesterone

Progesterone is another steroid hormone secreted by the ovary and placenta. It is the '*gestation hormone*' important in preparing for and maintaining pregnancy. Progesterone increases growth of the endometrium and breasts, causes changes in cervical mucus and inhibits uterine muscle activity.

Menstrual cycle

The **menstrual cycle** (**uterine cycle**) involves the changes occurring in the uterus as the endometrium responds to the secretion of ovarian hormones (*Figure 20.18*). It corresponds to the ovarian cycle and is repeated every 28 days (range 21–35 days) or so, except during pregnancy, from the menarche to the menopause. The menstrual cycle can be subdivided into three phases or stages: *proliferative, secretory* and *menstrual* phases. The whole point of the menstrual cycle is the preparation of the endometrium for possible implantation of the early embryo. Although the menstrual phase comes at the end of the cycle the first day of bleeding is, by convention, counted as day 1 of the menstrual cycle because it provides an obvious landmark.

Proliferative phase

The **proliferative phase** corresponds to the follicular phase of the ovarian cycle and starts around day 5 when menstrual bleeding has ceased. After menstruation only the stratum basalis of the endometrium remains. The release of oestrogen causes cell proliferation and regeneration of the stratum functionalis with its blood vessels and glands (see page 545 and 546). The proliferative phase ends with the maturation of a Graafian follicle and ovulation around day 14. At this point the endometrium is approximately 2mm thick. During the proliferative phase the cervical mucus changes from a thick plug blocking the cervix to profuse amounts of thin slippery mucus which the spermatozoa can penetrate.

Secretory phase

Commencing after ovulation, the **secretory phase** corresponds with the luteal phase of the ovarian cycle and lasts about 14 days. The oestrogen-primed endometrium is now influenced by progesterone, which causes the glands to enlarge and secrete glycogen, which would nourish the embryo during implantation. The spiral arteries of the endometrium increase in size and become more coiled. Endometrial thickness has increased to 5 mm. Cervical mucus becomes thick and again blocks the canal to protect the developing embryo if implantation occurs.

If fertilization does not occur the decline in hormones from the corpus luteum results in spiral artery spasm, which causes endometrial degeneration as it is deprived of nutrients. This leads, some 24 hours later, to menstruation. The spiral arteries dilate and bleed into the necrotic stratum functionalis, which starts to slough away. The secretion of prostaglandins by the endometrium also influences this part of the cycle.

Menstrual phase

It is worth reminding ourselves that the **menstrual phase** is the last

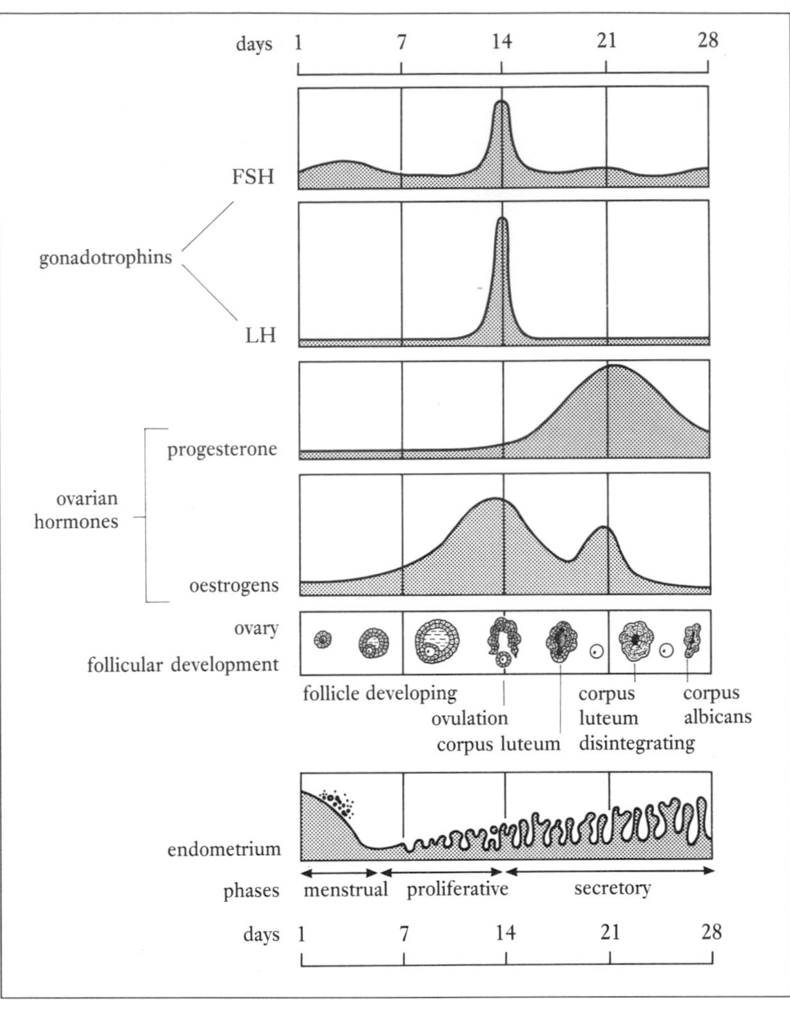

Figure 20.18. Menstrual cycle (with hormonal and ovarian cycle events).

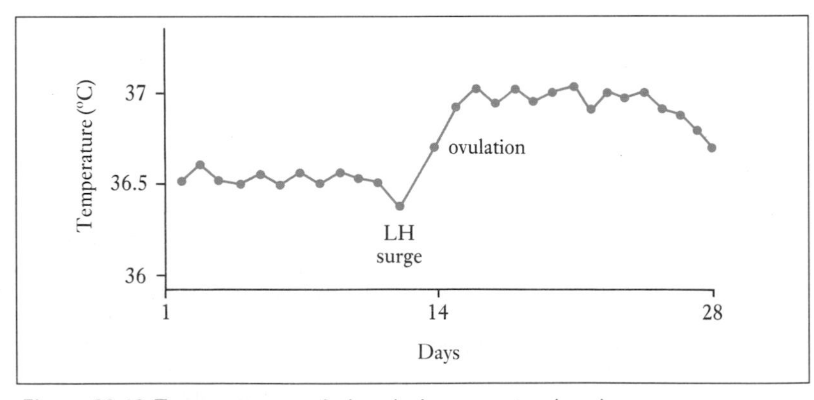

Figure 20.19. Temperature variation during menstrual cycle.

part of the menstrual cycle, in spite of being taken as day 1. The menstrual flow consists of blood and endometrial fragments, and lasts on average some 3–6 days. Blood loss at menstruation, which is usually 50–100 ml, varies considerably and is notoriously difficult to assess objectively. You might care to consider the amount of blood lost in a year where menstruation occurs 13 times and the loss is 100 ml each time: 13 x 100 = 1300 ml lost, which must be replaced – no wonder women need more iron than men during the reproductive years (see Chapter 9). Excessive menstrual loss (*menorrhagia*) is a common cause of iron deficiency anaemia.

Bleeding occurring between menstruations, e.g. after intercourse or examination, should always be investigated to exclude serious pathology such as cervical malignancy. Interestingly, menstrual blood is prevented from clotting within the uterus by the presence of anticoagulant enzymes. At the start of menstruation the uterus contracts, in response to prostaglandins, to expel the blood. These contractions result in the discomfort/pain, experienced by many women, known as *dysmenorrhoea.*

Body temperature may show variation during the menstrual cycle (*Figure 20.19*). There may be a fall in temperature which corresponds

Nursing Practice Application Premenstrual syndrome (PMS)

Premenstrual syndrome is sometimes called *premenstrual tension (PMT)* but either way it describes a collection of physical (many due to fluid retention) and emotional effects associated with the 10 days or so before the onset of menstruation. Most women experience some changes but for others the problems cause intense distress, disruption to their lives and relationships and in severe situations may have serious social implications. Dalton (1980) describes three serious criminal cases where women successfully pleaded diminished responsibility/mitigation due to PMS. *The features of PMS include:* (i) abdominal swelling; (ii) weight gain up to 3 kg; (iii) breast discomfort and heaviness; (iv) ankle oedema; (v) bloated feeling; (vi) headaches; (vii) nausea/vomiting; (viii) extreme tiredness; (ix) emotional lability with crying, depression and irritability; (x) insomnia; (xi) clumsiness; (xii) poor memory; (xiii) change in bowel habit.

Various studies have shown that women perform less well during the premenstrual and menstrual phases, e.g. driving, skilled tasks at work and examinations. During this time women are thought, by some authorities, more likely to commit crimes, have accidents, attempt suicide and require psychiatric treatment.

The aetiology of PMS is not fully understood but current theories include: (i) reduced progesterone in the secretory phase; (ii) elevated prolactin levels; (iii) a deficiency of pyridoxine (vitamin B_6); (iv) some, as yet unknown ovarian factor.

With such a diversity of presentation and possible causes the management of PMS needs to be considered on an individual basis, for what offers relief for one woman will be totally ineffective in another. Treatment regimens may include the administration of natural progesterone or synthetic progestogens. Pyridoxine may relieve PMS by its action on tryptophan (an amino acid), which is also a neurotransmitter concerned with mood. The use of pyridoxine supplements may also be helpful in reducing high prolactin levels. Diuretics (see Chapter 15) may be used to reduce oedema as may reducing sodium intake. Essential fatty acids, such as *gamma-linoleic acid (GLA)* found in the oil of evening primrose, may relieve some PMS symptoms. N.B. GLA is needed for the synthesis of prostaglandins.

General measures such as simple analgesia, distraction, exercise and a 'healthy' diet may all help women with PMS. Kirkpatrick *et al.* (1990) found that self-care measures, especially those related to nutrition and exercise, could be helpful. Women who are informed and understand what is happening each month can gain considerable benefit from open discussion with their partner, family or within a self-help group.

with the LH surge occurring about 24 hours before ovulation. Progesterone secreted by the corpus luteum then causes a rise in metabolic rate and body temperature of about 0.5°C that persists until menstruation occurs. These changes in temperature are a useful but rather crude guide to the timing of ovulation in some women. The knowledge that ovulation is occurring can be helpful for couples wishing to conceive; they may then time intercourse for the few days around ovulation when it is more likely to result in pregnancy. Conversely couples who wish to avoid pregnancy can use temperature changes as part of 'natural' family planning.

Other factors influencing the menstrual cycle are not, as yet, fully explained. Emotional factors are known to affect menstrual cycle regularity, e.g. grief, leaving home, starting a new course or job and extreme worry. Another interesting observation is the degree of synchrony in menstrual cycle patterns established by women living in close proximity to each other; this may be linked to the secretion of pheromones in sweat (see Chapter 19).

Sex differentiation and puberty in the female

The female genital tract develops from the same primitive embryonic ducts as the male tract (see page 3, 531 and 532), but this time it is the *Mullerian ducts* that develop. It is mainly the lack of testosterone which leads to the formation of female genitalia but there may be some oestrogenic effect from the fetal ovaries. The sex chromosomes determine the type of gonad formed. The genetically female fetus forms ovaries (which develop later than testis formation) because the male Y chromosome is absent. It is interesting to note that without testosterone all fetuses, genetic males included, will develop female external genitalia.

During childhood small amounts of adrenal sex hormones inhibit GnRH release but, as puberty nears, the hypothalamus becomes less sensitive and GnRH is released which in turn stimulates FSH, LH and ovarian hormone secretion.

At puberty the secretion of hormones (see page 555) activates the individual's reproductive potential, which in the case of the female is really a restart of oogenesis already commenced in fetal life, and development of the secondary sexual characteristics.

Puberty in the female may start any time between the ages of 9 and 13, but for most girls breast development and the growth of pubic and axillary hair occur at about 11 years. The menarche usually occurs between 11 and 17 but the mean age in developed countries is 12–13 years. The actual timing of the menarche appears to be linked with achievement of a minimum body weight body fat. This theory can be used to explain the fall in mean age of menarche over the last century in developed countries. It is possible that improved standards of health and nutrition have resulted in better growth rates which have triggered an earlier puberty. Body weight/fat influences are also demonstrated by the cessation of menstruation (*amenorrhoea*) following severe weight loss, e.g. 'crash diets' or exessive exercise.

Early menstrual cycles may occur at irregular intervals and, as already mentioned, the cycles are usually anovulatory. It comes as something of a suprise to a girl who has so-far had pain-free periods to experience the dysmenorrhoea which starts when ovulation becomes a regular event.

The onset of puberty in the female is generally earlier than in the male and is of shorter duration. Oestrogen accounts for most of the physical changes, which include:

- Growth and development of breasts.
- Development of internal genitalia.
- Oogenesis and follicle development.
- Vaginal mucosal changes and production of acid environment.
- Body hair.
- Development of female physique, e.g. pelvic shape (see Chapter 18), fat deposition, and 'growth spurts', which are usually complete by 15–17 years.
  NB males may continue to grow into their late teens/early 20s.

Climacteric and menopause

Female reproductive function and hormone production declines over a period of time known as the **climacteric**; this generally occurs between 46 and 55 years but may be earlier or later. The cessation of menstruation, known as the **menopause**, is a distinct episode occurring during the climacteric; the mean age for the menopause is 50 years in the UK. Often the term menopause is used, quite erroneously, to describe the climacteric.

Premenopausal women are often anxious that *hysterectomy* (removal of uterus) will cause the climacteric to occur. If their ovaries are left intact they will not experience the climacteric until the natural time (because ovarian hormone production will continue as normal). They will, however, stop menstruating and become infertile.

Obviously the climacteric and menopause bring to an end a woman's fertile life and the reduction in oestrogens produce important metabolic effects which range from short-term problems such as 'hot flushes' to lasting changes, e.g. raised serum cholesterol.

Basically the menopause occurs because the ovaries fail to respond to the gonadotrophins; there are still plenty of follicles but they no longer develop. Some women suddenly stop menstruating but for most the cycles become anovulatory, erratic and blood loss decreases. Eventually ovulation and menstruation cease and levels of oestrogen and progesterone decline. In response to this reduction in ovarian hormones the anterior pituitary secretes more and more FSH and LH. The physical and psychosocial effects of the climacteric are due either to declining oestrogen or elevated FSH and LH, and include:

- Atrophy of internal genitalia.
- Genital and bladder supports weaken, which in some women

leads to uterovaginal prolapse (see pages 547 and 548).

- Atrophy of labia and thinning of body hair.
- Breast atrophy
- Changes in vaginal mucosa, which thins and becomes dry. The loss of lactobacilli and the acid environment increases the vulnerability to infection. These changes of *atrophic (senile) vaginitis* can make intercourse painful or difficult (*dyspareunia*).
- 'Hot flushes' and night sweats caused by sudden vasodilation of skin vessels. Reducing oestrogen levels, linked in some way with LH surges, appear to be responsible for this vasomotor instability which occurs in some women.
- Mood changes such as depression, irritability, labile emotions and loss of confidence. These may also be linked with negative attitudes to the climacteric which many women still call the 'change of life' and to major life events occurring at that time such as children leaving home. Some women may fear the effects of the climacteric on their sexuality and relationships.
- Bone changes, which increase the incidence of osteoporosis in postmenopausal women (see Chapter 16).
- Loss of oestrogenic influence on serum cholesterol. After the menopause cholesterol and triglyceride levels rise and women develop the same risks of coronary heart disease as men of the same age (see Chapter 10).

Postmenopausal bleeding (PMB). Bleeding occurring a year after the menopause is considered abnormal and should always be investigated to exclude uterine malignancy. The bleeding may be due to vaginitis, which can be easily remedied with topical oestrogen preparations.

Hormone replacement therapy (HRT). For some women the administration of low-dose oestrogen and progesterone on a cyclical basis will help to alleviate the distressing effects of the climacteric, e.g. 'hot flushes'. The addition of progesterone prevents overstimulation of the endometrium. Women should be given information regarding the careful monitoring required during HRT to detect problems, which may include endometrial and breast malignancies (the incidence of breast cancer may increase with long-term therapy, i.e. some years). Also on the minus side, HRT will not prevent ageing or restore youth, in spite of the many anecdotal accounts to the contrary which appear in the popular press. Another effect of HRT which might not be considered by women is the fact that vaginal bleeding may occur during the treatment. HRT does, however, help many women cope better with the climacteric and it has obvious benefits in preventing osteoporosis and protecting against heart disease. Providing information about the climacteric and the options open to women is an important function of the health care services. Hughes (1992) describes the development of information meetings at a GP surgery and discusses the results.

Reproductive Physiology

After our separate considerations of the structure and functioning of the male and female reproductive systems it is highly appropriate to remind ourselves that neither system functions naturally in isolation. Even with the most sophisticated fertilization techniques it still requires the presence of an oocyte and a spermatozoon to produce a new life.

This part of the chapter looks at the sexual response, conception, pregnancy and parturition. In addition we will briefly discuss areas relevant to reproductive function, such as sexually transmitted diseases, contraception and infertility.

Male sexual response

The male sexual response consists of penile erection, which allows it entry into the female tract, and the ejaculation of semen within the vagina.

Erection takes place as the erectile tissue in the corpora cavernosa and corpus spongiosum fill with blood (see *Figure 20.4*). Arterioles in the erectile tissue are normally constricted and the penis flaccid, but during sexual excitement the arterioles dilate and the penis enlarges. These vascular changes occur through a spinal reflex and involve an example of parasympathetic-stimulated vasodilation, which is unusual. Mucus secretion by the bulbo-urethral glands is also stimulated by parasympathetic fibres. As the penis enlarges the venous return is impeded, which results in further engorgement.

Sexual excitement has a variety of triggers which include visual, auditory and olfactory stimuli, emotions and direct stimulation of the genitalia, e.g. touch receptors present in the glans penis.

The erection reflex operates through afferent pathways (pudendal nerve) which synapse in the sacral spinal cord and the efferent parasympathetic outflow from S2–S4 segments known as the *nervi erigentes* (see Chapter 6) which initiates vasodilation. There are also descending spinal pathways from the cerebrum, which can either expedite or prevent the erection reflex, e.g. thoughts alone can cause erection without any physical stimulation and anxiety or worry can prevent erection.

Impotence is the failure to achieve or maintain an erection. There are many physical and psychological causes which include excess alcohol, anxiety, certain drugs, vascular disease, nerve damage, e.g. during pelvic surgery, spinal injuries (Chapter 4), diabetes mellitus (Chapter 8) and multiple sclerosis (Chapter 3).

Ejaculation of semen also results from a spinal reflex and is accompanied by an intensely pleasurable sensation known as **orgasm** and physiological changes which include increased respiration, blood pressure and heart rate. The afferent pathway involved is the same as that for the erection reflex and when afferent impulses reach a critical point, as stimulation continues, the sympathetic efferents to the genitalia, mainly via L1 and L2, are stimulated. These sympathetic effects include contraction of smooth muscle in the accessory glands and ducts, which discharge their secretions into the urethra, closure of the internal urinary sphincter to prevent retrograde ejaculation of semen into the bladder or the leakage of urine, and vigorous contraction of the skeletal muscles at the base of the penis to expel semen.

Ejaculation is followed by a period of relaxation, known as the *latent period*, which varies from minutes to hours, during which further erection is impossible.

Female sexual response

The female sexual response is indicated by engorgement and erection of the clitoris and labia minora. This occurs through the same autonomic pathways as in the male. Vaginal lubrication is provided by the vestibular glands which increase secretion and the fluid which 'leaks' through the vaginal walls. There is breast enlargement and nipple erection with possible flushing of the skin of the chest and neck. Females may experience orgasm during coitus and as there is no latent or refractory period some women experience multiple orgasms. Although female orgasm is not essential for fertilization the rhythmic uterine contractions which occur may assist the passage of semen into the uterus.

As with the male, orgasm is accompanied by an increase in pulse, respiration and blood pressure.

Sexually transmitted diseases (STDs)

In the wake of massive publicity campaigns concerning AIDS (see Chapter 19) you could be forgiven for forgetting the existence of other STDs. The STDs include: *syphilis, gonorrhoea, chlamydia, genital herpes, trichomoniasis* and *candidiasis*. However, it must be stressed that many of these infections may also occur without sexual contact, e.g. chlamydial infection of an infant's eyes during birth.

It is not intended to explore each disease in depth but rather to concentrate on their effects within the reproductive system. Readers should refer to Further Reading for more specific information about individual infections.

Syphilis

Syphilis is caused by the spirochaete (bacterium) *Treponema pallidum*. Infection usually occurs during sexual contact where the organisms enter through the mucosa of the genitalia, anus or mouth. Syphilis may also be transmitted from mother to fetus via the placenta. The disease has well defined stages: primary, secondary, tertiary and quaternary. In primary syphilis painless lesions called *chancres* develop, which may affect the vulva, vagina, cervix, penis and other sites, including the anus and mouth. Secondary stage manifestations affecting the genitalia include wart-like *condylomata lata* on the penis or vulva, an associated lymphadenopathy and possibly a generalized body rash. The tertiary stage is characterized by skin ulcers called *gumma*, and quaternary syphilis has effects involving the nervous system, e.g. *tabes dorsalis* (degenerative changes in the spinal cord with sensory loss), or cardiovascular system, e.g. aortic aneurysm (see Chapter 10).

Gonorrhoea

Gonorrhoea is caused by the bacteria *Neisseria gonorrhoeae* which, being very fragile, does not survive for long outside the body. It infects the genital tract, urinary tract and the mucosa of the throat and anus. Gonorrhoea infection causes profound effects within the reproductive system, which can result in infertility. Men usually have *urethritis* with dysuria and discharge. Untreated this can lead to inflammation of the male duct system with stricture formation. Women may have urinary symptoms but unfortunately they may remain asymptomatic whilst severe damage occurs. Some women have vaginal discharge, abdominal pain and bleeding. The infection can cause *pelvic inflammatory disease (PID)* which leads to serious ill health and infertility if the delicate uterine tubes become blocked. Non-sexual spread may occur, e.g. an infant's eyes infected during birth or a prepubertal girl using contaminated items such as towels and flannels.

NB Syphilis and gonorrhoea are notifiable diseases which are classified legally as *venereal*. Generally in adults their spread is through sexual activity but this is certainly not exclusively so.

Chlamydia

Chlamydia is an increasingly common STD. It is caused by the organism *Chlamydia trachomatis* which has both bacterial and viral characteristics. In males it causes urethritis and testicular pain but may be asymptomatic. Women infected with chlamydia may have urethritis, vaginal discharge and irregular menstruation. It is a major cause of PID and can lead to serious eye and respiratory infections in infants exposed to the organism during birth.

Genital herpes

Genital herpes is commonly caused by the herpes simplex type 2 virus. The infection results in extremely painful blisters on the genitalia and the virus may infect the eyes. Both sexes become carriers as the virus lies dormant within their tissues, but 'flare-ups' or exacerbations of inflammation occur at intervals. Beardsley (1993) in a recent article outlines ways in which the spread of genital herpes may be reduced. The herpes virus may cause serious and possibly fatal infections in infants born to infected mothers. Also very worrying is the link between genital herpes and the development of cervical malignancy (see page 546).

Trichomoniasis

Trichomoniasis is caused by the protozoon *Trichomonas vaginalis*. Infected women have a foul, frothy, yellow vaginal discharge, vulval soreness and irritation. Men and sometimes women may be asymptomatic carriers of the organism.

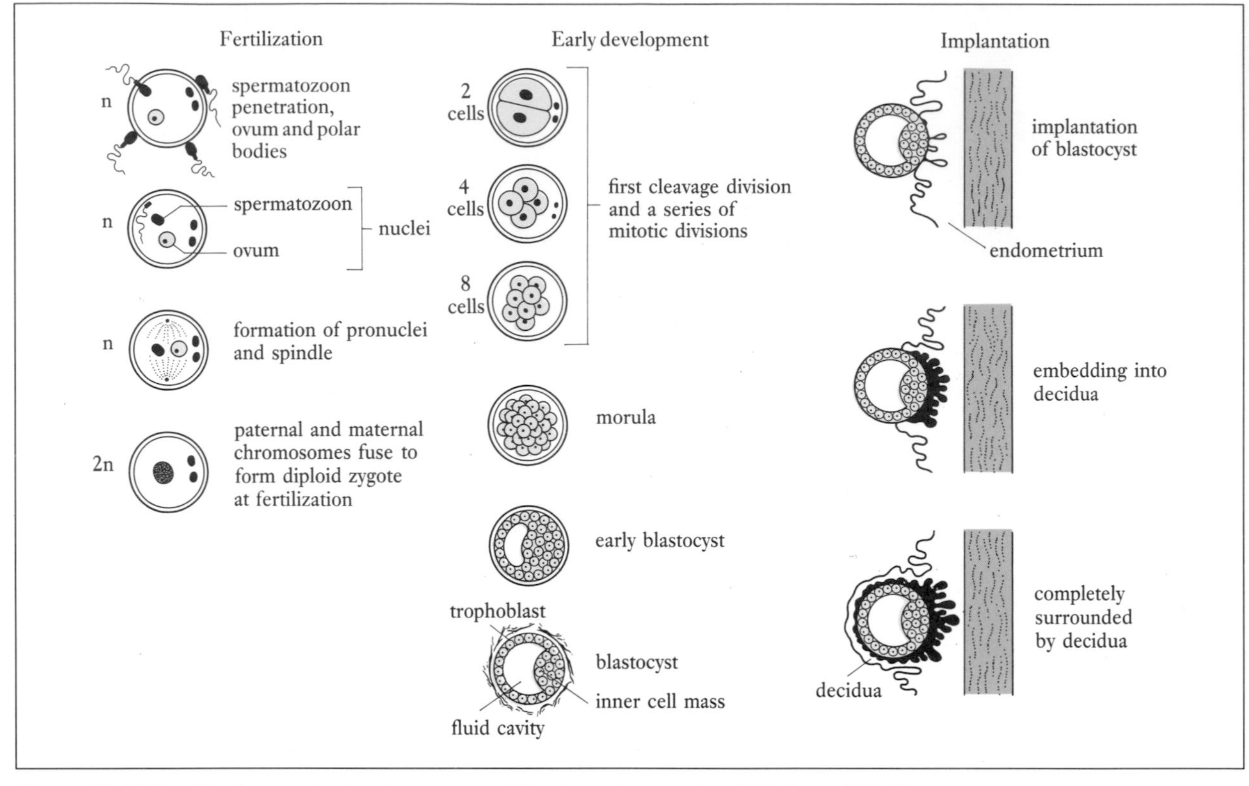

Figure 20.20. Fertilization, early development and implantation. n = haploid; $2n$ = diploid.

Candidiasis (thrush)

Candidiasis (thrush) is caused by the fungus *Candida albicans*. It causes intense pruritus and a thick creamy vaginal discharge. Although it may be sexually transmitted, candidiasis commonly affects women who are debilitated, where hormones have been modified, e.g. oral contraceptives, following antibiotic therapy and in diabetes mellitus due to the glycosuria (see Chapter 8). Candidiasis may cause balanitis (see page 533) in the male.

Conception and fertilization

Conception requires that a spermatozoon reaches and penetrates a secondary oocyte which completes meiosis II prior to the fusion of the two nuclei at fertilization (*Figure 20.20*). Time for fertilization is limited as the gametes have a finite life after which they deteriorate; the oocyte is viable for about 24 hours and spermatozoa for a longer period of 48 hours, but some may still be functional up to 72 hours.

After spermatozoa are deposited in the vagina during coitus they must make their way through the uterus and into the uterine tube and still need changes (see below) which allow them to penetrate the oocyte. Nature has ensured an abundance of spermatozoa which is just as well because millions are lost on the way – only a few thousand actually reach the ampulla of the uterine tube. Spermatozoa are motile, but they do have some help – they secrete

Special Focus **Fertility control and contraception**

Over the centuries human beings have devised many ways to control fertility. Some methods, such as wearing a 'magic' charm during the event, were destined to fail, as were the more active attempts to prevent the meeting of egg and sperm, e.g. packing the vagina with plant material or even jumping up and down after coitus!

The acceptance of fertility control and contraception took a long time – in Britain family planning services did not became generally available until the 1930s. Over the years there have been many developments in methods of contraception but the only sure way to avoid pregnancy is to abstain from intercourse. However, there are numerous ways in which couples may avoid pregnancy; some are highly reliable such as the 'pill' whilst others are nearly as risky as the magic charms.

Couples should be given sufficient information to allow them to choose the method which best suits them.

Coitus interruptus involves the withdrawal of the penis from the vagina before ejaculation occurs. This method is not at all reliable because semen may leak from the penis prior to ejaculation.

Natural methods where the fertile period is ascertained by cervical mucus changes (Billing's method), temperature changes (see page 559 and 560) or a combination of the two. This requires a high degree of motivation in both partners.

Barrier methods: (i) Spermicidal chemicals, e.g. nonoxinols, in the form of vaginal foams, gels, pessaries and impregnated film. They are not very effective when used without a vaginal diaphragm, cap or condom. (ii) Barrier methods such as condoms and vaginal diaphragms or caps are reliable when used properly, e.g. use of spermicides with a vaginal diaphragm and applying condoms carefully. Condoms have the additional advantage of reducing the spread of AIDS and other STDs. (iii) Other barrier methods include vaginal sponges and the fairly new female condom, a tube made of polyurethane which fits inside the vagina. Apart from contraception the female condom will also be useful in reducing the spread of STD's (Filshie and Guillebaud, 1989).

Intrauterine contraceptive devices (IUCD) made of polythene with or without copper are inserted through the cervix into the uterus. An IUCD or 'coil' probably exerts its contraceptive effect by preventing implantation (see page 571). An IUCD can be used as *postcoital contraception* if it is inserted within 5 days of unprotected intercourse.

Oral contraception or 'the pill' has been, so far, one of the major breakthroughs in fertility control. Either low dose oestrogen and progesterone combinations, which override the normal hypothalamic-pituitary-ovarian hormonal controls of the ovarian cycle, or a progesterone-only pill can be used. Oral contraceptives act in a variety of ways to prevent pregnancy. The combined oestogen–progesterone pill produces effects which include inhibition of FSH-release and failure of follicles to mature, no LH surge or ovulation, possible changes in uterine tube motility may affect the passage of an oocyte if ovulation did occur, endometrial proliferation is inhibited (making implantation unlikely if ovulation occurred) and the cervical mucus remains thick and impenetrable to spermatazoa. The combined pill is taken for 21 days followed by a break of 7 days, during which oestrogen-withdrawal bleeding occurs. The progesterone-only pill is less reliable because it does not always inhibit ovulation; its contraceptive effects are achieved mainly by its action on the cervical mucus, which remains thick and hostile to the passage of spermatozoa. Progesterone-only pills also affect endometrial proliferation and uterine tube motility. It is taken continuously, at the same time each day (no more than 3 hours late or its effects may be impaired), to provide its contraceptive effect.

Taken correctly the oral contraceptive offers very high reliability and, because the cycles are probably anovulatory they will be pain-free. If all this sounds too good to be true let us examine the minus points, which include some definite contraindications to their use, such as a history of breast cancer, ovarian cancer or thrombo-embolic disorders. There are side-effects which may occur e.g. nausea and headache, which are unacceptable to some women. Missed pills, gastrointestinal upsets or drugs such as antibiotics can all reduce the contraceptive effects. High-dose combined pills are used as postcoital contraception; the pills are administered, in two doses twelve hours apart, within 72 hours of unprotected intercourse. This type of therapy can be offered, for example, to rape victims or when contraception fails.

Various studies have suggested links between the oral contraceptive and the development of breast or cervical malignancy. There may well be a slight increase in the incidence of breast cancer in pill users but studies have produced conflicting findings. In other studies, women who had taken the combined pill appeared to show an increased incidence of cervical cancer.

Special Focus **Fertility control and contraception** (continued)

In the light of rather contradictory evidence it is important to remember that the risk factors are indeed very small and may well be acceptable to individual women in return for reliable contraception. Further studies suggest that some pills may actually reduce the risk of uterine or ovarian cancer. Also to be noted are the physical risks associated with pregnancy, even in a developed country, plus the emotional and social implications of an unwanted pregnancy. More research into the incidence of malignancy and its links with the pill is clearly needed.

Injections and subcutaneous implants of synthetic hormones which provide contraception for some weeks. There are many side-effects and they do not allow for a 'change of mind'. It is necessary to wait for their contraceptive effects to dissipate before pregnancy is again possible.

Sterilization surgery, which involves vasectomy (see pages 531 and 532) or tubal ligation. These surgical procedures should be considered to be permanent measures.

There are several possible developments for future fertility control which include: development of a 'male' pill, inhibition of FSH with substances such as inhibin and an immunological approach, which involves the production of short-term antibodies to the hormones required for successful implantation.

prostaglandins which cause contraction of the uterus and tubes.

Before spermatozoa can penetrate an oocyte they must undergo the processes of **capacitation** and **acrosome reaction**. You will remember that the spermatozoon carries enzymes which it will use to gain access to the oocyte. Obviously these proteolytic enzymes must remain intact on the journey but need to be available once the spermatozoa have located the oocyte. Capacitation involves structural changes to the acrosome which make the enzymes available for release. Many spermatozoa release their enzymes by the acrosome reaction, but only one will actually penetrate the secondary oocyte. There appear to be mechanisms operating to prevent the entry of further spermatozoa.

At last, the secondary oocyte which has wafted into the ampulla, and the spermatozoon are together. Once penetration by the spermatozoon head has occurred, the secondary oocyte will

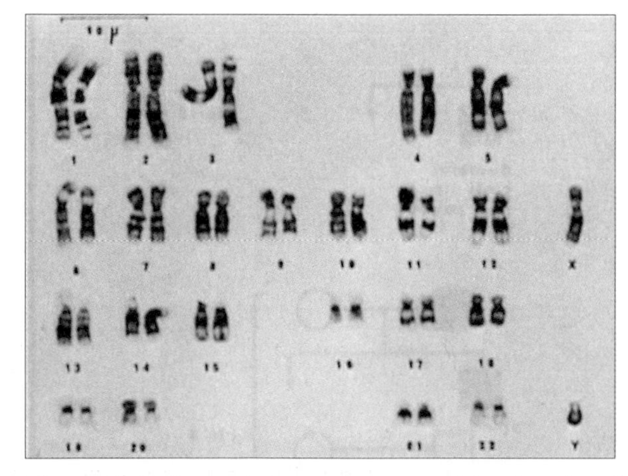

Figure 20.21. Normal chromosomes (male) - 22 pairs of autosomes plus one pair of sex chromosomes (XY) = 46 chromosomes. (Baraitser, M., Winter, R. *A Colour Atlas of Clinical Genetics.* Wolfe Medical Publications, Ltd. 1990. Reprinted with permission).

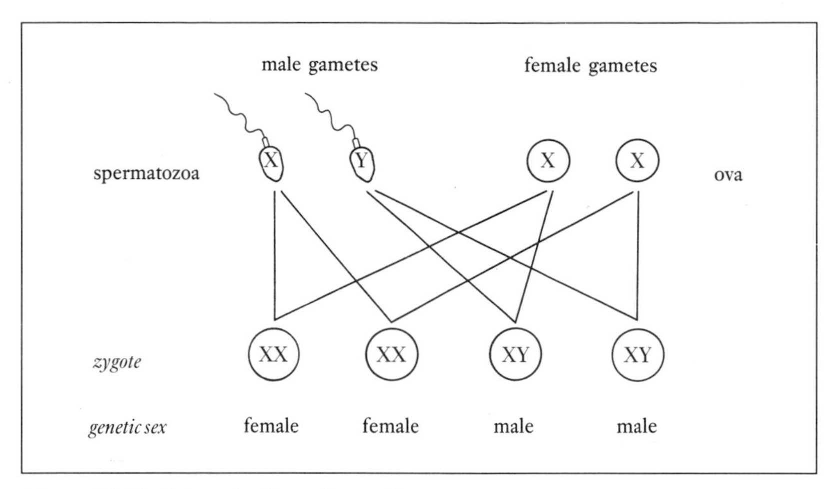

Figure 20.22. Determination of genetic sex.

complete meiosis II to produce the functional ovum and the second polar body. The haploid nuclei of both ovum and spermatozoon enlarge to form the **pronuclei**. A mitotic spindle forms between the two pronuclei and the maternal and paternal chromosomes combine to form the diploid **zygote** – the incredible process of fertilization is complete.

Chromosome complement and determination of genetic sex

It is appropriate here to discuss chromosomes and the determination of genetic sex. Normally the 23 chromosomes, 22 **autosomes** (a chromosome which is not a sex chromosome) + 1 **sex chromosome**, from both ovum and spermatozoon combine to give the zygote 46 chromosomes or 23 pairs (*Figure 20.21*). Human body (somatic) cells have 22 pairs of autosomes and one pair of sex chromosomes, either *XX (female)* or *XY (male)*. In females one of the X chromosomes becomes inactivated and appears in the somatic nuclei as a mass known as the *Barr body (sex chromatin)*.

Genetic sex is determined by the spermatozoon, which may carry either the X or Y sex chromosome, whereas the ovum will always carry the X chromosome. An ovum fertilized by the X spermatozoon produces a genetic female and that fertilized by the Y spermatozoon produces a genetic male (*Figure 20.22*). Inheritance of a sex chromosome only determines genetic sex; fetal sex hormones are required for genitalia differentiation and further hormones are required at puberty. Usually more males are born than females, the ratio in the UK is currently around M 105 : F 100.

Sometimes the complex meiotic divisions of **gametogenesis** (process by which the ova and spermatozoa are formed in the ovary or testis) go wrong and the resultant zygote has a chromosomal error either in number or structure. Many errors are incompatible with further development and the zygote is lost but in some situations an infant with the wrong number of chromosomes is born, e.g. *Down's syndrome* where the infant has 47 chromosomes – because there are three chromosomes in 'pair 21' it is also called *trisomy 21*. The only example of an error with 45 chromosomes that is compatible with life is *Turner's syndrome* where a female infant has only one X chromosome (XO). Both Down's syndrome and Turner's syndrome are attributed to *non-disjunction*, where chromosomes do not separate properly during meiosis (*Figure 20.5*) – one cell has 24 chromosomes and the other only 22.

Some errors produce abnormalities so severe that the infant is unable to survive, whereas other errors produce a range of abnormalities which depend upon the chromosome affected. The aetiology of chromosomal errors is not always known but one reason seems to be connected with the long resting stage of oogenesis; it is possible that agents such as radiation and viruses affect the oocyte during its long wait for maturation. It is certainly true that the incidence of Down's syndrome due to non-disjunction increases with maternal age at conception. Down's syndrome is the most common autosomal abnormality, with an overall incidence of about 1 in 700 live births. Women over 35 years show an increased risk and by 45 years the risk is 1 in 50. Another type of Down's syndrome occurs when translocation of a chromosome segment occurs between chromosomes 15 and 21 or between chromosomes 21 and 22.

Subfertility (infertility)

Before we continue with our discussion of early embryonic development it would be helpful to consider some of the obstacles to conception. By now you will be aware of reproductive complexity and it will be no surprise to learn that subfertility may be caused by a multitude of factors, including infrequent coitus, impotence (page 563), infrequent or anovulatory cycles, low spermatozoa counts, non-patent uterine tubes following ectopic pregnancy or PID, hostile cervical mucus and immunological causes where both sexes produce antibodies against spermatozoa (see Chapter 19).

The management of subfertility depends upon the cause, if it can be identified, and the wishes of the couple regarding investigation and certain treatments. Some people will find the intimate nature of investigations or treatments unacceptable, e.g. postcoital testing near to ovulation requires that the couple have intercourse 'to order' and the woman attends the clinic soon after. Couples may have moral objections to treatments such as *artificial insemination* using donated semen (*AID*) and others will feel unable to cope emotionally with the numerous visits, treatments and disappointments that may accompany *in vitro fertilization (IVF)*.

Subfertility and the techniques used to help childless couples are extremely complex areas, where many ideas challenging established attitudes have been developed. No attempt is made here to consider investigations and only a brief consideration of management is included. Readers requiring more information are referred to Further Reading.

Management

- Hyperprolactinaemia (increased prolactin in the blood) may be treated with the drug bromocriptine.
- Drugs to stimulate ovulation, e.g. clomiphene 'fertility drug' or

gonadotrophins. Treatment for anovulatory cycles is extremely effective with pregnancy occurring in the vast majority of couples.

- Men with low spermatozoa counts may have gonadotrophins or testosterone.
- Tubal surgery to remove adhesions.
- Artificial insemination (*AIH*) using the husband's/partner's semen may overcome problems with mucus or female anti-spermatozoa antibodies.
- AID, where the man produces few or no spermatozoa.
- IVF may be used for low spermatozoa counts, blocked uterine tubes or female anti-spermatozoa antibodies. In IVF the oocytes collected via a laparoscope are fertilized, using the partner's semen, in the laboratory. The fertilized ovum or ova are later introduced into the uterus with varying degrees of success. Ova or spermatozoa may be donated for IVF.
- *Gamete intrafallopian transfer (GIFT)*, a simpler technique used to overcome cervical mucus problems. Here the oocytes and spermatozoa are mixed together prior to being placed in the uterine tubes. The technique, which is much simpler than IVF, still requires the patency of one uterine tube (McLaughlin, 1989).
- Adoption is difficult with very few small babies available.
- Surrogacy, where another woman agrees to have a baby for an infertile couple, is an area for extreme caution with its potentially catastrophic legal, ethical and emotional problems.

Early embryonic development and implantation

The zygote, formed during fertilization, starts to divide by mitosis as it travels through the uterine tube towards the uterus (*Figure 20.20*). Earlier we mentioned non-identical (*dizygotic*) twins resulting from the fertilization of two oocytes (see page 555); in contrast, identical (*monozygotic*) twins result from changes after the fertilization of one oocyte. During early cell division two separate cell masses develop, which are destined to become two individuals who are genetically identical.

By the time the uterus is reached, 4–5 days after fertilization, a series of cell divisions has resulted in a mass of cells called the **morula**. The early embryo floats free within the uterine cavity for a further 2 days or so whilst progressing to the **blastocyst** stage. The blastocyst consists of a fluid-filled cavity surrounded by an outer layer of **trophoblast cells** and an **inner cell mass** protruding into the cavity. Trophoblast cells will become the **placenta** and the inner cell mass the embryo and its membrane (**amnion**).

Implantation of the blastocyst into the hormone-prepared endometrium commences about 6 or 7 days after fertilization. The trophoblast layer invades the endometrium which thickens to become the **decidua**. After a few days the blastocyst is completely enclosed within the decidua. At this stage the early embryo receives nutrients derived from the endometrial cell debris produced by the trophoblastic invasion.

During these very early days the corpus luteum is maintained by a

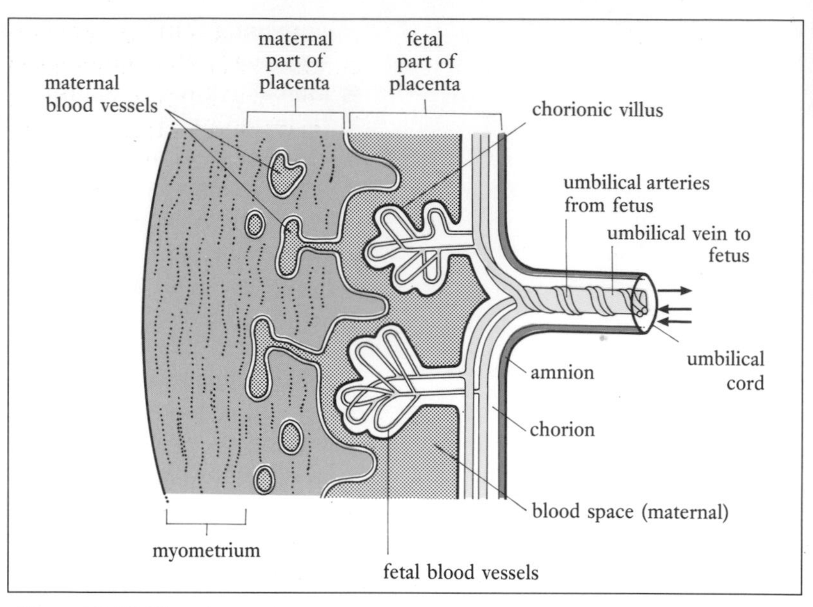

Figure 20.23. The placenta.

hormone, **human chorionic gonadotrophin** (**hCG**), secreted by the trophoblasts. This is essential to maintain the high levels of progesterone and oestrogen required to prevent menstruation and maintain the pregnancy. The detection of hCG in a woman's blood or urine forms the basis of pregnancy testing. It is possible to obtain a positive test as early as 2 weeks after conception but more reliable results are obtained 6–8 weeks after the last menstrual period (LMP). hCG levels reach their peak about 8–10 weeks after the LMP.

Coverage of the complexities of further embryonic development are not included here and readers are directed to Further Reading. However, we will consider briefly the placenta, fetal circulation and some of the hormones concerned with pregnancy.

Placental development and function

The **placenta** is a temporary structure which forms from the trophoblast layer (embryonic tissue) and the decidua (maternal tissue) (*Figure 20.23*). Early placental development includes:

- Formation of the **chorion** which will become the tough outer membrane surrounding the fetus inside its protective, fluid (**liquor amnii**) filled aminiotic membrane or 'bag of water'.
- Growth of vascular **chorionic villi**, from the chorion, into the decidua to form the close contacts with maternal blood vessels required for exchange of substances.

Meanwhile the blood vessels are forming in the embryo, which will eventually be linked to the placenta by three blood vessels (one vein and two arteries) which run in the **umbilical cord**. The placenta and vessels start to function within a few weeks and by two months the exchange system is fully operational. Soon the bag of membranes will completely fill the uterine cavity, with all the chorionic villi concentrated as the placenta.

The connection between mother and fetus ensures that oxygen, nutrients and protective antibodies can pass from the maternal blood across the placenta (normally there is no mixing of maternal and fetal blood) into the fetal blood. Fetal waste products including carbon dioxide pass in the opposite direction across the placenta into the maternal blood.

Unfortunately agents harmful to the fetus can also cross the placenta. These include:

- Viruses, such as rubella.
- Drugs, e.g. antibiotics, antithyroid, sedatives and many more.

NB all drugs should be avoided during pregnancy, especially in the first trimester, when organ development occurs – only drugs prescribed by a doctor should be taken.

- Alcohol.
- Chemicals from cigarettes.
- Harmful antibodies such as anti-D (see Chapters 9 and 19).

The placenta also forms part of the endocrine unit producing the hormones required to sustain the pregnancy (see pages 576 and 577), placental hormone production replaces that of the corpus luteum by 10–12 weeks.

Fetal circulation

During embryonic development the cardiovascular system is adapted for intrauterine life, a time when metabolic needs are met via the placenta. There are extra blood vessels and shunts which are converted at or soon after birth to an adult circulation pattern (see Chapter 10) which includes the pulmonary circulation.

By 4 weeks the fetal heart is beating and a blood vessel system is developing rapidly. Soon after this the embryo is connected to the placenta by the umbilical vessels, which remain the vital 'life-line' until birth and the infant's first breath.

Fetal blood, at 80% oxygen saturation, returns from the placenta within a large single **umbilical vein** to the liver (*Figure 20.24*). Most of the blood bypasses the liver and passes directly into the inferior vena cava via a shunt called the **ductus venosus**. The inferior vena cava takes blood, now mixed with the low oxygen content blood returning from the cells, to the right atrium.

From the right atrium, some of the blood is shunted through the **foramen ovale** (an opening in the septum), into the left atrium. This blood enters the left ventricle and leaves via the aorta to supply the head and upper limbs.

Blood returning via the superior vena cava to the right side of the heart and reaching the pulmonary artery is redirected to the aorta via the **ductus arteriosus**, a shunt between the pulmonary trunk and the aorta.

The foramen ovale and ductus arteriosus are shunts which enable most of the blood flow to bypass the non-functional lungs and

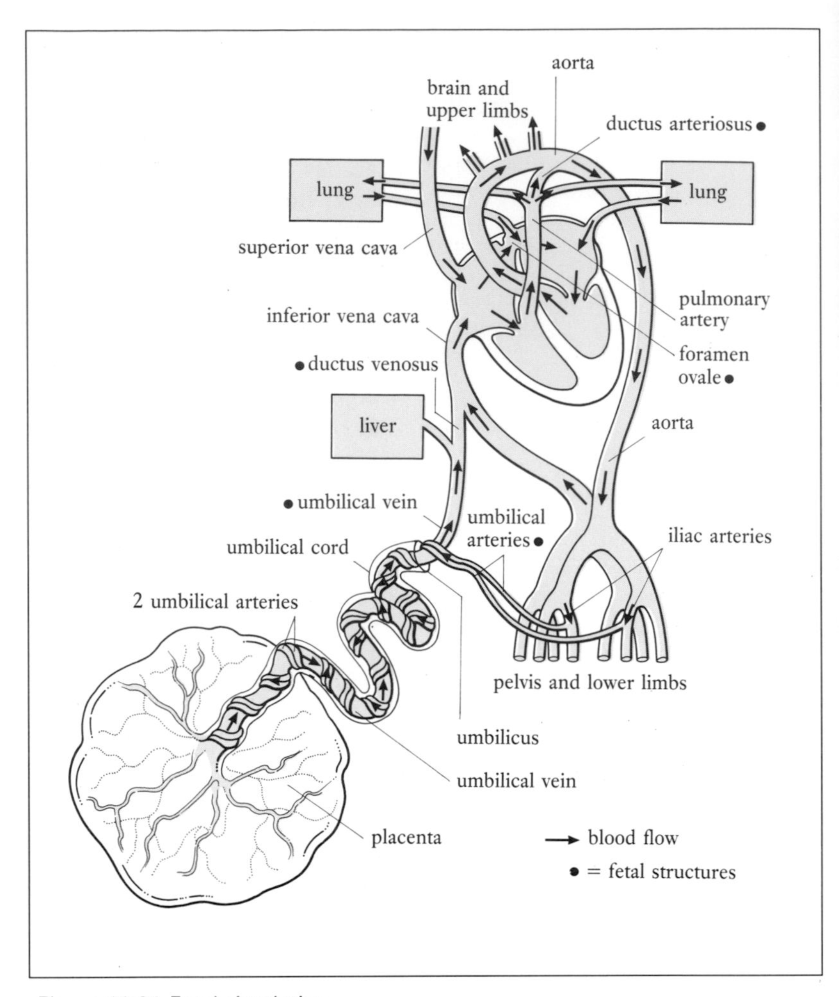

Figure 20.24. Fetal circulation.

pulmonary circulation. Blood flows through the shunts in the correct direction because pressure differentials are created by the minimal venous return from the lungs and blood flows from high to low pressure areas.

Blood with waste and a low oxygen content flows through the distal aorta to leave the fetus via the internal iliac arteries and the two **umbilical arteries** which carry the blood back to the placenta. Here fetal blood passes through the capillary network to exchange substances with the maternal blood prior to returning via the umbilical vein.

NB The liver, lungs and gastrointestinal tract are mostly bypassed because their functions are dealt with by maternal systems; they do, however, receive sufficient blood for their development.

Maternal changes during pregnancy

The numerous physical and psychological changes which occur in women during pregnancy may be attributed in part to major structural events, physiological adaptations to the pregnancy or

Special Focus **Screening for fetal abnormalities**

Investigations used during pregnancy can detect some, but not all, fetal abnormalities. Termination of pregnancy is an option for couples when serious fetal abnormality is diagnosed. Where appropriate the couple should be offered genetic counselling, where the risk of other pregnancies being similarly affected can be assessed, which allows the couple to make informed decisions about planning future pregnancies.

Recently a technique called *chorionic villus sampling (CVS)* has become available in some centres (*Figure 20.25 (a)*). Samples of fetal tissue can be obtained via the cervix for the detection of genetic abnormalities as early as 8–10 weeks of pregnancy. This has the great advantage of early diagnosis and if necessary a less traumatic first trimester termination. However, a recent evaluation (MRC, 1991) suggests that CVS is less safe than amniocentesis (which had a 1% miscarriage risk) and its decreased accuracy means that retesting may be required.

Many abnormalities can be diagnosed from a sample of the fetal cells shed into the amniotic fluid surrounding the fetus. The sample is obtained by *amniocentesis* where a wide-bore needle is passed into the amniotic sac via the abdominal wall (suitably anaesthetized) (*Figure 20.25 (b)*). Fetal cells obtained from the fluid can be grown and examined for chromosomal abnormalities such as Down's syndrome. The amniotic fluid may contain chemical markers for a particular abnormality, e.g. the presence of *alpha-fetoproteins* may indicate neural tube defects such as *spina bifida* (see Chapter 4).

The major disadvantage of amniocentesis is that it is not usually performed until 16–18 weeks of pregnancy and a further wait for the chromosome tests means that the couple are faced with the possibility of a very late termination. Earlier amniocentesis at 10–14 weeks may be developed as a possible alternative to CVS (MRC, 1991).

Tests where the chromosomes are examined also reveal the sex of the fetus which is relevant where a history exists of sex-linked genetic conditions such as haemophilia (see Chapter 9).

Both CVS and amniocentesis are invasive and carry an increased risk of spontaneous miscarriage; this should be explained to the couple. These tests are not routine and are normally offered only to: (i) women in their late 30s; (ii) couples with a history of previous fetal abnormality; or (iii) couples with a family history of genetic disorders.

Ultrasound scan (USS), a non-invasive technique, is used routinely to check the progress of normal pregnancies and assist in the diagnosis of fetal and placental abnormalities.

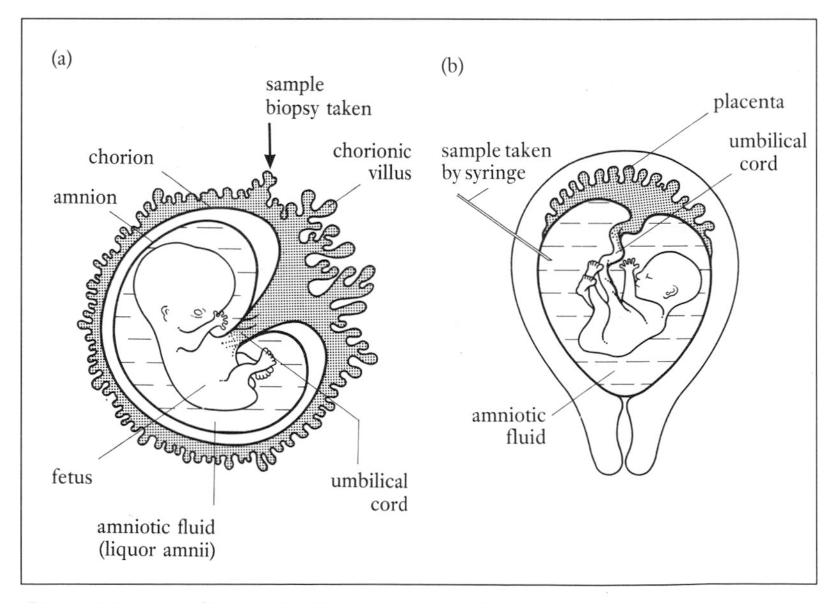

Figure 20.25. (a) Chorionic villus sampling; (b) amniocentesis.

hormonal effects (see below). For our purposes it is not appropriate to discuss all these changes in depth, but examples include:

- The uterus grows to fill most of the abdominal cavity and, by term (40 weeks), the fundus has reached the level of the xiphoid process. This enormously enlarged uterus will compress the abdominal structures, causing effects such as 'heartburn' where gastric acid is regurgitated into the oesophagus. Pressure on the bladder causes the frequency and urgency of micturition associated with the last weeks of pregnancy.
- During pregnancy blood volume increases by as much as 40% to cope with the extra demands. The increase necessitates changes in cardiovascular function which include an increased cardiac output.
- Changes in hormone levels have wide-ranging effects on metabolism. The nausea experienced by many women in the first 12 weeks is due to increasing progesterone and oestrogen secretion. Progesterone also reduces peristalsis, which results in constipation. There may be increased skin pigmentation affecting the nose and cheeks (*chloasma*), breast areola and a line from umbilicus to pubis may darken to form the *linea nigra*. Hormones are also responsible for the mood changes so characteristic of pregnancy. For some women the later stages of pregnancy are associated with a feeling of calm and placidity.

Hormones and pregnancy

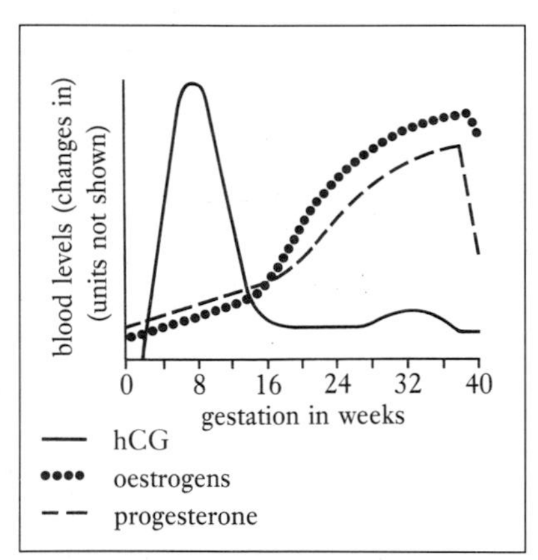

Figure 20.26. Hormone changes during pregnancy.

Progesterone (Figure 20.26)
Progesterone, produced initially by the corpus luteum and later by the placenta, is essential in sustaining the pregnancy. Its functions include maintenance of the endometrium, inhibition of uterine muscle contraction and helping to prepare the breasts for lactation.

Oestrogens (Figure 20,26)
Oestrogens, which are also produced by the corpus luteum and placenta, are concerned with uterine growth. Development of the breast duct system is stimulated by oestrogens and later in pregnancy oestrogens prepare the myometrium for parturition (see page 577).

Human chorionic gonadotrophin (Figure 20.26)
Human chorionic gonadotrophin (hCG) is initially secreted by the trophoblast cells and later by the chorion. hCG maintains the corpus luteum and its secretion of progesterone and oestrogens until placental production of these hormones is sufficiently advanced. The level of hCG peaks around 8–10 weeks, after which it declines suddenly to a low level which persists throughout the pregnancy. The presence of hCG in blood or urine is used to confirm pregnancy (see page 572).

Human placental lactogen
Human placental lactogen (hPL), also produced by the placenta, appears to stimulate growth and to be involved with breast preparation for lactation.

Oxytocin
Oxytocin, secreted by the posterior pituitary, is concerned with parturition (see below) and milk 'let down' during lactation (pages 579 and 580).

Relaxin
Relaxin, produced by the placenta, is concerned with preparing the pelvis for labour (see below).

Many other hormones are also concerned with the changes occurring during pregnancy, including growth hormone and other anterior pituitary hormones, thyroid hormones, corticosteroids and insulin.

Parturition

Parturition (giving birth) normally occurs after a gestation period lasting some 280 days (40 weeks) from the first day of the last menstrual period, or 266 days (38 weeks) from conception (assuming that cycles occurred regularly every 26–30 days). Most babies arrive within a fortnight either side of the expected date of delivery (EDD).

'Labour' is the term used to describe the processes required for parturition. The cervix dilates and the uterus contracts to expel the infant plus placenta and membranes, and afterwards prevents excess blood loss.

What actually initiates labour is not completely understood but several hormones are known to be involved. During the last weeks of pregnancy a change in the oestrogen/progesterone ratio appears to have two effects: the inhibitory effect of progesterone on uterine muscle is reduced and the myometrium becomes very sensitive to the posterior pituitary hormone oxytocin, which is known to stimulate myometrial contraction. Women experience irregular contractions, known as *Braxton Hicks* contractions, as pregnancy nears completion. The hormone relaxin increases flexibility of the pelvic ligaments, which allows for more 'give' as the fetus passes through the pelvis.

The secretion of *prostaglandins* ($F_{2\alpha}$) and high levels of oxytocin are known to initiate the contractions of 'true labour', but why the prostaglandins are released at a particular time is not clear. Both oxytocin and prostaglandins can be used to stimulate labour or termination of pregnancy. The use of antiprostaglandin drugs to inhibit premature labour can be taken as further evidence that prostaglandins are involved in labour.

Labour is divided into three stages (see *Family-centred Study*):

- First stage, which involves cervical dilatation.
- Second stage, during which the infant is expelled by powerful

Family-centred Study Ann, George and Harvey

Ann and George, expecting their second baby, were delighted when Ann was woken at 5am one morning (EDD + 5 days) by contractions in her lower abdomen. During the morning Ann had a 'show' of blood-stained mucus from the cervix and when the contractions, which had been every 10 minutes, started to increase they phoned the midwife, who came immediately. The plan was to have the baby at home and the midwife examined Ann to find that all was progressing well and that her cervix was thinning and was already 4 cm dilated.

Ann started using a transcutaneous electrical nerve stimulator (TENS) during contractions (see Chapter 4) and because she was getting good pain relief didn't feel that she needed any other analgesia. After checking with Ann and George about any last minute preparations and answering their questions the midwife arranged cover for her other work. Ann, who had formed a good relationship with the midwife during the pregnancy, felt confident and relaxed.

By 1 pm the contractions were coming every 3 minutes and Ann could feel that they were getting stronger. Although Ann was still coping well with the contractions she now had backache which was causing some distress.

Another examination revealed that the cervix was 8 cm dilated; Ann and George were delighted that the progress was so good. Ann started inhalation analgesia (nitrous oxide + oxygen) whilst George rubbed her back which really helped. At 3.30pm the membranes ruptured releasing the amniotic fluid (surrounding the baby) and Ann had a strong urge to 'bear down'. A quick examination revealed that the cervix was fully dilated and that the second stage had started.

The contractions were every 2 minutes and Ann was pushing with good effect. By 4 pm George and the midwife could see the baby's head which was a great boost to Ann who was working very hard. At 4.15 pm the baby, a boy (Harvey) was born, who cried (and breathed) at once. The umbilical cord was clamped and cut whilst Ann and George had a good look at their son, who looked quite 'perfect'.

Ann was given an injection of ergometrine and oxytocin to contract the uterus and the third stage was completed by 4.35 pm. The midwife confirmed that Ann's uterus was contracted and that there was no bleeding and then checked that the placenta and membranes (amnion and chorion) were complete (retained placental tissue could cause infection and haemorrhage).

Whilst all the checks and care were completed George collected their daughter, who had spent the afternoon with neighbours, so she could meet her brother.

contractions.

- Third stage, when the placenta and membranes are expelled

Changes to the fetal circulation at birth

Several changes must occur in the infant at birth so that it may adjust to extrauterine life, but the most immediate is the establishment of respiration. The infant will usually gasp and start breathing in response to the sudden exit from its cosy uterine surroundings.The first breath, which inflates the lungs, causes enormous circulatory changes. Blood flowing through the pulmonary circulation decreases pulmonary artery pressure and increases that in the left side of the heart and aorta. This reversal of pressure has two effects:

- The ductus arteriosus constricts and eventually becomes a fibrous cord.
- A flap occludes the foramen ovale, although closure will take some months to complete.

The umbilical vessels constrict and the ductus venosus empties. These structures will eventually become fibrosed but persist as ligaments, e.g. umbilical vein becomes the ligamentum teres of the liver (see Chapter 14).

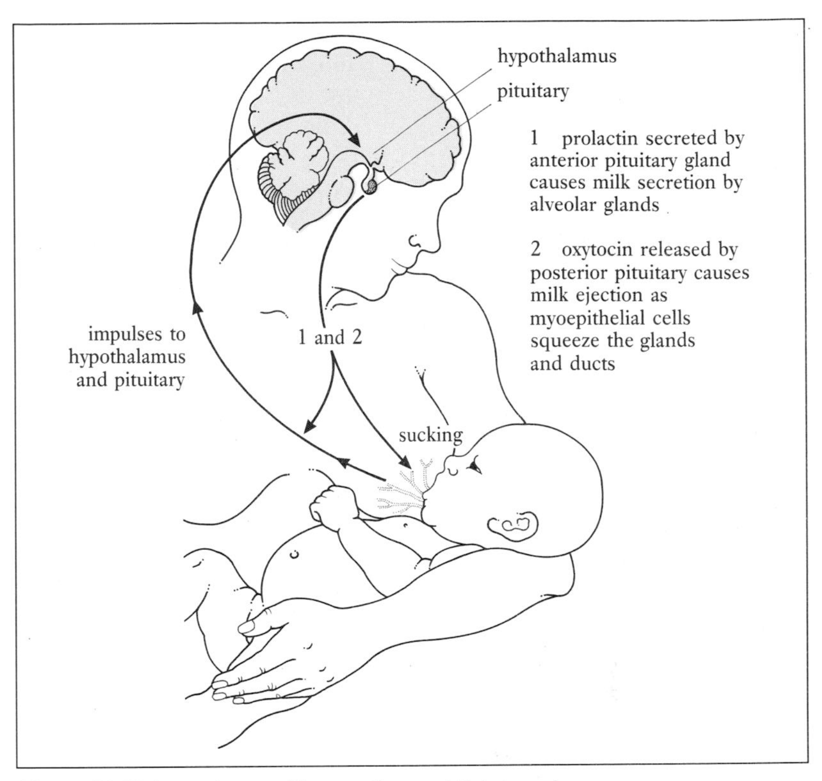

Figure 20.27. Lactation – milk secretion and 'let down'.

Congenital heart defects are sometimes caused by continued patency of foramen ovale or ductus arteriosus which, as you can imagine, cause considerable problems for the infant. These defects can usually be treated surgically.

Apgar Score

At birth, and again after 5 minutes, the infant's physical condition is assessed using the five physical signs of the *Apgar rating.* The infant is scored on a scale 0–2 for heart rate, respiratory effort, colour, reflexes and muscle tone (maximum score 10). Most healthy infants such as Harvey would be expected to score 8 or above.

Lactation

During pregnancy the secretion of hormones causes breast enlargement and development of the glandular secreting tissue and duct system. Late in pregnancy the breasts are already secreting a fluid called **colostrum**, which is also secreted during the 2–3 days following parturition.

Colostrum secretion is initiated by the hormones human placental lactogen (hPL; see page 577) and **prolactin** from the anterior pituitary (*Figure 20.27*). Prolactin release, which increases from early pregnancy, may be due to the suppression of prolactin inhibiting hormone or the presence of a hypothalamic releasing hormone (see Chapter 8). The colostrum, which is replaced by 'true' milk after 3 days, contains more protein, minerals and vitamins, but less lactose and fat than milk. Although prolactin starts off milk secretion it is the sucking by the infant which ensures continuity of supply. As the infant sucks, it initiates release of prolactin which in turns stimulates the milk for the next feed. Without the stimulation of sucking, prolactin levels fall and lactation is not established.

So far we have only discussed milk secretion, another hormone oxytocin is required for milk ejection or 'let down' reflex (*Figure 20.27*). Oxytocin release, which is also stimulated by the infant sucking, causes contraction of myoepithelial cells which surround the glandular tissue and milk is ejected. The release of oxytocin also causes uterine contraction, which helps the uterus to *involute* (return to prepregnant state); this explains the abdominal 'cramps' felt sometimes during breast-feeding.

Both colostrum and milk contain immunoglobulins (IgA) which may help to prevent gastrointestinal infections in the infant (see Chapter 19). Breast milk contains water, lactose, proteins, fats, minerals and vitamins in a form that the infant can most easily absorb and use.

During full lactation the prolactin levels are usually sufficient to prevent the return of ovarian cycles and menstruation. It is, however, unwise to rely upon this as contraception because ovulation can certainly occur during lactation.

Summary/Check List

Introduction

Male reproductive system – Structure. Testes. Testicular tumours. Duct system. Accessory glands. Penis, *Nursing Practice Application* – phimosis/paraphimosis. Functioning – meiosis. Spermatogenesis. Semen. Hormonal control, testosterone. Sex differentiation and puberty in the male. Changes with ageing.

Female reproductive system – Structure, ovaries, *Person-centred Study* – Lucy, Uterine tubes. Uterus, CIN/cervical cancer, *Nursing Practice Application* – cervical screening, uterine supports. Vagina. Vulva, *Nursing Practice Application* – episiotomy/perineal laceration. Mammary glands, *Person-centred Study* – Lucy (continued), breast cancer. Functioning – oogenesis. Ovarian cycle and hormonal control. Female steroid hormones. Menstrual cycle, *Nursing Practice Application* – PMS. Sex differentiation and puberty in the female. Climacteric/menopause, HRT.

Reproductive physiology – Sexual response. STD, *Special Focus* – contraception. Conception/fertilization. Chromosome complement and determination of genetic sex. Subfertility. Early embryonic development and implantation. Placental development and function. Fetal circulation, *Special Focus* – screening for fetal abnormalities. Maternal changes. Hormones and pregnancy. Parturition – labour, *Family-centred Study* – Ann, George and Harvey. Changes to fetal circulation. Lactation.

?

Self Test

1 Which of the following statements are true?
(a) The testes function best at body temperature.
(b) The epididymis is continuous with the vas deferens.
(c) Most of the fluid part of semen is produced by the prostate gland.
(d) The male urethra coveys both urine and semen.

2 Discuss ways in which gametogenesis differs between males and females.

3 Complete the following:
(a) Spermatogenesis occurs in the ____________ _______.
(b) A mature spermatozoon has a ____, ________ and ____.
(c) Enzymes are carried in the ________ of the spermatozoon.
(d) The gonadotrophins ___ and __ stimulate spermatogenesis and the release of the male hormone ____________.

4 Which of the following are part of the internal female genitalia and which are external genitalia?
(a) uterus (b) ovary (c) fourchette (d) fimbriae (e) cornua
(f) ovary (g) clitoris (h) external os (i) labia minora
(j) Bartholin's glands (k) endometrium (l) mons pubis.

5 Briefly describe the events of oogenesis.

6 Put the following in the correct chronological order:
(a) primary follicle (b) corpus luteum (c) primordial follicle
(d) corpus albicans (e) Graafian follicle
(f) maturing/secondary follicle.

7 Describe the phases of the menstrual cycle and the endometrial changes that occur.

8 Which of the following statements are true?
(a) Fertilization occurs in the ampulla.
(b) The zygote formed at fertilization is haploid.
(c) Implantation occurs around 6–7 days after fertilization.

9 Put the following in their correct pairs:
(a) prostaglandins and oxytocin (b) pregnancy testing
(c) progesterone (d) initiation of labour
(e) hCG (f) longest stage of labour
(g) cervical dilatation (h) inhibition of uterine contraction

10 What answers would you give to the following questions asked by Jane who is breast-feeding her first baby aged 2 days.
(a) Why do I have 'crampy' abdominal pains whilst feeding?
(b) How does colostrum differ from real milk?
(c) Why is the baby sucking so important to milk production?
(d) Is it true that I cannot get pregnant during lactation?

Answers
1. b, d. 2. See pages 536 and 554. 3. a, Seminiferous tubules; b, head, midpiece and tail; c, acrosome; d, FSH, LH and testosterone. 4. Internal – a, b, d, e, f, h, k. External – c, g, i, j, l. 5. See pages 552-554. 6. c, a, f, e, b, d. 7. See pages 557-560. 8. a, c. 9. a – d, c – h, e – b, g – f. 10. See pages 579-580.

References

Beardsley, J. (1993) Education to undermine a taboo. Understanding herpes simplex virus. *Professional Nurse*, **8** (5), 322-328.

Cancer Research Campaign (for the NHS Breast Screening Programme) (1991) *Be Breast Aware*. Oxford: Cancer Research Campaign.

Dalton, K. (1980) Cyclical criminal acts in premenstrual syndrome. *Lancet*, **ii**, 1070–1071.

Filshie, M. and Guillebaud, J. (1989) *Contraception Science and Practice*. London: Butterworth.

Gould, D. (1990) *Nursing Care of Women*. (Ch 10 and 13). London: Prentice Hall.

Health Education Authority (1989) *Can You Avoid Cancer? A Guide to Reducing Your Risks*. London: Health Education Authority.

Hiscock, E. Reece, G. (1988) Cytological screening for cervical cancer and human papillomavirus in general practice. *British Medical Journal*, **297**, 724–726.

Hughes, Y. (1992) Menopause – women of a certain age. *Nursing Times*, **88** (12), 36–37.

Kirkpatrick, M.K., Brewer, J.A., and Stocks, B. (1990) Efficacy of self-care measures for premenstrual syndrome (PMS). *Journal of Advanced Nursing*, **15** (3), 281–285.

McLaughlin, M. (1989) Gamete intrafallopian transfer (GIFT) – a treatment for infertility. *Midwives' Chronicle and Nursing Notes*, **102** (1212), 23.

MRC Working Party on the Evaluation of Chorionic Villus Sampling (1991) Medical Research Council European Trial of Chorionic Villus Sampling. *Lancet*, **i**, 1491–1499.

Thornhill, J.A., Conroy, R.M., Kelly, D.G., Walsh, A., Fennelly, J.J. and Fitzpatrick, J.M. (1986) Public awareness of testicular cancer and the value of self examination. *British Medical Journal*, **293**, 480–481.

Winston, R.M.L. (1986) *Infertility: A Sympathetic Approach*. London: Martin Dunitz.

Further Reading

Adler, M.W. (1986) *ABC of Sexually Transmitted Diseases*. London: British Medical Journal publications.

Edwards, R.G. (1980) *Conception in the Human Female*. London: Academic Press. (An advanced text.)

Edwards, R.G. and Steptoe, P.C. (1980) *A Matter of Life*. London: Hutchinson.

Emery, A.E.H. and Mueller, R.F. (1988) *Elements of Medical Genetics*, 7th edn. Edinburgh: Churchill Livingstone.

England, M.A. (1983) *A Colour Atlas of Life Before Birth – Normal Fetal Development*. London: Wolfe Medical Publications Limited.

Filshie, M. and Guillebaud, J. (1989) *Contraception Science and*

Practice. London: Butterworth.
Gould, D. (1990) *Nursing Care of Women.* London: Prentice Hall
Keele, C.A., Neil, E., and Joels, N. (1982 Reprinted with corrections 1984) *Samson Wright's Applied Physiology* (part XII), 13th edn. Oxford: Oxford University Press.
Masters, W. H. and Johnson, V.E. (1966) *Human Sexual Response.* Boston: Little, Brown and Company
Oakley, A., McPherson, A., and Roberts, H. (1984) *Miscarriage.* Glasgow: Fontana.
O'Brien, P.M.S. (1987) *Premenstrual Syndrome.* Oxford: Blackwell Scientific Publications.
Paterson, M.E.L. (1985) *The Menopause.* London: Update Publications Limited.
Sadow, J.I.D., Gulamhusein, A.P., Morgan, M.J., Naftalin, N.J. and Peterson, S.A. (1980) *Human Reproduction: An Integrated View.* London: Croom Helm.
Stevenson, J. (Ed) (1991) *Osteoporosis*, Update Postgraduate Centre Series.Guildford: Reed Healthcare Communications.
Warnock, M. (1985) A Question of Life – The Warnock Report on Human Fertilization and Embryology. Oxford: Basil Blackwell Limited.
Webb, C. (Ed) (1986) *Feminist Practice in Women's Health Care.* Chichester: Wiley.
Webb, C. (1985) *Sexuality, Nursing and Health.* Chichester: Wiley.
Williams, P.L., Wendell-Smith, C.P., and Treadgold, S. (1984) *Basic human embryology*, 3rd edn. London: Pitman Medical.

International System of Units (SI)

Introduction

The International System of Units (SI) or (*Systeme Internationale*) is a system of measurement used for medical, scientific, and technical purposes throughout most of the world. In the United Kingdom, SI units replaced those of the Imperial System, e.g. the kilogram is used for mass instead of the pound (N.B. in everyday situations, both mass and weight are measured in kilograms although weight, which varies with gravity, is really a measure of force).

There are seven base units and several derived units within the International System of Units. Each measurement unit has its own symbol and can be expressed as a decimal multiple or submultiple of the base unit by the use of prefixes (see Appendix B).

Writing Large Numbers and Decimals

Before we look at the details of SI, it would be helpful to establish the ground rules for writing large numbers and decimals. These have particular relevance for practice, e.g. prescriptions for medication and laboratory results, where it is essential to be clear and accurate so as to avoid confusion and error.

Large numbers should be written in groups of three digits (from right to left) with spaces, but without commas (which are used in some countries as a decimal point).

For example:

ten thousand	should be written as	10 000
one hundred thousand	should be written as	100 000

N.B. numbers with four digits are written without the space, e.g. one thousand — 1000.

Decimals should always have a zero (0) before the decimal point which should be positioned near the line, e.g. 0.25 and 0.005. Decimals with more than four digits should be written in groups of three digits (from left to right) with spaces, e.g. 0.000 25.

Base Units and Symbols

Measurement	SI Base Unit and Symbol
Mass	kilogram (kg)
Length	metre (m)
Temperature	kelvin (K)

Time	second (s)
Amount of substance	mole (mol)
Luminous intensity	candela (cd)
Electric current	ampere (A)

Derived Units (with nursing practice application)

Measurement	**SI Derived Unit and Symbol**
Celsius temperature	Celsius (°C)
Frequency	hertz (Hz)
Energy/quantity of heat/work	joule (J)
Absorbed dose of radiation	gray (Gy)
Radioactivity (radionuclide)	becquerel (Bq)
Dose equivalent	sievert (Sv)
Pressure	pascal (Pa)
Force	newton (N)
Electrical potential Potential difference Electromotive force	volt (V)
Power	watt (W)

Factors, Symbols and Prefixes for Decimal Multiples and Submultiples

Factor	**Prefix**	**Symbol**
10^9	giga	G
10^6	mega	M
10^3	kilo	k
10^2	hecto	h
10^1	deca	da
10^{-1}	deci	d
10^{-2}	centi	c
10^{-3}	milli	m
10^{-6}	micro	μ
10^{-9}	nano	n
10^{-12}	pico	p
10^{-15}	femto	f

N.B. As *kilogram* already contains a prefix, the convention is to use the prefixes with gram (g), e.g. milligram (mg).

Anomalies (encountered in clinical nursing practice)

Temperature — Celsius (°C) temperature is used in clinical practice for measuring temperature, rather than kelvin (K).

Time — although the second (s) is the SI base unit for time, it is acceptable to use minute (min), hour (h) and day (d), e.g. GFR = 125 ml/min.

Volume — the litre (l or L), a non-SI unit, is used clinically to measure volume of fluids and gases. The litre is based on the volume of a cube (10 cm x 10 cm x 10 cm). A smaller unit, the millilitre (ml), which forms one thousandth part of a litre, has numerous practice applications, e.g. recording fluid balance.

Pressure — the SI unit for pressure is the pascal (Pa), but in some areas, e.g. recording blood pressure, the old unit — millimetres of mercury pressure (mmHg) — is still widely used. The kilopascal (kPa) is increasingly used to measure blood gases and would replace mmHg pressure for blood pressure. Other clinical applications where the pascal is not used include: central venous pressure (CVP), which is measured in centimetres of water pressure (cm/H_2O); and the pressure of cerebrospinal fluid (CSF), which is measured in millimetres of water pressure (mm/H_2O).

Energy — the SI unit for energy is the joule (J) and for calculating energy composition or requirements the SI unit, kilojoule (kJ), is used. An old unit of measurement, the Calorie or kilocalorie (kcal), is however, still used. (N.B. 1 kcal = 4.2 kJ).

Amount of Substance — the SI base unit for amount of substance is the mole (mol), but some substances, e.g. enzymes in the serum, are measured in *International Units* (*IU* or *iu*).

The Components of Medical Words and Terms

Prefixes, Suffixes, Word Roots and Combining Forms

Medical words or terms are usually compound words which contain more than one component part. They are formed from the addition of a prefix or a suffix to a word root or combining form. An understanding of medical terms will be achieved more easily if you become familiar with the main prefixes, suffixes and combining forms. Once you learn the 'code,' it will often be possible for you to work out the meaning of new words when you enounter them.

Component	***Meaning***	***Example***
a/an-	without, lack of, not	*An*uria (lack of urine)
ab-	away from	*Ab*duction (move a limb away from the body)
-able	capable of	Vi*able* (capable of living or surviving)
acro-	extremity	*Acro*megaly (enlarged extremities - jaw, hands and feet)
ad-	towards	*Ad*duction (move a limb towards the body)
adeno-	glandular	*Adeno*ma (benign tumour arising in glandular tissue)
*-aemi*a	blood	An*aemia* (lack of blood - reduction in red cell numbers or haemoglobin)
aer-	air	*Aer*obic (requiring oxygen)
-aesthesia	sensation	Par*aesthesia* (disordered sensation - 'pins and needles')
-algia	pain	Proct*algia* (pain in the rectum)
ambi-	both	*Ambi*dextrous (equally skilled with both hands)
andro-	male	*Andro*gens (male hormones)

Component	Meaning	Example
angio-	blood vessel	*Angio*graphy (Xray picture taken after injection of opaque fluid into the vessel)
aniso-	unequal	*Aniso*cytosis (inequality in red blood cell size)
ante/antero-	before, in front	*Ante*natal (before birth)
anti-	against	*Anti*coagulant (substance which delays or prevents blood coagulation)
arthro-	joint	*Arthro*scopy (endoscopic joint examination)
- *ary*	associated, connected	Urin*ary* (associated with urine)
- asis/esis	state of	Cy*esis* (state of being pregnant)
- *ase*	enzyme	Lip*ase* (enzyme concerned with fat digestion)
auto -	self	*Auto*graft (graft taken from a person's own body)
bi/bis -	two	*Bi*cuspid (having two cusps)
bili -	bile	*Bili*rubin (a bile pigment)
bio -	life	*Bio*chemistry (the chemistry of cell life)
- *blast* -	germ, bud	Haemocyto*blast* (an immature stem cell which can develop into any blood cell)
bleph -	eyelid	*Bleph*aritis (inflammation of the eyelid)
brachi -	arm	*Brachi*al artery (an artery of the arm)
brachy -	short	*Brachy*cephaly (short skull)

Component	Meaning	Example
brady -	slow	*Brady*cardia (abnormally slow heart rate)
broncho -	bronchi	*Broncho*spasm (constriction of the bronchial muscle which narrows the air passages)
bucc -	cheek	*Bucc*inator (a cheek muscle)
carcino -	cancer	*Carcino*genic (agent which causes or predisposes to cancer)
cardio -	heart	*Cardio*myopathy (disease of heart muscle)
carpo -	wrist	*Carpo*metacarpal joints (joints between the wrist and hand bones)
cata -	down	*Cata*bolism (part of metabolism when substances are brokendown)
- *cele*	swelling/hollow	Hydro*cele* (fluid filled swelling associated with a testis)
cent -	hundred	*Cent*imetre (unit of length, 100th part of a metre)
- *centesis*	puncture	Amnio*centesis* (aspiration of amniotic fluid for diagnostic purposes)
cephal -	head	*Cephal*ometry (measurement of the head)
cerebro -	brain	*Cerebro*spinal fluid (fluid which surrounds the brain and spinal cord)
cervic -	cervix/neck	*Cervic*itis (inflammation of the uterine cervix)

Component	Meaning	Example
cheil/cheilo -	lip	*Cheilo*plasty (plastic operation on the lips)
chemo -	chemical	*Chemo*receptor (receptors able to respond to chemical stimuli, e.g.oxygen levels in the blood)
chol/chole -	bile	*Chole*cystokinin (hormone which stimulates gall-bladder contraction and bile release)
cholecyst-	gallbladder	*Cholecyst*ectomy (removal of the gallbladder)
choledocho -	bile ducts	*Choledocho*lithotomy (opening the bile duct to remove gallstones)
chondro -	cartilage	*Chondro*blast (an embryonic cartilage-producing cell)
chrom -	colour	Haemo*chrom*atosis (a condition where abnormal deposition of iron gives rise to skin pigmentation)
- cide	killing	Sui*cide* (self-destruction)
circum -	around	*Circum*duction (circular movement where the limb traces a cone in space, e.g. arm at the shoulder)
- cle/cule	small	Fasi*cle* /fasic*ule* (small bundle of fibres, e.g. of muscle)
co/con	together with	*Co*enzyme (substances, e.g. some vitamins, required for certain metabolic processes)
coli -	bowel	*Coli*form bacteria (organisms normally found in the bowel)

Component	Meaning	Example
colpo -	vagina	*Colpo*rrhaphy (repair of the vagina)
contra -	against	*Contra*ceptive (device or drug which prevents conception)
cost -	rib	Inter*cost*al muscles (muscles between the ribs)
cox -	hip	*Cox*algia (pain in the hip)
crani -	skull	*Crani*um (part of the skull enclosing the brain)
cryo -	cold	*Cryo*surgery (surgical technique which utilizes intense cold)
crypt -	hidden	*Crypto*menorrhoea (apparent amenorrhoea where menstrual blood is retained due to an imperforate hymen)
cyan -	blue	*Cyan*osis (blue discoloration of the skin and mucous membranes caused by poor oxygenation)
cysto -	bladder	*Cysto*scopy (endoscopic examination of the urinary bladder)
- cyte	cell.	Leuco*cyte* (white blood cell)
cyto -	cell	*Cyto*toxic (substance which is toxic to cells)
dacry -	tear	*Dacry*adenitis (inflammation of the tear (lacrimal) gland)
dactyl -	digit (usually fingers)	Poly*dactyl*y (presence of supernumerary digits)
de -	away.	*De*calcification (loss of calcium salts from bone)

Component	**Meaning**	**Example**
deci -	tenth	*Deci*litre (one tenth of a litre - 100 ml)
demi , *hemi*-	half	*Hemi*paraesis (weakness affecting one side of the body)
dent -	tooth	*Dent*ine (substance forming the bulk of a tooth)
derma -	skin	*Derma*titis (inflammation of the skin)
- *desis*	bind or fix together	Arthro*desis* (operation to fix a joint)
dextro -	right	*Dextro*cardia (congenital transposition of the heart to the right side)
di/diplo -	twos, double	*Di*saccharide (a sugar formed from two monosaccharide units, e.g. sucrose)
dia -	through	*Dia*pedesis (passage of white blood cells through blood vessel walls)
dis -	separation	*Dis*location (displacement of the articular surfaces of the bones forming a joint)
dors -	back	*Dors*al (relating to the back or the posterior part of an organ)
- *dynia*	pain	Pleuro*dynia* (pain felt in the intercostal muscles)
dys-	difficult, painful.	*Dys*menorrhoea (difficult or painful menstruation)
ec-	out from.	*Ec*topic (outside the normal place, e.g. a gestation where an embryo develops outside the uterus)

Component	Meaning	Example
- ectasis	dilation	Bronchi*ectasis* (a condition where the bronchi are abnormally dilated)
ecto -	outside	*Ecto*derm (outer germ layer of primitive embryonic tissue)
- ectomy	removal	Gastr*ectomy* (removal of the stomach)
electro -	electrical	*Electro*lyte (substance which dissociates in water to form electrically charged ions)
em -	in	*Em*pyema (collection of pus in a cavity)
- emesis	vomiting	Hyper*emesis* (excessive vomiting)
en/end/endo -	in, into, within	*Endo*metrium (inner layer of the uterus)
entero -	intestine	*Entero*kinase (intestinal enzyme which activates trypsinogen)
epi -	above, on, upon	*Epi*condyle (bony eminence above or upon a condyle)
erythr -	red	*Erythr*ocyte (red blood cell)
eu -	well, normal	*Eu*thyroid (having normal thyroid function)
ex/exo -	away from, out of	*Ex*ophthalmos (abnormal protrusion of the eyeball out of the orbit)
extra -	outside	*Extra*pyramidal (outside the pyramidal tracts)
faci -	face	*Faci*al (relating to the face)
- facient	making	Aborti*facient* (a drug which induces abortion)

Component	**Meaning**	**Example**
- ferent	carry	Af*ferent* (carry towards the centre)
ferri/ferro -	iron	*Ferri*tin (a storage form of iron)
feto -	fetus	Alpha-*feto*protein (a protein produced by the fetus)
fibro -	fibrous, fibre	*Fibro*sis (formation of fibrous tissue in areas of tissue damage)
flav/flavo -	yellow	*Flavo*proteins (coenzyme substances involved in metabolic processes)
fore -	before, infront of	*Fore*brain (front part of the brain)
- form	having the form of	Fili*form* (resembling a thread)
galact -	milk	*Galact*ose (a monosaccharide which combines with glucose to form lactose - milk sugar)
gastr/gastro -	stomach	*Gastr*in (local hormone produced by the stomach)
- genesis/ genetic	formation	Gluconeo*genesis* (formation of glucose from non-carbohydrate sources)
- genic	capable of causing	Pyro*genic* (causing fever)
genito -	genitals	*Genito*urinary (relating to the genital and urinary structures)
ger -	old age	*Ger*ontology (study of ageing)
gloss/glosso -	tongue	*Gloss*al (relating to the tongue)

Component	**Meaning**	**Example**
glyco -	sugar	*Glyco*lysis (series of reactions which breakdown sugar)
gnath -	jaw small jaw)	Micro*gnath*ia (abnormally
- *gogue*	increasing flow	Galacta*gogue* (agent which increases the flow of milk)
- *gram*	a tracing or drawing	Electroencephalo*gram* (tracing of brain wave patterns)
- *graph*	instrument for recording brain wave patterns)	Electroencephalo*graph* (instrument for recording
gynae -	female	*Gynae*comastia (male breast enlargement)
haem/haemo/ haemato -	blood	H*aemo*poiesis (formation of blood cells)
hemi -	half	*Hemi*anopia (loss of sight in half the visual field)
hepat/hepato -	liver	*Hepato*cytes (parenchymal cells of the liver)
hetero -	different	*Hetero*zygous (where the paired genes for a particular characteristic are different)
hexa -	six	*Hexa*gonal (having six sides, e.g. a liver lobule)
hist -	tissue	*Hist*iocytes (macrophages - phagocytic tissue cells)
homeo -	same, like, unchanging	*Homeo*thermic (warm-blooded; an animal which maintains the same core temperature range)
homo -	same	*Homo*zygous (where the paired genes for a particular characteristic are the same)

Component	Meaning	Example
hydro -	water	*Hydro*lysis (breakdown of complex substances by the addition of water)
hygro -	moisture	*Hygro*scopic (having the ability to absorb moisture)
hyper -	above, excessive	*Hyper*capnia (excess carbon dioxide in the blood)
hypno -	sleep	*Hypno*tic (drug which induces sleep)
hypo-	below, deficient	*Hypo*thalamus (area of the brain below the thalamus)
hyster -	uterus.	*Hyster*ectomy (removal of the uterus)
- *ia, iasis -*	state of, condition	Myop*ia* (condition of shortsightedness)
- *iatric(s)/iatry -*	healing, medical specialty	Ger*iatrics* (medical specialty which deals with the disorders of later life)
iatro -	physician	*Iatro*genic (condition caused by the physician)
- *ician*	person skilled in a particular field	Paediatr*ician* (person skilled in the treatment of children)
idio -	peculiar to an individual, self	*Idio*pathic (condition with no apparent cause)
ileo -	ileum	*Ileo*caecal valve (valve between the ileum and caecum)
ilio -	ilium	*Ilio*femoral (relating to the ilium and femur, e.g. iliofemoral ligament)
im , in	not, in, within	*Im*potent (not potent)

Component	Meaning	Example
immuno -	immunity	*Immuno*globulins (protein antibodies concerned with immunity)
inter -	between	*Inter*vertebral (between the vertebrae)
intra -	within	*Intra*uterine (within the uterus)
intro -	inward	*Intro*vert (an inward looking person)
ischio -	ischium	*Ischio*rectal (relating to the ischium and rectum, e.g. ischiorectal abscess)
- *ism*	condition/state	Hyperthyroid*ism* (condition where the thyroid gland is overactive)
iso -	same, equal	*Iso*topes (forms of the same element where atoms have a different number of neutrons)
- *itis*	inflammation.	Cys*itis* (inflammation of the bladder)
kerato -	keratin, horn, cornea	*Kerat*itis (inflammation of the cornea)
- *kin* -	movement	*Kin*etics (study of movement or change)
kypho -	rounded	*Kypho*sis (deformity of the thoracic spine with the formation of a rounded hump)
lacri -	tears	*Lacri*mal (relating to tears, e.g. lacrimal gland)
lact -	milk	*Lact*ation (process of milk production)

Component	Meaning	Example
laparo -	abdomen	*Laparo*tomy (opening the abdomen)
laryngo -	larynx	*Laryngo*spasm (spasm of the larynx)
later -	side	*Later*al (on the side, away from the midline)
leuco/leuko -	white	*Leuco*cyte (white blood cell)
lingua -	tongue	*Lingua*l (relating to the tongue, e.g. lingual artery)
lip/lipo -	fat	*Lip*ids (large group of fatty substances)
- lith/litho -	stones	*Litho*tripter (device which uses shock waves to disintegrate stones, e.g. kidney stones)
- lithiasis	presence of stones	Chole*lithiasis* (stones in the gallbladder)
- logy	science or study of	Cyto*logy* (study of cells)
- lysis	breakdown	Hydro*lysis* (breakdown of complex substances by the addition of water)
macul-	spot	*Macul*a lutea (yellow spot of the retina)
mal -	bad	*Mal*absorption (inability to absorb nutrients in sufficient quantities from the small intestine)
- malacia	softening	Osteo*malacia* (softening of the bones - adult rickets)
mamm/mast -	breast	*Mamm*ogram (radiographic examination of the breast)

Component	Meaning	Example
medi/meso -	middle	*Medi*an (in the middle, e.g. the median line an imaginary longtitudinal line which divides the body down the centre)
mega/megalo -	large	*Megalo*blastic (a type of anaemia characterized by abnormally large red cells)
- megaly	enlargement	Hepato*megaly* (enlarged liver)
melano -	black	*Melano*cytes (skin cells which produce the black pigment melanin)
meta -	after, beyond, between	*Meta*physis (the part of a long bone between the diaphysis and epiphysis)
- meter	measure	Thermo*meter* (device for measuring temperature)
metro -	uterus	*Metro*ptosis (prolapse of the uterus)
micro -	small	*Micro*cyte (abnormally small red blood cell)
milli -	thousand	*Milli*litre (thousandth part of a litre)
mio -	smaller	*Mio*sis (pupil constriction)
mono -	single, one	*Mono*nuclear (having a single nucleus)
- morph -/ morpho -	shape, form	Poly*morph*ous (having several forms)
muco -	mucus	*Muco*lytic (substance which reduces the viscosity of mucus)
multi -	many	*Multi*cellular (having many cells)

Component	Meaning	Example
myel-	marrow, spinal cord	*Myel*oblast (primitive bone marrow cell)
myo -	muscle	*Myo*metrium (uterine muscle)
narco -	stupor	*Narco*tic (drug which produces stupor and is used to relieve pain)
necr -	dead	*Necr*osis (death of tissue)
neo -	new	*Neo*plasm (new growth)
nephr/nephro -	kidney	*Nephr*on (functional unit of the kidney)
neuro -	nerves, nervous system	*Neuro*glia (non-excitable supporting cells of the nervous system)
noct/nyct -	night	*Noct*uria (passing urine at night)
normo -	normal	*Normo*cyte (a red blood cell of the correct size)
null/nulli -	none	*Nulli*parous (never had a child)
oculo -	eye.	*Oculo*motor (concerned with eye movement, e.g. oculomotor nerve)
odonto -	tooth	*Odonto*id (tooth-like, e.g. odontoid peg of the second cervical vertebra)
- ogen	precursor	Angiotensin*ogen* (the inactive precursor of angiotensin)
- oid	likeness	Sigm*oid* (shaped like the Greek letter sigma, e.g. sigmoid colon)

Component	**Meaning**	**Example**
oligo -	diminished	*Oligo*menorrhoea (infrequent or sparse menstruation)
- ology	science or study of	Gynaec*ology* (study of disorders affecting the female reproductive structures)
- oma	tumour	Sarc*oma* (a malignant tumour of connective tissue)
onco -	tumour, mass	*Onco*genic (an agent or substance which causes a tumour)
oo -	egg	*Oo*genesis (production of the egg or oocyte - the female gamete)
oophor -	ovary	*Oophor*itis (inflammation of the ovary)
ophthalmo -	eye	*Ophthalmo*scope (instrument used to examine the inside of the eye)
- opia	defect of vision or eye	Hypermetr*opia* (long sight)
orchido -	testis	*Orchid*ectomy (removal of the testis)
oro -	mouth	*Oro*pharynx (part of the pharynx behind the mouth - between the soft palate and the hyoid bone)
orth -	normal, straight	*Orth*optics (straightening or correction of visual problems such as strabismus)
os -	mouth, bone	*Os* calcis (bone of the heel also called the calcaneus)

Component	Meaning	Example
- osis	condition	Otoscler*osis* (condition of chronic thickening of middle ear ossicles with progressive hearing impairment)
oss, osteo -	bone	*Osteo*cyte (a bone cell)
- ostomy/stomy	opening	Trache*ostomy* (an opening into the trachea)
ot/oto -	ear.	*Oto*liths (tiny calcium deposits (ear stones) found in the utricle and saccule of the inner ear)
- otomy/tomy	incision	Oste*otomy* (incision into bone)
ovari -	ovary	*Ovari*an (relating to the ovary, e.g. ovarian artery)
ovi -	ovum, egg	*Ovi*duct (tube conveying ovum from ovary to uterus also called uterine or Fallopian tube)
oxy -	oxygen	*Oxy*haemoglobin (haemoglobin which is combined with oxygen)
pachy-	thick	*Pachy*dermia (thickening of the skin)
paed -	child	*Paed*iatrics (medical specialty dealing with the disorders of childhood)
pan -	all	*Pan*cytopenia (a reduction in the numbers of all blood cells)
para -	beside, near	*Para*median (close or near to the middle).
part -	birth	*Part*urition (the act of giving birth to a child)

Component	**Meaning**	**Example**
path -	disease	*Path*ogenicity (capacity to cause disease)
- pathy	disease	Nephro*pathy* (kidney disease)
- penia	lack of	Leuco*penia* (lack of white blood cells)
pent -	five	*Pent*ose (a five carbon sugar)
peps -	digest	*Peps*in (an enzyme which starts to digest protein)
per -	through	*Per*fusion (flow of fluid such as blood through an organ)
peri -	around	*Peri*natal (relating to the time around birth)
perineo -	perineum	*Perineo*rrhaphy (repair of the perineum)
- pexy	fixation	Orchido*pexy* (the operation to bring down an undescended testis and its fixation in the scrotum)
- phag -	ingest, swallow	Macro*phag*e (a tissue cell which ingests particles by phagocytosis)
- phagia	eating, swallowing	Dys*phagia* (difficulty in swallowing)
pharmac -	drug	*Pharmac*ology (science dealing with drugs and their effects)
pharyngo -	pharynx	*Pharyngo*tympanic tube (tube between the pharynx and middle ear)
- phasia	speech	Dys*phasia* (difficulty in speaking)

Component	**Meaning**	**Example**
- phil - *philo/philic -*	affinity for	Neutro*phil* (white blood cell which has an affinity for a neutral stain)
phlebo -	vein	Thrombo*phleb*itis (inflammation of a vein with thrombosis)
- phobia/phobe	fear of	Claustro*phobia* (fear of enclosed spaces)
phono -	voice, sound	*Phono*cardiography (recording of heart sounds and mumours using a phonocardiograph)
photo -	light	*Photo*therapy (treatment using light, e.g. for physiological jaundice in the newborn)
phren -	diaphragm, mind	*Phren*ic (relating to the diagphragm, e.g. phrenic nerves)
- phylaxis	protection	Pro*phylaxis* (measure taken for protection, e.g. immunization)
pilo -	hair	Arrector *pili* (muscle fibres around the hair follicle)
- plas -	form, grow	Hyper*plas*ia (grow larger by new cell production)
- plasty	reconstruct, plastic surgery	Rhino*plasty* (operation to reconstruct the nose)
- plegia	paralysis	Hemi*plegia* (paralysis down one side of the body)
pleur/pleuro -	pleura	*Pleur*isy (inflammation of the pleura)
pneumo -	air, lung	*Pneumo*nia (inflammation of the lungs)

Component	Meaning	Example
- pnoea	breathing	A*pnoea* (no breathing)
- poiesis	making	Erythro*poiesis* (making red blood cells)
poly -	many	*Poly*morphonuclear (many shaped nucleus)
post -	after	*Post*ganglionic (relating to the nerve fibre after or distal to a ganglion)
pre/pro -	in front, before	*Pre*ganglionic (relating to the nerve fibre before or proximal to a ganglion)
proct -	rectum, anus	*Proct*ology (study of disorders affecting the rectum and anus)
pseudo -	false	*Pseudo*cyesis (false pregnancy)
psycho -	mind	*Psycho*genic (originating in the mind)
pulmon -	lung	*Pulmon*ary (relating to the lungs, e.g: pulmonary artery)
pyelo -	renal pelvis	*Pyelo*lithotomy (operation to remove stone from the renal pelvis)
pyloro -	pylorus	*Pyloro*myotomy (operation to cut the muscle of the pylorus)
pyo -	pus	*Pyo*metra (pus in the uterus)
pyr -	fever, fire	*Pyr*ogen (substance causing fever)
quadri -	four	*Quadri*ceps (large four part muscle of the anterior thigh)

Component	**Meaning**	**Example**
radio -	radiation	*Radio*isotope (an unstable isotope which emits radiation)
re -	back, again	*Re*flux (flowing back, e.g. of stomach contents into the oesophagus)
ren -	kidney	*Ren*in (proteolytic enzyme produced by the kidney)
retro -	backward	*Retro*version (turned backwards, a displacement of the uterus)
rhin -	nose	*Rhin*opathy (disease affecting the nose)
- *rhythmia*	rhythm	Ar*rhythmia* (without rhythm, usually applied to a disturbance of cardiac rhythm)
- *rrhage/ rrhagia*	to burst, pour, excessive flow	Meno*rrhagia* (excessive menstrual flow)
- *rrhaphy*	to repair	Hernio*rrhaphy* (operation to repair a hernia)
- *rrhoea*	flow, discharge	Leuco*rrhoea* (white vaginal discharge)
rub -	red	*Rub*or (redness, a sign of inflammation)
sacro -	sacrum	*Sacro*iliac (relating to the sacrum and ilium, e.g. sacroiliac joints)
salpingo -	uterine/ Fallopian tube	*Salpingo*graphy (radiographic investigation to ascertain uterine tube patency)
sarco -	muscle, flesh	*Sarco*lemma (cell membrane enclosing a muscle fibre)

Component	Meaning	Example
sclero -	hard	*Sclero*sis (hardening, e.g atherosclerosis)
- *scope*	instrument for examining	Gastro*scope* (instrument used to examine the interior of the stomach)
- *scopy*	to examine, looking	Gastro*scopy* (endoscopic examination of the stomach)
semi -	half	*Semi*lunar (shaped like a half moon, e.g. semilunar cartilages or menisci of the knee)
sero -	serum	*Ser*ous (containing serum)
- *soma/somat* -	body	*Somat*ic (relating to the body, e.g. somatic nerves)
somni -	sleep	In*somni*a (not able to sleep)
- *sonic*	sound	Ultra*sonic* (high frequency sound beyond the range of the human ear)
sphygm -	pulse	*Sphygm*omanometer (instrument for measuring arterial blood pressure)
splen/spleno -	spleen	*Spleno*megaly (enlargement of the spleen)
spondyl/ spondylo -	vertebra	*Spondyl*itis (inflammation of the intervertebral joints)
- *stasis*	stand still, lack of movement	Haemo*stasis* (no bleeding - the arrest of haemorrhage)
steato -	fat	*Steato*rrhoea (passing undigested fat in the faeces)
steno -	narrow	*Steno*sis (abnormal narrowing, e.g. mitral valve stenosis)

Component	**Meaning**	**Example**
stern/sterno -	sternum	*Sterno*cleidomastoid (relating to the sternum and mastoid, e.g. sternocleidomastoid muscle)
sub -	below.	*Sub*arachnoid (below the arachnoid mater)
super/supra -	above	*Supra*condylar (above the condyles, e.g fracture of the elbow)
sym/syn -	together, union, with	*Syn*ergist (a muscle which works with other muscle groups
tachy -	fast	*Tachy*cardia (fast heart rate)
tars/tarso -	foot, eyelid	*Tars*als (bones of the ankle)
teno -	tendon	*Teno*synovitis (inflammation of tendon sheath)
tetra -	four	Fallot's *tetra*logy (congenital heart condition consisting of four defects)
thermo -	heat,temperature	*Thermo*graphy (investigation which measures and records heat production in different parts of an organ or the body)
thorac/thoraco-	chest	*Thora*cic (relating to the chest or thorax, e.g. thoracic vertebrae)
throm/thrombo -	clot	*Thrombo*plastin (clotting factor)
thyro -	thyroid	*Thyro*xine (a hormone produced by the thyroid)
tox -	poison	*Tox*ic (poisonous, e.g. toxic shock syndrome)

Component	Meaning	Example
trache/tracheo-	trachea	*Tracheo*-oesophageal (relating to the trachea and oesophagus, e.g. tracheo-oesophageal fistula)
trans -	across, through	*Trans*urethral (through the urethra, e.g. resection of prostate gland)
tri -	three	*Tri*cuspid (valve with three cusps)
trich -	hair	*Trich*omycosis (fungal disease of hair)
- tripsy	crushing	Litho*tripsy* (crushing stones, e.g. in the bladder)
- troph/trophy-	growth, nourishment	*Troph*oblast (cells covering the blastocyst, they invade the decidua and are concerned with the nutrition of the early embryo)
- trophic	changing, influencing	Adrenocortico*trophic* hormone (the hormone which influences the adrenal cortex)
ultra -	beyond, extreme	*Ultra*microscopic (particles too small to be viewed with a light microscope)
uni-	one	*Uni*lateral (on one side)
uretero-	ureter	*Uretero*vesical (relating to the ureter and bladder)
urethr/urethro-	urethra	*Urethr*itis (inflammation of the urethra)
uri/uro -	urine	*Uro*bilinogen (pigment excreted in the urine)
- uria	urine	Dys*uria* (pain or difficulty passing urine)

Component	Meaning	Example
uter/utero -	uterus	*Uter*ine (relating to the uterus, e.g. uterine tubes)
vas/vaso -	vessel, duct	*Vaso*constriction (contraction of blood vessels)
vene -	vein	*Vene*puncture (inserting a needle into a vein)
vesico -	bladder	*Vesico*vaginal (relating to the bladder and vagina)
viscer -	organs	*Viscer*al (relating to the internal organs especially those of the abdominal cavity)
xero -	dry	*Xero*derma (dry skin)
zoo -	animal	*Zoo*nosis (disease transmitted from animals to humans)

INDEX

A

B

C

D

E

J

K

L

M

N

O

P

T